	10	11	12	13	14	15	16	17	18
									2 He **4.00260** helium $1s^2$ $^1S_0$ 4.003 100% 0 3.016 10^{-4}% 1/2
				5 B **10.811** boron $[He]2s^22p^1$ $^2P_{1/2}$ 11.01 80.0% 3/2 10.01 20.0% 3	**6** C **12.011** carbon $[He]2s^22p^2$ $^3P_0$ 12.00 98.93% 0 13.00 1.07% 1/2	**7** N **14.0067** nitrogen $[He]2s^22p^3$ $^4S_{3/2}$ 14.00 99.64% 1 15.00 0.36% 1/2	**8** O **15.9994** oxygen $[He]2s^22p^4$ $^3P_2$ 15.99 99.76% 0 18.00 0.20% 0	**9** F **18.9984** fluorine $[He]2s^22p^5$ $^2P_{3/2}$ 19.00 100% 1/2 18.00 1.83h 1	**10** Ne **20.1797** neon $[He]2s^22p^6$ $^1S_0$ 19.99 90.48% 0 21.99 9.25% 0
				13 Al **26.9815** aluminum $[Ne]3s^23p^1$ $^2P_{1/2}$ 26.98 100% 5/2 25.99 710ky 5	**14** Si **28.0855** silicon $[Ne]3s^23p^2$ $^3P_0$ 27.98 92.23% 0 28.98 4.67% 1/2	**15** P **30.9738** phosphorus $[Ne]3s^23p^3$ $^4S_{3/2}$ 30.97 100% 1/2 32.97 25.3d 1/2	**16** S **32.065** sulfur $[Ne]3s^23p^4$ $^3P_2$ 31.97 95.0% 0 33.97 4.29% 0	**17** Cl **35.4527** chlorine $[Ne]3s^23p^5$ $^2P_{3/2}$ 34.97 75.77% 3/2 36.97 24.23% 3/2	**18** Ar **39.948** argon $[Ne]3s^23p^6$ $^1S_0$ 39.96 99.59% 0 35.97 0.34% 0
Co	**28** Ni **58.6934** nickel $[Ar]4s^23d^8$ $^3F_4$ 57.93 68.08% 0 59.93 26.22% 0	**29** Cu **63.546** copper $[Ar]4s^13d^{10}$ $^2S_{1/2}$ 62.93 69.2% 3/2 64.93 30.8% 3/2	**30** Zn **65.38** zinc $[Ar]4s^23d^{10}$ $^1S_0$ 63.93 48.6% 0 65.93 27.9% 0	**31** Ga **69.723** gallium $[Zn]4p^1$ $^2P_{1/2}$ 68.93 60.0% 3/2 70.93 40.0% 3/2	**32** Ge **72.64** germanium $[Zn]4p^2$ $^3P_0$ 73.92 36.5% 0 71.92 27.4% 0	**33** As **74.9216** arsenic $[Zn]4p^3$ $^4S_{3/2}$ 74.92 100% 3/2 72.92 80.3d 3/2	**34** Se **78.96** selenium $[Zn]4p^4$ $^3P_2$ 79.92 49.8% 0 77.92 23.5% 0	**35** Br **79.904** bromine $[Zn]4p^5$ $^2P_{3/2}$ 78.92 50.69% 3/2 80.92 49.31% 3/2	**36** Kr **83.798** krypton $[Zn]4p^6$ $^1S_0$ 83.91 57.0% 0 85.91 17.3% 0
Rh	**46** Pd **106.42** palladium $[Kr]4d^{10}$ $^1S_0$ 105.9 27.3% 0 107.9 26.7% 0	**47** Ag **107.87** silver $[Kr]5s^14d^{10}$ $^2S_{1/2}$ 106.9 51.83% 1/2 108.9 48.17% 1/2	**48** Cd **112.411** cadmium $[Kr]5s^24d^{10}$ $^1S_0$ 113.9 28.8% 0 111.9 24.0% 0	**49** In **114.818** indium $[Cd]5p^1$ $^2P_{1/2}$ 114.9 95.7% 9/2 112.9 4.3% 9/2	**50** Sn **118.710** tin $[Cd]5p^2$ $^3P_0$ 119.9 32.4% 0 117.9 24.3% 0	**51** Sb **121.760** antimony $[Cd]5p^3$ $^4S_{3/2}$ 120.9 57.3% 5/2 122.9 42.7% 7/2	**52** Te **127.60** tellurium $[Cd]5p^4$ $^3P_2$ 129.9 34.5% 0 127.9 31.7% 0	**53** I **126.904** iodine $[Cd]5p^5$ $^2P_{3/2}$ 126.9 100% 5/2 128.9 17My 7/2	**54** Xe **131.29** xenon $[Cd]5p^6$ $^1S_0$ 131.9 26.9% 0 128.9 26.4% 1/2
Ir	**78** Pt **195.08** platinum $[Xe]6s^14f^{14}5d^9$ $^3D_3$ 195.0 33.8% 1/2 194.0 32.9% 0	**79** Au **196.967** gold $[Xe]6s^14f^{14}5d^{10}$ $^2S_{1/2}$ 197.0 100% 3/2 195.0 186d 3/2	**80** Hg **200.59** mercury $[Xe]6s^24f^{14}5d^{10}$ $^1S_0$ 202.0 29.7% 0 200.0 23.1% 0	**81** Tl **204.383** thallium $[Hg]6p^1$ $^2P_{1/2}$ 205.0 70.50% 1/2 203.0 29.50% 1/2	**82** Pb **207.2** lead $[Hg]6p^2$ $^3P_0$ 208.0 52.4% 0 206.0 24.1% 0	**83** Bi **208.980** bismuth $[Hg]6p^3$ $^4S_{3/2}$ 209.0 100% 9/2 210 3.0My 9	**84** Po polonium $[Hg]6p^4$ $^3P_2$ 209.0 102y 1/2 208.0 2.9y 0	**85** At astatine $[Hg]6p^5$ $^2P_{3/2}$ 210.0 8h 5 211.0 7h 9/2	**86** Rn radon $[Hg]6p^6$ $^1S_0$ 222.0 4d 0 211.0 15h 1/2
Mt	**110** Ds darmstadtium 281 1.1m 280 7.5s	**111** Rg roentgenium 281 26s 280 3.6s	**112** Cn copernicium 285 11m 283 3m		**114** Fl flerovium 289 21s 288 6s		**116** Lv livermorium 292 52ms 290 29ms		

Dy	**67** Ho **164.930** holmium $[Xe]6s^24f^{11}$ $^4I_{15/2}$ 164.9 100% 7/2 162.9 4.6ky 7/2	**68** Er **167.26** erbium $[Xe]6s^24f^{12}$ $^3H_6$ 165.9 33.6% 0 167.9 26.8% 0	**69** Tm **168.934** thullium $[Xe]6s^24f^{13}$ $^2F_{7/2}$ 168.9 100% 1/2 170.9 1.92y 1/2	**70** Yb **173.054** ytterbium $[Xe]6s^24f^{14}$ $^1S_0$ 173.9 31.8% 0 171.9 21.9% 0	**71** Lu **174.967** lutetium $[Xe]6s^24f^{14}5d^1$ $^2D_{3/2}$ 174.9 97.41% 7/2 175.9 2.59% 7
Cf	**99** Es einsteinium $[Rn]7s^25f^{11}$ $^4I_{15/2}$ 254.1 276d 7 252.1 1.3y 5	**100** Fm fermium $[Rn]7s^25f^{12}$ $^3H_6$ 257.1 100d 9/2 253.1 3d 1/2	**101** Md mendelevium $[Rn]7s^25f^{13}$ $^2F_{7/2}$ 258.1 51.5d 8 260 32d	**102** No nobelium $[Rn]7s^25f^{14}$ $^1S_0$ 259.1 58m 9/2 255.1 3.1m 1/2	**103** Lr lawrencium $[Rn]7s^25f^{14}6d^1$** $^2D_{3/2}$ 262.1 3.6h 261.1 40m

*Multiple configurations contribute to the ground state.
**Electron configuration and term state assignment are tentative.

PHYSICAL CHEMISTRY

Quantum Chemistry

and Molecular Interactions

PHYSICAL CHEMISTRY

Quantum Chemistry

and Molecular Interactions

ANDREW COOKSY

PEARSON

Boston Columbus Indianapolis New York San Francisco Upper Saddle River
Amsterdam Cape Town Dubai London Madrid Milan Munich Paris Montréal Toronto
Delhi Mexico City São Paulo Sydney Hong Kong Seoul Singapore Taipei Tokyo

Editor in Chief: *Adam Jaworski*
Executive Editor: *Jeanne Zalesky*
Senior Marketing Manager: *Jonathan Cottrell*
Project Editor: *Jessica Moro*
Editorial Assistant: *Lisa Tarabokjia*
Marketing Assistant: *Nicola Houston*
Director of Development: *Jennifer Hart*
Development Editor: *Daniel Schiller*
Media Producer: *Erin Fleming*
Managing Editor, Chemistry and Geosciences: *Gina M. Cheselka*
Full-Service Project Management/Composition: *GEX Publishing Services*
Illustrations: *Precision Graphics*
Image Lead: *Maya Melenchuk*
Photo Researcher: *Stephanie Ramsay*
Text Permissions Manager: *Joseph Croscup*
Text Permissions Research: *GEX Publishing Services*
Design Manager: *Mark Ong*
Interior Design: *Jerilyn Bockorick, Nesbitt Graphics*
Cover Design: *Richard Leeds, BigWig Design*
Operations Specialist: *Jeffrey Sargent*
Cover Image Credit: *Tony Jackson/Getty Images*

Credits and acknowledgments borrowed from other sources and reproduced, with permission, in this textbook appear on the appropriate page within the text or in the back matter.

Library of Congress Cataloging-in-Publication Data
Cooksy, Andrew.
Physical chemistry : quantum chemistry and molecular interactions / Andrew Cooksy.
pages cm
Includes index.
ISBN-13: 978-0-321-81416-6
ISBN-10: 0-321-81416-9
1. Chemistry, Physical and theoretical--Textbooks. 2. Quantum chemistry--Textbooks. 3. Molecular dynamics--Textbooks. I. Title.
QD453.3.C655 2014
641--dc23
2012037314

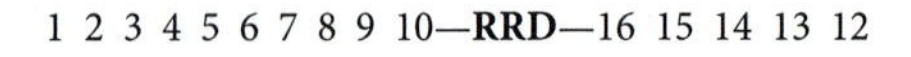
1 2 3 4 5 6 7 8 9 10—**RRD**—16 15 14 13 12

www.pearsonhighered.com

ISBN-10: 0-321-81416-9
ISBN-13: 978-0-321-81416-6

DEDICATION

To Mary, Wesley, and Owen

...our great creative Mother, while she amuses us with apparently working in the broadest sunshine, is yet severely careful to keep her own secrets, and, in spite of her pretended openness, shows us nothing but results.

—*Nathaniel Hawthorne* (1804–1864) The Birthmark

PHYSICAL CHEMISTRY BRIEF CONTENTS

Quantum Chemistry and Molecular Interactions

BRIEF TABLE OF CONTENTS FOR COOKSY'S THERMODYNAMICS, STATISTICAL MECHANICS, & KINETICS

Thermodynamics, Statistical Mechanics, & Kinetics

DETAILED CONTENTS

PHYSICAL CHEMISTRY

Quantum Chemistry and Molecular Interactions

TO THE READER

This book is intended to provide students with a detailed guide to the reasoning that forms the basis for physical chemistry—the framework that unites all chemistry. The study of physical chemistry gives us the opportunity to look at our science as an integrated whole, with each concept connected to the next. My goal has been to trace those connections, step-by-step whenever possible, to show how each new concept makes sense given its place in the framework.

Because its ideas build upon each other in this way, physical chemistry can serve as the foundation for an intuitive understanding of chemistry in all its forms, whether synthesizing new compounds, analyzing samples in a forensic laboratory, or studying the properties of novel materials. To that end, this book emphasizes the shared, fundamental principles of chemistry, showing how we can justify the form and behavior of complex chemical systems by applying the laws of mathematics and physics to the structures of individual particles and then extrapolating to larger systems. We learn physical chemistry so that we can recognize these fundamental principles when we run into them in our other courses and in our careers. The relevance of this discipline extends beyond chemistry to engineering, physics, biology, and medicine: any field in which the molecular structure of matter is important.

A key step toward cultivating an intuition about chemistry is a thorough and convincing presentation of these fundamentals. When we see not only what the ideas are, but also how they link together, those ideas become more discernible when we examine a new chemical system or process. The following features of this text seek to achieve that objective.

- My aim is to provide a rigorous treatment of the subject in a relaxed style. A combination of qualitative summaries and annotated, step-by-step derivations illuminates the logic connecting the theory to the parameters that we can measure by experiment. Although we use a lot of math to justify the theory we are developing, the math will always make sense if we look at it carefully. We take advantage of this to strengthen our confidence in the results and our understanding of how the math relates to the physics. Nothing is more empowering in physical chemistry than finding that you can successfully predict a phenomenon using both mathematics *and* a qualitative physical argument. The manifestation of atomic and molecular structure in bulk properties of materials is a theme that informs the unhurried narrative throughout the text.

- To illustrate how our understanding in this field continues to advance, we take the time to examine several tools commonly used in the laboratory ("Tools of the Trade"), while profiles of contemporary scientists ("Biosketches") showcase the ever-expanding frontiers of physical chemistry. Our intuition about chemistry operates at a deep level, held together by the theoretical framework, but these examples show how others are applying their understanding to solve real problems in the laboratory and beyond. They inspire us to think creatively about how the most fundamental chemical laws can answer our own questions about molecular structure and behavior.
- Our increasing appreciation and exploration of the interface between the molecular and the bulk scales has inspired a forward-looking coverage of topics that includes chapters dedicated to intermolecular interactions, nanoscale chemical structure, and liquid structure.

Acknowledgments

I thank Kwang-Sik Yun and Andrew P. Stefani for providing the original inspiration and encouragement to carry out this project. My love of this field owes much to my mentors—William Klemperer, Richard J. Saykally, and Patrick Thaddeus—and to the many students and colleagues who have patiently discussed chemistry with me. I am particularly grateful to my fellow physical chemistry faculty—Steve Davis and Kwang-Sik Yun at Ole Miss, David Pullman and Karen Peterson at SDSU—for their many insights and limitless forbearance, and to William H. Green and the late John M. Brown for kindly hosting my sabbatical work in their research groups. An early prospectus for this book formed part of the proposal for an NSF CAREER grant, and I thank the agency for that support.

For helping me see this through, I thank my friends and mentors at Pearson, especially Nicole Folchetti, Adam Jaworski, Dan Kaveney, Jennifer Hart, Jessica Moro, and above all Jeanne Zalesky. A great debt is also owed to Dan Schiller for his patient and extensive work editing the manuscript. Thanks to Mary Myers for much work on the original manuscript, and to many in the open-source community for the tools used to assemble it. Finally, I thank all of our faculty and student reviewers for their careful reading and thoughtful criticisms. A textbook author could strive for no higher goal than to do justice to the fascination that we share for this subject.

Reviewers

Ludwik Adamowicz
University of Arizona
Larry Anderson
University of Colorado
Alexander Angerhofer
University of Florida
Matthew Asplund
Brigham Young University
Tom Baer
University of North Carolina
Russ Baughman
Truman State University
Nikos Bentenitis
Southwestern University
John Bevan
Texas A&M University
Charles Brooks
University of Michigan
Mark Bussell
Western Washington University
Beatriz Cardelino
Spelman College
Donna Chen
University of South Carolina
Samuel Colgate
University of Florida
Stephen Cooke
Purchase College, the State University of New York
Paul Cooper
George Mason University
Phillip Coppens
University at Buffalo, the State University of New York
Biamxiao Cui
Stanford University
Alfred D'Agostino
University of Notre Dame
Paul Davidovits
Boston College
Borguet Eric
Temple University
Michelle Foster
University of Massachusetts
Sophya Garashchuk
University of South Carolina
Franz Geiger
Northeastern University
Kathleen Gilbert
New Jersey Institute of Technology
Derek Gragson
California Polytechnic State University
Hua Guo
University New Mexico
John Hagen
California Polytechnic State University
Cynthia Hartzell
Northern Arizona University
Bill Hase
Texas Tech University
Clemens Heske
University of Nevada, Las Vegas
Lisa Hibbard
Spelman College
Brian Hoffman
Northeastern University
Xiche Hu
University of Toledo
Bruce Hudson
Syracuse University
David Jenson
Georgia Institute of Technology
Benjamin Killian
University of Florida
Judy Kim
University of California, San Diego
Krzysztof Kuczera
University of Kansas
Joseph Kushick
University of Massachusetts, Amherst
Marcus Lay
University of Georgia, Athens
Lisa Lever
University of South Carolina
Louis Madsen
Virginia Polytechnic Institute and State University

Elache Mahdavian
Louisiana State University
Herve Marand
Virginia Polytechnic Institute and State University
Ruhullah Massoudi
South Carolina University
Gary Meints
Missouri State University
Ricardo Metz
University of Massachusetts
Kurt Mikkelson
University of Copenhagen
Phambu Nsoki
Tennessee State University
Jamiu Odutola
Alabama A&M University
Jason Pagano
Saginaw Valley State University
James Patterson
Brigham Young University
James Phillips
University of Wisconsin
Simon Phillpot
University of Florida
Rajeev Prabhakar
Miami University
Robert Quandt
Illinois State University
Ranko Richert
Arizona State University
Tim Royappa
University of West Florida
Stephen Sauer
University of Copenhagen
G. Alan Schick
Missouri State University
Charles Schmuttenmaer
Yale University
Rod Schoonover
California Polytechnic State University
Alexander Smirnov
North Carolina State University
J. Anthony Smith
Walla Walla University
Karl Sohlberg
Drexel University
David Styers-Barnett
Indiana University
James Terner
Virginia Commonwealth University
Greg Van Patten
Ohio University
John Vohs
University of Pennsylvania
Michael Wagner
George Washington University
Brian Woodfield
Brigham Young University
Dong Xu
Boise State University
Eva Zurek
University at Buffalo, the State University of New York

ABOUT THE AUTHOR

Andrew Cooksy

B.A., chemistry and physics, Harvard College, 1984;
Ph.D., chemistry, University of California, Berkeley, 1990;
Postdoctoral Research Associate, Harvard-Smithsonian Center for Astrophysics and Harvard University Department of Chemistry, 1990-1993;
Asst. and Assoc. Professor, University of Mississippi Department of Chemistry, 1993-1999. Asst. and Assoc. Professor, San Diego State University Department of Chemistry, 1999-2010. Professor, San Diego State University Department of Chemistry, 2010-.
Northrop-Grumman Excellence in Teaching Award, 2010
Senate Excellence in Teaching Award, SDSU College of Sciences, 2011

PHYSICAL CHEMISTRY is the framework that unites **all** chemistry—providing powerful insight into the discipline as an integrated series of connected concepts.

As an instructor and author, Andrew Cooksy helps students uncover these connections while showing how they can be expressed in mathematical form and demonstrating the power that derives from such expressions.

The text's lively and relaxed narrative illuminates the relationship between the mathematical and the conceptual for students. By formulating the fundamental principles of physical chemistry in a mathematically precise but easily comprehensible way, students are able to acquire deeper insight—and greater mastery—than they ever thought possible.

This innovative approach is supported by several exclusive features:

- **Split quantum and thermodynamics volumes** can be taught in either order for maximum course flexibility.
- **A discrete chapter** (Chapter A) included in each volume summarizes the physics and mathematics used in physical chemistry.
- **Chapter opening sections** orient the students within the larger context of physical chemistry, provide an overview of the chapter, preview the physical and mathematical relationships that will be utilized, and set defined chapter objectives.
- **Unique pedagogical features** include annotations for key steps in derivations and an innovative use of color to identify recurring elements in equations.

Uncovering connections between foundational concepts

Reflective of the author's popular lecture strategy, chapter opening and closing features ground each topic within the larger framework of physical chemistry and help students stay oriented as they follow the development of chapter concepts.

Learning Objectives outline the skills students should expect to acquire from their study of the chapter.

Visual Roadmaps help students see the relationship between the chapters in each part of the text and the topics in each chapter.

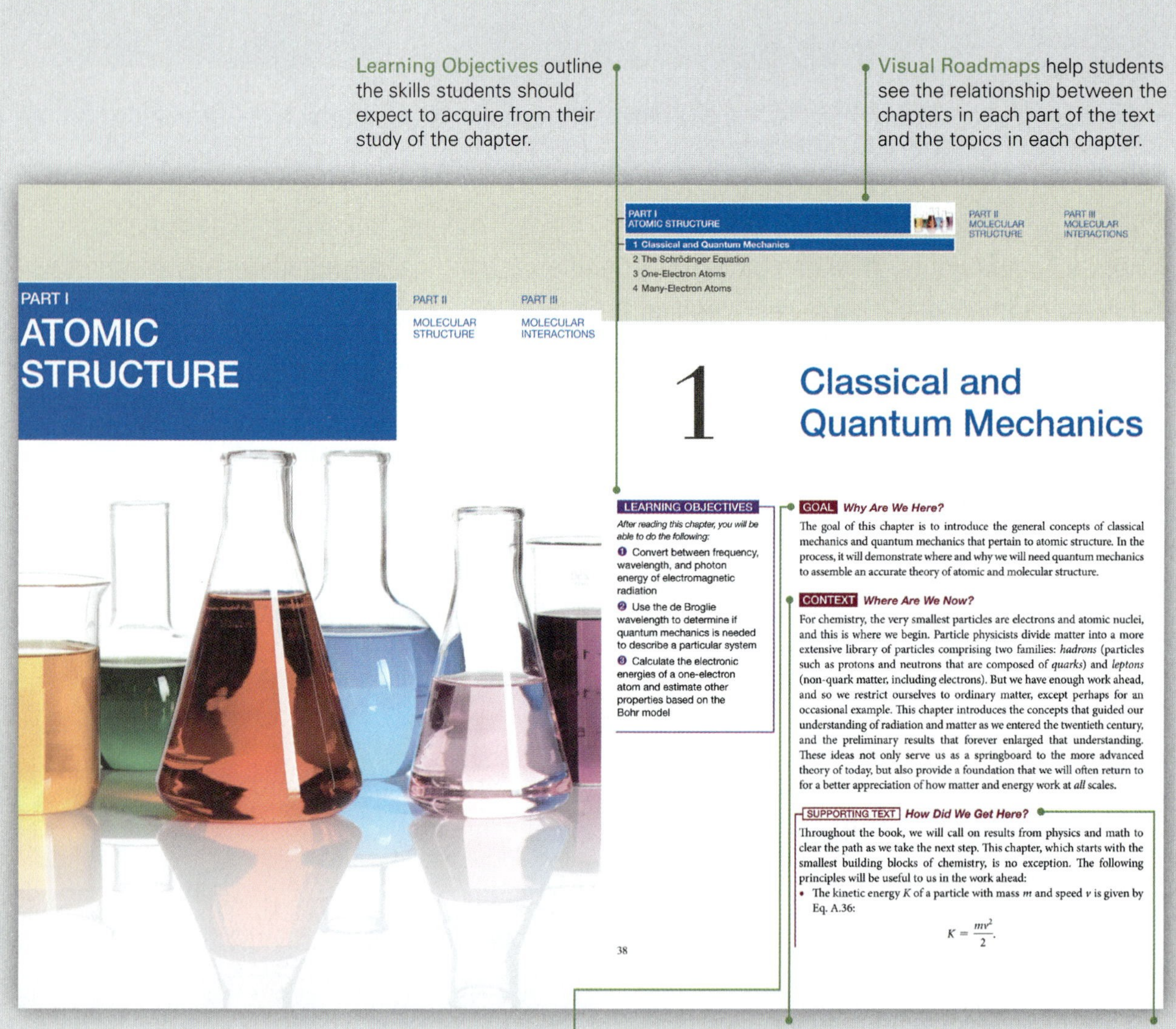

Context: Where Do We Go From Here? sections at the end of each chapter afford students a perspective on what they have just learned, and how it provides the foundation for the material explored in the next chapter.

Goal: Why Are We Here? chapter openers prepare students for the work ahead using one to two simple sentences.

Context: Where Are We Now? helps students understand how the chapter they are starting is related to what has come before and its place in the unfolding development of physical chemistry.

Supporting Text: How Did We Get Here? reviews previously introduced concepts, mathematical tools, and topical relationships that the new chapter will draw on.

Active research, tools, and techniques

Through learning about the instruments and methods of modern physical chemistry and meeting researchers at work today, students gain an appreciation for the practical applications of this science to many fields.

Tools of the Trade sections highlight the design and operation of commonly used experimental apparatuses and how they relate to the principles discussed in the chapter.

TOOLS OF THE TRADE Photoelectron Spectroscopy

One of the early keys to quantum theory was Albert Einstein's explanation of the **photoelectric effect.** In 1887, Heinrich Hertz discovered that the generation of electricity at a metal surface could be enhanced by ultraviolet light. From the results of subsequent experiments by Philipp Lenard, Einstein concluded that the energy of electromagnetic radiation is carried in units of photons, and that the surface abosorbs energy *one photon at a time*, with each interaction causing an energy change at the surface. If the photon energy surpasses a threshold value IE sufficient to expel an electron, any excess energy in the photon provides the kinetic energy of the ejected electron. Raising the intensity of the light merely provides more photons, increasing the number of interactions but not the energy of each interaction. If $h\nu$ is not high enough to ionize the sample, no electrons will be ejected. In a gas, the IE is the first ionization energy, the energy difference between the ion and the neutral atom or molecule M, $\text{IE} = E(\text{M}^+) - E(\text{M})$. In a solid IE is called the **work function.**

We continue to take advantage of this technique in order to measure electronic transition energies by using a known photon energy $h\nu$ and then measuring the kinetic energy of the electrons:

$$\Delta E = h\nu - \frac{m_e v^2}{2}.$$

This principle established the foundation for **photoelectron spectroscopy.**

What is photoelectron spectroscopy? In photoelectron spectroscopy, we strike the sample with a burst of ionizing photons, all at energy $h\nu$, and then measure the distribution in kinetic energies of the electrons that reach the detector in order to generate the spectrum of ΔE values.

Why do we use photoelectron spectroscopy? Photoelectron spectroscopy has three advantages over absorption spectroscopy: (*i*) the sensitivity is much higher because we can detect the tiny currents generated by a small number of electrons better than we can detect a tiny decrease in the intensity of radiation passing through a sample; (*ii*) ionizations are not subject to the $\Delta l = \pm 1$ selection rule (Section 3.3), because the ejected electron can carry away the necessary angular momentum; (*iii*) in a many-electron atom or molecule, both the neutral M and the ion M^+ have distinct quantum states, and the values of ΔE we measure tell us about the energy levels of the neutral and the ion in the same experiment.

Photoelectron spectroscopy is a zero-background technique, meaning that the detector sees nothing (except a weak noise signal) until there is an interaction between the radiation and the sample. In addition, our ability to measure charged particles such as electrons or ions is usually much better than our ability to detect photons. For one thing, we can use electric fields to accelerate charged particles before they reach the detector, so that they hit the detector with considerably greater energy than they had originally, and with greater energy than a typical photon in the experiment would have.

How does it work? In photoelectron spectroscopy, a pulse of laser light at a known photon energy ionizes the sample in an ultrahigh vacuum chamber. The electrons are directed by magnetic fields down a drift tube toward a detector to separate the different velocities. Just before the detector, the electrons are rapidly accelerated and focused to amplify the signal. The raw measurements consist of electron signals—electrical currents at the detector—tabulated as a function of the time after the laser pulse. From the drift times Δt and the known length d of the drift tube, we can calculate the electron speeds $v = d/\Delta t$ and convert the speeds into electron kinetic energies $mv^2/2$.

Gas-phase photoelectron spectroscopy is used chiefly for research into atomic and molecular energy levels. In addition, x-ray photoelectron spectroscopy is a common application of the technique, used on solid samples as a means of rapidly characterizing the elemental and molecular composition of materials or coatings.

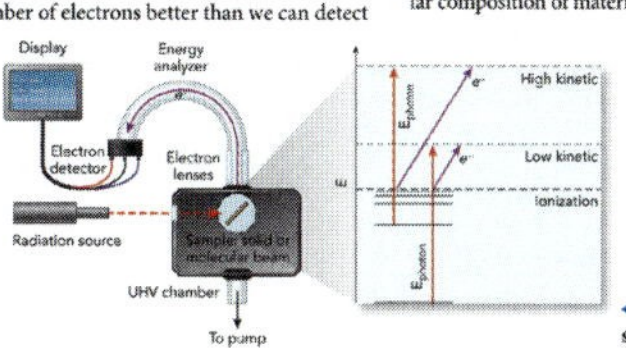

◀ **Photoelectron spectroscopy.**

Biosketches highlight a diverse array of contemporary scientists and engineers and their current research relating to physical chemistry.

BIOSKETCH Sylvia Ceyer

Sylvia Ceyer is the J. C. Sheehan Professor of Chemistry at MIT, where she and her research group investigate how molecules interact with solid surfaces. One of her goals has been a better understanding of the pressure-dependence of chemical reactions that occur on a surface. Surface chemistry is normally investigated under ultra-high vacuum conditions, at pressures of 10^{-13} bar or less, in order to allow methods like Auger spectroscopy (Example 4.3) and electron diffraction (Section 1.3) to characterize the reaction. These conditions make it difficult to study how pressure affects the reaction, however. The Ceyer group developed one technique that they christened "Chemistry with a Hammer." In this method, the reactant—methane, for example—is gently laid on the solid surface with too little energy to react. A high-speed beam of non-reactive noble gas atoms then strikes the surface, raising the effective temperature and pressure at the surface—simulating the reaction conditions the group wants to study, but only at the point at which the beam hits the surface. Overall, the pressure is still low enough to allow diagnostic tools like Auger spectroscopy and electron diffraction to function.

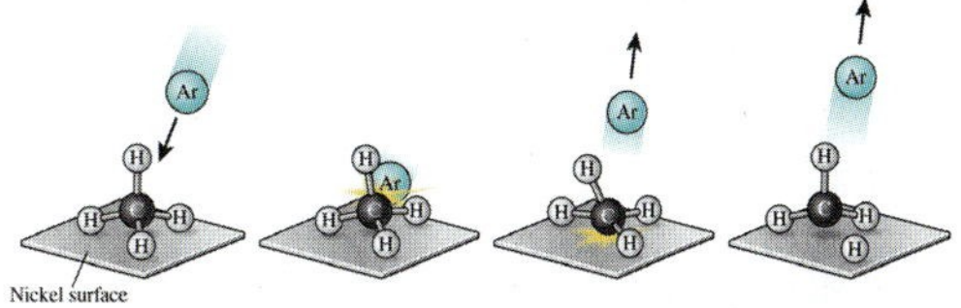

▲ Chemistry with a hammer. (After an image by Tom Dunne, *American Scientist* **87**, p. 21, 1999).

Conceptual Insight and Mathematical Precision in a Real World Context

A discrete summary of the prerequisite mathematics and physics adds flexibility and convenience by incorporating the necessary math tools in a single chapter.

TABLE A.5 Solutions to selected integrals. In these equations, a and b are constants, n is a whole number, and C is the constant of integration.

$\int x^n dx$	$= \frac{1}{n+1}x^{n+1} + C$	$\int a dx$	$= a(x + C)$
$\int \frac{1}{x} dx$	$= \ln x + C$	$\int e^x dx$	$= e^x + C$
$\int \ln x dx$	$= x \ln x - x + C$	$\int \frac{dx}{x(a+bx)}$	$= -\frac{1}{a}\ln\left(\frac{a+bx}{x}\right) + C$
$\int \sin x dx$	$= -\cos x + C$	$\int \cos x dx$	$= \sin x + C$
$\int \sin^2(ax) dx$	$= \frac{x}{2} - \frac{\sin(2ax)}{4a} + C$	$\int \cos^2(ax) dx$	$= \frac{x}{2} + \frac{\sin(2ax)}{4a} + C$
$\int [f(x) + g(x)]dx$	$= \int f(x) dx + \int g(x) dx$	$\int_a^b dx$	$= x\Big\|_a^b = b - a$
$\int_0^\infty x^n e^{-ax} dx$	$= \frac{n!}{a^{n+1}}$	$\int_0^\infty e^{-ax^2} dx$	$= \frac{1}{2}\left(\frac{\pi}{a}\right)^{1/2}$
$\int_0^\infty x e^{-ax^2} dx$	$= \frac{1}{2a}$	$\int_0^\infty x^2 e^{-ax^2} dx$	$= \frac{1}{4}\left(\frac{\pi}{a^3}\right)^{1/2}$
$\int_0^\infty x^{2n+1} e^{-ax^2} dx$	$= \frac{n!}{2a^{n+1}}$	$\int_0^\infty x^{2n} e^{-ax^2} dx$	$= \frac{[1\cdot3\cdot5\ldots(2n-1)]\sqrt{\pi}}{2^{n+1}a^{n+(1/2)}}$
$\int_0^s x^n e^{-ax} dx$	$= \frac{n!}{a^{n+1}} - e^{-as}\sum_{i=0}^{n}\frac{n! s^{n-i}}{a^{i+1}(n-i)!}$		

the value of C is lost. When we undo the derivative by taking the integral, we add an unknown constant of integration to the integrated expression. Omit this constant when solving definite integrals, because the limits of integration will determine its value.

3. The function being integrated is the **integrand,** and it is multiplied by the incremental change along the coordinates, called the volume element.

Most of the algebraic solutions to integrals that we need appear in Table A.5.

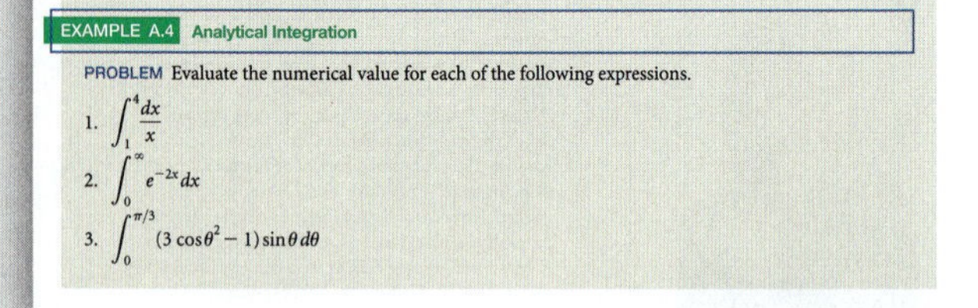
EXAMPLE A.4 Analytical Integration

PROBLEM Evaluate the numerical value for each of the following expressions.

1. $\int_1^4 \frac{dx}{x}$
2. $\int_0^\infty e^{-2x} dx$
3. $\int_0^{\pi/3} (3\cos\theta^2 - 1)\sin\theta\, d\theta$

SOLUTION These can be solved by substitution of the expressions in Table A.5.

(a) $\int_1^4 \frac{dx}{x} = \ln x\Big|_1^4 = \ln 4 - \ln 1 = 1.386 - 0 = 1.386$

(b) $\int_0^\infty e^{-2x} dx = -\frac{1}{2}e^{-2x}\Big|_0^\infty = -\frac{1}{2}(e^{-\infty} - e^0) = -\frac{1}{2}(0-1) = \frac{1}{2}$

(c) $\int_0^{\pi/3} (3\cos^2\theta - 1)\sin\theta\, d\theta = [-\cos^3\theta + \cos\theta]\Big|_0^{\pi/3}$

$= \left[-\left(\frac{1}{2}\right)^3 + \left(\frac{1}{2}\right)\right] - [-(1)^3 + (1)] = \frac{3}{8}.$

By the way, it is possible to apply rules of symmetry to extend some of the analytical solutions in Table A.5. For example, when the integrand is $x^{2n}e^{-ax^2}$, then the function is exactly the same from 0 to $-\infty$ as from 0 to $+\infty$ (Fig. A.3a). Therefore, the integral $\int_{-\infty}^{\infty} x^{2n}e^{-ax^2}dx$ is equal to 2 times $\int_0^\infty x^{2n}e^{-ax^2}dx$. However, if the power of x is odd, $2n + 1$, then the function is negative when $x < 0$ and positive when $x > 0$ (Fig. A.3b). The integral from $-\infty$ to 0 cancels the integral from 0 to $+\infty$, so $\int_{-\infty}^{\infty} x^{2n+1}e^{-ax^2}dx = 0$.

Numerical Integration

Not all integrals have algebraic solutions, and some have algebraic solutions only between certain limits (such as 0 and ∞). With suitable computers, any integral can be calculated without trying to cram it into some algebraic form. This is accomplished by going back to the definition in calculus,

$$\int_{x_1}^{x_2} f(x)\,dx = \lim_{\delta x \to 0}\left\{\sum_{i=1}^{N} f[x_1 + i\,\delta x]\right\}\delta x \quad (A.20)$$

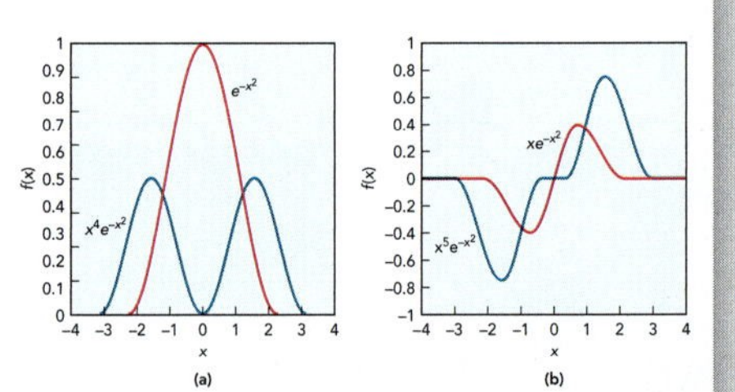

▶ **FIGURE A.3 Symmetry and definite integrals.** **(a)** If a function $f(x)$ is equal to $f(-x)$ for all values of x, then the integral from $-\infty$ to ∞ is equal to 2 times the integral from 0 to ∞. **(b)** If $f(x)$ is equal to $-f(-x)$, then the integral from $-\infty$ to ∞ is 0.

Chapter A provides a comprehensive summary of the physical laws and mathematical tools used to develop the principles of physical chemistry.

The distinctive use of color in the text's mathematical narrative allows students to identify important equation elements (such as the Hamiltonian operator) even as they take on different mathematical forms.

$$\begin{aligned}
\hat{L}^2\Theta(\theta) &= -\hbar^2\left[\frac{1}{\sin\theta}\frac{\partial}{\partial\theta}\sin\theta\frac{\partial}{\partial\theta} - \frac{m_l^2}{\sin^2\theta}\right]A_\theta \sin^k\theta && \text{Eqs. 3.5 and 3.10}\\
&= -A_\theta\hbar^2\left[\frac{1}{\sin\theta}\frac{\partial}{\partial\theta}\sin\theta\,(k\sin^{k-1}\theta\cos\theta) - m_l^2\sin^{k-2}\theta\right] && \text{take } \frac{\partial}{\partial\theta}\\
&= -A_\theta\hbar^2\left[\frac{1}{\sin\theta}\frac{\partial}{\partial\theta}(k\sin^k\theta\cos\theta) - m_l^2\sin^{k-2}\theta\right] && \times\sin\theta\\
&= -A_\theta\hbar^2\left[\frac{1}{\sin\theta}(k^2\sin^{k-1}\theta\cos^2\theta - k\sin^{k+1}\theta) - m_l^2\sin^{k-2}\theta\right] && \text{take } \frac{\partial}{\partial\theta}\\
&= -A_\theta\hbar^2[(k^2\sin^{k-2}\theta\cos^2\theta - k\sin^k\theta) - m_l^2\sin^{k-2}\theta] && \times\sin^{-1}\theta\\
&= -A_\theta\hbar^2[(k^2\sin^{k-2}\theta(1-\sin^2\theta) - k\sin^k\theta) - m_l^2\sin^{k-2}\theta]\\
& && \sin^2\theta + \cos^2\theta = 1\\
&= -A_\theta\hbar^2[(k^2\sin^{k-2}\theta - k^2\sin^k\theta - k\sin^k\theta) - m_l^2\sin^{k-2}\theta].
\end{aligned}$$

Thoughtful color-coding in key equations makes it easier for students to follow the development of complex derivations as well as recognize common mathematical elements that appear in the representation of different physical situations.

Derivations Demystified

$$\begin{aligned}
\hat{p}_x\psi_\pm(x) &= \frac{\hbar}{i}\frac{\partial}{\partial x}\left[\cos\left(\frac{2\pi x}{\lambda_{dB}}\right) \pm i\sin\left(\frac{2\pi x}{\lambda_{dB}}\right)\right]\\
&= \frac{\hbar}{i}\left[-\left(\frac{2\pi}{\lambda_{dB}}\right)\sin\left(\frac{2\pi x}{\lambda_{dB}}\right) \pm i\left(\frac{2\pi}{\lambda_{dB}}\right)\cos\left(\frac{2\pi x}{\lambda_{dB}}\right)\right] && \text{take } \frac{\partial}{\partial x}\\
&= \frac{2\pi\hbar}{\lambda_{dB}}\left[-\frac{1}{i}\sin\left(\frac{2\pi x}{\lambda_{dB}}\right) \pm \cos\left(\frac{2\pi x}{\lambda_{dB}}\right)\right] && \text{rearrange constants}\\
&= \frac{h}{\lambda_{dB}}\left[i\sin\left(\frac{2\pi x}{\lambda_{dB}}\right) \pm \cos\left(\frac{2\pi x}{\lambda_{dB}}\right)\right] && 2\pi\hbar = h,\ 1/i = -i\\
&= p_x\left[\pm\cos\left(\frac{2\pi x}{\lambda_{dB}}\right) + \sin\left(\frac{2\pi x}{\lambda_{dB}}\right)\right] && h/\lambda_{dB} = p_x,\ \text{switch terms}\\
&= \pm p_x\left[\cos\left(\frac{2\pi x}{\lambda_{dB}}\right) \pm i\sin\left(\frac{2\pi x}{\lambda_{dB}}\right)\right] && \text{factor out } \pm 1,\\
&= \pm|p_x|\,\psi_\pm(x). && (2.24)
\end{aligned}$$

Derivations are made transparent and comprehensible to students without sacrifice of mathematical rigor. Colored annotations provide crucial help to students by explaining important steps in key derivations.

DERIVATION SUMMARY The Angular Solution. We chose a reasonable guess for the angular wavefunction, leaving several free parameters undecided, and just operated on the thing with $\hat{L}^2$, requiring that we get an eigenvalue equation. That equation was only satisfied by wavefunctions with a squared angular momentum value L^2 of $\hbar^2 l(l+1)$, with l some whole number.

Summaries spell out the essential results of difficult derivations, making it easier to accommodate the needs of different courses, the preferences of different instructors, and the study and review habits of different students.

Supporting students' quest for deeper understanding

With numerous worked examples, robust review support, a wealth of end-of-chapter problems and a solutions manual written by the text's author, students have everything they need to master the basics of physical chemistry.

EXAMPLE 6.3 Point Group Operations

CONTEXT Problems in quantum mechanics can often be approached from different perspectives, and it becomes important to see when two processes, although described differently, are actually the same. For example, quantum mechanical tunneling has a dramatic impact on many chemical reactions that involve hydrogen transfer, because hydrogens are relatively light (which increases their tunneling probability). When there are several equivalent hydrogens in the same molecule, tunneling can also allow them to exchange places. In the 2-butyne molecule shown, it is possible for tunneling to exchange H atoms 1 and 2. A second tunneling exchange can then reverse the positions of atoms 2 and 3. Each of these exchanges is similar to a reflection of the methyl group through a mirror plane.

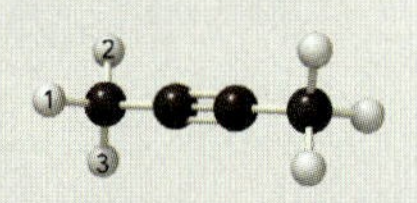

The combination of those two exchanges (switching atoms 1 and 2, and then switching 2 and 3) results in exactly the same arrangement as if the methyl group were rotated by 120°. The combination of two reflections in this case is equivalent to a single rotation by a third of a turn. This example illustrates in another way how we can determine that one combination of operations has the same result as another, single operation.

PROBLEM If $\hat{C}_2(z)$ indicates rotation by π about the z axis, $\hat{\sigma}_{xy}$ indicates reflection through the xy plane, and so on, then find the single operation that is identical to

$$\hat{\sigma}_{xy}\hat{C}_2(z)\hat{\sigma}_{yz}.$$

SOLUTION Remembering to carry out the operations from right to left, we have

$$\begin{aligned}\hat{\sigma}_{xy}\hat{C}_2(z)\hat{\sigma}_{yz}\psi(x,y,z) &= \hat{\sigma}_{xy}\hat{C}_2(z)\psi(-x,y,z)\\ &= \hat{\sigma}_{xy}\psi(x,-y,z) = \psi(x,-y,-z)\\ &= \hat{C}_2(x)\psi(x,y,z)\\ \hat{\sigma}_{xy}\hat{C}_2(z)\hat{\sigma}_{yz} &= \hat{C}_2(x).\end{aligned}$$

Therefore, a point group that contains $\hat{\sigma}_{xy}$, $\hat{C}_2(z)$, and $\hat{\sigma}_{yz}$, must also contain $\hat{C}_2(x)$.

Worked Examples provide students with context of the problem, clearly describe the parameters of the problem, and walk students step-by-step toward the solution.

KEY CONCEPTS AND EQUATIONS

KEY TERMS

OBJECTIVES REVIEW

PROBLEMS

End-of-chapter materials bring students full circle, helping them assess their grasp of current chapter concepts and synthesize information from prior chapters.

A **comprehensive online solutions manual**, written by author Andrew Cooksy, is filled with unique solution sets emphasizing qualitative results to help students move beyond the math to a deeper conceptual understanding.

MasteringChemistry® for Students

www.masteringchemistry.com

MasteringChemistry provides dynamic, engaging experiences that personalize and activate learning for each student. Research shows that Mastering's immediate feedback and tutorial assistance helps students understand and master concepts and skills—allowing them to retain more knowledge and perform better in this course and beyond.

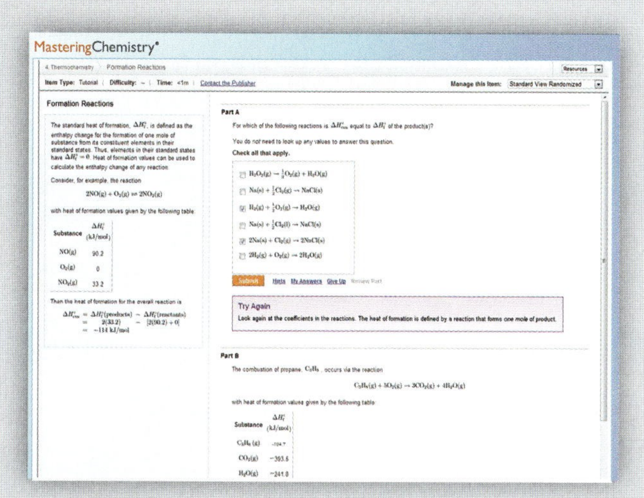

Student Tutorials

Physical chemistry tutorials reinforce conceptual understanding. Over 460 tutorials are available in MasteringChemistry for Physical Chemistry, including new ones on The Cyclic Rule and Thermodynamic Relation of Proofs.

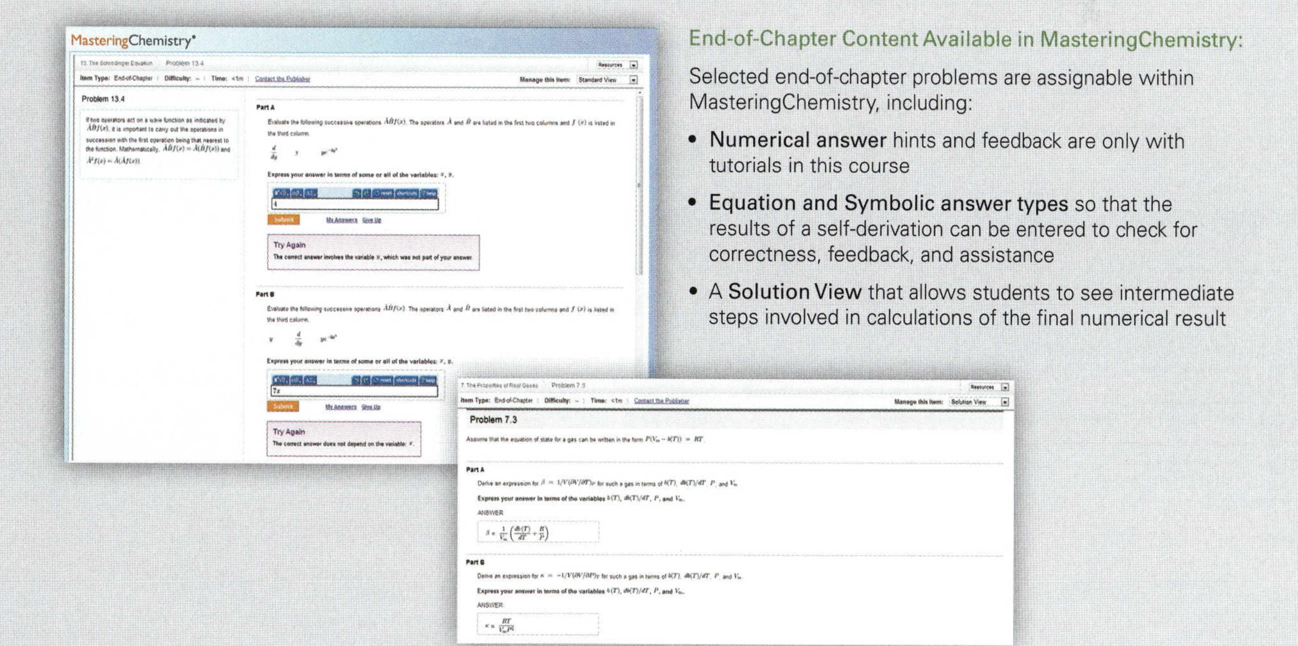

End-of-Chapter Content Available in MasteringChemistry:

Selected end-of-chapter problems are assignable within MasteringChemistry, including:

- **Numerical answer** hints and feedback are only with tutorials in this course
- **Equation and Symbolic answer types** so that the results of a self-derivation can be entered to check for correctness, feedback, and assistance
- A **Solution View** that allows students to see intermediate steps involved in calculations of the final numerical result

MasteringChemistry® for Instructors

www.masteringchemistry.com

Easy to get started. Easy to use.

MasteringChemistry provides a rich and flexible set of course materials to get you started quickly, including homework, tutorial, and assessment tools that you can use *as is* or customize to fit your needs.

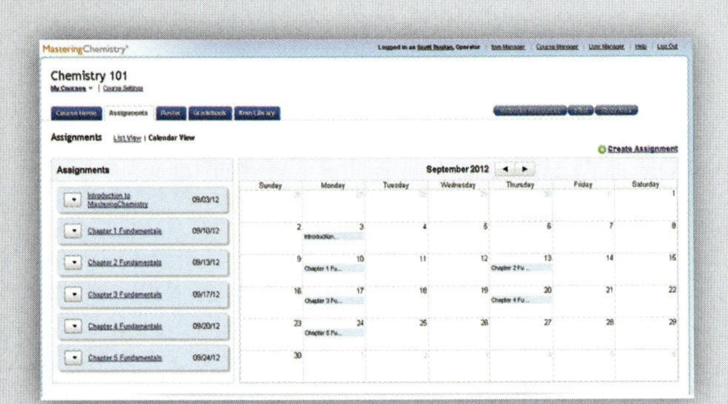

NEW! Calendar Features

The Course Home default page now features a **Calendar View** displaying upcoming assignments and due dates.

- Instructors can schedule assignments by dragging and dropping the assignment onto a date in the calendar. If the due date of an assignment needs to change, instructors can drag the assignment to the new due date and change the "available from and to dates" accordingly.
- The calendar view gives students a syllabus-style overview of due dates, making it easy to see all assignments due in a given month.

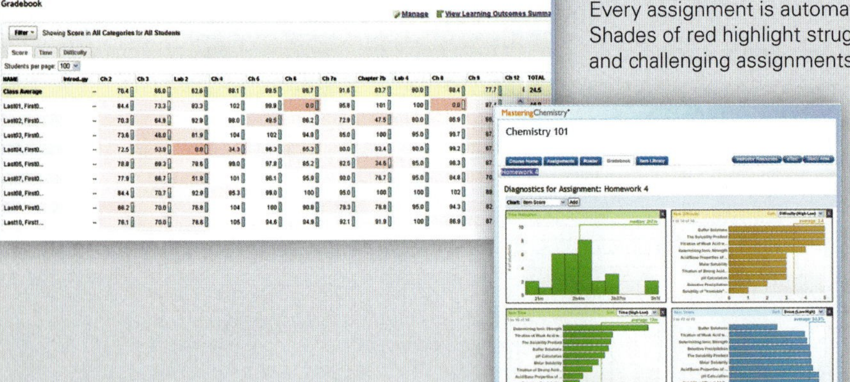

Gradebook

Every assignment is automatically graded. Shades of red highlight struggling students and challenging assignments at a glance.

Gradebook Diagnostics

This screen provides you with your favorite diagnostics. With a single click, charts summarize the most difficult problems, vulnerable students, grade distribution, and even score improvement over the course.

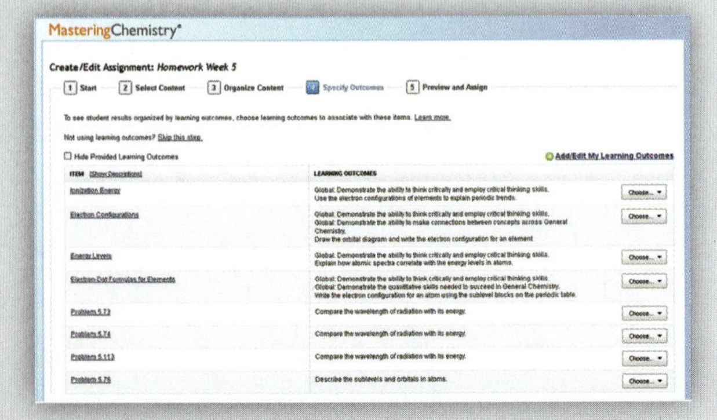

NEW! Learning Outcomes

Let Mastering do the work in tracking student performance against your learning outcomes:

- Add your own or use the publisher-provided learning outcomes.
- View class performance against the specified learning outcomes.
- Export results to a spreadsheet that you can further customize and share with your chair, dean, administrator, or accreditation board.

Introduction: Tools from Math and Physics

GOAL ***Why Are We Here?***

The goal of this textbook is a concise and elegant exposition of the theoretical framework that forms the basis for all modern chemistry. To accomplish this, we are going to draw regularly on your knowledge of algebra, geometry, calculus, mechanics, electromagnetism, and chemistry. Physical chemistry is both rewarding and challenging in this way.

Mathematics of several varieties is our most valuable tool, and in this text we shall be interested in it only as a tool. It is not necessary, for example, that you remember how to derive the algebraic solution to the integral $\int \ln x \, dx$, but it will help if you know that an algebraic solution exists and how to use it (because with it we will obtain a useful equation for diffusion).This chapter is a summary of the math and physics that serve as our starting point as we explore the theory of chemistry. If you are embarking on this course, you may wish to review any of the following topics that appear alarmingly unfamiliar at first glance.

A.1 Mathematics

Algebra and Units

Basic Formula Manipulations

The use of algebra in this text is similar to its use in introductory physics and chemistry courses. We will routinely encounter the basic manipulations of variables in equations, especially to solve for one unknown in terms of several known constants. A tough example would be to solve for n_B in the equation

$$T_B = T'_B \left[\frac{V_T - V_A}{V_T - V'_A} \right]^{-n_B R/C_B}$$

The key is to see that a solution must be available, because the variable we are solving for appears in only one place, and a series of operations will allow us to isolate it on one side of the equation. Once we recognize that, then we can methodically undo the operations on one side of the equation to leave n_B: divide both sides by T'_B, take the logarithm of both sides to bring n_B down to earth from the exponent, and finally divide both sides by the factor that leaves n_B alone on one side of the equation. Those steps eventually bring us to

$$n_B = -\frac{C_B}{R}\frac{\ln\left(\frac{T_B}{T'_B}\right)}{\ln\left(\frac{V_T - V_A}{V_T - V'_A}\right)}.$$

One issue that makes the algebra something of a challenge is the notation. To put it mildly, we will use a lot of algebraic symbols. In fact, with the exception of "O," which looks too much like a zero, we use the entire Roman alphabet at least twice, and most of the Greek.[1] The symbols have been chosen in hopes of an optimal combination of (a) preventing the same symbol from appearing with different meanings in the same chapter, (b) adherence to the conventional usage in the scientific literature, and (c) clarity of meaning. Unfortunately, these three aims cannot always be satisfied simultaneously. Physical chemistry is a synthesis of work done by pioneers in mathematics, physics, and chemistry, often without any intention that the results would one day become integrated into a general theory of chemistry. We bring together many fields that evolved independently, and the way these fields fit together is one of the joys of this course. Admittedly, the complexity of the notation is not.

The text provides guides to the notation used in long derivations and sample calculations to show how the notation is used. Please be aware, however, that no textbook gimmick can substitute for the reader's understanding of the parameters represented by these symbols. If you recognize the difference between the fundamental charge e and the base of the natural logarithm e, you are in no danger of confusing the two, even though they are both represented by the letter "e," sometimes appearing in the same equation.

Unit Analysis and Reasonable Answers

One of the most helpful tools for checking algebra and for keeping these many symbols under control is unit analysis. If a problem asks you to solve for the value of some variable Υ, and you're not certain what units you will get in the end, then it's likely that the meaning of Υ has not been made entirely clear. In many cases, including viscosities and wavefunctions, the units are not obvious from the variable's definition in words but are easily determined from an important equation in which the variable appears. Quick: how do you write the units for pressure in terms of mass and distance and time? If you recall the definition of the pressure as force per unit area

$$P = \frac{F}{A}$$

[1] If the lower case Greek letter upsilon (υ) didn't look so much like an italic "v" (v), there are at least two places it would have been used. It's bad enough that v and the Greek nu (ν) are so similar and sometimes appear in the same equation.

and know that force has units of mass times acceleration, then pressure must have units of

$$\frac{\text{force}}{\text{distance}^2} = \frac{\text{mass} \times \text{speed/time}}{\text{distance}^2} = \frac{\text{mass} \times \text{distance/time}^2}{\text{distance}^2}$$

$$= \frac{\text{mass}}{\text{distance} \times \text{time}^2} = \text{kg m}^{-1}\,\text{s}^{-2}. \quad \text{(A.1)}$$

It will not be worthwhile to attempt a problem before understanding the variables involved.

Unit analysis is also a useful guard against algebraic mistakes. An error in setting up an algebraic solution often changes the units of the answer, and a check of the answer's units will show the mistake. This does not protect against many other mistakes, however, such as dividing instead of multiplying by 10^{10} to convert a length from meters to angstroms. In such cases, there is no replacement for knowing what range of values is appropriate for the quantity. Recognizing a reasonable value for a particular variable is primarily a matter of familiarity with some typical parameters. The values given in Table A.1 are meant only to give common orders of magnitude for various quantities. Answers differing by factors of 10 from these may be possible, but not common.

TABLE A.1 Some typical values for parameters in chemical problems. These are meant only as a rough guide to expected values under typical conditions.

Parameter	Value (in typical units)
chemical bond length	1.5 Å
chemical bond energy	400 kJ mol^{-1}
molecular speed	200 m s^{-1}
mass density (solid or liquid)	1 g cm^{-3}

EXAMPLE A.1 Unreasonable Answers

PROBLEM Unit analysis and recognition of a reasonable value can prevent errors such as those that resulted in the following answers. Identify the problem with these results for the requested quantity:

Quantity	Wrong answer
the density of NaCl(*s*)	$1.3 \cdot 10^{-24}$ g cm^{-3}
the density of NaCl(*s*)	$3.3 \cdot 10^{7}$ g cm^{-1}
bond length of CsI	12.3 m
speed of a molecule	$4.55 \cdot 10^{11}$ m s^{-1}
momentum of electron	$5 \cdot 10^{-10}$ m s^{-1}

SOLUTION Each of those examples gives an answer of entirely the wrong magnitude (which could arise from using the wrong conversion factor, the wrong units, or both).

Quantity	Wrong answer	Why unreasonable
the density of NaCl(s)	$1.3 \cdot 10^{-24}$ g cm^{-3}	too small
the density of NaCl(s)	$3.3 \cdot 10^{7}$ g cm^{-1}	wrong units
bond length of CsI	12.3 m	too big
speed of a molecule	$4.55 \cdot 10^{11}$ m s^{-1}	too big (greater than speed of light)
momentum of electron	$5 \cdot 10^{-10}$ m s^{-1}	wrong units

In many problems, the units themselves require some algebraic manipulation because several units are products of other units. For example, the unit of pressure, 1 kg m^{-1} s^{-2}, obtained in Eq. A.1, is called the "pascal." We shall also encounter an equation

$$E_n = -\frac{Z^2 m_e e^4}{2(4\pi\varepsilon_0)^2 n^2 \hbar^2},$$

in which E_n has units of energy, Z and n are unitless, m_e has units of mass, e has units of charge, ε_0 has units of charge^2 energy^{-1} distance^{-1}, and $\hbar$ has units of energy $\times$ time. The units on each side of the equation must be identical, and this we can show by substituting in the appropriate units for mass, charge, and energy:

$$\begin{aligned} 1\,\text{J} &= 1\frac{(\text{kg})(\text{C})^4}{(\text{C}^2\,\text{J}^{-1}\,\text{m}^{-1})^2(\text{Js})^2} \\ &= 1\frac{(\text{kg})(\text{C})^4}{\text{C}^4\,\text{s}^2/\text{m}^2} \\ &= 1\,\text{kg}\,\text{m}^2\,\text{s}^{-2} = 1\,\text{J}. \end{aligned} \tag{A.2}$$

This may be a good place to remind you about that bothersome factor of $4\pi\varepsilon_0$ and some other aspects of the SI units convention.

SI Units

The accepted standard for units in the scientific literature is the Système International (SI), based on the meter, kilogram, second, coulomb, kelvin, mole, and candela.[2] It is acceptable SI practice to use combinations of these units and to convert up or down by factors of 1000. So, for example, the SI unit of force should have units of (mass $\times$ acceleration), or kg m s^{-2}, a unit commonly called the newton and abbreviated N. Energy has units of force $\times$ distance, so the SI unit is kg m^2 s^{-2}, also called the joule and abbreviated J. But the joule is inconveniently small for measuring, say, the energy released in a chemical reaction, so one could use the kilojoule (10^3 J) and remain true to the SI standard. We'll give special attention to energy units shortly.

A practical advantage of a single system for all physical units is that—if you're careful—the units take care of themselves. Allowing for the factors of 1000, if all the quantities on one side of an equation are in SI units, the value

[2]If you don't recall the candela, that's understandable. It's the unit of luminous intensity, and with that, makes its last appearance in this text.

on the other side will also be in SI units. If an object of mass 2.0 kg rests on a table, subject to the gravitational acceleration of 9.8 m s^{-2}, then I can calculate the force it exerts on the table by multiplying the mass and the acceleration,

$$F = ma = (2.0\ \mathrm{kg})(9.8\ \mathrm{m\ s^{-2}}) = 20\ \mathrm{N},$$

and I can be certain that the final value is in SI units for force, namely newtons.

Standardization of units takes time, however, and you can be certain that the chemical data you encounter in your career will not adhere to one standard. One formerly common set of units, now widely discouraged, is the **Gaussian** or **CGS system,** similar to SI except that it replaces the meter, kilogram, and coulomb with the centimeter, gram, and electrostatic unit, respectively. Another convention, now on the rise, is the set of atomic units, for which all units are expressed as combinations of fundamental physical constants such as the electron mass m_e and the elementary charge e.

The SI system, while having some features convenient to engineering, suffers from one inconvenience in our applications: elementary calculations that include electric charges or magnetic fields require the use of constants called the permeability μ_0 and permittivity ε_0 of free space. Although these constants originally appeared with a physical meaning attached, for our purposes they are merely conversion factors. In particular, the factor $4\pi\varepsilon_0$ converts SI units of coulomb squared to units of energy times distance, J • m. For example, the energy of repulsion between two electrons at a separation of $d = 1.0 \cdot 10^{-10}$ m is

$$\frac{e^2}{4\pi\varepsilon_0 d} = \frac{(1.602 \cdot 10^{-19}\ \mathrm{C})^2}{(1.113 \cdot 10^{-10}\ \mathrm{C^2\ J^{-1}\ m^{-1}})(1.0 \cdot 10^{-10}\ \mathrm{m})} = 2.306 \cdot 10^{-18}\ \mathrm{J}. \quad (A.3)$$

In contrast, the atomic and CGS units fold this conversion into the definition of the charge, and the factor of $4\pi\varepsilon_0$ would *not* appear in the calculation. For all equations in this text involving the forces between charged particles, we conform to the standards of the day and use SI units and the associated factor of $4\pi\varepsilon_0$.

In other cases, however, we will not adhere strictly to the SI standard. Even allowing for factors of 1000, I don't know any chemists who express molecular dipole moments in coulomb meters, a unit too large for its purpose by 30 orders of magnitude (not even prefixes like "micro-" and "nano-" are enough to save it). The conventional unit remains the debye, which is derived from CGS units (adjusted by 18 orders of magnitude, it must be said) and just the right size for measuring typical bond dipoles. The angstrom (Å) also remains in wide use in chemistry because it is a metric unit (1 Å = 10^{-10} m) that falls within a factor of 2 of almost any chemical bond length.

Of all the physical parameters, energy has the greatest diversity in commonly used scientific units. There are several ways to express energy, even after excluding all sorts of nonmetric energy units (such as the British thermal unit, kilowatt-hour, foot-pound, ton of TNT, and—most beloved of chemists—the calorie). Other conventions appear when discussing the interaction of radiation with matter, for which it is common to quantify energy in terms of the frequency (s^{-1}) or reciprocal wavelength (cm^{-1}) of the radiation. Under the proper assumptions, it may also be informative to convert an energy to a corresponding temperature, in units of kelvin. Typical laboratory samples of a compound have numbers of molecules in the range of 10^{20} or more, and molecular energies are therefore often given in terms of the energy per mole of the compound (e.g., $\mathrm{kJ\,mol^{-1}}$). These cases will be explained as they appear, and they are summarized in the conversion table for energies on this textbook's back endpapers.

Once these non-SI units are introduced, please make sure you are comfortable with the algebra needed to convert from one set of units to another. This one skill, mundane as it may seem, will likely be demanded of you in any career in science or engineering. Famous and costly accidents have occurred because this routine procedure was not given its due attention.[3]

Complex Numbers

Complex numbers are composed of a real number and an imaginary number added together. For our purposes, a complex number serves as a sort of two-dimensional number; the imaginary part contains data on a measurement distinct from the data given by the real part. For example, a sinusoidal wave that varies in time may be described by a complex number in which the real part gives the shape of the wave at the current time and the imaginary part describes what the wave will look like a short time later.

The imaginary part of any complex number is a real number multiplied by $i \equiv \sqrt{-1}$. (The symbol "≡" is used throughout this text to indicate a definition, as opposed to the "=" symbol, used for equalities that can be proved mathematically.) This relationship between i and -1 allows the imaginary part of a complex number to influence the real-number results of an algebraic operation. For example, if a and b are both real numbers, then $a + ib$ is complex, with a the real part and ib the imaginary part. The **complex conjugate** of $a + ib$, written $(a + ib)^*$, is equal to $a - ib$, and the product of any number with its complex conjugate is a real number:

$$(a + ib)(a - ib) = a^2 - iba + iba - i^2b^2 = a^2 + b^2. \tag{A.4}$$

Notice that the value of b—even though it was contained entirely in the imaginary parts of the two original complex numbers—contributes to the value of the real number quantity that results from this operation.

Many of the mathematical functions in the text are complex, but multiplication by the complex conjugate yields a real function, which can correspond directly to a measurable property. For that reason, we often judge the validity of the functions by whether we can integrate over the product f^*f. In this text, a well-behaved function f is single-valued, finite at all points, and yields a finite value when f^*f is integrated over all points in space. To be very well-behaved, the function and its derivatives should also be continuous functions, but we will use a few functions that are naughty in this regard.

---

[3] A prominent example is the loss in 1999 of the unmanned Mars Climate Orbiter, a probe that entered the Martian atmosphere too low and burned up because engineers were sending course correction data calculated using forces in pounds to an on-board system that was designed to accept the data in newtons.

EXAMPLE A.2 Complex Conjugates

PROBLEM Write the complex conjugate f^* for each of the following expressions f and show that the value of f^*f is real.

1. $5 + 5i$
2. $-x/i$
3. $\cos x - i\sin x$

SOLUTION

1. $f^* = 5 - 5i$

$$f^*f = (5 + 5i)(5 - 5i) = 25 + 25 = 50$$

2. First we would like to put this in the form $a + ib$, so we multiply by $\frac{i}{i}$ to bring the factor of i into the numerator:

$$f = -\frac{x}{i}\left(\frac{i}{i}\right) = -\frac{ix}{-1} = ix.$$

The real part of this function is zero, but for any complex conjugate, we change the sign on the imaginary term: $f^* = -ix$

$$f^*f = (ix)(-ix) = -i^2x^2 = x^2$$

3. $f^* = \cos x + i\sin x$

$$f^*f = \cos^2 x - i^2\sin^2 x = \cos^2 x + \sin^2 x = 1$$

Trigonometry

Elementary results from trigonometry play an important role in our equations of motion, and therefore you should know the definitions of the sine, cosine, and tangent functions (and their inverses) as signed ratios of the lengths of the sides of a right triangle. Using the triangle drawn in Fig. A.1, with sides of length y, x, and r, we would define these functions as follows:

$$\begin{aligned} \sin\phi &\equiv \frac{y}{r} & \csc\phi &\equiv \frac{1}{\sin\phi} = \frac{r}{y} \\ \cos\phi &\equiv \frac{x}{r} & \sec\phi &\equiv \frac{1}{\cos\phi} = \frac{r}{x} \\ \tan\phi &\equiv \frac{y}{x} & \cot\phi &\equiv \frac{1}{\tan\phi} = \frac{x}{y} \end{aligned} \tag{A.5}$$

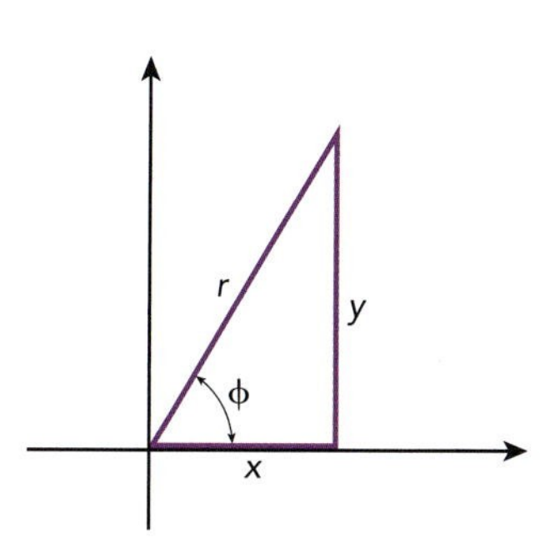

▲ FIGURE A.1 **Right triangle used to define trigonometric functions of the angle ϕ.**

The sign is important. If ϕ lies between 90° and 270°, then the x value becomes negative, so $\cos\phi$ and $\sec\phi$ would be less than zero. Similarly, $\sin\phi$ and $\csc\phi$ are negative for ϕ between 180° and 360°.

Please also make sure you are comfortable using the trigonometric identities listed in Table A.2. These are algebraic manipulations that may allow us to simplify equations or to isolate an unknown variable.

TABLE A.2 Selected trigonometric identities.

$\sin^2 x + \cos^2 x = 1$	$\sec^2 x - \tan^2 x = 1$
$\sin(x \pm y) = \sin x \cos y \pm \cos x \sin y$	$\cos(x \pm y) = \cos x \cos y \mp \sin x \sin y$
$\sin x \sin y = [\cos(x - y) - \cos(x + y)]/2$	$\cos x \cos y = [\cos(x + y) + \cos(x - y)]/2$
$\sin x \cos y = [\sin(x + y) + \sin(x - y)]/2$	
$\sin 2x = 2 \sin x \cos x$	$\cos 2x = 2\cos^2 x - 1$

Coordinate Systems

Mathematical functions are described by their variables, but we have some choice in deciding what those variables are. Rather than defining the function $f(x) = x^2$ as written, we could define it in terms of a new variable $y = 2x$, for which $f(y) = y^2/4$.

For functions that represent distributions in three-dimensional space, there are two common choices of variables: the **Cartesian coordinates,** (x, y, z); and the **spherical polar coordinates,** (r, θ, ϕ). The Cartesian coordinates can each vary from $-\infty$ to $+\infty$. The polar coordinates lie in the ranges

$$0 \leq r < \infty \qquad 0 \leq \theta < \pi \qquad 0 \leq \phi < 2\pi,$$

where π **radians** is equal to 180°, and the radian is the ratio of a circle's circumference to its diameter. Usually when we move between the two systems, we will take the angle θ as measured in any direction from the positive half of the z axis, and the angle ϕ as the angle measured parallel to the xy plane from the positive x axis towards the positive y axis. The distance r is always measured in any direction from the origin. These definitions are illustrated in Fig. A.2.

The Cartesian and spherical polar coordinate systems satisfy the fundamental requirements for a complete coordinate system in three-dimensional space—namely, that every point in space can be represented by some set of values for these coordinates, and every set of coordinates corresponds to only one point in space. Although the Cartesian coordinate representation of a single point may be easier for us to visualize than the representation in spherical coordinates, functions that have a lot of angular symmetry can be written and manipulated much more easily in spherical coordinates than in Cartesian coordinates.

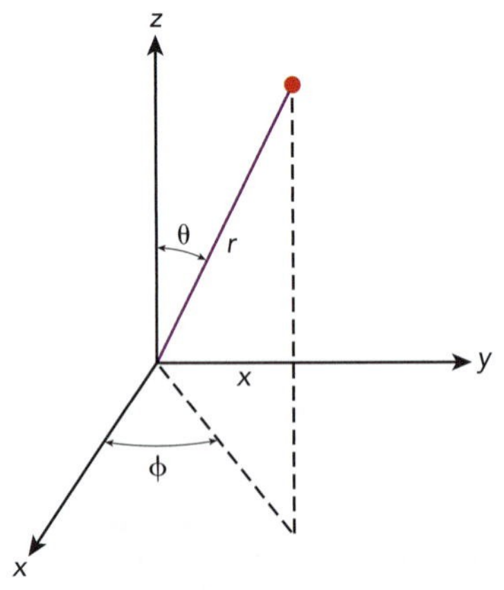

▲ **FIGURE A.2 The relation between spherical polar and Cartesian coordinates.**

Converting between Cartesian and spherical coordinates is straightforward but often tedious. The most crucial conversions between Cartesian and spherical coordinates have been done for us by someone else, and we should not be too shy to take advantage of all that hard work. Should it be necessary to convert between the two systems for a particular application, the following equations can be used:

$$\begin{aligned} x &= r \sin\theta \cos\phi & r &= (x^2 + y^2 + z^2)^{1/2} \\ y &= r \sin\theta \sin\phi & \theta &= \arccos\left(\frac{z}{r}\right) \\ z &= r \cos\theta & \phi &= \arctan\left(\frac{y}{x}\right) \end{aligned} \tag{A.6}$$

The most important conversion we will need is between the **volume elements,** abbreviated $d\tau$, that appear in all integrals. The volume element is so named because its integral, evaluated over some three-dimensional region, is the volume enclosed by that region. For an integral over three-dimensional space, the volume element is

$$d\tau \equiv dxdydz = r^2 dr \sin\theta \, d\theta \, d\phi. \tag{A.7}$$

Although this equation is not obvious at first glance, we can observe easily that $d\tau$ has units of volume as promised. For the Cartesian volume element, $dxdydz$ is the volume of a cube with sides of length dx, dy, and dz and has units of volume. The only spherical coordinate with units of distance, r, appears three times in the spherical volume element: twice in r^2 and once in dr (which has the same units as r), giving units of distance^3 or volume. The remaining terms, $\sin\theta \, d\theta \, d\phi$, are unitless.

EXAMPLE A.3 Cartesian and Polar Coordinates

PROBLEM Convert the following Cartesian expression into spherical coordinates and the spherical polar expression into Cartesian coordinates.

$$f(x,y,z) = ze^{-(x^2+y^2+z^2)/a^2}$$
$$g(r,\theta,\phi) = (3\cos^2\theta - 1)\tan\phi$$

SOLUTION We can directly substitute using the expressions in Eqs. A.6:

$$f(r,\theta,\phi) = (r\cos\theta)e^{-r^2/a^2}$$
$$g(x,y,z) = \left[3\left(\frac{z}{r}\right)^2 - 1\right]\tan\left[\arctan\frac{y}{x}\right]$$
$$= \left[3\left(\frac{z}{r}\right)^2 - 1\right]\frac{y}{x}$$

Linear Algebra

Linear algebra is so named because it grew out of methods for solving systems of linear equations. For our purposes, it is the branch of mathematics that describes how to perform arithmetic and algebra using vectors and matrices.

Vectors

Formally, a **vector** is a set of two or more variable values, but our use of the term will be restricted to *Euclidean* vectors, which are governed by the following definitions and rules:

1. A vector has direction, which can be specified by assuming one of the endpoints to be the origin and giving the coordinates of the other endpoint. As an example, the vector $(1,0,0)$ has one end at the origin and the other end at $x = 1$ on the x axis.
2. A vector $\overrightarrow{A} = (A_x, A_y, A_z)$ has a length or **magnitude,** indicated $|\overrightarrow{A}|$ or simply A, where

$$|\overrightarrow{A}| \equiv A = \sqrt{A_x^2 + A_y^2 + A_z^2}. \tag{A.8}$$

3. The **dot product** of two vectors $\vec{A} = (A_x, A_y, A_z)$ and $\vec{B} = (B_x, B_y, B_z)$ is a scalar quantity (*not* a vector) given by
$$\vec{A} \cdot \vec{B} \equiv A_x B_x + A_y B_y + A_z B_z. \tag{A.9}$$
4. The dot product of $\vec{A}$ and a **unit vector** (vector of length one) parallel to $\vec{B}$ is called the **projection** of $\vec{A}$ onto $\vec{B}$; this is often evaluated with $\vec{B}$ chosen to be one of the coordinate axes, such as
$$\vec{A} \cdot \hat{z} = A_z,$$
where $\hat{z} \equiv (0,0,1)$. This quantity gives the extent that the vector $\vec{A}$ stretches along the z direction, and is often called the z **component** of $\vec{A}$.
5. The **cross product** of two vectors is also a vector, given by
$$\vec{A} \times \vec{B} \equiv (A_y B_z - A_z B_y, A_z B_x - A_x B_z, A_x B_y - A_y B_x). \tag{A.10}$$
The cross product $\vec{A} \times \vec{B}$ is always perpendicular to the vectors $\vec{A}$ and $\vec{B}$.
6. The vector sum of $\vec{A}$ and $\vec{B}$ is given by
$$\vec{A} + \vec{B} = (A_x + B_x, A_y + B_y, A_z + B_z) \tag{A.11}$$
and is a vector with maximum magnitude $A + B$ (if the two vectors point in exactly the same direction) and minimum magnitude $|A - B|$ (if they point in exactly opposite directions).

Matrices

Although we will use vectors to represent physical quantities, such as position and angular momentum, to a mathematician a vector is any set of expressions that depend on some index. For example, the position vector $\vec{r}$ is the set of coordinate values r_i, where $r_1 = x$, $r_2 = y$, and $r_3 = z$. In that example, the index i lets us pick out one part of the vector. A **matrix** is a set of values or functions that depend on at least two different (and usually independent) indices. We will not encounter many matrices in this text, but there are a few places where they allow you to go one step farther in calculating important physical quantities in chemistry.

As an example, we may write the matrix **R** of values $r_i r_j$ for each i and j from 1 to 3:
$$\mathbf{R} = \begin{pmatrix} r_1 r_1 & r_1 r_2 & r_1 r_3 \\ r_2 r_1 & r_2 r_2 & r_2 r_3 \\ r_3 r_1 & r_3 r_2 & r_3 r_3 \end{pmatrix} = \begin{pmatrix} x^2 & xy & xz \\ yx & y^2 & yz \\ zx & zy & z^2 \end{pmatrix}.$$
This matrix gives all the possible combinations of x, y, and z with x, y, and z. The matrix **R** would be one short way to represent all the terms that would arise from expanding $(x + y + z)^2$:
$$(x + y + z)^2 = x^2 + y^2 + z^2 + 2xy + 2yz + 2xz.$$
It would also represent them in such a way that we could pick out any one of those terms—any single **matrix element** R_{ij}—by itself from the values of the two indices, as for example $R_{13} = r_1 r_3 = xz$.

There is an algebra for matrices. We can multiply a matrix by a constant:

$$c\begin{pmatrix} f \\ g \end{pmatrix} = \begin{pmatrix} cf \\ cg \end{pmatrix}. \tag{A.12}$$

We can also multiply a matrix and a vector $\overrightarrow{A}$, obtaining a new vector $\overrightarrow{B}$ according to the formula $B_i = \sum_j R_{ij} A_j$. For example, the product of any 2×2 matrix and a 2-coordinate vector is given by

$$\begin{pmatrix} r & s \\ t & u \end{pmatrix}\begin{pmatrix} f \\ g \end{pmatrix} = \begin{pmatrix} rf + sg \\ tf + ug \end{pmatrix}. \tag{A.13}$$

The product is a *new* vector. The multiplication just shown forms the basis for one of the most common applications of matrices in physics: changing a vector from one form to another. For example, start with the vector (a, b, c) where a, b, and c are constants giving the length of the vector along the x, y, and z axes, respectively. Now carry out the following multiplication:

$$\begin{pmatrix} 0 & 1 & 0 \\ 0 & 0 & 1 \\ 1 & 0 & 0 \end{pmatrix}\begin{pmatrix} a \\ b \\ c \end{pmatrix} = \begin{pmatrix} b \\ c \\ a \end{pmatrix}.$$

The result is a new vector with the same magnitude but pointing in a different direction, where a is now the length of the vector along the z axis instead of the x axis, and so on. The vector has been rotated by 90° around all three coordinate axes. What would be an awkward operation to carry out using trigonometry becomes relatively straightforward when we use matrix algebra. This example also illustrates how we can use a matrix to represent mathematically a real physical process, in this case the rotation of an object in space.

A second common application of matrix algebra is to solve a set of equations of the form

$$\begin{aligned} h_{11}ax + h_{12}by &= cax \\ h_{21}ax + h_{22}by &= cby. \end{aligned} \tag{A.14}$$

Here, the h_{ij}'s can be any coefficients, ax and by together form a vector in the xy plane, and c is some unknown constant that we want to find. Using our rules of matrix multiplication, these equations can be written as a single matrix equation:

$$\begin{pmatrix} h_{11} & h_{12} \\ h_{21} & h_{22} \end{pmatrix}\begin{pmatrix} ax \\ by \end{pmatrix} = c\begin{pmatrix} ax \\ by \end{pmatrix}. \tag{A.15}$$

Equation A.15 is an example of an eigenvalue equation, because after multiplying $\begin{pmatrix} ax \\ by \end{pmatrix}$ by $\begin{pmatrix} h_{11} & h_{12} \\ h_{21} & h_{22} \end{pmatrix}$ on the left, we get $\begin{pmatrix} ax \\ by \end{pmatrix}$ multiplied by a constant c on the right. (The eigenvalue equation is discussed in more detail in Section 2.1 of the *Quantum Mechanics* volume.) We can solve for the values of c that make Eq. A.15 true by a convenient feature of matrix algebra.

Say, for example, that we want to find the values of c that solve the two equations

$$\begin{aligned} 2ax + by &= cax \\ ax &= cby \end{aligned}$$

for any given values of a and b. Then, the matrix elements h_{ij} have the values:

$$\begin{pmatrix} h_{11} & h_{12} \\ h_{21} & h_{22} \end{pmatrix} = \begin{pmatrix} 2 & 1 \\ 1 & 0 \end{pmatrix}.$$

Then we find the values of c by **diagonalizing** the matrix. First, subtract the unknown value c from each value h_{ii} (these are the **diagonal elements** of the matrix):

$$\begin{pmatrix} 2-c & 1 \\ 1 & 0-c \end{pmatrix}.$$

Next, take the **determinant** of the matrix and set it equal to zero. The determinant is an algebraic combination of all the elements in a square matrix, with the following formulas for 2 × 2 and 3 × 3 matrices:

$$\begin{vmatrix} r & s \\ t & u \end{vmatrix} = ru - st \tag{A.16}$$

$$\begin{vmatrix} r & s & t \\ u & v & w \\ x & y & z \end{vmatrix} = r\begin{vmatrix} v & w \\ y & z \end{vmatrix} + s\begin{vmatrix} w & u \\ z & x \end{vmatrix} + t\begin{vmatrix} u & v \\ x & y \end{vmatrix} = rvz + swx + tuy - rwy - suz - tvx. \tag{A.17}$$

Using the 2 × 2 case, the determinant we need to set to zero in our example is

$$\begin{vmatrix} 2-c & 1 \\ 1 & -c \end{vmatrix} = (2 - c)(-c) - (1)(1) = c^2 - 2c - 1 = 0.$$

Solving for c with the quadratic formula, we obtain two solutions:

$$c = \frac{1}{2}[2 \pm \sqrt{4 + 4}] = 1 \pm \sqrt{2}.$$

There are two valid solutions to Eq. A.15, corresponding to the + and − signs. To show that they are solutions, substitute each result for c in Eqs. A.14:

$$\begin{aligned} 2ax + by &= (1 \pm \sqrt{2})ax && h_{11}ax + h_{12}by = cax \\ ax + (0)by &= (1 \pm \sqrt{2})\, by && h_{21}ax + h_{22}by = cby \\ ax &= (1 \pm \sqrt{2})by && \text{solve for } ax \\ 2(1 \pm \sqrt{2})by + by &= (1 \pm \sqrt{2})^2 by && \text{replace } ax \\ (3 \pm 2\sqrt{2})by &= (3 \pm 2\sqrt{2})by. \end{aligned}$$

The same method can be used to solve any number of related equations simultaneously, boiling the problem down to a single step: diagonalizing the matrix. Consequently, matrix diagonalization routines comprise a key element in computer programs designed to solve problems and simulate processes in virtually every realm of chemistry and physics.

Differential and Integral Calculus

If, like many of your classmates, you enjoyed everything about organic chemistry except its neglect of your calculus skills, rest assured that we won't make the same mistake in physical chemistry. Much of the problem-solving ahead of us involves taking a process that we understand on a tiny scale and expanding that description to a larger scale. That tiny-scale understanding will often be phrased mathematically using **derivatives,** which are an idealized version of how a property—such as electron position or chemical concentration—changes over a small step. Change makes *everything* interesting: how the colors of the leaves change with time, how the climate changes the closer we get to the coast, and how the taste of ice cream changes with the amount of vanilla added. For another example, we describe the interactions between particles in terms of the forces they exert on one another. Force is proportional to an acceleration, and acceleration is the derivative of the velocity with respect to time. A force describes where a particle is going to move right now. If we want to see a bigger picture, we can undo the derivative with **integration** and extract from the force law an idea of where the particle will be at different times. The force itself is a derivative (with respect to distance) of the energy, and integrating the force over distances can tell us how the energy of a system varies at different locations.

Another form of this extension from small scale to large scale requires us to calculate sums and averages—which are convenient ways to describe huge systems—from functions too detailed to bear patiently. For example, an understanding of the small-scale interaction between molecules and gravity leads us to predict that air is denser near sea level than at high altitudes. A clever equation even tells us how the air density varies with altitude. By integrating this equation over all altitudes, we can find the total amount of air present and drop all the information about the detailed interactions. It is this general approach of extrapolating from small to large that makes a journeyman command of calculus essential for the text.

Derivatives

Solutions to some standard derivatives appear in Table A.3. It does not hurt to know how to obtain derivatives and integrals, but we will be treating these aspects of calculus as just another kind of algebra. In other words, one may replace the derivative or integral expression by the correct algebraic expression, with the appropriate substitutions. This will suffice for almost all the calculus we encounter in the text.

When a function depends on more than one variable, then the derivative of the function with respect to one variable generally depends on the other variables as well. As one example, suppose that we have a variable P that depends on three other variables n, T, and V, and a constant R, such that

$$P = \frac{nRT}{V}.$$

TABLE A.3 Solutions to selected derivatives.

dx^c	$= cx^{c-1}dx\,(c \neq 0)$	$d(cx)$	$= cdx$
$d\ln x$	$= \frac{1}{x}dx$	de^x	$= e^x dx$
$d\sin x$	$= \cos x\,dx$	$d\cos x$	$= -\sin x\,dx$
$d[f(x)+g(x)]$	$= d[f(x)]+d[g(x)]$	$d[f(x)g(x)]$	$= f(x)d[g(x)]+g(x)d[f(x)]$
		$d[f(x)/g(x)]$	$= \frac{f(x)d[g(x)]-g(x)d[f(x)]}{g(x)^2}$

Then the derivative of P is related to the derivatives of the three variables, because small changes in n, in T, and in V will each contribute to the overall change in P. In general, derivatives of multivariable functions require knowing how all the variables depend on each other. In these instances, we will use the **partial derivative,** represented by the symbol ∂, which is simply the derivative of the function with respect to one variable *treating all the other variables as though they are constants.* The expression

$$\left(\frac{\partial P}{\partial V}\right)_{n,T}$$

represents the partial derivative of P with respect to V, treating n and T as though they were constants, just like R. Using the partial derivative, the total derivative of P may be written as a sum over the derivatives of the variables:

$$dP = \left(\frac{\partial P}{\partial n}\right)_{T,V} dn + \left(\frac{\partial P}{\partial T}\right)_{n,V} dT + \left(\frac{\partial P}{\partial V}\right)_{n,T} dV = \frac{RT}{V}dn + \frac{nR}{V}dT - \frac{nRT}{V^2}dV. \tag{A.18}$$

In the third partial derivative, for example, the variables n and T are treated as constants and factored out of the derivative. Hence the partial derivative simplifies to

$$\left(\frac{\partial P}{\partial V}\right)_{n,T} = \left(\frac{\partial(nRT/V)}{\partial V}\right)_{n,T} = nRT\left(\frac{\partial(1/V)}{\partial V}\right)_{n,T} = -\frac{nRT}{V^2}. \tag{A.19}$$

Table A.4 contains some useful relations involving partial derivatives.

Analytical Integrals

Please make sure that you understand the following terminology regarding integrals:

1. A **definite integral** is evaluated between **limits,** the quantities a and b in the expression $\int_a^b f(x)\,dx$. The integration of $f(x)$ in this case is only carried out from $x = a$ to $x = b$.
2. When the limits are not specified, the integral is an **indefinite integral.** The derivative of a constant C is zero. Therefore, when we take the derivative of a function $f(x) = g(x) + C$, all the information about

TABLE A.4 Relations involving partial derivatives.

reciprocal rule	$\left(\frac{\partial x}{\partial y}\right)_z \left(\frac{\partial y}{\partial x}\right)_z$	$= 1$
slope rule	$dz(x, y)$	$= \left(\frac{\partial z}{\partial x}\right)_y dx + \left(\frac{\partial z}{\partial y}\right)_x dy$
cyclic rule	$\left(\frac{\partial x}{\partial y}\right)_z$	$= -\left(\frac{\partial x}{\partial z}\right)_y \left(\frac{\partial z}{\partial y}\right)_x$
chain rule	$\left(\frac{\partial x}{\partial y}\right)_z$	$= \left(\frac{\partial x}{\partial w}\right)_z \left(\frac{\partial w}{\partial y}\right)_z$
	$\left(\frac{\partial x}{\partial y}\right)_z$	$= \left(\frac{\partial x}{\partial y}\right)_w + \left(\frac{\partial x}{\partial w}\right)_y \left(\frac{\partial w}{\partial y}\right)_z$

TABLE A.5 Solutions to selected integrals. In these equations, a and b are constants, n is a whole number, and C is the constant of integration.

$\int x^n dx$	$= \frac{1}{n+1} x^{n+1} + C$	$\int a dx$	$= a(x + C)$
$\int \frac{1}{x} dx$	$= \ln x + C$	$\int e^x dx$	$= e^x + C$
$\int \ln x dx$	$= x \ln x - x + C$	$\int \frac{dx}{x(a + bx)}$	$= -\frac{1}{a} \ln\left(\frac{a + bx}{x}\right) + C$
$\int \sin x dx$	$= -\cos x + C$	$\int \cos x dx$	$= \sin x + C$
$\int \sin^2(ax) dx$	$= \frac{x}{2} - \frac{\sin(2ax)}{4a} + C$	$\int \cos^2(ax) dx$	$= \frac{x}{2} + \frac{\sin(2ax)}{4a} + C$
$\int [f(x) + g(x)] dx$	$= \int f(x) dx + \int g(x) dx$	$\int_a^b dx$	$= x\vert_a^b = b - a$
$\int_0^\infty x^n e^{-ax} dx$	$= \frac{n!}{a^{n+1}}$	$\int_0^\infty e^{-ax^2} dx$	$= \frac{1}{2}\left(\frac{\pi}{a}\right)^{1/2}$
$\int_0^\infty x e^{-ax^2} dx$	$= \frac{1}{2a}$	$\int_0^\infty x^2 e^{-ax^2} dx$	$= \frac{1}{4}\left(\frac{\pi}{a^3}\right)^{1/2}$
$\int_0^\infty x^{2n+1} e^{-ax^2} dx$	$= \frac{n!}{2a^{n+1}}$	$\int_0^\infty x^{2n} e^{-ax^2} dx$	$= \frac{[1 \cdot 3 \cdot 5 \ldots (2n - 1)]\sqrt{\pi}}{2^{n+1} a^{n+(1/2)}}$
$\int_0^s x^n e^{-ax} dx$	$= \frac{n!}{a^{n+1}} - e^{-as} \sum_{i=0}^{n} \frac{n! s^{n-i}}{a^{i+1}(n - i)!}$		

the value of C is lost. When we undo the derivative by taking the integral, we add an unknown constant of integration to the integrated expression. Omit this constant when solving definite integrals, because the limits of integration will determine its value.

3. The function being integrated is the **integrand,** and it is multiplied by the incremental change along the coordinates, called the volume element.

Most of the algebraic solutions to integrals that we need appear in Table A.5.

EXAMPLE A.4 Analytical Integration

PROBLEM Evaluate the numerical value for each of the following expressions.

1. $\int_1^4 \frac{dx}{x}$
2. $\int_0^\infty e^{-2x} dx$
3. $\int_0^{\pi/3} (3 \cos\theta^2 - 1) \sin\theta \, d\theta$

SOLUTION These can be solved by substitution of the expressions in Table A.5.

(a) $\displaystyle\int_1^4 \frac{dx}{x} = \ln x \Big|_1^4 = \ln 4 - \ln 1 = 1.386 - 0 = 1.386$

(b) $\displaystyle\int_0^\infty e^{-2x}\,dx = -\frac{1}{2}e^{-2x}\Big|_0^\infty = -\frac{1}{2}(e^{-\infty} - e^0) = -\frac{1}{2}(0 - 1) = \frac{1}{2}$

(c) $\displaystyle\int_0^{\pi/3} (3\cos^2\theta - 1)\sin\theta\, d\theta = \left[-\cos^3\theta + \cos\theta\right]\Big|_0^{\pi/3}$

$\displaystyle = \left[-\left(\frac{1}{2}\right)^3 + \left(\frac{1}{2}\right)\right] - [-(1)^3 + (1)] = \frac{3}{8}.$

By the way, it is possible to apply rules of symmetry to extend some of the analytical solutions in Table A.5. For example, when the integrand is $x^{2n}e^{-ax^2}$, then the function is exactly the same from 0 to $-\infty$ as from 0 to $+\infty$ (Fig. A.3a). Therefore, the integral $\int_{-\infty}^{\infty} x^{2n}e^{-ax^2}dx$ is equal to 2 times $\int_0^\infty x^{2n}e^{-ax^2}dx$. However, if the power of x is odd, $2n + 1$, then the function is negative when $x < 0$ and positive when $x > 0$ (Fig. A.3b). The integral from $-\infty$ to 0 cancels the integral from 0 to $+\infty$, so $\int_{-\infty}^{\infty} x^{2n+1}e^{-ax^2}dx = 0$.

Numerical Integration

Not all integrals have algebraic solutions, and some have algebraic solutions only between certain limits (such as 0 and ∞). With suitable computers, any integral can be calculated without trying to cram it into some algebraic form. This is accomplished by going back to the definition in calculus,

$$\int_{x_1}^{x_2} f(x)\,dx = \lim_{\delta x \to 0}\left\{\sum_{i=1}^{N} f[x_1 + i\,\delta x]\right\}\delta x \qquad \text{(A.20)}$$

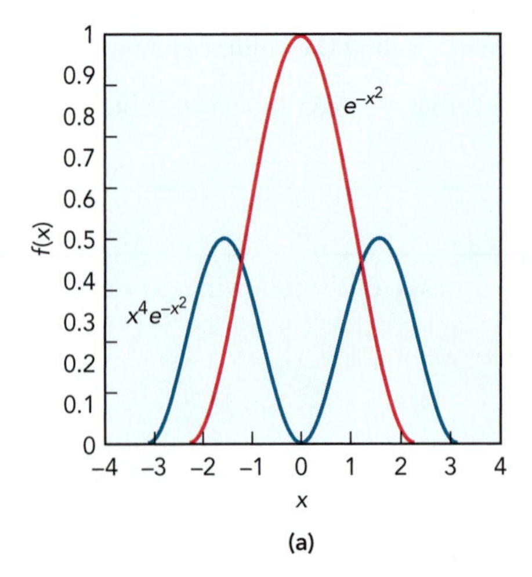

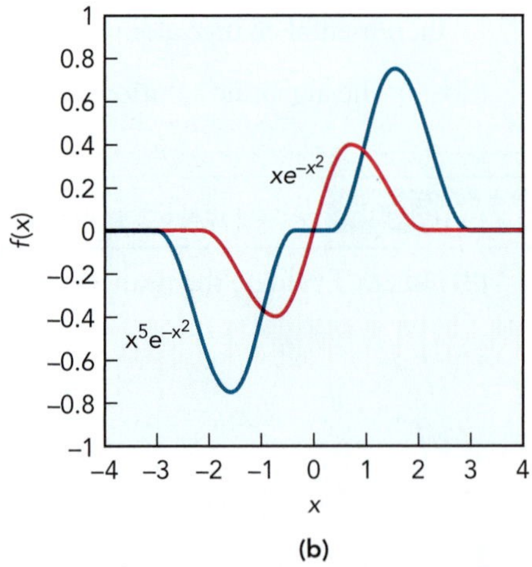

▶ **FIGURE A.3 Symmetry and definite integrals.** **(a)** If a function $f(x)$ is equal to $f(-x)$ for all values of x, then the integral from $-\infty$ to ∞ is equal to 2 times the integral from 0 to ∞. **(b)** If $f(x)$ is equal to $-f(-x)$, then the integral from $-\infty$ to ∞ is 0.

where

$$\delta x = \frac{x_2 - x_1}{N}.$$

One increment of this sum is illustrated in Fig. A.4. If the computer has $f(x)$ stored not as an algebraic expression but as a list of N points, it can carry out that sum very quickly. This is a simple **numerical** (rather than analytical) method of solving the integral. A thousand points ($N = 1000$) would take much less than a second for any modern computer.

It's straightforward to carry out such an integration with a number of elegant programs. Let's say we need to evaluate the integral

$$\int_0^2 f(r)\, r^2\, dr = \int_0^2 \exp(-\pi^{1/3} r^2) r^2\, dr.$$

The appropriate commands for some selected programs are given in Table A.6. For each program, the result is obtained in less time than it takes to enter the command.

The problem is that in many scientific problems we want to integrate a function of several coordinates, and the number of points we need to sample increases as the *power* of the number of coordinates. As an example, in computational chemistry we often integrate an energy equation over the x, y, and z coordinates of every electron in a molecule. If this has to be done numerically, then we may want to sample 10 points along each coordinate. Even if the molecule has only 10 electrons, then a brute force approach would require that we check the value of the function at 10 points along each of 30 coordinates (x, y, and z for each electron), which would require 10^{30} calculations, clearly too demanding even by today's standards. In practice, we reduce the problem to as few coordinates as possible for what we're trying to study, often by factoring the integral into as many independent, low-dimensional integrals as possible. Clever spacing of the points (where the function changes faster, we want more points per unit x) and other tricks can make this even more efficient. The search for more efficient numerical integrators for many-dimensional functions has fueled research projects in mathematics for decades.

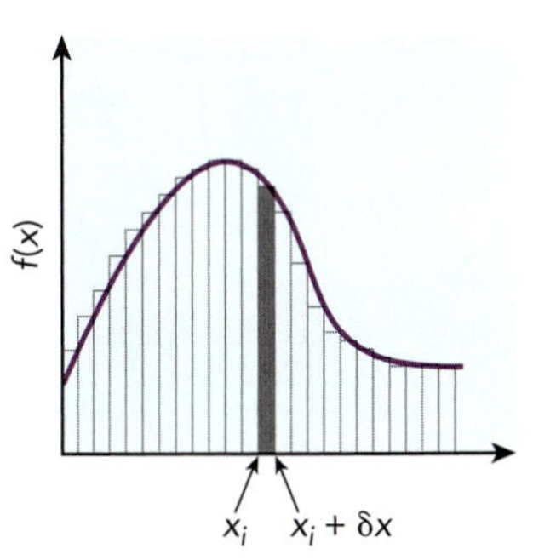

▲ **FIGURE A.4 Numerical integration.** In the simplest numerical integration of a function (solid line), the area of a rectangle of height $f(x_i)$ between two points, x_i and $x_i + \delta x$, is calculated, and the areas of all such adjacent rectangles are summed to obtain an approximate value for the integral.

TABLE A.6 Syntax for numerical integration of the expression $\int_0^2 \exp(-\pi^{1/3} r^2) r^2\, dr$ for common symbolic math programs. The solution is 0.2479.

Program	Command(s)
Maple™	int(exp(-P^(1/3)*r^2)*r^2, r=0..2)
Mathematica®	Integrate[Exp[-Pi^(1/3)*r^2]*r^2, {r,0,2}]
MATLAB®	int(exp(-P^(1/3)*r^2)*r^2, r,0,2)
Octave	function xr = x(r); xr = r^2 * exp(-(pi^(1/3))*r^2); endfunction quad("x",0,2)

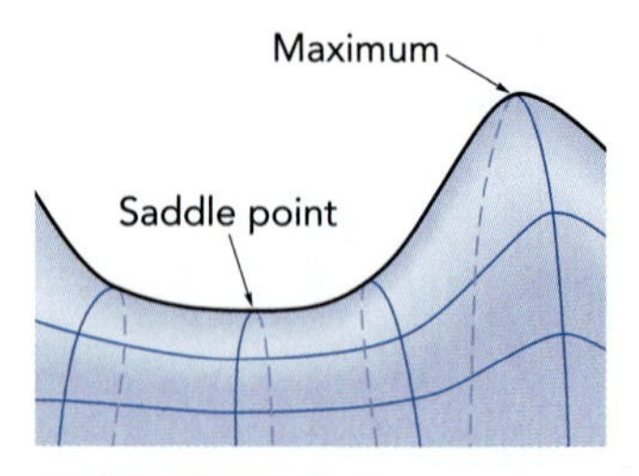

▲ FIGURE A.5 **Stationary points on a function of two variables.**

Volumes and Stationary Points

We are going to consider several different kinds of algebraic functions, including those that describe the energy of the reactants at different stages of a chemical reaction, or the distribution of the electrons in a molecule. These can be so complicated that even the gross features of the function may be difficult to visualize.

One way calculus can help us manage a complicated function is by identifying **stationary points,** regions where the slope of the function is zero. If the function has only one coordinate, such as the time elapsed since a chemical reaction began, then the stationary points are local minima or maxima along the curve. With more dimensions, such as the electron distribution of a molecule in three-dimensional space, the stationary points can also be **saddle points** on the surface of the function, locations where some coordinates reach local maxima while the others are at local minima. Figure A.5 shows a function that could represent the potential energy of a molecule as it changes from one conformation to another, with different stationary points occurring along the surface.

These stationary points are identified by solving for the first derivative of the function and finding the values of the coordinates for which the first derivative—the slope—is zero. For example, $df(x,y,z)/dx$ gives the slope of $f(x,y,z)$ along the x direction. At a minimum, maximum, or saddle point of a three-dimensional function, $df(x,y,z)/dx = df(x,y,z)/dy = df(x,y,z)/dz = 0$.

Another task for which we employ calculus is the computation of areas or volumes contained by functions. These can be obtained by taking the integral of the function between the appropriate limits. Two examples follow.

EXAMPLE A.5 First Derivatives

PROBLEM Find the maximum value of the function $3\cos^2\theta - 1$ where $0 \leq \theta < \pi$.

SOLUTION Take the derivative with respect to the variable, in this case θ:

$$\frac{d}{d\theta}(3\cos^2\theta - 1) = -6\cos\theta\sin\theta.$$

This derivative is zero when either $\sin\theta$ or $\cos\theta$ is zero, in other words at $\theta = 0$ or $\pi/2$. To determine which of these values corresponds to the maximum, we can substitute the two values back into the original function, obtaining

$$3\cos^2(0) - 1 = 3\cdot(1)^2 - 1 = 2,$$

$$3\cos^2(\pi/2) - 1 = 3\cdot(0)^2 - 1 = -1.$$

The maximum value of $3\cos^2\theta - 1$ is 2.

EXAMPLE A.6 Integration

PROBLEM Find the area of a rectangle with length a along the x axis and width b along the y axis.

SOLUTION This can be obtained from the integral

$$\int_0^a f(x)dx = \int_0^a b\,dx = b\int_0^a dx = b(a-0) = ab.$$

Alternatively, one could evaluate the double integral

$$\int_0^a dx \int_0^b dy = (a-0)(b-0) = ab.$$

EXAMPLE A.7 Triple Integrals

PROBLEM Prove the equation for the volume of a sphere of radius R.

SOLUTION In spherical polar coordinates, the volume of a sphere with radius R is found from the triple integral:

$$\int_0^{2\pi} d\phi \int_0^{\pi} \sin\,\theta d\theta \int_0^R r^2\,dr = (\phi)\Big|_0^{2\pi} (-\cos\theta)\Big|_0^{\pi} \frac{1}{3}r^3\Big|_0^R$$

$$= (2\pi - 0)[-(-1) - (-1)]\left(\frac{1}{3}R^3 - 0\right) = \frac{4\pi R^3}{3}.$$

Fourier Transforms

While we're on the subject of integrals, here's a related mathematical manipulation that sees wide application in the sciences, but that you may not have seen before. An integral examines the values of a continuous function at each step along some coordinate, like position or time, and then adds up all the values. The resulting integral doesn't depend on the value of the coordinate any more, because *all* of the coordinate values over the range of integration are used. But if the integrand depends on a second coordinate as well, the integral can also be used to **transform** the function's dependence from one variable to another.

The **Fourier transform** allows us to take any well-behaved function and rewrite it as an integral over sine and cosine functions. If we have a function $f_t(t)$ that varies with (as an example) the time t, then we can rewrite this as a "new" function $f_\omega(\omega)$, where ω is a coordinate with units of $1/t$. The function $f_\omega(\omega)$ is not really a new function but a splitting of the original $f_t(t)$ into many pieces, where each piece is a sine or cosine function with frequency ω.

The simplest case is for a *periodic* function, which starts at $t = 0$, carries on for a period of time τ, and then repeats over and over again. There is *always* a way to write exactly the same function as a sum of sines and cosines:

$$f_t(t) = \frac{a_0}{2} + \sum_{n=1}^{\infty}\left(a_n \cos\frac{2\pi nt}{\tau} + b_n \sin\frac{2\pi nt}{\tau}\right)$$

where n is an integer. The values of the coefficients a_n and b_n are the amplitudes of the cosine and sine functions that we add together to make $f_t(t)$, where each of those functions has an oscillation frequency $\omega = 2\pi n/\tau$. Therefore, we can think of the a_n and b_n coefficients as the values of a function $f_\omega(\omega)$ that tells us how much cosine or sine we must add together to obtain the original $f_t(t)$. The transform is the process of finding the values of those coefficients and, as you were warned, it is an integration:

$$a_n = \frac{2}{\tau}\int_{-\tau/2}^{\tau/2} f_t(t)\cos\frac{2\pi n t}{\tau}\,dt = \frac{2}{\tau}\int_{-\tau/2}^{\tau/2} f_t(t)\cos\omega t\,dt$$

$$b_n = \frac{2}{\tau}\int_{-\tau/2}^{\tau/2} f_t(t)\sin\frac{2\pi n t}{\tau}\,dt = \frac{2}{\tau}\int_{-\tau/2}^{\tau/2} f_t(t)\sin\omega t\,dt.$$

In order to keep track of both the sine and cosine components simultaneously, we can make $f_\omega(\omega)$ a complex function, in which the real part gives the cosine component and the imaginary part gives the sine component:

$$f_\omega(\omega) = a_n + ib_n.$$

For a periodic function, $\omega = 2\pi n/\tau$ is a discrete variable, meaning that it is limited to only certain values. For a non-periodic function, we can still carry out a Fourier transform, but we have to extend the period τ to infinity. As τ becomes infinite, the gaps between values of the frequency ω become smaller and smaller, until ω is a continuous variable, like the time t. In that limit, the Fourier transform is written

$$\mathcal{F}[f_t(t)] \equiv f_\omega(\omega) = \frac{1}{\sqrt{2\pi}}\int_{-\infty}^{\infty} f_t(t)\,[\cos(\omega t) + i\sin(\omega t)]dt. \quad \text{(A.21)}$$

Notice that the original function depended only on t, but integrating over all values of t removes the t-dependence, giving a function that depends only on ω. We can reclaim the original function from its Fourier transform by carrying out the inverse Fourier transform:

$$\mathcal{F}^{-1}[f_\omega(\omega)] = f_t(t) = \frac{1}{\sqrt{2\pi}}\int_{-\infty}^{\infty} f_\omega(\omega)\,[\cos(\omega t) - i\sin(\omega t)]d\omega. \quad \text{(A.22)}$$

In short, this particular Fourier transform shows us how to rewrite a function of *time* as the sum of many sine waves having different *frequencies*. This procedure can be used, for example, to find the frequencies in an audio recording that carry the data we want to keep, and then design an electronic filter that will remove the high-frequency noise we don't want.

Fourier transforms are not limited to conversions between time and frequency. Clever imaging techniques in astronomy and chemistry employ Fourier transforms involving functions of position, for example.

Differential Equations

Equations in which the variables of interest appear with different orders of their derivatives are called **differential equations.** Examples are $x + (dx/dy) = 0$ and $(d^2x/dy^2) + y = 5(dx/dy)$. The solution to these is essentially a problem in integration, and (like integration) need not have an algebraic solution. Some of

EXAMPLE A.8 Fourier Transform

PROBLEM Find the Fourier transform $f_k(k)$ of a step function $f_x(x)$ (Fig. A.6a), for $f_x(x)$ equal to f_0 in the range $-a \le x < a$ and zero everywhere else.

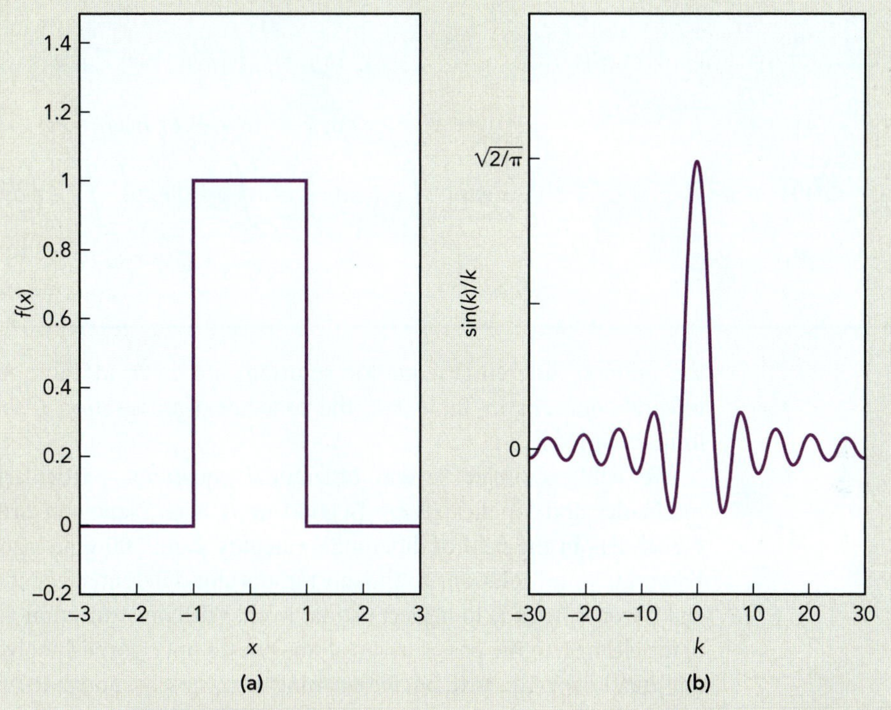

▲ FIGURE A.6 **The step function (a) $f_x(x)$ for f_0 and $a = 1$, and (b) its Fourier transform $f_k(k)$.**

SOLUTION We'll need to use one of the results from Table A.5 and remember that we can break the integral up into different regions:

$$\begin{aligned}
\mathcal{F}[f_x(x)] &= \frac{1}{\sqrt{2\pi}}\int_{-\infty}^{\infty} f_x(x)[\cos(kx) + i\sin(kx)]\,dx \\
&= \frac{1}{\sqrt{2\pi}}\left\{\int_{-\infty}^{-a}(0)\,dx + \int_{-a}^{a} f_0[\cos(kx) + i\sin(kx)]\,dx + \int_{a}^{\infty}(0)\,dx\right\} \\
&= \frac{1}{\sqrt{2\pi}} f_0\left[\frac{1}{k}\sin(kx) - \frac{i}{k}\cos(kx)\right]_{-a}^{a} \\
&= \frac{1}{\sqrt{2\pi}} f_0\left[\frac{2}{k}\sin(ka) - 0\right] \\
&= \frac{1}{\sqrt{2\pi}}\frac{2f_0}{k}\sin(ka) \equiv f_k(k).
\end{aligned}$$

We can plot this new function, $f_k(k)$, versus k and get the graph in Fig. A.6b. This tells us that the step function $f_x(x)$ can be formed by adding together sine waves with wavelengths $1/k$, with the amplitude or height of each sine wave given by the values in Fig. A.6b.

TABLE A.7 **Solutions to selected differential equations.**

Differential equation	Solution
$\frac{dx}{dy} = \frac{g(y)}{f(x)}$	$\int f(x)dx = \int g(y)dy + C$
$\frac{dx}{dy} = h(w), w = x/y$	$\ln y = \int [h(w)-w]^{-1}\, dw + C, h(w) \neq w$
	$\ln x = \ln y + C, h(w) = w$
$\frac{dx}{dy} + g_1(y)x = g_2(y)$	$x \exp\left(\int g_1(y)dy\right) = \int g_2(y) \exp\left(\int g_1(y)dy\right)dy + C$
$\frac{d^2y}{dx^2} = a^2 y$	$y = C_1 e^{ax} + C_2 e^{bx}$

the simplest differential equation solutions are given in Table A.7. As with the integral solutions in Table A.5, the constant of integration C is used when no limits are given.

We will encounter several differential equations, particularly in quantum mechanics and kinetics. There is good news here. Napoleon firmly planted the French flag in the field of differential calculus about 200 years ago. His Académie Française found solutions to the most important differential equations in our field, and it would be rude to neglect all that work. With one exception—the distribution of the electron in the one-electron atom—we do not rigorously solve those differential equations in this text, but we examine them closely enough to find the origins of important physical effects. Some exercises and problems also take advantage of the solutions available in Table A.7 in order to be a little more general or interesting.

In any case, we are not excused from knowing what a differential equation is. If one of our French friends hands us the solution, we should be able to show that it does in fact solve the equation.

EXAMPLE A.9 Differential Equations

PROBLEM Prove that the first three examples in Table A.7 predict the same solution for x at $y = 2$ given the differential equation

$$\frac{dx}{dy} = -\frac{2x}{y},$$

where $x = 2$ at $y = 1$.

SOLUTION Our equation is in the proper form to apply the first solution in Table A.7, with $g(y) = -2/y$ and $f(x) = 1/x$. We rewrite "x" in the integrand as a dummy variable x' to avoid confusing it with x:

$$\int_2^x f(x')dx' = \int_1^2 g(y)dy$$

$$\int_2^x \frac{dx'}{x'} = -2\int_1^2 \frac{dy}{y}$$

$$\ln\left(\frac{x}{2}\right) = -2\ln 2$$
$$x = \frac{1}{2}.$$

To apply the second solution in Table A.7, we would set $h(w) = -2w$ and solve:

$$\ln y\Big|_1^2 = \int_2^{x/2} [h(w) - w]^{-1} dw$$
$$\ln 2 = -\frac{1}{3}\int_2^{x/2} \frac{dw}{w}$$
$$\ln 2 = -\frac{1}{3}\ln\left(\frac{x}{4}\right)$$
$$\ln\left(\frac{1}{8}\right) = \ln\left(\frac{x}{4}\right)$$
$$x = \frac{1}{2}.$$

Note that the limits for the integral over w are the limits of $w = x/y$: 2/1 at $x = 2$ and $y = 1$, and $x/2$ at $y = 2$.

For the third solution in Table A.7, set $g_1(y) = 2/y$ and $g_2(y) = 0$:

$$x\exp\left(\int_1^y g_1(y)dy\right) = \int_1^2 g_2(y)\exp\left(\int_1^2 g_1(y)dy\right)dy + C$$
$$x\exp\left(\int_1^y \frac{2dy}{y}\right) = C$$
$$x\exp(2\ln y) = C$$
$$xy^2 = C \qquad \text{for } x=2, y=1 \text{ we find } C=2$$
$$x(2)^2 = 2 \qquad \text{now set } y=2$$
$$x = \frac{1}{2}.$$

Power Series

On many occasions in the text, we will find it convenient to express a mathematical function as a sum over an infinite number of terms. We shall often employ the **power series** expansion, which has the form

$$\sum_{n=0}^{\infty} a_n x^n = a_0 + a_1 x + a_2 x^2 + a_3 x^3 + \ldots, \tag{A.23}$$

where the a_i's are constants. This is one way of representing an unknown function in some mathematical form on our way to solving its equation.

Another use appears when the later terms in the expansion are much smaller than the leading terms. If x is near 0, then the terms on the right-hand side decrease precipitously as the exponents get larger, and the series eventually converges to some finite value. The advantage to us is that it allows a sophisticated function to be rewritten in a more approachable, but approximate, form. For example, the function $\sin(2x)^{1/2}$ is not easy to integrate on paper, but when x is small, the power series expansion allows this function to be set roughly equal to $1 - x^2$, which integrates easily to $x - (x^3/3) + C$.

More importantly, approximations such as this make the equations more intuitively meaningful, allowing us to better predict how matter will behave *qualitatively.* An approximate equation may sound like a poor replacement for the exact result, but this course is likely to enhance your appreciation of a simple concept over precision. As satisfying as a ten-significant-digit answer may be, the most highly prized advances in physical chemistry are those that reveal a new insight, and insights are hard to gain from final equations that take up a page or more. (Just wait; you'll see for yourself.) Approximations are essential tools in our work ahead, because they provide some algebraic answers that are unattainable otherwise, and they greatly simplify others that would just be too cumbersome to be informative. Many of our approximations will be based on the power series given next.

Trigonometric and Exponential Series

The power series expansion we will use most frequently is the one for the exponential function:

$$e^x = \sum_{n=0}^{\infty} \frac{x^n}{n!} = 1 + x + \frac{x^2}{2} + \frac{x^3}{6} + \frac{x^4}{24} + \ldots. \tag{A.24}$$

This leads to a common approximation when the magnitude of x is small:

$$\text{for } |x| \ll 1: \quad e^x \approx 1 + x. \tag{A.25}$$

Two other important power series are used for the trigonometric functions $\sin x$ and $\cos x$ (with x in units of radians, not degrees):

$$\sin x = \sum_{n=0}^{\infty} (-1)^{n+1} \frac{x^{2n+1}}{(2n+1)!} = x - \frac{x^3}{6} + \frac{x^5}{120} - \ldots \tag{A.26}$$

$$\cos x = \sum_{n=0}^{\infty} (-1)^n \frac{x^{2n}}{(2n)!} = 1 - \frac{x^2}{2} + \frac{x^4}{24} + \ldots \tag{A.27}$$

Of all the basic math equations that we draw on in the text, these must be two of the most amazing. The functions $\sin x$ and $\cos x$ never return values outside the range -1 to 1, no matter how gargantuan a value of x you plug in. To look at the equation for $\sin x$, summing over x and x^3 and x^5 and beyond, it seems ridiculous that the value should ever converge if we set $x = 1001$, for example, but it always does.

Even if you're not impressed by that, notice that these equations prove the small angle approximations

$$\text{for } |x| \ll 1: \quad \sin x \approx x, \quad \cos x \approx 1.$$

The various trigonometric identities, such as $\cos(-x) = \cos x$, can also be tested using these expansions:

$$\cos(-x) = \sum_{n=0}^{\infty} (-1)^n \frac{(-x)^{2n}}{(2n)!}$$

$$= \sum_{n=0}^{\infty}(-1)^n(-1)^{(2n)}\frac{(x)^{2n}}{(2n)!}$$

$$= \sum_{n=0}^{\infty}(-1)^n\frac{(x)^{2n}}{(2n)!} = \cos x.$$

And better still, we can combine the power series for $\sin x$ and $\cos x$ (see Problem A.9) to obtain the Euler equation,

$$e^{ix} = \cos x + i\sin x \tag{A.28}$$

which will play a major part in our interpretation of the atomic orbital wavefunctions.

The Taylor Series

The power series given previously for e^x, $\sin x$, and $\cos x$ are special cases of a general power series equation that allows any single function $f(x)$ to be expressed in terms of its derivatives at some point x_0:

$$f(x) = \sum_{n=0}^{\infty}\frac{1}{n!}\left(\frac{d^nf(x)}{dx^n}\right)\bigg|_{x_0}(x - x_0)^n$$

$$= f(x_0) + \left(\frac{df(x)}{dx}\right)\bigg|_{x_0}(x - x_0) + \frac{1}{2}\left(\frac{d^2f(x)}{dx^2}\right)\bigg|_{x_0}(x - x_0)^2$$

$$+ \frac{1}{6}\left(\frac{d^3f(x)}{dx^3}\right)\bigg|_{x_0}(x - x_0)^3 + \ldots, \tag{A.29}$$

where the subscript x_0 after the derivative means the derivative is evaluated at the point x_0. This power series is called the **Taylor series,** and its use is confined to those cases when only the leading terms (the lowest order derivatives) are important—in other words, for slowly varying functions. The first derivative of a function is the function's slope, the second derivative is the slope of the slope, and so on. If a function doesn't have rapid oscillations or sharp peaks, its higher derivatives tend to converge toward zero, and the Taylor series becomes a useful approximation.

One common application of the Taylor series expansion is to the natural logarithm $\ln x$. If we choose $x_0 = 1$, since $\ln(1) = 0$, then we get the following:

$$0 < x < 2:\ \ln x = \ln(1) + \left(\frac{d\ln x}{dx}\right)\bigg|_{x=1}(x - 1) + \frac{1}{2}\left(\frac{d^2\ln x}{dx^2}\right)\bigg|_{x=1}(x - 1)^2$$

$$+ \frac{1}{6}\left(\frac{d^3\ln x}{dx^3}\right)\bigg|_{x=1}(x - 1)^3 + \ldots$$

$$= 0 + \frac{1}{x}\bigg|_{x=1}(x - 1) + \frac{1}{2}\left(-\frac{1}{x^2}\right)\bigg|_{x=1}(x - 1)^2 + \frac{2}{6}\left(\frac{1}{x^3}\right)\bigg|_{x=1}(x - 1)^3 + \ldots$$

$$= (x - 1) - \frac{1}{2}(x - 1)^2 + \frac{1}{3}(x - 1)^3 + \ldots = \sum_{n=1}^{\infty}(-1)^{n+1}\frac{1}{n}(x - 1)^n. \tag{A.30}$$

The logarithm function is undefined for nonpositive numbers, so $x > 0$. In addition, the Taylor series for the logarithm **diverges** if $x - 1 > 1$, because the denominators are increasing only as n. Consequently, this expansion is useful only over the interval $0 < x < 2$.

In another case, the Taylor expansion is applied to the function $1/(1 + x)$, where $|x| < 1$. Choosing $x_0 = 0$ in the Taylor expansion, we find

$$|x| < 1: \frac{1}{1+x} = 1 - x + x^2 - x^3 + \ldots = \sum_{n=0}^{\infty}(-1)^n x^n. \tag{A.31}$$

The range on this expansion is also restricted because the series diverges when $|x| \geq 1$.

A third example in which we use the Taylor series is the expansion of the function $(1 + x)^{1/m}$, where $|x| \ll 1$. Again we choose $x_0 = 0$ and expand:

$$|x| < 1: (1 + x)^{1/m} = 1 + \frac{1}{m}(1)^{(1/m)-1}x + \left(\frac{1}{2}\right)\left[\left(\frac{1}{m}\right)\left(\frac{1}{m} - 1\right)\right](1)^{(1/m)-2}x^2 + \ldots$$

$$= 1 + \frac{1}{m}x - \frac{m-1}{2m^2}x^2 + \ldots. \tag{A.32}$$

This is one form of the **binomial series.** As one example, we will use Eq. A.32 to approximate the square root function, when $m = 2$:

$$|x| < 1: (1 + x)^{1/2} = 1 + \frac{x}{2} - \frac{x^2}{8} + \ldots. \tag{A.33}$$

EXAMPLE A.10 Taylor Series

PROBLEM Use the Taylor series expansion to find an approximate equation for $1/x$. Let $x_0 = 1$ and keep terms up to order $(x - x_0)^2$.

SOLUTION Drawing on Eq. A.29,

$$\frac{1}{x} = \frac{1}{x_0} + (x - x_0)(-x_0^{-2}) + \frac{(x - x_0)^2}{2}(2x_0^{-3}) + \frac{(x - x_0)^3}{2 \cdot 3}(-2 \cdot 3x_0^{-4}) + \ldots$$

$$= \frac{1}{x_0} - \frac{(x - x_0)}{x_0^2} + \frac{(x - x_0)^2}{x_0^3} - \frac{(x - x_0)^3}{x_0^4} + \ldots$$

$$= 1 - (x - 1) + (x - 1)^2 - (x - 1)^3 + \ldots$$

$$\approx 2 - x + (x - 1)^2.$$

In the last two steps we set $x_0 = 1$ and then dropped the higher order terms, which is only valid for values of x close to 1. For $x = 1.1$, this equation predicts $1/x = 0.910$, very near the correct value of 0.909. However, for $x = 1.6$, the correct value is 0.625 and this equation predicts 0.760.

There is an important result from this, by the way. The variable in a power series, such as the x in e^x or the θ in $\cos\theta$, must not have physical units, strictly speaking. That is because there is no way to add quantities that represent different physical properties. If we set $x = 2\,\text{cm}$, then e^x is nonsense, because we could then write $e^x = 1 + 2\text{ cm} + 2\text{ cm}^2 + \ldots$, and there is no single quantity in the real world that can represent "distance plus area." For this reason, the text also gives most angles in radians (for example, $\pi/4$ instead of 45°) but leaves the units implied instead of writing them down.[4] Having said all that, this text violates the rule against units in power series in the intermediate steps of some derivations, when it is convenient to split the logarithm function into different pieces. It is okay to write $\ln([\text{A}]/[\text{A}]_0)$, with both $[\text{A}]$ and $[\text{A}]_0$ being concentrations, because the units cancel before we take the logarithm. We can replace this, as a mathematical exercise, with $\ln[\text{A}] - \ln[\text{A}]_0$, but please be aware that the individual terms $\ln[\text{A}]$ and $\ln[\text{A}]_0$ do not have physically meaningful values until we somehow get rid of the concentration units, perhaps by combining these terms with other concentrations in the problem.

A.2 Classical Physics

Classical or Newtonian physics describes nature on the **macroscopic** scales of time, mass, and energy—measured in seconds, kilograms, and joules—to which we are most accustomed. Quantum mechanics and relativity describe deviations from classical mechanics, but they operate more subtly in our experience because their effects are strongest at energy scales much smaller (quantum) or much larger (relativity) than we normally perceive with our own senses. Our interest in this volume is at the **microscopic** scale, which we will take to mean the scale of individual atoms and molecules: distances of a few nanometers or less, masses less than 1000 atomic mass units, and energies of no more than about 10^{-18} J. Nevertheless, Isaac Newton's laws of motion for macroscopic bodies are often indispensable in visualizing the motions of **microscopic** entities, such as individual electrons, atoms, and molecules, sometimes with no adjustment at all. Therefore, it may be useful to review a few topics from classical physics that will show up in the text.

Force and Energy

In the absence of forces, a moving object with mass m will travel in a straight line with velocity vector $\vec{v}$. The vector has a magnitude (the speed) and a direction (the trajectory). In order to change the object's speed, its trajectory, or both, the object must undergo an **acceleration** $d\vec{v}/dt$. A train leaving a station may accelerate to increase its speed along a straight track (so fixed trajectory), and the moon is subject to an acceleration that continuously changes its trajectory, to maintain a roughly circular orbit around the earth, at roughly constant speed. Acceleration to change a velocity $\vec{v_1}$ to a new velocity $\vec{v_2}$ requires a force F, but the amount of force depends on exactly how the acceleration is applied: all at once, continuously over a long time, or in a series of jerks. The force is minus one times the energy expended per unit distance:

$$-\frac{dE}{dx} \equiv F_x. \tag{A.34}$$

[4]Strictly speaking, the radian is a unit (we measure the size of angles with it) but has no dimensions (because it is effectively the ratio of two distances: a circle's circumference divided by its diameter). Thus, it's still okay to put it in a power series.

Newton's law in turn relates the force to the acceleration:

$$F_x = m\frac{dv_x}{dt}, \quad \text{or} \quad \vec{F} = m\frac{d\vec{v}}{dt}. \tag{A.35}$$

For motion in a circular orbit of radius r at constant speed v, a **centripetal acceleration** of v^2/r is necessary to keep changing the direction of the velocity vector. Acceleration is caused by a force, and centripetal acceleration is caused by a **centripetal force** of mv^2/r, where m is the mass of the orbiting object.

Conservation of Mass and Energy

At all times before, during, and after any chemical process, the overall mass and overall energy must each remain constant. In chemistry, we can ignore the exceptions to this rule.[5] Whether the process is the collision of two atoms, the combustion of 30 kg of fuel, or the binding of an electron to a proton, we shall enforce both of these constraints—the overall mass is constant, and the overall energy is constant.

In the absence of external forces, the overall linear momentum $\vec{p} = m\vec{v}$ is also conserved. This is especially important in mechanics because it vastly simplifies the equations of motion for a system with many colliding bodies. We will use it in our discussions of molecular collisions.

Kinetic, Potential, and Radiant Energy

Energy is a parameter of great importance in chemistry, and one of the parameters that we will be following from beginning to end of the text. It is a convenience to separate energy into different forms, as, for example, the different ways that a molecule can store energy. However, energy can be converted from one form to another, and sometimes the distinction between two forms of energy may suddenly become unclear. There are nevertheless three forms for which we will impose fairly rigid definitions:

1. **Kinetic energy** K: any energy due to the motion of an object with mass m:

$$K = \frac{mv^2}{2}, \tag{A.36}$$

   where v is the speed of the object

2. **Potential energy** U: any energy due to the interaction of an object with fields of the fundamental forces:

$$U = -\int F ds, \tag{A.37}$$

   where ds is a distance derivative and F is the force arising from gravitational fields, electromagnetic fields, or the fields of the nuclear forces (but don't expect to see that last set often in this text)

3. **Radiant energy**: any energy present in the form of electromagnetic radiation

[5]The exceptions include the conversions between mass and energy accompanying reactions in particle physics, the postulated formation of virtual particles, and—for the most part—the scaling of mass and energy at relativistic speeds. This last exception becomes important when treating the motions of core electrons in heavy atoms.

Kinetic energy will be involved when there is any motion of matter: motions of electrons within atoms, atoms within molecules, and molecules within massive solids or gases or liquids. Potential energy in our work will almost always result from the **Coulomb force,** the force on a particle with charge q_1 due to an electric field $\vec{\mathcal{E}}$:

$$\vec{F}_{\text{Coulomb}} = q_1 \vec{\mathcal{E}}. \tag{A.38}$$

A charged particle generates an electric field in the surrounding space,

$$\vec{\mathcal{E}} = \frac{q_1 \vec{r}}{4\pi\varepsilon_0 r^3}, \tag{A.39}$$

where q_1 is the particle's charge and $\vec{r}$ is the vector connecting the particle to the point where the field is measured. For two particles with charges q_1 and q_2, therefore, the Coulomb force acting on each particle is

$$\vec{F}_{\text{Coulomb}} = \frac{q_1 q_2 \vec{r}_{12}}{4\pi\varepsilon_0 r_{12}^3}, \tag{A.40}$$

where $\vec{r}_{12}$ is the vector connecting the two charges. Often we want only the force along the axis connecting the two charges, in which case this force law may be written

$$F_{\text{Coulomb}} = \frac{q_1 q_2}{4\pi\varepsilon_0 r_{12}^2}. \tag{A.41}$$

This force will result in some motion, and therefore some kinetic energy K, unless some canceling force is present.

There is also a potential energy U due to the interaction of the particles with the electric field, whether or not there is any motion. One way of looking at this is as follows. Force is the derivative with respect to position of the energy taken *from the field* and used to accelerate the particle. The potential energy is the energy *still available in the field* to accelerate the particle, before the particle acquires it. The energy consumed would be the integral of the force, and therefore the potential energy is the *negative* integral of the force:

$$U_{\text{Coulomb}} = \int_0^{U(r_{12})} dU = -\int_\infty^{r_{12}} F_{\text{Coulomb}}\, dr = -\int_\infty^{r_{12}} \frac{q_1 q_2}{4\pi\varepsilon_0 r^2}\, dr = \frac{q_1 q_2}{4\pi\varepsilon_0 r_{12}}. \tag{A.42}$$

The lower limit of the integral in r is chosen to be infinity because that is where U is zero.

Throughout this textbook we will be studying the Coulomb force interactions of various particles: atomic nuclei and electrons, atoms and other atoms, molecules and other molecules. From the potential energy function and the total energy, we can in principle determine *all the possible results* of these interactions. Because the total energy is conserved, the potential energy becomes the key to many central problems throughout physical chemistry. Keep an eye on it.

Both kinetic and potential energy are measured relative to some reference energy. The observable results that we predict cannot depend on which reference points we select, as long as we are careful to stick with them throughout the problem. The kinetic energy depends on the speed v, but this value depends in turn on the observer's own speed. Similarly, the integral $\int F ds$ that gives the potential energy has to be integrated from some origin (mathematically, the lower limit of the integral), and we are free to choose whatever origin we like. We usually set zero potential energy to correspond to some convenient physical state, such as an ionized atom or the most stable geometry of a molecule.

We will usually measure positions and speeds relative to the **center of mass** of our system, The center of mass is the location $\vec{r}_{\text{COM}}$ such that

$$\sum_{i=1}^{N} m_i(\vec{r}_i - \vec{r}_{\text{COM}}) = 0, \tag{A.43}$$

where the sum is over all the particles i, each with mass m_i and position $\vec{r}_i$. The center of mass makes a convenient origin for our coordinate system because most of the motions that we want to study are the *relative* motions of particles, such as the influence of a positively charged nucleus on a nearby electron, rather than the overall motion of the system.

To see how the relative and overall motion can be separated, let's take the case of two particles with masses m_1 and m_2, position vectors $\vec{r}_1$ and $\vec{r}_2$, and velocity vectors $\vec{v}_1$ and $\vec{v}_2$. The center of mass is at position $\vec{r}_{\text{COM}}$ such that

$$m_1(\vec{r}_1 - \vec{r}_{\text{COM}}) + m_2(\vec{r}_2 - \vec{r}_{\text{COM}}) = 0.$$

We convert this to a relationship among the velocities by taking the derivative of both sides with respect to time t:

$$m_1\left(\frac{d\vec{r}_1}{dt} - \frac{d\vec{r}_{\text{COM}}}{dt}\right) + m_2\left(\frac{d\vec{r}_2}{dt} - \frac{d\vec{r}_{\text{COM}}}{dt}\right)$$
$$= m_1(\vec{v}_1 - \vec{v}_{\text{COM}}) + m_2(\vec{v}_2 - \vec{v}_{\text{COM}}) = 0.$$

Now we can solve for the center of mass velocity vector,

$$\vec{v}_{\text{COM}} = \frac{m_1\vec{v}_1 + m_2\vec{v}_2}{m_1 + m_2}.$$

This velocity tells us how fast and in what direction the center of mass of the system is moving, no matter what the individual motions of the two particles. There is a kinetic energy $mv^2/2$ associated with this motion, where the speed is the magnitude of the $\vec{v}_{\text{COM}}$ vector and the mass is the combined mass of both particles:

$$\begin{aligned} K_{\text{COM}} &= \frac{1}{2}(m_1 + m_2)\,\vec{v}_{\text{COM}}^{\,2} \\ &= \frac{1}{2}(m_1 + m_2)\frac{(m_1v_1)^2 + 2m_1m_2(\vec{v}_1 \cdot \vec{v}_2) + (m_2v_2)^2}{(m_1 + m_2)^2} \\ &= \frac{1}{2}\frac{m_1^2v_1^2 + 2m_1m_2(\vec{v}_1 \cdot \vec{v}_2) + m_2^2v_2^2}{m_1 + m_2}. \end{aligned} \tag{A.44}$$

This value is not the same as the *total* kinetic energy, however, because it ignores any motions of the particles *relative* to the center of mass, and therefore relative to one another. (By motions "relative to one another," we mean changes in the distance between the particles, or in the direction of one from the other.)

The total kinetic energy K_{tot} is the sum of $mv^2/2$ for both particles. But the number we need to calculate much of the time is the kinetic energy K_{rel} for the relative motion, so we find that by subtracting K_{COM} from K_{tot},

$$\begin{aligned}
K_{rel} &= K_{tot} - K_{COM} \\
&= \frac{1}{2}(m_1 v_1^2 + m_2 v_2^2) - \frac{1}{2}\frac{m_1^2 v_1^2 + 2m_1 m_2(\vec{v_1}\cdot\vec{v_2}) + m_2^2 v_2^2}{m_1 + m_2} && \text{by Eq. A.44} \\
&= \frac{1}{2}\left[m_1 v_1^2\left(1 - \frac{m_1}{m_1 + m_2}\right) - \frac{2m_1 m_2(\vec{v_1}\cdot\vec{v_2})}{m_1 + m_2} + m_2 v_2^2\left(1 - \frac{m_2}{m_1 + m_2}\right)\right] && \text{combine } mv^2 \text{ terms} \\
&= \frac{1}{2}\left[m_1 v_1^2\left(\frac{(m_1 + m_2) - m_1}{m_1 + m_2}\right) - \frac{2m_1 m_2(\vec{v_1}\cdot\vec{v_2})}{m_1 + m_2} + m_2 v_2^2\left(\frac{(m_1 + m_2) - m_2}{m_1 + m_2}\right)\right] && 1 = (m_1 + m_2)/(m_1 + m_2) \\
&= \frac{1}{2}\left[m_1 v_1^2\left(\frac{m_2}{m_1 + m_2}\right) - \frac{2m_1 m_2(\vec{v_1}\cdot\vec{v_2})}{m_1 + m_2} + m_2 v_2^2\left(\frac{m_1}{m_1 + m_2}\right)\right] \\
&= \frac{1}{2}\frac{m_1 m_2}{m_1 + m_2}[v_1^2 - 2(\vec{v_1}\cdot\vec{v_2}) + v_2^2] && \text{factor out } m_1 m_2/(m_1 + m_2) \\
&= \frac{1}{2}\frac{m_1 m_2}{m_1 + m_2}(\vec{v_1} - \vec{v_2})^2 \\
&\equiv \frac{1}{2}\mu(\vec{v_1} - \vec{v_2})^2 && \text{(A.45)}
\end{aligned}$$

In the last step, we define the **reduced mass** $\mu = m_1 m_2/(m_1 + m_2)$. The reduced mass is the effective mass that appears in the equation for the relative kinetic energy of the two particles. Notice that the velocity that appears in that expression, $\vec{v_1} - \vec{v_2}$, is the velocity difference between the two particles, so K_{rel} depends only on the relative motion of the two particles.

When we cover the motions of electrons around atomic nuclei and the motions of atoms bound together in a molecule, we will take for granted this breakdown of the kinetic energy into a component for the center of mass and a separate component for the relative motion. In the relative motion term, the masses of the particles will appear in terms of the reduced mass defined in Eq. A.45.

Angular Momentum

Angular momentum is among the most challenging concepts in first-year physics. It is an important concept, however, for we will find that the energies of electrons, atoms, and molecules are often stored in some form of angular motion, and nearly all magnetic properties in chemistry arise from angular motion. The value of angular momentum lies in its being conserved in the absence of external forces, just as energy, mass, and linear momentum are. Angular momentum and all its effects result from changing the *direction* of motion of an object. This requires a force and involves an acceleration. That does not necessarily mean a change in kinetic energy, because circular motion at fixed speed requires a constant acceleration but no change in v nor in $K = mv^2/2$.

The angular momentum $\vec{L}$ of an object with linear momentum $\vec{p} = m\vec{v}$ and position vector $\vec{r}$ relative to the system's center of mass is defined to be the vector cross-product

$$\vec{L} = \vec{r} \times \vec{p} = m\vec{r} \times \vec{v}. \tag{A.46}$$

The angular momentum is also a vector; it has a magnitude $|\vec{L}| = L$, and it has a direction that is *perpendicular* to the plane of motion. For circular motion, the vectors $\vec{r}$ and $\vec{p}$ are always perpendicular, and the magnitude of the cross product simplifies to

$$\textit{circular motion:} \quad L = rp = mvr. \tag{A.47}$$

You may well ask what the angular momentum represents physically. The easiest answer is that it is a purely mathematical expression of great convenience. Its conservation law may suggest that it has a fundamental physical meaning, but the conservation of angular momentum is a direct consequence of conservation of the linear momentum and energy, not a new basic principle. Angular momentum is by no means unique in this way, however, because nearly all the parameters that we study in this text are similar mathematical constructs—including, for example, mass, charge, force, and energy. There is very little that we can claim to measure directly. The astonishing thing is how successfully we can convert those few direct measurements—and a great deal of abstract interpretation—into predictions for how matter will behave. It is that modeling of the behavior of matter, the extrapolation from theory to real observations, that is the subject from this point on.

PROBLEMS

For this chapter, only a few representative and illustrative problems are given, for the benefit of students interested in using some of the skills and concepts that we will draw on in the main text.

Unit Analysis and Reasonable Values

A.1 If the pK_a is $-\log_{10} K_a$, and K_a is equal to $\exp[-\Delta G/(RT)]$, write an equation for the pK_a in terms of $\Delta G/(RT)$.

A.2 Which of the following results is the most reasonable value for the average speed of an electrical signal traveling through a copper wire?

a. $2 \cdot 10^{10}$ m s^{-1} b. $2 \cdot 10^{5}$ m s^{-1} c. 2 m s^{-1}

A.3 Which of the following results is the most reasonable value for the number of atoms in a sample of crystal with dimensions 5.0 Å × 5.0 Å × 5.0 Å?

a. $8 \cdot 10^{10}$ b. $8 \cdot 10^{5}$ c. 8

A.4 Are the following values reasonable for the quantities indicated?

a. $25 \cdot 10^{-8}$ m for the C—O bond length in H_2CO.

b. 78 amu for the mass of one 6-carbon hydrocarbon molecule.

A.5 What are the correct units for the constants indicated in the following equations?

a. k in $-d[\mathrm{A}]/dt = k[\mathrm{A}][\mathrm{B}]$, if $[\mathrm{A}]$ and $[\mathrm{B}]$ are in units of mol L^{-1}, and t is in units of s

b. k_B in $k = Ae^{-E_a/(k_B T)}$, if E_a is in units of J, T is in units of K, and A is in units of s^{-1}

c. K_c in $K_c = [\mathrm{C}][\mathrm{D}]/([\mathrm{A}][\mathrm{B}])$, if the concentrations [X] are all in units of mol L^{-1}

d. k in $\omega = \sqrt{k/\mu}$, if ω has units of s^{-1} and μ has units of kg

Mathematics

A.6 Solve for all possible values of x that satisfy the equation

$$-\frac{(2x+1)^2 e^{-ax^2}}{a^2} = 0,$$

where a is a finite and non-zero constant.

A.7 Find the complex conjugates of the following:

a. $x - iy$ b. ix^2y^2

c. $xy(x + iy + z)$ d. $(x + iy)/z$

e. e^{ix} f. 54.3

A.8 For the vectors $\vec{A} = (1,0,0)$, $\vec{B} = (1,0,1)$, and $\vec{C} = (0,2,1)$, find the following:

a. $|\vec{C}|$ b. $\vec{A} + \vec{B}$

c. $\vec{A} \cdot \vec{B}$ d. $\vec{A} \cdot \vec{C}$

e. $\vec{A} \times \vec{B}$

A.9 Prove the Euler formula,

$$e^{ix} = \cos x + i \sin x,$$

using the Taylor series expansions for these functions.

A.10 Solve V_m to three significant digits in the following equation using a math program or successive approximation, given the values $P = 1.000$ bar, $a = 3.716$ L mol^{-2}, $b = 0.0408$ L mol^{-1}, $R = 0.083145$ L mol^{-1} K^{-1}, and $T = 298.15$ K:

$$\frac{\left(P - \frac{a}{V_m^2}\right)(V_m - b)}{RT} = 1.$$

A.11 Evaluate the derivatives with respect to x for the following functions.

a. $f(x) = (x + 1)^{1/2}$

b. $f(x) = [x/(x + 1)]^{1/2}$

c. $f(x) = \exp[x^{1/2}]$

d. $f(x) = \exp[\cos x^2]$

A.12 Evaluate the following integrals:

a. $\int_0^{\infty} e^{-ax} dx$

b. $\int_1^5 x^2 dx$

c. $\int_1^5 x^{-3/2} dx$

d. $r^2 \int_0^{2\pi} d\varphi \int_0^{\pi} \sin\theta d\theta$

Physics

A.13 Calculate the force in N between two electrons separated by a distance of 1.00 Å.

A.14 Express the kinetic energy K of an object in terms of its momentum p and its mass m.

A.15 Let the gravitational force obey the law $F_{\text{gravity}} = mg$, where the constant g, equal to 9.80 m s^{-2}, is the acceleration due to gravity at the earth's surface. What is the gravitational potential energy as a function of the height r and the mass m?

A.16 In dealing with molecular forces, gravity is usually ignored. Show that this is justified by comparing the Coulomb force between the electron and proton at a distance of 0.529 Å to the gravitational force on the hydrogen atom exerted by the earth near the earth's surface (earth's gravitational force $F_{\text{gravity}} = mg$, where $g = 9.80$ m s^{-2}).

A.17 One method for solving Newton's equations of motions for a mechanical system employs a function called the **Lagrangian** L, commonly defined as the kinetic energy K minus the potential energy U:

$$L \equiv K - U.$$

For the mechanics problems we encounter, the potential energy depends *only* on position, and the kinetic energy depends *only* on speed. (For example, if two identical particles pass through the same point in space at different times, they have the same potential energy even if one of them is traveling faster.) Using this, show that the Lagrangian has the feature that

$$\frac{\partial L}{\partial x} = \frac{d}{dt}\frac{\partial L}{\partial v_x}.$$

A.18 Two particles with masses m_1 and m_2 are moving with initial speeds v_1 and v_2, respectively. They collide with each other, and then continue along. None of the energy is lost to other kinds of motion, and there are no external forces working on the particles (i.e., the potential energy can be set to zero). Use conservation of energy and linear momentum to find the speeds v_1' and v_2' of the particles after the collision.

A.19 Calculate the angular momentum L of an electron (assuming that the velocity and position vectors are perpendicular to each other) when the electron is in a state with $K = -U$ at a distance of 1.0 Å from a proton. The electron's kinetic energy is K, and U is its potential energy due to the Coulomb force. Use the proton as the center of mass of the system.

A.20 This is a basic mechanics problem that includes some trigonometry, some vector algebra, and some mechanics. You may solve any or all of the individual pieces; together they add up to a verification of conservation of angular momentum for a very simple collision problem. The collision system is illustrated in the accompanying figure. Two classical, spherical particles (labeled 1 and 2) of equal mass m and equal diameter d approach each other, each with speed v_0 but traveling in opposite directions parallel to the z axis. They initially travel with constant x coordinates, such that $0 \leq x_1 < d/2$ and $x_2 = -x_1$. This guarantees that the two particles will eventually collide, changing their velocity vectors. When the particles collide, a line can be drawn between their centers of mass, and we'll designate as θ the angle that this line makes with the z axis, where $0 \leq \theta < \pi/2$.

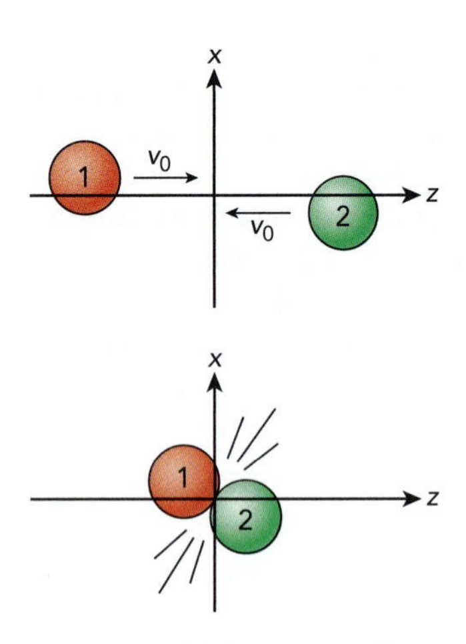

a. Show that when the collision occurs, the particle centers of mass are at the positions (relative to the center of mass of the entire system) $\vec{r}_1^{(0)} = ((d/2)\sin\theta, 0, -(d/2)\cos\theta)$ and $\vec{r}_2^{(0)} = (-(d/2)\sin\theta, 0, (d/2)\cos\theta)$.

b. Set the time when the collision occurs to $t = 0$, so t is negative before the collision and positive after the collision. The velocity vectors before the collision are $\vec{v}_1'' = v_0(0,0,1)$ and $\vec{v}_2'' = v_0(0,0,-1)$. Show that the velocity vectors after the collision are $\vec{v}_1' = v_0(\sin 2\theta, 0, -\cos 2\theta)$ and $\vec{v}_2' = v_0(-\sin 2\theta, 0, \cos 2\theta)$.

c. The position vectors are functions of time and may be written $\vec{r}_i = \vec{r}_i^{(0)} + \vec{v}_i t$, where i is 1 or 2, using the velocity vectors $\vec{v}_i''$ or $\vec{v}_i'$ appropriate for t negative or positive, respectively. Show that the angular momentum vector for this system is $\vec{L} = -mdv_0(0, \sin\theta, 0)$ when calculated before *and* after the collision.

A.21 This is a demonstration of how the magnetic field is generated by the motion of a charged particle. Start with two charged particles 1 and 2 at Cartesian coordinates $(-x_1,0,0)$ and $(0,y_2,0)$, respectively. We generally think of these two particles as interacting via their electric fields, which exert forces along field lines emanating from the charged particle as in (a) of the accompanying figure.

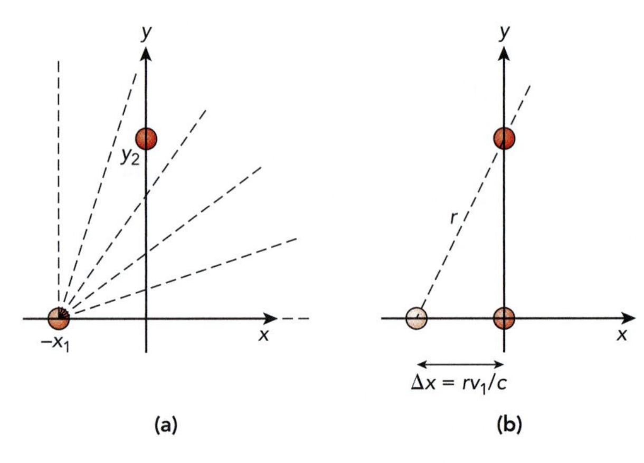

We now move particle 1 along the x axis at a constant speed v_1 and check the forces when particle 1 reaches the origin, as shown in (b) of the accompanying figure. The field lines of particle 1 move as well, and this changes the electrical force at particle 2. However, that change is transmitted at the speed of light, not instantaneously. The time required for the field of particle 1 at time t to reach particle 2 is r/c, where r is the separation between the two particles at time t. In the meantime, particle 1 has traveled a distance rv_1/c.

a. Write an equation for the electric field vector $\vec{\mathcal{E}}_1$ generated by particle 1 at the location of particle 2. Write the vector in its Cartesian form, showing the x- and y-components, as for example $\vec{\mathcal{E}}_1 = c(x_0,y_0,0)$. The electric field vector can be defined by its relation to the force: $\vec{F}_{\text{elec}} = q_2\vec{\mathcal{E}}_1$.

b. Now let's define the magnetic flux density by the relation $\vec{B} = \frac{1}{c^2}\vec{\mathcal{E}}_1 \times \vec{v}_1$. Write $\vec{B}$ in vector form.

c. Next, find the vector representing the magnetic force exerted on particle 2 in the field generated by particle 1, defining that force to be $\vec{F}_{\text{mag}} = q_2\vec{v}_1 \times \vec{B}$.

d. Finally, we want to show that this magnetic force is the difference between the actual Coulomb force and the classical Coulomb force that would be present if the electric field were transmitted instantaneously. Write the magnitude of the actual Coulomb force F (using the distance r) and the classical force F' (using the distance y_2). Calculate the difference between these, and simplify the result assuming $v_1 \ll c$ to show that you get the same value as predicted by the equations.

PART I

ATOMIC STRUCTURE

PART II

MOLECULAR STRUCTURE

PART III

MOLECULAR INTERACTIONS

1 Classical and Quantum Mechanics

LEARNING OBJECTIVES

After reading this chapter, you will be able to do the following:

❶ Convert between frequency, wavelength, and photon energy of electromagnetic radiation

❷ Use the de Broglie wavelength to determine if quantum mechanics is needed to describe a particular system

❸ Calculate the electronic energies of a one-electron atom and estimate other properties based on the Bohr model

GOAL *Why Are We Here?*

The goal of this chapter is to introduce the general concepts of classical mechanics and quantum mechanics that pertain to atomic structure. In the process, it will demonstrate where and why we will need quantum mechanics to assemble an accurate theory of atomic and molecular structure.

CONTEXT *Where Are We Now?*

For chemistry, the very smallest particles are electrons and atomic nuclei, and this is where we begin. Particle physicists divide matter into a more extensive library of particles comprising two families: *hadrons* (particles such as protons and neutrons that are composed of *quarks*) and *leptons* (non-quark matter, including electrons). But we have enough work ahead, and so we restrict ourselves to ordinary matter, except perhaps for an occasional example. This chapter introduces the concepts that guided our understanding of radiation and matter as we entered the twentieth century, and the preliminary results that forever enlarged that understanding. These ideas not only serve us as a springboard to the more advanced theory of today, but also provide a foundation that we will often return to for a better appreciation of how matter and energy work at *all* scales.

SUPPORTING TEXT *How Did We Get Here?*

Throughout the book, we will call on results from physics and math to clear the path as we take the next step. This chapter, which starts with the smallest building blocks of chemistry, is no exception. The following principles will be useful to us in the work ahead:

- The kinetic energy K of a particle with mass m and speed v is given by Eq. A.36:

$$K = \frac{mv^2}{2}.$$

- Two point charges q_1 and q_2 separated by a distance r_{12} interact through the Coulomb force,

$$F_{\text{Coulomb}} = \frac{q_1 q_2}{4\pi\varepsilon_0 r_{12}^2},$$

which is –1 times the derivative of the Coulomb potential energy $U_{Coulomb}$ (Eq. A.42):

$$U_{\text{Coulomb}} = \frac{q_1 q_2}{4\pi\varepsilon_0 r_{12}}.$$

- When a point mass moves in a circular orbit of radius r at speed v, the force that causes the constant change in the particle's direction is defined to be the centripetal force $F_{\text{cent}} = mv^2/r$. The particle will also have a conserved angular momentum L (Eq. A.47):

$$L = mvr.$$

- Energy in our system is also conserved, if we're careful to look everywhere (Section A.2). We will be looking at a system in this chapter that combines potential energy (from the interaction of charges), kinetic energy (from the motion of an electron), and radiant energy (from a beam of light). Energy is transferred between these different forms, but if we add it all up, the total energy in all these forms keeps the same value. In particular, when radiation is absorbed by an atom, the energy of the radiation is added to the energy of the atom.

- Fourier transforms (Section A.1) allow us to write a function of one variable, call it $f_x(x)$, as a function of the inverse variable k, using a summation of sine and cosine waves. The new variable k has units of $1/x$, and for every point x, there is a point $k=1/x$. In other words, a function $f_x(x)$ may always be written instead as $\sum_k a_k \cos(kx) + b_k \sin(kx)$, where the values of a_k and b_k are determined by the function $f_k(k)$, so that the same function can be represented by either $f_k(k)$ or $f_x(x)$.

- We can save ourselves a lot of trouble in manipulating equations if we recall that all the trigonometric functions depend on one another, through identities given in Table A.2.

1.1 Introduction to the Text

Physical chemistry is the foundation on which all chemistry rests. In physical chemistry we apply fundamental laws of mathematics and physics to obtain a description of molecular structure and interactions. A successful description will accurately predict the structures and interactions of as yet undiscovered molecules, and of known molecules in untested environments, allowing us eventually to predict the mechanism and outcome of *any* chemical process. We will call this description the **chemical model,** and this text is intended as a guide to the model's underlying reasoning and to its application.

The crux of the chemical model is this: chemical behavior depends on molecular structure. In order to predict the structure and the chemical activity of matter under any conditions, we must understand the structure and motions of individual molecules. We begin with the laws that govern the simplest single atoms, and on these atomic principles we build an explanation of the structure of single molecules. The model we develop for the behavior of matter on the microscopic scale we can then extend to the structure of **bulk matter,** groups of molecules large enough to be handled by traditional laboratory equipment such as tongs, beakers, or gas cylinders. That extension is one of the major topics in the companion volume, *Physical Chemistry: Thermodynamics, Statistical Mechancis, and Kinetics (TK).*

When we know how to apply the chemical model to the *structure* of matter in this volume, we can then apply it to the motions or *dynamics* of matter. In *TK*, we average over variations in the properties of the molecules across our sample, ultimately keeping track of only those variables that describe the entire sample. Then we use these results to investigate processes such as heating or expansion, which change these bulk properties without affecting the molecular structure. We extend this application to processes that grow more strongly connected to the molecular nature of matter, including phase transitions and solvation, until at last we study chemical reactions. The chemical reaction is the most fascinating and complex of all the processes we examine in physical chemistry, and it takes a grasp of the molecular and the bulk perspectives to appreciate the depth of that topic.

The great strength of the chemical model is the startling range of its applicability. We do confine ourselves to **ordinary matter**—matter composed of protons, electrons, and neutrons at temperatures low enough that these particles coalesce into atoms and molecules. But within that constraint, the scope of the chemical model is enormous. We use it to describe the interactions between individual molecules softly vibrating against one another at the base of a glacier, the binding of oxygen to hemoglobin in the blood coursing through our veins, and the countless chemical reactions that occur throughout an interstellar nebula more than a light year across. All of these examples appear as the finest details in a single vast but cohesive picture. Everything we claim to understand about molecular behavior in organic synthesis, industrial chemical processing, pharmaceutics, materials design, biochemistry, and any other discipline is based on the predictive power of this one theoretical framework.

The aim of physical chemistry is to test and extend the chemical model.

Our daily experience with matter involves masses of kilogram quantities, sizes on the order of meters, and energies on the order of joules, give or take three orders of magnitude. At this scale, matter obeys the laws of motion that Isaac Newton formulated, the laws of **classical mechanics.** Our first step in this text, however, is to describe individual atoms. These have masses, volumes, and energies some 20 orders of magnitude smaller than those to which we are accustomed. At that scale, classical mechanics is no longer accurate, and we turn to **quantum mechanics,** the set of laws devised initially to apply to matter at the tiny scale of single atoms. Quantum mechanics is the backbone of the next eight chapters of this text.

Quantum mechanics not only presents an exquisitely detailed depiction of individual atoms and molecules, it even tells us exactly how well we can know those details. This can get to be too much information, and we will fall back on approximations to quantum mechanics more and more as the molecules get bigger and more numerous. All the same, we can see the effects of quantum mechanics well into the world of benchtop chemistry, if we know where to look.

1.2 The Classical World

Quantum mechanics employs several mathematical tools and physical concepts that may be unfamiliar. Some of these concepts deal with radiation rather than matter because the interaction between matter and radiation is critical to many applications of physical chemistry. To smooth the way a bit before we discuss quantum mechanics, let's first briefly summarize the fundamental properties of matter and radiation as viewed from a classical perspective.

Classical Matter

A classical object has mass and occupies space. Mass is the characteristic of an object that determines (a) how much energy is required to accelerate the object from one velocity to another, regardless of what kind of force is applied, and (b) how much force is exerted on the object by a gravitational field. The volume of space occupied by a classical object cannot be simultaneously occupied by another object.

In classical mechanics, every parameter associated with an object is a continuous variable. The mass, volume, and kinetic energy may each have any positive and finite value. This is consistent with everyday experience. Say, for example, that we have a toy truck that we're going to roll down a playground slide (Fig. 1.1). The truck will accelerate as it descends, reaching the bottom of the slide at some kinetic energy. If we start at the bottom of the slide, we get a minimum kinetic energy—zero, if the slide levels off at the bottom. If we start at the top of the slide, we get a maximum kinetic energy. According to classical mechanics, we may adjust the starting point of the truck to any point between those two and get *any* kinetic energy between the minimum and maximum values. The final kinetic energy of the truck is a *continuous* variable.

Classical Radiation

The classical theory of electromagnetic radiation, attributed primarily to James Clerk Maxwell, describes radio waves, microwaves, infrared and visible and ultraviolet light, x-rays, and γ-rays all as combinations of oscillating

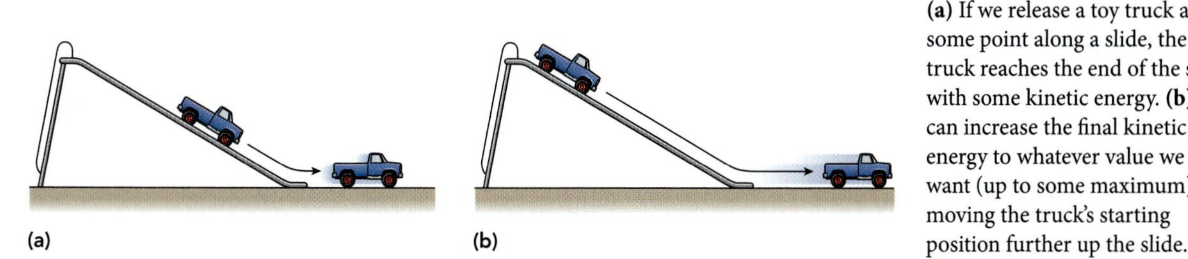

◀ **FIGURE 1.1 Classical energy is a continuous parameter.** **(a)** If we release a toy truck at some point along a slide, the truck reaches the end of the slide with some kinetic energy. **(b)** We can increase the final kinetic energy to whatever value we want (up to some maximum) by moving the truck's starting position further up the slide.

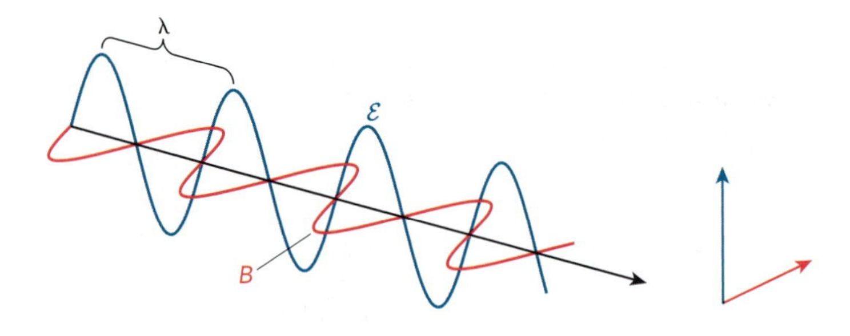

▶ FIGURE 1.2 **Maxwell's model for electromagnetic radiation.** In this model, a single ray of light is composed of perpendicular electric ($\mathcal{E}$) and magnetic (B) fields oscillating in phase and at the same frequency. The spacing between two adjacent maxima in the electric or magnetic field is the wavelength λ.

electric and magnetic fields, acting in accordance with the laws of wave mechanics formulated by Christiaan Huygens (a Dutch contemporary of Newton's). A common representation of this model is shown in Fig. 1.2.

Within a single ray of electromagnetic radiation, there is an electric field vector $\vec{\mathcal{E}}$ that oscillates up and down in a plane containing the axis of the ray. Perpendicular to the plane of the electric field is an oscillating magnetic field vector $\vec{B}$.[1] If we measure the two fields at some observation point in space, we find that they oscillate sinusoidally at a rate (in cycles per unit time, or simply reciprocal time) known as the **frequency,** ν. The frequency is on the order of $10^6\ \mathrm{s}^{-1}$ for radio waves, $10^{14}\ \mathrm{s}^{-1}$ for visible light, and so on. If, on the other hand, we freeze the ray in time, the distance between the peaks in each field is the **wavelength** of the radiation, λ. For the oscillation at our observation point to occur at frequency ν, we need the waves to travel a distance λ in time $1/\nu$, so the speed c of the whole package is related to the frequency and wavelength,

$$\lambda\nu = c, \tag{1.1}$$

where the speed of light in a vacuum is a constant given by $c = 2.998 \cdot 10^8\ \mathrm{m\,s}^{-1}$. The magnitude of the oscillations in the electric and magnetic field, from trough to peak, determines the **amplitude** of the radiation field. In our classical picture, the amplitude can have any value, and it corresponds to how much energy the radiation carries: greater oscillations mean more energy because the fields can push charged particles farther and faster.

The correspondence between these frequencies, their wavelengths, and the regions of the electromagnetic spectrum is shown in Fig. 1.3. Our eyes turn out to be sensitive to a relatively narrow range of wavelengths, centered at about 600 nm (a frequency of $5 \cdot 10^{14}\,\mathrm{s}^{-1}$) and spanning only about a factor of two, from roughly 400 nm to 800 nm. In contrast, what we call the infrared,

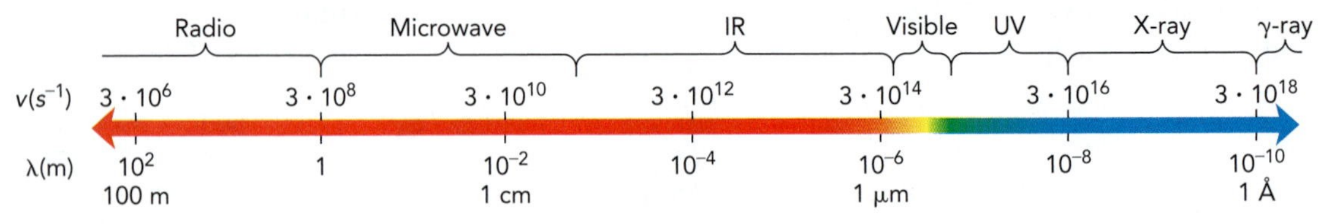

▲ FIGURE 1.3 **The electromagnetic spectrum.**

[1]We shall call the magnitude B of the vector $\vec{B}$ the magnetic field, although properly this is the symbol for the magnetic flux density. Because we are going to be concerned only with the action of magnetic fields across a vacuum, B and the magnetic field intensity H are identical for our purposes. This helps avoid confusion with the Hamiltonian, $\hat{H}$, which appears shortly.

microwave, and radio frequencies each cover some three orders of magnitude in wavelength, and the ultraviolet and x-ray regions each cover more than a factor of ten. The colors that make up the visible part of the spectrum represent a tiny fraction of what we may find as we explore these other regions. For example, warm molecules glow in the infrared in such a way that each chemical structure has its own infrared "color," a topic we explore in Chapter 8.

Unlike our experience of objects, it is not at all strange if two or more waves occupy the same space. When that happens, the two waves **interfere,** and the amplitude of the combined oscillation at any point is the sum of the individual wave amplitudes. If the values of the two waves at some point have the same sign, then **constructive interference** gives the combined wave a greater magnitude (whether positive or negative) than the individual component waves (Fig. 1.4a). Where the individual waves have the opposite sign, the magnitudes cancel (at least partially), an effect we call **destructive interference** (Fig. 1.4b).

If the interfering waves are **coherent waves,** they have a constant phase difference. This requires that they have the same frequency and that their points of origin are well-defined. For coherent waves, a pattern emerges over several wave-cycles. This **interference pattern** is composed of constructive and destructive interactions between the component waves as shown in Fig. 1.5. The alternating regions of constructive and destructive interference are called **interference fringes.** Interference patterns form at the overlap of any type of waves, including sound waves, light, and waves of water.

(a) In phase—
constructive interference

(b) Out of phase—
destructive interference

▲ FIGURE 1.4 **Superposition of waves.** When two waves overlap, we add the amplitudes. **(a)** Where the original amplitudes have the same sign, the interference is constructive, increasing the overall amplitude. **(b)** Where the original amplitudes have opposite phase, the interference is destructive, reducing the overall amplitude or even cancelling it completely.

SAMPLE CALCULATION **Convert between Frequency and Wavelength.**

Calculate the frequency ν of electromagnetic radiation with a wavelength of 25.4 mm. Using Eq. 1.1, we find

$$\nu = \frac{c}{\lambda} = \frac{2.998 \cdot 10^{8}\ \mathrm{m\,s^{-1}}}{0.0254\ \mathrm{m}} = 1.18 \cdot 10^{10}\ \mathrm{s^{-1}}.$$

This lies in the radio frequency range of the spectrum, near the boundary with the microwave region.

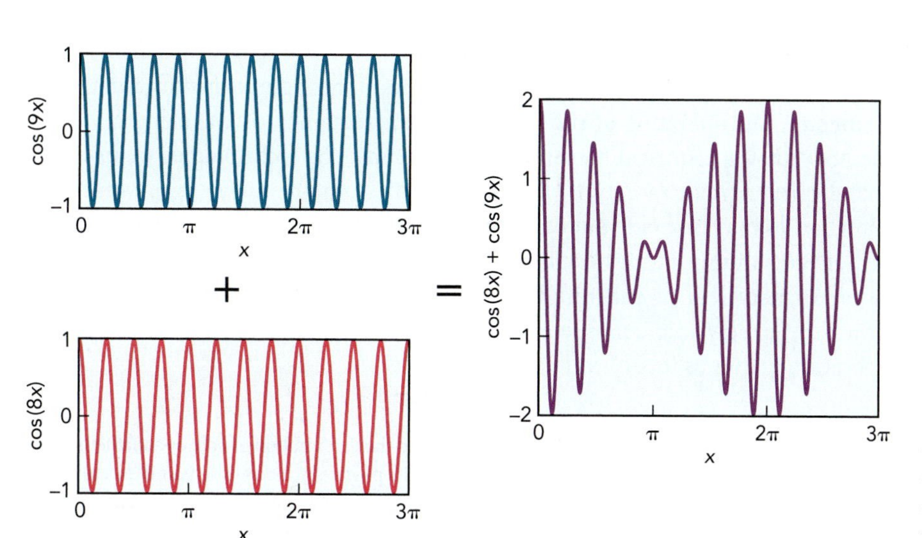

◄ FIGURE 1.5 **The interference pattern of two waves with unequal frequencies, or equal frequency but traveling along different trajectories.** In some regions, the interference is constructive and the wave amplitude is greater than that of either component wave; in other regions, destructive interference occurs and the wave appears weaker than its components.

1.3 The Quantum World

As our experimental methods improved, and as we thought more carefully about observations already made, it became apparent that the classical laws of radiation and matter did not form a complete picture of the natural world. Our understanding of radiation *and* matter underwent a rapid and dramatic evolution in the first third of the twentieth century, resulting in a new and fundamentally integrated understanding of both phenomena.

Quantum Radiation

In 1901, Max Planck shattered the classical picture of radiation by postulating that the energy in a field of electromagnetic radiation was absorbed or emitted only in quantities called **photons.**[2] The energy contained in a single photon is given by the frequency of the radiation ν or by the wavelength λ:

$$E_{\text{photon}} = h\nu = \frac{hc}{\lambda}. \tag{1.2}$$

This equation introduces a new fundamental constant, **Planck's constant,** h, equal to $6.626 \cdot 10^{-34}$ J s. Like the speed of light in a vacuum, c, Planck's constant is a number for which we have no derivation, and which quantifies the fundamental phenomena that shape and move our universe.

Let's get back to the photon. In a beam of red light with frequency $\nu \approx 5.0 \cdot 10^{14}\ \text{s}^{-1}$, for example, an atom can absorb only $h\nu = 3.3 \cdot 10^{-19}$ J of energy at a time—not $1.1 \cdot 10^{-19}$ J or $3.3 \cdot 10^{-20}$ J or any other amount. This violates the classical view of radiation, in which the energy of a beam of light is a continuous variable. It suggests instead that radiation has some characteristics traditionally associated with particles—that it behaves like separate chunks of energy rather than a single, continuous wave.

EXAMPLE 1.1 Planck's Law

CONTEXT Our different names for the regions of the electromagnetic spectrum grew out of the experiments in which they were originally observed. In some cases it was not understood that these forms of radiation were precisely the same phenomenon, but operating at different scales of frequency or wavelength. The various names persist, however, not only for historical reasons but also because the technologies we use to generate, manipulate, and detect electromagnetic radiation vary enormously from one wavelength range to another. For microwave radiation, lenses may be made out of Teflon, (polytetrafluoroethlyene) whereas for x-rays, lenses may be composed of pure aluminum or beryllium. Similarly, the natural processes that give rise to electromagnetic radiation span orders of magnitude in characteristic energies. The energy of photons emitted by a supernova—the cataclysmic collapse of a star—is vastly greater than the energy of photons emitted by room temperature matter as it cools off at night.

[2]The history is more complicated. Briefly, Planck proposed the quantization of light as a mathematical model, but it was Albert Einstein in 1905 who asserted its physical validity. The name "photon" was bestowed on these quanta in 1926 by G. N. Lewis, better known to general chemistry veterans for his dot structures.

PROBLEM 1. Infrared cameras are available that are sensitive over many different ranges within the IR region of the spectrum. One popular camera covers the 8–14 micrometer ($8 \cdot 10^{-6}$ m–$14 \cdot 10^{-6}$ m) range. How much energy does one photon at the high end of this wavelength range carry?

2. The γ-ray bursts emitted by supernovae generally consist of photons with energies less than about $3.0 \cdot 10^5$ electron volts (eV). Convert this energy to SI units, and calculate the wavelength and frequency of the radiation.

SOLUTION 1. We need Planck's law:

$$E_{\text{photon}} = \frac{hc}{\lambda} = \frac{(6.626 \cdot 10^{-34}\,\text{J s})(2.998 \cdot 10^8\,\text{m s}^{-1})}{14 \cdot 10^{-6}\,\text{m}} = 1.4 \cdot 10^{-20}\,\text{J}.$$

This is a typical order of magnitude for the energies we will encounter in this microscopic realm of individual atoms, molecules, and photons.

2. Looking up the appropriate conversion factor in the endpapers, we calculate

$$E_{\text{photon}} = (3.0 \cdot 10^5\,\text{eV})(1.000\,\text{eV}/1.602 \cdot 10^{-19}\,\text{J})^{-1} = 4.8 \cdot 10^{-14}\,\text{J}.$$

Now the units let us swiftly determine how to obtain λ and ν from Planck's law:

$$\nu = \frac{E_{\text{photon}}}{h} = \frac{4.8 \cdot 10^{-14}\,\text{J}}{6.626 \cdot 10^{-34}\,\text{J s}} = 7.3 \cdot 10^{19}\,\text{s}^{-1}$$

$$\lambda = \frac{c}{\nu} = \frac{2.998 \cdot 10^8\,\text{m s}^{-1}}{7.3 \cdot 10^{19}\,\text{s}^{-1}} = 4.1 \cdot 10^{-12}\,\text{m} = 0.041\,\text{Å}.$$

Quantum Matter

We base our quantum version of radiation on one relationship, Planck's law (Eq. 1.2): $E_{\text{photon}} = h\nu$. The quantum mechanics of matter also grow out of a single equation, one that ascribes a property to matter completely alien to the classical view: a *wavelength*. At the tiny distances over which atoms and subatomic particles interact, matter exhibits the characteristics of waves, consistent with a **de Broglie wavelength,**

$$\lambda_{\text{dB}} = \frac{h}{mv} = \frac{h}{p}, \tag{1.3}$$

where m is the mass of the particle, v is its speed, and the product mv gives the momentum p. By this expression, conceived by Louis de Broglie in 1924, lightweight and slow-moving particles have a long de Broglie wavelength.

Long wavelengths make interference patterns (like the pattern in Fig. 1.5) measurable because the interference fringes are widely spaced. The distance between fringes of constructive interference in two overlapping waves is roughly proportional to the wavelengths of the original waves, so long wavelengths spread out the interference pattern. Interference is a phenomenon that does not appear in the classical description of matter, so observation of interference fringes in matter provides a critical test of de Broglie's hypothesis. But the values

EXAMPLE 1.2 De Broglie Wavelength and the Correspondence Principle

CONTEXT This is the fundamental application of de Broglie's hypothesis to chemistry. At the bulk scale, where we carry our compounds in beakers and measure volumes by eye, classical mechanics is perfectly accurate. However, when we deal with the structure within *individual* atoms and molecules, the relevant masses and distances are too small for classical mechanics to accurately describe the system.

PROBLEM Calculate the de Broglie wavelengths of the following:

1. A 0.20 kg vial of acid thrown at a speed of 10.0 m s^{-1} toward a lovesick organist, soon to become the legendary Phantom of the Opera
2. An electron of mass $m_e = 9.109 \cdot 10^{-31}$ kg traveling in an atom with a speed of $v \approx 1.0 \cdot 10^{7}$ m s^{-1}

SOLUTION 1. The acid vial has a de Broglie wavelength of

$$\lambda_{\text{dB}} = \frac{h}{mv} = \frac{6.626 \cdot 10^{-34}\,\text{Js}}{(0.20\,\text{kg})(10.0\,\text{m}\,\text{s}^{-1})} = 3.3 \cdot 10^{-34}\,\text{m}.$$

This distance is 24 orders of magnitude smaller than a single atom and is too small for us to measure. Wave properties of the acid vial, such as destructive interference with another vial, likewise cannot be measured; the conditions needed to see destructive interference, for example, would require matching the positions of the two vials to much less than a single nuclear diameter. The mass and speed of the acid vial are large enough that the vial behaves like a classical object; its wavelike properties are undetectable.

2. In contrast, the electron in (2) has a de Broglie wavelength of

$$\lambda_{\text{dB}} = \frac{6.626 \cdot 10^{-34}\,\text{J}\,\text{s}}{(9.109 \cdot 10^{-31}\,\text{kg})(1.0 \cdot 10^{7}\,\text{m}\,\text{s}^{-1})} = 7.3 \cdot 10^{-11}\,\text{m}.$$

Yes, this is still a small distance, but it is almost 1 Å and roughly as large as the atom itself. Similarly, the de Broglie wavelength of an atom in a molecule ($m \approx 10^{-26}$ kg, $v \approx 10^{2}$ m s^{-1}) is on the order of 1 Å, the same size as the bonds in which the atoms move.

Based on these results, we don't need help from anyone but Newton in predicting the motion of our acid vial, but we'll have to learn some quantum mechanics to deal with the motion of the electron in an atom and the motion of an atom in a molecule.

work out so that only the tiniest particles have measurable wavelengths. Massive or fast-moving particles have short de Broglie wavelengths, which render the wavelike properties—such as interference—undetectable.

The significance of λ_{dB} unfolds over this and the next eight chapters, but it's worth pointing out now that we won't encounter any simpler or more important equation in this volume.

Being so simple, Eq. 1.3 wields enormous power as a qualitative tool in quantum mechanics. Here's how. At every step in our study of physical chemistry, we consider a particular system: usually a group of particles with forces acting on and between them that tend to limit their motions to some region of space.

The distance that the particle can travel within the system is the particle's **domain.** If λ_{dB} for a particle is much smaller than the particle's domain, then classical mechanics should be fine. But should the particle be trapped within a region no bigger than a few λ_{dB}'s, the classical equations will become inaccurate. This one calculation of λ_{dB} tells us whether our system is in the quantum or classical regime: *use quantum mechanics when the de Broglie wavelength is comparable to the particle's domain.*

Another, qualitative way to state the de Broglie equation is known as the **correspondence principle:** the quantum-mechanical solution to a problem must approach the classical solution as the system grows in mass, distance (more precisely, domain), or energy. The correspondence principle clarifies that the quantum and classical worlds are not completely distinct. Depending on how much work we're willing to do and how accurate we want our answers, we may be able to choose which perspective is the best for any particular problem.

De Broglie's theory was substantiated in 1927, three years after it was published, by the electron diffraction experiments of Clinton Davisson and Lester Germer. In these experiments, a beam of electrons was reflected off different layers of a nickel crystal surface, giving in effect two coherent electron beams that could then interfere with one another. As shown in Fig. 1.6, the electrons that bounce off the lower layer travel a slightly longer distance than the electrons bouncing off the top layer, with the difference equal to roughly $2d \cos\theta$, where d is the spacing between layers and θ is the electron beam's angle of incidence. When the two beams meet at the detector, they may have different phases, and the phase difference can be controlled by adjusting the angle θ. In order to see maximum current at the detector, we require constructive interference between the two beams, which means that the phase difference has to be equal to some whole number n of de Broglie wavelengths:

$$n\lambda_{dB} = 2d \cos\theta, \tag{1.4}$$

with n an integer.

By varying the detector angle, Davisson and Germer showed that the electrons did indeed behave like waves, destructively interfering at some angles and constructively interfering at other angles. Moreover, using a previously measured value of 2.15 Å for the spacing between layers of the nickel crystal, they were able

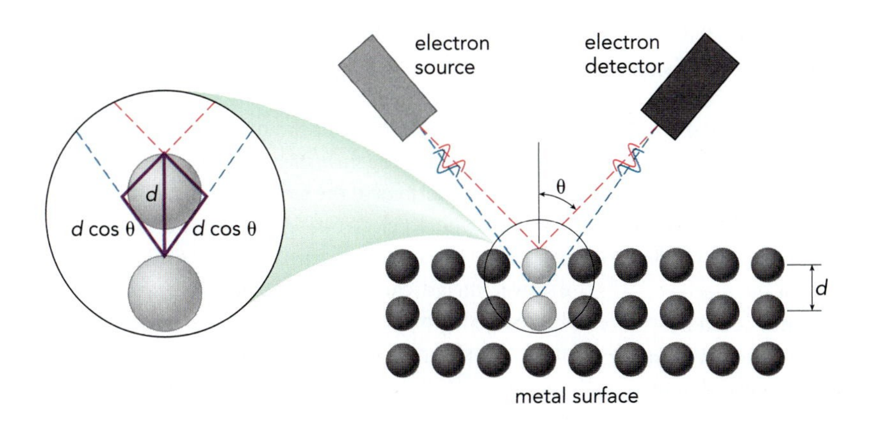

◀ **FIGURE 1.6 The Davisson–Germer electron diffraction experiment.** A beam of slow electrons has two possible paths between source and detector, via reflection off the first or second layer of atoms. The difference between the pathlengths is equal to roughly $d \cos\theta$ for the incident and for the reflected beams, or $2d \cos\theta$ overall.

to establish an effective wavelength of 1.65 Å for their electrons when they used a beam with electron kinetic energy K of 54 eV. De Broglie's equation predicted

$$\begin{aligned}\lambda_{\mathrm{dB}} &= \frac{h}{p} = \frac{h}{m_e v} \\ &= \frac{h}{\sqrt{2m_e K}} \qquad K = m_e v^2/2 \\ &= \frac{6.626 \cdot 10^{-34}\,\mathrm{J\,s}}{\sqrt{2(9.109 \cdot 10^{-31}\,\mathrm{kg})(54\,\mathrm{eV})(1.602 \cdot 10^{-19}\,\mathrm{J/eV})}} \\ &= 1.67 \cdot 10^{-10}\,\mathrm{m} = 1.67\,\text{Å},\end{aligned}$$

in excellent agreement with Davisson and Germer's experiment.

TOOLS OF THE TRADE | Low Energy Electron Diffraction

Low energy electron diffraction (LEED) is the extension of Clinton Davisson and Lester Germer's 1927 experiment at Bell Labs into a practical tool for characterizing the surfaces of clean solids. The experiments agreed so convincingly with de Broglie's predictions of a few years earlier that particles and waves could never again be completely separated in the minds of physicists.

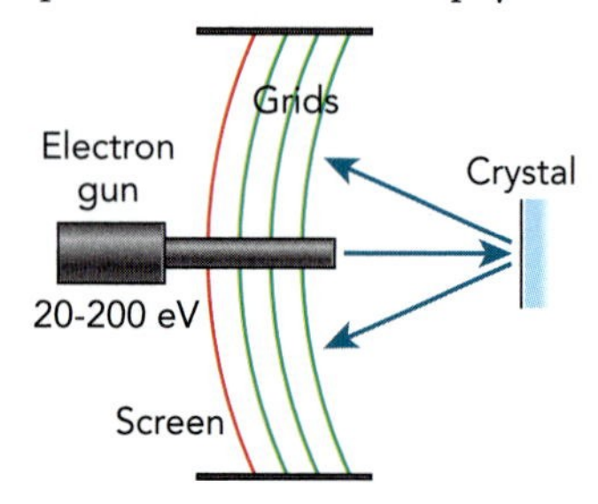

Why do we use LEED? LEED probes the structure of a surface by means of electron beams. The cleaner and flatter the surface, the less complex are the paths that the electrons take on their way to the detector, and the simpler the interference pattern—the LEED signal—that the detector records. The LEED signal allows researchers to rapidly assess the surface quality, and to quantify the effects of physical and chemical processes on the surface.

We want to study surfaces because they support numerous reactions in research and industry. The use of metal surfaces to catalyze reactions dates back to 1796, and the field continued to expand through the 1800s. By the time of Davisson's experiments, reactive surfaces were a major tool in industrial chemical synthesis.

But virtually nothing was known about the mechanism for such reactions, much less how they could be improved upon. To study those reactions in detail, it was necessary to isolate the surface from the many different substances present in an industrial reactor and to determine the structure and cleanliness of the surface. The surfaces of crystals can be quite flat, but unless a special effort is made, the crystal surface tends to show pits and scratches, which may profoundly affect its activity as a catalyst. A way to determine how the surface structure could influence the chemistry would require being able to observe the surface without damaging it. Electron diffraction provided a means for doing just that.

However, electron diffraction was originally a difficult and costly process because it was essential to carry out the measurements in very high vacuum (meaning very low pressure, less than 10^{-7} bar). The technology for achieving such low pressures became widely available only in the 1950s. This development opened the door for studies of solid surfaces and their chemistry, using LEED to determine before the experiment how clean and flat the surface was.

How does it work? LEED uses an electron gun to generate roughly a microamp flow of electrons at some well-defined energy, usually between 20 and 200 eV. The electron gun is a filament—a wire that emits electrons when current passes through it—placed near charged rings and plates that direct and regulate the flow of the electrons. The beam of electrons diffracts off the surface and is imaged onto a hemispherical screen. Some of the electrons bounce straight off the surface and are *elastically scattered,* but many are *inelastically scattered* and lose energy to the motions of the surface atoms. Only the elastically scattered electrons will generate a clean diffraction pattern, so a series of charged grids prevents the slower, inelastically scattered electrons from reaching the screen. The image that appears on the screen is a Fourier transform of the actual distribution of atoms on the surface. The more regular the pattern of atoms on the surface, the simpler the LEED image.

1.4 One-Electron Atoms

The Hydrogen Atom Spectrum

Let's step back a hundred years from de Broglie and his colleagues, to the observation that led to the need for a new theory of matter.

The sun emits electromagnetic radiation over a very wide range of wavelengths, from the long-wavelength radio waves to the shorter wavelengths of visible light and the even shorter-wavelength γ-rays. The component wavelengths that make up broadband radiation of this sort can be separated by various means, including dispersion of the visible light through a prism.

The field of **spectroscopy** was born in 1814 when the optics manufacturer Joseph von Fraunhofer fashioned the highest quality prisms attainable at the time, capable of dispersing visible light into its spectrum of different wavelengths with a clarity never before seen. Fraunhofer observed, in the diffracted light from the sun, tiny dark bands at certain wavelengths (Fig. 1.7). The sun generates light at all visible wavelengths, but some light, at specific wavelengths, is partially absorbed by gases in the solar atmosphere (Fig. 1.8). The absorption dims the light at those wavelengths, resulting in a pattern of dark bands in the dispersed light. It was subsequently found that, for light passing through a gas of any given chemical composition, the absorptions always occurred at the same set of wavelengths, and the set of absorptions came to be called the **spectrum of the gas.** Seventy years after Fraunhofer's discovery, Johann Balmer and others measured a series of these spectral lines for atomic hydrogen and found the

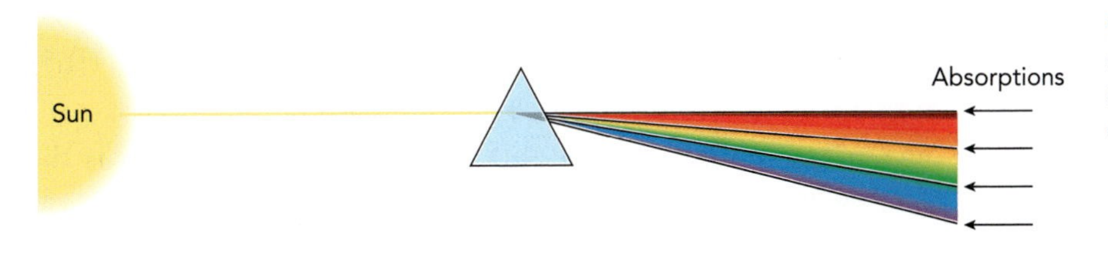

◀ FIGURE 1.7 **Wavelength separation of sunlight reveals absorptions at certain wavelengths.**

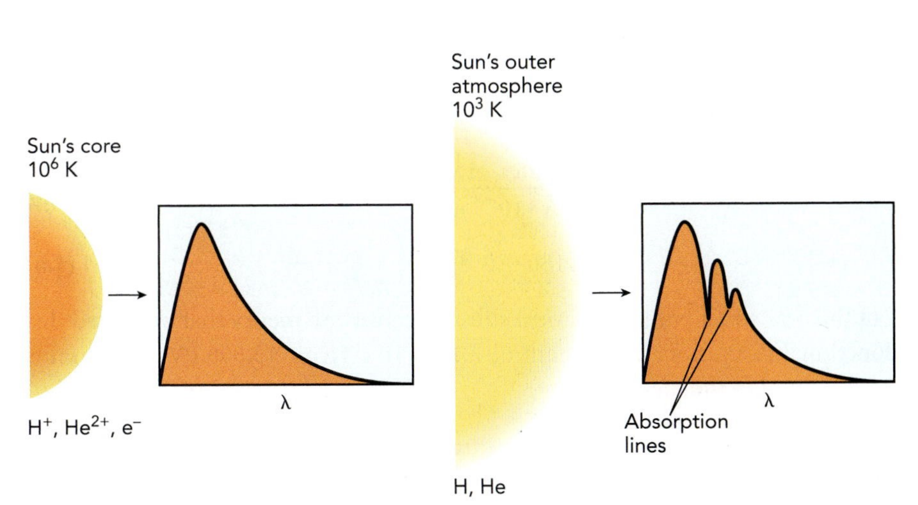

◀ FIGURE 1.8 **Light from the sun's interior is absorbed by atoms in its outer atmosphere.**

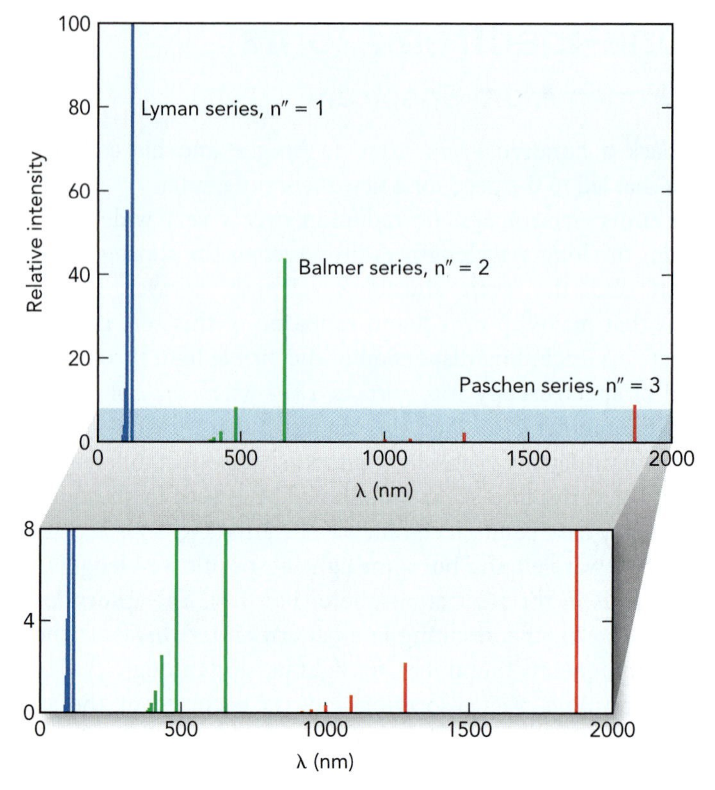

▶ FIGURE 1.9 **Simulated absorption spectrum of the hydrogen atom at photon wavelengths below 2000 nm.** The vertical axis represents the intensity of the absorption. The transitions appear in groups, with the $n'' = 1$ transitions in the ultraviolet named the Lyman series, the $n'' = 2$ transitions in the visible being the Balmer series, and the $n'' = 3$ transitions in the infrared being the Paschen series.

pattern shown in Fig. 1.9. Balmer's analysis for a few absorptions was extended by Johannes Rydberg to all the spectral lines of the hydrogen atom that were observable at the time. The series of wavelengths obeyed the **Rydberg equation,**

$$\lambda = (9.113 \cdot 10^{-8}\,\mathrm{m})\left(\frac{1}{n''^2} - \frac{1}{n'^2}\right)^{-1}, \tag{1.5}$$

where n' and n'' are positive integers, n' is greater than n'', and λ is the wavelength of the light absorbed. This equation was based purely on observation, and no successful interpretation was given for nearly thirty years, a full century after Fraunhofer's discovery.

With the discovery of Planck's Law, investigators saw that they could relate the wavelengths in the H atom spectrum to the photon energies:

$$\begin{aligned} E_{\text{photon}} &= \frac{hc}{\lambda} && \text{by Eq. 1.2} \\ &= \frac{hc}{9.113 \cdot 10^{-8}\,\mathrm{m}}\left(\frac{1}{n''^2} - \frac{1}{n'^2}\right) && \text{by Eq. 1.5} \\ &= (2.180 \cdot 10^{-18}\,\mathrm{J})\left(\frac{1}{n''^2} - \frac{1}{n'^2}\right), && (1.6) \end{aligned}$$

but this was still a purely empirical rule. There was no theory that predicted this functional form, nor the constant of $2.180 \cdot 10^{-18}$ J (or 13.606 eV, to use units more typical to the measurement).[3]

[3]The eV is the kinetic energy gained by one electron accelerated through a region where the electrical potential changes by 1 volt. It is a convenient unit for experiments in which electrons are used to carry energy into or out of the system, as in electron impact or ionization studies. It is not an SI unit, but it has made enough of the right friends to stick around.

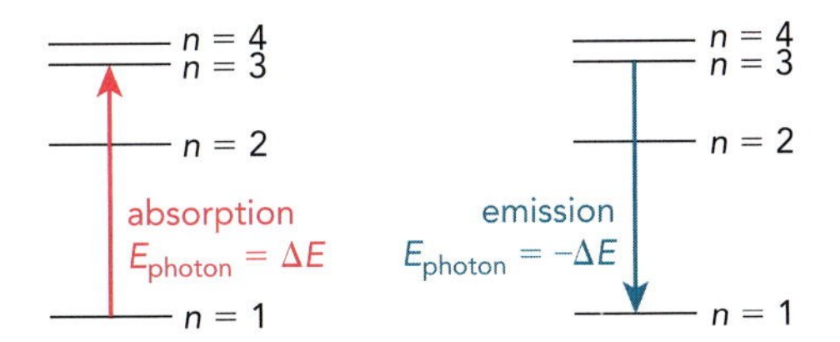

◀ **FIGURE 1.10 Transition Energies.** An atom may move from one level to a higher energy level by absorption of a photon with energy E_{photon} equal to the transition energy ΔE, or to a lower energy level by emission of a photon that carries away an energy E_{photon} equal to $-\Delta E$ (where ΔE is negative).

Equation 1.6 tells us that when atomic hydrogen is exposed to radiation, it absorbs the radiation *only at certain photon energies.* And if atomic hydrogen *emits* radiation, those photon energies are also given by Eq. 1.6. Scientists concluded that the energies of atomic hydrogen are themselves limited to certain values.

A common illustration of the concept is shown in Fig. 1.10, the first of several **energy level diagrams** that we will need. The vertical axis is the energy of the system, in this case a single hydrogen atom. Each horizontal line represents one of the possible energy values available to the system, an **energy level.** The system begins in an *initial* energy level and undergoes a transition to another, *final* level. Energy is conserved during these transitions, and in spectroscopy the energy gained or lost by the atom is balanced by the energy of a photon. If the initial energy is lower than the final energy, a photon is absorbed, and we say that the atom is **excited** into a higher energy level. If the final energy is lower than the initial energy, a photon is emitted and we say that the atom has **relaxed.** A hydrogen atom absorbs or emits a photon only if E_{photon} is equal to the energy gap ΔE between the initial and final levels of the atom. The energy difference ΔE is the **transition energy** for the process.

EXAMPLE 1.3 The Rydberg Equation

CONTEXT Atomic hydrogen makes up much of the observable matter in galaxies, and the spiral shape of our own galaxy was first mapped by taking images of the sky using telescopes set to detect the emission wavelengths of the hydrogen atom spectrum. Energy is transferred into the diffuse gas of interstellar space by the intense light of stars, by shock waves, and by cosmic rays. This energy excites electronic states of hydrogen, which then radiate to release the energy again, but at the photon energies specific to the hydrogen atom spectrum. The transitions in hydrogen with $n''=1$, the Lyman series, are perhaps the best known wavelengths in astronomy. In the same way, we can observe the presence of other elements in space by their distinctive spectra. The relative strengths of these atomic emission transitions are how we first began to determine the cosmic abundances of the different elements.

PROBLEM Predict the wavelength of the radiation generated by the hydrogen atom when $n'=3$ and $n''=1$. What region of the spectrum is this in? What are the longest and shortest wavelengths possible if $n''=1$?

SOLUTION From the Rydberg equation, Eq. 1.5:

$$\lambda = (9.113 \cdot 10^{-8}\,\mathrm{m})\left(\frac{1}{1^2} - \frac{1}{3^2}\right)^{-1} = 1.025 \cdot 10^{-7}\,\mathrm{m} = 102.5\,\mathrm{nm}.$$

This wavelength is in the ultraviolet. From the equation we find that when $n''=1$, the longest wavelength radiation is obtained when $n'=2$ and $\lambda = 122\,\mathrm{nm}$, whereas the shortest wavelength is for $n' \rightarrow \infty$ (so that $1/n'^2 \approx 0$), when $\lambda = 91\,\mathrm{nm}$. If $n''=1$, then for all values of n', the emitted radiation is ultraviolet.

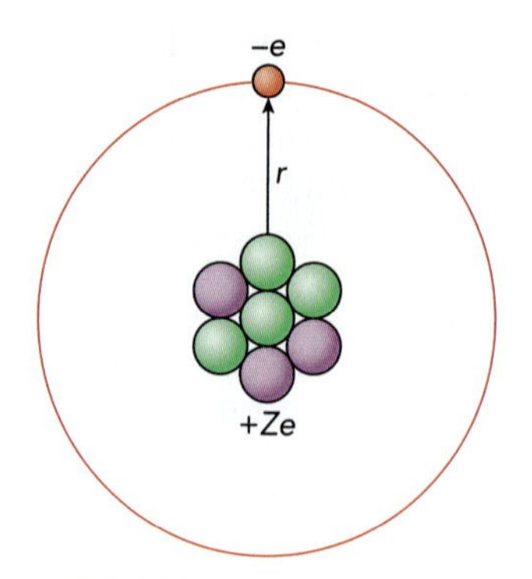

▲ FIGURE 1.11 **The Bohr model of the one-electron atom.** The single electron with charge $-e$ follows a circular orbit about the nucleus with charge $+Ze$.

But what determines the energies of the atom? That question opened the door to quantum chemistry.

Spectroscopy remains the routine experimental application of quantum mechanics. A typical spectroscopy experiment uses large numbers of photons to probe the properties of individual atoms and molecules. Each photon interacts with only one molecule at a time, bringing the experiment to the nanometer distances—the quantum scale—of single molecules. By measuring how a beam of photons is affected by its interaction with matter, we can now deduce the quantum properties of the substance. As we develop the theory of atomic and molecular quantum mechanics over the next several chapters, we will keep coming back to the practical issue of how our results can be supported by laboratory spectroscopy.

The Bohr Model

To explain the Rydberg equation, Eq. 1.5, Niels Bohr proposed the model of the atom pictured in Fig. 1.11. This was an irresistibly simple proposal that answered the fundamental question raised by the Rydberg equation: what determines the energy levels of the hydrogen atom? The model would turn out to be valuable, but imperfect. It was viewed with some suspicion from the beginning, for it rested on unfounded—and apparently unjustifiable—assumptions.

Earlier experiments had established that atoms consisted of electrons, each with a charge of $-e$, surrounding a small, dense nucleus, with charge $+Ze$, where Z is the atomic number. The Coulomb force, governing the attraction between the nucleus and each electron, could therefore be written (Eq. A.41):

$$F_{\text{Coulomb}} = \frac{q_1 q_2}{4\pi\varepsilon_0 r^2} = -\frac{Ze^2}{4\pi\varepsilon_0 r^2}.$$

Bohr set out to derive the energy levels of atoms that had only one electron. He hypothesized first that (I) *the electron travels in a circular orbit* around the nucleus, requiring that the centripetal force of $-mv^2/r$, needed to maintain a stable orbit, is equal to the Coulomb force

$$F_{\text{cent}} = F_{\text{Coulomb}}$$

$$-\frac{m_e v^2}{r} = -\frac{Ze^2}{4\pi\varepsilon_0 r^2}, \tag{1.7}$$

where m_e is the mass of the electron and v is its speed (not to be confused with frequency ν).[4] The minus signs indicate that the force is working to pull the electron towards the nucleus.

If we treat the nucleus and electron as point charges, something is already wrong. The nucleus and electron pair establish an **electric dipole moment,** a net change in charge along an axis. The dipole moment gives rise to an electric field.

[4]When one particle is said to orbit another, the center of the rotation is actually the center of mass of both particles, not the center of one of the particles. For a nucleus of mass m_{nuc}, that means that the centripetal force actually depends on the reduced mass, $\mu = m_e m_{\text{nuc}}/(m_e + m_{\text{nuc}})$, of the electron-nucleus pair in this equation instead of m_e. For simplicity, we ignore that for now because the electron is always at least a factor of 1000 less massive than the nucleus, and μ is equal to m_e to within 0.1%.

As the electron orbits the nucleus, this dipole moment and its electric field rotate. The trouble is that James Maxwell had shown that any oscillating dipole electric field must radiate energy. This arises from Maxwell's definition of light as an oscillating electromagnetic field traveling through space. When an electric field changes direction, the change propagates through space at the speed of light. Therefore, an oscillating atomic dipole creates an oscillating electric field (and, it turns out, an accompanying magnetic field) traveling at the speed of light. This is the same way we use a changing electric current in a radio antenna to broadcast radio waves: the oscillating electric and magnetic fields *are* what make up electromagnetic radiation. Bohr's atom, according to what was then known, should be emitting photons and losing energy.

If the atom were radiating energy constantly, then conservation of energy would require the orbit to decay, the electron eventually colliding with the nucleus. The Bohr model ignores this paradox on purpose, just to reproduce the experimental observations. Bohr's second assumption is (II) *the orbit of the electron does not radiate energy.*

Bohr's final critical assumption was necessary to get the pattern of absorptions described by Rydberg's equation. Bohr stated that (III) the **orbital angular momentum,** L (the angular momentum resulting from the circular motion of the electron), *is restricted to integer multiples of* $\hbar$, where $\hbar$ is Planck's constant divided by 2π:

$$L_n = \frac{nh}{2\pi} \equiv n\hbar. \tag{1.8}$$

The subscript n in L_n indicates that L is now a function of the integer n.

The angular momentum also obeys the classical equation for a circular orbit (Eq. A.47),

$$L_n = m_e v r = n\hbar,$$

and this lets us find the effect of Bohr's assumptions on the force law Eq. 1.7:

$$-\frac{m_e v^2}{r} = -\frac{Ze^2}{4\pi\varepsilon_0 r^2}$$

$$m_e v^2 r^2 = \frac{Ze^2 r}{4\pi\varepsilon_0} \qquad \text{multiply by } -r^3$$

$$m_e^2 v^2 r^2 = \frac{Ze^2 r m_e}{4\pi\varepsilon_0} \qquad \text{multiply by } m_e$$

$$n^2\hbar^2 = \frac{Ze^2 r m_e}{4\pi\varepsilon_0} \qquad \text{by Eq. 1.8 and Eq. A.47}$$

$$r = \frac{4\pi\varepsilon_0 n^2\hbar^2}{Ze^2 m_e} \qquad \text{solve for } r \tag{1.9}$$

$$v = \frac{L}{m_e r} \qquad \text{by Eq. A.47}$$

$$= \frac{n\hbar}{m_e}\left(\frac{Ze^2 m_e}{4\pi\varepsilon_0 n^2\hbar^2}\right) = \frac{Ze^2}{4\pi\varepsilon_0 n\hbar} \qquad \text{substitute for } L \text{ from Eq. A.47 and } r \text{ from Eq. 1.9} \tag{1.10}$$

In order to explain the spectrum, what we're really after are the energies of the electron. Now we can use Eq. 1.10 for v to write the kinetic energy K (Eq. A.36), our equation for r to write the Coulomb potential energy U (Eq. A.42), and add the two together to get the overall energy E:

CHECKPOINT By equating the Coulomb and centripetal forces on the electron and imposing a quantized angular momentum, the Bohr model arrives at quantized values for the radius, speed, and energy of the one-electron atom. We will find that a wavelike electron justifies some of the assumptions that got us here.

$$E = K + U = \tfrac{1}{2}m_e v^2 - \frac{Ze^2}{4\pi\varepsilon_0 r} \qquad \text{by Eqs. A.36 and A.42} \qquad (1.11)$$

$$= \frac{1}{2}m_e\left(\frac{Ze^2}{4\pi\varepsilon_0 n\hbar}\right)^2 - \frac{Ze^2}{4\pi\varepsilon_0}\left(\frac{Ze^2 m_e}{4\pi\varepsilon_0 n^2\hbar^2}\right) \qquad \text{by Eqs. 1.9 and 1.10}$$

$$= \frac{Z^2e^4 m_e}{(4\pi\varepsilon_0)^2 n^2\hbar^2}\left(\tfrac{1}{2} - 1\right) = -\frac{Z^2e^4 m_e}{2(4\pi\varepsilon_0)^2 n^2\hbar^2}. \qquad (1.12)$$

As a result of limiting the values of L, all the other characteristics of the orbit become discrete (rather than continuous) variables, indexed by the value of n:

$$r_n = \frac{4\pi\varepsilon_0 n^2\hbar^2}{Ze^2 m_e} \equiv \frac{n^2}{Z}a_0 \qquad (1.13)$$

$$v_n = \frac{Ze^2}{4\pi\varepsilon_0 n\hbar} \qquad (1.14)$$

$$E_n = -\frac{Z^2e^4 m_e}{2(4\pi\varepsilon_0)^2 n^2\hbar^2} \equiv -\frac{Z^2}{2n^2}E_h. \qquad (1.15)$$

These equations introduce two convenient new units of measurement, both defined by combinations of fundamental physical constants: a distance unit a_0 known as the **Bohr radius** and an energy unit E_h named the **hartree:**

$$1a_0 \equiv \frac{4\pi\varepsilon_0\hbar^2}{m_e e^2} = 5.292 \cdot 10^{-11}\,\text{m} \qquad 1\,E_h \equiv \frac{m_e e^4}{(4\pi\varepsilon_0)^2\hbar^2} = 4.360 \cdot 10^{-12}\,\text{J}. \quad (1.16)$$

Equations 1.8 and 1.13–1.15 are in stark opposition to the classical picture of matter: now the energy, speed, radius, and angular momentum of the orbit can take on only certain values. These values together define a **quantum state** of the system. Each quantum state differs from any other quantum state by the value of at least one parameter such as these. Rather than label each quantum state by the precise values of each parameter, we can instead identify the quantum state by its *quantum numbers*. For example, in the Bohr model we can find the values of r_n, v_n, and E_n once we know the **principal quantum number,** n. The orbital radius r would be a continuous variable in a classical system, but in the Bohr atom r has become quantized to values $r_n = n^2 a_0/Z$, equal for example to $a_0 = 0.5292\,\text{Å}$ for the $n = 1$ state of the H atom and $4a_0 = 2.117\,\text{Å}$ for the $n = 2$ state. The quantization of properties such as size and energy, which are continuous in classical systems, is what gives quantum mechanics its name.

Atomic and molecular physicists commonly express energies in units of E_h and distances in units of a_0, because they absorb all the values of the fundamental constants $\hbar$, m_e, and e. These are examples of **atomic units,** units derived from multiplicative combinations of the fundamental constants. Table 1.1 shows how these and other quantities can be obtained by appropriate combinations.

EXAMPLE 1.4 Properties of the Bohr Atom

CONTEXT We will find that, while the Bohr model has its limits, it predicts the correct values for the kinetic and potential energies of the one-electron atom. Oddly enough, this is not the same as predicting the correct values of the speed (used to calculate the kinetic energy) or the radius of the orbit (used to calculate the potential energy), and in fact we have to rethink what we mean by these parameters in the picture of the atom that we use today. We can take the Bohr model's values for v and r as easily obtained first approximations. As an example, this calculation of electron speed is suitable for determining when the motion of core electrons in heavier atoms becomes relativistic. At atomic numbers between 20 and 30, this estimated v becomes significant compared to the speed of light c, and more careful calculations require special relativity to correctly explain the structure of the atom.

PROBLEM Calculate the speed (in m/s), the orbital radius (in Å), and the energy (in J and in E_h) of the electron, according to Bohr's model of the atom, in the He^+ ion when the electron is in the energy level for $n = 5$.

SOLUTION From the list of fundamental constants given in the inside back cover, substitute the appropriate values into Eqs. 1.13–1.15, remembering to set $Z = 2$ for He^+:

$$r_n = \frac{n^2}{Z}a_0 = \frac{25}{2}a_0 = 6.61 \cdot 10^{-10}\text{m} = 6.615\text{Å}$$

$$v_n = \frac{Ze^2}{4\pi\varepsilon_0 n\hbar} = \frac{2(1.602 \cdot 10^{-19}\text{C})^2}{(1.113 \cdot 10^{-10}\text{C}^2\text{J}^{-1}\text{m}^{-1})5(1.055 \cdot 10^{-34}\text{J s})} = 8.743 \cdot 10^5 \text{ m s}^{-1}$$

$$E_n = -\left(\frac{Z^2}{2n^2}\right)E_h = -\left(\frac{2^2}{2 \cdot 5^2}\right)(4.360 \cdot 10^{-18}\text{J}) = -3.488 \cdot 10^{-19}\text{ J}.$$

Notice that this speed is already more than a thousandth the speed of light, suggesting correctly that a precise treatment of electrons in atoms with high atomic number Z requires relativistic corrections.

TABLE 1.1 Atomic units for common physical quantities, and their SI equivalents.

quantity	atomic unit	SI value
mass	m_e	$9.1094 \cdot 10^{-31}$ kg
distance	$a_0 \equiv \frac{4\pi\varepsilon_0\hbar^2}{m_e e^2}$	$5.2918 \cdot 10^{-11}$ m
time	$\frac{\hbar}{E_h} = \frac{(4\pi\varepsilon_0)^2\hbar^3}{m_e e^4}$	$2.4189 \cdot 10^{-17}$ s
charge	e	$1.6022 \cdot 10^{-19}$ C
speed	$\frac{e^2}{4\pi\varepsilon_0\hbar}$	$2.1877 \cdot 10^{6}$ m s^{-1}
force	$\frac{m_e^2 e^6}{(4\pi\varepsilon_0)^3\hbar^4}$	$8.2387 \cdot 10^{-8}$ N
energy	$E_h \equiv \frac{m_e e^4}{(4\pi\varepsilon_0)^2\hbar^2}$	$4.3597 \cdot 10^{-18}$ J
angular momentum	$\hbar$	$1.0546 \cdot 10^{-34}$ J s

We measure the potential energy of a system with respect to some reference point, and that reference point may vary depending on what is convenient. Often, we'll set the minimum possible potential energy equal to zero, for example, so that the energies in the problem are always positive. For the one-electron atom, however, we can't measure the potential energy from its minimum value, because $-Ze^2/(4\pi\varepsilon_0 r)$ approaches $-\infty$ as r approaches zero. Instead, we set the potential energy to zero when r becomes infinite. At finite r values, the potential energy is a negative number, increasing in magnitude as the electron approaches the nucleus. Letting the kinetic energy keep its usual definition, $mv^2/2$, the total energy $E = K + U$ is zero when the electron has exactly enough kinetic energy to break free of the nucleus. Positive energies correspond to ionized electrons, and negative energies to bound electrons. The Coulomb attraction between nucleus and electron makes the atom more stable (i.e., have lower energy) than a separated nucleus and electron, so the electron energy in Eq. 1.15 is negative. The energy levelsfor the hydrogen atom ($Z = 1$), according to the Bohr theory, are plottedin Fig. 1.12.

Equation 1.15 not only gives us the means of calculating transition energies in the one-electron atom, it also predicts the **ionization energy** (IE) of the atom, the minimum energy necessary to remove the electron from the lowest energy state of the atom. The atom is ionized when we reach $E = 0$, because we defined $E = 0$ to be the minimum energy of the atom when the electron and nucleus are completely separated (Fig. 1.13, p. 57). To reach $E = 0$ from the $n = 1$ state of the hydrogen atom, we need to put in an amount of energy equal to $\Delta E = 0 - E_{n=1} = E_h/2$ or 13.606 eV. To put it more simply, the IE of atomic hydrogen is 13.606 eV.

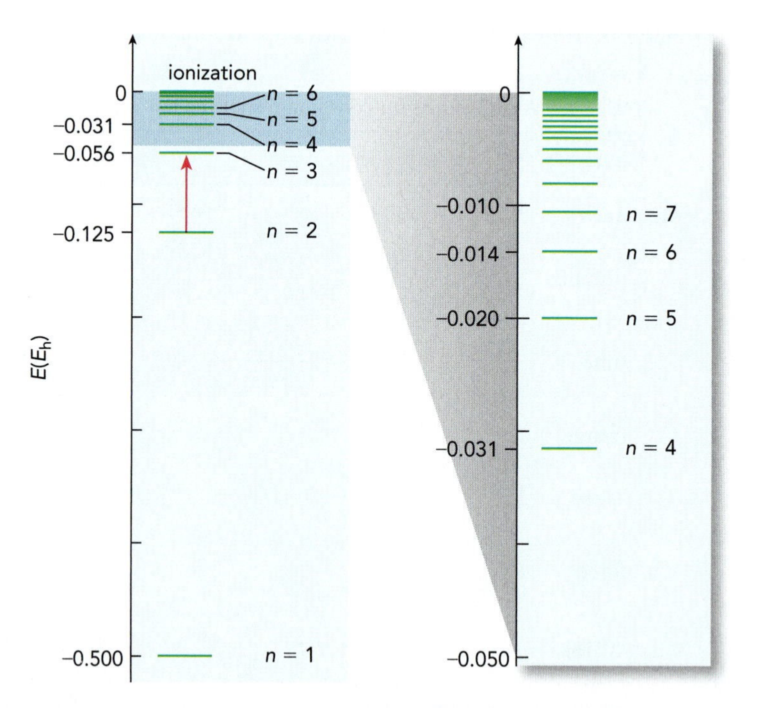

▲ FIGURE 1.12 **Energy levels of the Bohr H atom.** The $n = 2 \rightarrow 3$ transition is shown with a red arrow. The range from -0.050 to 0 Eh is expanded to show detail.

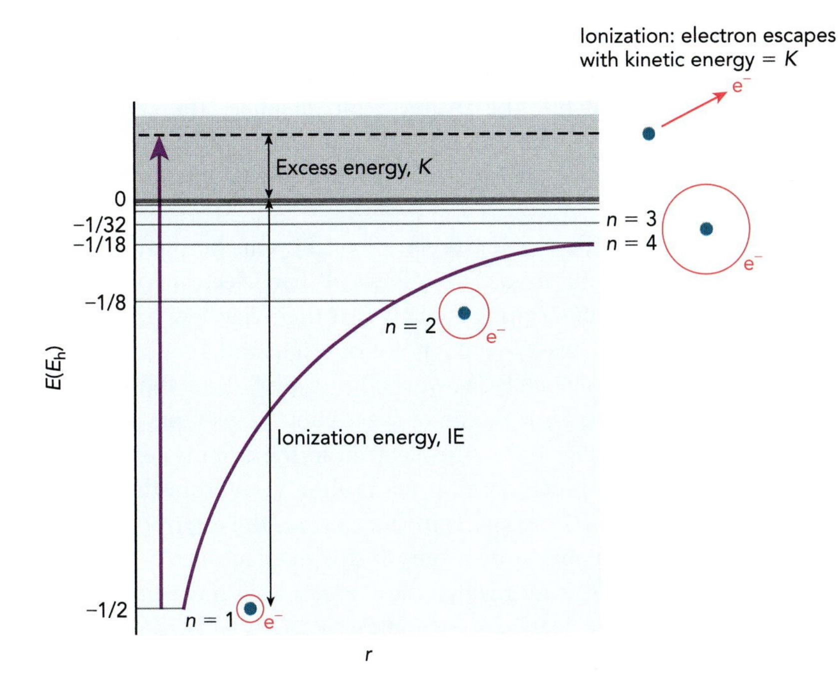

◀ FIGURE 1.13 **Ionization of atomic hydrogen.** Energy added to the ground state $n = 1$ atom promotes the system beyond the other negative energy bound states to the shaded area of continuous free particle energy levels where the electron is no longer bound to the nucleus. The minimum energy required for this is the ionization energy IE, and any extra input energy is converted to kinetic energy K of the electron and nucleus. The higher energy bound states, such as $n = 2$ and $n = 3$, are indicated by Bohr atoms with larger electron orbits.

SAMPLE CALCULATION **One-Electron Atom Ionization Energies.** To calculate the ionization energy in E_h of the Be^{3+} ion, we find the magnitude of the electron orbital energy when $n = 1$ and $Z = 4$ using Eq. 1.15:

$$E_n = -\frac{Z^2}{2n^2} E_h = \frac{4^2}{2 \cdot 1^2} E_h = 3.49 \cdot 10^{-17}\,\text{J}.$$

1.5 Merging the Classical and Quantum Worlds

Bohr's original model of the atom relied on a completely classical version of matter (with point charges for the nucleus and electron) and a completely quantum version of the radiation (with energy quantized into photons). The assumptions he had to make suggested that the picture remained incomplete, but this problem of how to complete the picture motivated the subsequent work to bring together the worlds of classical and quantum physics. In this section we see de Broglie's own justification of Bohr's assumptions, and the ramifications of the wavelike nature of particles on our ability to measure their properties.

De Broglie Waves in the Bohr Atom

De Broglie's hypothesis can be used to justify two of Bohr's major assumptions, advancing the atomic model toward the picture we hold of the atom today. Bohr's assumption (II), that the electron does not radiate, is supported

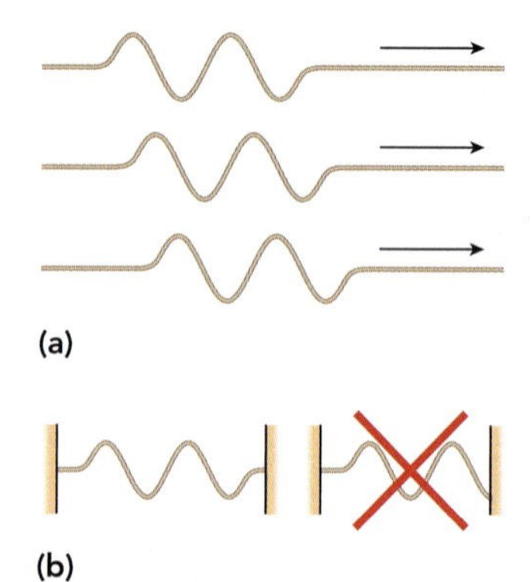

▲ FIGURE 1.14 **Traveling waves and standing waves.** **(a)** Traveling waves, such as water waves approaching a beach, can have any wavelength. **(b)** Standing waves, such as a vibrating piano wire strung between two posts, can sustain oscillations only at specific wavelengths in order to avoid destructive interference.

by our recasting the electron as a spread-out wave, rather than as a tiny localized particle. The electron does not move from one side of the atom to the other in a circle, but fills the entire "orbit" at once. The charge of the electron is equally distributed along this wave, so the atom has no net dipole moment and does not lose energy. This justification of assumption (II) is correct in its essentials.

Bohr's assumption (III), that $L_n = n\hbar$ (Eq. 1.8), can be justified for the circular orbits of the Bohr model as follows. When electrons travel freely through space, quantum mechanics predicts that their wavelike properties are those of **traveling waves**—waves with a point of origin and a trajectory. When the electron is bound to a nucleus, however, it behaves like a **standing wave.** Standing waves occupy the same region of space continuously, having no point of origin or net motion (Fig. 1.14). The electron in the atom is constrained to remain near the nucleus; it can have no net motion away from the nucleus if the atom is to keep from falling apart. In such a case, the electron wave overlaps itself and can be a stable system only if this overlap does not cause any destructive interference. This means that the electron wave must exactly retrace itself from one orbit to the next, as shown in Fig. 1.15.

The circumference of the orbit, $2\pi r_n$ for a particular state n, must be equal to some integer number n of electron de Broglie wavelengths:

$$2\pi r_n = n\lambda_{\mathrm{dB}} = \frac{nh}{m_e v_n}. \tag{1.17}$$

This restriction on the allowed values of wavelength for a standing wave is called a **boundary condition.** Dividing through by 2π and multiplying by $m_e v_n$ gives the angular momentum for the state n:

$$L_n = m_e r_n v_n = \frac{nh}{2\pi} = n\hbar. \tag{1.18}$$

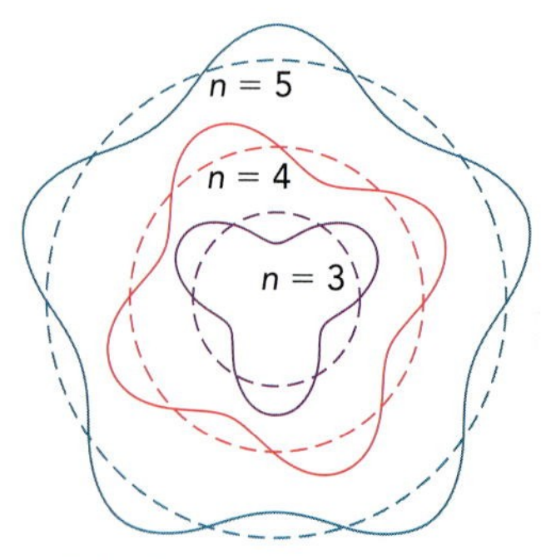

▲ FIGURE 1.15 **Application of de Broglie's theorem to the circular orbits of the Bohr model.** Only certain orbits have the right combination of circumference (determined by r) and electron de Broglie wavelength (determined by v) to satisfy the standing wave boundary condition. Note the longer λ_{dB} at larger n, consistent with the lower speed predicted by the Bohr model (Eq. 1.17).

This justifies Bohr's assumption of quantized angular momentum. Therefore, each quantum state of the Bohr model corresponds to a different number of wavelengths of the electron at the orbital radius. Bohr's model is extraordinarily tempting in its simplicity and its power to predict the atomic spectra. The hidden strength of the model is that it accurately portrays the relationship between the electron energy and its characteristic v and r values. It is all the more important, therefore, that we be aware of its shortcomings.

We have constructed an improved model of the electron in the one-electron atom, a model in which a wavelike electron occupies a circular orbit about the nucleus, justifying Bohr's second and third assumptions. While it is no accident that this model correctly predicts the energy levels of the one-electron atom, we've ignored the Bohr model's one great error: the first assumption (I) that the electron is distributed wholly along a circular orbit, confined to a single plane. A proper application of de Broglie's hypothesis to the electron rules this out, because the electron is a three-dimensional particle and should behave like a wave along *all three dimensions.* The electron distribution must be three-dimensional. The Bohr atom is utterly unable to predict the typical distributions of the electron around the nucleus. The angular momentum values are incorrect as well, because the orbital angular momentum vector $\vec{l}$ depends on the angular motion of the electron in all three dimensions.[5] We see in Chapter 3 how the Bohr model of the atom was transformed into our present-day model with the formulation of quantum mechanics in the late 1920s.

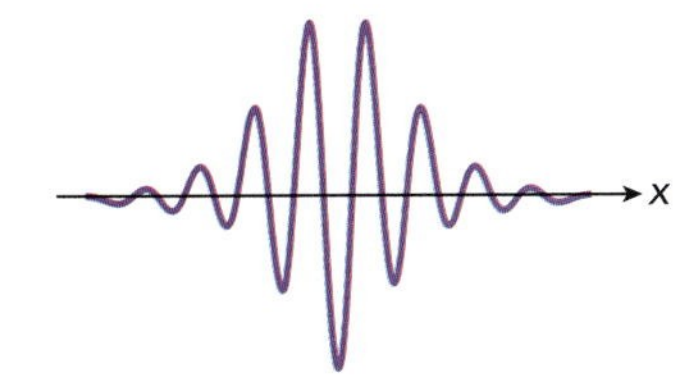

▲ FIGURE 1.16 **A wavepacket.** An electron wavepacket is a localized but oscillating function of mass and charge. A photon, by contrast, is a combination of oscillating but localized electric and magnetic fields.

The Wavepacket and the Uncertainty Principle

If you're uncomfortable with the counter-intuitive nature of the quantum models for matter and radiation—in particular the wavelike nature of the electron discussed previously—then you may sympathize with the many respectable physicists who still debate the proper interpretation of these models. One picture that combines the wave and particle aspects of quantum systems is the **wavepacket,** with an example drawn in Fig. 1.16. This is a function that oscillates, just as a classical wave does, but the significant amplitude of the oscillation is restricted to a narrow region. The size of the wave/particle is the size of this region where the amplitude is big enough to be detected.

The wavepacket provides a dependable mental image of the electron and other quantum-mechanical systems, useful in familiarizing us with their behavior in real chemistry. For example, the wavepacket picture provides a straightforward illustration of a fundamental law in quantum mechanics that has no analog in classical mechanics: we cannot simultaneously know exact values for *all* the physical parameters of a quantum system.

The most famous application of this principle is to the position and momentum of a particle. Notice that for our wavepacket in Fig. 1.16 there is no single value of x that defines its location. A wave cannot be tied to a single, fixed point in space, and neither can a particle whose position is defined by a wavepacket. If we carry out an experiment that reports a position for a particle wavepacket, we may get any of several answers. From our standpoint, the position of the particle is now an uncertain parameter because we can't be sure what value the measurement will report. However, we are skilled experimentalists, and we can become more certain of the particle's location by making the wavepacket more and more narrowly confined. But as we do that, there is less room for the oscillations within the wavepacket, and the wavelength of those oscillations is more poorly determined.

These oscillations define the particle's de Broglie wavelength, and therefore its momentum. If we wanted to know the momentum exactly, we would need a perfect sine wave that extended forever; consequently we would not know anything about its position. If we wanted to know its position exactly, the wavepacket would become infinitely narrow along the x axis and the wavelength (and hence momentum) would become an unknowable parameter. As the position of a quantum particle wavepacket becomes more certain, its momentum becomes more uncertain, and vice versa. We cannot know exact values of both the position and momentum of the particle at the same time. That is the qualitative version of the **Heisenberg uncertainty principle.** More precisely, the uncertainty

[5]There is an awkward overlap in notation because the angular momenta share symbols with, but *are not equal to,* the corresponding quantum numbers. We use $\vec{l}$ for the orbital angular momentum vector of one electron and $\vec{L}$ when we have several electrons. The magnitude of $\vec{l}$ will be written as L to avoid confusing it with the unitless quantum number l.

principle states that if measurements of position x and momentum p_x come with uncertainties δx and δp_x, respectively, then the product $\delta x \delta p_x$ is limited by

$$\delta x\, \delta p_x \geq \frac{\hbar}{2}. \tag{1.19}$$

If we measured the values of x and p_x over and over again on some quantum system with the best possible equipment, unavoidable sources of error would typically cause the probability of measuring any particular value of x to be given by a Gaussian function,

$$\mathcal{P}(x) = A_x^2\, e^{-x^2/(2\delta x^2)}, \tag{1.20}$$

where A_x is some constant we won't need to know and δx is the standard deviation of the x measurement, proportional to the width of the Gaussian. For simplicity, we will assume the average value of x is zero. This $\mathcal{P}(x)$ represents the intensity of the wavepacket. In the same way that the intensity of light is the square of the amplitude of the radiation field, this probability is proportional to the square of the wavepacket amplitude. Therefore, we can write our function as

$$F_x(x) = \left[\mathcal{P}(x)\right]^{1/2} = A_x\, e^{-x^2/(4\delta x^2)}. \tag{1.21}$$

Similarly, the probability of different momentum values could be expressed in terms of a momentum function,

$$F_p(p_x) = A_p\, e^{-p_x^2/(4\delta p_x^2)}. \tag{1.22}$$

Now we use a *Fourier transform*, $\mathcal{F}$, which allows us to rewrite any wavelike function $f(q)$ in terms of sine and cosine waves of different wavelengths in a variable $1/q$ (Section A.1). What we are doing now is an ideal application of the Fourier transform. The wavepacket, which is a function of x, can be expressed as a sum of sinusoidal waves, each with a different de Broglie wavelength, and therefore with different momenta. Conveniently, the Fourier transform $\mathcal{F}$ of one Gaussian function yields another Gaussian function, specifically

$$\mathcal{F}\left[e^{-x^2/(4\delta x^2)}\right] = A\, e^{-\pi^2(4\delta x^2)/\lambda_{\text{dB}}^2} = A\, e^{-\pi^2(4\delta x^2)p_x^2/h^2}. \tag{1.23}$$

CHECKPOINT This derivation applies just as well to the wavepacket picture of electromagnetic radiation: the more strongly localized the radiation field of a photon, the more poorly defined is the photon's frequency, and therefore its energy. Because the speed of light is a constant, a more localized photon corresponds to a burst of radiation over a shorter time δt, and this form of the uncertainty principle is written

$$\delta E_{\text{photon}}\, \delta t \geq \frac{\hbar}{2}.$$

This exponential must be the same as the distribution of p_x values we would measure for the particle, so we can set the terms in the argument of the exponentials in Eqs. 1.22 and 1.23 equal to each other:

$$-\frac{\pi^2(4\delta x^2)p_x^2}{h^2} = -\frac{p_x^2}{4\delta p_x^2} \qquad \text{combine Eq. 1.22 and Eq. 1.23}$$

$$\frac{\pi^2(4\delta x^2)}{h^2} = \frac{1}{4\delta p_x^2}$$

$$\delta p_x^2 \delta x^2 = \frac{h^2}{(4\pi)^2}$$

$$\delta p_x \delta x = \frac{h}{4\pi} = \frac{\hbar}{2}. \tag{1.24}$$

The equality in Eq. 1.24 holds only if the quality of our measurements is the best possible. If the quality is worse, then the value of $\delta p_x \delta x$ is higher, and the principle is usually written as $\delta p_x \delta x \geq \hbar/2$ (Eq. 1.19).

The Heisenberg uncertainty principle brings us to the brink of the challenging questions that tie quantum mechanics to measurement and perception. We will pause only for a moment at this brink, before turning right around and going back to chemistry, because the uncertainty principle will appear implicitly in all our applications of quantum mechanics. In classical mechanics, one could ask at any time for exact values of a particle's parameters such as position, velocity, energy, angular momentum, direction of the angular momentum, and time spent in motion. In quantum mechanics, all of these parameters still have the same meaning, and are still measurable. The difference is that if we want to know exact and fixed values of one or two of these parameters, we will end up with other parameters that are *not* fixed. The wavepacket description is one tool for visualizing why this is so.

The concept of the wavepacket also elucidates a remarkable *simplification* of quantum over classical physics: matter and radiation are distinct but similar phenomena because both can be represented by wavepackets. To some extent, a wavelength and location can be specified for a particle of matter and for a photon. But these wavepackets carry different properties: both matter and radiation carry energy, but a matter wavepacket always has mass and sometimes has charge, while a photon wavepacket has neither.

One final note about the wavepacket: the Heisenberg uncertainty principle prevents us from knowing exact values of all the parameters of our particle at the same time, but it doesn't force us to pick which particular set of parameters we do know. For example, we will usually *choose* to require our quantum states to have exact values of the total energy, the total angular momentum L, and the z-component of the angular momentum, L_z. If we had chosen a different set of parameters (say, L_x instead of L_z), we would obtain a different set of quantum states. It annoys our instincts that the very thing we want to characterize seems to change depending on just what parameter we decide to measure. The explanation comes back to the wavepacket description: we can force our particle to adopt a single value of L_z by measuring that parameter, but in exchange for narrowing the wavepacket to define L_z, we spread out the wavepacket in other directions. In this example, we would find that if we know L_z, we cannot measure a unique value of L_x.

CHECKPOINT In a quantum system, the Heisenberg uncertainty principle prevents us from knowing exact values of certain combinations of parameters (such as momentum and position) for a particle at the same time. This is because waves, unlike classical particles, do not have exact values for position or other parameters, and in the quantum regime, particles act a lot like waves.

CONTEXT *Where Do We Go from Here?*

Although we are building a framework for describing systems when classical mechanics fails, we are not abandoning classical mechanics. In fact, throughout this text it will be valuable to think of these microscopic particles classically, keeping in mind the results from quantum mechanics that must be obeyed. Those results will become clearer as we work through other examples.

In this chapter, we've examined a semi-classical version of the simplest atoms that predicts how the energy in these microscopic systems comes to be a quantized parameter. Now we need to develop a more rigorous and more general approach, starting from de Broglie's hypothesis but with fewer preconceptions about the nature of the particles that make up the atom.

KEY CONCEPTS AND EQUATIONS

1.3 The Quantum World.

- Electromagnetic radiation is a combination of oscillating electric and magnetic fields, which interacts with matter in units of energy determined by Planck's law

$$E_{\text{photon}} = h\nu = \frac{hc}{\lambda}. \quad (1.2)$$

- Quantum mechanics, rather than classical mechanics, provides a more accurate prediction of a system's motions as the de Broglie wavelength λ_{dB} approaches the size of the domain, where

$$\lambda_{\text{dB}} = \frac{h}{mv} = \frac{h}{p}. \quad (1.3)$$

1.4 One-Electron Atoms.

- The electronic transition energies in the hydrogen atom were found by experiment to obey the equation

$$E_{\text{photon}} = (2.180 \cdot 10^{-18}\,\text{J})\left(\frac{1}{n''^2} - \frac{1}{n'^2}\right), \quad (1.6)$$

where n' and n'' are positive integers.

- The Bohr model of the atom correctly predicts the energies of individual electronic states:

$$E_n = -\frac{Z^2 m_e e^4}{2(4\pi\varepsilon_0)^2 n^2 \hbar^2} \equiv -\frac{Z^2}{2n^2}E_{\text{h}}. \quad (1.15)$$

1.5 Merging the Classical and Quantum Worlds.

- De Broglie's representation of wavelike particles creates a limitation in our ability to define simultaneously the location and the momentum of a particle, expressed by the Heisenberg uncertainty principle

$$\delta x\, \delta p_x \geq \frac{\hbar}{2}. \quad (1.19)$$

KEY TERMS

- **Interference** is the phenomenon that describes how overlapping waves can reinforce one another (**constructive interference**) or cancel (**destructive interference**), in whole or in part. This is considered a key characteristic of waves, distinguishing wavelike behavior from particle-like behavior.
- **Radio, microwave, infrared, visible, ultraviolet, x-ray,** and **γ -ray** are all forms of electromagnetic radiation, differing only in the frequency (and therefore wavelength) of the radiation.
- A **photon** is the smallest unit of electromagnetic radiation, carrying an energy E_{photon} proportional to the radiation frequency ν, such that $E_{\text{photon}} = h\nu$.
- The **correspondence principle** states that in the limit that the system has large mass, large kinetic energy, and/or large travel distance (**domain**), the results predicted by quantum mechanics are the same as the results predicted by classical mechanics.
- The **Heisenberg uncertainty principle** asserts that quantum mechanics places limits on the precision with which we can simultaneously know the values of certain pairs of parameters, especially position and momentum.
- The **transition energy** is the energy difference between the upper and lower states of a change in quantum state.
- A **spectrum** is a set of transition energies (often expressed in related units, such as frequency or wavelength) determined for a particular chemical sample.
- **Atomic units** are a set of measurement units based on combinations of fundamental parameters such as $\hbar$, e, and m_e. The atomic unit for distance is the **Bohr radius** a_0, and the atomic unit for energy is the **hartree** E_{h}.

OBJECTIVES REVIEW

1. *Convert between frequency, wavelength, and photon energy of electromagnetic radiation.*
   Find the frequency and photon energy of radiation at wavelength 1.0 mm.
2. *Use the de Broglie wavelength to determine if quantum mechanics is needed to describe a particular system.*
   Hydrogen atoms can move along metal surfaces to catalyze chemical reactions. Determine whether quantum mechanics is likely to be essential in describing the motions of ^{1}H atoms over distances of about 1.0 μm, assuming a kinetic energy of about $4.0 \cdot 10^{-21}$ J.
3. *Calculate the electronic energies of a one-electron atom and estimate other properties based on the Bohr model.*
   Find the total energy and the potential energy of the electron in the 2*s* orbital of the He^+ ion.

PROBLEMS

The problems at the end of each chapter are intended to cover most of the chapter's material, ranging from straightforward numerical calculations to more difficult derivations. Discussion problems emphasize a qualitative understanding of the topics in the chapter, and are often motivated by apparent paradoxes or possible misconceptions. In later chapters, problems may take the opportunity to combine results from different chapters.

Discussion Problems

1.1 If we rewrite Planck's law in terms of the wavelength, $\lambda = hc/E_{\text{photon}}$, it has the form of de Broglie's equation $\lambda_{\text{dB}} = h/p$ but is written for radiation instead of matter. We've used de Broglie's equation to devise a correspondence principle to tell us when we need to use quantum mechanics to describe a system. Does this form of Planck's law suggest a version of the correspondence principle appropriate for radiation?

1.2 The operation of a radio transmitter can be imagined in classical terms as the electromagnetic equivalent of dropping a rock in a pond, with waves emanating outward from a single location. The amplitude of the waves diminishes *continuously* as they recede from the source, so a weaker signal is detected at great distances. On the other hand, if we think of the transmitter as emitting individual photons, like pebbles thrown outward from the middle of the pond, then the photons have to travel in specific directions. Eventually, as we pull the receiver away from the transmitter, won't the photons start to miss the receiver, so that at some distance the signal *suddenly* vanishes?

1.3 In order for Davisson and Germer's experiment to work, the electrons arriving at the detector from each crystal plane must be coherent. They used accelerating voltages to force the electrons to a specific energy, and therefore a common λ_{dB}, yet they used no special technique to force the electrons to have the same *phase*. How could they be sure that the phases at the detector would not be randomly distributed?

1.4 In Davisson and Germer's experiment, the waveforms of an electron beam reflected off different layers of a crystal destructively interfere at some angles, and no current is measured at the detector. What has happened to the mass and the charge on the electrons? Is mass not conserved for quantum particles?

1.5 I get a particle moving at 10 m s^{-1}. Then I put a tiny and very dedicated student on another particle, moving at exactly the same speed and in the same direction. I measure the speeds of the two particles and find that they are the same. Knowing the masses of the particles and the student, I can therefore calculate the de Broglie wavelengths of the two bodies. But when the student on particle 2 measures the speed of particle 1, the speed of particle 1 appears to be zero, because particle 2 is moving with exactly the same velocity. Wouldn't this mean that the student observes the de Broglie wavelength of particle 1 to be infinite? If so, how can I ever measure a finite de Broglie wavelength in my lab?

1.6 The Bohr model of the atom successfully predicts which of the following?

a. the spectrum of the neutral helium atom
b. the distribution of the electron in the neutral hydrogen atom
c. the 2nd ionization energy of the helium atom (the ionization energy of He^+)

Planck's Law and the Bohr Model

1.7 What is the energy in J absorbed during the transition $n = 1 \rightarrow 2$ in one He^+ ion?

1.8 Calculate the wavelength in m of the radiation emitted when the He^+ ion drops from the state $n = 101$ to $n = 100$. What region of the electromagnetic spectrum (radio, microwave, infrared, etc.) is this in?

1.9 A Li^{2+} ion is initially in the state $n = 1$. It absorbs radiation of wavelength 10.4 nm, and immediately emits a photon with wavelength 828 nm to land in an excited final state. What is the n value of this final state?

1.10 Calculate the transition energy in E_h of the longest wavelength absorption transition from the $n = 2$ state of He^+.

1.11 In a spectrum of highly ionized atoms, a physicist observes absorptions at the following energies: 1600 eV, 1613 eV, 1621 eV, and 1626 eV. Assume that these are due to transitions from the ground state of a one-electron atom and identify the element. THINKING AHEAD ▶ [Why are these energy values so close together?]

1.12 Find F_{Coulomb} for an electron in the $n = 1$ state of atomic hydrogen according to the Bohr model.

1.13 Evaluate the potential energy in J, according to the Bohr model, of the electron in the $n = 1$ state of the He^+ atom.

1.14 Write an equation for the kinetic energy of the electron in the Bohr atom in terms of the atomic number Z and principal quantum number n.

1.15 The following table lists properties of the $n=2$ state of the He^+ electron predicted by the Bohr model. Fill in the corresponding values for the $n=3$ state of Li^{2+}.

	$He^+ n = 2$	$Li^{2+} n = 3$
r_n	1.06 Å	
v_n	$2.19 \cdot 10^6$ m s^{-1}	
U_n	$-4.36 \cdot 10^{-18}$ J	

1.16 Two dipole moments μ_A and μ_B can attract each other according to a force law

$$F_{\text{d-d}} = -\frac{A\mu_A\mu_B}{r^4},$$

where A is a constant. Imagine a system analogous to the Bohr model of the atom, but replace the electron orbiting the nucleus by μ_A orbiting μ_B. Using the force law given and Bohr's central assumptions, find an equation for r_n of this system.

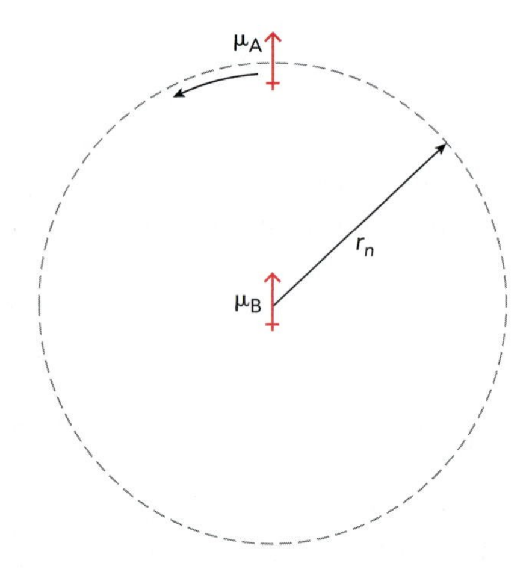

1.17 Find an expression, in terms of n and Z and the fundamental constants, for the centripetal force of the electron in the Bohr model of the atom.

1.18 Values for some properties of the $n = 1$ state of the Bohr model of the hydrogen atom are given in the following table. Write the value of the same parameter (in the same units) for the $n = 2$ state.

parameter	$n = 1$	$n = 2$
momentum (kg m s^{-1})	$1.99 \cdot 10^{-24}$	
de Broglie wavelength (nm)	0.333	
kinetic energy (E_h)	0.500	
transition energy to $n = 3$ (E_h)	0.444	

1.19 Find the lowest value of Z for which the ground state electron of the one-electron atom is predicted by the Bohr model to have a speed faster than one tenth the speed of light. At this point, we would certainly expect relativistic terms to be crucial to a precise description of the system.

1.20 Find the quantum numbers for the $\Delta n = 1$ transition in He^+ for which the energy is 0.045 E_h. THINKING AHEAD ▶ [How does this transition energy compare to the ionization energy?]

1.21 Calculate the linear momentum p (in kg m s^{-1}) predicted by the Bohr model for the electron in the $n = 2$ state of the He^+ ion.

1.22 If the Bohr atom *was* radiating as the electron orbited, what would be the radiation frequency in the $n = 4$ state of Li^{+2}?

1.23 Absorption energies from the ground state of Be^{+3} are at 7.50 E_h and 7.68 E_h. Find the upper state quantum numbers.

1.24 If the microwave range of the spectrum extends exactly from a wavelength of 1.0 mm to 100.0 cm, find the *minimum* value of n such that the $n \rightarrow \infty$ transition (this would be the transition for ionization) in He^+ occurs in the microwave.

1.25 A spectroscopist measuring the spectrum of the Li^{+2} ion observes an absorption signal for radiation at a photon energy of $1.74 \cdot 10^{-17}$ J. What are the n quantum numbers of the observed transition?

1.26

a. Find an equation for the rotational frequency (the number of orbits per second) of the electron in the Bohr model.

b. Calculate the change in this rotational frequency for the $n = 4 \rightarrow 5$ transition in atomic hydrogen.

c. Calculate the frequency of the radiation necessary to induce the same transition.

1.27 Identify the one-electron atom or ion for which 25.6 nm is the wavelength of the $n = 1 \rightarrow 3$ transition.

1.28 If the proton and electron in the Bohr model of hydrogen were replaced by uncharged particles with the same masses, there would still be a very small gravitational attraction between the particles. What is the ionization energy of this atom in J? The equation for the attractive force between two masses is

$$F_{\text{grav}} = -\frac{m_1 m_2 G}{r^2},$$

where the gravitational constant G is equal to $6.67 \cdot 10^{-11}\,\text{m}^3\,\text{kg}^{-1}\text{s}^{-2}$.

1.29 Assuming the Bohr model, identify the one-electron ion for which an electron with an orbital radius of $1.286a_0$ has a speed of $2.333e^2/(4\pi\varepsilon_0\hbar)$.

1.30 Use the Bohr model to calculate the de Broglie wavelength λ_{dB} in m of the muon in the ground state "muonium" atom composed of a proton and a muon (charge $= -e$, mass $= 1.884 \cdot 10^{-28}$ kg). Assume the center of mass is still at the proton.

Photons, de Broglie Wavelengths, and Uncertainty

1.31 If a lightbulb emits $10\text{ W} = 10\text{ J s}^{-1}$ of light with average wavelength of 590 nm, roughly how many photons are emitted per second?

1.32 The photon that excites the $n = 1 \rightarrow 10$ transition in He^+ has the same energy necessary to excite $n = 5 \rightarrow 6$ in what other one-electron ion?

1.33 A YAG laser (based on neodymion-doped yttrium aluminum garnet) is a powerful source of infrared radiation at 1064 nm. To get more energetic photons, this radiation is often processed by a crystal that triples the frequency. Find the photon energy in J of a frequency-tripled YAG laser.

1.34 Difference frequency lasers mix two beams of radiation, usually in the visible, at frequencies ν_1 and ν_2 to generate infrared radiation at the frequency $\nu_3 = |\nu_2 - \nu_1|$. Find the wavelength in microns of the difference frequency laser radiation generated by mixing beams with wavelengths 551 nm and 438 nm.

1.35 A stationary detector directly measures the intensity of free particles called *neutrinos* as a function of time. We position the detector to monitor a beam of neutrinos at some constant speed v. Assume that the wavelike oscillations of the neutrinos all have the same phase and that the neutrino mass is $m = 1.0 \cdot 10^{-37}$ kg.

a. Sketch what the detector sees when $v_1 = 0.010c$, where c is the speed of light. Give numbers and units for the t axis, but not the vertical axis.

b. Do the same for $v_2 = 0.020c$, using the same scale for the t axis.

1.36 An Ar^+ laser emits radiation at a wavelength of 514 nm, and a power of 8 watts ($1\text{ W} = 1\text{ J s}^{-1}$). Determine how many argon ions at a minimum must undergo the emission per second in order to maintain this power output.

1.37 Find the kinetic energies in eV of (a) a free electron and (b) an ion of ^{56}Fe, if both have de Broglie wavelengths equal to the wavelength of visible light with frequency $5.25 \cdot 10^{14}\text{ s}^{-1}$.

1.38 One manifestation of particle-wave duality is that photons carry momentum, even though they have no mass. The same relationship between wavelength and momentum that applies to matter holds for radiation.

a. Calculate the momentum in SI units of one photon of wavelength 450 nm.

b. Find the total energy in a beam of 450 nm wavelength radiation if the beam has exactly enough momentum to accelerate a 1.0 g marble by 0.010 m s^{-1}.

1.39 Find the average de Broglie wavelengths of an electron and a proton in a hydrogen atom in the $n = 1$ state when the net speed of the entire atom is $2 \cdot 10^3\text{ m s}^{-1}$.

1.40 Find the de Broglie wavelength of an electron traveling through free space with speed $6 \cdot 10^5\text{ m s}^{-1}$, the speed after acceleration from rest through a 1 V potential.

1.41 A question of long-standing interest has been the motion and reactions of atomic oxygen along the surface of aluminum crystals. Assume that a typical speed for the O atom is $8 \cdot 10^5\text{ cm s}^{-1}$, and the typical distance traveled along the surface is 40 Å. Use the de Broglie wavelength to show whether or not quantum mechanics is critical to this analysis.

1.42 Like the Bohr model, our version of the de Broglie equation is simplified to neglect special relativity.

a. According to our version of the equation, what is the range of possible wavelengths for the electron, permitting speeds between 0 and c?

b. In atoms with high atomic number, relativistic effects in the motion of the lowest energy electrons are detectable. The momentum p with the relativistic correction is $mv/\sqrt{1 - (v/c)^2}$. Does the relativistic correction make the energy levels farther apart (more quantum) or closer together (more classical) than we would calculate otherwise?

1.43 The center of mass of the Bohr hydrogen atom lies between the nucleus and the electron, a distance $m_e r/(m_{\text{nuc}} + m_e)$ from the nucleus, where m_{nuc} is the mass of the nucleus. What is the ratio of the de Broglie wavelengths of the nucleus and the electron in this atom, λ_{dB} (nucleus)/λ_{dB} (electron)? THINKING AHEAD ▶ [Based on the mass alone, do you expect λ_{dB} to be bigger or smaller for the nucleus than for the electron? Next, how will the speed of the nucleus compare to that of the electron, and how will *this* effect the final λ_{dB}?]

1.44 If the uncertainty in position of an electron is $\delta x = 1.0\,\text{Å}$ and its average speed is $3.0 \cdot 10^6\,\text{m s}^{-1}$, find the electron's de Broglie wavelength and the *minimum uncertainty* in the de Broglie wavelength. The relationship between the uncertainties is given by $\delta p/\delta\lambda_{\text{dB}} = |dp/d\lambda_{\text{dB}}|$.

1.45 We can rewrite the Heisenberg uncertainty principle for the position and distance of a particle in terms of the de Broglie wavelength,

$$\delta x\, \delta\left(\frac{h}{\lambda_{\text{dB}}}\right) \geq \frac{\hbar}{2}.$$

Replace λ_{dB} by the photon wavelength λ to derive a similar inequality in terms of δE_{photon} (the uncertainty of the photon energy for electromagnetic radiation) and one other uncertainty.

2 The Schrödinger Equation

LEARNING OBJECTIVES

After reading this chapter, you will be able to do the following:

1. Find the eigenvalue, if any, of a function under a given operation.
2. Use the average value theorem to calculate properties of a system with a known wavefunction.
3. Devise a Schrödinger equation for simple systems in Cartesian coordinates
4. Determine the energies, wavefunctions, and other properties of a particle in a one- or three-dimensional box

GOAL *Why Are We Here?*

The goal of this chapter is to introduce the basic methods of quantum mechanics, which we will use to tackle questions and gain insight about the structures and motions of matter on the scale of individual atoms and molecules. We base our approach on the Schrödinger equation, so we begin by analyzing the components of this equation and then apply what we've learned to three important, fundamental sample systems.

CONTEXT *Where Are We Now?*

This chapter introduces the mathematical tools of quantum mechanics, which extend the laws of mechanics to the tiny sizes and masses of atoms and molecules. The structure and motions of these particles are the essence of chemistry, and quantum mechanics provides our only means of accurately predicting their experimental properties when studied one at a time. Once we have a little practice writing and employing the Schrödinger equation, we will use the same approach as we investigate the detailed structure of the atom in Chapters 3 and 4, and then carry those results forward to the study of molecular properties in Chapters 5–9. Our results from quantum mechanics remain valuable beyond the microscopic scale, forming the key that we use to answer critical questions in chemical thermodynmics and kinetics as well.

SUPPORTING TEXT *How Did We Get Here?*

- We can save ourselves a lot of trouble in manipulating equations if we recall that all the trigonometric functions depend on one another, through identities given in Table A.2. Also, some simple differential equations can be solved using the expressions in Table A.7.
- The de Broglie wavelength (Eq. 1.3) $\lambda_{dB} = h/p$ defines the wavelike nature of a particle with mass m and momentum $p = mv$. This

wavelength accurately predicts, among other phenomena, the constructive and destructive interference of beams of electrons, establishing that the electron cannot always be treated as a point mass or point charge. Instead, the electron is distributed through space. We will have to take this property into account as we construct a new model of the electron in the atom.

- The Bohr model of the atom accurately predicts the quantized energies of the one-electron atom (Eq. 1.15):

$$E_n = -\frac{Z^2 e^4 m_e}{2(4\pi\varepsilon_0)^2 n^2 \hbar^2} \equiv -\frac{Z^2}{2n^2} E_{\mathrm{h}}.$$

  The energy in this equation is measured relative to the ionization energy of the atom, which means that the lower energy states in which the electron and nucleus are bound together correspond to *negative* energies. The Bohr model is limited by simplifying approximations, which we will correct, but this predicted energy remains valid for one-electron atoms.
- A complex number is the sum of a real number and an imaginary number (Section A.1), with the imaginary number being a real number times i. The complex conjugate of a complex number is the same pair of terms but with the opposite sign of the imaginary component. So for every complex number $a + ib$, there is a complex conjugate $(a + ib)^* = (a - ib)$. The complex conjugate is designed to guarantee that the product $(a + ib)(a + ib)^* = a^2 + b^2$ will be a non-negative real number. We use complex numbers in this chapter, and in what follows, to keep track of the direction of motion.

2.1 Mathematical Tools of Quantum Mechanics

As with classical mechanics, the goal of quantum mechanics is to describe the motions of interacting particles. Between 1925 and 1927, Werner Heisenberg, Erwin Schrödinger, and Paul Dirac formulated different mathematical approaches to this problem, all of which remain in use today. Schrödinger's method offers the simplest mathematics for the quantum mechanics that we will need.

The Schrödinger Equation

In the 1920s, Schrödinger was fascinated by the prospects for general theories that would unify different areas of physics. In 1925, following a presentation by de Broglie, he was inspired to attempt a derivation of the hydrogen atom spectrum, starting from de Broglie's insight into the wavelike nature of small particles. Schrödinger relied more heavily on principles of classical mechanics than did Heisenberg and Dirac, bringing concepts from Newtonian physics to bear on a decidedly non-Newtonian problem. Unlike Bohr's model of the atom, the solution that Schrödinger arrived at explicitly treated the electron by the laws of wave mechanics, and allowed the electron to occupy all three dimensions.

Deceptively concise in its simplest form, the **Schrödinger equation** is written:

$$\hat{H}\psi = E\psi. \tag{2.1}$$

We will use this equation as our starting point as we predict the structures and motions in atomic and molecular systems.

The first hurdle in using the Schrödinger equation is identifying the symbols, but there are only three: the **Hamiltonian operator,** $\hat{H}$; the **wavefunction,** ψ; and the energy, E. The energy is the sum of the potential and kinetic energies of the particles in the system, such as the energy of the electron in the one-electron atom. The wavefunction may be a vector or a matrix or an algebraic expression that describes the distribution of the particles in space or time or some combination. The Hamiltonian is an example of an **operator,** a mathematical process such as multiplying by 3 or taking a derivative, that acts on a function to leave a (usually different) mathematical function behind. *The Hamiltonian is specifically the operator that allows us to solve for the possible energies and wavefunctions of the system, starting only from physical properties of the system such as the forces at work and the masses of the particles.*

CHECKPOINT Here we introduce a color code for the three components of the Schrödinger equation: the Hamiltonian $\hat{H}$ (blue), the wavefunction ψ (black), and the energy E (red). The colors serve two purposes: to distinguish among these expressions as we discuss each in turn, and (later) to illustrate qualitatively how the Hamiltonian extracts the energy from the wavefunction.

Note that the Hamiltonian is not a coefficient multiplying the wavefunction. Instead it is a series of mathematical steps, including derivatives. The derivatives make the Schrödinger equation a differential equation. Unlike an algebraic equation, we don't need to know two of the three terms in the equation to find the third. If we know everything about the Hamiltonian *or* the set of possible wavefunctions *or* the set of possible energies, we can find the other two terms.

It will take the next several pages to make sense out of the wavefunction and the Hamiltonian, but we can paint a crude picture of how the Schrödinger equation works by referring to some classical physics. The standard problem in classical mechanics runs along these lines: you start with a bunch of particles at given initial positions and velocities and interacting through some given force law, and you have to solve for the positions and velocities of the particles at some later time. The solution can be obtained using Newton's laws of physics and a lot of integration. William Hamilton devised an approach to this problem that breaks the energy of the particles into kinetic energy and potential energy contributions, in effect rewriting Newton's laws—which are expressed in terms of forces—so that they depend instead on these energies. Although the solution is the same either way, Hamilton's method yields insights that might be missed otherwise, because—unlike forces—the total energy of the system is a conserved quantity.

Hamilton's method begins by constructing a Hamiltonian expression, which summarizes what we know about the system: how many particles are present, what their masses are, and what forces govern their interactions. From the Hamiltonian, we can then predict what the system will look like at a later time. But the method yields a *family* of solutions, rather than one specific solution, because the final positions and velocities of the particles depend on what energies the particles have initially. We get different solutions for the locations of the particles—the *distribution* of the particles—for different values of the total energy E.

The Hamiltonian represents our *theoretical* understanding of the system, because we can assemble it from a few system parameters and fundamental formulas. From the Hamiltonian, we can predict and relate the *experimental* observables of the system: its energy, and the distribution of the particles. Similarly, if we were to measure in the laboratory the final energy and distribution of the particles, we could work backward to find the Hamiltonian. This would tell us, for example, what the masses of the particles are and how they were originally distributed.

The Schrödinger equation applies the same logic. If we wish to apply our theoretical understanding to predict experimental results—perhaps to forecast the molecular structure of a new solid fuel—then we set about the problem in the following way. We first construct a Hamiltonian operator $\hat{H}$ based on what we know about the system—how many atomic nuclei, their charges and masses, how many electrons—and based also on the force law for the interactions between the particles. We then solve the Schrödinger equation—starting from only the Hamiltonian—by integration. The solution yields a family of results: different energies E of the molecule lead to different possible distributions of the electrons and nuclei, the molecular structures we wanted to find. In the case of quantum mechanics, the distribution of particles is given by the product of the wavefunction and its complex conjugate, $|\psi|^2 \equiv \psi^*\psi$.

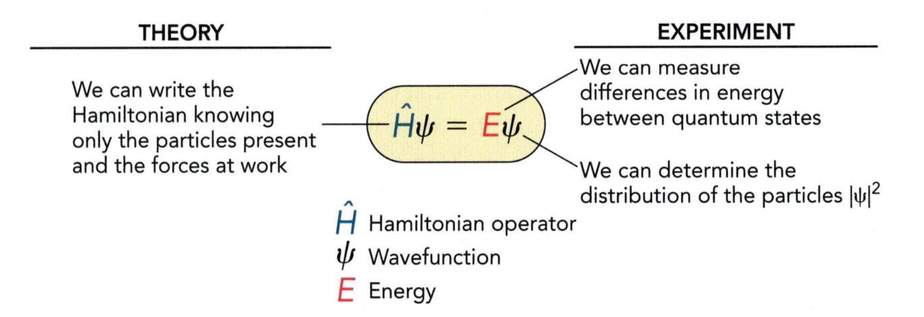

Let's now examine the wavefunction and the quantum-mechanical Hamiltonian in some depth.

The Wavefunction

Experimental measurements of atomic and molecular properties typically probe one of two parameters: how much energy is present in a particular type of motion, and where the particles—electrons and/or nuclei—can be found. Spectroscopy determines energies, and a variety of experiments determine location. Conductivity measurements, for example, can provide one measure of where electron charge is distributed within a molecule.

But how shall we define the *location* of atomic and molecular particles in the quantum world, where these particles have properties we'd usually associate with waves? We can no longer ask for a precise location. In classical mechanics, we could imagine the behavior of a point mass, or a point charge, but there's no such thing as a "point wave." A wave must extend over a distance, because it is defined by the difference in amplitude between one point and another point. This is where we need the first piece of our Schrödinger equation.

Mathematical and Physical Definitions

Instead of specifying a single location, we describe the position of a quantum particle in terms of its **probability density.** There are two common interpretations of the probability density:

1. If we make a measurement that can only tell us whether or not the particle is at some point x, then the probability density at x gives the fraction of times that we will detect the particle there. For example, if the particle is an electron, and we position a tiny wire at x with a switch that will close (to allow the electrons through) only at certain times, then the probability density at x tells us what chance we have of seeing the electron appear at the other end of the wire when we close the switch. This interpretation treats the electron like a classical particle, with a well-defined location.
2. If instead we measure some quantitative property of the particle, such as its mass or charge, then the probability density at point x tells us what fraction of the parameter's total value will be found at point x. In this case, imagine that we now connect a charge meter at point x so that we continuously monitor how much charge from the electron is at that location. In this experiment, if the probability density where we make our measurement is 0.1, then we will measure a total charge of $-0.1e$. This interpretation treats the electron more like a wave, with its properties distributed over space rather than concentrated in one location.

This brings us at last to the wavefunction ψ that appears in the Schrödinger equation. The probability density of a particle (a molecule or electron, for example) at any point is given by the the wavefunction multiplied by its complex conjugate (Section A.1), the **square modulus** $|\psi|^2 \equiv \psi^*\psi$, at that point.

If we define the wavefunction in terms of the probability density, and it's the probability density that determines what we measure in the laboratory, then why do we need the wavefunction at all? The wavefunction is a mathematical convenience that allows us not only to predict the properties of individual quantum states, but also to predict results that stem from combinations of quantum states. It is the wavefunction, not the probability density, that solves the Schrödinger equation, and—like a wave—the wavefunction can undergo constructive and destructive interference with other wavefunctions to accurately predict the resulting probability densities.

To illustrate this last point, if we let two wavefunctions ψ_1 and ψ_2 overlap, forming a total wavefunction $\psi = \psi_1 + \psi_2$, then the probability density of the result is

$$|\psi_1 + \psi_2|^2 = (\psi_1 + \psi_2)^*(\psi_1 + \psi_2). \tag{2.2}$$

This is different from the sum of the two probability densities, $|\psi_1|^2 + |\psi_2|^2$, and correctly describes the way electrons and other quantum-mechanical particles interact. However, we must emphasize that our experiments never directly measure the wavefunction. In fact, it has been proven that the wavefunction is not *needed* to calculate properties of our particles (we'll see more on this under Density Functional Theory in Section 7.3), but it gets us those results more directly and with greater accuracy than any other method.

Wavefunctions are often complex expressions (having imaginary components), and for our purposes the imaginary part may be thought of as carrying information about how the system changes with time. Multiplying the wavefunction by its complex conjugate ψ^* ensures that the probability density—which is what we can actually measure—has no imaginary components.

Any wavefunction for a real system obeys the following rules, common to well-behaved functions that describe physical phenomena:

1. It has only one value at a given point in time and space.
2. It is finite and continuous at all points, and so are its first and second derivatives with respect to distance.
3. The function $|\psi|^2$ has a finite integral over all space.

(Having said all that, we will examine some idealized systems that violate rules 2 and 3, but those are limited to this chapter.)

In our notation, Ψ will often be used to describe the "complete" wavefunction, which may include time-dependent or magnetic terms. Conveniently, we are more often interested only in the terms that describe the distribution of the system in space—the spatial wavefunction—for which we will use the lower case ψ.

To explore an example further, let's take two wavefunctions that happen to be discussed in the next chapter and see what happens when we add them. For now, we'll use only one coordinate—the angle θ measured from the Cartesian z axis. The wavefunction for a $2s$ electron is similar to a sphere, having no angle-dependence,

$$\psi_{2s}(\theta) = \sqrt{\frac{1}{2}}, \tag{2.3}$$

while the wavefunction for a $2p$ electron has the form

$$\psi_{2p}(\theta) = \sqrt{\frac{3}{2}}\cos\theta. \tag{2.4}$$

These functions are shown in Fig. 2.1.

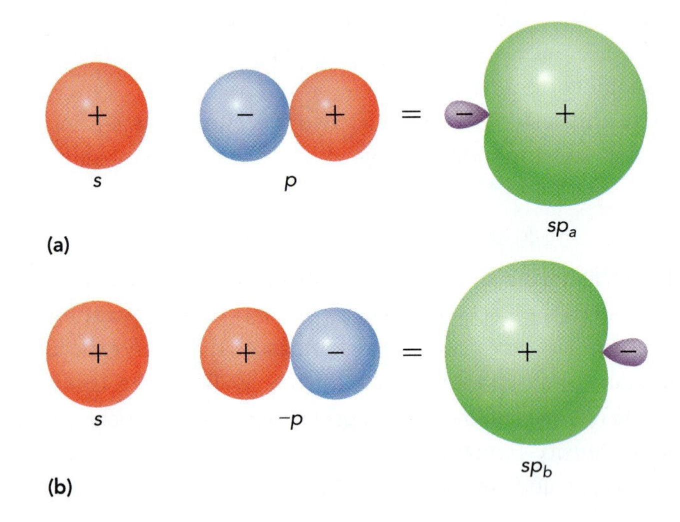

▶ FIGURE 2.1 **The $2s$ wavefunction and $2p$ wavefunction can be combined to form two distinct sp wavefunctions.** Each diagram shows a surface along which the wavefunction has a constant value.

The probability of finding the 2*s* electron at some position does not depend on the angle. The probability of finding the 2*p* electron at a given angle θ is proportional to $\cos^2\theta$. If we combine these two, we obtain an *sp* hybrid wavefunction described by the equation

$$\psi_a(\theta) = \psi_{2s}(\theta) + \psi_{2p}(\theta) = \sqrt{\frac{1}{2}} + \sqrt{\frac{3}{2}}\cos\theta, \tag{2.5}$$

with a probability density given by $|\psi_a(\theta)|^2$, represented to the right of the equal sign in Fig. 2.1a. If we had written instead

$$\psi_b(\theta) = \sqrt{\frac{1}{2}} - \sqrt{\frac{3}{2}}\cos\theta, \tag{2.6}$$

we would have had a distribution oriented in the opposite direction, the mirror image of the first result, as shown in Fig. 2.1b.

The **phase** of a wavefunction is any coefficient that multiplies the whole wavefunction but has no effect on the probability density. Mathematically, that means that the phase is any number a that has a square modulus of 1:

$$|a|^2 \equiv a^*a = 1. \tag{2.7}$$

For example, if my wavefunction is $\psi(\theta) = \sin\theta$, then multiplying ψ by -1 changes the value of the wavefunction at most points but leaves the probability density $\psi^2(\theta) = \sin^2\theta$ unchanged. The value -1 in this example is a phase factor. The square root of -1, i, is also a phase factor:

$$\begin{aligned}|i\psi(\theta)|^2 &= [i\psi(\theta)]^*[i\psi(\theta)] = [-i\psi(\theta)]\,[i\psi(\theta)] \\ &= -i^2\psi^2(\theta) = -(-1)\psi^2(\theta) = \psi^2(\theta).\end{aligned}$$

Multiplying a wavefunction by i does not affect the probability density.

Why then does the phase of the wavefunction matter, if all we can measure is the probability density? The overall phase of the wavefunction does *not* matter, but its phase *relative to another wavefunction* does affect the predicted outcomes. The difference between $\psi_a(\theta)$ and $\psi_b(\theta)$ in the previous example is that we reversed the phase of the 2*p* term *relative* to the 2*s* term, which reversed the direction of the *sp* wavefunction.

If we worked only with real functions, our phase factors would be 1 or -1. However, wavefunctions can have imaginary components, and so can the phase factors. A general way to represent the phase that allows for complex values is by a factor $e^{i\phi}$, where ϕ varies from 0 to 2π. The **Euler formula** (see Problem A.9) allows us to write this as

$$e^{i\phi} = \cos\phi + i\sin\phi. \tag{2.8}$$

For $\phi = 0$, $e^{i\phi}$ is one; for $\phi = \pi/2$, $e^{i\phi}$ is i; for $\phi = \pi$, $e^{i\phi}$ is -1; for $\phi = \pi/4$, $e^{i\phi}$ is $(1 + i)/\sqrt{2}$; and so on. No matter what value of ϕ you select, the square modulus $e^{i\phi}e^{-i\phi}$ is equal to one. The functions $\psi(\theta)$, $-\psi(\theta)$, $i\psi(\theta)$, and $(i - 1)\psi(\theta)/\sqrt{2}$ all have different phase factors, but they have the same square modulus, and their measurable properties, such as where the mass and charge of the particle are located, will be identical. Effectively, phase factors merely shift the coordinate system. A change in the absolute phase cannot affect what the wavefunction predicts we will measure in the lab. Multiplying a wavefunction by -1 is like deciding that you want the z axis to point in the opposite direction when you measure the location of a particle. That decision changes the sign of z in all the

EXAMPLE 2.1 General Form of Wavefunctions

CONTEXT As we will see in subsequent chapters, we often solve the Schrödinger equation by assuming at the outset that we know what kind of functional form—say, exponential or sine wave—a particular wavefunction will have. The question then is, of all the functions possible in our coordinate system, which is the best guess to start from? We simplify that problem by taking advantage of the constraints on the wavefunction: it must be finite and continuous, for example. We look for functions that are also easy to work with, but most of all they must meet these minimum requirements to help us predict measurable parameters.

PROBLEM For each of the following functions, state whether or not it could be a rigorously valid *wavefunction.* If not, state why not.

1. $f(x) = x^2 - 5x + 2$
2. $f(x) = \pm e^{-x^3}$
3. $f(x) = \sin x$
4. $f(x) = e^{-x^2}$

SOLUTION 1. No. The function diverges, becoming infinite as x approaches infinity, and therefore this cannot be a valid wavefunction:

$$\lim_{x \rightarrow \infty} (x^2 - 5x + 2) = \lim_{x \rightarrow \infty} x^2 \rightarrow \infty.$$

There are other ways to state this: the probability density at large x is infinite; the normalization constant is zero.

2. No, because $f(x)$ has two different values for most values of x (the function is not single valued); and furthermore, it also diverges at large negative x values.

3. Strictly speaking, no, because $|f(x)|^2$ does not have a finite integral over all values of x. Even though the value of $\sin x$ always lies between -1 and 1, the function continues oscillating all the way to $x = \pm\infty$, and the integral of the probability density adds up to infinity:

$$\int_{-\infty}^{\infty} \sin^2 x \, dx = \infty.$$

4. Yes. This function is finite, continuous, and single valued at all values of x, and it converges to zero as x approaches either ∞ or $-\infty$, which means $|f(x)|^2$ will have a finite integral over all space.

expressions that describe the system. If the particle starts at $z = 0$ and travels until it reaches $z = 10$ in the original coordinate system, changing the sign of z means we would say that the particle travels from 0 to $z = -10$ but the particle has the same physical location at the end of the trip in either case. The *physical* results of an experiment are not affected by the choice of absolute phase, but the numerical results may be affected, as in this case.

Normalization

With few exceptions, we will work with **normalized** wavefunctions, which are simply wavefunctions multiplied by the proper coefficient such that the probability density over all space totals one. In other words, if we look everywhere, we must have one chance in one (i.e., a 100% chance) of finding the particle described by that wavefunction. Even if the particle is distributed here and there as a de Broglie wave, by the time we have counted up all the charge and all the mass in every location, we must have exactly the mass and charge of the particle. Mathematically, we write this as

$$\int_{\substack{\text{all}\\\text{space}}} |\psi|^2 d\tau = 1, \tag{2.9}$$

where $d\tau$ is the volume element appropriate to the coordinate system being used (e.g., $dxdydz$ for Cartesian coordinates), and the integral is over every point in that coordinate system.

Normalizing the wavefunction is a separate problem from solving the Schrödinger equation. If we found a wavefunction ψ that satisfied the Schrödinger equation $\hat{H}\psi = E\psi$, then any other wavefunction $\psi' = A\psi$ (where A is some constant) would also satisfy the equation, because A would multiply both sides. However, there is only one value of $|A|^2$ such that the square modulus $|\psi'|^2$ gives the correct probability density. Normalizing the wavefunction thus removes any ambiguity as to the magnitude of ψ. Now, if we integrate over only *part* of the space, then the integral $\int |\psi|^2 d\tau$ for a normalized wavefunction gives the fractional probability density of the system within the limits of the integral.

For example, suppose we make our system a classical ball rolling back and forth at constant speed along a flat floor between two walls, one wall at $x = 0$ and one at $x = a$ (Fig. 2.2). The wavefunction is a trivial one, a constant $\psi(x) = A$, indicating that the ball has an equal chance of being found anywhere between the two walls. The fractional probability of finding the ball between the two walls is 1, meaning that the ball is always somewhere between the walls. Put mathematically,

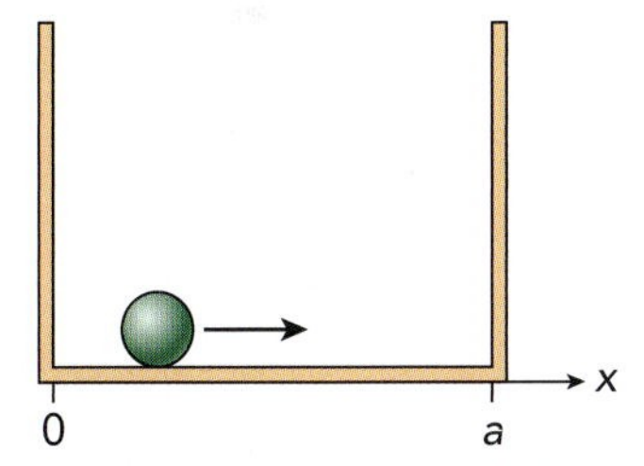

▲ **FIGURE 2.2 A classical ball rolling back and forth between two walls.**

$$\int_0^a |A|^2\, dx = |A|^2 \int_0^a dx = |A|^2 a = 1.$$

The constant A may be equal to $a^{-1/2}$, or $-a^{-1/2}$, or any $e^{i\phi}a^{-1/2}$. The overall phase factor is irrelevant to the normalization; normalization fixes the *magnitude* of the wavefunction, not its phase. If we wanted to show that the ball spent half its time between $x = 0$ and $x = a/2$, we could integrate over that region of the floor, now using the value of A we determined from the normalization step:

$$\int_0^{a/2} |A|^2\, dx = \int_0^{a/2} \frac{dx}{a} = \left.\frac{x}{a}\right|_0^{a/2} = \frac{1}{2}.$$

The result, 1/2, means that the ball spends half of its time between $x = 0$ and $x = 1/2$. To be more accurate, we've shown that if we measured the mass present in the region between $x = 0$ and $x = 1/2$ over many round trips of the ball, we'd get an average mass equal to half the mass of the ball. If the wavefunction were not properly normalized, we would not get this correct answer. For example, if the distance between the walls is 10 m, and I choose the wavefunction to be a constant $\psi'(x) = 1$, the integral over half the distance gives me

$$\int_0^{5\,\text{m}} \psi'(x)^2 dx = \int_0^{5\,\text{m}} dx = 5\,\text{m},$$

EXAMPLE 2.2 Normalization and Probability Density

CONTEXT Once we've solved a Schrödinger equation to get the wavefunction for the electrons in an atom or molecule, we can calculate the distribution of the electrons. This tells us, for example, how much charge from the electron is at each point in the system, allowing us to predict dipole moments or nuclear magnetic resonance spectra (such as the spectrum below; more on this topic in Section 5.5). In this example, we show how the wavefunction for a $2p$ electron, if properly normalized, can tell us how much electron density lies close to the plane through the middle of the orbital.

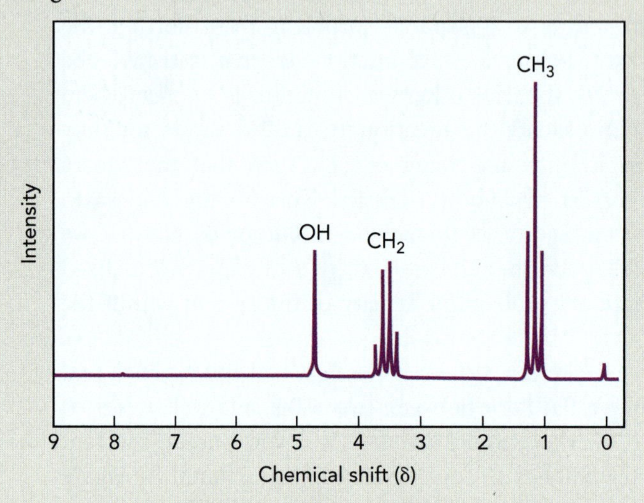

PROBLEM (a) Prove that the $2p$ angular function $\psi_{2p}(\theta) = A\cos\theta$ has a normalization constant $A = \sqrt{3/2}$, if $0 \leq \theta < \pi$.

(b) In this $2p$ wavefunction, what fraction of the electron charge will be found between $\theta = 60^\circ$ ($\pi/3$) and $\theta = 120^\circ$ ($2\pi/3$)?

SOLUTION (a) To test the normalization, the following parts of the problem need to be addressed:

- Our unnormalized wavefunction ψ is $\cos\theta$.
- The complex conjugate ψ^* is $\cos\theta$ (taking the complex conjugate of a real function leaves the function unchanged).
- The square modulus $|\psi|^2 = \psi^*\psi$ is therefore $\cos^2\theta$.
- This wavefunction is a function only of θ. Therefore, the volume element is $\sin\theta\, d\theta$, and the range "all space" in this coordinate falls between the limits 0 and π.

Now we're all set:

$$\int_0^{\pi} A^2\cos^2\theta \sin\theta\, d\theta = \frac{1}{3}A^2(-\cos^3\theta)\Big|_0^{\pi} = \frac{2}{3}A^2 = 1.$$

So for the $2p$, $A = \sqrt{\frac{3}{2}}$.

(b) We integrate the probability density $|\psi|^2$ over $\pi/3 \leq \theta < 2\pi/3$:

$$\int_{\frac{\pi}{3}}^{\frac{2\pi}{3}} A^2\cos^2\theta \sin\theta\, d\theta = \frac{1}{3}\left(\frac{3}{2}\right)(-\cos^3\theta)\Big|_{\pi/3}^{2\pi/3} = \frac{1}{2}\left(\frac{1}{8} + \frac{1}{8}\right) = \frac{1}{8} = 0.125.$$

We're looking over one third of the range of θ, and if we integrate $\sin\theta\, d\theta$ alone we'll find that this range is half of the volume around the atom. Yet the electron probability density in this range is only one-eighth of the total. For this wavefunction, the probability density lies close to the z axis and approaches zero near the xy plane.

which does not tell me what I want to know. Instead, using the *normalized* wavefunction $\psi(x) = (10\,\text{m})^{-1/2}$, this integral gives me

$$\int_0^{5\,\text{m}} \psi(x)^2 dx = \frac{1}{10\,\text{m}} \int_0^{5\,\text{m}} dx = \frac{5}{10} = \frac{1}{2},$$

which is the fractional probability density in the first half of the path.

Orthogonality and Basis Sets

The algebraic form of a wavefunction can be very complicated, and any method of describing wavefunctions that provides some immediate qualitative insight into their nature is a valuable tool. The most common approach to a complicated wavefunction is to write it as a sum over the members of a **basis set** of relatively simple wavefunctions.

Let's say, for example, that we have to find an expression for an unnormalized wavefunction $\psi(x)$ between $x = 0$ and $x = 10$, and we know that at $x = 0$ the wavefunction has a value $\psi(0) = 2$. Then we could start with a guess that our wavefunction can be written $\psi(x) = 2$. Next we learn that a second point on the wavefunction has the value $\psi(1) = 3$, so we improve our guess wavefunction to write $\psi(x) = 2 + x$, which has the correct value at both points. If we measure a third point to be $\psi(2) = -2$, then the function $\psi(x) = 2 + 4x - 3x^2$ is consistent with all three points. We are building up an nth order polynomial, and we can just keep adding terms to the polynomial until our $\psi(x)$ matches every point in the system where we have a measurement. In this example, our basis functions—which we'll call φ_i — are the polynomial terms $\varphi_1 = x^0 = 1$ (the constant term), $\varphi_2 = x^1$, $\varphi_3 = x^2$, and so on. We wrote our wavefunction as a **linear combination** of the basis functions φ_i:

$$\psi = \sum_{i=1}^{\infty} c_i \varphi_i.$$

The c_i in Eq. 2.10 are the numerical coefficients that multiply the basis functions, so in our expression $\psi(x) = 2 + 4x - 3x^2$, we would set $c_1 = 2$, $c_2 = 4$, and $c_3 = -3$. The basis set of all x^n constitutes a **complete** set because no matter what the shape of our wavefunction, some polynomial—some combination of x^n—can be assembled that has that shape. The basis sets we will use are complete, and they get that way by containing an infinite number of functions. Happily, we will only use the simplest functions in each set.

However, unlike the x^n basis set, the basis sets that we will use are constructed from independent functions. By independent functions, we mean that one function (for example, the coordinate x) can be set at any value without affecting the value of a second function (such as y). In contrast, it is not possible to adjust the value of x without also changing the value of x^2. In mathematics, we would say that our independent basis functions are always **orthogonal,** meaning that the functions have been chosen so that, for any two basis functions φ_i and φ_j,

$$\int_{\substack{\text{all}\\ \text{space}}} \varphi_i^* \varphi_j d\tau = 0, \text{if } i \neq j.$$

If the basis functions are all normalized and orthogonal to each other, then we have an **orthonormal** basis set, and we can combine both features into a single equation,

$$\int_{\substack{\text{all}\\\text{space}}} \varphi_i^* \varphi_j d\tau = \delta_{ij},$$

where the **Kronecker delta function** δ_{ij} is 1 if $i = j$ and 0 otherwise. Orthogonality ensures that the basis set does not have any functions that can be made by simple addition of other functions in the basis set. In other words, an orthogonal basis set contains the *minimum* number of functions for a complete set. An orthogonal set of functions will result naturally every time we solve the Schrödinger equation, and we will see an example of how they are obtained by the end of this chapter.

EXAMPLE 2.3 Orthogonality

CONTEXT Quantum mechanics often involves solving a lot of integrals, and it's a big help if we can set up a problem so that we *know* in advance that a whole class of integrals over complicated functions evaluate to zero. We usually select the basis functions that we will use to write wavefunctions so that they are mutually orthogonal. By clever planning along these lines, the daunting mathematics of quantum mechanics became sufficiently tractable during the 1980s that we could start reliably predicting geometries of molecules based on only the fundamental physics. Computational quantum mechanics is now a major tool in the development of new drugs and new materials.

PROBLEM Prove that the θ-dependent s and p wavefunctions in Eqs. 2.3 and 2.4 are orthogonal.

SOLUTION We need only integrate the product $\psi_{2s}(\theta)\psi_{2p}(\theta)$ over all space (between 0 and π for θ) and show that this is zero. For integration over θ alone, our volume element is $\sin\theta\, d\theta$, which is the θ-dependent part of the volume element in spherical coordinates:

$$\begin{aligned}\int_0^\pi \psi_{2s}(\theta)\psi_{2p}(\theta)\, d\tau &= \int_0^\pi \left(\sqrt{\frac{1}{2}}\right)\left(\sqrt{\frac{3}{2}}\cos\theta\right)\sin\theta\, d\theta \\ &= \sqrt{\frac{3}{4}}\int_0^\pi \cos\theta \sin\theta\, d\theta \\ &= -\sqrt{\frac{3}{4}}\left(\frac{1}{2}\right)(\cos^2\theta)\,\Big|_0^\pi \\ &= \sqrt{\frac{3}{4}}\left(\frac{1}{2}\right)(1-1) = 0.\end{aligned}$$

Because the answer is zero, the two functions are indeed orthogonal.

We needn't have done the math to show that the integral is zero if we recognized that $\cos\theta\sin\theta$ between 0 and π is divided equally into positive and negative regions. Integrating $\cos\theta\sin\theta$ over this region must give zero, since the positive and negative regions of the graph exactly cancel. Multiplying the function by any constant only changes the amplitude of the wave; the integral will still be zero. This kind of symmetry analysis is often useful for simplifying the calculus we encounter in this text.

EXAMPLE 2.4 Orthogonality by Symmetry

PROBLEM Which pairs, if any, of the following three functions can we expect to be orthogonal over the interval from 0 to π, given the symmetry properties apparent in the accompanying graphs?

$$f_1(x) = \sin x \sin(2x) \qquad f_2(x) = \sin x \cos(3x) \qquad f_3(x) = \sin(2x)\cos(3x)$$

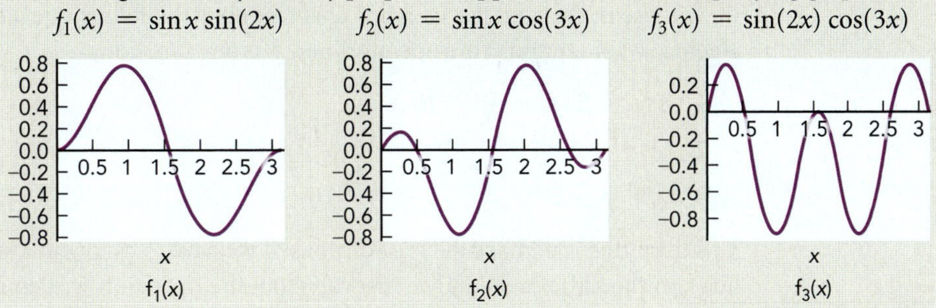

SOLUTION Between 0 and π, $f_1(x)$ and $f_2(x)$ are odd functions because $f_1(x) = -f_1(\pi - x)$ and $f_2(x) = -f_2(\pi - x)$. On the other hand, $f_3(x)$ is even in that $f_3(x) = f_3(\pi - x)$. That means that the products $f_1(x)f_3(x)$ and $f_2(x)f_3(x)$ are also odd functions and will integrate to zero, whereas $f_1(x)f_2(x)$ will be even and in general will not integrate to zero. These products are the integrands $f_i(x)f_j(x)$ that we would need when checking orthogonality by evaluating the integrals $\int_0^\pi f_i(x)f_j(x)\,dx$. The integrals over all space of odd integrands vanish. If we examine these product functions, we can see that $f_1(x)f_3(x)$ and $f_2(x)f_3(x)$ have canceling areas of positive and negative value, and those are the orthogonal pairs.

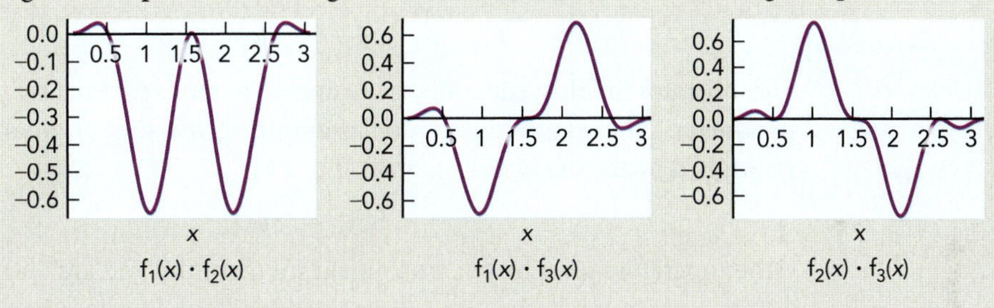

BIOSKETCH Lene Vestergaard Hau

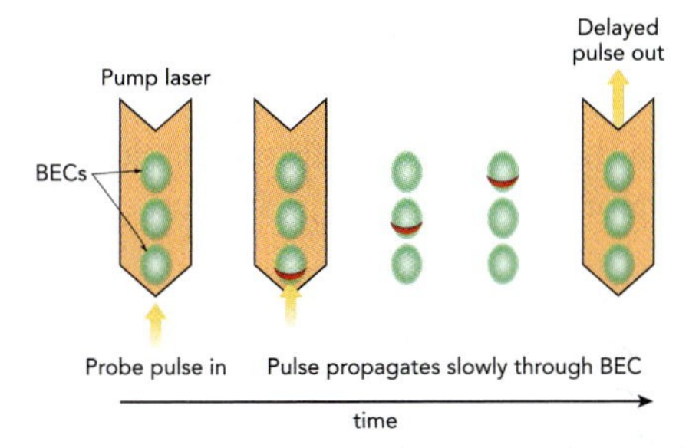

Lene Vestergaard Hau is the Mallinckrodt Professor of Physics and of Applied Physics at Harvard University. Her group has been working on the interaction of light with cold matter. The constant $c = 2.998 \cdot 10^8$ m s^{-1} that we use throughout this book is the speed of light in a vacuum, but light travels slower than this when it passes through matter. The Hau group has studied ways of slowing laser light as it propagates through a very cold gas known as a Bose–Einstein Condensate, or BEC (See the companion volume of this text, *Physical Chemistry: Statistical Mechanics, Thermodynamics, and Kinetics*, Chapter 4). In a sample of atomic sodium gas cooled to roughly 10^{-6} K, Hau and her coworkers slowed the photons to a speed of only 17 m s^{-1}. This braking of the light can be switched on and off by a second laser called the *pump*. With the pump laser off, a burst of light from the first laser (the *probe*) travels through the gas near its vacuum speed. Turning on the pump laser mixes together two quantum states of the sodium atoms in a cleverly planned interaction that prevents the light from the probe laser from being absorbed but forces it to interact so strongly with the sample that the light slows down to a ten-millionth of its typical speed. The effect is used to study the nature of the BEC as well as the light that it affects so dramatically.

The Hamiltonian and Other Operators

The Hamiltonian $\hat{H}$ is essentially defined by the Schrödinger equation to be the mathematical operation that acts on our wavefunction to determine the energy of that wavefunction's quantum state. The operators, including the Hamiltonian, that we will use in this and the next few chapters obey the following rules of **operator algebra**: where $\hat{\alpha}$ and $\hat{\beta}$ are operators and f is any function,

1. if $\hat{\alpha} = \hat{\beta}$, then $\hat{\alpha}f = \hat{\beta}f$;
2. $(\hat{\alpha} + \hat{\beta})f = \hat{\alpha}f + \hat{\beta}f$;
3. $\hat{\alpha}\hat{\beta}f = \hat{\alpha}(\hat{\beta}f)$.

This third rule is perhaps the least obvious: if we have to carry out a series of operations on the same function, we first carry out the operation written *immediately to the left* of the function. If we change the order of operations in a problem, in general we won't get the same result. Operators include multiplication factors, addends, and derivatives, and there are others that we will use to test symmetry.

The operators that we'll use operate on wavefunctions. As discussed earlier, we have to decide which parameters we want to be able to measure exactly and which parameters the uncertainty principle (Eq. 1.19) will prevent us from knowing. When possible, we choose our wavefunctions so that the result of the operation is the original wavefunction multiplied by some coefficient:

$$\hat{\alpha}\psi = a\psi.$$

The operator in this equation is $\hat{\alpha}$, and a is an experimentally determinable parameter. The Hamiltonian, for example, gives the original wavefunction multiplied by the energy of the state (Eq. 2.1):

$$\hat{H}\psi = E\psi.$$

Other operators give the position, linear momentum, and angular momentum of a state, and we can derive what these operators look like based on the nature and number of the particles in the system. We will see a couple of examples of that in the rest of this section. How the operators extract these parameters from the wavefunction depends on the operator, but as an example let's take a sine wave with frequency ν:

$$\psi(x) = \sin(2\pi\nu x).$$

There is an operator $\hat{\alpha}$ in this case that will extract from this wavefunction the square of the frequency. That operator is $\hat{\alpha} = -\frac{1}{(2\pi)^2}\frac{d^2}{dx^2}$, because

$$\hat{\alpha}\,\psi(x) = -\frac{1}{(2\pi)^2}\frac{d^2}{dx^2}\sin(2\pi\nu x) = -\frac{2\pi\nu}{(2\pi)^2}\frac{d}{dx}\cos(2\pi\nu x)$$
$$= -\frac{(2\pi\nu)^2}{(2\pi)^2}[-\sin(2\pi\nu x)] = \nu^2\sin(2\pi\nu x) = \nu^2\psi(x).$$

However, the uncertainty principle tells us that we cannot know all the parameters of the particle simultaneously. In the language of these operators, that means that we can never choose one single wavefunction that will be unchanged after operation of, for example, *both* the position and momentum operators. Instead, if we find a wavefunction that is unchanged after operating on it with, say, $\hat{\alpha}$ as shown earlier, then when we operate on that same wavefunction with another operator we may instead get a result,

$$\hat{\beta}\psi = b\psi',$$

where the shapes of the functions ψ and ψ' are different.

If the original wavefunction is left unaltered after an operation except for multiplication by some constant, such that $\hat{\alpha}\psi = a\psi$, then the coefficient a is an **eigenvalue** of the operator $\hat{\alpha}$ and the wavefunction is its **eigenfunction** (also called an *eigenstate* or *eigenvector)*.

The function $f(x) = x^2$, for example, is an eigenfunction of the operator $\hat{\alpha} = x(d/dx)$, because

$$\hat{\alpha}f(x) = x\left(\frac{d}{dx}\right)x^2 = x(2x) = 2x^2 = 2f(x).$$

In this case, $f(x)$ is an eigenfunction of $\hat{\alpha}$ and the eigenvalue is 2.

On the other hand, the function $g(x) = \ln x$ is not an eigenfunction of $\hat{\alpha}$, because

$$\hat{\alpha}g(x) = x\left(\frac{d}{dx}\right)\ln x = x\left(\frac{1}{x}\right) = 1.$$

This time, the result of the operation does not contain the original function $\ln x$. We would say that $g(x)$ is not an eigenfunction of $\hat{\alpha}$, and there is no eigenvalue.

The Schrödinger equation is an eigenvalue equation in which ψ is an eigenfunction of $\hat{H}$ and E is the eigenvalue.

Operators corresponding to quantities that can be physically measured are always **linear,** meaning that

$$\hat{\alpha}[f(x) + g(x)] = \hat{\alpha}f(x) + \hat{\alpha}g(x).$$

They are also **Hermitian,** meaning that for any two wavefunctions ψ_i and ψ_j,

$$\int \psi_i^* \hat{H}\psi_j \, d\tau = \int \psi_j^* \hat{H}\psi_i \, d\tau.$$

The only operators we will deal with at first involve only multiplication and differentiation, and are therefore linear.

However, it is not necessarily the case that our operators will commute. Commuting operators $\hat{\alpha}$ and $\hat{\beta}$ obey the relation

$$\hat{\alpha}\hat{\beta}f(x) - \hat{\beta}\hat{\alpha}f(x) = 0.$$

For example, if $\hat{\alpha}f(x) = 3f(x)$, and $\hat{\beta}f(x) = 2f(x)$, then

$$\hat{\alpha}\hat{\beta}f(x) - \hat{\beta}\hat{\alpha}f(x) = 3[2f(x)] - 2[3f(x)] = 6f(x) - 6f(x) = 0.$$

Multiplication is a commutative operation. But not all operators commute. For example, if we change our operators to $\hat{\alpha}f(x) = df(x)/dx$, and $\hat{\beta}f(x) = 2xf(x)$, then

$$\hat{\alpha}\hat{\beta}f(x) - \hat{\beta}\hat{\alpha}f(x) = \frac{d}{dx}[2xf(x)] - 2x\frac{d}{dx}f(x)$$

$$= 2f(x) + 2x\frac{df(x)}{dx} - 2x\frac{df(x)}{dx} = 2f(x) \neq 0.$$

These two operators do not commute.

When two operators commute, they share a set of eigenfunctions. The Hamiltonian commutes with one of the angular momentum component operators, for example $\hat{L}_z$, but not with the total angular momentum operator $\hat{L}$. This means that the same set of eigenfunctions ψ that obey the Schrödinger equation will also give $L_z\psi$ when operated on by $\hat{L}_z$, but they will *not* give $L\psi$ when operated on by $\hat{L}$.

EXAMPLE 2.5 Commuting Operators

CONTEXT If we have two operators whose eigenvalues are experimentally measured quantities, such as energy and angular momentum, then we can determine both eigenvalues simultaneously only if the two operators commute. If the operators do not commute, then some form of the Heisenberg uncertainty principle (Eq. 1.19) applies: no quantum state can be prepared in which both quantities can be known exactly at the same time. Our work on the atom is constrained by sets of non-commuting operators that determine which parameters we can measure simultaneously for any given state of the electron.

Another application of the same principle appears in the study of molecules over very short time scales. Lasers exist that emit light in bursts lasting only a few femtoseconds (10^{-15}s). If we wish to know the photon energy in that burst, however, we run into the Heisenberg uncertainty principle again. It turns out that the energy operator (the Hamiltonian) does not commute with the time operator (multiplication by t), and so we have another form of the uncertainty principle that reads $\delta E \delta t \geq \frac{\hbar}{2}$. As a result, although we can probe molecules over very short timescales, the photon energy of the radiation becomes very poorly defined. We can tell how long it takes for the molecule to change from one structure to another, for example, but we can't measure the energy of that structure change very precisely. An uncertainty in time of 10^{-15}s corresponds to an energy uncertainty of 30% at visible wavelengths. This fundamental limitation in our knowledge of the system can be traced back to the non-commutation of the operators for energy and time.

PROBLEM (a) Determine whether or not operators $\hat{P}$ and $\hat{Q}$ and operators $\hat{Y}$ and $\hat{Z}$ commute when

1. $\hat{P}f(x) = e^{f(x)}$ and $\hat{Q}f(x) = f(x)^2$;

2. $\hat{Y}f(x) = x\frac{d}{dx}f(x)$ and $\hat{Z}f(x) = \frac{d}{dx}[xf(x)]$.

(b) Show that functions of the form $f(x) = ax^{-2}$ (with a some constant) are eigenfunctions of the operators $\hat{Y}$ and $\hat{Z}$ and find the eigenvalues.

SOLUTION (a) We need to determine whether or not $\hat{P}\hat{Q}f(x) = \hat{Q}\hat{P}f(x)$ and $\hat{Y}\hat{Z}f(x) = \hat{Z}\hat{Y}f(x)$ for all functions $f(x)$. For the first case,

$$\hat{P}\hat{Q}f(x) = \hat{P}f(x)^2 = e^{f(x)^2}$$
$$\hat{Q}\hat{P}f(x) = \hat{Q}e^{f(x)} = (e^{f(x)})^2 = e^{2f(x)}.$$

It is not generally the case that $f(x)^2 = 2f(x)$, so these two operators do not commute. But for the second case,

$$\hat{Y}\hat{Z}f(x) = x\frac{d}{dx}\left\{\frac{d}{dx}[xf(x)]\right\} = x\frac{d}{dx}\left\{f(x) + x\frac{df(x)}{dx}\right\}$$
$$= x\left\{\frac{df(x)}{dx} + \frac{df(x)}{dx} + x\frac{d^2f(x)}{dx^2}\right\} = 2x\frac{df(x)}{dx} + x^2\frac{d^2f(x)}{dx^2}$$

$$\hat{Z}\hat{Y}f(x) = \frac{d}{dx}\left[x^2\frac{df(x)}{dx}\right] = 2x\frac{df(x)}{dx} + x^2\frac{d^2f(x)}{dx^2} = \hat{Y}\hat{Z}f(x),$$

so these two operators do commute.

(b) Insert this $f(x)$ in front of the operator, and look for $f(x)$ to reappear after the operation. Any new coefficient multiplying $f(x)$ is the eigenvalue:

$$\hat{Y}f(x) = x\frac{d}{dx}(ax^{-2}) = x(-2ax^{-3}) = -2ax^{-2} = -2f(x)$$

$$\hat{Z}f(x) = \frac{d}{dx}(x\,ax^{-2}) = \frac{d}{dx}(ax^{-1}) = -ax^{-2} = -f(x).$$

The eigenvalues are -2 for $\hat{Y}$ and -1 for $\hat{Z}$.

If $\hat{A}$ corresponds to a physical property of the system, whether or not ψ is an eigenfunction of $\hat{A}$, the average value of that property is given by the **expectation value** of the operator,

$$\langle A \rangle = \int_{\substack{\text{all}\\\text{space}}} \psi^*(\hat{A}\psi)d\tau, \tag{2.10}$$

CHECKPOINT For any wavefunction that gives the distribution of the particle *in space,* there is no single value for the position, and we don't expect to be able to write an eigenvalue equation for the position operators such as x and r.

where the wavefunction ψ *must* be normalized. This is the **quantum average value theorem.** The uncertainty principle prevents us from obtaining exact values for all the physical parameters of a system simultaneously. Some of the parameters will be intrinsically uncertain, meaning that a series of measurements will yield a range of values for that parameter. However, if we know the system's wavefunction, then Eq. 2.10 extracts the average we would get from that series of measurements.

This is the quantum version of the **classical integrated average.** If a classical system has a coordinate x and the probability of finding the system at any point x is given by the probability distribution function $\mathcal{P}(x)$, then the average value of a parameter $A(x)$ is given by the average taken by integrating over all possible values of x, weighted by the probability function $\mathcal{P}(x)$:

$$\langle A \rangle = \int_{-\infty}^{\infty} \mathcal{P}(x)A(x)dx. \tag{2.11}$$

In this expression the classical probability distribution function takes the place of our quantum-mechanical probability distribution $|\psi|^2$. As in the quantum mechanical case, this equation requires that the probability function $\mathcal{P}(x)$ has been normalized so that $\int_{-\infty}^{\infty} \mathcal{P}(x)dx = 1$.

The alert reader will notice that we're near the end of the section and a lot has been said about operators in general and precious little about the Hamiltonian in particular. That's because, starting with the next section of the text, we will be tailoring the equations for our Hamiltonians to each specific application. What makes the Hamiltonian important is that it represents a *general* expression, representing the fundamental physics of the system. In trying to predict what an

EXAMPLE 2.6 Quantum Average Value Theorem

CONTEXT We use the average value theorem not only to predict experimentally measurable quantities from a wavefunction we know, but also as the basis for a common approach to solving the Schrödinger equation in the first place (shown in Section 4.2). Here, let's look at how the average value theorem can give us an expectation value for the position of a particle that is distributed across a small domain, and then at an even simpler problem: what happens to the dipole moment in the Bohr model of the atom.

PROBLEM (a) The normalized wavefunction $f(x)$ is equal to $\sqrt{3k}(1 - kx)$ for x between 0 and $1/k$ and zero everywhere else. Find the average position of the particle.
(b) In the Bohr model of the atom, the electron travels in a circular orbit through an angle ϕ at a distance r from the nucleus. If we transform this to a de Broglie wave satisfying the condition $n\lambda = 2\pi r$, then the wavefunction for the $n = 1$ electron may be written as $\psi(\phi) = (1/\sqrt{\pi})\sin\phi$, where $0 \leq \phi < 2\pi$. Find the average value of the dipole moment μ if the dipole moment operator is $\hat{\mu} = er\cos\phi$.

SOLUTION (a) We cannot find the position using an eigenvalue equation because $f(x)$ is not an eigenfunction of multiplication by x. The average value theorem allows us to evaluate the average position by integrating:

$$\begin{aligned}\langle x \rangle &= \int_0^{1/k} f(x)^2\, x\, dx = (3k)\int_0^{1/k} (1 - kx)^2\, x\, dx \\ &= (3k)\int_0^{1/k} (x - 2kx^2 + k^2x^3)\, dx = (3k)\left[\frac{x^2}{2} - \frac{2kx^3}{3} + \frac{k^2x^4}{4}\right]_0^{1/k} \\ &= (3k)\left[\frac{1}{2k^2} - \frac{2}{3k^2} + \frac{1}{4k^2}\right] = (3k)\left[\frac{1}{12k^2}\right] = \frac{1}{4k}.\end{aligned}$$

The average position is 1/4 of the way from the origin to the x-intercept.
(b) To find the expectation value of μ, we solve

$$\begin{aligned}\langle \mu \rangle &= \int_0^{2\pi} \psi^* \hat{\mu} \psi\, d\phi \\ &= \frac{1}{\pi}\int_0^{2\pi} \sin^2\phi\,(er\cos\phi)\, d\phi \\ &= -\frac{er}{\pi}\left(\frac{1}{3}\right)\sin^3\phi\,\Big|_0^{2\pi} = 0.\end{aligned}$$

This is almost a trivial case, but it shows how we would verify mathematically that the dipole moment averages to zero when we evenly distribute the electron around the nucleus.

experiment will tell us about a new quantum-mechanical system, we usually determine the Hamiltonian and then find the energies and wavefunctions that make the Schrödinger equation true. On the other hand, if we have the experimental results in hand, we can work through the Schrödinger equation in the *other* direction, deriving the Hamiltonian and learning about the fundamental properties of the system, such as the masses of the particles and the nature of the forces between them.

Why is the Schrödinger equation based on the Hamiltonian operator, as opposed to the position or angular momentum operators? The Hamiltonian's eigenvalue is the energy, and we focus on energy for several reasons, both theoretical and practical. For one, the correspondence principle tells us that the quantum nature of our system decreases as we increase the mass, energy, and domain. Of these, energy is a parameter that we can usually adjust in our system, whereas the mass and size are less convenient variables to a chemist. If you work with fluorine gas, you're stuck with molecules of 38 atomic mass units and fixed molecular size. Furthermore, spectroscopy remains the most widespread laboratory application of quantum mechanics, and the one direct measurement that a spectrum provides is the energy gap between quantum states. Energy is also a crucial parameter in determining chemical reactivities and stabilities, so it makes sense in the long run to start keeping track of it at this early stage.

One perspective on the typical interplay between experiment and theory in quantum mechanics is summarized in Fig. 2.3. Experimental approaches to quantum mechanics are usually based on spectra (which allow us to determine relative quantum state energies) or other measurements, such as density or conductivity (which are sensitive to the electron distribution and therefore the probability density). From these, we can determine the form of the Hamiltonian that must solve the Schrödinger equation. In particular, the potential energy function for the molecule describes almost everything we might want to know about the geometry and motions of the molecule. A more typical problem for the theoretician is to predict the properties of the system knowing at first only the particles that are present and the forces that act between them. The masses and number of particles determine the kinetic energy contribution to the Hamiltonian, and the forces determine the potential energy. So the theoretician more often knows $\hat{H}$ and solves the Schrödinger equation to obtain the energies and probability densities. We use the Schrödinger equation as the gateway between the two approaches.

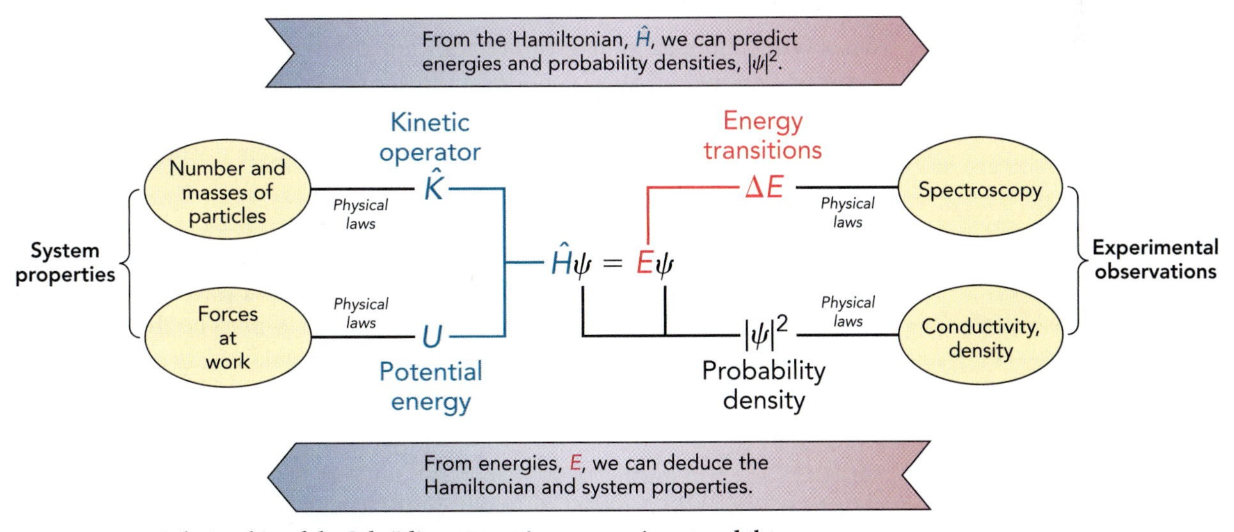

▲ **FIGURE 2.3 Relationship of the Schrödinger equation to experiment and theory.** Quantum mechanics provides a theory that we can use to predict spectra and other experimentally measurable properties. Similarly, from experimental measurements of energies and electron distributions, we can unlock the potential energy function U of a chemical system, which tells us about the structure and forces of the system, including, for example, the distribution of chemical bonds in a molecule.

2.2 Fundamental Examples

The mathematical challenge in quantum mechanics takes a common form: we know the masses, charges, and force laws involved in our system from the outset (which is enough to write the Hamiltonian), but we don't know the possible energies and we don't know the wavefunctions for the system. The Hamiltonian involves a kinetic energy operator that takes a second derivative of the wavefunction, which means that the Schrödinger equation is always a *second order differential equation.* Solving a second-order differential equation is a matter of effectively integrating both sides of the equation twice over, to undo the two levels of derivatives. In doing so, we obtain the energies and wavefunctions simultaneously. We will see that in this section as we apply the Schrödinger equation to three idealized systems. These examples illustrate how we use quantum mechanics to describe matter and introduce three basic quantum-mechanical systems that we shall refer to often in the remaining text.

The general procedure for solving the Schrödinger equation is as follows:

1. Select an appropriate coordinate system and any approximations that you may need.
2. Write the Hamiltonian explicitly, adding the kinetic energy operator to the potential energy function for your system.
3. Then solve the second-order differential equation $\hat{H}\psi = E\psi$. You know $\hat{H}$, but you need to find E and ψ. Some differential equations have exact algebraic solutions we can look up, and—if we have the time and patience—we can always resort to numerical solutions on a computer if analytical solutions aren't available.

The Free Particle

One of the simplest and most important Hamiltonians gives the energy of a free particle, a particle that is not influenced by any forces whatsoever. For this particle, the potential energy U is equal to zero everywhere. Because this is our first Hamiltonian, we must begin by deriving the kinetic energy operator $\hat{K}$.[1]

The Kinetic Energy Operator

The operator we're looking for should satisfy an eigenvalue equation $\hat{K}\psi = K\psi$, where ψ for now is chosen to be an eigenfunction of $\hat{K}$, a wavefunction with a single, exact value for the kinetic energy. The operator $\hat{K}$ must be able to find the numerical value of the kinetic energy within the expression for ψ and pull it out as a factor, leaving ψ unharmed. What is it in the wavefunction that carries that information?

[1] In physics, the symbol $\hat{T}$ is usually used for this operator, and V is used for the potential energy. We will use $\hat{K}$ and U instead to avoid confusion when the temperature T and volume V appear in the same equations as these operators. While we're on the subject of notation, the potential energy operator is always a multiplicative function in our systems, and it contains no derivative operators. Therefore, it is typical to drop the operator notation, and we will write U rather than $\hat{U}$ for the potential energy.

The reason we've turned to quantum mechanics is that we want a precise treatment of atoms and molecules that takes account of their *wavelike* character. We've defined that wavelike character using the de Broglie wavelength, $\lambda_{dB} = h/p$, and it turns out that the momentum p is just the foothold we need to calculate the kinetic energy K:

$$K = \frac{mv^2}{2} = \frac{(mv)^2}{2m} = \frac{p^2}{2m} = \left(\frac{\hbar^2}{2m}\right)\left(\frac{p}{\hbar}\right)^2 = \left(\frac{\hbar^2}{2m}\right)\left(\frac{h/\lambda_{dB}}{h/2\pi}\right)^2 = \left(\frac{\hbar^2}{2m}\right)\left(\frac{2\pi}{\lambda_{dB}}\right)^2. \tag{2.12}$$

CHECKPOINT The wavefunction of a particle can be thought of as a sum of sine waves, each oscillating with wavelength equal to λ_{dB}. The kinetic energy operator in quantum mechanics uses a second derivative to pull that de Broglie wavelength out of the sine functions without changing the sine waves themselves. The de Broglie wavelength in turn gives the momentum of the particle, and that's enough for us to get the kinetic energy.

If we can extract the value of λ_{dB} from the wavefunction, then we can find the kinetic energy.

Only one kind of function has a constant value for the wavelength: a pure sinusoidal wave (which includes the functions sine, cosine, and e^{ix}). Therefore, the only states that are *eigenfunctions* of $\hat{K}$ are waves of the form $\sin(kx + \phi)$, where the **wavenumber** k is inversely proportional to the wavelength,

$$k \equiv \frac{2\pi}{\lambda_{dB}}, \tag{2.13}$$

and where ϕ is some phase angle that allows the whole sine wave to be shifted along the x axis.

From this, we can work out the kinetic energy operator. We want to find $\hat{K}$ such that

$$\hat{K}\sin(kx + \phi) = \left(\frac{\hbar^2}{2m}\right)\left(\frac{2\pi}{\lambda_{dB}}\right)^2 \sin(kx + \phi) = \left(\frac{\hbar^2}{2m}\right)k^2\sin(kx + \phi). \tag{2.14}$$

The operator that pulls out k^2 while leaving $\sin(kx + \phi)$ unchanged is the second derivative operator:[2]

$$\frac{\partial^2}{\partial x^2}\sin(kx + \phi) = \frac{\partial}{\partial x}[k\cos(kx + \phi)] = -k^2\sin(kx + \phi). \tag{2.15}$$

To complete the kinetic energy operator, we just include the factor of $\hbar^2/2m$:

$$\hat{K}\sin(kx + \phi) = -\left(\frac{\hbar^2}{2m}\right)\frac{\partial^2}{\partial x^2}\sin(kx + \phi) = \left(\frac{\hbar^2}{2m}\right)k^2\sin(kx + \phi) = K\sin(kx + \phi). \tag{2.16}$$

To summarize,

$$\hat{K} = -\left(\frac{\hbar^2}{2m}\right)\frac{\partial^2}{\partial x^2} \tag{2.17}$$

for motion of a mass m along the x axis. Equation 2.17 turns out to be one of the fundamental equations of quantum mechanics. This operator is present, and takes essentially this form in the Hamiltonian, for any quantum-mechanical system. The only thing we assumed was that the kinetic energy of the particle was a constant (which, as we'll see later, doesn't have to be true).

[2]We are using partial derivatives even though there is only one coordinate, x. We could use normal derivatives here, but we have used the partials to be consistent with the multi-dimensional examples that follow.

Wavefunction Phase

Actually, this derivation of Eq. 2.17 includes one other assumption, which was not mentioned because it doesn't affect $\hat{K}$. We used the unnormalized function $\sin(kx + \phi)$ as the general form of a wavefunction with well-defined kinetic energy. There's a hidden assumption here that the wavefunction has no complex values, that it is a pure real function.

When we report numbers we can measure in the laboratory, we do expect them to be pure real numbers, but we never *directly measure* wavefunctions. We allow wavefunctions to be complex functions, having both real and imaginary components, because the imaginary component can carry information about the interference properties of wavefunctions. This is just a more sophisticated treatment of the phase angle ϕ. A more general way to write a function that has a well-defined kinetic energy uses Euler's equation (Eq. 2.8),

$$Ae^{ikx} = A\cos(kx) + Ai\sin(kx),$$

in which we can select the phase of the real component of the wave by adjusting the constant A. If A is a pure real number, then the real part of Ae^{ikx} depends only on the cosine term; if A is pure imaginary, then the real part of Ae^{ikx} is a pure sine wave. Complex values of A will give phases in between the pure sine and pure cosine functions. But no matter what the value of A or k, this function still satisfies the eigenvalue equation for the kinetic energy because it's just the sum of two sinusoidal waves:

$$\begin{aligned}\hat{K}(Ae^{ikx}) &= -\left(\frac{\hbar^2}{2m}\right)A\frac{\partial^2(e^{ikx})}{\partial x^2} = -\left(\frac{\hbar^2}{2m}\right)(ik)^2Ae^{ikx} \\ &= \left(\frac{k^2\hbar^2}{2m}\right)Ae^{ikx} = \frac{p^2}{2m}(Ae^{ikx}).\end{aligned} \tag{2.18}$$

Solving the Schrödinger Equation

The Hamiltonian operator for the free particle in one dimension is just the kinetic energy operator, because there are no forces present to generate potential energy:

$$\hat{H} = \frac{\hat{p}_x^2}{2m} = -\frac{\hbar^2}{2m}\frac{\partial^2}{\partial x^2}. \tag{2.19}$$

We can solve the Schrödinger equation for this Hamiltonian using the corresponding equation in Table A.7:

$$\hat{H}\psi = -\frac{\hbar^2}{2m}\frac{\partial^2\psi}{\partial^2} = E\psi; \tag{2.20}$$

$$-\frac{\hbar^2}{2mE}\frac{\partial^2\psi}{\partial x^2} = \psi \tag{2.21}$$

$$\psi(x) = c\,e^{ikx} = c[\cos(kx) + i\sin(kx)],\ k = \frac{\sqrt{2mE}}{\hbar}. \tag{2.22}$$

This solution is verified by substituting the wavefunction ψ into Eq. 2.20. The Schrödinger equation yields an infinite set of wavefunctions, corresponding to the infinite set of possible kinetic energies E imparted to the particle. The wavefunctions consist of sinusoidal waves with wavelength $2\pi/k = h/\sqrt{2mE}$.

As the kinetic energy increases, the wavelength decreases in agreement with de Broglie's hypothesis. De Broglie, by postulating the particle wavelength λ_{dB}, had stocked the toolbox for Schrödinger's wave mechanics, in particular providing the basis for the kinetic energy operator that appears in all of our quantum mechanics problems. The possible energy values are continuous, completely covering the range $0 \leq E < \infty$, which is consistent with classical physics, unlike the electronic energies in the Bohr model. The cosine component of these wavefunctions is graphed at several energy levels in Fig. 2.4.

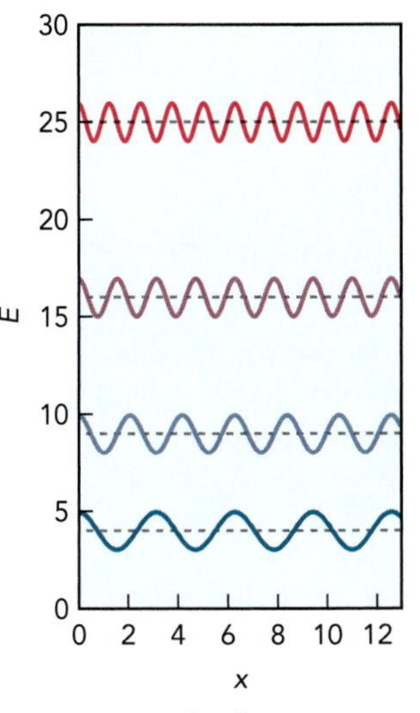

▲ FIGURE 2.4 **Real parts $\cos(kx)$ of the free particle wavefunctions at selected energies.**

From the $\hat{p}_x^2$ operator we can also obtain the momentum operator $\hat{p}_x$, because the square of an operator is simply the result of applying that operator twice. Therefore, we may write:

$$\hat{p}_x = \frac{\hbar}{i}\frac{\partial}{\partial x}, \tag{2.23}$$

where $1/i = i/i^2 = -i$. As with square roots of ordinary numbers, we have a choice as to phase, and here we could have chosen a phase factor of i instead of $1/i$. In this case, the phase reflects the fact that the momentum is a vector, with direction; the particle could move in either the $+x$ $(p_x > 0)$ or in the $-x$ direction $(p_x < 0)$ and still have the same value of p_x^2. Our choice here is such that a free particle wavefunction

$$\psi_{\pm}(x) = e^{\pm ikx} = \cos\left(\frac{2\pi x}{\lambda_{dB}}\right) \pm i\sin\left(\frac{2\pi x}{\lambda_{dB}}\right)$$

will have positive or negative momentum as determined by the $\pm$ sign:

$$\begin{aligned}
\hat{p}_x\psi_{\pm}(x) &= \frac{\hbar}{i}\frac{\partial}{\partial x}\left[\cos\left(\frac{2\pi x}{\lambda_{dB}}\right) \pm i\sin\left(\frac{2\pi x}{\lambda_{dB}}\right)\right] \\
&= \frac{\hbar}{i}\left[-\left(\frac{2\pi}{\lambda_{dB}}\right)\sin\left(\frac{2\pi x}{\lambda_{dB}}\right) \pm i\left(\frac{2\pi}{\lambda_{dB}}\right)\cos\left(\frac{2\pi x}{\lambda_{dB}}\right)\right] && \text{take } \frac{\partial}{\partial x} \\
&= \frac{2\pi\hbar}{\lambda_{dB}}\left[-\frac{1}{i}\sin\left(\frac{2\pi x}{\lambda_{dB}}\right) \pm \cos\left(\frac{2\pi x}{\lambda_{dB}}\right)\right] && \text{rearrange constants} \\
&= \frac{h}{\lambda_{dB}}\left[i\sin\left(\frac{2\pi x}{\lambda_{dB}}\right) \pm \cos\left(\frac{2\pi x}{\lambda_{dB}}\right)\right] && 2\pi\hbar = h,\ 1/i = -i \\
&= p_x\left[\pm\cos\left(\frac{2\pi x}{\lambda_{dB}}\right) + \sin\left(\frac{2\pi x}{\lambda_{dB}}\right)\right] && h/\lambda_{dB} = p_x, \text{ switch terms} \\
&= \pm p_x\left[\cos\left(\frac{2\pi x}{\lambda_{dB}}\right) \pm i\sin\left(\frac{2\pi x}{\lambda_{dB}}\right)\right] && \text{factor out } \pm 1, \\
&= \pm|p_x|\,\psi_{\pm}(x).
\end{aligned} \tag{2.24}$$

We have taken the trouble here to define λ_{dB} as $h/|p_x|$, to ensure that it is a positive number. This choice of phase defines the relationship between the momentum operator and the Cartesian axes for everything that follows.

The free particle has the continuous energies of a classical system because the flat potential energy function means that the domain of the particle is infinite, larger than any de Broglie wavelength. To introduce the energetics of a

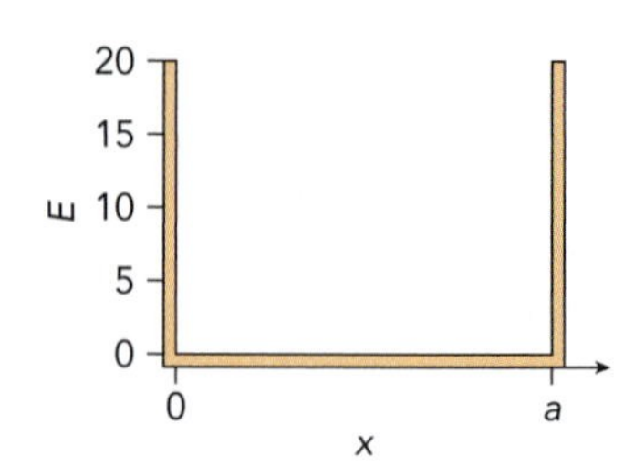

▲ FIGURE 2.5 **The one-dimensional box potential energy function.**

quantum-mechanical system, we need a potential energy function that puts up walls, limiting the particle's domain and thus allowing λ_{dB} to become comparable to the domain. Our next consideration, therefore, is the solution to a Schrödinger equation with a more interesting potential function.

The Particle in a One-Dimensional Box

Among the simplest quantum mechanical problems is the motion of a particle on the potential curve drawn in Fig. 2.5, defined by:

$$U(x \leq 0) = \infty \qquad U(0 < x \leq a) = 0 \qquad U(x > a) = \infty,$$

which is called the **one-dimensional box potential.** Solving the Schrödinger equation for this potential surface will allow us to predict the properties of a particle trapped along a linear path between two impenetrable walls but otherwise free from any external forces. This is similar to the ball rolling back and forth between two walls in Fig. 2.2, or to a π-bond electron in conjugated chain molecules such as butadiene.

First, let us consider the classical solution to our problem. In the region $0 < x < a$, the potential energy is zero, so the kinetic energy is equal to the total energy: $E = mv^2/2$, where v is the speed of the particle. Higher energies correspond to higher speeds. The particle cannot be found outside the walls, in the regions $x \leq 0$ or $x > a$, because any place where the potential energy becomes greater than the total energy, the particle must turn around. Because the system has energy $E = K + U$ and the kinetic energy $K = mv^2/2$ is always positive, a classical system can never have an energy less than the potential energy at any given point:

$$\text{classical system:} \quad E \geq U. \tag{2.25}$$

Any point at which the potential energy becomes greater than E is called the **classical turning point.** Energy is conserved, so once the system is given some energy E, the potential curve restricts motion to a region between the classical turning points. The classical particle travels between the two walls at constant speed, reversing direction whenever it hits either wall. At any given time, there is an equal chance of finding the particle at any value of x between 0 and a. The probability distribution for the classical particle is independent of x.

Now do the quantum mechanics. If we consider first the region where U is infinite, the Hamiltonian (including both kinetic and potential energy terms) is

$$\hat{H} = -\frac{\hbar^2}{2m}\frac{\partial^2\psi}{\partial x^2} + \infty. \tag{2.26}$$

The Schrödinger equation is therefore

$$\hat{H}\psi = -\frac{\hbar^2}{2m}\frac{\partial^2\psi}{\partial x^2} + \infty\,\psi = E\psi. \tag{2.27}$$

Regardless of the kinetic term, the Schrödinger equation has only two kinds of solutions: the energy E is infinite, which we reject as unattainable, or the wavefunction is zero throughout this region. Our conclusion: a quantum-mechanical object cannot exist in a region where the potential energy is infinite:

$$\psi(x \leq 0) = \psi(x > a) = 0.$$

This is nothing new; the classical solution in this region is the same. Conveniently, though, this establishes boundary conditions for our solution inside the box. The wavefunction must be a well-behaved (in this case meaning continuous) function, and therefore ψ must be zero at $x = 0$ and at $x = a$.

We only need to solve the Schrödinger equation in the middle region, where the potential is zero. In this region, we need the free particle Hamiltonian,

$$\hat{H}\psi = -\frac{\hbar^2}{2m}\frac{\partial^2\psi}{\partial x^2} = E\psi, \tag{2.28}$$

which we can rewrite

$$\frac{\partial^2\psi}{\partial x^2} = -\frac{2mE}{\hbar^2}\psi. \tag{2.29}$$

The general solution to this second-order differential equation is the same as for the free particle,

$$\psi_{\text{free particle}}(x) = c\,e^{ikx} = c[\cos(kx) + i\sin(kx)] \quad k = \sqrt{2mE/\hbar^2},$$

with an important exception: we must also apply the boundary conditions for the box potential. Because $\sin(0) = 0$ and $\cos(0) = 1$, the cosine term in our free particle solution must be dropped, and we retain only the sine component. The wavefunction is zero at $x = 0$ if it has the form

$$\psi(x) = c\sin(kx). \tag{2.30}$$

The wavefunction must also be 0 at $x = a$, which requires that $ka = n\pi$ where n is any positive integer. Substituting our Eq. 2.22 for k and solving for the energy, we find

$$E_n = \frac{k^2\hbar^2}{2m} = \frac{n^2\pi^2\hbar^2}{2ma^2} \quad n = 1, 2, 3, \ldots. \tag{2.31}$$

(We neglect the trivial case $n = 0$, for which $\psi = 0$.) We have our first genuinely quantum-mechanical energy expression. Like the energies of Bohr's one-electron atom, only discrete values of the energy are possible.

Furthermore, unlike the free particle, no state exists for which $E = 0$. Even the lowest energy quantum state, the **ground state,** has some kinetic energy. The difference between the energy of the ground state and the minimum value of the potential energy is called the **zero-point energy,**

$$E_{\text{zero-point}} = E_{\text{gnd}} - U_{\text{min}}, \tag{2.32}$$

and it is present in any quantum-mechanical system for which the potential energy limits the particle's domain.[3] For the particle in a box, the zero-point energy is $E_1 = \pi^2\hbar^2/(2ma^2)$.

At the same time we find the energies of the particle in a box, we also obtain solutions for the wavefunctions by substituting $k = n\pi/a$ and normalizing:

$$\psi_n(x) = \sqrt{\frac{2}{a}}\sin\left(\frac{n\pi x}{a}\right), \tag{2.33}$$

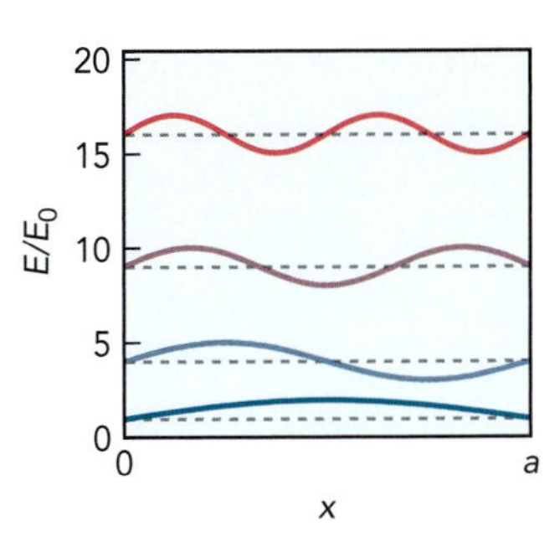

▲ FIGURE 2.6 **Energy levels and wavefunctions for the four lowest states of the particle in a one-dimensional box.** The energy levels are given in units of $E_0 = \pi^2\hbar^2/(2ma^2)$.

[3]The particle's de Broglie wavelength requires this, because if the particle could have zero kinetic energy, then λ_{dB} would become infinite. The only way for the particle to be at rest is for it to occupy all space. We will see, however, in Section 9.1 that, in the right coordinate system, this does not require the particle itself to be infinitely large.

where $c_1 = \sqrt{2/a}$ has been substituted to normalize the wavefunction. These wavefunctions are plotted in Fig. 2.6. The non-physical units of $m^{-1/2}$ for the wavefunction remind us that all of the observable parameters predicted by ψ are based on the probability density $|\psi|^2$, which has more reasonable units of m^{-1}, the probability per unit distance.

These wavefunctions are qualitatively similar to the Bohr electron orbits. The energies and wavefunctions are discrete functions of an integer n, and each wavefunction has $n-1$ **nodes,** the points at which the wavefunction *crosses* zero. We emphasize that the wavefunction crosses zero at the nodes to distinguish nodes from the regions outside the box, where the wavefunction stays at zero. Because the probability density is given by $|\psi|^2$, a node in the wavefunction corresponds to a region where the probability density also reaches zero. The mass, the charge, all the measurable quantities we associate with the particle—all of these are zero at a node.

The correspondence principle holds for mass and size here. Because E_n is inversely proportional to mass and to a^2, the energies of the states collapse as the mass increases or as the length of the box increases, approaching a classical (continuous) distribution of values. The dependence of the energy on n is much different this time, though, with allowed energy levels getting farther and farther apart as n increases, rather than closer together as was the case for the electron energies in the Bohr model.

The particle in a box is the first non-trivial system we have examined using the mathematics of quantum mechanics, and it demonstrates when the classical picture breaks down. Whereas the classical particle has an equal chance of being observed at any value of x, the quantum particle in the lowest energy state, $n = 1$, has a higher probability of being detected near the *middle* of the box than near the walls, even though it has constant kinetic energy. Try explaining that to Isaac Newton.

We can also see where classical mechanics becomes useful again. A particle of mass 1.0 g trapped in a one-dimensional box of length 1.0 cm would have energies of $5.5 \cdot 10^{-61} n^2$ J. Suppose we could measure energies to within $1.0 \cdot 10^{-50}$ J, the incredibly small kinetic energy of one electron moving at a speed of 1 Å s^{-1}. We would still be ten orders of magnitude away from measuring the spacing between the $n = 1$ and $n = 2$ states. For our purposes, the energy levels would be too close together for the gaps to be seen, and the energy would appear to be a continuous variable. At any reasonably large energy, say corresponding to a temperature of about 100 K, we would expect to find the 1 g particle excited to roughly the state $n = 10^{20}$, at which point the spacing between nodes would be too small to measure and the probability density would be the classical result, i.e., equal probability anywhere inside the box.

If we go to high energy, the spacing between the levels becomes bigger. At very high values of n (around $5 \cdot 10^9$), the energy level spacing would become measurable in the earlier example, but that is an unusual feature of this potential surface brought about by two characteristics: the steepness of the potential walls and the restriction to one dimension. In three dimensions, our next example, such a massive particle behaves classically at all energies.

EXAMPLE 2.7 Particle in a Box Energies

CONTEXT The energies of the particle in a one-dimensional box are a surprisingly good model for the energies of the electronic states in polyene dyes, such as the diethylcyanine shown. The electrons are restricted to move primarily along the conjugated bonds that connect the two nitrogen atoms, so the motion is primarily one dimensional, with steep increases in the potential energy at the ends of the conjugated chain. The change in absorption wavelength from one dye to another, as these chains are lengthened or shortened, can be predicted by the particle in a box model to within about 10%.

PROBLEM Find the energy in E_h of the $n = 1$ state of an electron in a one-dimensional box of length $2a_0 = 1.06$ Å.

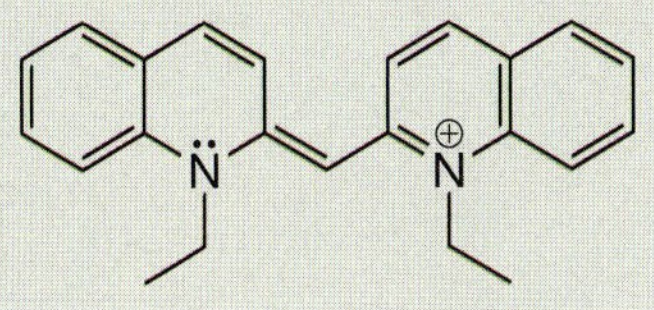

SOLUTION We can use the general solution, Eq. 2.31, and plug in the values for n, m, and a. The mass of the electron is a constant, and the problem gives the values for n and a:

$$\begin{aligned} E &= \frac{n^2\pi^2\hbar^2}{2m_e a^2} \\ &= \frac{\pi^2(1.055\cdot 10^{-34}\,\mathrm{J\,s})^2}{2(9.109\cdot 10^{-31}\,\mathrm{kg})[(1.06\,\text{Å})(10^{-10}\,\mathrm{m}\,\text{Å}^{-1})]^2} && \text{put values in SI units} \\ &= 5.36\cdot 10^{-18}\,\mathrm{J} && \text{convert to } E_h \\ &= 1.23\,E_h. \end{aligned}$$

(With sharp eyes at the second step, we could also use the definition of a_0 as $4\pi\varepsilon_0\hbar^2/(m_e e^2)$ to convert the energy directly to $\pi^2 m_e e^4/(8(4\pi\varepsilon_0)^2\hbar^2)$, which is equal to exactly $\pi^2/8$ hartree.) The energies of an electron in a one-dimensional box are comparable to those in the one-electron atom, if the box is about the same size as the diameter of the Bohr orbit.

The Particle in a Three-Dimensional Box

Now we try the three-dimensional box potential, such that

$$\begin{aligned} &U(x \leq 0) = \infty \quad && U(0 < x \leq a) = 0 \quad && U(x > a) = \infty, \\ &U(y \leq 0) = \infty && U(0 < y \leq b) = 0 && U(y > b) = \infty, \\ &U(z \leq 0) = \infty && U(0 < z \leq c) = 0 && U(z > c) = \infty. \end{aligned}$$

In words, the particle is trapped in a box with rectangular walls, having lengths a, b, and c along the x, y, and z coordinates, respectively.

Energies and Wavefunctions

First we need the correct Hamiltonian. Because this is our first example with more than one dimension, we should check our kinetic energy operator, which was derived from the operator $\hat{p}^2$. In three dimensions, this operator can be written as

$$\begin{aligned} \hat{p}^2 &= \hat{p}_x^2 + \hat{p}_y^2 + \hat{p}_z^2 \\ &= -\hbar^2\left(\frac{\partial^2}{\partial x^2} + \frac{\partial^2}{\partial y^2} + \frac{\partial^2}{\partial z^2}\right) \\ &= -\hbar^2\nabla^2, \end{aligned} \tag{2.34}$$

where the operator

$$\frac{\partial^2}{\partial x^2} + \frac{\partial^2}{\partial y^2} + \frac{\partial^2}{\partial z^2} \equiv \nabla^2 \tag{2.35}$$

is the **Laplacian operator** in Cartesian coordinates. The Hamiltonian again has only a kinetic term, which we now write as

$$\hat{H} = \hat{K} = -\frac{\hbar^2}{2m}\nabla^2. \tag{2.36}$$

There are no terms in the Hamiltonian that couple the coordinates x, y, or z, meaning that there are no terms in which different coordinates appear at the same time, such as $\partial^2/(\partial x\,\partial y)$. In any such case, we can simplify the problem by **separation of variables**: the eigenstate can be written as a product of functions that each depend on only *one* coordinate. In this case, if we let

$$\psi = \psi_x(x)\,\psi_y(y)\,\psi_z(z), \tag{2.37}$$

then

$$\hat{H}\psi = -\frac{\hbar^2}{2m}\left[\frac{\partial^2}{\partial x^2}\psi_y(y)\psi_z(z)\psi_x(x) + \frac{\partial^2}{\partial y^2}\psi_z(z)\psi_x(x)\psi_y(y) + \frac{\partial^2}{\partial z^2}\psi_x(x)\psi_y(y)\psi_z(z)\right]$$

$$\begin{aligned}
\hat{H}\psi &= -\frac{\hbar^2}{2m}\left[\psi_y(y)\psi_z(z)\frac{\partial^2\psi_x(x)}{\partial x^2} + \psi_z(z)\psi_x(x)\frac{\partial^2\psi_y(y)}{\partial y^2} + \psi_x(x)\psi_y(y)\frac{\partial^2\psi_z(z)}{\partial z^2}\right] \\
&= \psi_y(y)\psi_z(z)E_x\,\psi_x(x) + \psi_z(z)\psi_x(x)E_y\,\psi_y(y) + \psi_x(x)\psi_y(y)E_z\,\psi_z(z) \\
&= E_x\,\psi_y(y)\psi_z(z)\psi_x(x) + E_y\,\psi_z(z)\psi_x(x)\psi_y(y) + E_z\,\psi_x(x)\psi_y(y)\psi_z(z) \\
&= (E_x + E_y + E_z)\,\psi_x(x)\psi_y(y)\psi_z(z) \\
&= E\,\psi_x(x)\psi_y(y)\psi_z(z).
\end{aligned} \tag{2.38}$$

The part of $\hat{H}$ that contains $\partial/\partial x$ treats $\psi(y)$ and $\psi(z)$ like constants, so they can be factored out of the derivatives; the analogous treatments hold for the $\partial/\partial y$ and $\partial/\partial z$ terms as well. The total energy E ends up being the sum over the energies E_x, E_y, and E_z for motion along each coordinate axis. We have simply solved three separate one-dimensional box problems simultaneously, and the solution for the wavefunction just puts our one-dimensional wavefunction at each coordinate and multiplies them all together:

$$\psi_{n_x,n_y,n_z}(x,y,z) = \left(\frac{8}{abc}\right)^{1/2}\sin\left(\frac{n_x\pi x}{a}\right)\sin\left(\frac{n_y\pi y}{b}\right)\sin\left(\frac{n_z\pi z}{c}\right), \tag{2.39}$$

where the wavefunction has already been normalized. The n_x, n_y, and n_z quantum numbers are again limited to positive integers. Similarly, the energy is the sum of contributions from each of the three coordinates:

$$E_{n_x,n_y,n_z} = \frac{\pi^2\hbar^2}{2m}\left(\frac{n_x^2}{a^2} + \frac{n_y^2}{b^2} + \frac{n_z^2}{c^2}\right) \tag{2.40}$$

$$\begin{aligned}
&\approx \frac{\pi^2\hbar^2}{2m(abc)^{2/3}}(n_x^2 + n_y^2 + n_z^2) \\
&= \frac{h^2}{8mV^{2/3}}n^2,
\end{aligned} \tag{2.41}$$

where $n^2 \equiv n_x^2 + n_y^2 + n_z^2$ and where $V = abc$ is the volume of the box. Equation 2.40 is valid for any rectangular box. The approximation used to reach Eq. 2.41 is valid if a, b, and c are roughly equal, i.e., if the box is cubical in shape. For simplicity, let's assume a cubical box unless otherwise stated. To simplify the notation, let's define an energy unit ε_0 that absorbs all the constants for a given system so that when we want a short form of Eq. 2.41 we can just write

$$E_{n_x,n_y,n_z} = n^2\varepsilon_0 \text{ where } \varepsilon_0 \equiv \frac{h^2}{8mV^{2/3}}. \quad (2.42)$$

SAMPLE CALCULATION **Energy of a Particle in a Three-Dimensional Box.** To find the ground state energy of a hydrogen atom in a cubical box 5 nm on a side, we first set $n_x,n_y,n_z = 1,1,1$. Using Eq. 2.42, we would write

$$\varepsilon_0 = \frac{h^2}{8mV^{2/3}} = \frac{(6.626 \cdot 10^{-34}\,\text{J s})^2}{8(1.008\,\text{amu})(1.661 \cdot 10^{-27}\,\text{kg amu}^{-1})\left[(5\,\text{nm})^3(10^{-9}\,\text{m/nm})^3\right]^{2/3}}$$

$$= 1.3 \cdot 10^{-24}\,\text{J}$$

$$E_{1,1,1} = (1^2 + 1^2 + 1^2)\varepsilon_0 = 3\varepsilon_0 = 4 \cdot 10^{-24}\,\text{J}.$$

Equations 2.39 and 2.40 yield these important lessons:

1. If a Hamiltonian of several coordinates is separable into a sum of one-coordinate terms, the total wavefunction is a product of the individual wavefunctions for each coordinate, and the total energy is the sum of the independent coordinate energies.
2. There is a zero-point energy for each coordinate. As long as the box is bound by walls along x, the lowest energy state has some non-zero value of E_x. The same holds for the y and z axes.
3. It is possible to have several states with different quantum numbers, and therefore different wavefunctions, but with the same energy.

This last characteristic is a new wrinkle among the examples we've considered. It becomes a central player in our description of macroscopic systems, so let's give it some special attention here.

Degeneracy and the Three-Dimensional Box

Table 1.2 lists the values of n^2 for all of the quantum states of a cubical box up to $n^2 = 59$, and shows that there may be several ways of arriving at the same value. States with the same energy are called **degenerate,** or **isoenergetic,** and the total number of states sharing that energy is the degeneracy g of that energy level. Assuming a cubical box, the 4 states in Table 2.1 with $n^2 = (n_x^2 + n_y^2 + n_z^2) = 27$ are degenerate. It is not even necessary for the states to be equivalent, as the $(n_x,n_y,n_z) = (5,1,1)$, $(1,5,1)$, $(1,1,5)$ states are, for degeneracy to exist.[4]

[4]The word *equivalent* here has the same meaning as when we apply it to chemical structures: equivalent items are the same except for some change in location or orientation. For example, the two nuclei in the molecule $^1H_2$ are equivalent. *Equivalent* here does not mean "equal," because once we set our coordinate system, there is a physical difference between the locations of the x and y axes, so $\sin(n\pi x/a)$ and $\sin(n\pi y/a)$ are equivalent but distinct. To compare, $\sin(n\pi x/a)$ and $\sin[2n\pi x/(2a)]$ are *equal*—no measurement will distinguish between them.

TABLE 2.1 Lowest energy states of a particle in a cubical box. The table lists values of n^2 (equal to the energies in units of ε_0) and degeneracies g for a particle with quantum numbers (n_x,n_y,n_z). Permutations are omitted; for example, at level $n^2 = 6$ the state (2,1,1) is shown, but not (1,2,1) and (1,1,2).

n^2	g	States	n^2	g	States
3	1	(1,1,1)	35	6	(5,3,1)
6	3	(2,1,1)	36	3	(4,4,2)
9	3	(2,2,1)	38	9	(5,3,2)(6,1,1)
11	3	(3,1,1)	41	9	(4,4,3)(6,2,1)
12	1	(2,2,2)	42	6	(5,4,1)
14	6	(3,2,1)	43	3	(5,3,3)
17	3	(3,2,2)	44	3	(6,2,2)
18	3	(4,1,1)	45	6	(5,4,2)
19	3	(3,3,1)	46	6	(6,3,1)
21	6	(4,2,1)	48	1	(4,4,4)
22	3	(3,3,2)	49	6	(6,3,2)
24	3	(4,2,2)	50	6	(5,4,3)
26	6	(4,3,1)	51	6	(5,5,1)(7,1,1)
27	4	(3,3,3)(5,1,1)	53	6	(6,4,1)
29	6	(4,3,2)	54	12	(5,5,2)(6,3,3)(7,2,1)
30	6	(5,2,1)	56	6	(6,4,2)
33	6	(4,4,1)(5,2,2)	57	6	(5,4,4)(7,2,2)
34	3	(4,3,3)	59	9	(5,5,3)(7,3,1)

The change from one coordinate (x) to three coordinates (x, y, and z) has brought the particle in a box more in keeping with the correspondence principle, for now the energy levels approach more classical behavior—becoming harder to distinguish from one another—at higher energy. The energy becomes a continuous, classical variable.

This is a little difficult to see, looking at individual energy levels, where the degeneracies hop around from, for example, $E = 53\varepsilon_0$ (6 states) to $E = 54\varepsilon_0$ (12 states) to $E = 55\varepsilon_0$ (0 states). The approach to a continuously variable energy becomes plainer when we examine the **density of quantum states** $W(E)$, the number of quantum-mechanical states encountered per unit energy. This is a useful parameter because, once the system gets large, ε_0 gets too small for us to measure, let alone control, and adjacent energy levels begin to blur together. Instead of being sensitive to the number of states at a specific energy, our results begin to depend instead on the average number of states over a particular *range* of energies, and this is what we get from $W(E)$.

The energy levels for the particle in a three-dimensional box are shown in Fig. 2.7, using units of ε_0. So, for example, in the range of energies where $n^2 < 20$ (covering an energy range of $20\varepsilon_0$), there are 26 states. That gives an average quantum state density of $W = 26/(20\varepsilon_0) = 1.30\varepsilon_0^{-1}$. In the range $n^2 = 20$ to 39,

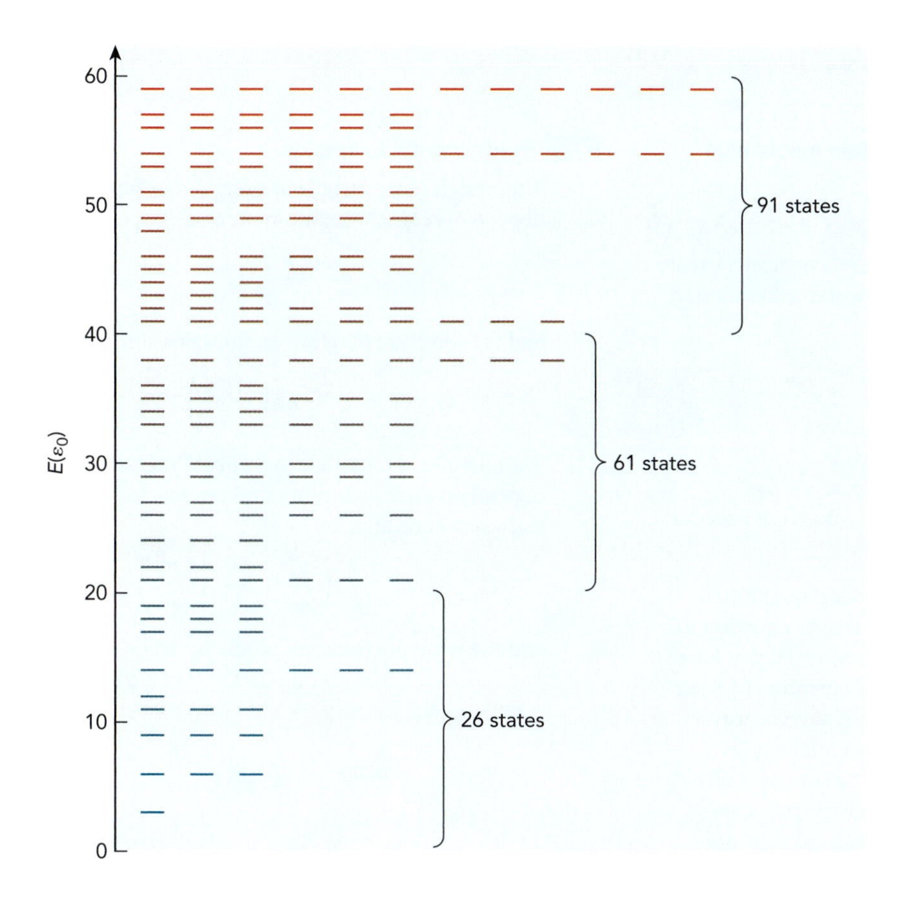

◀ **FIGURE 2.7 The energy levels of the three-dimensional box.** The number of states at each energy level between 0 and 60 (in units of ε_0) is shown.

there are 61 states ($W = 3.05\varepsilon_0^{-1}$), and in the range 40–59 there are 91 states ($W = 4.55\varepsilon_0^{-1}$). The trend continues, with the density of states increasing at higher energy to form an essentially continuous distribution of energies.

The total energy E, the number of independent coordinates or **degrees of freedom** N_{dof}, the potential energy function U, and the degeneracy g are four parameters of a system that will interest us for *every* system we study throughout this text. In essence, all of our problems boil down to this: we want to describe the motion of particles with some energy E along the N_{dof} coordinates of a potential energy function U, as they sample the g degenerate quantum states of the system. The degeneracy and the potential energy function become more difficult to describe accurately as the systems grow larger and more sophisticated. These simpler systems can be especially helpful as we familiarize ourselves with the influence these parameters have on what we can measure.

CONTEXT *Where Do We Go from Here?*

In this chapter, we've examined the quantum mechanics of systems with the simplest possible potential energy functions. Now we are ready to apply the principles of quantum mechanics in three dimensions to the structure of the one-electron atom. This will nearly complete the mathematical model that all modern chemists use to visualize individual atoms, and how they interact to form molecules.

KEY CONCEPTS AND EQUATIONS

2.1 Mathematical tools of quantum mechanics.

- The Schrödinger equation,
$$\hat{H}\psi = E\psi, \tag{2.1}$$
relates the energy of a quantum-mechanical particle to its wavefunction ψ, a measure of the distribution of the particle in space.

- The average value theorem,
$$\langle A \rangle = \int_{\substack{\text{all}\\\text{space}}} \psi^*(\hat{A}\psi)d\tau, \tag{2.10}$$
allows us to calculate expectation values for parameters, even if the parameter does not have the same value at all points in space.

- The standard problem in theoretical quantum mechanics is to solve the Schrödinger equation to get energies and wavefunctions when all you have to start with is the Hamiltonian operator. To write the Hamiltonian, we combine a kinetic energy operator $\hat{K}$ with the classical potential energy U. The kinetic energy operator is the sum over all the Cartesian coordinates and all the particles in the system of the following one-particle, one-dimensional operator:
$$\hat{K} = -\left(\frac{\hbar^2}{2m}\right)\frac{\partial^2}{\partial x^2}, \tag{2.17}$$
where m is the mass of the particle.

2.2 Fundamental examples.

- The energies of a particle in a one-dimensional box depend on a single quantum number n,
$$E_n = \frac{n^2\pi^2\hbar^2}{2ma^2}, \tag{2.31}$$
and the wavefunctions are sections of a sine wave,
$$\psi_n(x) = \sqrt{\frac{2}{a}}\sin\left(\frac{n\pi x}{a}\right). \tag{2.33}$$

- The energies of a particle in a *three*-dimensional box depend on *three* quantum numbers, one for each Cartesian coordinate,
$$E_{n_x,n_y,n_z} = \frac{\pi^2\hbar^2}{2m}\left(\frac{n_x^2}{a^2} + \frac{n_y^2}{b^2} + \frac{n_z^2}{c^2}\right), \tag{2.40}$$
and the wavefunctions are products of sine waves,
$$\psi_{n_x,n_y,n_z}(x,y,z) = \left(\frac{8}{abc}\right)^{1/2}\sin\left(\frac{n_x\pi x}{a}\right)\sin\left(\frac{n_y\pi y}{b}\right)\sin\left(\frac{n_z\pi z}{c}\right). \tag{2.39}$$

KEY TERMS

- A **wavefunction** is a representation of the wavelike character of a quantum mechanical system, used to predict properties of the system, but it is not directly measurable itself.
- **Probability density** is the square modulus of the wavefunction and is the function that describes (depending on the experiment) (a) how much of a particle (i.e., how much mass, charge, etc.) will be found at a particular position, or (b) what fraction of the time the particle will be detected at a particular position.
- A **basis set** is a set of (usually idealized) wavefunctions that can be combined to form new, often more realistic, wavefunctions.
- An **operator** is an expression that performs a manipulation on a function. This may be as simple as multiplying the function by a number, but many of the operators we use also take derivatives of the function.
- In an **eigenvalue equation,** an operator acts on a function to yield only the original function (the **eigenstate**) times a number called the **eigenvalue,** where the eigenvalue does not depend on the coordinates that describe the eigenstate.
- The **Hamiltonian** is the operator whose eigenvalue is the energy of the system. The exact form of the Hamiltonian varies from system to system.
- A **Hermitian** operator obeys the relation
$$\int \psi_i^*\hat{H}\psi_j\,d\tau = \int \psi_j^*\hat{H}\psi_i\,d\tau.$$
- The **expectation value** is the observed value of a parameter for a particular wavefunction, effectively the value of the parameter at each point in space weighted by the probability density at that point.

OBJECTIVES REVIEW

1. *Find the eigenvalue, if any, of a function under a given operation.*
Determine whether or not (a) $f(x) = 3xe^{2x}$ or (b) $g(x) = 3e^{2x^2}$ is an eigenfunction of the operator
$$\hat{\alpha} = \frac{1}{x}\frac{d}{dx}.$$
2. *Use the average value theorem to calculate properties of a system with a known wavefunction.*
Find the average value of x^4 for the wavefunction
$$\psi(x) = \sqrt{\frac{2}{a}}\sin\left(\frac{2\pi x}{a}\right)\ 0 < x < a.$$
3. *Devise a Schrödinger equation for simple systems in Cartesian coordinates.*
Write the Schrödinger equation for a particle of mass m traveling in a potential with a constant slope, $U(x) = U_0 x$.
4. *Determine the energies, wavefunctions, and other properties of a particle in a one- or three-dimensional box.*
Calculate the energy of a proton in the (100,1,1) state of a cubical box with a volume of $1.0 \cdot 10^{-18}\ \text{m}^3$.

PROBLEMS

Discussion Problems

2.1 The phase function e^{ikx} was given as the solution to the free particle wavefunction in Eq. 2.22, but e^{ax} (with a a real number) was not mentioned, although it can also satisfy the Schrödinger equation (Eq. 2.21),
$$-\frac{\hbar^2}{2m}\frac{\partial^2 e^{ax}}{\partial x^2} = E\,e^{ax},$$
because
$$\frac{\partial^2 e^{ax}}{\partial x^2} = a^2\,e^{ax}.$$
Why then is e^{ax} not a correct solution to the free particle wavefunction?

2.2 A particle moves between two walls. The correspondence principle predicts that the system will behave more *classically* when we increase which of the following parameters?
a. the mass of the particle
b. the spacing between the walls
c. the de Broglie wavelength of the particle
d. the speed of the particle

Wavefunctions and Operators

2.3 Normalize the wavefunction
x, y, z positive: $\psi(x,y,z) = e^{-(x^2+y^2+z^2)}\sqrt{xyz}$
otherwise: $\psi(x,y,z) = 0$.

2.4 Normalize the following wavefunction:
$$\psi(r,\theta,\phi) = e^{-r/(2a_0)}\left[\left(\frac{r}{a_0}\right)\left(\cos\theta + \frac{1}{\sqrt{2}}e^{-i\phi}\sin\theta - \frac{1}{\sqrt{2}}e^{i\phi}\sin\theta\right) + 2\left(1 - \frac{r}{2a_0}\right)\right].$$

THINKING AHEAD ▶ [Will the normalization constant depend on r, θ, or ϕ? Also, the wavefunction is complex. Does that mean the normalization constant is complex too?]

2.5 Normalize the wavefunction
$$\psi(x) = A(x+1)^{-a/2}\sqrt{\ln(x+1)},$$
$$\text{where } x > 0, a > 1.$$

2.6 For the following wavefunction, calculate the probability that the particle will be found between $r = 0$ and $r = a_0$. The wavefunction is not normalized.
$$\psi(r) = \left(\frac{r}{a_0}\right)e^{-2(r/a_0)^5}$$

2.7 Prove that any function $A\sin(ax)$ is orthogonal to any function $B\cos(bx)$ over the interval from $-x_0$ to x_0, where a, b, A, B, and x_0 are constants. It is not necessary to actually solve any integrals.

2.8 Example 2.2 tells us that the electron in our $2p$ wavefunction,
$$\psi_{2p}(\theta) = \sqrt{\frac{3}{2}}\cos\theta,$$
has a probability density equal to 0.125 between $\theta = 60^\circ$ and 120°. If an experiment measures the amount of charge in the region of space between $\theta = 0^\circ$ and 60°, how much charge (in coulombs) will it find for one electron having this $2p$ wavefunction?

2.9 Find the probability density of a particle between $x = 0$ and $x = 2$ if the particle has the complex wavefunction
$$\psi(x) = \sqrt{\frac{1}{21}}(x - 2i) \qquad \text{for } 0 \leq x < 3$$
$$\psi(x) = 0 \qquad \text{for x} < 0, 3 \leq x.$$

(Obtain a numerical answer, but you do not need to simplify it. Don't worry about the normalization; that has been taken care of.)

2.10 An electron has a normalized wavefunction

$$\psi(x) = \sqrt{\frac{30}{a^5}}[(a/2)^2 - x^2] \quad -a/2 < x \leq a/2$$

$$\psi(x) = 0 \quad x \leq -a/2, x > a/2,$$

which is not an eigenfunction of the kinetic energy operator. Find the expectation value of the kinetic energy in terms of the constant a.

2.11 The wavefunction $\psi(x) = (0.805)\exp(-mx^k)$ is an eigenfunction of the operator $\hat{\alpha} = \frac{2}{x^5}\frac{d}{dx}$, with eigenvalue equal to -6. Find the values of m and k, which are both integers.

2.12 Using each operator exactly *once*, find one combination of the three operators

$$\hat{\alpha} = \left(\frac{d^2}{dx^2} - \frac{1}{x^2}\right) \qquad \hat{\beta} = 2x \qquad \hat{\gamma} = x$$

such that $x + \frac{1}{x^2}$ is an eigenfunction of the combination. Give the combined operator (such as $\hat{\alpha}\hat{\beta}\hat{\gamma}$) and the eigenvalue.

2.13 Find the eigenvalue of the function $x^4y^2/3$ when acted on by the operator

$$\hat{A} = 2x\frac{\partial}{\partial x} + 2y\frac{\partial}{\partial y}.$$

2.14 Other than multiplying by 4, find any combination of differentiation (i.e., derivatives) and multiplication that gives an operator such that the eigenfunction $\psi(\phi) = e^{2i\phi}$ has an eigenvalue of 4.

2.15 Find which of the following functions are eigenstates of the operator (d^2/dx^2), given that a and b are constants. For any that are an eigenstate, give the eigenvalue.

a. $e^{-ax}e^{-bx}$

b. $b\sin(ax)\cos(ax)$

c. $\cos(ax)e^{-ax}$

2.16 Show that the function

$$\left(\sqrt{\frac{3}{4}} + \frac{i}{\sqrt{2}}\right)e^{ikx}$$

has a real part equal to $\sin(kx + \phi)$ and find the phase shift ϕ.

2.17 Find the momentum of a particle with wavefunction $\psi(x) = 2\cos(3kx) + 2i\sin(3kx)$, where k is a constant.

2.18 Show that the operators $\hat{x} = x$ and $\hat{p}_x = -i\hbar\frac{\partial}{\partial x}$ do not commute when acting on a function $f(x)$, and find the value of the operator $\hat{x}\hat{p}_x - \hat{p}_x\hat{x}$. (This operator is called the commutator for the operators $\hat{x}$ and $\hat{p}_x$ and is written $[\hat{x},\hat{p}_x]$.)

2.19 Find an integral expression for $\langle[\hat{A},\hat{B}]\rangle$ in terms of the wavefunction ψ; its derivatives ψ', ψ''; and so on, where $\hat{A} = \frac{d^2}{dx^2}$ and $\hat{B} = x\frac{d}{dx}$. In this expression, $[\hat{A},\hat{B}]$ is the commutator $\hat{A}\hat{B} - \hat{B}\hat{A}$.

2.20 Verify that the wavefunction

$$\psi(x) = Ae^{-(k\mu)^{1/2}x^2/(2\hbar)}$$

(where A, μ, k are all constants) is a solution to the Schrödinger equation

$$\left[-\frac{\hbar^2}{2\mu}\frac{d^2}{dx^2} + \frac{1}{2}kx^2\right]\psi(x) = E\psi(x),$$

and find the energy in terms of μ and k.

2.21 Allow the wavepacket of some quantum particle to have the general form

$$\psi(x) = f(x)\cos(2\pi x/\lambda_0).$$

In this expression, $f(x)$ is an arbitrary, well-behaved function that gives the overall shape of the wavepacket and that varies slowly enough so that several oscillations of the cosine wave are possible within the wavepacket. If we operate with $\hat{p}^2$ at one point along the wavefunction, we will define the local squared momentum *at that point only* to be

$$p(x)^2 \equiv -\frac{\hbar^2}{\psi(x)}\frac{d\psi(x)^2}{dx^2}.$$

For example, if $f(x)$ were a constant, then the wavefunction would be a pure cosine wave with a constant de Broglie wavelength of λ_0 and a constant square momentum $p_0^2 = h^2/\lambda_0^2$. However, the position of the particle would be completely unknown because the cosine wave would extend infinitely over all values of x. Now, we shrink the domain of $f(x)$ by making $f(x) = 0$ everywhere except over a finite range of x. As $f(x)$ becomes more localized, the position becomes more certain, but the local momentum is no longer a constant; i.e., the value of the momentum becomes more uncertain. To show this, consider the points $x_n = 2n\lambda_0/\pi$, with n an integer, where the pure cosine wave reaches its maximum amplitude. Find a general equation for the squared momentum at any x_n in terms of p_0, $f(x)$, and the second derivative $f''(x) = d^2f(x)/dx^2$.

The Particle in a One-Dimensional Box

2.22 Find an equation for the de Broglie wavelength of the particle in a one-dimensional box as a function only of the energy of the particle E and the particle's mass m.

2.23 How big is a one-dimensional box containing an electron if $1/\lambda$ for the $n = 2 \rightarrow 3$ transition is $1.0\ \text{cm}^{-1}$?

2.24 If an electron in a one-dimensional box has a lowest energy state with $E_1 = 1.3 \cdot 10^{-19}$ J, what is the electron's energy when its de Broglie wavelength is exactly one half the length of the box? **THINKING AHEAD ▶** [Based on the correspondence principle, is the new state at high n or low n?]

2.25 There are many subatomic particles besides electrons, protons, and neutrons. Suppose one of these particles, the π-meson, is trapped in a one-dimensional box of length 3.00 Å. The *lowest energy* transition for the system absorbs photons of energy $7.60 \cdot 10^{-21}$ J. What is the mass of the π-meson?

2.26 Write the wavefunctions that solve the Schrödinger equation for a single particle of mass m traveling with energy E along one coordinate x, when the potential energy has the form

$$U(x < 0) = \infty, \quad U(x \geq 0) = 0.$$

Do not try to normalize the wavefunctions.

2.27 It is possible to construct molecular-scale tubes out of networks of carbon atoms (see Section 11.2). To a particle traveling along one such narrow tube, very little motion may be possible perpendicular to the walls. The potential function then can be approximated using the particle in a box, with the length of the box equal to the length of the tube. Find the transition frequency (the transition energy divided by h) between the lowest two quantum states of a sodium ion Na^+ inside one of these nanotubes 2.0 cm long.

2.28 Find an equation for the root mean square speed $\langle v^2 \rangle^{1/2}$ of a particle with mass m in a one-dimensional box of length a and quantum state n.

2.29 Imagine that we take the one-dimensional box potential and add an inifinitely thin wall in the middle, at $x = a/2$. At $x = a/2$, the potential energy is infinite, but the particle is still allowed to occupy both halves of the box.

a. Write the kinetic energy operator for a particle of mass m anywhere along x.
b. Write the boundary condition for the wavefunction at $\psi(x = 0)$, $\psi(x = a)$, and $\psi(x = a/2)$. (In other words, what is the value of ψ at those points?)
c. Write an equation for the normalized wavefunctions that satisfy all of these boundary conditions.

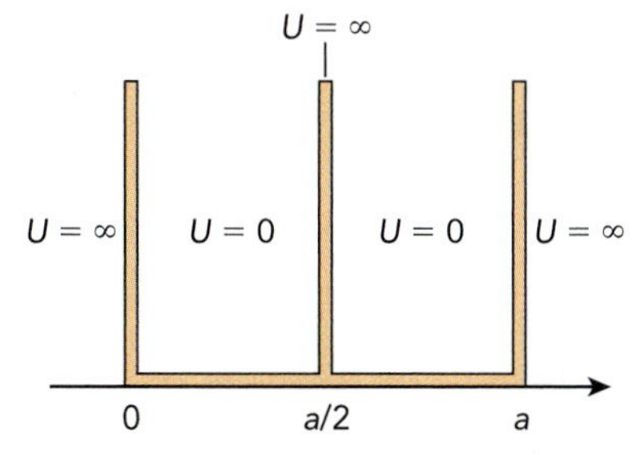

2.30 What is the eigenvalue when the $\hat{p}^2$ operator operates on the $n = 2$ state of a proton in a box of length 2.0 Å?

2.31 Calculate the probability densities of the particle in a one-dimensional box over the interval from 0 to $a/4$ in the states $n = 1$ and $n = 21$. What is the classical value?

2.32 Calculate the fraction of time the particle in the $n = 1$ state of a one-dimensional box can be found in each fifth of the box, i.e., $0 < x < a/5$, $a/5 < x < 2a/5$, $2a/5 < x < 3a/5$, $3a/5 < x < 4a/5$, and $4a/5 < x < a$.

2.33 What are the wavefunctions and energies of the particle in a one-dimensional box that extends from $-a/2$ to $+a/2$? THINKING AHEAD ▶ [If you are a particle in this box, what properties of the box can you tell have changed, compared to Fig. 2.6?]

2.34 The wavefunctions of the particle in a one-dimensional box are *all* orthogonal. Prove the $n = 1$ and $n = 2$ states are orthogonal.

2.35 Find $\langle x \rangle$ for the particle in a one-dimensional box at *any* n. THINKING AHEAD ▶ [What is the answer for the classical system?]

2.36 Evaluate the rms (root mean square) momentum of the particle in a one-dimensional box, $\langle p_x^2 \rangle^{1/2}$, as a function of the quantum number n and the length of the box a.

2.37 Write the integral necessary to evaluate the average of the product xp^2, where p is the momentum, for a particle in the $n = 3$ state of a one-dimensional box of length a. It is not necessary to solve the integral.

2.38 Write the Hamiltonians needed in region A (0 to 2 Å) and in region B (2 Å to 4 Å) for a particle of mass m trapped in a one-dimensional funnel, according to the following potential energy curve (the solid line).

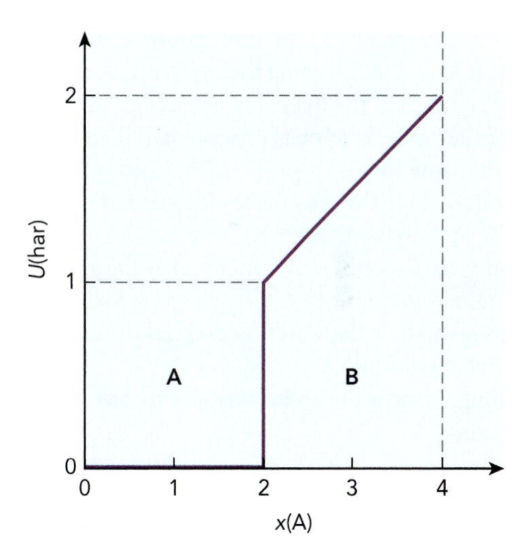

Particles in Two and Three Dimensions

2.39 Verify that $(2/a)^{3/2}$ is the normalization constant for the wavefunctions of a cubical box.

2.40 Check the following entry in the list of energy levels (in units of ε_0) for the particle in a three-dimensional box. Cross out any states that do not belong, and correct the degeneracy if necessary. No new states need to be added.

E	$g(E)$	states
89	18	(9,2,2)(8,4,3)(7,6,2)(7,5,5)

2.41 A particle of mass m in a three-dimensional box has quantum states indicated by the quantum numbers (n_x,n_y,n_z). The box is cubic and lies in the region $x = 0$ to a, $y = 0$ to a, $z = 0$ to a, where $a = 4.0$ Å. You do not need to solve any integrals for this problem.

a. What is the value of $|\psi|^2$ in the center of the box for the state (1,1,1)?
b. What is the value of $|\psi|^2$ at the point $x,y,z = 0, a, a$ for the state (1,1,1)?
c. What is the value of the probability density *integrated* from the origin to $x,y,z = a/2,a/2,a/2$ for the state (1,1,1)?
d. What are the x,y,z coordinates at *each* position where the state (2,1,1) reaches its maximum probability density?

2.42 Find the general solution for the wavefunction of the free particle in three dimensions. It is not possible to normalize these wavefunctions; please don't try.

2.43 Two three-dimensional boxes each have sides a and b of equal length, $a = b = 10.0$ nm. However, box A is a cubical box with $c = 10.0$ nm, while box B is not a cube. Find the value of c in box B if a proton in box B has the same ground state energy as an electron in box A.

2.44 A particle is in the $(n_x,n_y,n_z) = (3,2,1)$ state of a three-dimensional box with dimensions $a = 2.0$ Å, $b = 4.0$ Å, $c = 8.0$ Å along the x, y, and z axes respectively. For each of the following properties, identify the appropriate axis (x, y, or z) or write "all" if all three axes have the same value.

a. Motion along this axis makes the greatest contribution to the total kinetic energy.
b. Along this axis the wavefunction has the greatest number of nodes.
c. Along this axis the wavefunction has the greatest de Broglie wavelength.
d. Along this axis the *probability density* has exactly two maxima.

2.45 A particle in a three-dimensional cubical box of volume 8.0 Å^3 is in the (2,2,2) state with a total kinetic energy of $6.0 \cdot 10^{-20}$ J. Give the following:

a. the zero-point energy (in J),
b. λ_{dB} (in Å) of the particle along any axis,
c. the new energy in the (2,2,2) state if we double the length of each side of the box, and
d. the mass (in kg) of the particle.

2.46 What is the degeneracy of the lowest excited energy level for a particle in a box with dimensions $a = b = 4.0$ nm, c $= 2.0$ nm?

2.47 Find all the states (n_x,n_y,n_z) with energy $7\pi^2\hbar^2/(ma^2)$ for a particle of mass m in a three-dimensional box with sides of length a, $b = a$, and $c = 2a$.

2.48 What are the quantum numbers for the wavefunction in a three-dimensional box that has one node along x, three nodes along y, and no nodes along z?

2.49 Find the two lowest energy sets of *degenerate* quantum states of an electron in a three-dimensional box with dimensions $a = 2$ Å, $b = 2$ Å, $c = 4$ Å. Indicate each state by its quantum numbers (n_x,n_y,n_z). (You do *not* have to calculate the energy itself.)

2.50 A first guess at the wavefunction of a non-polar molecule on the square face of a metal crystal could be that of a particle in a *two*-dimensional box of the same size as the crystal face. Give an expression for the energy levels in such a system, clearly identifying all the symbols used. Give the quantum numbers for the lowest set of degenerate states.

2.51 Derive an equation for the degeneracy and density of states of a *two-dimensional* box with area A in the limit of large n, based on the derivation of Eq. 2.40.

2.52 For an electron trapped in a cubic box with each side of length a_0, the energy in hartrees can be written $E(E_h) = \frac{\pi^2}{2}(n_x^2 + n_y^2 + n_z^2)$. What is the fastest speed v_z (in cm/s) along the z axis that an electron can have if it is in this box with an energy of $12\pi^2$ hartrees?

2.53 Consider a three-dimensional box of lengths $a = 2.0$ Å, $b = 4.0$ Å, and $c = 1.0$ Å. Identify the quantum numbers n_x,n_y,n_z for the wavefunction of a particle in this box when the spacings between the nodes along the x axis, the y axis, and the z axis are all equal, and when there is exactly one node along the z axis.

2.54 What size does the box have to be in order for the three-dimensional box potential to yield the same electronic transition energy for the $(n_x,n_y,n_z) = (1,1,1) \rightarrow (2,1,1)$ transition as the $n = 1 \rightarrow 2$ transition in the Bohr atom with atomic number Z?

2.55 Extend Table 2.1 to $E = 65\pi^2\hbar^2/(2mV^{2/3})$.

2.56 Make a table similar to Table 2.1 for *two* particles with distinguishable quantum numbers in a three-dimensional cubical box, giving energies, degeneracies, and quantum numbers for the lowest five energy values. The lowest energy state, for example, will be (1,1,1)(1,1,1).

2.57 Assume that the conduction electrons in a 8 cm^3 cubical block of metal obey the quantum mechanics for a particle in a three-dimensional box. Estimate the separation (to within a factor of three) between energy levels of these electrons in units of cm^{-1}.

3 One-Electron Atoms

LEARNING OBJECTIVES

With the material in this chapter, you will be able to do the following:

❶ Write the electronic wavefunction for any quantum state of any one-electron atom.

❷ Use the wavefunction to evaluate properties of the electron in that quantum state.

❸ Identify properties of various orbital wavefunctions based on the quantum numbers.

GOAL *Why Are We Here?*

The goal of this chapter is to demonstrate how we arrive at the mathematical model used by modern chemists to describe one-electron atoms. In other words, we seek to solve the Schrödinger equation for the one-electron atom, to find the energies (eigenvalues) and wavefunctions (eigenstates). At the heart of our reasoning will be a derivation of many little steps but enormous implications. The major results from that derivation will be a simple expression for the energies and two tables, based on a new set of quantum numbers, from which we may assemble the wavefunction for the electron in any one-electron atom.

CONTEXT *Where Are We Now?*

We need to build up a reliable description of atoms because atoms are the building blocks we will use to construct molecules. Chapters 1 and 2 present basic rules of quantum mechanics, using the semi-classical Bohr model and the idealized particle-in-a-box problems as fundamental examples. But the Bohr model is not a three-dimensional model, and our atoms—and the chemical bonds that they form—*are* three-dimensional. Therefore, we return to the structure of atoms, but now, with the tools of quantum mechanics at our disposal, we will describe them more accurately.

To keep the math manageable, we limit ourselves to atoms with only one electron for now. This is a stepping stone between the semiclassical Bohr model and the many-electron atoms that follow. For the one-electron atom, we can solve our Schrödinger equation exactly. That capability will last us only until we get to the next chapter, so let's enjoy it while we can!

SUPPORTING TEXT **How Did We Get Here?**

The main *qualitative* concept we draw on for this chapter is that the electron in an atom behaves enough like a wave that quantum mechanics restricts its energy and other parameters to specific values. For a more detailed background, we will draw on the following equations and sections of text to support the ideas developed in this chapter:

- In Section 2.2 we learned how to develop a Hamiltonian that combines the kinetic and potential energy contributions to the system. The first step we will take in solving the Schrödinger equation for the atom, in order to predict its properties based on the particles present and the forces at work, is to develop a Hamiltonian $\hat{H}$ that extracts the energy from a wavefunction ψ. The forces determine the potential energy function, which we normally include in the Hamiltonian as a classical, algebraic expression. The kinetic energy is not expressed classically, but it reliably has the form (Eq. 2.36)

$$\hat{K} = -\frac{\hbar^2}{2m}\nabla^2$$

for each particle, where (Eq. 2.35)

$$\nabla^2 \equiv \frac{\partial^2}{\partial x^2} + \frac{\partial^2}{\partial y^2} + \frac{\partial^2}{\partial z^2}$$

is the Laplacian operator. The Laplacian finds the curvature of the wavefunction by taking the second derivative. The greater the curvature, the faster the wavefunction oscillates and the shorter the de Broglie wavelength (Eq. 1.3),

$$\lambda_{\text{dB}} = \frac{h}{mv} = \frac{h}{p}.$$

The de Broglie wavelength, in turn, relates the wavefunction to the momentum p. In this way, the Laplacian effectively extracts the kinetic energy, $K = p^2/(2m)$, from the wavefunction.
- Section 1.4 described the Bohr model of the one-electron atom, a semiclassical explanation of how the atom works. Although it incorrectly treats the electron as a point particle moving in a circular orbit, the Bohr model provides a helpful point of entry into the quantum mechanical picture because it correctly divides the energy into its kinetic and potential contributions and combines these to obtain the correct value of the total, quantized energy.

3.1 Solving the One-Electron Atom Schrödinger Equation

We're going to get to the math soon enough. First, let's look at the problem qualitatively and ask what the general form of the electron distribution will turn out to be.

What to Expect

We can anticipate much of what the calculus will reveal about a system if we apply the *qualitative* concepts of quantum mechanics, given the forces that define the potential energy function.

To describe the three-dimensional nature of the electron's wavelike properties, we now liberate the electron from a circular orbit and allow it to have density anywhere: along a circular path, out of the plane of that circle, and even at different distances from the nucleus — all while it's in a single quantum state. Let the circular orbit in the Bohr model correspond to motion along an angle ϕ, measured from the x axis in the xy plane. Now we need to introduce two *new* coordinates to accommodate the new motions permitted to the electron. We will call the new coordinates θ (the angle measured in any direction from the z axis) and r (the distance from the center of mass, which is roughly at the nucleus). Fig. 3.1 provides definitions for these spherical coordinates. (These are the conventional definitions for θ and ϕ in physics and chemistry, but unfortunately they are the reverse of the definitions often used in mathematics courses.)

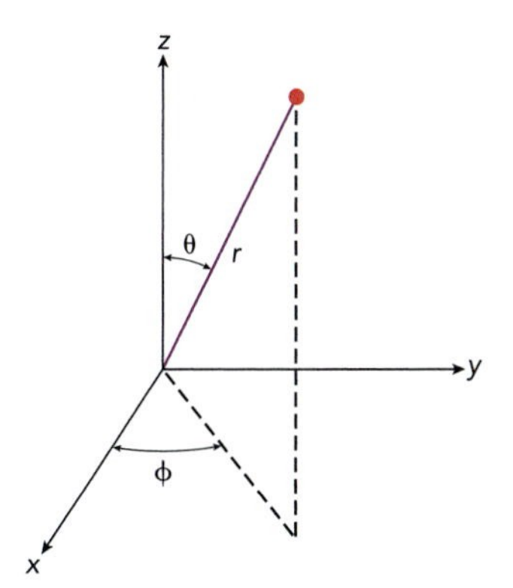

▲ FIGURE 3.1 **Relation between spherical and Cartesian coordinates.**

Each independent coordinate along which the electron can move is an electronic **degree of freedom,** a different way that the electron can store its energy. In the semiclassical Bohr model, the electron could travel only along the coordinate ϕ, tracing out a circular path. Now, with motion allowed in all three directions, the energy may be divided among the different degrees of freedom: r, θ, and ϕ (Fig. 3.2).

When we apply the de Broglie electron/wave to Bohr's circular orbit (Sec. 1.5), we find some similarities to the particle-in-a-box wavefunction. Those similarities remain, now that we extend the electron's motion to r and θ. For example,

- The electron is *not* a point particle in this system. It behaves instead like a cloud of mass and charge, with its density diminishing as the potential energy climbs at distances far from the nucleus.
- As we put more energy into one degree of freedom, the number of nodes along that coordinate increases. There may be nodes at particular distances r from the nucleus, as well as at certain angles ϕ or θ. As we saw for the particle in a three-dimensional box, some wavefunctions will have nodes along all three coordinates.
- Also as with the three-dimensional box, we expect that as we increase the energy in the system, there will be ever more states to choose from, leading to a greater degeneracy g.

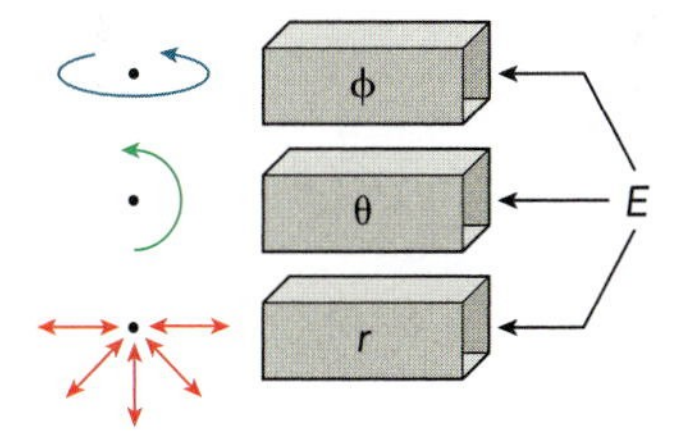

▲ FIGURE 3.2 **Energy is expressed as motion along different coordinates.** Real electrons exist in three dimensions and can move along ϕ, θ, or r. Each coordinate provides a different degree of freedom for the electron, like a set of different boxes in which the energy can be stored. In general, some of the energy may go into each of the three degrees of freedom.

The system will have some new features as well. For example—here's a surprise—the electron doesn't need to orbit the nucleus at all, meaning that the angular momentum can be zero. That is because we now have **radial** kinetic energy for motion along r, and that motion alone is sufficient to keep the electron in a stable distribution about the nucleus. If we imagine the Coulomb potential as a bowl with the nucleus at the bottom, Bohr's orbits correspond to rolling the electron like a marble along the lip of the bowl at some fixed speed and always the same distance from the nucleus, so that only the angle ϕ changes. But we can also allow the

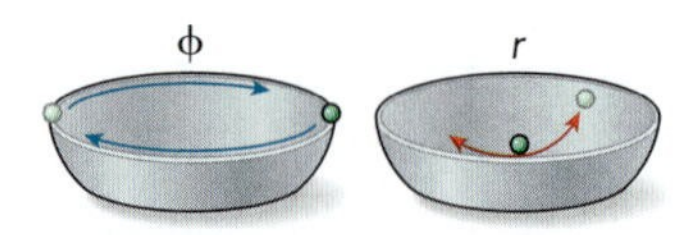

▲ FIGURE 3.3 **Angular and radial motion.** For a marble rolling around the rim of a bowl along angle ϕ, we can measure the values of the kinetic energy and the angular momentum. If we instead roll the marble from one side of the bowl through the middle and to the other side, then the motion is along the distance r from the center of the bowl. Both motions may have the same kinetic energy, but the radial motion has no angular momentum.

marble to roll from one end of the bowl to the other, across the middle; there is no angular motion in this case, only radial motion (Fig. 3.3). We can put the same kinetic energy into the radial motion, but the angular momentum is now zero. In the same way, the electron can oscillate along r without motion along the angular coordinates θ or ϕ, although most of the quantum states we find will involve some combination of radial and angular motion.

The One-Electron Atom Schrödinger Equation

Now the math. To apply the methods of quantum mechanics to atoms with one electron, we solve the Schrödinger equation for an electron and an atomic nucleus interacting through the Coulomb potential. First we need the Hamiltonian for this system, which we obtain by adding together the appropriate kinetic and potential energy operators.

The kinetic energy operator for motion in three dimensions, Eq. 2.36, can be applied to both the nucleus and electron. If the nucleus has mass m_{nuc} and Laplacian ∇^2_{nuc}, and the electron has mass m_e and Laplacian ∇^2_{elec}, then we can obtain a total kinetic energy operator

$$\begin{aligned}\hat{K} &= \hat{K}_{\text{nuc}} + \hat{K}_{\text{elec}} \\ &= -\frac{\hbar^2}{2m_{\text{nuc}}}\nabla^2_{\text{nuc}} - \frac{\hbar^2}{2m_e}\nabla^2_{\text{elec}}.\end{aligned}$$

This kinetic energy operator treats all the motions of the two particles (the electron and the nucleus). We're specifically interested in the motion of the electron *relative* to the nucleus, but not the overall motion of the whole atom from, say, one side of the room to the other side of the room. A common transformation in physics (Eq. A.45) allows us to rearrange the kinetic energy terms for the nucleus and the electron into a *reduced mass* term for the relative motions, with the reduced mass $\mu = m_{\text{nuc}}m_e/(m_{\text{nuc}} + m_e)$, and a *center of mass* term for the overall motion, with mass $m_{\text{nuc}} + m_e$:

$$\hat{K} = -\frac{\hbar^2}{2\mu}\nabla^2 - \frac{\hbar^2}{2(m_{\text{nuc}} + m_e)}\nabla^2_{\text{COM}}.$$

We will ignore the center of mass term for now to focus on the relative motions of the particles. Furthermore, the nucleus is always at least 1800 times more massive than the electron, and even a carbon nucleus is already four orders of magnitude more massive than the electron. So the center of mass—to a good approximation—is at the nucleus. That means that the motions relative to the center of mass are the motions of the electron, and the reduced mass can be approximated by the mass of the electron:

$$\mu = \frac{m_{\text{nuc}}m_e}{m_{\text{nuc}} + m_e} \approx \frac{m_{\text{nuc}}m_e}{m_{\text{nuc}}} = m_e. \tag{3.1}$$

Therefore, our kinetic energy operator starts off in this form:

$$\hat{K} = -\frac{\hbar^2}{2m_e}\nabla^2. \tag{3.2}$$

We have to make one major adjustment before proceeding: we have to figure out the best way to write the Laplacian, ∇^2. For atomic systems, spherical coordinates are much more convenient to use than Cartesian coordinates.

The kinetic energy operator is most easily written in terms of x, y, and z because motion is linear when no forces are at work. But in this chapter, we do have a force at work: the Coulomb attraction of the electron for the nucleus — a force that depends only on the distance r, a spherical coordinate. So we have a choice: the kinetic energy operator is expressed most simply in Cartesian coordinates, and the potential energy in spherical coordinates. The coordinates convenient for the potential function win out. The motions of the electron, and therefore the shapes of its wavefunctions, are determined by the Coulomb force, and by using the coordinates convenient for U we will obtain wavefunctions in the most convenient coordinates as well.

With a lot of trigonometry, you could show that the Laplacian in spherical coordinates is

$$\begin{aligned}\nabla^2 &= \frac{\partial^2}{\partial x^2} + \frac{\partial^2}{\partial y^2} + \frac{\partial^2}{\partial z^2} \\ &= \frac{1}{r^2}\frac{\partial}{\partial r}r^2\frac{\partial}{\partial r} + \frac{1}{r^2\sin\theta}\frac{\partial}{\partial\theta}\sin\theta\frac{\partial}{\partial\theta} + \frac{1}{r^2\sin^2\theta}\frac{\partial^2}{\partial\phi^2},\end{aligned} \tag{3.3}$$

with angles defined as usual for spherical coordinates. Be careful how you read Eq. 3.3: according to our rules of operator algebra, we carry out the operations from right to left:

$$\frac{1}{r^2}\frac{\partial}{\partial r}r^2\frac{\partial}{\partial r}\psi = \frac{1}{r^2}\left\{\frac{\partial}{\partial r}\left[r^2\left(\frac{\partial\psi}{\partial r}\right)\right]\right\}.$$

The coordinate conversion doesn't affect the physics that ∇^2 represents. We can see in Eq. 3.3 that the Laplacian is still a sum over three second derivatives, and each term still has units of (distance)$^{-2}$. Combining Eqs. 3.2 and 3.3, we obtain the kinetic energy operator in spherical coordinates:

$$\begin{aligned}\hat{K} &= -\frac{\hbar^2}{2m_e}\left[\frac{1}{r^2}\frac{\partial}{\partial r}r^2\frac{\partial}{\partial r} + \frac{1}{r^2\sin\theta}\frac{\partial}{\partial\theta}\sin\theta\frac{\partial}{\partial\theta} + \frac{1}{r^2\sin^2\theta}\frac{\partial^2}{\partial\phi^2}\right] \\ &= -\frac{\hbar^2}{2m_e}\frac{1}{r^2}\frac{\partial}{\partial r}r^2\frac{\partial}{\partial r} + \frac{1}{2m_e r^2}\hat{L}^2(\theta,\phi),\end{aligned} \tag{3.4}$$

where

$$\hat{L}^2(\theta,\phi) = -\hbar^2\left[\frac{1}{\sin\theta}\frac{\partial}{\partial\theta}\sin\theta\frac{\partial}{\partial\theta} + \frac{1}{\sin^2\theta}\frac{\partial^2}{\partial\phi^2}\right]. \tag{3.5}$$

The $\hat{L}^2$ term contains all the dependence on the angular motion of the electron. In classical mechanics, we express the kinetic energy of a particle as $mv^2/2$, and we can break the contributions to the velocity up into a radial speed v_r and an angular speed $v_{\theta,\phi}$, which allows us to obtain the kinetic energy in a form similar to the quantum kinetic energy operator in Eq. 3.4:

$$\begin{aligned}K &= \frac{mv^2}{2} = \frac{mv_r^2}{2} + \frac{mv_{\theta,\phi}^2}{2} \\ &= \frac{mv_r^2}{2} + \frac{(mrv_{\theta,\phi})^2}{2mr^2} = \frac{mv_r^2}{2} + \frac{L^2}{2mr^2},\end{aligned} \qquad L = mrv_{\theta,\phi} \tag{3.6}$$

where L is the angular momentum. By analogy between Eqs. 3.4 and 3.6, let us claim that the eigenvalue of the operator $\hat{L}^2(\theta,\phi)$ is indeed L^2, the square of the angular momentum. [Problem 3.12 steps you through a more rigorous proof that the eigenvalue of $\hat{L}^2(\theta,\phi)$ is L^2.]

CHECKPOINT The goal of this chapter is to solve the Schrödinger equation for the one-electron atom, and to do that we first have to write the Hamiltonian, $\hat{H}$. We just obtained the kinetic energy operator we need. Now we need the potential energy, and that will complete $\hat{H}$.

So much for the kinetic energy. The potential energy is the Coulomb energy for the interaction between two point charges (Eq. A.42), where we set the charges equal to Ze for the nucleus and $-e$ for the electron:

$$U = \frac{q_1 q_2}{4\pi\varepsilon_0 r_{12}} = -\frac{Ze^2}{4\pi\varepsilon_0 r}. \tag{3.7}$$

This potential energy curve, drawn in Fig. 3.4, leads to a dramatic departure from the results we find for the idealized examples of the free particle or particle in a box.

Combining the kinetic (Eq. 3.4) and potential (Eq. 3.7) energy operators, the Hamiltonian for the one-electron atom becomes

$$\hat{H} = -\frac{\hbar^2}{2m_e}\frac{1}{r^2}\frac{\partial}{\partial r}r^2\frac{\partial}{\partial r} + \frac{1}{2m_e r^2}\hat{L}^2(\theta, \phi) - \frac{Ze^2}{4\pi\varepsilon_0 r}. \tag{3.8}$$

Plan of Attack

With any multivariable math problem, it helps if we can break the problem into individual parts for each variable. This Hamiltonian does not have so straightforward a separation of variables as we have in the three-dimensional box problem (Eq. 2.38), because the three variables (r, θ, and ϕ) do not each appear alone in separate terms of the Hamiltonian. Instead, the derivative with respect to ϕ is multiplied by a function of θ, and the θ- and ϕ-dependent parts of the Hamiltonian are then together multiplied by a function of r. The problem is still separable, but to a lesser degree. Rather than solving the three parts independently, the way we did with the three-dimensional box, we solve this Schrödinger equation by a specific sequence of steps. At each step, we replace an operator that depends on a specific coordinate (like θ or ϕ) with the operator's eigenvalue, which has a numerical value for a particular quantum state. We say that we replace the *explicit* dependence of the Hamiltonian on θ and ϕ by an *implicit* dependence, where θ and ϕ have determined certain numerical values in the Hamiltonian but no longer appear directly in the expression themselves. Here are the steps:

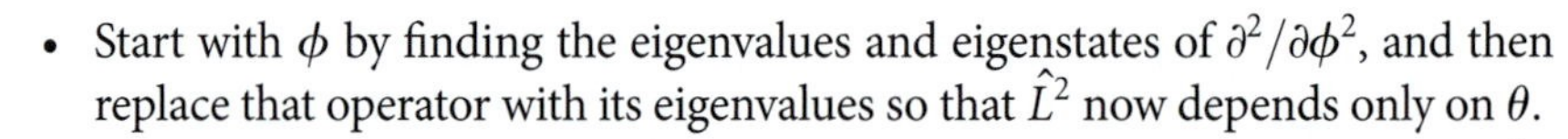

- Start with ϕ by finding the eigenvalues and eigenstates of $\partial^2/\partial\phi^2$, and then replace that operator with its eigenvalues so that $\hat{L}^2$ now depends only on θ.
- Next solve the θ-part by finding the eigenvalues and eigenstates of $\hat{L}^2$, and replace that operator with its eigenvalues so that $\hat{H}$ now depends only on r.
- Finally, solve the differential equation in r to get the radial wavefunctions and energies of the electron.

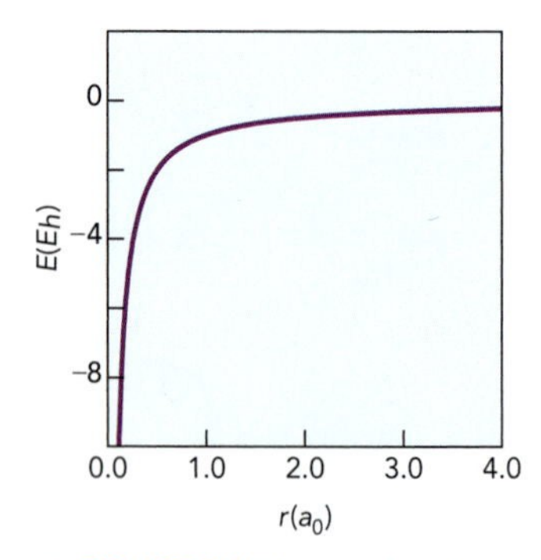

▲ FIGURE 3.4 **The Coulomb potential energy of the hydrogen atom.** The electrical potential energy E (in units of hartrees, E_h) is given as a function of distance r between the nucleus and electron (in units of the Bohr radius, a_0).

Only the last step actually solves the Schrödinger equation, with the final eigenvalues being the energies, but the eigenstates we find along the way will each contribute to the final wavefunctions.

We will focus on this Schrödinger equation more closely than on any other, because these energies and wavefunctions are *the* fundamental results from quantum mechanics that we use in chemistry. The angular part of the atomic wavefunction is predominantly responsible for the geometries of covalently bonded molecules. The radial part of the wavefunction effectively sets the size of the atom, and consequently controls the onset of steric interactions, among many other properties of molecules.

That's the big picture. If you want the specifics, get comfortable and stay tuned. To skip the math and get to the solution, jump to the **DERIVATION SUMMARIES** that appear after Eqs. 3.20 and 3.32.

The Angular Solution

CHECKPOINT This is the hard work of solving a partial differential equation in three variables. Problems like this continue to keep mathematicians the world over happily occupied. The approach here may seem haphazard—we make up trial functions and then show that they work out—but this is a typical approach to solving differential equations. At the end of this process, we will have the wavefunctions and energies of the one-electron atom.

The angular part of the Schrödinger equation has been absorbed into a single operator, $\hat{L}^2$ (Eq. 3.5),

$$\hat{L}^2(\theta,\phi) = -\hbar^2\left[\frac{1}{\sin\theta}\frac{\partial}{\partial\theta}\sin\theta\frac{\partial}{\partial\theta} + \frac{1}{\sin^2\theta}\frac{\partial^2}{\partial\phi^2}\right],$$

and the eigenvalues of this operator will be the square of the angular momentum for the electron. We can separate the angular problem further into the individual contributions from the two angles, θ and ϕ.

We can't begin with θ, because part of the θ-dependence is locked up in the second term $(1/\sin^2\theta)\partial^2/\partial\phi^2$ and can't be solved until we have the second derivative with respect to ϕ. Also, the part of our wavefunction that depends on ϕ—let's call it $\Phi(\phi)$—is easier to solve because the operator $\partial^2/\partial\phi^2$ is similar to the $\partial^2/\partial x^2$ operator we saw in the free particle Schrödinger equation (Section 2.2). The free particle has a Hamiltonian (Eq. 2.19)

$$\hat{H} = \frac{\hat{p}_x^2}{2m} = -\frac{\hbar^2}{2m}\frac{\partial^2}{\partial x^2},$$

and the eigenfunctions of this Hamiltonian are the wavefunctions that solve the Schrödinger equation:

$$\psi = ce^{ikx} = c\left[\cos(kx) + i\sin(kx)\right].$$

Now we have another operator, $\partial^2/\partial\phi^2$, that depends entirely on the second derivative of the coordinate, and it has a similar set of eigenfunctions:

$$\Phi(\phi) = A_\phi e^{im_l\phi}. \tag{3.9}$$

If we take the second derivative of this $\Phi(\phi)$, we find the eigenvalues

$$\frac{\partial^2\Phi(\phi)}{\partial\phi^2} = -A_\phi m_l^2 e^{im_l\phi} = -m_l^2\,\Phi(\phi). \tag{3.10}$$

The eigenvalue of $\partial^2/\partial\phi^2$ is $-m_l^2$. If we also require that $\Phi(\phi)$ corresponds to a standing wave, such that $\Phi(\phi) = \Phi(\phi + 2\pi)$, then,

$$e^{im_l\phi} = e^{im_l(\phi+2\pi)} = e^{im_l\phi}e^{im_l2\pi},$$

which in turn requires that

$$e^{im_l2\pi} = \cos(2\pi m_l) + i\sin(2\pi m_l) = 1.$$

This is true only if m_l is any integer, including negative values and zero.

Before we go any further, it is reasonable to ask what these eigenvalues mean physically. Recall that the angular momentum $\overrightarrow{L}$ is a vector defined by the cross product of the position and momentum vectors (Eq. A.46):

$$\overrightarrow{L} = \overrightarrow{r} \times \overrightarrow{p}.$$

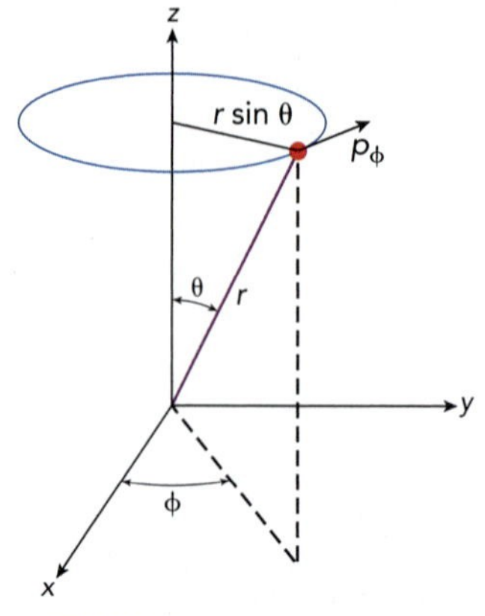

▲ **FIGURE 3.5 Motion along ϕ corresponds to circular motion about the z axis.**

Since the cross product is always perpendicular to the two vectors from which it's formed, $\vec{L}$ is perpendicular to $\vec{r}$ and $\vec{p}$. Motion along the angle ϕ is a circular motion parallel to the xy plane (see Fig. 3.5), and therefore it has an angular momentum $\vec{L}$ directed along the z axis. In other words, there is some non-zero projection L_z of $\vec{L}$ onto the z axis, which depends on how much of the angular motion is along angle ϕ (i.e., in the xy plane). The magnitude of L_z is the product of the linear momentum p_ϕ along ϕ and the distance of the electron from the z axis, which is equal to $r \sin\theta$, giving

$$L_z = p_\phi r \sin\theta. \tag{3.11}$$

The linear momentum can be rewritten in terms of the de Broglie wavelength for this motion,

$$p_\phi = \frac{h}{\lambda_\phi}.$$

But now let's again require (as in Fig. 1.15) that, to avoid destructive interference, some whole number m_l wavelengths fit exactly into the circumference of the circuit. The circumference of the orbit is $2\pi r \sin\theta$, and therefore

$$\lambda_\phi = \frac{2\pi r \sin\theta}{m_l}.$$

Combining these expressions gives a formula for the z component of the angular momentum in terms of our new quantum number, the integer m_l:

$$L_z = \frac{hr \sin\theta}{\lambda_\phi} = \frac{hr \sin\theta}{2\pi r \sin\theta / m_l} = m_l \hbar. \tag{3.12}$$

This value is also obtained if we construct a quantum mechanical operator $\hat{L}_z$ from the linear momentum operator $\hat{p}$. A complete circuit of ϕ from 0 to 2π covers a distance $2\pi r \sin\theta$, and traveling any fraction $d\phi$ of that corresponds to a distance

$$2\pi r \sin\theta \frac{d\phi}{2\pi} = r \sin\theta \, d\phi.$$

In Section 2.2, we learn that the momentum operator for motion along x is (Eq. 2.23) $\hat{p}_x = \frac{\hbar}{i}\frac{d}{dx}$. For motion along ϕ, we use the same formula for the momentum operator, except that we replace the distance increment dx by the distance traveled when the angle changes by an increment $d\phi$,

$$\hat{p} = \frac{\hbar}{i} \frac{1}{r \sin\theta} \frac{\partial}{\partial \phi},$$

and the angular momentum component L_z is given by the operator

$$\hat{L}_z = r \sin\theta \hat{p}_\phi = \frac{\hbar}{i} \frac{\partial}{\partial \phi}. \tag{3.13}$$

Returning now to the larger problem of the angular Hamiltonian, we can replace the second derivative of ϕ in $\hat{L}^2$ by the eigenvalue $-m_l^2$. Now we have to find the θ-dependent part of the wavefunction, $\Theta(\theta)$. This is more difficult, and we take a shortcut by assuming a reasonable, general form for the wavefunction. We need a function that can suffer differentiation with respect to θ, suffer multiplication and division by $\sin\theta$, and still look the way it did when we started.

Two functions that could help are sine and cosine. Second derivatives leave sine and cosine within a multiplicative constant of where they started. Let's use that property to assume initially that $\Theta(\theta)$ takes the form $A_\theta \sin^k \theta$ and see where it takes us:

$$\begin{aligned}
\hat{L}^2\Theta(\theta) &= -\hbar^2\left[\frac{1}{\sin\theta}\frac{\partial}{\partial\theta}\sin\theta\frac{\partial}{\partial\theta} - \frac{m_l^2}{\sin^2\theta}\right]A_\theta sin^k\theta && \text{Eqs. 3.5 and 3.10}\\
&= -A_\theta\hbar^2\left[\frac{1}{\sin\theta}\frac{\partial}{\partial\theta}\sin\theta\,(k\sin^{k-1}\theta\cos\theta) - m_l^2\, sin^{k-2}\theta\right] && \text{take } \frac{\partial}{\partial\theta}\\
&= -A_\theta\hbar^2\left[\frac{1}{\sin\theta}\frac{\partial}{\partial\theta}(k\sin^k\theta\cos\theta) - m_l^2\sin^{k-2}\theta\right] && \times \sin\theta\\
&= -A_\theta\hbar^2\left[\frac{1}{\sin\theta}(k^2\sin^{k-1}\theta\cos^2\theta - k\sin^{k+1}\theta) - m_l^2\sin^{k-2}\theta\right] && \text{take } \frac{\partial}{\partial\theta}\\
&= -A_\theta\hbar^2\left[(k^2\sin^{k-2}\theta\cos^2\theta - k\sin^k\theta) - m_l^2\sin^{k-2}\theta\right] && \times \sin^{-1}\theta\\
&= -A_\theta\hbar^2\left[(k^2\sin^{k-2}\theta(1-\sin^2\theta) - k\sin^k\theta) - m_l^2\sin^{k-2}\theta\right]\\
& && \sin^2\theta + \cos^2\theta = 1\\
&= -A_\theta\hbar^2\left[(k^2\sin^{k-2}\theta - k^2\sin^k\theta - k\sin^k\theta) - m_l^2\sin^{k-2}\theta\right].
\end{aligned}$$

CHECKPOINT The $\hat{L}^2$ operator is part of the angular term in the Hamiltonian, and its eigenvalue contributes to the energy obtained when we solve the Schrödinger equation. To illustrate that here, we color code the $\hat{L}^2$ operator in blue and the factors that form its eigenvalue in red. This is our first major step in solving the Schrödinger equation.

If the angular momentum is well defined for this wavefunction, we should obtain an eigenvalue equation such that our original wavefunction $\Theta(\theta)$ appears in the last line of math shown previously. But $\Theta(\theta)$ depends only on terms in $\sin^k\theta$, not $\sin^{k-2}\theta$. To get the $\sin^{k-2}\theta$ terms to cancel, we must set $m_l^2 = k^2$, and then we have indeed obtained an eigenvalue equation:

$$\begin{aligned}
\hat{L}^2\Theta(\theta) &= -A_\theta\hbar^2\left[(-k^2\sin^k\theta - k\sin^k\theta)\right] && m_l^2 = k^2\\
&= A_\theta\hbar^2(k^2+k)\sin^m\theta = k(k+1)\hbar^2\Theta(\theta). && (3.14)
\end{aligned}$$

What we have found is one set of valid solutions to the angular wavefunction, having an eigenvalue of $\hat{L}^2$ equal to $k(k+1)\hbar^2$. This is an eigenvalue equation only if we equate the exponent k of the sine function to $|m_l|$, where m_l is our quantum number in $e^{im_l\phi}$. Therefore, k must be an integer like m_l, but (unlike m_l) k cannot cannot be negative if the wavefunction is defined at all points, because $\sin^{-1}\theta$ is infinite at $\theta = 0$.

We started with a function that had pure $\sin\theta$ character. These functions keep the angular motion restricted to the xy plane as much as possible (because $\sin\theta$ reaches a maximum when $\theta = \pi/2$, which is where the xy plane is). Therefore, the functions $\Theta(\theta) = \sin^m\theta$ are the angular functions with the maximum value of L_z. To examine other possibilities, with the angular momentum in some other direction, we need to incorporate some $\cos\theta$ character into our wavefunctions. But these in general have to be polynomials of $\cos\theta$. With a polynomial, it is possible to cancel unwanted terms so that we can get the original wavefunction back after the operation, which is what we will need in order to have an eigenvalue equation.

CHECKPOINT At this point, we've found out that the wavefunction has a factor $e^{im_l\phi}$, where m_l must be an integer. We've also shown that the eigenvalue of $\hat{L}^2$ has the form $k(k+1)\hbar^2$, and this will lead to the angular momentum quantum number l.

So let our general function $\Theta(\theta)$ be given by

$$\Theta(\theta) = A_\theta \sin^m\theta \sum_{j=0}^{k} a_j \cos^j\theta, \qquad (3.15)$$

where we set $m \equiv |m_l|$ to mark that the m_l value determines the power of $\sin\theta$, without having to carry absolute value bars around. Now we run through $\hat{L}^2\Theta(\theta)$ again, with this more general function. The approach is identical to that used to obtain Eq. 3.14:

$$
\begin{aligned}
\hat{L}^2\Theta(\theta) &= -\hbar^2\left\{\frac{1}{\sin\theta}\frac{\partial}{\partial\theta}\sin\theta\frac{\partial}{\partial\theta} - \frac{m^2}{\sin^2\theta}\right\}A_\theta\sin^m\theta\sum_{j=0}^{k}a_j\cos^j\theta && \text{Eqs. 3.5 and 3.15} \\
&= -A_\theta\hbar^2\left\{\frac{1}{\sin\theta}\frac{\partial}{\partial\theta}\sin\theta\left[m\sin^{m-1}\theta\sum_{j=0}^{k}a_j\cos^{j+1}\theta\right.\right. \\
&\quad \left.- \sin^{m+1}\theta\sum_{j=0}^{k}a_j j\cos^{j-1}\theta\right] - m^2\sin^{m-2}\theta\sum_{j=0}^{k}a_j\cos^j\theta\Bigg\} && \text{take } \frac{\partial}{\partial\theta} \\
&= -A_\theta\hbar^2\left\{\frac{1}{\sin\theta}\frac{\partial}{\partial\theta}\left[m\sin^m\theta\sum_{j=0}^{k}a_j\cos^{j+1}\theta - \sin^{m+2}\theta\sum_{j=0}^{k}a_j j\cos^{j-1}\theta\right]\right. \\
&\quad \left. - m^2\sin^{m-2}\theta\sum_{j=0}^{k}a_j\cos^j\theta\right\} && \times\sin\theta \\
&= -A_\theta\hbar^2\left\{\frac{1}{\sin\theta}\left[m^2\sin^{m-1}\theta\sum_{j=0}^{k}a_j\cos^{j+2}\theta - m\sin^{m+1}\theta\sum_{j=0}^{k}a_j(j+1)\cos^j\theta\right.\right. \\
&\quad \left. - (m+2)\sin^{m+1}\theta\sum_{j=0}^{k}a_j j\cos^j\theta + \sin^{m+3}\theta\sum_{j=0}^{k}a_j j(j-1)\cos^{j-2}\theta\right] \\
&\quad \left. - m^2\sin^{m-2}\theta\sum_{j=0}^{k}a_j\cos^j\theta\right\} && \text{take } \frac{\partial}{\partial\theta} \\
&= -A_\theta\hbar^2\left\{m^2\sin^{m-2}\theta\sum_{j=0}^{k}a_j\cos^{j+2}\theta - m\sin^m\theta\sum_{j=0}^{k}a_j(j+1)\cos^j\theta\right. \\
&\quad - (m+2)\sin^m\theta\sum_{j=0}^{k}a_j j\cos^j\theta + \sin^{m+2}\theta\sum_{j=0}^{k}a_j j(j-1)\cos^{j-2}\theta \\
&\quad \left. - m^2\sin^{m-2}\theta\sum_{j=0}^{k}a_j\cos^j\theta\right\} && \times\sin^{-1}\theta \\
&= -A_\theta\hbar^2\left\{m^2\sin^{m-2}\theta\cos^2\theta\sum_{j=0}^{k}a_j\cos^j\theta - m\sin^m\theta\sum_{j=0}^{k}a_j(j+1)\cos^j\theta\right. \\
&\quad - (m+2)\sin^m\theta\sum_{j=0}^{k}a_j j\cos^j\theta + \sin^m(1-\cos^2\theta)\sum_{j=0}^{k}a_j j(j-1)\cos^{j-2}\theta \\
&\quad \left. - m^2\sin^{m-2}\theta\sum_{j=0}^{k}a_j\cos^j\theta\right\} && \sin^2\theta = 1-\cos^2\theta \\
&= -A_\theta\hbar^2\left\{m^2\sin^{m-2}\theta(\cos^2\theta - 1)\sum_{j=0}^{k}a_j\cos^j\theta - m\sin^m\theta\sum_{j=0}^{k}a_j(j+1)\cos^j\theta\right. \\
&\quad - (m+2)\sin^m\theta\sum_{j=0}^{k}a_j j\cos^j\theta + \sin^m\theta\sum_{j=0}^{k}a_j j(j-1)\cos^{j-2}\theta \\
&\quad \left. - \sin^m\theta\sum_{j=0}^{k}a_j j(j-1)\cos^j\theta\right\} && \text{combine terms with } \cos^j\theta
\end{aligned}
$$

$$= -A_\theta \hbar^2 \Big\{ -m^2 \sin^m\theta \sum_{j=0}^{k} a_j \cos^j\theta - m \sin^m\theta \sum_{j=0}^{k} a_j (j+1)\cos^j\theta$$
$$-(m+2)\sin^m\theta \sum_{j=0}^{k} a_j j \cos^j\theta + \sin^m\theta \sum_{j=0}^{k} a_j j(j-1)\cos^{j-2}\theta$$
$$- \sin^m\theta \sum_{j=0}^{k} a_j j(j-1)\cos^j\theta \Big\} \qquad \cos^2\theta - 1 = -\sin^2\theta$$
$$= A_\theta \hbar^2 \sin^m\theta \Big\{ \sum_{j=0}^{k} a_j [m^2 + m(j+1) + j(m+2) + j(j-1)] \cos^j\theta$$
$$- \sum_{j=0}^{k} a_j j(j-1)\cos^{j-2}\theta \Big\} \qquad \text{factor out } \cos^j\theta$$

and shift the values of j in the second sum by 2 (we'll see why in a moment):

$$= A_\theta \hbar^2 \sin^m\theta \Big\{ \sum_{j=0}^{k} a_j [m^2 + mj + m + mj + 2j + j^2 - j] \cos^j\theta$$
$$- \sum_{j'=-2}^{k-2} a_{j'+2}(j'+2)(j'+1)\cos^{j'}\theta \Big\}$$
$$= A_\theta \hbar^2 \sin^m\theta \sum_{j=0}^{k} \Big\{ a_j [(m+j)^2 + (m+j)] - a_{j+2}(j+1)(j+2) \Big\} \cos^j\theta. \tag{3.16}$$

We make the shift in the j summation index[1] so that we may combine the factors of $\cos^j\theta$. In order for Eq. 3.16 to be an eigenvalue equation, we need

$$A_\theta \hbar^2 \sin^m\theta \sum_{j=0}^{k} \Big\{ a_j [(m+j)^2 + (m+j)] - a_{j+2}(j+1)(j+2) \Big\} \cos^j\theta.$$
$$= L^2 A_\theta \sin^m\theta \sum_{j=0}^{k} a_j \cos^j\theta. \tag{3.17}$$

The equation must be true for all values of θ. That requirement in turn means that the equation must be satisfied individually for each power j of $\cos^j\theta$. In other words, the equality holds just between the terms proportional to $\cos\theta$, just between the terms proportional to $\cos^2\theta$, and so on. Let's start with the simplest case by setting $j = k$, which knocks out the a_{j+2} term from Eq. 3.17 (because there is no a_j where $j > k$), leaving

$$\hbar^2 a_k [(m+k)^2 + (m+k)] = L^2 a_k$$
$$L^2 = \hbar^2[(m+k)^2 + (m+k)] \equiv \hbar^2 l(l+1). \tag{3.18}$$

CHECKPOINT In Eq. 3.18 we solve an eigenvalue equation for the squared angular momentum L^2, rather than for the energy E. We are still using the red color coding here, however, because value of L^2 will give us the angular contribution to the total energy, as shown by the corresponding angular momentum operator in Eq. 3.8.

We have introduced a new quantum number l, equal to the sum of m and k, which are the maximum powers of $\sin\theta$ and $\cos\theta$, respectively, in the wavefunction. Because the eigenvalue L^2 depends only on this quantum number, we will

[1] That shift may look suspicious and even more so the merging of two sums in Eq. 3.16, even though j and j' have different lower and upper bounds. The factor of $(j'+2)(j'+1)$ in the previous step guarantees that the first two terms, $j' = -2$ and $j' = -1$, in the sum are zero, so the sum over j' can begin at $j' = 0$. Similarly, as long as we stipulate that $a_j = 0$ for any j greater than k, the sum over j' will be zero for values of j' greater than $k - 2$, so the upper bound for that sum may be changed to k.

use l to identify the angular momentum of our wavefunctions in the one-electron atom. We required $m = |m_l|$ and k to be positive integers, and therefore l must also be a positive integer.

We can take the more general case, $j < k$, in Eq. 3.17 and use that to find the coefficients in the polynomial. Replacing $L^2/\hbar^2$ by $l(l+1)$, we have

$$\frac{a_j}{a_{j+2}} = \frac{(j+1)(j+2)}{[(m+j)^2 + (m+j)] - l(l+1)} \tag{3.19}$$

for *all* the a_j's. To satisfy this equation, we need either all odd or all even values of j (i.e., all odd or all even powers of $\cos\theta$); the others may be set to zero.

As a more specific example, let's find the expression for $\Theta(\theta)$ that corresponds to $m = 0, k = 2, l = m + k = 2$:

$$\frac{a_0}{a_2} = \frac{(1)(2)}{(0)^2 + (0) - (2)(3)} = -\frac{2}{6}.$$

These results tell us that $a_2 = -3a_0$, and therefore we may write $\Theta(\theta)$ as $A_\theta(3\cos^2\theta - 1)$. The polynomials that satisfy these conditions are called the **Legendre polynomials**[2] $\mathcal{P}_l^{m_l}(\theta)$.

Combining the two angular parts, $\Theta(\theta)$ and $\Phi(\phi)$, we obtain the eigenfunctions of $\hat{L}^2(\theta,\phi)$, which are a class of functions called the **spherical harmonics,**

$$Y_l^{m_l}(\theta,\phi) \equiv \Theta(\theta)\Phi(\phi) = A_{lm_l}\mathcal{P}_l^{m_l}(\theta)e^{im_l\phi}. \tag{3.20}$$

Given values of l and m_l, we can find $\mathcal{P}_l^{m_l}(\theta)$ by using Eq. 3.19, or better yet by just looking it up, for example in Table 3.1. Polar plots of these functions in the xz plane are given in Fig. 3.6. We have shown that the eigenvalue for the operator $\hat{L}^2(\theta,\phi)$ is $\hbar^2 l(l+1)$, the square of the orbital angular momentum. Hence, the angular momentum L has a magnitude equal to $\hbar\sqrt{l(l+1)}$.

DERIVATION SUMMARY The Angular Solution. We chose a reasonable guess for the angular wavefunction, leaving several free parameters undecided, and just operated on the thing with $\hat{L}^2$, requiring that we get an eigenvalue equation. That equation was only satisfied by wavefunctions with a squared angular momentum value L^2 of $\hbar^2 l(l+1)$, with l some whole number.

We also showed by qualitative argument that the value of L_z is given entirely by the ϕ component of the momentum (Eq. 3.12) and has a value of $m_l\hbar$. The maximum value of m_l is l, so the maximum value of L_z is $l\hbar$. Looking at our earlier eigenvalue of $\hat{L}^2$, however, we find that the angular momentum itself, L, is always *greater* than $l\hbar$. That means that you cannot get the angular momentum vector to line up exactly with the z axis (or any other axis). We could have anticipated this: in order for L to be parallel to the z axis, any angular motion of the electron must be parallel to the xy plane — meaning the momentum p_z is exactly zero. But if we know the p_z value exactly, the Heisenberg uncertainty principle mandates that we cannot know z at all, so the electron cannot be restricted to our atom. Similarly, the electron cannot be kept in a circular orbit in the xy plane. In that case,

[2]Specifically these are the *associated Legendre polynomials of the first kind* and are usually written as functions of $\cos\theta$ rather than θ. They are named after Adrien-Marie Legendre (1752–1833), who discovered them as a general family of solutions to differential equations in spherical coordinates while he was working on a mathematical description of the motions of stars. His colleague, Simon-Pierre Laplace (1749–1827), then drew on the Legendre polynomials to formulate the three-dimensional spherical harmonics.

TABLE 3.1 **The angular wavefunctions $Y_l^{m_l}(\theta, \phi)$ of the one-electron atom.**

l	m_l	$Y_l^{m_l}(\theta,\phi)$
0	0	$\sqrt{\frac{1}{4\pi}}$
1	0	$\sqrt{\frac{3}{4\pi}}\cos\theta$
1	± 1	$\sqrt{\frac{3}{8\pi}}\sin\theta e^{\pm i\phi}$
2	0	$\sqrt{\frac{5}{16\pi}}(3\cos^2\theta - 1)$
2	± 1	$\sqrt{\frac{15}{8\pi}}\sin\theta\cos\theta e^{\pm i\phi}$
2	± 2	$\sqrt{\frac{15}{32\pi}}\sin^2\theta e^{\pm 2i\phi}$
3	0	$\sqrt{\frac{7}{16\pi}}\cos\theta(5\cos^2\theta - 3)$
3	± 1	$\sqrt{\frac{21}{64\pi}}\sin\theta(5\cos^2\theta - 1)e^{\pm i\phi}$
3	± 2	$\sqrt{\frac{105}{32\pi}}\sin^2\theta\cos\theta e^{\pm 2i\phi}$
3	± 3	$\sqrt{\frac{35}{64\pi}}\sin^3\theta\, e^{\pm 3i\phi}$

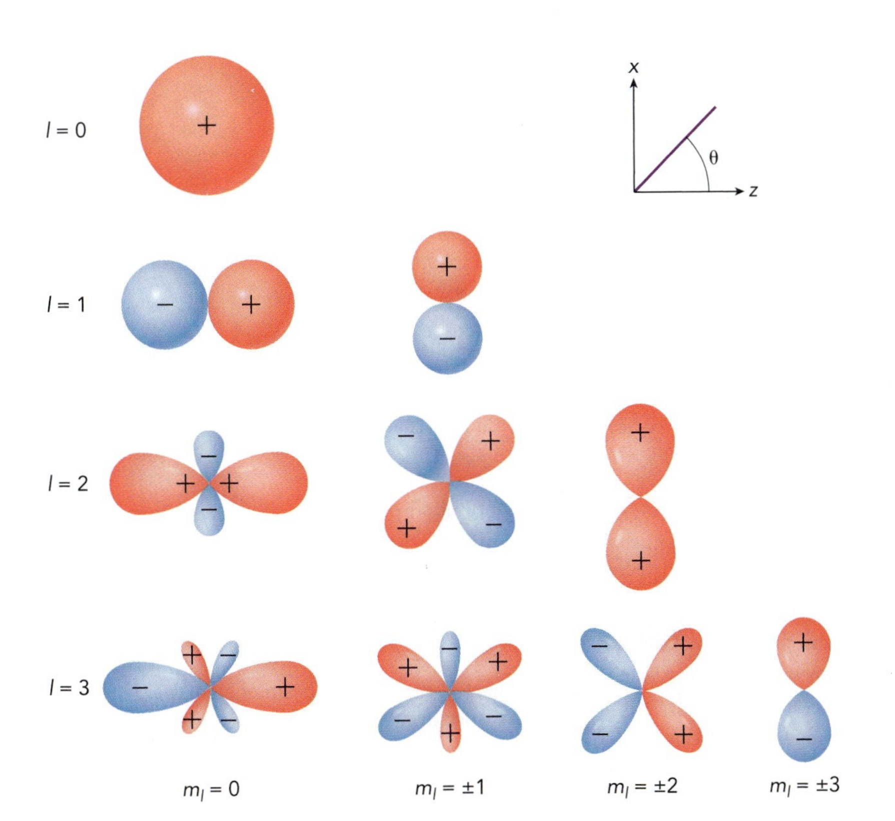

◀ **FIGURE 3.6** **The θ-dependent term $\mathcal{P}_l^{m_l}(\theta)$ of the angular wavefunctions for the electron in a one-electron atom.** These are cross-sections taken of the angular wavefunction in the *xz* plane.

the z value would be known to be exactly zero, and so p_z would be completely uncertain. But if p_z is uncertain, we can't require the electron to be stationary, and if the electron moves along z, then its z value is no longer zero. The values of z and p_z are both somewhat indefinite, and this in turn restricts how much we can know about the angular momentum.

The Radial Solution

Replacing $\hat{L}^2$ by its eigenvalue $\hbar^2 l(l+1)$ in Eq. 3.18, the Schrödinger equation is now

$$\left[-\frac{\hbar^2}{2m_e}\frac{1}{r^2}\frac{\partial}{\partial r}r^2\frac{\partial}{\partial r}+\frac{\hbar^2 l(l+1)}{2m_e r^2}-\frac{Ze^2}{4\pi\varepsilon_0 r}\right]R(r)\,Y_l^{m_l}(\theta,\phi)=ER(r)Y_l^{m_l}(\theta,\phi), \tag{3.21}$$

where $\psi(r,\theta,\phi)$ has been divided into a radial wavefunction $R(r)$ and an angular function $Y_l^{m_l}(\theta,\phi)$.

Because the operator in Eq. 3.21 does not depend on θ or ϕ, we can divide both sides by $Y_l^{m_l}(\theta,\phi)$, leaving only the r-dependent part of the equation. As we did with the angular solution, let's spare some math by first invoking a qualitative argument for the general form of the wavefunctions and then getting to the specifics. Again, we will simply guess what the radial wavefunction should look like, writing it in terms of several unknown parameters, and then we will solve the Schrödinger equation to get the values of those parameters. In this case, we need

BIOSKETCH **Peter Beiersdorfer**

Peter Beiersdorfer is a group leader in the Physics Division at Lawrence Livermore National Laboratory, and he heads the team using the Electron Beam Ion Trap (EBIT) to generate *very* highly charged ions, such as those produced in the intense heat of stars. EBIT employs a dense beam of accelerated electrons, with currents as high as 5000 A/cm^2, to strip some or all of the bound electrons from gas-phase atoms. Highly charged ions have been known in physics for over a century, beginning with Ernest Rutherford's discovery that He^{2+} ions (α-particles) are emitted by the radioactive decay of uranium. Since then, the greatest ionic charges have been produced in the particle accelerators that deliver bare nuclei of the heaviest elements. However, most natural environments include electrons as well as nuclei, and EBIT allows, for example, the study of one-electron ions of uranium and other elements. These studies simulate the hottest plasmas, where enormous electric and magnetic fields direct the complex interactions among ions and electrons. This work also provides a direct test of our understanding of special relativity by making it possible to obtain spectra of core electrons in heavy atoms, in the absence of complications from the valence electrons.

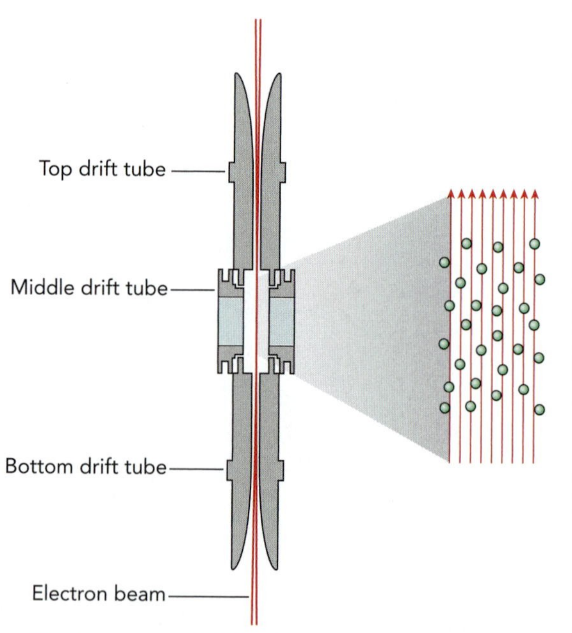

▲ **The heart of the EBIT.** A beam of high-energy electrons intersects three drift tubes, where the ions are formed.

wavefunctions that describe the radial distribution of the electron. Unlike the angles, r extends from zero to infinity, rather than repeating after a cycle of 0 to 2π, for example. This causes the boundary conditions to be different: the angular function had to be periodic because at $\phi = 2\pi$ the function had to be the same as at $\phi = 0$, but that is not a requirement for functions of r. Instead, we need the wavefunction to vanish at large r so that the probability density is finite. The convergence to zero at large r is satisfied by the exponential function, e^{-br}—where b is some real, positive constant—and, unlike other functions that converge to zero, the exponential is already an eigenfunction of the derivative operator.

Our Hamiltonian is more complicated than mere differentiation, however. Simplifying the notation a little, we need to find ψ and E that satisfy the equation

$$(\hat{H} - E)R(r) = \left[-\frac{c_1}{r^2}\frac{\partial}{\partial r}r^2\frac{\partial}{\partial r} + \frac{c_1 l(l+1)}{r^2} - \frac{c_2}{r} - E\right]R(r) = 0, \qquad (3.22)$$

with

$$c_1 \equiv \frac{\hbar^2}{2m_e} \quad \text{and} \quad c_2 \equiv \frac{Ze^2}{4\pi\varepsilon_0}.$$

The Hamiltonian includes terms with different powers of r, and in solving the Schrödinger equation we will need all those powers of r to cancel, leaving us with only the energy times our wavefunction. That suggests that in addition to the exponential term e^{-br}, we will also need a polynomial in r, with the coefficients set so that any unwanted terms vanish. The general form for our radial wavefunction $R(r)$is therefore:

$$R(r) = A_r e^{-br}(a_0 + a_1 r + a_2 r^2 + \ldots) = A_r e^{-br}\sum_{i=0}^{k} a_i r^i, \qquad (3.23)$$

with A_r the normalization constant. As with the angular functions, the polynomial allows us to get unwanted terms to cancel. The polynomial will also allow the radial function to pick up nodes as the energy increases, a general feature of wavefunctions, while the exponential will keep the function finite over the domain $0 \leq r < \infty$.

Now substitute $R(r)$ into the Schrödinger equation and see what the constraints are on the energy E and on the coefficients b and a_i, which will determine the wavefunctions. As a first step, let's just deal with the radial kinetic term, the one with the derivatives:

$$-\frac{c_1}{r^2}\frac{\partial}{\partial r}r^2\frac{\partial}{\partial r}R(r) = -\frac{c_1}{r^2}\frac{\partial}{\partial r}r^2\frac{\partial}{\partial r}\left[A_r e^{-br}\sum_{i=0}^{k} a_i r^i\right]$$

$$= -\frac{c_1}{r^2}\frac{\partial}{\partial r}r^2 A_r e^{-br}\left\{-b\sum_{i=0}^{k} a_i r^i + \sum_{i=1}^{k} a_i i r^{i-1}\right\} \qquad \text{take } \frac{\partial}{\partial r}$$

$$= -\frac{c_1}{r^2}\frac{\partial}{\partial r}A_r e^{-br}\left\{-b\sum_{i=0}^{k} a_i r^{i+2} + \sum_{i=1}^{k} a_i i r^{i+1}\right\} \qquad \text{move factor } r^2 \text{ inside brackets}$$

$$= -\frac{c_1}{r^2}A_r e^{-br}\left\{-b\left[-b\sum_{i=0}^{k} a_i r^{i+2} + \sum_{i=1}^{k} a_i i r^{i+1}\right]\right. \qquad \text{take } \frac{\partial}{\partial r}$$

$$\left. + \left[-b\sum_{i=0}^{k} a_i(i+2)r^{i+1} + \sum_{i=1}^{k} a_i i(i+1)r^i\right]\right\}$$

CHECKPOINT In this derivation, we begin with the radial part of the Hamiltonian (in blue) operating on the wavefunction (in black). As we carry out the sequence of operations, the terms that will eventually comprise the energy expression (red) become increasingly apparent. In each of these first few steps, the Hamiltonian is extracting information about the energy from the wavefunction. After the fourth line, the operation is complete, and the remaining work is to reorganize the expression to separate the energy and the wavefunction.

$$= -c_1 A_r e^{-br}\left\{-b\left[-b\sum_{i=0}^{k} a_i r^i + \sum_{i=1}^{k} a_i i r^{i-1}\right]\right.$$ move factor r^{-2} inside brackets

$$\left.+\left[-b\sum_{i=0}^{k} a_i(i+2)r^{i-1} + \sum_{i=1}^{k} a_i i(i+1)r^{i-2}\right]\right\}$$

$$= -c_1 A_r e^{-br}\left\{b^2\sum_{i=0}^{k} a_i r^i - b\sum_{i=1}^{k} a_i i r^{i-1}\right.$$

$$\left.- b\sum_{i=0}^{k} a_i(i+2)r^{i-1} + \sum_{i=1}^{k} a_i i(i+1)r^{i-2}\right\}$$ combine terms of like powers of r

$$= -c_1 A_r e^{-br}\left\{b^2\sum_{i=0}^{k} a_i r^i - b\sum_{i=0}^{k} a_i(2i+2)r^{i-1} + \sum_{i=1}^{k} a_i i(i+1)r^{i-2}\right\}. \quad (3.24)$$

Now we add in the other terms from our Schrödinger equation (Eq. 3.22), obtaining

$$0 = (\hat{H} - E)R(r)$$
$$= A_r e^{-br}\left\{-c_1 b^2\sum_{i=0}^{k} a_i r^i + c_1 b\sum_{i=0}^{k} a_i(2i+2)r^{i-1} - c_1\sum_{i=1}^{k} a_i i(i+1)r^{i-2}\right.$$
$$\left.+ c_1 l(l+1)\sum_{i=0}^{k} a_i r^{i-2} - c_2\sum_{i=0}^{k} a_i r^{i-1} - E\sum_{i=0}^{k} a_i r^i\right\}. \quad (3.25)$$

We can divide out the factors of $A_r\, e^{-br}$. This equation needs to be true at all values of r, so let's compare equal powers of r. Each term in Eq. 3.25 involves a sum over i, beginning with $i - 0$. If we change the value of i where each sum starts, we can rewrite this sum so that each term is a sum over r^i, making it easier to compare the terms directly. We are not changing the contributions to the sums, only how we label them:

$$0 = -c_1 b^2\sum_{i=0}^{k} a_i r^i + c_1 b\sum_{i=-1}^{k-1} a_{i+1}(2i+4)r^i - c_1\sum_{i=-1}^{k-2} a_{i+2}(i+2)(i+3)r^i$$
$$+ c_1 l(l+1)\sum_{i=-2}^{k-2} a_{i+2} r^i - c_2\sum_{i=-1}^{k-1} a_{i+1} r^i - E\sum_{i=0}^{k} a_i r^i.$$

$$0 = -c_1 b^2\sum_{i=0}^{k} a_i r^i - E\sum_{i=0}^{k} a_i r^i + c_1 b\sum_{i=-1}^{k-1} a_{i+1}(2i+4)r^i - c_2\sum_{i=-1}^{k-1} a_{i+1} r^i$$ regroup terms by coefficient

$$-c_1\sum_{i=-1}^{k-2} a_{i+2}(i+2)(i+3)r^i + c_1 l(l+1)\sum_{i=-2}^{k-2} a_{i+2} r^i. \quad (3.26)$$

With a little patience, we can pull out some remarkable results. For example, if we set $i = k$, only the first two terms are non-zero, giving

$$0 = -c_1 b^2 a_k - E a_k \text{ and } E = -c_1 b^2. \quad (3.27)$$

Now check the terms when $i = k - 1$:

$$0 = -c_1 b^2 a_{k-1} - E a_{k-1} + c_1 b a_k(2k+2) - c_2 a_k. \quad (3.28)$$

According to Eq. 3.27 the first two terms cancel, leaving

$$c_2 = c_1 b(2k + 2)$$

$$b = \frac{c_2}{(2k + 2)c_1}$$

$$= \frac{Ze^2/(4\pi\varepsilon_0)}{2(k + 1)\hbar^2/(2m_e)} = \frac{Zm_e e^2}{(k + 1)(4\pi\varepsilon_0)\hbar^2} \qquad \text{by Eq. 3.22}$$

$$= \frac{Z}{(k + 1)a_0}. \qquad a_0 = 4\pi\,\varepsilon_0\,\hbar^2/(m_e e^2) \qquad (3.29)$$

The value $k + 1$ becomes so important that we'll assign it to a new quantum number, n. The exponential term on our wavefunctions is then of the form $e^{-br} = e^{-Zr/(na_0)}$. We can also use this value of b to find the energies, using Eq. 3.27:

$$E = -c_1 b^2 = -\frac{\hbar^2}{2m_e}\left(\frac{Z}{na_0}\right)^2 = -\frac{Z^2}{2n^2}E_h. \qquad (3.30)$$

This is the correct energy equation, the same as predicted by the Bohr model, but this time we have accurate wavefunctions that describe the three-dimensional electron distribution.

Taking a look at those wavefunctions, if we return to Eq. 3.26 and take the general case where $i < k - 1$, we can solve for the values of all the coefficients in the wavefunction:

$$0 = -c_1 b^2 a_i - Ea_i + c_1 b a_{i+1}(2i + 4) - c_2 a_{i+1}$$
$$-c_1(i + 2)(i + 3)a_{i+2} + c_1 l(l + 1)a_{i+2}$$
$$= c_1 b a_{i+1}(2i + 4) - c_2 a_{i+1} \qquad E = -c_1 b^2$$
$$-c_1(i + 2)(i + 3)a_{i+2} + c_1 l(l + 1)a_{i+2}$$
$$= [2(i + 2)c_1 b - c_1 b(2k + 2)]a_{i+1} \qquad c_2 = c_1 b(2k + 2)$$
$$+c_1[(i + 2)(i + 3) - l(l + 1)]a_{i+2}[1]$$
$$= [2ic_1 b - c_1 b(2k + 2)]a_{i-1} + c_1[i(i + 1) - l(l + 1)]a_i \qquad \text{shift } i \text{ by 2}$$
$$\frac{a_{i-1}}{a_i} = \frac{l(l + 1) - i(i + 1)}{2ib - b(2k + 2)} = \left(\frac{1}{2b}\right)\frac{l(l + 1) - i(i + 1)}{i - (k + 1)}$$
$$= -\left(\frac{na_0}{2Z}\right)\frac{l(l + 1) - i(i + 1)}{n - i}. \qquad n = k + 1,\ b = Z/(na_0) \qquad (3.31)$$

This final equation tells us the rest about the wavefunction. If $l = i$, then the ratio a_{i-1}/a_i vanishes, meaning that $a_{i-1} = a_{l-1}$ must be zero. And if a_{l-1} is zero, then all the coefficients a_i with $i < l$ are zero also, according to Eq. 3.31. Therefore, the terms in the polynomial extend from r^l up to r^{n-1}, and the terms of order $l-1$ or less are all zero. That also means that the maximum value of l is $n-1$, because otherwise there are no terms in the polynomial and the wavefunction is zero.

CHECKPOINT One benefit of following this derivation through is seeing how the math leads naturally to the rules for the quantum numbers that you may have encountered in general chemistry: m_l can be any integer between $-l$ and $+l$, n can be any positive integer, and here we learn why l must be less than n.

Once again, the polynomials are well-known functions found in tables or automatically generated in computer programs from Eq. 3.31. The general radial wavefunction may be written

$$R_{n,l}(r) = A_{n,l}e^{-Zr/(na_0)}\left(\frac{Zr}{a_0}\right)^l \mathcal{L}_{n-l-1}(r), \qquad (3.32)$$

where the $\mathcal{L}$'s are the associated **Laguerre polynomials.**[3] The indices n and l must again be integers, with $n \geq 1$ and $0 \leq l < n$. See Table 3.2 and Fig. 3.7 for representations of the radial part of the wavefunction.

DERIVATION SUMMARY The Radial Solution. Again, we started from a reasonable guess, this time for the radial wavefunction with several free parameters, and operated on it to see what had to happen in order to satisfy the eigenvalue equation. The eigenvalue equation in this case was the complete Schrödinger equation, and we obtained (*i*) the same energies as the Bohr model, and (*ii*) radial wavefunctions that required l to be a positive integer less than n.

TABLE 3.2 The radial wavefunctions $R_{n,l}(r)$ of the one-electron atoms.

n	l	$R_{n,l}(r)$
1	0	$2\left(\frac{Z}{a_0}\right)^{3/2} e^{-Zr/a_0}$
2	0	$\frac{1}{\sqrt{2}}\left(\frac{Z}{a_0}\right)^{3/2}\left(1 - \frac{Zr}{2a_0}\right)e^{-Zr/(2a_0)}$
2	1	$\frac{1}{\sqrt{24}}\left(\frac{Z}{a_0}\right)^{3/2}\left(\frac{Zr}{a_0}\right)e^{-Zr/(2a_0)}$
3	0	$\frac{2}{\sqrt{27}}\left(\frac{Z}{a_0}\right)^{3/2}\left(1 - \frac{2Zr}{3a_0} + \frac{2Z^2r^2}{27a_0^2}\right)e^{-Zr/(3a_0)}$
3	1	$\frac{4\sqrt{2}}{27\sqrt{3}}\left(\frac{Z}{a_0}\right)^{3/2}\left(\frac{Zr}{a_0}\right)\left(1 - \frac{Zr}{6a_0}\right)e^{-Zr/(3a_0)}$
3	2	$\frac{2\sqrt{2}}{81\sqrt{15}}\left(\frac{Z}{a_0}\right)^{3/2}\left(\frac{Zr}{a_0}\right)^2 e^{-Zr/(3a_0)}$
4	0	$\frac{1}{4}\left(\frac{Z}{a_0}\right)^{3/2}\left(1 - \frac{3Zr}{4a_0} + \frac{Z^2r^2}{8a_0^2} - \frac{Z^3r^3}{192a_0^3}\right)e^{-Zr/(4a_0)}$
4	1	$\frac{\sqrt{5}}{16\sqrt{3}}\left(\frac{Z}{a_0}\right)^{3/2}\left(\frac{Zr}{a_0}\right)\left(1 - \frac{Zr}{4a_0} + \frac{Z^2r^2}{80a_0^2}\right)e^{-Zr/(4a_0)}$
4	2	$\frac{1}{64\sqrt{5}}\left(\frac{Z}{a_0}\right)^{3/2}\left(\frac{Zr}{a_0}\right)^2\left(1 - \frac{Zr}{12a_0}\right)e^{-Zr/(4a_0)}$
4	3	$\frac{1}{768\sqrt{35}}\left(\frac{Z}{a_0}\right)^{3/2}\left(\frac{Zr}{a_0}\right)^3 e^{-Zr/(4a_0)}$

[3]The notation here is simplified to emphasize the r-dependence of the polynomials. However, in the conventional notation, these are associated Laguerre polynomials and would be written as $\mathcal{L}_{n-l-1}^{2l+1}(\rho)$, where $\rho = 2Zr/(na_0)$ is a unitless coordinate proportional to r. Edmond Laguerre (1834–1886) was studying approximation methods when he came across these solutions to the radial differential equation. At first, these functions were merely an exercise in interesting but abstract mathematics, and their appearance in these wavefunctions remains their principal application.

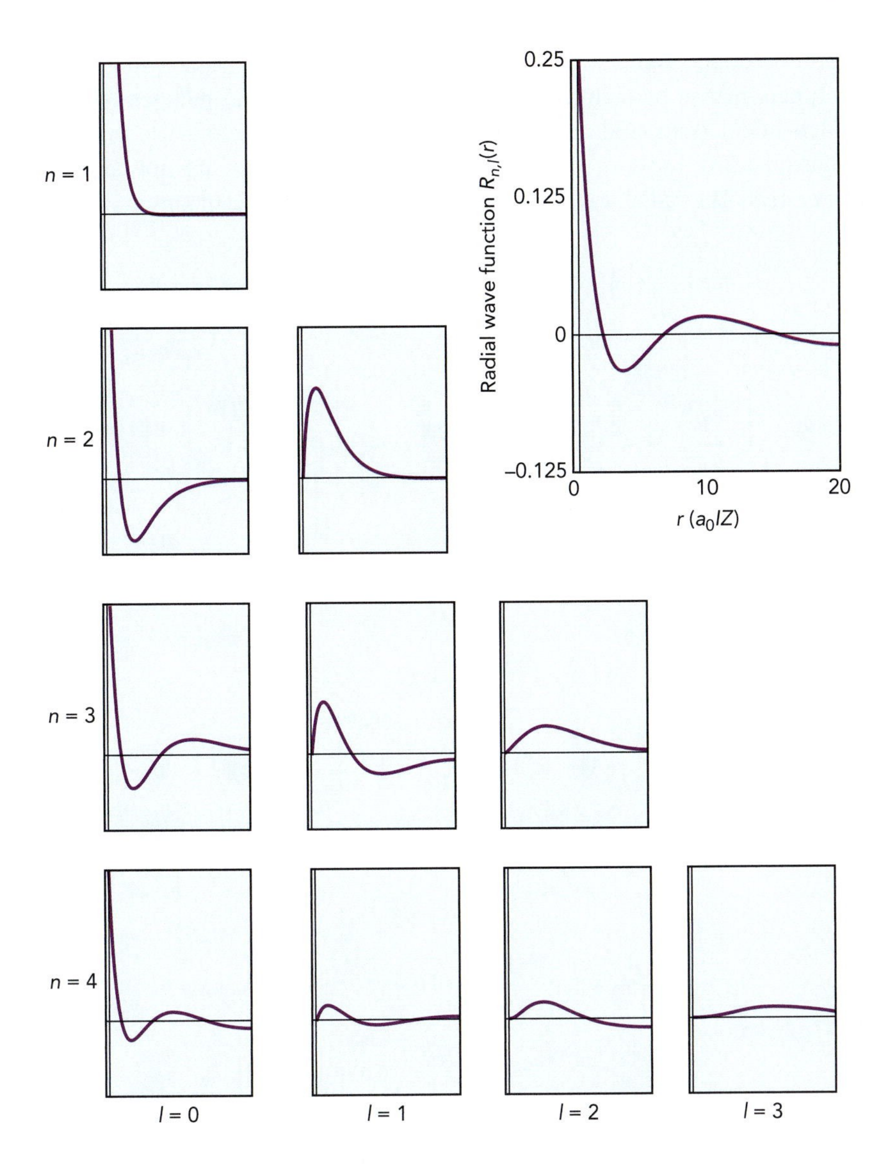

FIGURE 3.7 The radial wavefunctions $R_{n,l}(r)$ for the electron in a one-electron atom. The inset in the upper right corner gives the vertical scale in units of $(a_0/Z)^{-3/2}$ and horizontal scale in a_0/Z.

3.2 The One-Electron Atom Orbital Wavefunctions

The combined wavefunction, with a lot of help from 19th-century French mathematicians, is

$$\psi_{n,l,m_l} = R_{n,l}(r)Y_l^{m_l}(\theta,\phi) = A_{n,l,m_l}e^{-Zr/(na_0)}\left(\frac{Zr}{a_0}\right)^l \mathcal{L}_{n-l-1}(r)\mathcal{P}_l^{m_l}(\theta)e^{im_l\phi} \quad (3.33)$$

with energies (Eq. 3.30)

$$E_n = -\frac{Z^2}{2n^2}E_{\mathrm{h}}.$$

As promised, the wavefunctions are still separable to a degree, in that each wavefunction may be factored into three distinct terms: an r-dependent term, a θ-dependent term, and a ϕ-dependent term. Unlike the particle in a three-dimensional box wavefunctions, though, those three terms are not completely independent. The radial wavefunction $R_{n,l}(r)$ depends on the quantum number l,

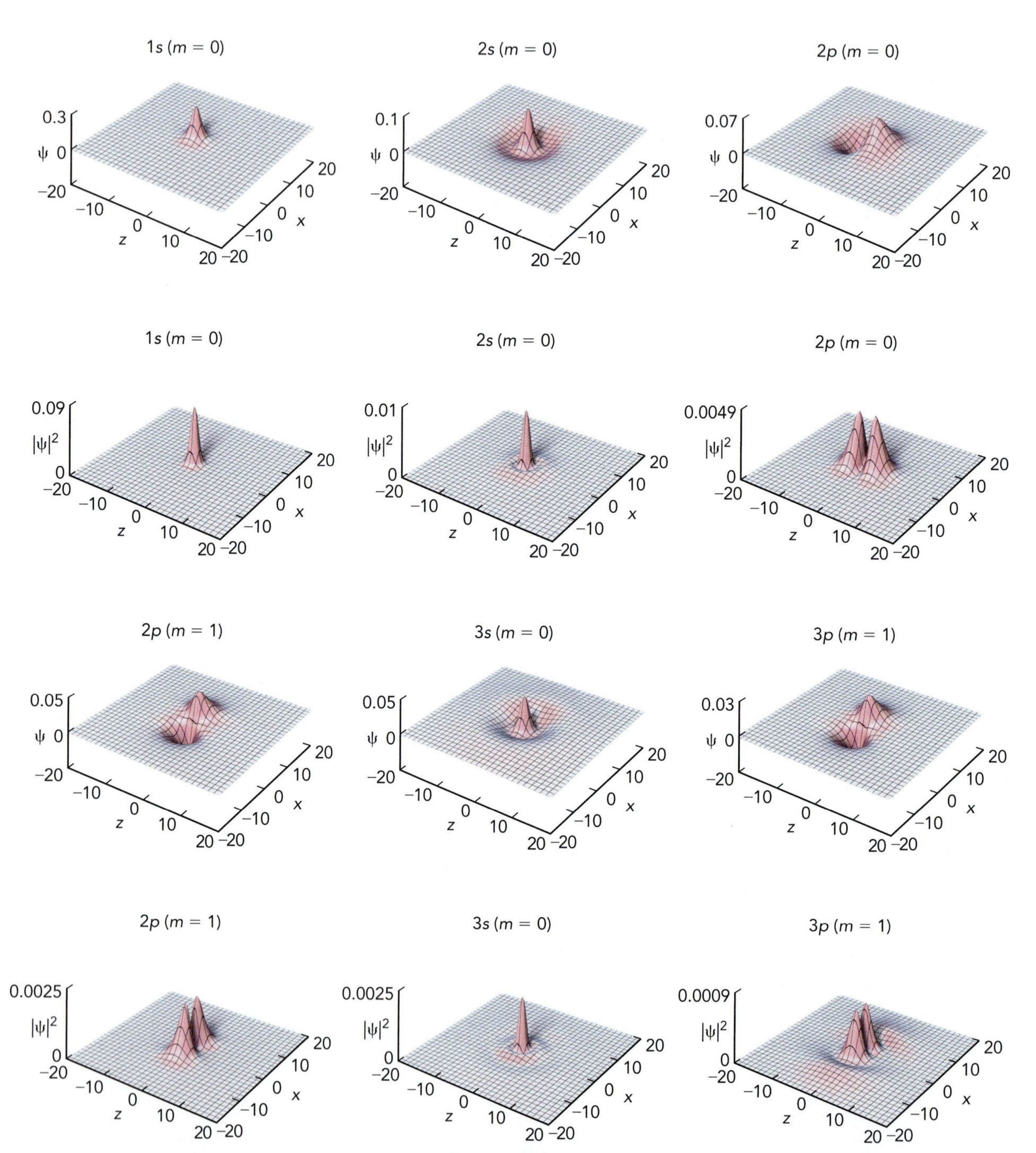

▲ **FIGURE 3.8 Cross-sections in the *xz* plane of the wavefunctions and probability densities for selected one-electron orbitals.** Only the real part of the wavefunction is graphed. Axis values are in atomic units (a_0 for x and z, $a_0^{-3/2}$ for ψ, a_0^{-3} for $|\psi|^2$). The horizontal axes are of equal scale in all graphs, but the vertical axes are adjusted for clarity.

which is the order of the polynomial in the θ-dependent part of the wavefunction, and the θ-part itself is determined partly by the value of the quantum number m_l, which is set by the ϕ-dependent part. It can be difficult to visualize these functions in all three dimensions. Fig. 3.8 looks at two dimensions simultaneously by graphing the amplitude of the wavefunctions and probability densities in the xz plane, which offers a look at both the r- and θ-dependence of these remarkable functions. For example, the one radial node and the one angular node are both visible for the $n = 3, l = 1$ ($3p$) wavefunction.

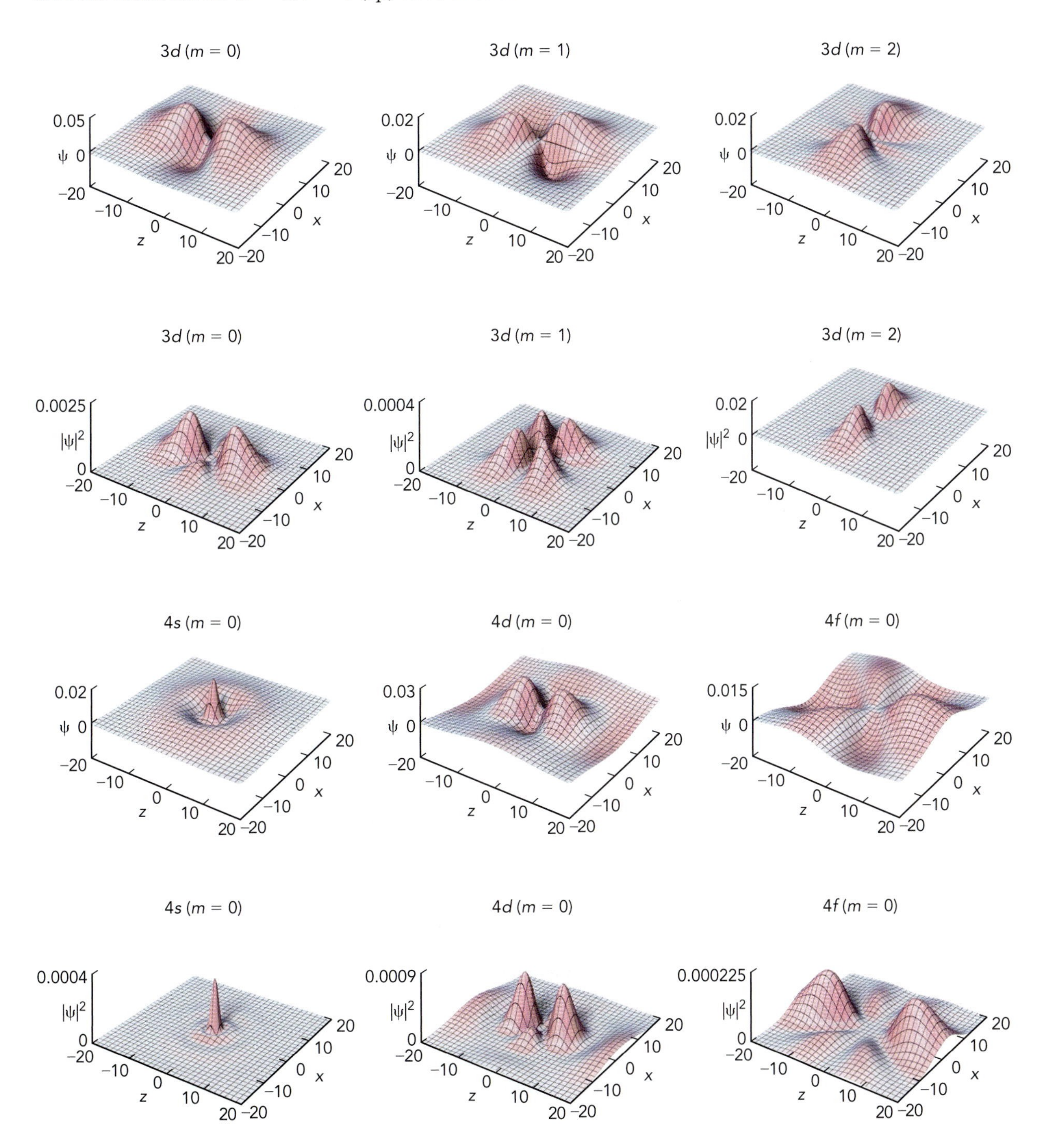

These wavefunctions form a complete and orthogonal basis set: complete because any three-dimensional function can be written as a sum over these wavefunctions, and orthogonal because

$$\int_{\text{all space}} \psi^*_{n,l,m_l} \psi_{n',l',m_{l'}} d\tau = 0$$

unless all three quantum numbers (n, l, and m_l) are the same for both wavefunctions. Yes, we now have three quantum numbers, and they quantize three parameters of the atom: the energy, the angular momentum, and the z-component of the angular momentum,

$$E_n = -\frac{Z^2}{2n^2}E_h \quad L = \hbar\sqrt{l(l+1)} \quad L_z = \hbar m_l.$$

The solution for the energy is the same as for the Bohr model, which is good because Bohr got that part right. But now, with the explicit wavefunction, we can draw the probability distributions of the electron and see that they differ considerably from the classical, circular orbits.

Let's pause for a brief digression while we catch our breath. Just how does the Bohr model succeed in getting the right energies, even though the circular orbits badly misrepresent the electron distribution? Or, to put it another way, why do the solutions to the correct but complicated Schrödinger equation have the same energies as our simple de Broglie wave in a circular orbit (Fig. 1.15)? Looking at it one way, Bohr's solution is based on a fiendishly clever manipulation of the system that greatly simplifies, but does not invalidate, the energy problem. We found that the correct angular momentum term, proportional to $l(l + 1)$, does not have to be known to solve the equations (Eqs. 3.27 and 3.29) that give the energy. Bohr used the correct equation for the potential energy and then put all of the kinetic energy into motion through the angle ϕ. The energy—and the *quantization* of the energy—are independent of how the energy is divided between angular and radial motion, so this is a terrific way to get the correct answer to the energy problem. As we've seen, the easiest part of the quantum mechanics is the solution for the ϕ-dependent part, and what Bohr does is artificially make this the *only* part of the quantum mechanics. At this point, his solution is perfectly correct: if motion is restricted to the xy plane, then there is no uncertainty in the orientation of the angular momentum. The value of L is the same as L_z, and that is quantized to values of $m_l\hbar$, with m_l some integer—Bohr's findings exactly. What's most interesting about this is that Bohr's model reveals a striking simplicity to the energy problem, one that's hard to dig out of our several pages of quantum mechanics. The correct energies, in all their three-dimensional, wavefunctional glory, must still satisfy the constraints of the humble circular orbit illustrated in Fig. 1.11.

Interpretation of the Quantum Numbers

We can see by now that quantum numbers are nothing but convenient numerical labels for quantum states, often related to the index of a sum, or the exponent in a power series, or both. In any quantum system, distinct quantum numbers are used for each degree of freedom, each coordinate into which energy can be channeled. Each quantum number is a unitless number, and the higher its value, the more energy there is in the motion along its corresponding degree of freedom. We have three degrees of freedom in the one-electron atom—r, θ, and ϕ—and three quantum numbers:

1. The **principal quantum number,** n, determines the total number of nodes (equal to $n - 1$) in the real part of the wave function, and the corresponding energy of the state. We can associate n roughly with the coordinate r because increases in the energy cause the electron to move further from the nucleus on average, regardless of any change in the angular motion. We shall call the group of quantum states that share the same value of n the nth **shell.**
2. We name l the **azimuthal quantum number** (the first and last time that term appears in this book), and it distinguishes among states with different orbital angular momenta, as given by the eigenvalue $\hbar^2 l(l + 1)$ of $\hat{L}^2$. Because it describes the overall orbital angular momentum, l is associated with excitation along both θ and ϕ. We shall call the group of quantum states that share the same values of both n and l a **subshell.** The quantum number l determines the letter symbol in the customary subshell designations: s ($l = 0$), p ($l = 1$), d ($l = 2$), f ($l = 3$), g ($l = 4$), and so on.
3. The **magnetic quantum number,** m_l, distinguishes among wavefunctions with the same angular momentum but different *projections* of the angular momentum onto the z axis. We showed in Eq. 3.12 that the eigenvalue of the operator $\hat{L}_z$ is $\hbar m_l$. Because L_z is determined by the motion along ϕ, the ϕ-dependent part of the wavefunction determines the value of m_l. The state that corresponds to a particular value of n, l, and m_l will be called an **atomic orbital.**

If we want to specify the subshell, we will write the n value and the symbol for the l value; for example, the $3d$ subshell has $n = 3$ and $l = 2$. For convenience, we will add the m_l value as a subscript if we want to specify the exact orbital in the subshell (e.g., $3d_{-1}$ indicates the $m_l = -1$ orbital of the $3d$ subshell), but this is not a standard notation.

The Angular Part

With the whole thing solved, let's go back and look just at the shapes of the wavefunctions along various coordinates. If we take a little time to find the patterns within the expressions in Table 3.1, each expression becomes a little more comprehensible, and we can more readily identify errors that may crop up.

General features of the angular wavefunctions represented in Table 3.1 and Fig. 3.6 include these:

1. Knowing the pattern of nodes can be helpful in understanding and anticipating the shapes of the orbital wavefunctions. The $Y_l^{m_l}$'s have nodes at $l - |m_l|$ values of θ, where $0 \leq \theta < \pi$. (It is not a node if the wavefunction goes to zero at $\theta = 0$, because θ can't be less than 0, and so the wavefunction does not *cross* zero at that point.)

2. The real part of the $e^{im_l\phi}$ term is equal to $\cos m_l\phi$, and it would contribute $|m_l|$ additional nodal planes. Consequently, the total number of angular nodes in the *real* part of ψ is equal to l. Each angular node corresponds to a surface (a plane for nodes in ϕ, a cone for nodes in θ) on which the electron wavefunction vanishes. For example, the d_0 wavefunction has nodes at $\theta = 54.7°$ and 125.3°, and all the points at each value of θ form a cone centered on the z axis. On the other hand, the real part of the p_1 wavefunction has a node at $\phi = 90°$, which defines the yz plane. The nodal planes along ϕ vanish when $Y_l^{m_l}$ is multiplied by its complex conjugate, and therefore the probability density $|Y_l^{m_l}|^2$ has no nodes along ϕ, but retains the $l - |m_l|$ nodes along θ.

EXAMPLE 3.1 Spherical Harmonics and Angular Momentum

CONTEXT The spherical harmonics were devised to help solve problems in celestial mechanics, but they have found application in many other forms. In Chapter 10, we find that the electric field of a molecule can be written as a sum over spherical harmonics. Spherical harmonics also provide an efficient set of integrands for computing the simulated lighting of three-dimensional spaces in computer games and other digital imaging. As we study atomic structure, what interests us most about spherical harmonics is that they are eigenfunctions of the square angular momentum operator $\hat{L}^2$. The eigenvalues of this operator differentiate the shapes of s, p, and d orbitals.

PROBLEM Verify that Y_1^0 is an eigenfunction of the operator $\hat{L}^2$ and find its eigenvalue, where

$$\hat{L}^2 = -\hbar^2\left[\frac{1}{\sin\theta}\frac{\partial}{\partial\theta}\sin\theta\frac{\partial}{\partial\theta} + \frac{1}{\sin^2\theta}\frac{\partial^2}{\partial\phi^2}\right].$$

SOLUTION We obtain the spherical harmonic Y_1^0 from Table 3.1 and then carry out the operation $\hat{L}^2$ on that function:

$$Y_1^0(\theta,\varphi) = \sqrt{\frac{3}{4\pi}}\cos\theta;$$

$$\hat{L}^2 Y_1^0(\theta,\varphi) = -\hbar^2\sqrt{\frac{3}{4\pi}}\left[\frac{1}{\sin\theta}\frac{\partial}{\partial\theta}\sin\theta\frac{\partial}{\partial\theta}(\cos\theta) + \underbrace{\frac{1}{\sin^2\theta}\frac{\partial^2}{\partial\phi^2}(\cos\theta)}_{=0}\right] \qquad \text{no } \phi \text{ in } Y_1^0$$

$$= -\hbar^2\sqrt{\frac{3}{4\pi}}\left[\frac{1}{\sin\theta}\frac{\partial}{\partial\theta}(-\sin^2\theta)\right] \qquad \text{take } \frac{\partial}{\partial\theta}$$

$$= -\hbar^2\sqrt{\frac{3}{4\pi}}\left[-\frac{1}{\sin\theta}(2\sin\theta\cos\theta)\right] \qquad \text{take } \frac{\partial}{\partial\theta}$$

$$= 2\hbar^2\sqrt{\frac{3}{4\pi}}\cos\theta$$

$$= 2\hbar^2 Y_1^0(\theta,\varphi).$$

So Y_1^0 has an eigenvalue $2\hbar^2 = \hbar^2 l(l+1)$ and is an eigenfunction of $\hat{L}^2$.

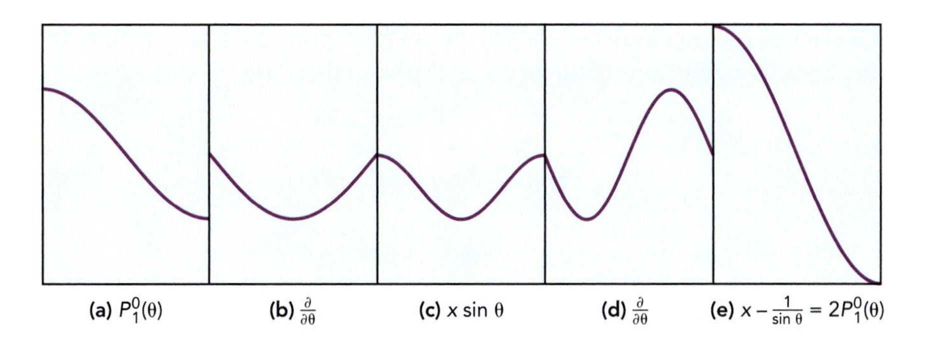

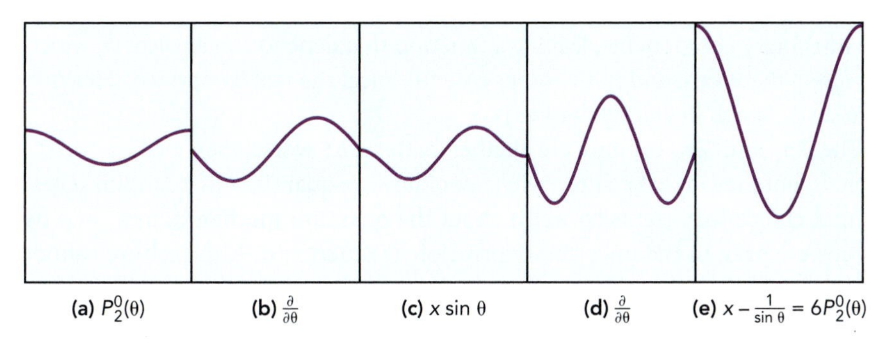

◀ FIGURE 3.9 **Visual guide to the operation of $\hat{L}^2$.** The sequence of results is graphed as we carry out each component operation of the θ-term in $\hat{L}^2$ on the $P_1^0(\theta)$ and $P_2^0(\theta)$ functions, showing that the final function (e) equals the eigenvalue $l(l + 1)$ times the original function (a).

The angular momentum operator $\hat{L}^2$ fell in our laps as a result of changing from Cartesian to spherical coordinates, but the first term, with the θ derivatives, is daunting. As an exercise in visual calculus, its operation on the $P_1^0(\theta)$ and $P_2^0(\theta)$ Legendre polynomials is shown as a sequence of steps in Fig. 3.9. These functions are proportional to the p_z and d_{z^2} orbital wavefunctions, respectively, and the ϕ-dependent part of $\hat{L}^2$ gives zero. If you know the shapes of the sine and cosine functions, it is possible to sort out how this operator works on its eigenstates. In essence, what it does is use the derivative operators to count the nodes in the function between zero and π, and the multiplication and division by $\sin\theta$ to restore the wavefunction. The first derivative in Fig. 3.9a magnifies the function by the number of nodes, a factor of 1 for $P_1^0(\theta)$ and a factor of 2 for $P_2^0(\theta)$. The second derivative, taken in Fig. 3.9(d), magnifies the function again, and the final division in Fig. 3.9(e) brings back the original shape of the function, now multiplied by the eigenvalue.

Cartesian Orbitals

Here's one reason some people don't like quantum mechanics: after all that effort to solve the Schrödinger equation, the angular wavefunctions listed in Table 3.1 could be written in a *different*—and equally valid—form.

The angular part of the $2p_0$ orbital is a cosine function with no imaginary component. Cosine has its maximum magnitude at $\theta = 0$ and $\theta = \pi$, which correspond to the z axis, and goes to zero away from the z axis, so we also call this orbital the $2p_z$ orbital. However, our other $2p$ angular wavefunctions are complex expressions, each containing an imaginary component that defines the direction of angular motion, so we cannot define them as lying along one axis or the other.

Or can we? A closer look reveals that we can form pure real wavefunctions by taking linear combinations of the $m_l = \pm 1$ orbitals like this:

$$\begin{aligned} \psi_{2,1,0} &= 2p_z \\ \frac{1}{\sqrt{2}}(\psi_{2,1,1} + \psi_{2,1,-1}) &= 2p_x \\ \frac{i}{\sqrt{2}}(\psi_{2,1,1} - \psi_{2,1,-1}) &= 2p_y. \end{aligned} \tag{3.34}$$

The explicit math is saved for Problem 3.41, but a summary of how we accomplish this is shown in Fig. 3.10. We combine the two orbitals in one case to cancel the imaginary components, leaving a function that depends on $\sin\theta\cos\phi$, which lies along the x axis, and in the other case to cancel the real components, leaving $\sin\theta\sin\phi$, which lies along the y axis.

The $2p_x$ and $2p_y$ orbitals are neither better nor worse than our $m_l = \pm 1$ orbitals, but they reflect a different *choice* of how we quantize our quantum states. It's that choice that seems so weird about the quantum mechanics here, and its origins are back in the uncertainty principle (Section 1.5). Although we cannot know exact values for all the parameters of our atom simultaneously, we can choose which ones we do want to know. When we use the $m_l = \pm 1$ angular wavefunctions, we have states for which the z-component of the angular motion is well defined. When we use $2p_x$ and $2p_y$, we have built **Cartesian atomic orbitals** that are pure real and that correspond to well-defined positions in space.

These two ways of looking at it are mutually exclusive: the Cartesian orbitals $2p_x$ and $2p_y$ are mixtures of states with different m_l values, so m_l is no longer a good quantum number. And the $m_l = \pm 1$ states line up with neither the x nor the y axes, because in these functions the electron distribution orbits the z axis. But notice that in both cases the energy and the angular momentum of the $2p$ orbitals is the same. That's because we have only mixed states of different m_l to form the Cartesian orbitals. The orbitals we combined share the same values of the n and l quantum numbers. Since n determines the energy, and l determines the magnitude of the angular momentum, neither of those parameters is affected when we mix and match the orbitals within a subshell.

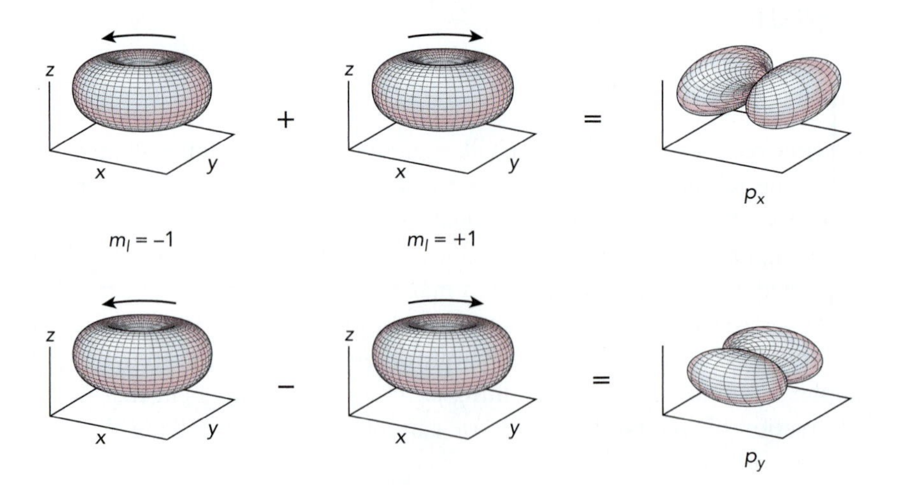

▶ FIGURE 3.10 **Cartesian 2p orbitals.** We are able to combine the two $m_l = \pm 1$ $2p$ orbitals to get pure real orbitals that line up along either the x or y axis.

The way we set up the Schrödinger equation makes it easiest to solve for the angular momentum wavefunctions in Table 3.1 first, but the Cartesian orbitals will be more useful when we get to molecules, where the orientation of the electron density along different Cartesian axes becomes important.

The Radial Part

An important distinction arises here between Cartesian and spherical coordinates: the probability density can be inconsistently defined if we're not cautious. The probability density of a wavefunction $R(r)$ is $|R(r)|^2$, but this gives the probability per unit volume of the particle being at some particular value of r. Often, however, we are not interested in any specific direction from the nucleus, and we want to know the density at some value of r *added up over all the angles.* In that case, the function we want is the **radial probability density,** equal to $R_{nl}(r)^2r^2$ because the volume sampled by the radial wavefunction increases, like the surface area of a sphere, as r^2. Looking at it another way, when we integrate over the probability density, we add a factor of r^2 in the volume element. That is the same factor we include when we write the radial probability density. As shown in Fig. 3.11, the nodes that appear in $R_{n,l}(r)$ remain in the radial probability density, but the amplitude increases more clearly as r increases.

As we did for the angular part, let's go over some general features of the radial wavefunctions.

1. The Laguerre polynomials are of order $n - l - 1$ in r and have $n - l - 1$ nodes. For every value of $r > 0$ for which $\mathcal{L}_{n-l-1}(r) = 0$, there is a nodal sphere centered at the nucleus on which the electron has no probability density. For a given value of l, the number of such nodes increases as n increases.
2. $\mathcal{L}_{n-l-1}(0) \neq 0$ only when $l = 0$, so only s functions have non-zero amplitude at $r = 0$, which is at the nucleus. (It may seem impossible that the nucleus and the electron could simultaneously occupy the same space, but the nucleus and the electron both have significant wave character, and waves can overlap.) The p, d, f, and other functions all go to zero at the nucleus. They must, because the magnitude of the angular momentum, $|\vec{r} \times \vec{p}|$, is a constant for our wavefunctions, and that constant must be zero if the wavefunction is allowed at $r = 0$. Therefore $l \neq 0$ states, which have non-zero angular momentum, vanish at the nucleus.
3. For $l = n - 1$, the Laguerre polynomial is a constant, and the radial wavefunction has the shape $(Zr/a_0)^n e^{-Zr/(na_0)}$. These functions—the 1$s$, 2$p$, 3$d$, and so on—have no radial nodes.
4. There is an overall exponential envelope $e^{-Zr/(na_0)}$, which falls off more slowly as n increases. This is consistent with Bohr's result that highly excited states correspond to a greater average distance of the electron from the nucleus.

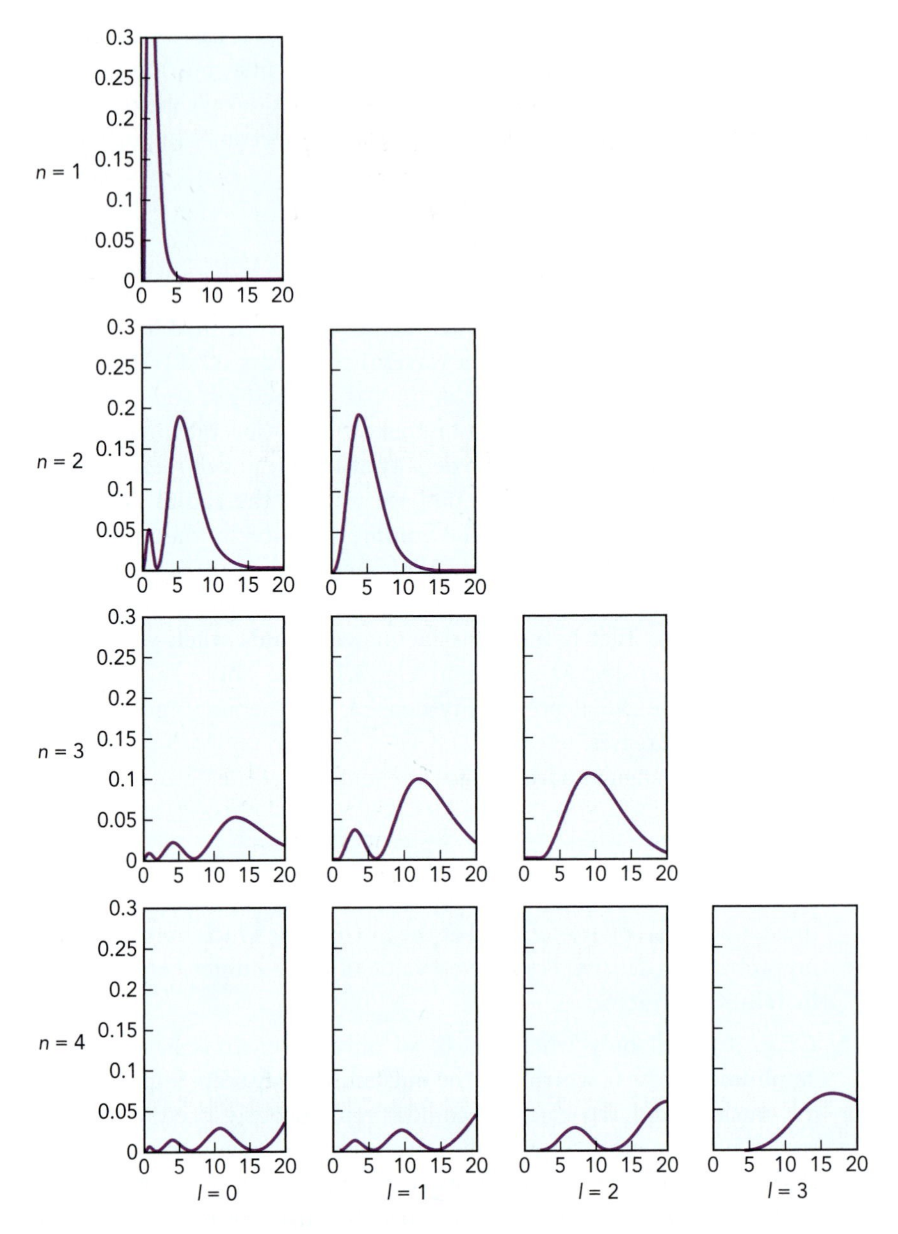

▶ **FIGURE 3.11 The radial probability densities $R^2_{n,l}(r)r^2$.** These are drawn for the same orbitals as appear in Fig. 3.7. The vertical scale is in units of $(a_0/Z)^{-1}$, and the horizontal scale is in a_0/Z.

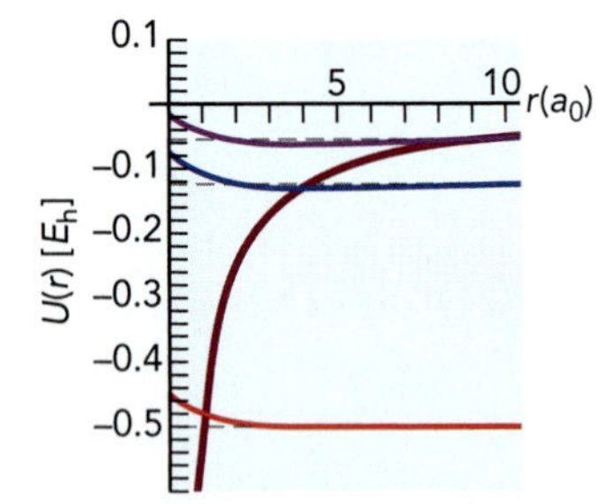

▲ **FIGURE 3.12 The 1*s*, 2*s*, and 3*s* radial wavefunctions.** The wavefunctions and their energies are plotted with the potential energy curve $U(r) = -Ze^2/r$. Tunneling allows the wavefunctions to remain non-zero even when $E < U(r)$.

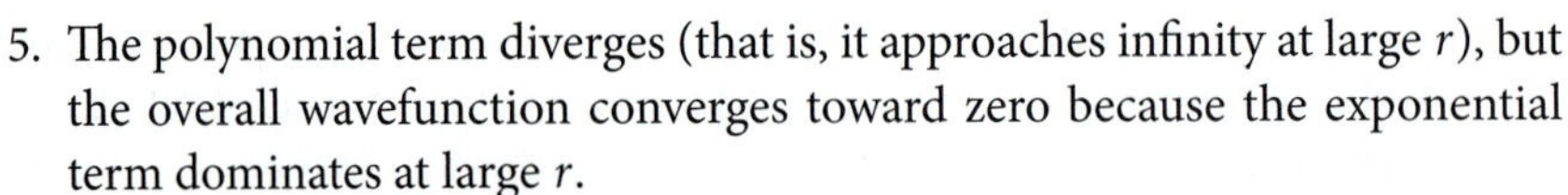

5. The polynomial term diverges (that is, it approaches infinity at large r), but the overall wavefunction converges toward zero because the exponential term dominates at large r.
6. But, incredibly, the wavefunction never quite reaches zero, even at values of r where the energy of the state is *less* than the potential energy. This is a non-classical effect called **tunneling.**

Tunneling must be the oddest of the unexpected effects that we encounter in our look at the quantum world. We can represent the wavefunctions and energy levels of the electron by drawing them on the potential energy surface, as shown in Fig. 3.12. Where the energy of the state equals the potential energy, the wavefunction reaches its classical turning points. Beyond the classical turning point, the energy E is less than U. If $K = E - U$, then this region corresponds to *negative kinetic energy,* an impossibility for a classical kinetic energy of $mv^2/2$. Why then is this possible in quantum mechanics?

The critical distinction between a classical particle and a quantum-mechanical particle is wavelike character, and one property that defines a wave is its distribution over space. A wave is never located at precisely one point. Therefore, a wavelike particle that has a specific energy E can extend over a region of space in which the potential energy varies. The particle experiences an *average* potential energy, given by the expectation value $\langle U \rangle$. The particle's kinetic energy must be averaged over the same region, and that average kinetic energy, $\langle K \rangle = \langle E - U \rangle$, is always positive. Although we don't measure bizarre kinetic energies for tunneling particles, tunneling does allow particles to travel through barriers that classical particles would never penetrate. There is always a chance that a wavelike particle will escape through a wall, for example. But the probability of tunneling through a barrier decreases rapidly as we increase the mass of the particle, the length of the barrier, or the energy difference $U - E$, all as predicted by the correspondence principle.

SAMPLE CALCULATION Atomic Orbital Wavefunctions. To write the electronic wavefunction for the $3d_{-1}$ state of the Li^{+2} ion, we start by finding what quantum numbers we have. The $3d$ subshell corresponds to $n = 3$, $l = 2$. For lithium, the atomic number Z is 3, and this is a factor in the radial wavefunction. Therefore, we combine the angular wavefunction for $l, m_l = 2, -1$ with the radial wavefunction for $n, l = 3,2$ and $Z = 3$:

$$\begin{aligned} \psi_{3,2,-1}(r,\theta,\phi) &= R_{3,2}(r) Y_2^{-1}(\theta,\phi) \\ &= \left(\frac{Z}{a_0}\right)^{3/2} \left(\frac{2\sqrt{2}}{81\sqrt{15}}\right) \left(\frac{\sqrt{15}}{\sqrt{8\pi}}\right) \left(\frac{Zr}{a_0}\right)^2 e^{-Zr/(3a_0)} \sin\theta \cos\theta \, e^{-i\phi} \\ &= \left(\frac{3}{a_0}\right)^{3/2} \frac{1}{81\sqrt{\pi}} \left(\frac{3r}{a_0}\right)^2 e^{-r/a_0} \sin\theta \cos\theta \, e^{-i\phi}. \end{aligned}$$

EXAMPLE 3.2 Atomic Orbital Characteristics

CONTEXT In subsequent chapters, we will find that the sizes and shapes of the atomic orbital wavefunctions determine the bond lengths and bond angles of molecules, among other properties. The various placements of angular nodes in different atomic orbitals allow hybrid orbitals to distribute bonding electron pairs around a central atom so as to minimize electron–electron repulsion while increasing the opportunities for chemical bonds to form. The radial distribution of the electron determines how close atoms can easily approach each other, forming the basis for our understanding of steric effects in organic chemistry. The angular momentum of the electrons is partly responsible for the soft glow of stellar nebulae and the aurora borealis. A little intuition about the form of these wavefunctions carries us a long way, because everything we want to describe in the rest of this book is made from these wavefunctions.

PROBLEM Answer the following questions about the form of a $4p$ atomic orbital.

1. How many times does the wavefunction change sign between $r = 0$ and $r = \infty$, keeping θ and ϕ fixed?
2. Does the number of angular nodes increase, decrease, or stay the same as the nuclear charge increases? Does the number of radial nodes change?
3. Does the integrated probability density between $r = 0$ and $r = a_0$ increase, decrease, or stay the same as the nuclear charge increases?
4. What is the numerical value for the orbital angular momentum of an electron in this orbital?

SOLUTION

1. The radial wavefunction crosses zero 2 times. See the curve for $n = 4, l = 1$ in Fig. 3.7.
2. The numbers of angular and radial nodes stay the same. The angular part of the wavefunction does not depend on nuclear charge at all. As for the radial part, the nuclear charge does determine the distances at which the radial nodes appear, but not the order of the Laguerre polynomial in Eq. 3.32. The radial nodes are the roots of this polynomial, and therefore the nuclear charge does not affect the number of radial nodes.
3. The probability density near the nucleus increases as we increase the nuclear charge, Ze. Increasing Z makes the exponential term $e^{-Zr/(na_0)}$ in Eq. 3.32 decay faster, so the probability density near the nucleus goes up.
4. The orbital angular momentum is given by $\hbar\sqrt{l(l+1)} = \sqrt{2}\hbar = 1.5 \cdot 10^{-32}\,\mathrm{kg\,m^2\,s^{-1}}$.

EXAMPLE 3.3 Atomic Orbital Probability Densities

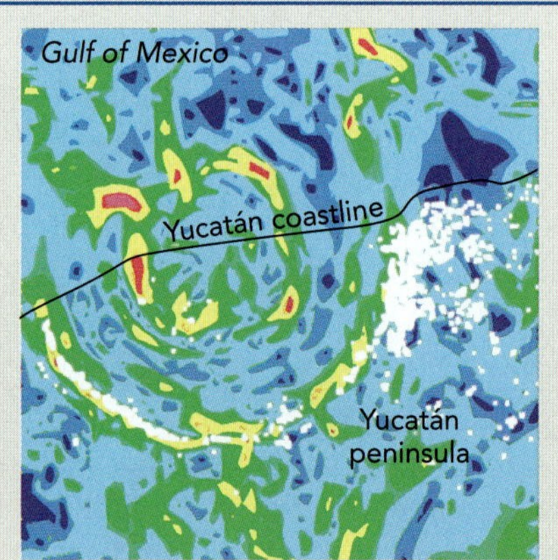

The crater of an ancient asteroid or comet impact lies long the Mexican coast.

CONTEXT Formal charges in molecules, NMR chemical shifts, and chemical bond orders are all determined by the probability density of the electron in specific regions of space. Even in individual atoms, we can use the probability density close to the nucleus to distinguish one element from among many others in a rich mixture of compounds, such as soil. The interaction between the nucleus and the electron density near the nucleus is one probe that has been used to find evidence of prehistoric comet impacts on Earth and their possible role in mass extinctions. The Cretaceous-Tertiary (K-T) boundary is a visible layer in the earth's crust that divides the fossil record into the times before and after the dinosaurs. Roughly 65 million years ago, the K-T boundary was formed, and after that, no more dinosaurs. Physicist Luis Alvarez in 1980 suggested that the extinction of the dinosaurs was caused by aftereffects of a tremendous comet or asteroid impact on Earth, based on his discovery that the K-T boundary held elevated levels of iridium, an element that is 10,000 times more abundant in comets and asteroids than in the earth's crust. Scientists were slow to accept the hypothesis, until geologists presented mounting evidence that a massive impact *had* taken place at that time. One key piece of evidence was the structure of minerals at that time, including the abundant mineral quartz, which is composed of SiO_2. When quartz has been subjected to a shock, such as a comet impact, the structure of the crystal is compressed, and this shifts the electron density. The nuclear magnetic resonance spectrum (Section 5.5) of silicon-29 in shocked quartz is shifted relative to the spectrum of unshocked quartz, because the shocked crystals have a perceptibly different electron distribution near the silicon nucleus. The presence of shocked quartz in the K-T boundary further supports Alvarez's theory. An impact is also known to have occurred at about the right time, near Yucatán in Mexico. The crater has been covered by sedimentary rock over the millions of years since, but a circular pattern of compression in the rock is revealed by gravity measurements along the Yucatán coast.

PROBLEM Find the fraction of the electron density that lies *inside* the radial node of a 2*s* orbital.

SOLUTION The 2*s* radial wavefunction is

$$\frac{1}{\sqrt{2}}\left(\frac{Z}{a_0}\right)^{3/2}\left(1 - \frac{Zr}{2a_0}\right)e^{-Zr/(2a_0)}.$$

The node for this function occurs at the value of r where the polynomial term $1 - \frac{Zr}{2a_0}$ is zero, and that's at $r_1 = 2a_0/Z$. We need to find the probability density for $r < r_1$, so we take the square modulus of the wavefunction, multiply by the volume element, and integrate from $r = 0$ to $r = r_1$. Because this question depends only on the r dependence of the wavefunction, we could include the angle dependence, but we don't have to. If we put in $Y_0^0(\theta,\phi)$ and integrated over θ and ϕ (as well as r), the angular part of the wavefunction would just integrate to 1. That's because $Y_0^0(\theta,\phi)$ is normalized, and we'd just be integrating $|Y_0^0(\theta,\phi)|^2$ over all of the angular space. The radial part is different because we're integrating over only *part* of the range of r values:

$$\begin{aligned}
\int_0^{r_1} R_{2,0}(r)^2 r^2 dr &= \frac{1}{2}\left(\frac{Z}{a_0}\right)^3 \int_0^{r_1}\left[1 - \frac{Zr}{a_0} + \frac{Z^2r^2}{4a_0^2}\right]e^{-Zr/(a_0)}r^2 dr \\
&= \frac{1}{2}\left(\frac{Z}{a_0}\right)^3\left[\int_0^{r_1} r^2 e^{-Zr/(a_0)}dr - \frac{Z}{a_0}\int_0^{r_1} r^3 e^{-Zr/(a_0)}dr + \frac{Z^2}{4a_0^2}\int_0^{r_1} r^4 e^{-Zr/(a_0)}dr\right] \\
&= \frac{1}{2}\left(\frac{Z}{a_0}\right)^3\left[\left(\frac{a_0}{Z}\right)^3(2 - 10e^{-2}) - \frac{Z}{a_0}\left(\frac{a_0}{Z}\right)^4(6 - 38e^{-2}) + \frac{Z^2}{4a_0^2}\left(\frac{a_0}{Z}\right)^5(24 - 168e^{-2})\right] \\
&= \frac{1}{2}\left[(2 - 6 + 6) - (10 - 38 + 42)e^{-2}\right] = 1 - 7e^{-2} = 0.0527.
\end{aligned}$$

This is consistent with the small radial probability density in the inner lobe of the 2*s* function in Fig. 3.11.

EXAMPLE 3.4 Average Properties of the One-Electron Atom

CONTEXT Electron distributions can be complicated functions, but often we want only a single number, such as the average distance or the average kinetic energy of the electron. These expectation values help distill the entire distribution into more easily digestible individual quantities, but it takes a bit of calculus to obtain them. One parameter that depends on the average position of the electron is the **atomic radius,**

an approximate measure of the size of the atom. The electron density around the nucleus does not suddenly stop at some particular distance. Instead, it just gets weaker at large distances as the exponential function $e^{-Zr/(na_0)}$ steadily decreases. Nevertheless, we use some effective atomic radius to estimate chemical bond lengths, possible steric interactions, and other chemical properties. Tabulated values of atomic radii are often based on experimental measurements of bond lengths, but they can also be calculated from the atomic orbital wavefunctions. The average value of r is one such theoretical estimate of the atomic radius.

PROBLEM Find an equation for the average value of r of the 1s wavefunction of the one-electron atom. The final expression should be in units of a_0 and a function of Z.

SOLUTION We do not need the angular momentum part of the wavefunction, because we are examining only the r-dependence of the wavefunction. Get the radial wavefunction from Table 3.2. You can also take advantage of general solutions to the exponential integral from Table A.5:

$$\int_0^\infty x^n e^{-ax} dx = \frac{n!}{a^{n+1}}.$$

$$R_{1,0}(r) = 2\left(\frac{Z}{a_0}\right)^{3/2} e^{-Zr/a_0}$$

$$\langle r \rangle = \int_0^\infty R_{1,0}(r)^2 r^2 dr = \frac{4Z^3}{a_0^3}\int_0^\infty e^{-2Zr/(a_0)} r^3\, dr = \frac{4Z^3}{a_0^3}\left(\frac{3!}{(2Z/a_0)^4}\right)$$

$$= \frac{4 \cdot 6Z^3 a_0^4}{16Z^4 a_0^3} = \frac{3a_0}{2Z}.$$

In the Bohr model, the $n = 1$ electron is at a distance of a_0/Z. The average distance (the expectation value of r) in the real one-electron atom is one and a half times that.

Degeneracy and the Imaginary Part of the Wavefunction

Despite the differences in angular momentum, the energies of the one-electron wavefunctions depend only on n. For each n there are n values of l, each with $2l + 1$ values of m_l. For example, $n = 2$ has one $l = 0$ state and three $l = 1$ states, yielding $1 + 3 = 4$ quantum states, each with a distinct spatial wavefunction but all with the same energy. This number of states for a given value of n is the degeneracy, g, of the spatial wavefunctions and is equal to

$$g = \sum_{l=0}^{n-1}(2l + 1) = n^2. \tag{3.35}$$

We'll find in Section 3.4 that this is not the whole story, but Eq. 3.35 is a good starting point.

The degeneracies in these states come from two sources: different magnitudes of the orbital angular momentum $L = \hbar\sqrt{l(l + 1)}$ and different projections $L_z = \hbar m_l$ of that angular momentum onto the z axis. When $l = 0$, there is no angular momentum, and all the kinetic energy is in radial motion. As l increases for a particular n value, say as we go from the 3s orbital to 3p and then to 3d, more and more of the kinetic energy is transferred into angular motion, but the total kinetic energy remains the same.

Once the value of l is set, different orientations of the angular momentum correspond to the available values of m_l. It makes sense that the energy is the same for each value of m_l if we are only changing the orientation of the angular motion, but then why does the m_l-dependence come with an imaginary part? What does it mean to change the wavefunction by a phase factor $e^{im_l\phi}$? One qualitative explanation, looking at it classically, is that the phase factor allows us to couch some time-dependence in our ostensibly time-independent wavefunction. Time is not a variable in our calculation, because we are interested in identifying only those quantum states that are stable indefinitely. But time has an important ramification for the number of these states: an electron wave orbiting clockwise is in a different state from the same wave orbiting counterclockwise, but the two cases may have exactly the same energy and the same time-averaged probability distribution. So we need a way to write the electron wavefunction that incorporates not only *where* the electron is found, but in *what direction* it moves.

One way to think of the relation between phase and time-dependence is as follows. If we set $m_l\phi = x$, then the Euler formula (Eq. 2.8) states

$$e^{ix} = \cos x + i\sin x.$$

A moving wave may look like $\cos x$ at time $t = 0$, but when the wave has moved forward by one quarter wavelength (or $\pi/2$ radians), it looks identical to $\sin x$ when examined against its original coordinate axes (Fig. 3.13). If instead the wave moves *backward* by $\pi/2$, then it takes the form $-\sin x$. The state $\cos x + i\sin x$ looks like $\cos x$ at $t = 0$, but $\sin x$ after moving $\pi/2$. The state $\cos x - i\sin x$ will become $-\sin x$ after traveling $\pi/2$. These are wavefunctions moving in opposite directions, and that is why the angular momentum projections for $m_l = \pm 2$, for example, are opposite in sign: the electron distributions are rotating in opposite directions and therefore have angular momentum vectors with opposite signs.

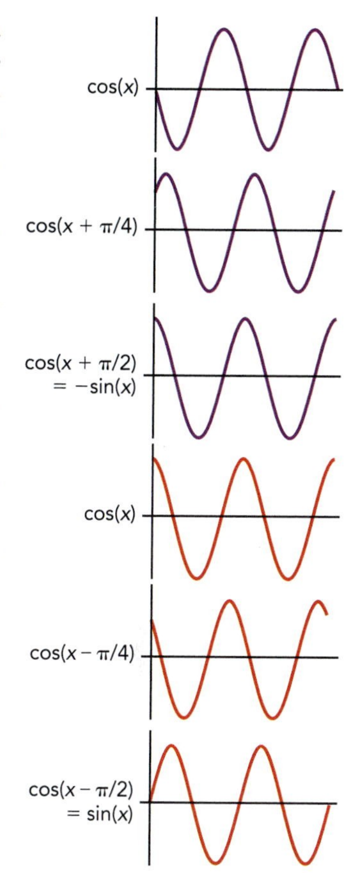

▲ FIGURE 3.13 **Phase in the angular wavefunction.** A wavefunction $\cos x$ traveling in the $+x$ direction becomes $\sin x$ after a phase shift of $\pi/2$. If it travels the same amount in the $-x$ direction, the cosine wave becomes $-\sin x$. The imaginary term in $e^{im_l\phi}$ can be thought of as indicating the direction of rotation of the wavefunction.

3.3 Electric Dipole Interactions

The electric dipole moment μ is a measure of the separation of positive and negative charges along an axis. All of our atomic orbitals, as intricate as their shapes may be, distribute the charge evenly on opposite sides of the nucleus. For this reason, the atom normally possesses no electric dipole moment. However, an external electric field $\mathcal{E}$ will exert opposing forces on the positively charged nucleus and the negatively charged electrons, changing the distribution—at least temporarily. The **polarizability,** α, measures the extent to which that distortion *induces* a dipole moment in the atom:

$$\mu_{\text{induced}} = \alpha\mathcal{E}, \tag{3.36}$$

where μ_{induced} is the magnitude of the induced dipole moment. A high polarizability means the electrons are like water sloshing about in a bucket: easily pushed around by an external force.

Now imagine that our electron distribution is completely wavelike, oscillating up and down on either side of the nucleus at some constant rate. If we introduce an oscillating electric field, it is reasonable that we can change the rate of the electron's oscillation up and down, without conferring a dipole moment on the atom. This is a loose description of the principal method by which radiation interacts with matter to result in an absorption transition. Similarly, during an emission transition, the periodic motion of the electron itself releases energy in the form of an oscillating electromagnetic field, provided that there is a suitable quantum state at lower energy for it to land in.

These are the simplest processes in spectroscopy. The principles of spectroscopy will be a recurring theme as we probe the microscopic structure of many-electron atoms and molecules, because spectroscopy remains the most precise and adaptable tool for controlling and measuring the quantum mechanical characteristics of a chemical system. From spectroscopy comes our most precise molecular geometries and successful theories of chemical bonding, as well as many of our most powerful analytical techniques.

The probability that absorption or emission will actually occur depends on several factors, including the nature of the initial and final quantum states of the atom. One useful rule is that in its strongest interactions with matter, the photon behaves like a particle with one unit of angular momentum,[4] and, therefore, when a photon is absorbed or emitted the angular momentum changes by one unit in the atom: $\Delta l = \pm 1$ (Fig. 3.14). This means that transitions of the type $s \leftrightarrow p$ and $p \leftrightarrow d$ are strong or **allowed,** whereas $s \leftrightarrow s$, $p \leftrightarrow p$, or $s \leftrightarrow d$, $p \leftrightarrow f$ are weak or **forbidden.** An allowed transition is one that is likely to take place; a much greater number of atoms will normally absorb or emit photons for an allowed transition than for a forbidden transition. The rule $\Delta l = \pm 1$ is one of the **selection rules** for electric dipole transitions, one of the rules determining what changes in the quantum numbers allow a strong transition.

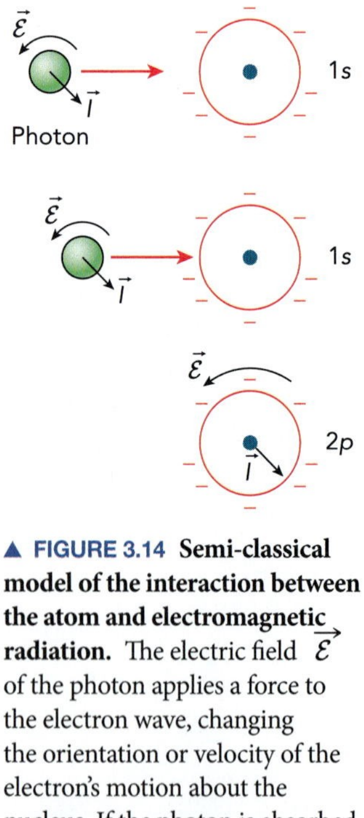

There are also selection rules for m_l, the quantum number determined by the motion along the angle ϕ, around our chosen z axis. Light polarized so that its electric field is *parallel* to that z axis cannot change the angular motion *perpendicular* to that axis (along ϕ), so for that polarization m_l cannot change and $\Delta m_l = 0$. However, light polarized perpendicular to the z axis *will* affect the motion perpendicular to that axis, and for that polarization we have the selection rule $\Delta m_l = \pm 1$. For unpolarized light, which is a combination of these different polarizations, $\Delta m_l = 0, \pm 1$.

Magnetic dipole transitions that violate these selection rules may occur by interaction with the magnetic field of the radiation, connecting quantum states with different orientations and/or magnitudes of the magnetic field. However, the forces arising from magnetic field interactions are generally about 1000 times weaker than those in electric field interactions, and the transitions are correspondingly weak.

▲ FIGURE 3.14 **Semi-classical model of the interaction between the atom and electromagnetic radiation.** The electric field $\vec{\mathcal{E}}$ of the photon applies a force to the electron wave, changing the orientation or velocity of the electron's motion about the nucleus. If the photon is absorbed by the atom, a decrease in the energy of the radiation is observed, and the electron is left in a higher energy state with a corresponding change in angular momentum $\vec{l}$.

[4]That will be good enough for us, but there's much more to the story. The photon picture of radiation has its uses, but our version is not sufficient to explain all the properties of the radiation field. Photons can have both spin and orbital angular momentum, and single-photon electric quadrupole transitions take place where the angular momentum changes by two units instead of one.

TOOLS OF THE TRADE | Atomic Absorption Spectroscopy

The lack of light from the sun at certain wavelengths was first observed and characterized by Wollaston and Fraunhofer before 1820. But it was Gustav Kirchoff and Robert Bunsen (of Bunsen burner fame) who in 1859 developed our modern understanding that a spectrum results from the wavelength-dependent interaction of light with different chemical materials. This principle was immediately applied to chemical analysis, and Bunsen's group even discovered several new elements by analyzing the emission spectra of flames. (The important aspect of Bunsen's burner design in this work was that the flame generated a lot of heat but little light, allowing emission from the atomized samples to be easily visible.) Nevertheless, it was nearly another century before chemists began using spectroscopy as a means of routine chemical analysis.

What is atomic absorption spectroscopy? In atomic absorption spectroscopy (AAS), the elements that make up a sample are identified by reducing the sample to atoms (as completely as practical) and then matching the absorption spectrum of the resulting gas against known spectra.

Why do we use atomic absorption spectroscopy? Before the development of AAS, individual elements were identified by a complex series of isolations and individual chemical tests. In general, it was necessary to know what to look for in order to carry out an effective analysis. In contrast, AAS allows a single probe—a beam of light—to yield results that can rapidly finger any of some 70 different elements.[5] With AAS, it first became possible to identify all the elements in chemically diverse samples of food or soil, for example, without having to do any extensive sample preparation.

How does it work? A typical AA spectrometer schematic is shown. The system requires a way of atomizing the sample, a radiation source to probe the atoms, and a way to detect the wavelength-dependent absorptions.

Although AAS is a remarkably general tool for elemental analysis in principle, that broad applicability requires that the sample be reduced to atoms. (Each compound also gives rise to a distinct spectrum, but the number of possible compounds is unlimited, and the spectra are much more complex.) Therefore, the atomization of the sample forms a key step.

The atomizer has been an evolving component of these spectrometers over the last few decades. Atomizers have long used flames, with the **analyte** (the substance being studied) dissolved in a solution and injected into the flame. The flame is generated in air/acetylene or nitrous oxide/acetylene mixtures above a long, narrow nozzle. The long nozzle maximizes the path length of the radiation through the flame, increasing the signal. Graphite furnaces, which are becoming more standard than flames, are capable of starting from solid samples as well as solutions, and they also tend to atomize the analyte more efficiently than the flame, allowing smaller concentrations to be detected.

The rest of the spectrometer usually mimics the experiments by Fraunhofer and others 200 years ago. Lamps are the most common radiation source, providing light over a broad range of wavelengths simultaneously, much like the sun. Light of specific wavelengths is absorbed by the sample, and we separate the light into its component wavelengths in order to figure out where those absorptions took place. A monochromator with a diffraction grating carries out this task, similar to the way that Fraunhofer's prism separated the wavelengths of light in his work.

A competing method for elemental analysis is inductively coupled plasma emission spectroscopy (ICP-ES). The ICP is capable of much higher temperatures, roughly 10,000 K versus less than 3000 K in the flame, and this allows it to atomize a greater range of elements effectively. The ICP is also less susceptible than the AAS furnace to chemical reaction. However, the plasma generates its own spectra, which limits its sensitivity. Furnace AAS instruments are generally able to detect smaller concentrations of analytes than ICP.

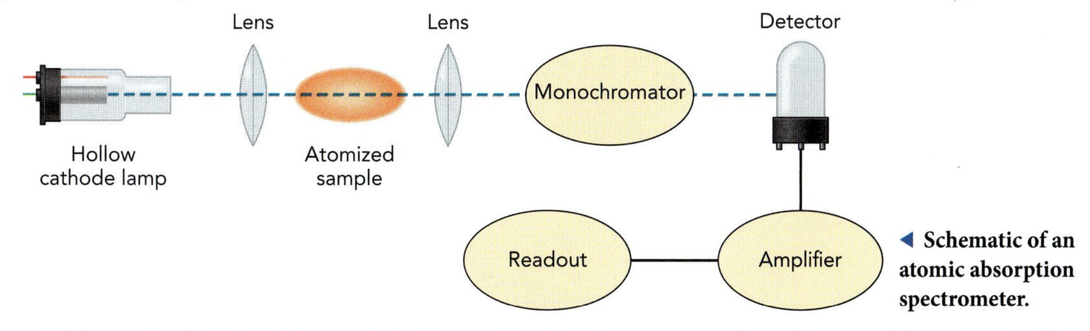

◀ **Schematic of an atomic absorption spectrometer.**

[5]A typical commercial AAS will not be programmed to identify especially rare or unstable elements.

3.4 Magnetic Dipole Interactions

Even when regarded as a wave, the electron has charge and kinetic energy, and any charged mass in motion gives rise to a magnetic field. That magnetic field provides another mechanism for interactions with electromagnetic radiation, and a mechanism for interactions with other magnetic fields.

Orbital Magnetic Moment

Stealing a result derived from one of Maxwell's equations, we can show that an electron with orbital velocity $\vec{v}$ at a position $\vec{r}$ measured relative to the nucleus must generate an orbital magnetic field $\vec{B}_l$ parallel to the angular momentum vector $\vec{l}$:

$$\begin{aligned}
\vec{B}_l &= \frac{1}{c^2}\vec{\mathcal{E}} \times \vec{v} && \text{From Maxwell's laws} \\
&= \frac{1}{c^2}\mathcal{E}\frac{\vec{r}}{r} \times \vec{v} && \vec{r}/r \text{ gives vector direction} \\
&= \frac{1}{c^2}\frac{-e}{(4\pi\varepsilon_0)r^2}\frac{\vec{r}}{r} \times \vec{v} && \mathcal{E} = q/(4\pi\varepsilon_0 r^2) \\
&= -\frac{e}{(4\pi\varepsilon_0)m_e c^2 r^3}\vec{r} \times (m_e\vec{v}) && \times\, m_e/m_e \\
&= -\frac{e}{(4\pi\varepsilon_0)m_e c^2 r^3}\vec{l} && \vec{l} = \vec{r} \times \vec{p} = \vec{r} \times (m\vec{v}) \qquad (3.37)
\end{aligned}$$

A one-electron atom in a state with $l > 0$ generates an orbital magnetic field.

Equation 3.37 only gives us the magnetic field near the electron. The distribution of magnetic field lines is more complicated, but the field is often represented in calculations by a single vector, parallel to the angular momentum, called the **magnetic dipole moment, $\vec{\mu}$**. If we set a tiny circular loop of wire in the electron's orbital magnetic field, perpendicular to the field lines near the particle (Fig. 3.15), then the field will cause a current I to flow through the wire, and the orbital magnetic moment $\vec{\mu}_l$ is defined such that

$$\vec{\mu}_l \equiv -IA\frac{\vec{l}}{|\vec{l}|}, \tag{3.38}$$

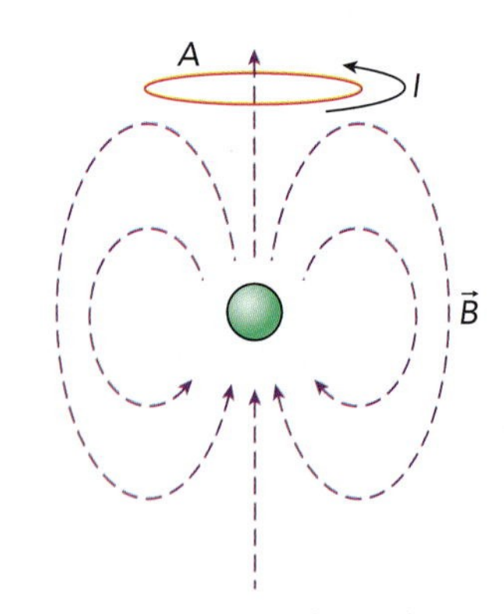

▲ **FIGURE 3.15 A charged particle with angular momentum generates a magnetic field $\vec{B}$.** The magnetic field, in turn, can generate a current I in a small wire loop of area A. The induced current is proportional to the magnetic moment.

where the wire loop encloses an area A and $\vec{l}/|\vec{l}|$ is a unit vector that points in the same direction as the $\vec{l}$ angular momentum vector. The minus sign accounts for the field pointing in the opposite direction from the angular momentum vector. The current is the rate at which charge passes through a given point in the wire, and the area is determined by the radius r of the loop. These relations let us express the magnetic moment in terms of the charge and mass of the electron:

$$\begin{aligned}
\tau &= \frac{2\pi r}{v} && \text{period} = \text{circumference/speed} \\
I &= \frac{q}{\tau} = \frac{ev}{2\pi r} && \text{current} = \text{charge/time} \qquad (3.39) \\
\vec{\mu}_l &\equiv -IA\frac{\vec{l}}{|\vec{l}|} && \text{definition of } \mu
\end{aligned}$$

$$= -\frac{ev}{2\pi r}(\pi r^2)\frac{\vec{l}}{|\vec{l}|} \qquad \text{by Eq. 3.39, } A = \pi r^2$$

$$= -\frac{evr}{2}\frac{\vec{l}}{|\vec{l}|} = -\frac{el}{2m_e}\frac{\vec{l}}{|\vec{l}|} \qquad l = mvr$$

$$= -\frac{e}{2m_e}\vec{l}. \tag{3.40}$$

Equation 3.40 establishes the critical relationship: the magnetic moment $\vec{\mu}_l$ is proportional and parallel to the angular momentum vector $\vec{l}$. We will need this relationship later when we examine electronic states of atoms and molecules, and when we discuss nuclear magnetic resonance spectroscopy in Section 5.5.

CHECKPOINT Our solutions to the Schrödinger equation in Section 3.1 deal with the major contributions to the energy of an electron in an atom. However, smaller contributions to the energy come from the interactions between different magnetic fields within the atom. We've found that angular momentum of the electron produces a magnetic field, and next we will see how this affects the energy of the electron.

We rewrite Eq. 3.40 to introduce two new constants of proportionality,

$$\vec{\mu}_l = -\frac{g_l \mu_B}{\hbar}\vec{l}. \tag{3.41}$$

In this expression, the **Bohr magneton** μ_B combines fundamental constants,

$$\mu_B = \frac{e\hbar}{2m_e} = 9.27402 \cdot 10^{-24}\,\mathrm{J\,T^{-1}}, \tag{3.42}$$

to provide a useful conversion factor between magnetic field strength (in tesla) and the energy (in joules) when two fields interact. The second constant, g_l, is the unitless **orbital gyromagnetic factor.** Gyromagnetic factors provide any remaining multiplicative correction that might be necessary. For the orbital motion of the electron, that correction arises mainly from our assumption that the nucleus is at the center of mass:

$$g_l = 1 - \frac{m_e}{m_{\mathrm{nuc}}}, \tag{3.43}$$

where m_{nuc} is the nuclear mass. The magnetic dipole moment of the electron determines how the energy of the electron will be affected by other magnetic fields. The magnetic moment is a convenient parameter for magnetic properties because it is independent of position, whereas the magnetic field strength of a particle varies with location.

An external magnetic field $\vec{B}$ exerts a force on a magnetic moment $\vec{\mu}$ that tries to line the two vectors up. The strength of the force depends on the relative orientation of the two vectors. Therefore, the energy that corresponds to that force may be written

$$E_{\mathrm{mag}} = -\langle \vec{\mu} \cdot \vec{B} \rangle, \tag{3.44}$$

where the expectation value allows for averaging over fields that change with position, and the minus sign ensures that the lowest energy is when the magnetic moment and magnetic field vectors point in the same direction. Therefore, putting a one-electron atom into a magnetic field should give different energies for different orientations of the orbital magnetic moment μ_l within the field. We can always define the magnetic field axis to be the z axis,

$$\vec{B} = (0, 0, B_0),$$

in which case the magnetic interaction energy is a constant (so we don't need the expectation value brackets) equal to

$$E_{\mathrm{mag},l} = -\vec{\mu}_l \cdot \vec{B} = -\mu_{l,z} B_0$$

$$= g_l \mu_B l_z B_0/\hbar$$

$$E_{\mathrm{mag},l} = g_l \mu_B m_l B_0. \tag{3.45}$$

We call m_l the magnetic quantum number because it plays this role in determining the energy of the atom in a magnetic field. If there were no other magnetic fields to worry about, then Eq. 3.45 would be all we need to describe the response of an atom to a laboratory magnetic field. However, there is always at least one other source of magnetism available, which we get to next.

Spin Magnetic Moment

Otto Stern and Walter Gerlach demonstrated in 1922 that ground state silver atoms, which have an unpaired electron but no orbital angular momentum, still exhibit an internal magnetic field, and the field appears with only two possible orientations. Although it would be some years before the experiments would be correctly interpreted,[6] they had shown that all electrons have an intrinsic angular momentum, unrelated to their motion around a nucleus. Even free electrons have this angular momentum, and it was lastingly misnamed **electron spin** because a particle spinning about a central axis also has an angular momentum independent of its motion through space. That picture will work well enough for us, but, for the record, the origin of an electron's spin angular momentum is taken to be an intrinsic property of the electron, like its mass or charge, that is independent of its position or its classical motion through space. The property of spin in elementary particles such as the electron can be derived from solving the Schrödinger equation with a Hamiltonian that takes into account special relativity.

We will return briefly to the origin of the electron spin at the end of this section, but, for better or worse, in chemistry we needn't aim for a such a deep understanding. The math we will use cares nothing about the origin of a particle's angular momentum: any angular momentum $\vec{j}$ is quantized by a quantum number j such that the magnitude of the angular momentum is $\hbar\sqrt{j(j+1)}$ and its z-components are $\hbar m_j$, where the magnetic quantum number m_j may be any value from $-j$ to $+j$, changing in increments of one. This is shown for several values of j in Fig. 3.16. The two orientations of the spin magnetic field observed

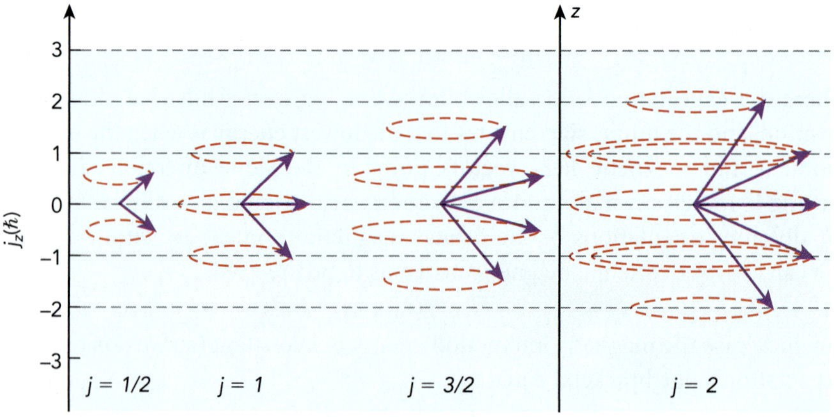

▶ FIGURE 3.16 **Quantized angular momentum vectors.** Vectors are drawn for quantum numbers j equal to 1/2, 1, 3/2, and 2. In each case, the magnitude of the angular momentum is $\hbar\sqrt{j(j+1)}$ and the z component of the vector is given by $\hbar m_j$. Because we have chosen to know the z component, the uncertainty principle forbids our knowing the x or y components of the angular momentum, so we cannot require the vectors to be fixed in the page as drawn. They are free to rotate around the z axis.

[6] Stern and Gerlach believed at first that they had added support to Bohr's model by showing that the *orbital* angular momentum of the electron in a circular orbit could have only two orientations, corresponding to clockwise and counterclockwise rotation.

by Stern and Gerlach correspond to the two possible values of the **spin magnetic quantum number,** m_s. This is quite different from orbital angular momentum in one respect: the number of m_l values for a given atomic subshell is always an odd number, $2l + 1$, because there are l positive values plus l negative values, plus one for $m_l = 0$. But Stern and Gerlach observed only two orientations of the electron spin, two values of m_s. If we invent a spin quantum number s such that m_s can take all the values from $-s$ to s, changing in units of one, then s must be $\frac{1}{2}$, with m_s then equal to $-\frac{1}{2}$ or $\frac{1}{2}$. Unlike the orbital angular momentum quantum numbers, which are integers, the spin quantum numbers are **half-integers,** numbers equal to some integer plus $\frac{1}{2}$.

The magnetic moment in this case is given by

$$\vec{\mu}_s = -\frac{g_s \mu_B}{\hbar}\vec{s}. \tag{3.46}$$

The gyromagnetic factors, g, are obtained from classical electromagnetism, in which they represent the ratio of the magnetic moment (in units of μ_B) to the angular momentum (in units of $\hbar$) that causes the magnetic field. The resulting **spin gyromagnetic factor** g_s is then equal to roughly $(1/s) = 2$, with additional corrections yielding a more precise value of 2.00232.

We may write an equation analogous to Eq. 3.45 for the energy of the interaction between the electron spin and an external magnetic field:

$$E_{\text{mag},s} = g_s \mu_B m_s B_0. \tag{3.47}$$

If atomic hydrogen in the ground $1s$ configuration ($n = 1$, $l = m_l = 0$, $m_s = \pm\frac{1}{2}$) is placed in the field of a strong laboratory magnet, the energies of the $m_s = +\frac{1}{2}$ and $m_s = -\frac{1}{2}$ states separate by an amount proportional to the strength B_0 of the applied field.

Unlike the orbital angular momentum, spin plays only an indirect role in determining where the electron is. We define a new coordinate ω, which is internal to the electron, for the extent of the spin motion. To a good approximation, the spin motion is specified by some spin wavefunction $\chi_{m_s}(\omega)$, which is completely independent of the electron's spatial wavefunction, $\psi_{n,l,m_l}(r,\theta,\phi)$. The combined spin-spatial wavefunction is then the product of those two pieces:

$$\Psi_{n,l,m_l,m_s}(r,\theta,\phi,\omega) = \psi_{n,l,m_l}(r,\theta,\phi)\ \chi_{m_s}(\omega). \tag{3.48}$$

Because there are only two possible values for m_s, there are only two possible spin wavefunctions. We skip trying to write $\chi(\omega)$ in any kind of algebraic form and instead simply label the two possible spin wavefunctions as follows:

$$\chi_{m_s=+\frac{1}{2}}(\omega) \equiv \alpha, \qquad \chi_{m_s=-\frac{1}{2}}(\omega) \equiv \beta.$$

A spin-spatial wavefunction for one electron may be written, for example, as $2p_0\beta$, which specifies the four quantum numbers ($n = 2$, $l = 1$, $m_l = 0$, $m_s = -\frac{1}{2}$) that we would need to determine all the parts of the wavefunction.

The half-integer spin has a rather spooky physical implication. Let's assume that the spin *does* correspond to some kind of classical turning motion through an angle ω. Then the spin magnetic quantum number m_s has the same physical interpretation as m_l and differs only in that it corresponds to some hidden degree of freedom in the electron. The form of the spin wavefunction $\chi(\omega)$, which quantizes m_s, should then resemble the $e^{-im_l\phi}$ term that quantizes m_l in our spherical harmonics:

$$\chi_{m_s}(\omega) = A\ e^{im_s\omega} = A\ e^{\pm i\omega/2}, \tag{3.49}$$

where A is some constant. Now change ω by 2π, a full rotation. It doesn't matter what the original value of ω was—the result is that the function changes sign:

$$\begin{aligned}
\chi(\omega + 2\pi) &= A\ e^{\pm i(\omega+2\pi)/2} \\
&= A\ e^{\pm i\omega/2}\ e^{\pm i\pi} && e^{ab} = e^a e^b \\
&= Ae^{\pm i\omega/2}\Big[\underbrace{\cos(\pm\pi)}_{=-1} + i\underbrace{\sin(\pm\pi)}_{=0}\Big] && e^{i\phi} = \cos\phi + i\sin\phi \quad \text{by Eq. 1.25} \\
&= -\chi(\omega + 2\pi)\,. && (3.50)
\end{aligned}$$

It appears that a full rotation doesn't bring us back to the original function! The de Broglie wavelength for the electron along this coordinate is equal to *two times* the circumference, and it is necessary to rotate the system *twice* to get back to the original function. This occurs because our treatment of the electron spin as an angular momentum is sufficient to explain many of its features, but the spin is not correctly described by the sort of classical rotational motion just described. Dirac's solution to the Schrödinger equation expresses the spin wavefunction $\chi(\omega)$ using 2×2 matrices called **spinors,** obtained partly by combining time and space into a symmetric set of coordinates. Spinors take the math a step further than we'd like, and we'll avoid them in this text. The cost is that you will have to forgive one more seemingly magical property of particle spins that we apply in Section 4.3 to obtain the Pauli exclusion principle.

Spin-Orbit Interaction

Before we get too complacent separating spin and spatial wavefunctions, we examine one important exception. A hydrogen atom in its 1s ground state has an electron spin magnetic moment but no orbital magnetic moment. Once we excite the atom into any of the $l > 0$ states, the spin magnetic moment can interact with the magnetic field due to $\vec{l}$. The orbital magnetic moment of the electron $\vec{\mu}_l$ gives the magnitude and direction of the magnetic field $\vec{B}_l$ generated by the orbital motion, and therefore

$$E_{\text{spin-orbit}} = -\langle \vec{\mu}_s \cdot \vec{B}_l \rangle \equiv \frac{A_{\text{so}}\langle \vec{l} \cdot \vec{s} \rangle}{\hbar^2}, \qquad \text{by Eqs. 3.37 and 3.46} \tag{3.51}$$

$$A_{\text{so}} \approx \frac{e\hbar\mu_B g_s}{(4\pi\varepsilon_0)m_e c^2}\langle r^{-3}\rangle$$

where the **spin-orbit constant** A_{so} has units of energy. An improved value of A_{so} includes a factor of two in the denominator, the **Thomas precession correction,** which corrects for having to rotate the spin wavefunction by 4π to restore the original state:

$$A_{\text{so}} = \frac{e\hbar\mu_B g_s}{2(4\pi\varepsilon_0)m_e c^2}\langle r^{-3}\rangle. \tag{3.52}$$

SAMPLE CALCULATION The Spin-Orbit Constant. The expectation value $\langle r^{-3}\rangle$ of the $2p$ state of a one-electron atom is given by $Z^3/(24a_0^3)$. Equation 3.52 then estimates the value of A_{so} for $2p$ excited state hydrogen to be

$$\begin{aligned} A_{so} &= \frac{e\hbar\mu_B g_s}{2(4\pi\varepsilon_0)m_e c^2}\langle r^{-3}\rangle \\ &= \frac{(1.602\cdot 10^{-19}\,\mathrm{C})(1.055\cdot 10^{-34}\,\mathrm{J\,s})(9.27402\cdot 10^{-24}\,\mathrm{J\,T^{-1}})(2.00232)}{2(1.113\cdot 10^{-10}\,\mathrm{C^2\,J^{-1}\,m^{-1}})(9.109\cdot 10^{-31}\,\mathrm{kg})(2.998\cdot 10^{8}\,\mathrm{m\,s^{-1}})^2} \\ &\quad\times\left(\frac{1}{24(5.292\cdot 10^{-11}\,\mathrm{m})^3}\right) \\ &= 4.841\cdot 10^{-24}\,\mathrm{J} = 1.110\cdot 10^{-6}\,E_h. \end{aligned}$$

This value gives a typical order of magnitude for the spin-orbit interaction energy, in hydrogen almost 500,000 times smaller than the ionization energy but still a large effect in certain kinds of spectroscopy.

The possible values of the spin-orbit energy therefore depend on the possible values of the dot product $\vec{l}\cdot\vec{s}$. Keeping in mind that the orbital angular momentum establishes a magnetic field, the spin can still take only two possible orientations (for $m_s = \pm\frac{1}{2}$) relative to the orbital angular momentum, and therefore the one-electron atom with $l > 0$ has two different **spin-orbit states.** The two states are labeled by the possible values of the overall angular momentum, the vector sum of $\vec{l}$ and $\vec{s}$ (Fig. 3.17):

$$\vec{j} = \vec{l} + \vec{s}. \tag{3.53}$$

Similarly, the two values of the quantum number j are given by adding s to l or subtracting s from l:

$$\text{one} - \text{electron atom:}\quad j = l \pm \tfrac{1}{2}. \tag{3.54}$$

For a one-electron atom, j must be a half-integer (an integer plus $\frac{1}{2}$), because $s = \frac{1}{2}$ and l is an integer. Like l and s, it has a projection quantum number, m_j, which can have values between $-j$ and j, changing in units of one:

$$m_j = -j, -j+1, -j+2, \ldots, j.$$

Because $\vec{j}$ is the vector sum of $\vec{l}$ and $\vec{s}$, its projection onto the z axis, $m_j\hbar$, must be equal to the sum of the z projections of $\vec{l}$ and $\vec{s}$. Therefore,

$$m_j = m_l + m_s. \tag{3.55}$$

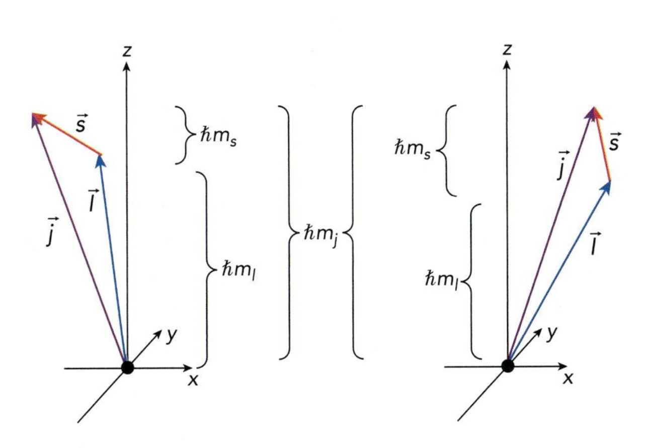

◀ FIGURE 3.17 **The vector sum $\vec{l} + \vec{s}$ yields the total angular momentum $\vec{j}$.** The magnitudes of $\vec{l}$, $\vec{s}$, $\vec{j}$, and m_j may all be constant, but m_l and m_s may vary.

EXAMPLE 3.5 **Spin-Orbit States of 2*p* H**

CONTEXT Perhaps the most fundamental parameter that limits computer processing speed is the size of the processor circuit. The processor circuit involves, among many other components, *bits* where voltages are set, detected, and altered as part of a single computational process. The smaller the circuit, the less time it takes for charge to travel to and from the bits, and the faster we can add or remove charge from the bit. Contemporary chip-making methods involve circuit dimensions of tens of nanometers, which impinges on the distance scale where quantum mechanics becomes important. In the quantum limit, electrical circuits function very differently, for example because the charge can no longer be strongly localized and can even tunnel between adjacent circuit elements. This sounds at first like a disaster for building electrical circuits, but a different approach to processor design seizes on the intrinsic separation of energy levels in a quantum system as an advantage. The quantum version of the bit, called a **qubit,** consists of some two-state quantum mechanical system. One of the candidates for a suitable qubit is the spin-orbit interaction involving a single unpaired electron in nanometer-sized wires. The energy changes by a measurable extent when the electron spin angular momentum alters its orientation relative to the orbital angular momentum, flipping between the two possible j states.

PROBLEM What are the values of m_j, and what is the degeneracy for each spin-orbit state of atomic hydrogen in the $2p$ state?

SOLUTION For the H atom in a $2p$ orbital, possible values of m_l are -1, 0, and 1, while $m_s = \frac{1}{2}$ or $-\frac{1}{2}$. Examining all the possible combinations of $m_l + m_s$, we obtain these values of m_j:

$$\begin{array}{lll} -1 - \frac{1}{2} = -\frac{3}{2} & 0 - \frac{1}{2} = -\frac{1}{2} & 1 - \frac{1}{2} = \frac{1}{2} \\ -1 + \frac{1}{2} = -\frac{1}{2} & 0 + \frac{1}{2} = \frac{1}{2} & 1 + \frac{1}{2} = \frac{3}{2}. \end{array}$$

There are two values of j, $l - s = \frac{1}{2}$ and $l + s = \frac{3}{2}$, and each of these should have values of m_j that run from $-j$ to j, changing in units of one. For $j = \frac{1}{2}$, the m_j values should be $-\frac{1}{2}$ and $\frac{1}{2}$, giving a degeneracy of 2. For $j = \frac{3}{2}$, the m_j values should be $-\frac{3}{2}, -\frac{1}{2}, \frac{1}{2}$, and $\frac{3}{2}$, giving a degeneracy of 4. These account for all the m_j states we found by adding m_l and m_s.

The $j = \frac{1}{2}$ and $j = \frac{3}{2}$ levels correspond to two different spin-orbit energies, separated by the different orientation of the spin $\vec{s}$ with respect to the orbital angular momentum $\vec{l}$. The energy differences between spin-orbit levels from the same electron configuration is sometimes called **fine structure** because it was a small ("fine") effect in early spectra.

Choice of Angular Momentum Quantum Numbers

Here's a question that may be bothering you. We now have four coordinates to describe the motion of our one electron: r, θ, ϕ, and the spin coordinate ω. Therefore, we have four quantum numbers (n, l, m_l, and m_s) to parametrize the excitation energy along these coordinates. How then, do we suddenly have two new quantum numbers, j and m_j, when we haven't actually added any new coordinates? The answer, though a little subtle, ushers in an issue that will be critical in our analysis of many-electron atoms. What we've done is take two quantum numbers we had already—m_l and m_s—and refashion them into the new quantum numbers j and m_j. This is another example of what we did in forming the *sp* wavefunctions in Eqs. 2.5 and 2.6. We used one set of simple, easy-to-work-with wavefunctions as a basis set, from which we created a set of new wavefunctions, more accurate but more cumbersome. In Example 3.5 the conversion went as follows:

$$\begin{aligned}
&m_l, m_s = -1, -\tfrac{1}{2} \rightarrow j, m_j = \tfrac{3}{2}, -\tfrac{3}{2} \\
&\left.\begin{aligned} m_l, m_s &= -1, \tfrac{1}{2} \\ m_l, m_s &= 0, -\tfrac{1}{2} \end{aligned}\right\} \rightarrow \left\{\begin{aligned} j, m_j &= \tfrac{1}{2}, -\tfrac{1}{2} \\ j, m_j &= \tfrac{3}{2}, -\tfrac{1}{2} \end{aligned}\right. \\
&\left.\begin{aligned} m_l, m_s &= 1, -\tfrac{1}{2} \\ m_l, m_s &= 0, \tfrac{1}{2} \end{aligned}\right\} \rightarrow \left\{\begin{aligned} j, m_j &= \tfrac{1}{2}, \tfrac{1}{2} \\ j, m_j &= \tfrac{3}{2}, \tfrac{1}{2} \end{aligned}\right. \\
&m_l, m_s = 1.\tfrac{1}{2} \rightarrow j, m_j = \tfrac{3}{2}, \tfrac{3}{2}.
\end{aligned} \tag{3.56}$$

The sum $m_l + m_s = m_j$ remains a constant as we impose the spin-orbit coupling on our basis states, but the specific values of m_l and m_s are no longer known for the $m_j = \pm\frac{1}{2}$ states. The $m_j = \frac{1}{2}$ states were obtained by combining together the $m_l, m_s = 0,\frac{1}{2}$ and $1,-\frac{1}{2}$ basis states. When this sort of coupling occurs, quantum numbers that remain eigenvalues of the wavefunction throughout (such as l, s, j, m_j in this example) are called **good quantum numbers.** An important feature of this process, as we form a new set of quantum states by mixing our basis functions, is that we form exactly as many distinct new states as we had basis states initially. We started with six basis states on the left side of Eq. 3.56 and end with six states on the right.

The possible orientations of $\vec{s}$ relative to $\vec{l}$ limit the values of j such that $|l - s| \leq j \leq l + s$. For $l = 0$ states, only $j = \frac{1}{2}$ satisfies that condition; there is no spin-orbit splitting in the energies because there is no orbital angular momentum. In one-electron atoms, the spin-orbit energy increases with j. For example, the $j = \frac{1}{2}$ level has lower energy than $j = \frac{3}{2}$ in $2p$ H atom. The transition between these states is a magnetic dipole transition (it requires changing the orientation of the spin magnetic moment) and is therefore quite weak. For $2p$ H, the spin-orbit splitting is $1.66 \cdot 10^{-6} E_h$ (0.365 cm^{-1}), compared to $0.375 E_h$ for the $1s \rightarrow 2p$ spacing. Spectroscopy to measure the spin-orbit interaction has been a valuable tool in the analysis of atoms and molecules with unpaired electrons, particularly metal-containing molecules whose chemistry depends on the number of unpaired electrons at the metal atom.

The spin-orbit interaction splits many of the energies of the states that we called degenerate for the purposes of Eq. 3.35. However, for one-electron atoms with relatively low atomic number, the spin-orbit splitting is small compared to the energy gap between states with different n values. In this case, we can assume that Eq. 3.35 still represents the number of *spatial* wavefunctions with (roughly) the same energy, but the overall degeneracy of the electronic states is greater by a factor of two to account for the two possible *spin* states of the electron:

$$g = 2\sum_{l=0}^{n-1}(2l + 1) = 2n^2. \tag{3.57}$$

Magnetic Resonance

In the laboratory, we sometimes apply our own magnetic fields in order to take advantage of the resulting interaction with these magnetic moments intrinsic to the atom. In the simplest case, a very high magnetic field $\vec{B}$, aligned along an axis Z fixed in the laboratory,[7] interacts with the different m_l and m_s states according to the same law that gave us Eq. 3.51:

$$\begin{aligned} E_{\text{Zeeman}} &= -(\vec{\mu}_l + \vec{\mu}_s)\cdot\vec{B} \\ &= g_l\,\mu_B\,\vec{l}\cdot\vec{B} + g_s\,\mu_B\,\vec{s}\cdot\vec{B} \\ &= g_l\,\mu_B\,m_l\,B_0 + g_s\,\mu_B\,m_s\,B_0. \end{aligned} \tag{3.58}$$

This leads to a separation of the quantum states with different m_l and m_s into different energies and splits the spectroscopic transition energies as well, a phenomenon called the **Zeeman effect.**

We have chosen the Z axis of our magnetic field so that m_l and m_s give the projections of $\vec{l}$ and $\vec{s}$ onto that axis. Consequently $\vec{l}\cdot\vec{B}$ is just equal to $m_l B_0$. Notice that the energy increases as B_0 increases for states where m_l and m_s are positive, and it decreases with B_0 when m_l and m_s are negative. That is why m_l and m_s are called "magnetic" quantum numbers; the different m_l or m_s states are separated in energy only by the presence of another magnetic field.

At low magnetic field strengths, the interaction between $\vec{l}$ and $\vec{s}$ is more important than interactions with the external B_Z field. Here the energies obey the equation

$$E_{\text{Zeeman}} = g_j \mu_B m_j B_Z. \tag{3.59}$$

In Fig. 3.18a, we see two j levels with energies given by Eq. 3.59, shifting in proportion to $m_j B$, where m_j has values $\pm\frac{1}{2}$ and $\pm\frac{3}{2}$ for $j = \frac{3}{2}$ and $\pm\frac{1}{2}$ for $j = \frac{1}{2}$. As the applied field becomes stronger than the magnetic fields generated by the electron (Fig. 3.18b), the spin and angular momenta try to line up with it and stop trying to line up with one another. Beyond that point, the energies are better described by Eq. 3.58. Approximating g_s by 2 and g_l by 1, that leads to the five energy levels at high field, with $m_l, m_s = 1, -\frac{1}{2}$ and $-1, \frac{1}{2}$ having roughly the same energy.

Transitions among these states are also magnetic dipole transitions and quite weak, but they are the basis of **electron paramagnetic resonance** or **electron spin resonance** spectroscopy, used to characterize inorganic compounds based on the rate at which the m_j energy levels shift with increased magnetic field. Why is it helpful to complicate the atom by introducing yet another magnetic field? One strength of this experiment is its ability to identify a quantum state by counting the number of magnetic components: an energy level that splits into four m_j values, for example, should correspond to a j value of $\frac{3}{2}$ (so that $m_j = -\frac{3}{2}, -\frac{1}{2}, \frac{1}{2}, \frac{3}{2}$). By far the most common application of the Zeeman effect

[7]We will label the coordinate system of the laboratory by capital letters X,Y,Z, to distinguish from the coordinates x,y,z of the atom.

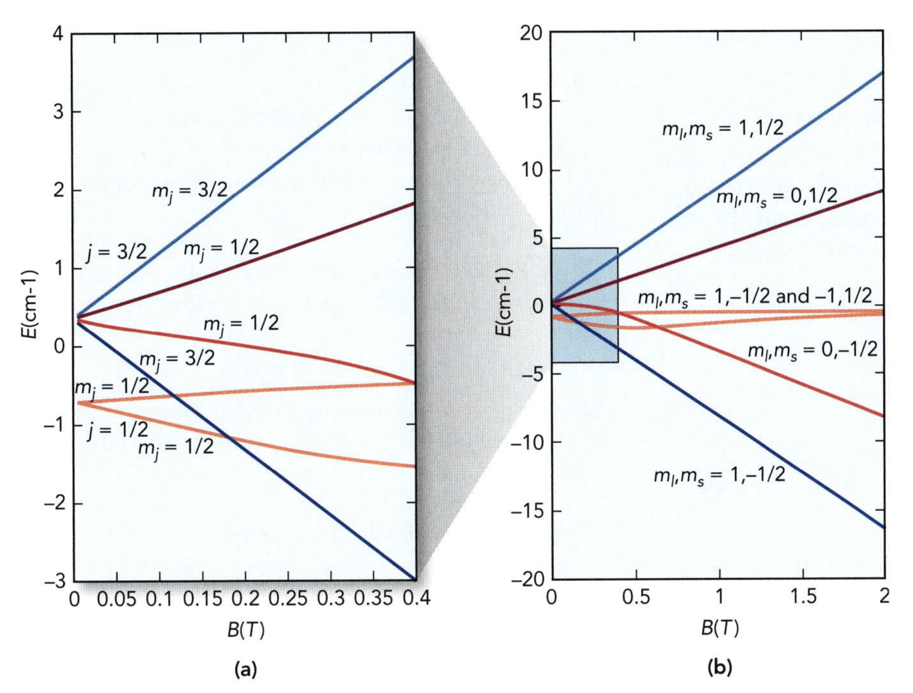

◀ **FIGURE 3.18 The Zeeman effect.** The electronic energies in $2p$ H are shown at **(a)** low magnetic field, where the spin and orbital moments of the electron couple to one another to form j, and at **(b)** high field, where the orbital and spin moments couple to the external magnetic field. Note the change in energy scale from (a) to (b).

is to the splitting of the nuclear magnetic energy levels. Because the nuclear magnetic field is so small, the splitting of nuclear spin states by interaction with the electrons is difficult to observe. The introduction of high external magnetic fields amplifies these energy differences for applications such as nuclear magnetic resonance spectroscopy, where the magnified energy difference in the magnetic field allows us to distinguish between nuclear spin states with slightly different chemical environments. We will expand on that principle in Section 4.5.

As we add electrons to the atom, we complicate this picture by increasing the number of available magnetic fields and the number of ways they can couple to one another. Because these interactions remain important in understanding the electronic states of many-electron atoms, we will return to this topic in Section 4.4.

CONTEXT *Where Do We Go from Here?*

Having put together a one-electron atom, we now have to move on to many-electron atoms if we have any hope of dealing with most of the periodic table. Adding more electrons to the picture of the atom we have now is a bigger step than you might think, because it takes the mechanics beyond what we can solve exactly. But it also opens the door to some fascinating physics that control the structure of the periodic table and the structure of chemical bonds.

KEY CONCEPTS AND EQUATIONS

3.1 Solution of the Schrödinger equation.

a. The Schrödinger equation for the one-electron atom may be written as the sum of three terms: a radial (r-dependent) kinetic energy term, an angular kinetic energy term, and a potential energy term:

$$\hat{H} = -\frac{\hbar^2}{2m_e}\frac{1}{r^2}\frac{\partial}{\partial r}r^2\frac{\partial}{\partial r} + \frac{1}{2m_e r^2}\hat{L}^2(\theta,\phi) - \frac{Ze^2}{4\pi\varepsilon_0 r}, \quad (3.8)$$

where

$$\hat{L}^2(\theta,\phi) = -\hbar^2\left[\frac{1}{\sin\theta}\frac{\partial}{\partial\theta}\sin\theta\frac{\partial}{\partial\theta} + \frac{1}{\sin^2\theta}\frac{\partial^2}{\partial\phi^2}\right] \quad (3.5)$$

is the operator whose eigenvalue is the square of the orbital angular momentum:

$$L^2 = \hbar^2 l(l+1). \quad (3.18)$$

One contribution to the $\hat{L}^2$ operator comes from the operator $\hat{L}_z$, which has eigenvalue

$$L_z = m_l\hbar, \quad (3.12)$$

that gives the projection of the orbital angular momentum onto the z axis.

b. The solution to this Schrödinger equation yields energies that agree with the predictions of the Bohr model,

$$E = -\frac{\hbar^2}{2m_e a_0^2}\left(\frac{Z}{n}\right)^2 = -\frac{Z^2}{2n^2}E_h, \quad (3.30)$$

and wavefunctions that factor into a radial term $R_{n,l}(r)$ and an angular term $Y_l^{m_l}(\theta,\phi)$:

$$\psi_{n,l,m_l} = R_{n,l}(r)Y_l^{m_l}(\theta,\phi)$$
$$= A_{n,l,m_l}e^{-Zr/(na_0)}\left(\frac{Zr}{a_0}\right)^l L_{n-l-1}(r)P_l^{m_l}(\theta)e^{im_l\phi}. \quad (3.33)$$

The radial part depends on quantum numbers n and l, where n is any integer greater than zero, and l is any integer between 0 and $n-1$, inclusive. The angular part of the wavefunction depends on the same value of l, and also on the integer quantum number m_l, which must lie between $-l$ and l, inclusive. The energy depends only on the principal quantum number, n.

3.3 Electric dipole interactions. In the presence of an external electric field, the density of the electron will distort in proportion to its polarizability:

$$\mu_{\text{induced}} = \alpha\mathcal{E}. \quad (3.36)$$

This response to an applied field is what allows the electron to be excited by electromagnetic radiation, to induce a change from one quantum state to another.

3.4 Magnetic dipole interactions.

a. Any moving charge generates a magnetic field. The orbital motion of electrons in the $l \neq 0$ states generates a magnetic field with a magnetic dipole moment given by

$$\vec{\mu}_l = -g_l\mu_B\vec{l}/\hbar. \quad (3.41)$$

b. Electrons in $l = 0$ states do not have orbital magnetic fields, but any electron, whether it's part of an atom or not, generates a spin magnetic field. The motion that creates this spin is independent (to a good approximation) of any other motion of the electron, such as the motion that accounts for which orbital the electron occupies, as determined by the quantum numbers n, l, and m_l. Therefore, we usually assume that we can cleanly factor the spin-spatial wavefunction into two parts:

$$\Psi = \psi_{n,l,m_l}(r,\theta,\phi)\,\chi_{m_s}(\omega). \quad (3.48)$$

The spin also results in a magnetic moment,

$$\vec{\mu}_s = -g_s\mu_B\vec{s}/\hbar. \quad (3.46)$$

We can measure the effects of the orbital and spin motions, because if the electron possesses a net magnetic moment $\vec{\mu}$, then the energy of the electron will shift in an external magnetic field by an amount

$$E_{\text{mag}} = -\langle\vec{\mu}\cdot\vec{B}\rangle. \quad (3.44)$$

KEY TERMS

- **Laguerre polynomials** are one factor in the radial part of the one-electron atom electronic wavefunctions; the other factor is the exponential envelope $e^{-Zr/(na_0)}$.
- **Legendre polynomials** give the θ-dependence of the **spherical harmonics**, which turn out to represent the angular part of the electronic wavefunction in a one-electron atom.

- The **radial probability density** is equal to $R(r)r^2$ and gives the probability density of the electron at any given distance r from the nucleus, regardless of direction.
- **Tunneling** is the property of quantum mechanical particles that allows them to penetrate regions of space where the potential energy is greater than the particle's total energy, a result that would be impossible for a classical particle. The probability of tunneling through such a barrier decreases as we increase the mass of the particle, the distance through the barrier, or the height of the barrier.
- **Polarizability** measures how easily the distribution of the electrons around an atom or molecule can be deformed by an applied electric field.
- **Absorption** and **emission** transitions in spectroscopy are quantum state changes that take in or release photons, respectively. An absorption corresponds to an increase in the energy of the atom or molecule; emission corresponds to a decrease.
- The **principal quantum number** n determines the energy of an electron in the one-electron atom and the approximate size of the electron distribution. All the quantum states with a particular value of n comprise a **shell.**
- The **azimuthal quantum number** l determines the orbital angular momentum of the electron. All the quantum states with the same values of n and l make up a **subshell.**
- The **orbital magnetic quantum number** m_l gives the orientation of the electron's angular motion. The values of n, l, and m_l specify an **orbital.**
- The **spin magnetic quantum number** m_s gives the orientation of the electron's intrinsic magnetic field. For a single electron, the only allowed values of m_s are $\pm 1/2$.

OBJECTIVES REVIEW

1. *Write the electronic wavefunction for any quantum state of any one-electron atom.*
   Write out the normalized wavefunction for the $3p$, $m_l = -1$ orbital of Li^{2+}.
2. *Use the wavefunction to evaluate properties of the electron in that quantum state.*
   Write the integral for finding the average distance from the nucleus of the $3p$, $m_l = -1$ electron in Li^{2+}. Evaluate the integral if you can.
3. *Identify properties of various orbital wavefunctions based on the quantum numbers.*
   How many angular and radial nodes does the $3p$, $m_l = -1$ orbital wavefunction have?

PROBLEMS

Discussion Problems

3.1 For the particle in a three-dimensional box, the three kinetic energy contributions from x and y and z were completely independent of each another. In the kinetic energy operator for the one-electron atom as given in Eq. 3.4, the radial kinetic energy term (which has the derivatives with respect to r) does not depend on angle. Why then does the kinetic energy operator for motion along θ depend on r (varying as $1/r^2$), and why does the operator for motion along ϕ depend on both θ and r, having a factor $1/(r^2 \sin^2\theta)$?

3.2 An accelerating charged particle 1 loses energy because it releases electromagnetic radiation that can do work on some other charged particle 2, but what about a stationary charge? The Coulomb force of a *stationary* particle 1 with charge q_1 acts on another particle with charge q_2 according to $F = -q_1 q_2/(4\pi\varepsilon_0 r^2)$. The force on q_2 causes particle 2 to accelerate, so work is done. Does the stationary particle 1 lose energy?

3.3 Based on the correspondence principle, would you expect the electron to behave more like a classical particle in the $2s$ or in the $5s$ subshell?

3.4 For the $4d$ subshell of an atom,

a. what is the value of l?
b. what are all the possible values of m_l?
c. what are all the possible values of m_s?
d. how many orbitals are there?

3.5 For the $6p$ subshell,

a. what are the values of n and l, and what are the possible values of m_l?
b. how many radial nodes does the wavefunction have?
c. how many angular nodes does the wavefunction have?
d. how many orbitals are there in the subshell?
e. what is the probability density at $r = 0$?

3.6 The kinetic energy operator for the one-electron atom (Eq. 3.2) is taken directly from the operator used for the particle in a three-dimensional box (Eq. 2.36). All of a sudden, angular momentum appears as we solve the atomic Schrödinger equation, whereas we didn't use angular momentum at all for the particle-in-a-box of one or three dimensions. Do *any* wavefunctions for the particle in a box have any angular momentum?

3.7 The ϕ-dependent wavefunctions $e^{im_l\phi}$ appear to offer something quite new when compared to the particle-in-a-box wavefunctions of Eq. 2.39. For the first time, we see a pair of wavefunctions with exactly the same probability density but counted as separate because they move in opposite directions, that is, one corresponding to clockwise motion and the other to counterclockwise motion. Why, then, are the particle-in-a-box wavefunctions non-degenerate, when classically the particle could move in either the $+x$ or $-x$ direction?

3.8 The hyperfine and spin-orbit interactions are both magnetic moment interactions involving the electron spin, but the hyperfine energy is typically less than 10^{-4} times the spin-orbit energy. Is this consistent with the correspondence principle?

Angular Momentum and the One-Electron Atom Hamiltonian

3.9 If the electron actually did travel in a fixed, circular orbit in the xy-plane about the nucleus, as described by the Bohr model, it could still be described by a quantum mechanical Hamiltonian. Write that Hamiltonian. THINKING AHEAD ▶ [What coordinate or coordinates best parameterizes this motion?]

3.10 Based on the Hamiltonians we've seen for the particle in a three-dimensional box and the one-electron atom, write the Hamiltonian using cylindrical coordinates r, z, ϕ for an electron travelling near a charged wire where the potential energy is $U = -C/r$, where C is a constant. In cylindrical coordinates, z is the same as the Cartesian z coordinate, r is the distance from the z axis, and ϕ has the same definition as for spherical polar coordinates.

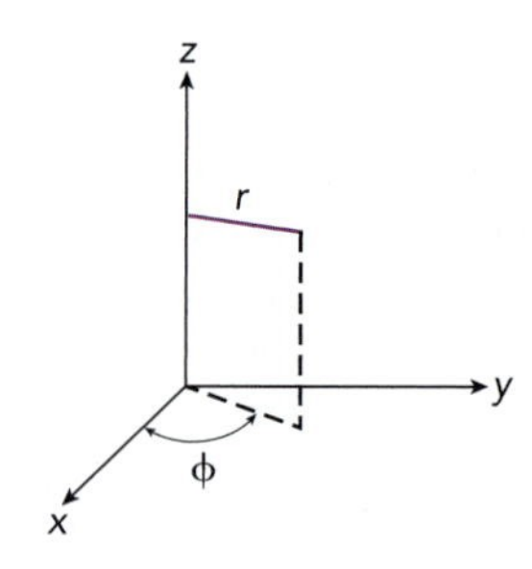

3.11 Determine whether or not the state $\psi_{n,l,m_l} = \psi_{2,1,1}$ is an eigenstate of the operator ∇^2, and if so, give the eigenvalue.

3.12 The kinetic energy term of our one-electron atom Hamiltonian is resolved into an angular term, $\hat{L}^2/(2mr^2)$, and a radial term in Eq. 3.4. The classical kinetic energy in Cartesian coordinates can be written

$$K = \frac{mv^2}{2} = \frac{m}{2}(v_x^2 + v_y^2 + v_z^2) = \frac{m}{2}(\dot{x}^2 + \dot{y}^2 + \dot{z}^2),$$

where $\dot{q} = dq/dt$ for $q = x, y, z$. Rewrite this classical kinetic energy equation in spherical coordinates, using $\dot{r}$, $\dot{\theta}$, and $\dot{\phi}$, and show that in the classical case the angular motion also contributes $L^2/(2mr^2)$ to the kinetic energy. THINKING AHEAD ▶ [How will I recognize the angular motion in my expression for the kinetic energy operator?]

3.13 Section 3.2 uses the uncertainty principle to stipulate that the orbital angular momentum vector $\vec{L}$ cannot be exactly parallel to the z axis. Start with the Bohr model of a point electron in a circular orbit, and assume that the uncertainty principle requires this orbit to be tilted out of the xy plane by some minimum angle β. The vector $\vec{L}$ must be exactly perpendicular to the plane of the orbit, and so it will be at an angle β from the z axis. As the electron orbits the nucleus in this tilted plane, its z value then varies between $z = -r\sin\beta$ and $z = r\sin\beta$, and the projection of the linear momentum p onto the z axis varies from $p_z = -p\sin\beta$ to $p_z = p\sin\beta$. The root mean square values of these parameters—which we will use to estimate their uncertainties—can be shown by integration $\left(\int_0^{2\pi}(r\sin\phi)^2 d\phi / \int_0^{2\pi} d\phi\right)$ to be equal to $\delta z = r/\sqrt{2}$ and $\delta p_z = p/\sqrt{2}$, respectively. With the angle β being the minimum possible angle of tilt, the system will have the maximum possible value of $L_z \equiv l\hbar$. Use the uncertainty principle to estimate the value of L in terms of l. The binomial theorem should simplify the result.

3.14 It is traditional to represent electron spins with arrows that point straight up and down. However, we've learned that the direction of the angular momentum vector cannot be known perfectly. The electron spin is the simplest quantum case for non-zero angular momentum. The spin operators $\hat{s}^2$ and $\hat{s}_z$ have eigenvalues $s(s+1)\hbar^2$ and $m_s\hbar$, respectively (i.e., they are analogous to the orbital angular momentum operators $\hat{L}^2$ and $\hat{L}_z$). Calculate the angle that the electron spin angular momentum vector makes with the z axis.

3.15 The angular momentum vector is given by $\vec{r} \times \vec{p}$, which has a z component equal to $xp_y - yp_x$. Find the eigenvalue of $(x\hat{p}_y - y\hat{p}_x)$ for the $2p_{m_l=1}$ wavefunction, replacing the angular terms with Cartesian coordinates:

$$\sin\theta = \frac{\sqrt{x^2+y^2}}{\sqrt{x^2+y^2+z^2}}$$
$$\cos\phi = \frac{x}{\sqrt{x^2+y^2}}$$
$$\sin\phi = \frac{y}{\sqrt{x^2+y^2}}.$$

3.16 Find the value of $\hat{L}_z\hat{H} - \hat{H}\hat{L}_z$, where the eigenvalue of $\hat{L}_z$ is m_l.

3.17 For the energy levels of the one-electron atom within any given shell n, the energy is the same for all the orbitals but may be distributed differently among the radial and angular coordinates. Write short equations for the minimum and maximum values of the average *radial* kinetic energy of an electron in a one-electron atom. The equations should be in units of E_h and in terms of n and Z. You can take advantage of the correct result from the Bohr model that $K = -E_n$ for the one-electron atom (Problem 1.14). You may also take advantage of the normalization constant for the $l = n-1$ radial wavefunctions in Table 3.2 being

$$A_{n,l=n-1} = \left(\frac{2^{2n}}{n^{2n+2}(2n-1)!}\right)^{1/2}\left(\frac{Z}{a_0}\right)^{3/2}.$$

The One-Electron Atom Wavefunctions

3.18 Show that if $R_{n,l}(r)$ and $Y_l^{m_l}(\theta,\phi)$ are normalized, then the one-electron wavefunction $\psi_{n,l,m_l}(r,\theta,\phi)$ is normalized.

3.19 Verify that R_{21} is an eigenfunction of

$$\hat{H} = \hat{H}_r + \frac{\hbar^2 l(l+1)}{2m_e r^2},$$

and find the corresponding eigenvalue.

3.20 At what distance in Å from the nucleus does the 2s He^+ electron have a radial node?

3.21 At what value of r does the 2p wavefunction in He^+ reach its maximum magnitude?

3.22 Find the values of θ for the two angular nodes in the $l = 2$, $m_l = 0$ function (which corresponds to the d_{z^2} orbital).

3.23 Find the values of n, l, $|m_l|$, and Z for the electronic wavefunction whose real part has zero derivatives at (and only at) $r = 0$, $r = a_0$, $r = \infty$, $\theta = \pi/4$, $\theta = 3\pi/4$, $\phi = 0$, and $\phi = \pi$. THINKING AHEAD ▶ [How many nodes does this imply there are along each coordinate?]

3.24 Identify the quantum numbers n, l, and the absolute value of m_l for the wavefunction of the He^+ ion that has exactly one radial node, at $6a_0$, and with exactly one angular node in θ.

3.25 The radial distribution function $R_{n,l}(r)^2 r^2$ describes the probability density of the electron as a function of r, added over all angles. Similarly, we may write a **polar distribution function,** which gives the probability density of the electron as a function of the polar angle θ, summing over all values of r and ϕ. Sketch a polar graph of the polar distribution functions for the 3s, 3p_0, and 3d_0 orbitals. Include the proper normalization.

3.26 In Fig. 3.6, the angular wavefunction $Y_3^3(\theta,\phi)$ is graphed as a function of θ and appears to have only one node, at $\theta = 0$. Sketch the graph of the *real* part of $Y_3^3(\theta,\phi)$ as a function of ϕ after setting $\theta = \pi/2$. What is the *total* number of angular nodes for the real part of this function?

3.27 One of the valid one-electron wavefunctions is

$$\psi_{n,l,m_l} = A_{n,l,m_l}(Zr/a_0)^4 e^{-Zr/(5a_0)}\sin^3\theta\cos\theta e^{3i\phi}.$$

What are its quantum numbers?

3.28 Write the wavefunction for any 5g atomic orbital in terms of r, θ, and ϕ, and identify the n, l, and m_l quantum numbers. Don't worry about the normalization constant.

3.29

a. Calculate the distance from the nucleus in Å that begins the tunneling region for the electron in 2p $m_l = 1$ He^+. THINKING AHEAD ▶ [What is the condition for tunneling, and how can we relate one of the parameters in that condition to the distance?]

b. Call the answer to part (a) r_t, and write an integral expression (including correct limits, normalization constant, and volume element) that gives the total mass of the electron in the tunneling region.

3.30 Identify by the quantum numbers n,l,m_l,m_s any atomic quantum state with two angular nodes and one radial node.

3.31 Find the values of r or θ or ϕ for all the nodes in the real part of the $4d_{-1}$ wavefunction of the He^+ ion.

3.32 Write the explicit expression in terms of r, θ, and ϕ for the one-electron atomic wavefunction for the $3p$ state $n = 3, l = 1, m_l = -1$.

3.33 Write the explicit, normalized expression in terms of r, θ, and ϕ for any one-electron atomic wavefunction with exactly one radial node and exactly one angular node.

3.34 Verify the normalization constant for the $2p_z$ one-electron atom wavefunction $\psi_{2,1,0}$.

3.35 Write the complete wavefunction for the $3p_{m_l=0}$ state of He^+, and identify the values of the coordinates for all the nodes.

3.36 Find the values of θ and ϕ that correspond to the angular nodes in the real part of the $4f_{m_l=1}$ wavefunction.

3.37 For the $7d$ state of Li^{2+} with $m_l = -1$, calculate the following in SI units:

a. the energy
b. the value of the orbital angular momentum
c. the projection of the orbital angular momentum onto the z axis

3.38 Find the angle between the xy plane and the angular momentum vector $\vec{L}$ of an electron in the $4f$ $m_l = 2$ orbital.

3.39 Find an equation for the r value of the maximum radial probability density in any orbital where $l = n - 1$, as a function of Z and n.

3.40 Give the m_l value *or values* for the orbitals in the $4f$ subshell with the following characteristics:

a. having ϕ-component equal to $e^{-i\phi}$
b. having a pure real wavefunction
c. having a complex conjugate equal to the $m_l = 2$ wavefunction
d. having the greatest density in the xy plane
e. having the greatest number of nodes along θ

3.41 For the $2p$ subshell, the $m_l = 0$ orbital is identical to the $2p_z$ orbital. However, the $2p_x$ and $2p_y$ orbitals are composed by mixing together the $m_l = 1$ and $m_l = -1$ orbitals. For the following wavefunctions ψ_a and ψ_b,

$$\psi_a = \psi_{2,1,1} + \psi_{2,1,-1}$$
$$\psi_b = i(\psi_{2,1,1} - \psi_{2,1,-1})$$

show that the starting wavefunctions $\psi_{2,1,1}$ and $\psi_{2,1,-1}$ are orthogonal, find the normalization constants for ψ_a and ψ_b, and determine which corresponds to the $2p_x$ and which to the $2p_y$ Cartesian orbitals. Do not evaluate any complicated integrals for the normalization.

3.42 The traditional labels for the spatial d orbitals are d_{z^2}, d_{xy}, d_{xz}, d_{yz}, and $d_{x^2-y^2}$. As in the case of the labels p_x, p_y, and p_z, the subscripts given to these d orbitals are such that the orbital has the same nodes and phases (signs) as the function in the subscript. Find a normalized expression for each of these in terms of the spherical harmonic functions $Y_l^{m_l}(\theta,\phi)$ in Table 3.1.

3.43 Prove that any two atomic orbital wavefunctions with different m_l are orthogonal.

3.44 For a one-dimensional system, the Schrödinger equation can be rewritten to solve for the second derivative of the wavefunction $\psi(x)$:

$$\frac{\partial^2\psi(x)}{\partial x^2} = -\frac{2m}{\hbar^2}(E - U)\psi(x).$$

In a tunneling region, the wavefunction has the form

$$\psi(x) = A\,e^{-ax}.$$

a. Find an equation for a in this wavefunction in terms of m, E, and U.
b. How can you tell that this wavefunction is invalid everywhere *except* in the tunneling region?

3.45 Identify the locations of all the radial and angular nodes in the real part of the wavefunction $\psi_{3,1,-1}$ for He^+.

Integrals to Calculate Properties of the Orbitals

3.46 Write the explicit integral expression in terms of r, θ, and ϕ, including limits of integration and normalization, necessary to evaluate the probability of finding the electron in the $3d$ state $n = 3, l = 2, m_l = -2$ between the angles $\theta = \pi/2$ and $\theta = \pi$.

3.47 Write the integral, including limits, that is necessary to find the probability density of the $3d$ He^+ electron within a distance $r = 2.00\,\text{A}$ of the nucleus. Evaluate the integral.

3.48 The one-electron atomic wavefunctions given in Tables 3.1 and 3.2 are all mutually orthogonal. Write the explicit integral that you would need to demonstrate that the $2s$ and $3d(m_l = 2)$ wavefunctions are orthogonal to each other.

3.49 Write the explicit integral expression in terms of r, θ, and ϕ, including limits of integration and normalization, necessary to evaluate the average value of L_x for the $3p$ wavefunction with $m_l = -1$, where

$$\hat{L}_x = i\hbar\left(\sin\phi\frac{\partial}{\partial\theta} + \cot\theta\,\cos\phi\frac{\partial}{\partial\theta}\right).$$

3.50 Write the integral, including limits, normalization constant, and volume element, necessary to find the average kinetic energy of the electron in the $3p_{m_l=1}$ H atom.

3.51 Motion of the electron can be expressed as a combination of radial and angular motion, and the electron's total kinetic energy can be separated into radial and angular contributions. Write an integral equation for finding the average *radial* kinetic energy of the electron, $\langle K_r \rangle$, in the $2p_1$ state of He^+.

3.52 The hyperfine effect is a very small splitting in atomic or molecular energy levels proportional to

$$\left\langle \frac{\mu_s \mu_I (3\cos^2\theta - 1)}{r^3} \right\rangle,$$

where μ_s and μ_I are constants (the electronic and nuclear spin magnetic moments). Write (but do not evaluate) the complete integral you could use to calculate this average for a Li^{+2} ion in the state $n, l, m_l = 2,1,1$.

3.53 Find the probability density of the electron at distances from the nucleus less than the Bohr radius, $\mathcal{P}(r < a_0)$, for the H atom in the 1*s* ground state.

3.54 Find the value of r_0 such that half of the charge on the electron in the H 1*s* orbital is found at $r \leq r_0$ and half is at $r > r_0$.

3.55 Use the average value theorem to calculate the average value of r for the normalized wavefunction

$$R(r) = \frac{16\sqrt{2}}{27\sqrt{5a_0^3}} \left(\frac{12r}{a_0}\right)^2 e^{-4r/a_0}.$$

3.56 Identify the n, l, and m_l quantum numbers of the electron and the atomic number Z of the one-electron atom being analyzed by the following integral, and identify the property being evaluated:

$$-\frac{128\hbar^2}{3\pi a_0^5} \int_0^\infty \int_0^\pi \int_0^{2\pi} r\left(1 - \frac{r}{a_0}\right) e^{-2r/a_0}$$
$$\sin\theta\, e^{-i\phi} \left\{ \nabla^2 \left[r\left(1 - \frac{r}{a_0}\right) e^{-2r/a_0} \right.\right.$$
$$\left.\left. \sin\theta\, e^{i\phi} \right] \right\} r^2 dr \sin\theta\, d\theta\, d\phi.$$

3.57 Write the integral that you would use to find the probability density of an electron in the $3p_z$ orbital of He^+ in the region where x, y, and z are all positive. Then reason or calculate the solution to the integral.

3.58 To convert from the Cartesian coordinate x to spherical coordinates, we use the relation $x = r \sin\theta \cos\phi$. Write the expression we would need to solve in order to find the root mean square value of x, equal to $\sqrt{\langle x^2 \rangle}$, of the electron in the $4f_{m_l=-3}$ state of atomic hydrogen. Factor your final answer into single-integral terms (meaning, one integral for each coordinate). Provide all the information needed for a computer to calculate the value.

3.59 One can show that the speed of the electron in Bohr's model of the atom, v_{Bohr}, is not (in general) equal to the expectation value $\langle v \rangle$ of the electron in our quantum wavefunctions, but instead

$$v_{\text{Bohr}} = \langle v^2 \rangle^{1/2}.$$

Write a similar equation showing how the radius of the electron's orbit in the Bohr model, r_{Bohr}, is related to one of the averaged properties of the one-electron wavefunction. **THINKING AHEAD ▶** [What does the Bohr model predict correctly, and how could that be related to the distance r?]

3.60 Find $\langle r^2 \rangle^{1/2}$ for the 2*s* state of H.

3.61 The dipole moment of a system with two charges $+e$ and $-e$ separated by a distance z is ez. Similarly, the dipole moment operator is $ez = er\cos\theta$. For an electric dipole transition, $\psi_1 \rightarrow \psi_2$, due to radiation polarized along the z axis, the expression $\left| \int \psi_2^* \, r\cos\theta \, \psi_1 \, d\tau \right|^2$ gives the transition intensity, proportional to the amount of light absorbed or emitted during the transition. Using this integral, show that the transitions $(n, l, m_l) = (2,1,0) \rightarrow (2,1,1)$ and $(2,1,0) \rightarrow (3,1,0)$ are forbidden. We only need to show that the integral along any one of the coordinates is zero. Don't worry about normalizing the wavefunctions.
THINKING AHEAD ▶ [The coordinates of these wavefunctions are separable. Could that simplify the work needed here, the way it often simplifies the integrals for calculating probabilities or expectation values?]

3.62 Write the expression in terms of r, θ, and ϕ necessary to determine whether or not the $1s \rightarrow 2p_{-1}$ atomic transition is forbidden, given that the **transition moment operator** μ_t is $er\cos\theta\cos\phi$. Include normalization constants but do not evaluate the expression.

3.63 Show that the transitions $(2,1,0) \rightarrow (1,0,0)$ and $(2,1,0) \rightarrow (2,0,0)$ are allowed, using the expression for the transition intensity given in the previous problem. Which is more intense? To compare transition intensities, the wavefunctions must be normalized.

Magnetic Field Interactions

3.64 For the $4d$ group of wavefunctions in a one-electron atom,

a. list the possible values of m_l.
b. list the possible values of j.
c. write the number of radial nodes.

3.65 What is the total number of states, including spin-orbit and hyperfine splittings, within the one-electron subshell $3d$ when the nuclear spin $I = 1$? Do not label the states; just find the total.

3.66 The spin-orbit constant A_{SO} decreases as the size of the electron orbital increases. Arrange the following one-electron systems in order of increasing maximum spin-orbit energy $A_{SO} l \cdot s$: (a) the $2p_z$ Be^{+3} atom; (b) the $2p_z$ H atom; (c) the $2s$ H atom; (d) the $8p_z$ H atom.

Many-Electron Atoms

LEARNING OBJECTIVES

After this chapter, you will be able to do the following:

❶ Write the Hamiltonian for any atom.

❷ Find a zero-order wavefunction (the electron configuration) and zero-order energy of any atom.

❸ Use the Hartree-Fock orbital energies to estimate the effective atomic number of the orbital.

❹ Determine the permutation symmetry of a many-particle function.

❺ Determine the term symbols and energy ordering of the term states resulting from any electron configuration.

GOAL *Why Are We Here?*

The principal goal of this chapter is to introduce the Schrödinger equation for the many-electron atom, together with some new physics related to spin, and show how we can solve that Schrödinger equation by approximations to the energies and wavefunctions. We need the energies in order to predict the energies of chemical bonds; we need the wavefunctions because the distribution of the electrons will determine the geometry of the molecules we build.

CONTEXT *Where Are We Now?*

Chapter 3 describes—in considerable detail—a two-body problem: finding the energies and distributions of one electron bound to an atomic nucleus. Chemistry is largely determined by where electrons go, and if we're to get closer to real chemistry (where most electrons come in pairs), we'll need more than one electron. In this chapter, we add the other electrons. That may sound simple enough, but the *many*-body problem cannot be solved exactly. Much of our effort in this chapter must be directed toward finding approximations that help us approach the right answer, while maintaining some *intuitive* understanding of the structure of the atom. We can also expect new intricacies, both in the electric field interactions as we add more charges and in the magnetic field interactions because each new electron introduces at least one new magnetic moment. But once we have built a comprehensive model of the many-electron atom, we'll have the entire periodic table at our disposal as we move on to assembling molecules.

SUPPORTING TEXT *How Did We Get Here?*

The main *qualitative* concept we draw on for this chapter is that the electron in an atom is sufficiently wavelike that quantum mechanics restricts its energy and other parameters to specific values. For a more detailed background, we will draw on the following equations and sections of text to support the ideas developed in this chapter:

- Section 2.1 explained how we can use operators, such as a combination of multiplication and differentiation, to extract information from a function. In particular, the Hamiltonian $\hat{H}$ operates on a wavefunction ψ to return the energy E times the wavefunction. We call this the Schrödinger equation (Eq. 2.1):
$$\hat{H}\psi = E\psi.$$
- However, the Schrödinger equation works as written only if the wavefunction is an eigenstate of the Hamiltonian. We will find in this chapter that one useful approximation uses a time-independent wavefunction ψ that is *not* an eigenstate of $\hat{H}$, and in that case we can use the average value theorem to find an expectation value of the energy (Eq. 2.10):
$$\langle A \rangle = \int_{\substack{\text{all}\\\text{space}}} \psi^*(\hat{A}\psi)d\tau.$$
- Section 2.2 introduced the particle in a one-dimensional box, a fundamental problem in quantum mechanics in which the Schrödinger equation can be solved analytically to give the wavefunctions (Eq. 2.33)
$$\psi_n(x) = \sqrt{\frac{2}{a}}\sin\left(\frac{n\pi x}{a}\right)$$
and the energies (Eq. 2.31)
$$E_n = \frac{n^2\pi^2\hbar^2}{2ma^2}.$$
- The one-dimensional box depends only on a single Cartesian coordinate. The atom, however, has a potential energy function that can be more naturally described in spherical coordinates, because the Coulomb attraction between the electron and the nucleus depends only on r, the distance between them. As a result, the solutions to the Schrödinger equation are also easier to express in spherical coordinates. To convert between the two coordinate systems, we use Eqs. A.6:
$$x = r\sin\theta\cos\phi \qquad r = (x^2 + y^2 + z^2)^{1/2}$$
$$y = r\sin\theta\sin\phi \qquad \theta = \arccos\left(\frac{z}{r}\right)$$
$$z = r\cos\theta \qquad \phi = \arctan\left(\frac{y}{x}\right).$$
These relationships also allow us to convert the Cartesian volume element $dx\,dy\,dz$ to the volume element in spherical coordinates, $r^2dr\sin\theta\, d\theta d\phi$, which is an essential component of any integrals we may need to carry out.
- The Schrödinger equation for the one-electron atom can also be solved analytically, yielding the energies (Eq. 3.30)
$$E_n = -\frac{Z^2 m_e e^4}{2(4\pi\varepsilon_0)^2 n^2\hbar^2} \equiv -\frac{Z^2}{2n^2}E_h.$$
We also obtain the wavefunctions as products of a radial wavefunction $R_{n,l}(r)$ (given in Table 3.2) and an angular wavefunction $Y_l^{m_l}(\theta,\phi)$ (given in Table 3.1): $\psi_{n,l,m_l} = R_{n,l}(r)Y_l^{m_l}(\theta,\phi)$.

- One tool we will turn to for a critical step is the Taylor expansion (Eq. A.29),

$$f(x) = \sum_{n=0}^{\infty} \frac{(x - x_0)^n}{n!} \left(\frac{d^n f(x)}{dx^n} \right) \Bigg|_{x_0},$$

which lets us rewrite any algebraic expression as a power series. The trick with power series is to set them up so that each higher-order term is smaller than the last term, so that you can ignore all but the first few terms. In this chapter, we use a Taylor series expansion of the Schrödinger equation itself to find approximate solutions to the energies and wavefunctions.

4.1 Many-Electron Spatial Wavefunctions

Neglecting some relatively small effects,[1] we can solve the Schrödinger equation of the hydrogen atom to get algebraic expressions for the energies and wavefunctions. Once we add a second electron to an atom, that becomes impossible. Although many numerical methods work well, the exact solution cannot be written in general form. We will go as far as we can in the space of one chapter, which is pretty far.

Electron Configurations

CHECKPOINT As in Chapter 3, we are color-coding the math related to the Schrödinger equation to help distinguish terms in the Hamiltonian operator (blue) from contributions to the scalar energy (red), and to help illustrate the transformation of one to the other.

First, let's write the Hamiltonian for a two-electron atom as an extension of the Hamiltonian we used for the one-electron atom:

$$\hat{H} = -\frac{\hbar^2}{2m_e}\nabla(1)^2 - \frac{\hbar^2}{2m_e}\nabla(2)^2 - \frac{Ze^2}{4\pi\varepsilon_0 r_1} - \frac{Ze^2}{4\pi\varepsilon_0 r_2} + \frac{e^2}{4\pi\varepsilon_0 r_{12}}, \quad (4.1)$$

where

$$\nabla(i)^2 = \left(\frac{\partial^2}{\partial x_i^2} + \frac{\partial^2}{\partial y_i^2} + \frac{\partial^2}{\partial z_i^2} \right), \quad r_i = (x_i^2 + y_i^2 + z_i^2)^{1/2};$$

x_i, y_i, and z_i are the Cartesian coordinates for electron i; and r_{12} is the distance between electrons 1 and 2. These coordinates are defined in Fig. 4.1. The potential energy terms are now written relative to the energy of the *completely* ionized atom; in other words, the potential energy is zero when r_1, r_2, and r_{12} are all infinite. For the more general case of the atom with N electrons, we sum up the kinetic energies of all the electrons and their potential energies of Coulomb attraction to the nucleus. Then, for *each pair* of electrons we add the repulsion energy:

$$\hat{H} = \sum_{i=1}^{N} \left[-\frac{\hbar^2}{2m_e}\nabla(i)^2 - \frac{Ze^2}{4\pi\varepsilon_0 r_i} + \sum_{j=1}^{i-1} \frac{e^2}{4\pi\varepsilon_0 r_{ij}} \right]. \quad (4.2)$$

What's new and disturbing in the many-electron atom Hamiltonians is the electron–electron repulsion term, $e^2/(4\pi\varepsilon_0 r_{12})$, which prevents the electron 1 and electron 2 terms of the Hamiltonian in Eq. 4.1 from being separable.

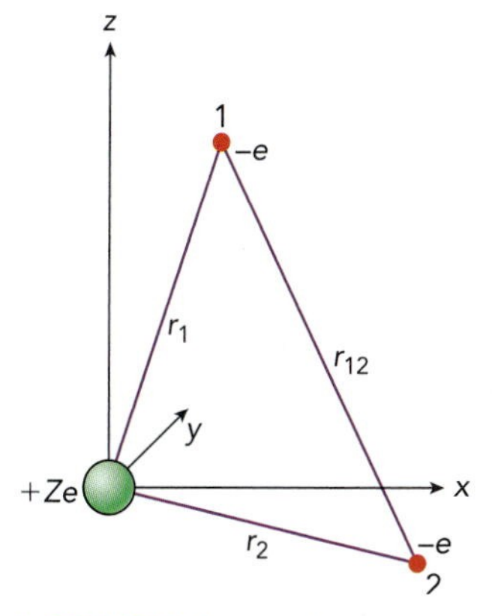

▲ **FIGURE 4.1 Coordinate definitions for the two-electron atom.**

[1] Magnetic interactions, finite mass of the nucleus, special relativity, to name a few.

EXAMPLE 4.1 Many-Electron Atom Hamiltonians

CONTEXT The element beryllium doubles as one of the most useful and frightening metals available to us. It is the lightest of all the structurally useful metals (neglecting lithium, which is soft and highly reactive), and also among the most hazardous. It has a strength comparable to steel, but at a fraction of the mass. Beryllium-based products were widely produced in the United States and Europe in the decades prior to World War II, but then it became gradually clear that the beryllium in factory dust was sickening and killing workers, sometimes years after exposure. Chronic beryllium disease resembles pneumonia, and can be contracted by repeated exposure to 0.5 mg/m^3 or more of beryllium in the air. The mechanism for its biological activity is only partly understood, but it apparently combines with a protein and then deposits itself primarily in lung tissue, triggering an over-reaction by the body's immune system—a response called *hypersensitivity* that leads to destructive inflammation of the lungs. The high toxicity has led to severe restrictions on its use in manufacturing, but its properties are still tempting. By accurately simulating the electronic properties of beryllium on a computer, we may identify the most valuable applications of this element at minimal risk.

PROBLEM Write the Hamiltonian you would need to calculate the electronic wavefunctions of the neutral beryllium (Be) atom. Leave the Laplacian for each electron in the form ∇^2.

SOLUTION With four electrons, there would be one kinetic energy term for each, one nuclear attraction term for each, and one repulsion term for each possible pair of electrons. We use $Z=4$ for beryllium:

$$\hat{H} = -\frac{\hbar^2}{2m_e}\left(\nabla^2(1) + \nabla^2(2) + \nabla^2(3) + \nabla^2(4)\right) - \frac{4e^2}{4\pi\varepsilon_0}\left(\frac{1}{r_1} + \frac{1}{r_2} + \frac{1}{r_3} + \frac{1}{r_4}\right)$$
$$+ \frac{e^2}{4\pi\varepsilon_0}\left(\frac{1}{r_{12}} + \frac{1}{r_{23}} + \frac{1}{r_{34}} + \frac{1}{r_{13}} + \frac{1}{r_{24}} + \frac{1}{r_{14}}\right).$$

Nevertheless, we will often neglect the coupling between the electrons and describe the electronic state of the many-electron atom by its **electron configuration,** which describes the many-electron wavefunction by associating each electron with a particular subshell of the one-electron atom. In other words, the one-electron or **hydrogenic** orbitals will become a basis set for describing the states of many-electron atoms. Once an electron is assigned to a particular hydrogenic orbital, such as the 3*s* orbital, that orbital is said to be **occupied.** The electron configuration is written as the product of all the occupied subshells, each with a superscript indicating the number of electrons assigned to it. For example, the ground state electron configuration of the H atom is $1s^1$, indicating that the single electron has a 1*s* orbital wavefunction. The ground state He atom places two electrons in the 1*s* orbital, and its electron configuration is written $1s^2$. The presence of additional electrons changes the radial wavefunctions of those orbitals significantly, so the wavefunctions of the two 1*s* electrons in ground state He are not identical to the 1*s* wavefunction of He^+. However, to a good approximation the angular wavefunction of each orbital is still given by one of the spherical harmonics $Y_l^{m_l}(\theta,\phi)$.

The electron configuration is essentially a first-guess, inaccurate spatial wavefunction for the many-electron atom. It is generated by the product of a number of one-electron wavefunctions, the atomic orbitals. We can write this wavefunction explicitly, but we must remember that the coordinates of each electron are distinct. Knowing that each spatial wavefunction ψ_{n,l,m_l} depends on the r, θ, and ϕ coordinates for that electron, let's now just write $\psi_{n,l,m_l}(1)$ to indicate the spatial wavefunction for electron 1, a function of the coordinates r_1, θ_1, and ϕ_1. For example, the spatial part of the wavefunction for the Li atom configuration with two electrons in the $1s$ orbital and one in the $2s$ could be approximated:

$$\begin{aligned} 1s^2 2s^1 &\approx \psi_{1,0,0}(1)\,\psi_{1,0,0}(2)\,\psi_{2,0,0}(3) \\ &= \left(\frac{1}{\sqrt{\pi}}\right)\left(\frac{3}{a_0}\right)^{3/2} e^{-3r_1/a_0}\left(\frac{1}{\sqrt{\pi}}\right)\left(\frac{3}{a_0}\right)^{3/2} e^{-3r_2/a_0} \qquad \text{Tables 3.1, 3.2} \\ &\times \left(\frac{1}{\sqrt{8\pi}}\right)\left(\frac{3}{a_0}\right)^{3/2}\left(1 - \frac{3r_3}{2a_0}\right)e^{-3r_3/(2a_0)}. \end{aligned} \tag{4.3}$$

For now, we neglect the effect of one electron on the orbital of another, and simply stack the one-electron wavefunctions on top of one another, as illustrated in Fig. 4.2. Notice that the electron configuration does not contain the spin wavefunctions α and β and therefore describes only the spatial part of the many-electron wavefunction.

To arrive at the electron configuration of the atom's ground state, we employ the ***Aufbau* principle.** (Literally, *Aufbau* means "building up," although it also translates appropriately as "structure.") We count the number of electrons in the atom and fill the available orbitals with them, starting from the lowest-energy orbital and working up until all the electrons are accounted for. We limit the number of electrons in each shell to the degeneracy of the shell, $2(2l + 1)$. As each shell is completely filled, it contributes to the set of **core electrons,** which are more tightly bound to the nucleus and less likely to budge in chemical processes than the outermost **valence electrons** in unfilled shells.

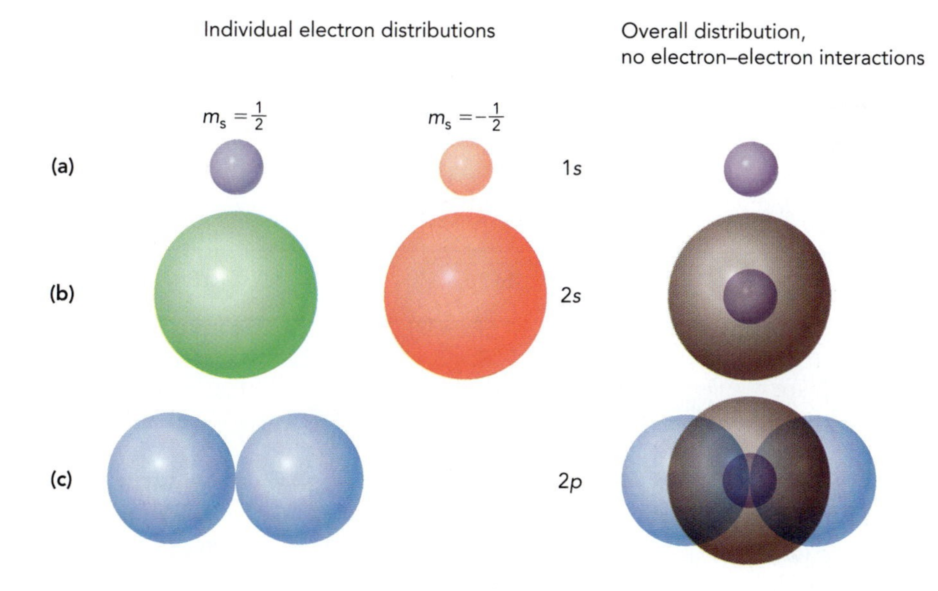

◀ FIGURE 4.2 **A zero-order representation of the many-electron atom.** This picture is obtained by putting up to two electrons in each orbital. In the case of ground state boron, for example, we combine **(a)** two $1s$ electrons (one with each value of m_s), **(b)** two $2s$ electrons, and **(c)** one electron in a $2p$ orbital. We neglect any interaction between the orbitals for now.

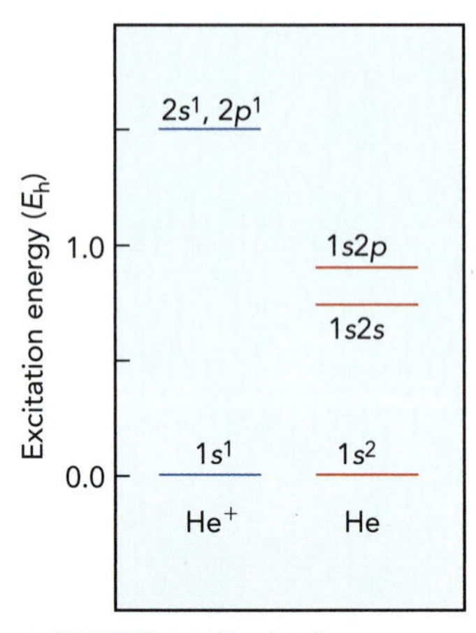

▲ **FIGURE 4.3 Excitation energies of the lowest quantum states of He^+ and He.** (Note that this graph gives excited state energies relative to the ground states—*not* relative to the ionization limit, which we normally use.) The 2*s* and 2*p* subshells have the same energy in a one-electron atom such as He^+, but the energies separate in a multi-electron atom such as He.

Electron–Electron Repulsion and Shielding

In the one-electron atom, all the states with the same value of n had the same energy. That is no longer the case. With every additional electron in the atom, the nuclear charge experienced by the other electrons is reduced, an effect called **electron shielding.** This is just another way of looking at the electron–electron repulsion contribution to the Hamiltonian $e^2/(4\pi\varepsilon_0 r_{12})$ in Eq. 4.1. Electrons in low-n orbitals that are tightly bound to the nucleus reduce the average charge of the nucleus experienced by higher energy electrons.

The 2*s* and 2*p* orbitals of the one-electron atom He^+ have the same energy, neglecting relatively small effects such as spin-orbit interaction. However, the degeneracy of the 2*s* and 2*p* subshells is broken once we add a second electron to make neutral He (Fig. 4.3). The 1*s*2*s* excited state of helium is lower in energy than the 1*s*2*p* excited state by more than 3000 cm^{-1}. Keep in mind that the average of the potential energy, $\langle Ze^2/(4\pi\varepsilon_0 r)\rangle$, is the same for all orbitals in the same n level of the one-electron atom, so we cannot claim, say, that the 2*p* electron is simply distributed farther from the nucleus than the 2*s* electron. In fact, the average distances $\langle r \rangle$ are $6a_0/Z$ for the 2*s* and $5a_0/Z$ for the 2*p*, so a 2*p* electron is *closer* on average. Instead, the effect must arise somehow from the relative positions of the excited 2*s* or 2*p* electron and the 1*s* electron.

From the plots of $R_{n,l}(r)^2 r^2$ in Fig. 4.4, we see that as n increases, the probability density at large r increases. Orbitals at larger r are stabilized less by the nucleus, and so the energy increases with n. The difference between the subshells *within* a particular n shell is more complicated, and it isn't obvious yet why the 2*p* should be less stable than the 2*s*. But one clear difference is that as l decreases within a shell, the electron density is increasingly divided by nodes at various values of r, with the region of largest density appearing at high r and the smallest density at low r, next to the nucleus.

The 2*p* orbital in Fig. 4.4b has a single, broad distribution along r, whereas the 2*s* orbital distribution (Fig. 4.4a) is divided into two distinct regions: a little lump near the nucleus and a large lump farther out. Although the smallest peak in

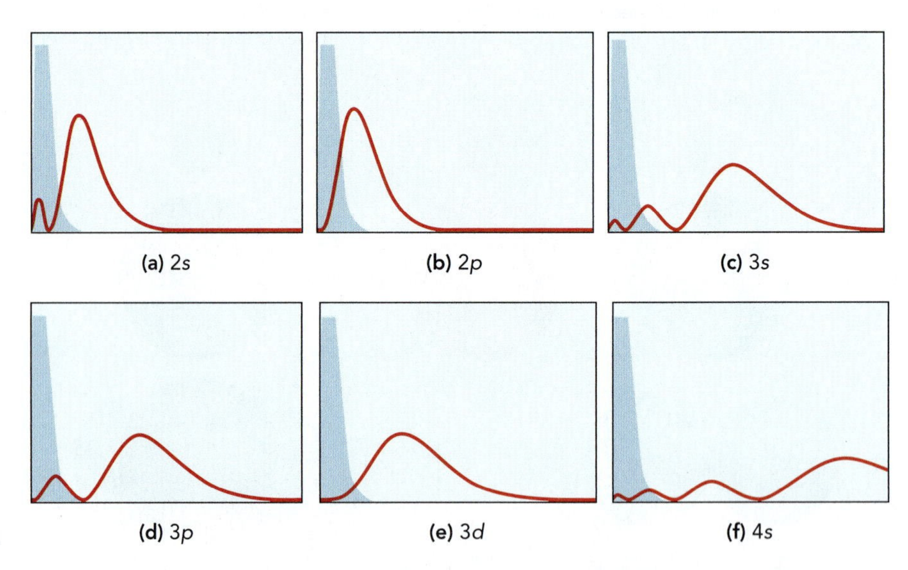

▶ **FIGURE 4.4 Electron shielding.** Shown are the radial probability densities $R_{nl}(r)^2 r^2$ of the 2*s*, 2*p*, 3*s*, 3*p*, 3*d*, and 4*s* orbitals, overlapped with the radial probability density of the 1*s* orbital probability density (shaded area). The shorter the distance between the probability density and the shaded area, the greater the repulsion by the 1*s* orbital and the greater the shielding from the nuclear charge.

the $2s$ distribution falls well within the shaded area of the $1s$ electron, most of the $2s$ electron density lies at larger distances. The overlap between the $1s$ and $2p$ distributions, on the other hand, is much greater, and the peak of the $2p$ distribution is closer to the region occupied by the $1s$ core.

Consequently, the $1s$ and $2p$ electrons are, on average, closer together than the $1s$ and $2s$ electrons. This causes greater repulsion between the $1s$ and $2p$ electrons than between the $1s$ and $2s$, effectively shielding the $2p$ subshell from the nuclear charge.[2] Comparing the separation between the $1s$ core and the peak probabilities of the $n=3$ subshells in Fig. 4.4, the same trend is observable: the average distance to the core electrons diminishes as l increases from $3s$ to $3p$ to $3d$, and the repulsion energy climbs. Consequently, in many-electron atoms, the orbitals for a given value of n increase in energy with l:

$$E_{ns} < E_{np} < E_{nd} < E_{nf}.$$

At values of n greater than 2, the effect of electron shielding is so pronounced that it is comparable to the energy separation between states with different n. Figure 4.4 shows, for example, that the shielding experienced by the $4s$ electrons is actually going to be less than that experienced by $3d$ electrons, because the $3d$ electrons suffer greater repulsion by the core electrons. The energy ordering of the subshells in typical ground state atoms is roughly as follows:

$$1s, 2s, 2p, 3s, 3p, (4s, 3d), 4p, (5s, 4d), 5p, (6s, 4f, 5d), 6p, 7s, (5f, 6d).$$

The groups contained in parentheses are subshells that lie close enough in energy that the ordering varies with the number of electrons, as spin-orbit energies and other effects become significant. Different orbitals of each subshell have the same energy, for our purposes. A more quantitative, but still approximate, illustration of the many-electron atom energy levels appears in Fig. 4.5.

This ordering justifies the common form of the periodic table of the elements. Each period (row) of the table begins with the elements whose highest-energy electrons are in the ns orbital. After filling the $1s$ orbital (H and He), we move on to the $n=2$ shell and start the next period, which contains the eight elements (Li–Ne) with ground state electron configurations that end either in the two $2s$ orbitals or in the six $2p$ orbitals. However, we then find that the third period also contains only eight elements (Na–Ar), because after filling the $3s$ and $3p$ orbitals, the next lowest-energy subshell is not the $3d$ but the $4s$. We start a new shell, so we start a new period. This begins a row of 18 elements (K–Kr), including the first row of the **transition metals,** which are characterized by partial filling of the highly degenerate d subshell. The $5s$ valence electrons inaugurate a row of 18 more elements (Rb–Xe), including a second row of transition metals, now filling the $4d$ orbitals. And the $6s$ and $7s$ orbitals mark the beginnings of rows on the periodic table containing both transition metals and the **rare earths**—elements with

[2]A more common perspective on this effect points out that lower l orbitals penetrate the core electron density more effectively than high l orbitals (e.g., the $2s$ distribution has a small peak near the nucleus that the $2p$ distribution lacks). However, that neglects the much larger $1s-2s$ electron–electron repulsion that also occurs in that region. Contributions to the overall energy evaluated as a function of r would argue instead that it is the other part of the $2s$ distribution—more distant from the nucleus but also from the $1s$ core—that contributes most to the stability of the $2s$ relative to $2p$ (Problem 4.26).

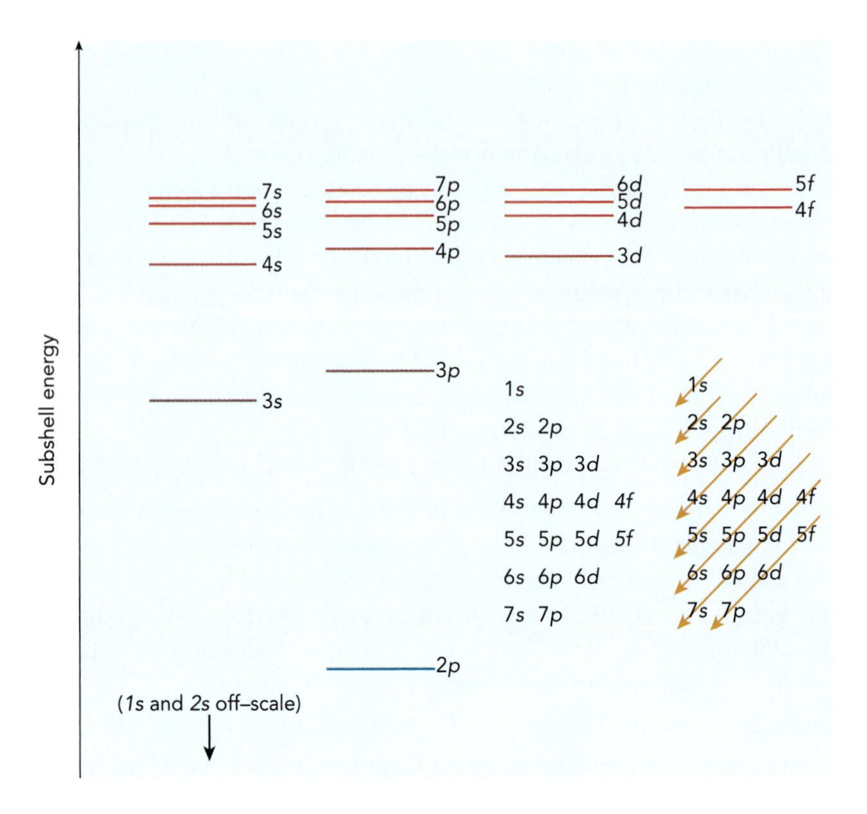

▶ **FIGURE 4.5 Approximate relative energies of the subshells in the many-electron atom, reflecting the shifts due to nuclear shielding.** The 1*s* and 2*s* orbital energies are too low to appear in this plot. Actual relative energies will vary with number and excitation of electrons. The inset shows a common and effective visual mnemonic for the typical energy ordering of the subshells.

partially filled *f* subshells. So, for an 11-electron atom, such as Na or Mg^+, we would construct the electron configuration as follows: first put two electrons in the 1*s* orbital (one for each value of m_s), two in the 2*s* orbital, six in the 2*p* (which has three values of m_l to use up and two values of m_s for each), and we begin filling the 3*s* before we have used up all the electrons. This configuration is written $1s^2 2s^2 2p^6 3s^1$, where the superscripts indicate the number of electrons in each subshell that has more than one electron. An ionized sodium atom has ten electrons, and in its ground state these electrons also obey the *Aufbau* principle, yielding the electron configuration $1s^2 2s^2 2p^6$. To obtain the ground state configurations of atomic ions, one simply adds to or removes from the neutral atom configuration the appropriate number of electrons.

Normally, the electron configuration offers an incomplete description of the electronic state because it does not specify the m_l and m_s values, much less the spin-orbit states that result from these. Nevertheless, it is usually an excellent starting point, and not just for ground states. Excited electron configurations are formed by violating the *Aufbau* principle, skipping an unfilled subshell as we add electrons. The lowest-energy excited electron configuration can often be obtained by placing the highest-energy ground state electron in the lowest-energy *unoccupied* atomic subshell. The lowest excited electron configuration of Na, for example, is obtained by moving the 3*s* electron to the 3*p* subshell, giving a configuration $1s^2 2s^2 2p^6 3p$. On the other hand, nearly degenerate subshells may lead to low-lying excited configurations where an electron is transferred from one occupied subshell to another occupied level. The lowest excited configuration of carbon, for example, is $1s^2 2s\, 2p^3$ rather than $1s^2 2s^2 2p\, 3s$.

TOOLS OF THE TRADE | Photoelectron Spectroscopy

One of the early keys to quantum theory was Albert Einstein's explanation of the **photoelectric effect.** In 1887, Heinrich Hertz discovered that the generation of electricity at a metal surface could be enhanced by ultraviolet light. From the results of subsequent experiments by Philipp Lenard, Einstein concluded that the energy of electromagnetic radiation is carried in units of photons, and that the surface abosorbs energy *one photon at a time*, with each interaction causing an energy change at the surface. If the photon energy surpasses a threshold value IE sufficient to expel an electron, any excess energy in the photon provides the kinetic energy of the ejected electron. Raising the intensity of the light merely provides more photons, increasing the number of interactions but not the energy of each interaction. If $h\nu$ is not high enough to ionize the sample, no electrons will be ejected. In a gas, the IE is the first ionization energy, the energy difference between the ion and the neutral atom or molecule M, IE $= E(M^+) - E(M)$. In a solid IE is called the **work function.**

We continue to take advantage of this technique in order to measure electronic transition energies by using a known photon energy $h\nu$ and then measuring the kinetic energy of the electrons:

$$\Delta E = h\nu - \frac{m_e v^2}{2}.$$

This principle established the foundation for **photoelectron spectroscopy.**

What is photoelectron spectroscopy? In photoelectron spectroscopy, we strike the sample with a burst of ionizing photons, all at energy $h\nu$, and then measure the distribution in kinetic energies of the electrons that reach the detector in order to generate the spectrum of ΔE values.

Why do we use photoelectron spectroscopy? Photoelectron spectroscopy has three advantages over absorption spectroscopy: (*i*) the sensitivity is much higher because we can detect the tiny currents generated by a small number of electrons better than we can detect a tiny decrease in the intensity of radiation passing through a sample; (*ii*) ionizations are not subject to the $\Delta l = \pm 1$ selection rule (Section 3.3), because the ejected electron can carry away the necessary angular momentum; (*iii*) in a many-electron atom or molecule, both the neutral M and the ion M^+ have distinct quantum states, and the values of ΔE we measure tell us about the energy levels of the neutral and the ion in the same experiment.

Photoelectron spectroscopy is a zero-background technique, meaning that the detector sees nothing (except a weak noise signal) until there is an interaction between the radiation and the sample. In addition, our ability to measure charged particles such as electrons or ions is usually much better than our ability to detect photons. For one thing, we can use electric fields to accelerate charged particles before they reach the detector, so that they hit the detector with considerably greater energy than they had originally, and with greater energy than a typical photon in the experiment would have.

How does it work? In photoelectron spectroscopy, a pulse of laser light at a known photon energy ionizes the sample in an ultrahigh vacuum chamber. The electrons are directed by magnetic fields down a drift tube toward a detector to separate the different velocities. Just before the detector, the electrons are rapidly accelerated and focused to amplify the signal. The raw measurements consist of electron signals—electrical currents at the detector—tabulated as a function of the time after the laser pulse. From the drift times Δt and the known length d of the drift tube, we can calculate the electron speeds $v = d/\Delta t$ and convert the speeds into electron kinetic energies $mv^2/2$.

Gas-phase photoelectron spectroscopy is used chiefly for research into atomic and molecular energy levels. In addition, x-ray photoelectron spectroscopy is a common application of the technique, used on solid samples as a means of rapidly characterizing the elemental and molecular composition of materials or coatings.

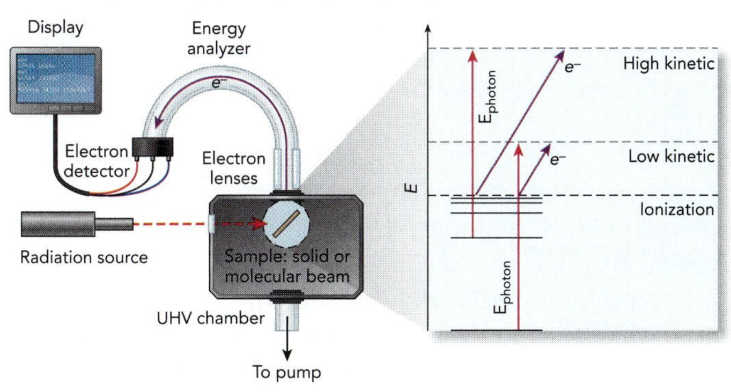

◀ **Photoelectron spectroscopy.**

4.2 Approximate Solution to the Schrödinger Equation

So far we've seen the many-electron Hamiltonian but only discussed the qualitative implications of the electron–electron repulsion term. That term prevents us from finding an exact, algebraic solution to the Schrödinger equation. Instead, we resort to approximate methods, and even these solutions are beyond what most of us would willingly solve on paper. The two methods described next are the basis for nearly all the computational tools used to model the quantum mechanics of atoms and molecules. The first method (using *perturbation theory*) writes the Schrödinger equation as a power series, allowing us to increase the accuracy of the calculation by adding more (and increasingly complicated) high-order terms. The second method (the *variational method*) uses the full Schrödinger equation, but with an approximate wavefunction that is adjusted to improve the value of the predicted energy.

Some problems with only one or two coordinates (including several important systems) are tractably solved by these techniques, but it becomes increasingly necessary to present results based on computations carried out elsewhere. Our job as chemists then includes comparing these results against our qualitative understanding of the system. If there are discrepancies, we have the pleasure of improving either our own understanding or the accuracy of the computer model.

Perturbation Theory

The electron configuration can be viewed as an approximate wavefunction that neglects the electron–electron interaction. In effect, we arrive at this by solving the Schrödinger equation while leaving out the hard part, the $e^2/(4\pi\varepsilon_0 r_{ij})$ term. It may sound illegitimate, but this is a common approach to daunting problems in physics: if the problem is easy except for one very troublesome contribution, then just ignore the hard part for a start, and later bring the hard part back a little at a time. That is the idea behind **perturbation theory,** illustrated by Fig. 4.6. We start from the easy part, in our case the uncoupled terms in the Hamiltonian. Each electron i has a kinetic energy and a potential energy for attraction to the nucleus:

$$\hat{H}_0(i) = -\frac{\hbar^2}{2m_e}\nabla(i)^2 - \frac{Ze^2}{4\pi\varepsilon_0 r_i}. \tag{4.4}$$

We add these terms up for all N of the electrons to obtain our **zero-order Hamiltonian:**

$$\hat{H}_0 = \sum_{i=1}^{N}\hat{H}_0(i). \tag{4.5}$$

The part of the Hamiltonian that we have left out, the hard part, is the perturbation:

$$\hat{H}' = \sum_{i=2}^{N}\sum_{j=1}^{i-1}\hat{H}'(i,j) = \sum_{i=2}^{N}\sum_{j=1}^{i-1}\frac{e^2}{4\pi\varepsilon_0 r_{ij}}. \tag{4.6}$$

The accuracy of perturbation theory improves as the perturbation becomes small compared to the zero-order contribution. In our case, the electron–electron repulsion energy must be less than the other terms in the Hamiltonian if the atom is to be stable, so we are off to a good start.

The coordinates for each electron are separable in our zero-order Hamiltonian, and therefore the eigenfunctions of the zero-order Schrödinger equation—the

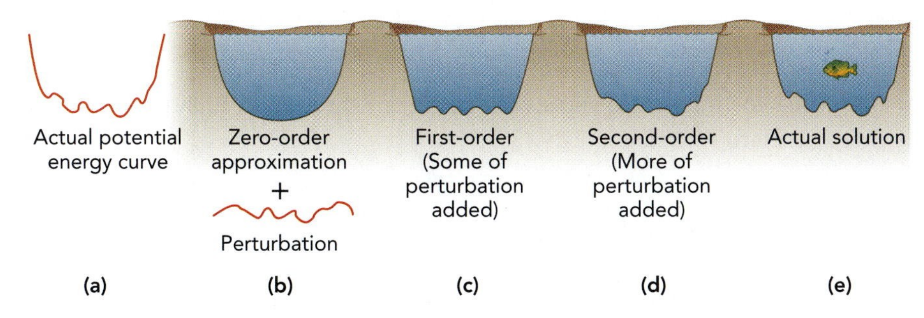

◀ **FIGURE 4.6 Perturbation theory used to find an approximate formula for the shape of water at the bottom of a pond.** **(a)** The pits and mounds in the earth at the pond bottom establish a potential energy function, in which points at lower height have lower potential energy. **(b)** We break the function up into two pieces: a zero-order piece that has the right general shape and obeys a simple formula—for example, the formula for a parabola—and a perturbation, which contains all the complicated parts. For the zero-order part, we can solve the problem exactly (the distribution of the water in this limit is known, but not very accurate). **(c)** Then we add some of the perturbation, to get a first-order solution (the distribution of the water becomes more accurate). **(d)** As we go to higher order, we get closer and closer to including the full perturbation, approaching **(e)** the distribution of water in the real system.

zero-order wavefunctions—can be written as products of one-electron wavefunctions, and the total energy can be written as a sum over the one-electron energies. For example, the solution to our two-electron $1s^2$ zero-order Hamiltonian can be obtained from the one-electron results:

$$\begin{aligned}\psi_0 &\equiv \psi_{1s}(1)\psi_{1s}(2)\\ \hat{H}_0\psi_0 &= \hat{H}_0(1)\psi_{1s}(1)\psi_{1s}(2) + \hat{H}_0(2)\psi_{1s}(1)\psi_{1s}(2)\\ &= E_{1s}\psi_{1s}(1)\psi_{1s}(2) + E_{1s}\psi_{1s}(1)\psi_{1s}(2)\\ &= 2E_{1s}\psi_{1s}(1)\psi_{1s}(2) \equiv E_0\psi_0.\end{aligned} \tag{4.7}$$

For the ground state He atom ($Z = 2$, $n = 1$, two electrons), the zero-order energy is

$$\begin{aligned}1s^2\text{ He:}\quad E_0 = E_{1s} + E_{1s} &= \left(-\frac{Z^2}{2n_1^2} - \frac{Z^2}{2n_1^2}\right)E_h\\ &= \left(-\frac{2^2}{2\cdot 1^2} - \frac{2^2}{2\cdot 1^2}\right)E_h = -4E_h.\end{aligned} \tag{4.8}$$

Now we want to improve this result by including the electron–electron repulsion term, while still keeping the problem solvable. We introduce the perturbation by turning a dial that continuously increases its value. We do this by multiplying $\hat{H}'$ by a constant λ that we vary from zero (no perturbation) to one (the full perturbation). At very small values of λ, the wavefunctions and energies that solve the Schrödinger equation must satisfy the Taylor series expansion (Eq. A.29) about the zero-order solution:

$$\begin{aligned}\psi &= \psi_0 + \left(\frac{\partial\psi}{\partial\lambda}\right)_{\lambda=0}\lambda + \frac{1}{2}\left(\frac{\partial^2\psi}{\partial\lambda^2}\right)_{\lambda=0}\lambda^2 + \dots\\ &= \psi_0 + \lambda\psi_1' + \lambda^2\psi_2' + \dots\end{aligned} \tag{4.9}$$

$$\begin{aligned}E &= E_0 + \left(\frac{\partial E}{\partial\lambda}\right)_{\lambda=0}\lambda + \frac{1}{2}\left(\frac{\partial^2 E}{\partial\lambda^2}\right)_{\lambda=0}\lambda^2 + \dots\\ &= E_0 + \lambda E_1' + \lambda^2 E_2' + \dots,\ \text{with}\end{aligned} \tag{4.10}$$

$$\psi_1' \equiv \left(\frac{\partial\psi}{\partial\lambda}\right)_{\lambda=0} \qquad \psi_2' \equiv \frac{1}{2}\left(\frac{\partial^2\psi}{\partial\lambda^2}\right)_{\lambda=0}$$

$$E_1' \equiv \left(\frac{\partial E}{\partial\lambda}\right)_{\lambda=0} \qquad E_2' \equiv \frac{1}{2}\left(\frac{\partial^2 E}{\partial\lambda^2}\right)_{\lambda=0}$$

In this notation, E_0 is our zero-order energy, and the primes indicate corrections to be added to the zero-order value. Thus, E_1' is the *first-order correction* to the energy, E_2' is the *second-order correction*, and the second-order energy, overall, is $E_2 = E_0 + E_1' + E_2'$. Similarly, ψ_0 is our initial wavefunction, and ψ_1' is the first-order correction to the wavefunction.

We can substitute these expansions into the Schrödinger equation to get the general result to second order in λ:

$$(\hat{H}_0 + \lambda \hat{H}')\psi = E\psi$$

$$(\hat{H}_0 + \lambda \hat{H}')(\psi_0 + \lambda\psi_1' + \lambda^2\psi_2' + \ldots) = (E_0 + \lambda E_1' + \lambda^2 E_2' + \ldots) \times (\psi_0 + \lambda\psi_1' + \lambda^2\psi_2' + \ldots). \quad (4.11)$$

Expanding Eq. 4.11 and dropping terms beyond second order, we have[3]

$$\hat{H}_0\psi_0 + \lambda(\hat{H}_0\psi_1' + \hat{H}'\psi_0) + \lambda^2\hat{H}'\psi_1' = E_0\psi_0 + \lambda(E_0\psi_1' + E_1'\psi_0) + \lambda^2(E_0\psi_2' + E_1'\psi_1' + E_2'\psi_0). \quad (4.12)$$

CHECKPOINT The goal of perturbation theory is to find an approximate solution to the many-electron Schrödinger equation. Here we use the parameter λ to divide the problem into corrections that can be added to the zero-order solution. Each added correction improves the accuracy of the solution, but contributes less and is harder to calculate than the previous correction.

Here's the trick. This way of writing the Schrödinger equation does not depend on the value of λ, and it must be correct (to second order) for *any* value of λ. If we change λ, then terms that are proportional to λ change at a different rate than terms that are proportional to λ^2. The only way the equation can be true at all values of λ is for each set of terms proportional to the same power of λ to satisfy their own equation. As an example, if we know that

$$a_1x^2 + b_1x + c_1 = a_2x^2 + b_2x + c_2$$

is true at *all* values of x, then a_1 must be equal to a_2. You see, $b_1x + c_1$ and $b_2x + c_2$ can only describe straight lines, so the curvature of these two functions must come from exactly the x^2 term. But then if $a_1x^2 = a_2x^2$, we can subtract those terms from both sides of the equation, and obtain an equation for the straight lines: $b_1x + c_1 = b_2x + c_2$. For these two lines to be equal at all x, they must have the same slopes b_1 and b_2 and the same intercepts c_1 and c_2. When two polynomials are equal at all values of the variable, then each term in one polynomial must be equal to the term of the same power in the other polynomial.

In Eq. 4.12, we have two polynomials of λ that we are requiring to be true at all values of λ. This lets us separate Eq. 4.12 into individual equations, each containing the terms for a particular power of λ. For example, the term that is zero-order in λ is the zero-order Schrödinger equation that we've already seen:

$$\hat{H}_0\psi_0 = E_0\psi_0.$$

Next we separate out the terms that are proportional to the first power of λ:

$$\lambda(\hat{H}_0\psi_1' + \hat{H}'\psi_0) = \lambda(E_0\psi_1' + E_1'\psi_0)$$

$$\hat{H}_0\psi_1' + \hat{H}'\psi_0 = E_0\psi_1' + E_1'\psi_0 \qquad \text{divide by } \lambda$$

$$\hat{H}_0\psi_1' - E_0\psi_1' = E_1'\psi_0 - \hat{H}'\psi_0 \qquad \text{group by } \psi_0, \psi_1'$$

$$\psi_0^*\hat{H}_0\psi_1' - \psi_0^*E_0\psi_1' = \psi_0^*E_1'\psi_0 - \psi_0^*\hat{H}'\psi_0 \qquad \times\ \psi_0^*$$

$$\int \psi_0^*\hat{H}_0\psi_1' d\tau - \int \psi_0^*E_0\psi_1' d\tau = \int \psi_0^*E_1'\psi_0 d\tau - \int \psi_0^*\hat{H}'\psi_0 d\tau \qquad \text{integrate both sides}$$

$$\int \psi_0^*\hat{H}_0\psi_1' d\tau - E_0\underbrace{\int \psi_0^*\psi_1' d\tau}_{=0,\ \text{orthogonal}} = E_1'\underbrace{\int \psi_0^*\psi_0 d\tau}_{=1,\ \text{normalized}} - \int \psi_0^*\hat{H}'\psi_0 d\tau$$

---

[3]We will choose ψ_0 to be normalized, but ψ_1' and ψ_2' are small corrections and are *not* normalized. However, it is guaranteed that ψ_1' and ψ_2' must be orthogonal to ψ_0, as long as they are all constructed using an orthonormal basis set such as the one-electron wavefunctions of Chapter 3.

$$\int \psi_0^* \hat{H}_0 \psi_1' d\tau = E_1' - \int \psi_0^* \hat{H}' \psi_0 d\tau. \tag{4.13}$$

The Hamiltonian is a Hermitian operator[4] (Section 2.1). This allows us to switch the wavefunctions in the first integral shown, giving

$$\int \psi_1'^* \hat{H}_0 \psi_0 \, d\tau = E_1' - \int \psi_0^* \hat{H}' \psi_0 \, d\tau$$

$$E_0 \underbrace{\int \psi_1'^* \psi_0 \, d\tau}_{=0,\text{ orthogonal}} = E_1' - \int \psi_0^* \hat{H}' \psi_0 \, d\tau$$

$$E_1' = \int \psi_0^* \hat{H}' \psi_0 \, d\tau. \tag{4.14}$$

Our final value for the first-order correction is the average value theorem (Eq. 2.10) applied to the perturbation Hamiltonian $\hat{H}'$ and integrated over the zero-order wavefunction ψ_0.

This entire process is first-order perturbation theory. To get the total energy to first order, we add this correction to the zero-order energy:

$$E_1 = E_0 + E_1' = E_0 + \int \psi_0^* \hat{H}' \psi_0 d\tau. \tag{4.15}$$

Returning to the $1s^2$ He atom, our Eq. 4.14 from first-order perturbation theory gives

$$E_1' = \int \psi_0^* \hat{H}' \psi_0 \, d\tau = \int \psi_0^* \frac{e^2}{4\pi\varepsilon_0 r_{12}} \psi_0 \, d\tau. \tag{4.16}$$

For any given electronic state, the first-order correction for the electron–electron repulsion term can be found, but a general solution is not available. For both electrons in the $1s$ orbital (the ground state), the integral we need to solve to get E_1' is actually a six-dimensional integral over the coordinates r, θ, and ϕ for each electron. We will normally abbreviate this as a double-integral (to indicate that we're integrating over the coordinates of two different electrons):

$$\int \psi_0^* \frac{e^2}{4\pi\varepsilon_0 r_{12}} \psi_0 \, d\tau = \frac{1}{\pi^2}\left(\frac{Z}{a_0}\right)^6 \int\int \frac{e^2}{4\pi\varepsilon_0 r_{12}} (e^{-Zr_1/a_0} e^{-Zr_2/a_0})^2 \, d\tau_1 d\tau_2. \tag{4.17}$$

Much of the r-, θ-, and ϕ-dependence is hidden in Eq. 4.17 for simplicity, but we could bring it out by writing r_{12} in terms of the one-electron coordinates (Fig. 4.7). If we start from the Cartesian coordinates

$$r_{12} = [(x_1 - x_2)^2 + (y_1 - y_2)^2 + (z_1 - z_2)^2]^{1/2}, \tag{4.18}$$

this becomes (with the help of Eqs. A.6)

[4]We haven't proven this. Briefly, the potential energy term must be Hermitian because it is a multiplicative operator, and the order of multiplication in the integrand does not affect the integral. For the kinetic energy, we show that the derivative operator $\frac{\partial}{\partial x}$ is Hermitian as follows. Integration by parts yields

$$\int_{x_1}^{x_2} \psi_0^* \frac{\partial \psi_1'}{\partial x} dx = \int_{x_1}^{x_2} \psi_1' \frac{\partial \psi_0^*}{\partial x} - [\psi_0^* \psi_1']_{x_1}^{x_2},$$

and x_1 and x_2 can be chosen to be $\pm\infty$, where ordinary wavefunctions must vanish, so the second term goes to zero. If the derivative operator is Hermitian, then the one-dimensional momentum operator $\hat{p}_x = -i\hbar\frac{\partial}{\partial x}$ is also Hermitian, and by extension, $\hat{p}_x^2$, the kinetic energy operator, and the Hamiltonian are also all Hermitian operators.

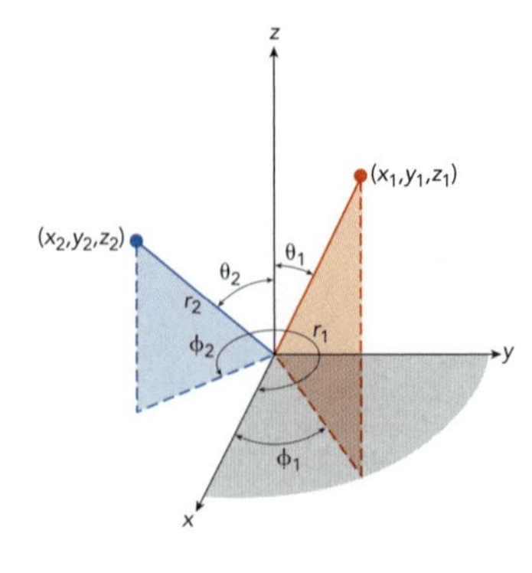

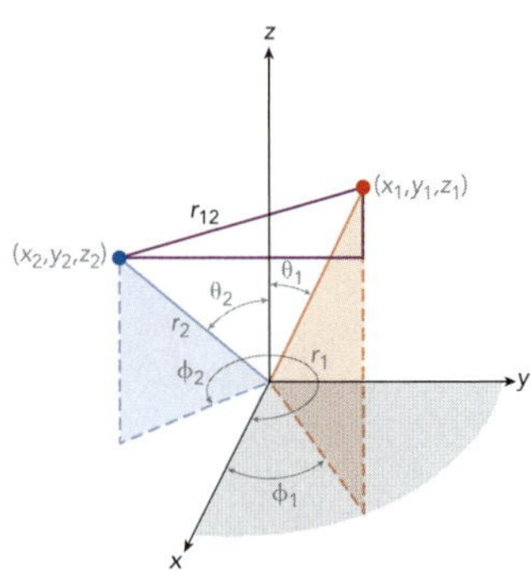

▲ **FIGURE 4.7 Calculating the distance between two electrons.** We can identify the positions of electrons 1 and 2 by the spherical coordinates (r_1,θ_1,ϕ_1) and (r_2,θ_2,ϕ_2) and then use trigonometry to convert each of those points to Cartesian coordinates (x_1,y_1,z_1) and (x_2,y_2,z_2). The distance r_{12} between the two electrons is then given by the Pythagorean relation Eq. 4.18.

$$r_{12} = [(r_1 \sin\theta_1 \cos\phi_1 - r_2 \sin\theta_2 \cos\phi_2)^2 + (r_1 \sin\theta_1 \sin\phi_1 - r_2 \sin\theta_2 \sin\phi_2)^2 + (r_1 \cos\theta_1 - r_2 \cos\theta_2)^2]^{1/2}. \quad (4.19)$$

Substituting this explicit expression for r_{12} in Eq. 4.17 gives an integral we evaluate numerically, and it comes to $1.25\,E_h$ for $1s^2$ He. For ground state He, the two-electron zero-order energy is $E_0 = -4.00\,E_h$, so the total energy to first-order is

$$1s^2\ He:\quad E_1 = E_0 + E_1' = (-4.00 + 1.25)\,E_h = -2.75\,E_h.$$

Before we compare this to the right answer, note that we can be sure our first-order energy of $-2.75\,E_h$ will be too high (i.e., the correct value should be more negative). That's because we have included the electron–electron repulsion term but with a wavefunction ψ_0 that is certain to keep the electrons too close together. The repulsion term, after all, will cause the electrons to repel one another. So the electrons will take on new wavefunctions that put the electrons on average farther away from one another, reducing the repulsion and leading to a more stable (lower energy) system.

The experimental value for $E(1s^2\ \text{He})$ is $-2.90\,E_h$. Thus, the zero-order energy ($-4.00\,E_h$) brings us only to within 38% of the correct energy of the state, while first-order perturbation theory ($-2.75\,E_h$) brings us to within 5.2%.

The next step in the perturbation theory would be to solve for ψ_1, and the complete Schrödinger equation can be solved to arbitrary precision through a cycle whereby improved, high-order wavefunctions and energies are used to calculate the next higher-order wavefunctions and energies, always returning to Eq. 4.11 and extracting the next power of λ. Our first-order wavefunction ψ_1 is obtained from the following expression:

$$\psi_1 = \psi_0 + \psi_1' = \psi_0 + \sum_{k=1}^{\infty} \frac{\int \psi_0^* \hat{H}' \psi_{0,k}\, d\tau}{E_0 - E_{0,k}} \psi_{0,k}, \quad (4.20)$$

where $\psi_{0,k}$ and $E_{0,k}$ are the set of zero-order wavefunctions and their energies that we would calculate for all the excited state electron configurations, such as $1s^1 2s^1$ and $2s^2$. The first-order wavefunctions then participate in the solution to the second-order energy correction:

$$E_2' = \int \psi_0^* \hat{H}' (\psi_0 - \psi_1)\, d\tau = \sum_{k=1}^{\infty} \frac{\left[\int \psi_0^* \hat{H}' \psi_{0,k}\, d\tau\right]^2}{E_0 - E_{0,k}}. \quad (4.21)$$

We can be sure that the second-order correction to a ground state energy, such as the He $1s^2$ ground state energy, will always *stabilize* the system, regardless of the perturbation. That's because the numerator in Eq. 4.21 is always positive, but the denominator is always negative (because for the ground state, E_0 is always less than $E_{0,k}$). If we tally up the sum for the He $1s^2$ ground state, we get a correction of $-0.05\,E_h$, giving a total energy of $-2.80\,E_h$ and bringing the value to within 3.5% of the experimentally observed energy.

In the case at hand—the many-electron atom—we can justify the qualitative behavior of the first- and second-order energy corrections as follows. The first-order correction must be positive, because it is the average of the electron–electron repulsion. However, because that repulsion is averaged over the densities of *non-interacting* electrons (using the zero-order wavefunctions), it overestimates the repulsion. Once we turn on the interaction, the electrons push away from each other, as shown by the radial probability distributions in Fig. 4.8. Although the *peak* of the first-order distribution stays at about $0.5a_0$—the same distance as in the zero-order wavefunction—much of the electron distribution shifts to larger r. Because the electrons are more separated in first-order, the repulsion energy drops a little in second-order. That's why the second-order correction to the energy is negative but smaller in magnitude than the first-order correction (Fig. 4.9).

As you can see, the computations necessary for these corrections become increasingly complicated as we go to higher-order. When perturbation theory works well, the relatively little effort required for the first- and second-order corrections accounts for most of the perturbation. We can see that perturbation theory converges fairly quickly for the two electrons in ground state helium because the second-order correction to the energy is much smaller than the first-order correction. Similarly, Fig. 4.10 shows that the first-order, two-electron wavefunction is about 67% $1s^2$ and about 30% $1s^1 2s^1$, with much smaller contributions from other states.

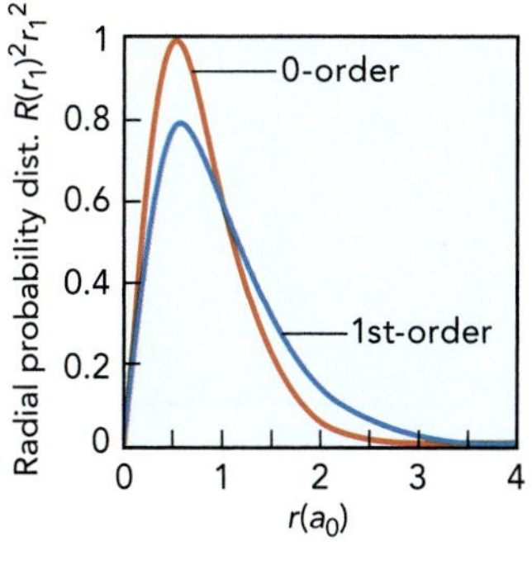

▲ FIGURE 4.8 **Radial distribution functions for the perturbation theory solution to helium.** The average radial distribution functions for each of the two electrons in the zero-order and first-order wavefunctions are shown. The zero-order case is the same distribution as occupied by the single electron in $1s^1$ He^+, because we have turned off the electron–electron repulsion. Once we turn on the repulsion (by adding the first-order correction to the energy), each electron pushes the other away. As a result, the electron density expands to a larger average distance from the nucleus.

DERIVATION SUMMARY Perturbation Theory. The Hamiltonian was broken up into a zero-order part that could be solved exactly and a perturbation. By expanding the Schrödinger equation in a Taylor series about the zero-order solution, a series of equations could be obtained that provides a systematic way of improving the energies and wavefunctions. Our principal results are Eqs. 4.14 and 4.21 for the first- and second-order energy corrections.

One particular strength of perturbation theory is its intuitive simplicity. That may not be apparent when you've just emerged from two pages of calculus, but the first- and second-order corrections to the energy are very useful conceptually. As we've seen, the first-order correction is essentially the quantum average value theorem applied to the perturbation Hamiltonian $\hat{H}'$. In many cases,

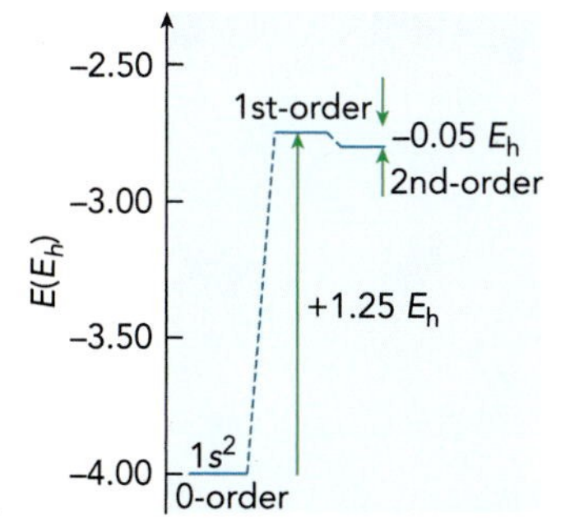

▲ FIGURE 4.9 **Energies calculated at different orders of perturbation theory.** The zero-order energy of ground state He is the energy of two electrons in the 1s orbital not interacting with each other. Once we turn the interaction on, the repulsion pushes the energy up by 1.25 E_h. If we then correct the wavefunction to increase the average distance between electrons caused by the repulsion, the repulsion is less, so the energy drops slightly.

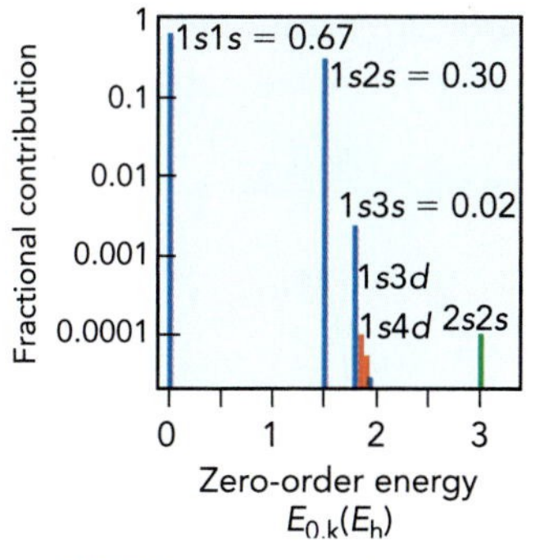

▲ FIGURE 4.10 **Contributions to the first-order ground state He wavefunction.** The first-order wavefunction is assembled from different zero-order, two-electron states, plotted here as strength of contribution versus zero-order energy. The vertical scale is a log scale to show some of the weaker contributions.

however, we are able to select a zero-order wavefunction in which the first-order correction is zero. In that case, we turn to the second-order correction. All the terms in Eq. 4.21 have a positive numerator and a denominator $E_0 - E_{0,k}$, which is negative when $E_0 < E_{0,k}$. This tells us that the second-order correction always works to push the energies of the interacting states away from one another.

EXAMPLE 4.2 Second-Order Perturbation Theory

CONTEXT The electron–electron repulsion in a many-electron atom is an interaction involving the x, y, and z coordinates of all the electrons, so it is tough to visualize. But we can take lessons from a simpler problem. Let's start from an electron in a one-dimensional box and add a repulsion to one side of the box, representing the region occupied by core electrons. What is the effect on the distribution of our electron? The repulsion is represented by an increase in the potential energy in the second half of the box.

PROBLEM Use perturbation theory to find the first-order wavefunction and second-order energy for the ground state of a particle in a one-dimensional box that has an added repulsion in the second half of the box, a perturbation $\hat{H}' = \pi^2\hbar^2/(2ma^2)$ from $x = a/2$ to $x = a$. Include only the contributions from the $n = 2$ state to keep it manageable.

SOLUTION We have a zero-order case, the particle in a regular one-dimensional box, to which we already have the solutions (Eqs. 2.33 and 2.31):

$$\psi_{0,n}(x) = \sqrt{\frac{2}{a}}\sin\left(\frac{n\pi x}{a}\right) \quad E_{0,n} = \frac{n^2\pi^2\hbar^2}{2ma^2}.$$

In order to find the first-order energy correction, we obtain the average of the perturbation over the zero-order ground state function $\psi_{0,n=1}(x)$ (Eq. 4.14):

$$E'_{1,n=1} = \int \psi^*_{0,n=1}\hat{H}'\psi_{0,n=1}dx = \frac{2}{a}\int_{a/2}^{a}\left(\frac{\pi^2\hbar^2}{2ma^2}\right)\sin^2\left(\frac{\pi x}{a}\right)dx = \frac{\pi^2\hbar^2}{4ma^2} = \frac{1}{2}E_{0,n=1}.$$

The integral from $x = 0$ to $x = a/2$ is zero because $\hat{H}'$ is zero over that range. The first-order wavefunction will include a contribution from the $n = 2$ excited state (Eq. 4.20):

$$\psi^*_{1,n=1} = \psi_{0,n=1} + \frac{\int\psi^*_{0,n=1}\hat{H}'\psi_{0,n=2}\,dx}{E_{0,n=1} - E_{0,n=2}}\psi_{0,n=2}$$

$$= \psi_{0,n=1} + \frac{\frac{2}{a}\int_{a/2}^{a}\left(\frac{\pi^2\hbar^2}{2ma^2}\right)\sin\left(\frac{\pi x}{a}\right)\sin\left(\frac{2\pi x}{a}\right)dx}{\frac{\pi^2\hbar^2}{2ma^2} - \frac{4\pi^2\hbar^2}{2ma^2}}\psi_{0,n=2}$$

$$= \psi_{0,n=1} + \frac{\frac{2}{a}\left(\frac{\pi^2\hbar^2}{2ma^2}\right)\left(-\frac{2a}{3\pi}\right)}{-\frac{3\pi^2\hbar^2}{2ma^2}}\psi_{0,n=2} = \psi_{0,n=1} + \frac{4}{9\pi}\psi_{0,n=2}.$$

This leads to the second-order energy correction (Eq. 4.21):

$$E'_{2,n=1} = \frac{\left[\int\psi_{0,n=1}\hat{H}'\psi_{0,n=2}\,d\tau\right]^2}{E_{0,n=1} - E_{0,n=2}}$$

$$= \frac{\left[\frac{2}{a}\left(\frac{\pi^2\hbar^2}{2ma^2}\right)\left(-\frac{2a}{3\pi}\right)\right]^2}{-\frac{3\pi^2\hbar^2}{2ma^2}}$$

$$= -\frac{8\hbar^2}{27ma^2} = -\frac{16}{27\pi^2}E_{0,n=1}.$$

This predicts that adding the step to the second half of the box shifts the ground state wavefunction toward the first half of the box and increases the energy from $E_{0,n=1} = \pi^2\hbar^2/(2ma^2)$ to

$$E_{0,n=1} + E'_{1,n=1} + E'_{2,n=1} = \left(1 + \frac{1}{2} - \frac{16}{27\pi^2}\right)E_{0,n=1} = 1.44\,E_{0,n=1}.$$

As we might expect, by adding a positive potential energy to the system, the first-order energy correction raises the average energy of the state. The second-order correction then reduces the energy slightly, taking into account how the wavefunction $\psi_{1,n=1}$ shifts toward the first half of the box, away from the barrier. The unnormalized probability density $\psi^2_{1,n=1}$ is compared in the following figure to the zero-order $\psi^2_{0,n=1}$ density, showing the shift toward the lower energy half of the box.

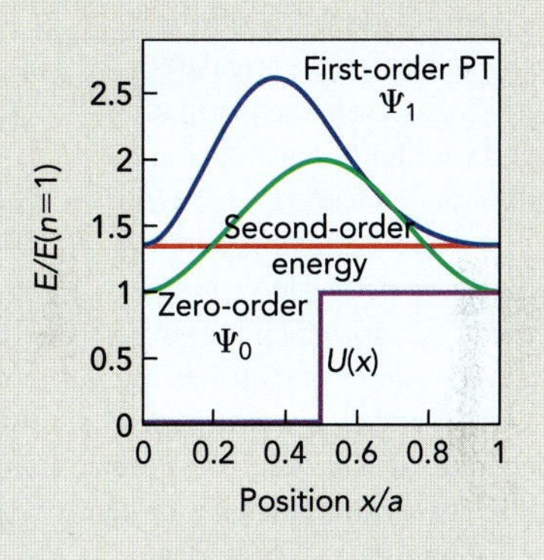

We can expect similar results for a valence electron when we turn on the interactions with the core electrons. The repulsion from the core electrons will push the valence electron distribution to a larger average distance from the nucleus and will raise the energy.

Variation Theory

The difficulty with perturbation theory is primarily with the initial assumption: if the perturbation is not small compared to the zero-order Hamiltonian, convergence to an accurate set of energies and wavefunctions can be an excruciating process. Another option, not quite as sensitive to the complexities of the Hamiltonian, employs the **variational principle:** *the correct ground state wavefunction of any system is the wavefunction that yields the lowest possible value of the energy.* To rephrase it from a practical perspective, we start off with a guess wavefunction, and adjust that wavefunction to get the lowest-energy we can.

The variation method has an appealingly intuitive justification, namely that the forces at work on the electrons can always push them into the lowest-energy distribution possible (see Fig. 4.11). All we have to do is try to find that minimum energy. A general approach to solving a Schrödinger equation variationally works like this:

1. We write a zero-order wavefunction ψ_0 to start from, but we include in it parameters that can be adjusted.
2. Our ψ_0 does not solve the Schrödinger equation, which means that it is not an eigenvalue of the Hamiltonian. However, we can calculate the expectation value of the energy using the average value theorem (Eq. 2.10), $\langle E \rangle = \int \psi_0^* \hat{H} \psi_0 d\tau$.
3. Adjust the parameters of ψ_0 to get a new wavefunction ψ_1.

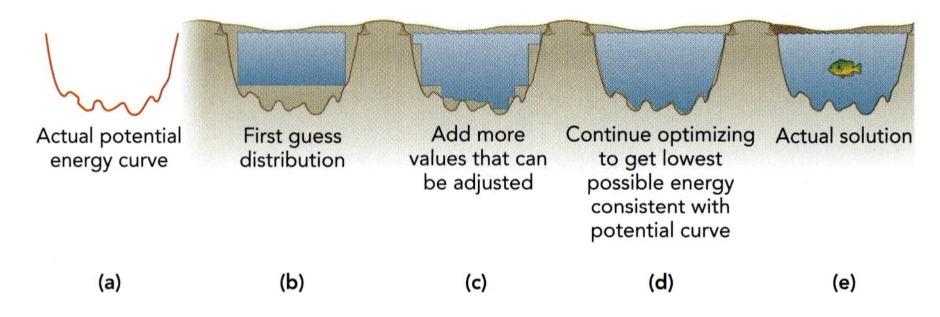

▶ **FIGURE 4.11 The variational principle applied to find an approximate formula for the shape of water at the bottom of a pond.** **(a)** The pits and mounds in the earth at the pond bottom establish a potential energy function, in which points at lower height have lower potential energy. **(b)** We can attempt to represent the distribution of water along the pond bottom with a series of rectangles of a fixed width, with each rectangle just touching the pond bottom. Then we can calculate the total potential energy based on this distribution. **(c)** If we change our function to allow diagonal edges, we have a more flexible function, and we adjust it to get the lowest total energy we can, which is lower than the energy we calculated from (b) and presumably more accurate. **(d)** Our goal is to get as close as we can to the distribution of water in the real system, in which the force of gravity works on the water molecules to attain by nature the lowest-energy distribution possible.

4. If $\langle E \rangle$ is now lower than the previous value, then keep the new wavefunction. Otherwise return to the previous wavefunction.
5. Repeat from step 3 until $\langle E \rangle$ cannot get any lower.

We start off with a trial wavefunction ψ_0—just a guess—and calculate the variational energy E^{var} by the average value theorem (Eq. 2.10, $\langle A \rangle = \int \psi^* \hat{A} \psi d\tau$):

$$E_0^{\text{var}} = \frac{\int \psi_0^* \hat{H} \psi_0 \, d\tau}{\int \psi_0^* \psi_0 d\tau}. \tag{4.22}$$

For our $1s^2$ He problem, ψ_0 is a two-electron wavefunction. We continue adjusting ψ to minimize E^{var}. When no change in the wavefunction lowers E^{var}, we should have the correct wavefunction for the lowest state with that angular momentum. The concept is straightforward: given some angular momentum in the atom that must be conserved, nature allows the wavefunction to find its most stable (lowest-energy) form. This means that inexact wavefunctions always give an energy that is too high.

One way to look at the difference between the variational method and perturbation theory is to compare the zero-order Schrödinger equations that they each start from. When we used perturbation theory, the zero-order Schrödinger equation used the right wavefunction to accompany the wrong Hamiltonian because we left the electron–electron repulsion out of $\hat{H}$. As we gradually add the perturbation to the Hamiltonian (by going to higher-order), we also have to adjust the wavefunction to keep pace. In the variational approach, we use the right Hamiltonian in zero order—with all the terms included—but the wrong wavefunction. The Hamiltonian stays the same, and all our effort is devoted to improving the wavefunction. In short, the difference between our two approaches is that perturbation theory finds the wavefunctions and energies starting from an inexact Hamiltonian, whereas variation theory uses the correct Hamiltonian throughout but starts from an inexact wavefunction.

How should we adjust this particular wavefunction? Still using ground state $1s^2$ He atom as the example, we can reason that electron shielding causes a net reduction in the effective charge of the nucleus, Ze. Our zero-order wavefunction is

$$\psi_0 = Ae^{-Zr_1/a_0}e^{-Zr_2/a_0},$$

and we change that now so that the **effective atomic number** Z_{eff} is a variable:

$$\psi_1 = Ae^{-Z_{\text{eff}}r_1/a_0}e^{-Z_{\text{eff}}r_2/a_0}. \tag{4.23}$$

In Problem 4.68, we find that the energy for the one-electron terms can be written

$$\langle E_0 \rangle = -2Z_{\text{eff}}\left(Z - \frac{Z_{\text{eff}}}{2}\right)E_{\text{h}}. \tag{4.24}$$

The electron–electron repulsion energy for this wavefunction is given by

$$\left\langle \frac{e^2}{4\pi\varepsilon_0 r_{12}} \right\rangle = 0.62 Z_{\text{eff}}\, E_{\text{h}}.$$

To find the value of Z_{eff} that minimizes the energy, we take the derivative of the total energy and find the value of Z_{eff} where this derivative vanishes:

$$E_1 = \langle E_0 \rangle + \left\langle \frac{e^2}{4\pi\varepsilon_0 r_{12}} \right\rangle$$

$$= -2Z_{\text{eff}}\left(Z - \frac{Z_{\text{eff}}}{2}\right) + 0.62 Z_{\text{eff}}$$

$$\frac{\partial}{\partial Z_{\text{eff}}}\langle E_1 \rangle = -2Z + 2Z_{\text{eff}} + 0.62 = 0$$

$$Z_{\text{eff}} = Z - \frac{0.62}{2} = 2.00 - 0.31 = 1.69.$$

The optimum effective atomic number $Z_{\text{eff}} = 1.69$ minimizes the energy predicted by the Schrödinger equation and gives the result $E_1(1s^2\ \text{He}) = -2.85\,E_{\text{h}}$, with the radial electron distribution shown in Fig. 4.12. This answer is reasonable: $Z_{\text{eff}} < Z$, as predicted, but the second electron does not completely cancel the charge of one of the protons, so $Z_{\text{eff}} > 1$.

CHECKPOINT The variational method provides another avenue for solving the many-electron Schrödinger equation, this time by systematically improving an initial approximate wavefunction. Although it doesn't have the rapidly increasing complexity that perturbation theory has, the variational method doesn't offer a formula for how to write the wavefunction. The success of the method relies on our ability to choose a good set of variational parameters.

While we cannot perform these calculations explicitly here, this method lends itself well to computation. The computer does not even need an algebraic expression for ψ; it can calculate ψ as a grid of points, evaluating the integrals numerically. A weakness of the variational method in some applications is that we directly solve only for the lowest-energy state of any given symmetry, where the symmetry in this case is given by the number of electrons with each value of l.[5] For example, the $1s^1 2s^1$ excited state of He, which has a spherically symmetric wavefunction, will be adjusted in a variational calculation to give the lower energy, spherically symmetric $1s^2$ wavefunction. However, we can preserve symmetry during the calculation to find the lowest-energy excited states that have different symmetry from the ground state, such as $1s^1 2p^1$ He.

Variation theory may seem difficult because it often requires solving integrals that don't have algebraic solutions, but the solution of numerical integrals by computer (as described in Section A.1) is a routine practice. The trouble is that the computer time goes up exponentially with the number of dimensions. The electronic wavefunction for a two-electron atom has six coordinates: r_1, θ_1, ϕ_1, r_2 θ_2, and ϕ_2. To sample the two-electron wavefunction with 100 points in each coordinate means a wavefunction array of $100^6 = 10^{12}$ points (Fig. 4.13). This is already a lot of points, and the variational method requires calculating the integrals several times for different parameter values, so algebraic solutions to the integrals are preferred, especially when the system has many dimensions.

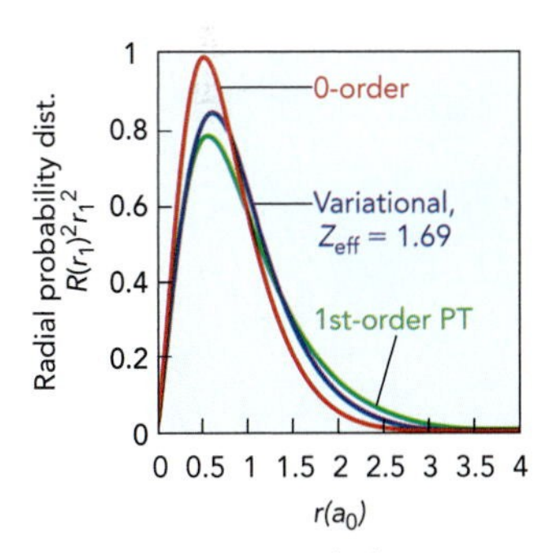

▲ **FIGURE 4.12 Radial probability density for the variational solution to helium.** Shown is the density for one electron based on the wavefunction in Eq. 4.23 with $Z_{\text{eff}} = 1.69$. Like the radial probability density of the first-order wavefunction from perturbation theory, the distribution is shifted to larger average distance relative to the density of the pure $1s$ orbital.

[5] Linear variation calculations escape this limitation by starting with a finite, orthonormal basis set and requiring that all the excited state wavefunctions at the end of the variational calculation are orthogonal to the ground state.

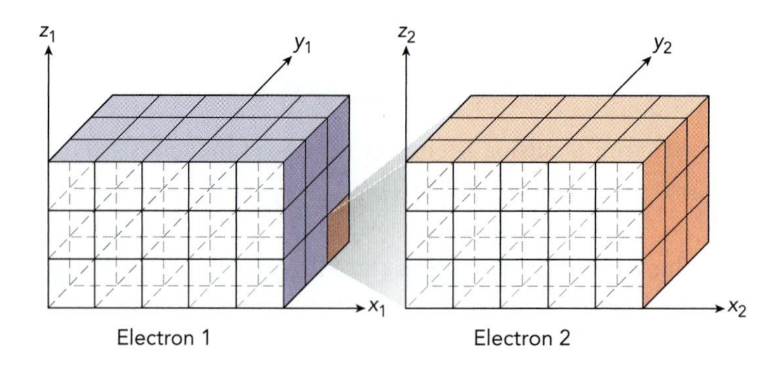

▶ FIGURE 4.13 **Numerical integration over the coordinates of two electrons.** We evaluate the function over many individual blocks of space in the coordinates of electron 1 (shown here as Cartesian coordinates for simplicity), but for *each* one of those blocks we must also integrate over all three coordinates of electron 2.

Hartree-Fock Calculations

The success of our earlier sample variational calculation conceals a haphazard approach. In order for the variational method to be useful to a broad range of chemical systems, we need a systematic way of constructing and optimizing the guess wavefunctions. We attain this by choosing a basis set of the one-electron orbitals formulated in Tables 3.1 and 3.2, and then adding these together to form a **linear combination of atomic orbitals** (LCAO):

$$\begin{aligned}\varphi(i) &= c_{1s}\psi_{1s}(i) + c_{2s}\psi_{2s}(i) + c_{2p_x}\psi_{2p_x}(i) + \ldots \\ &= \sum_{n=1}^{n_{\max}} \sum_{l=0}^{n-1} \sum_{m_l=-l}^{l} c_{nlm_l}\psi_{nlm_l}(i)\,, \end{aligned} \tag{4.25}$$

where the $\psi_{nlm_l}(i)$'s are the pure one-electron orbitals. This is a variational wavefunction, in which the variational parameters are the coefficients c_{nlm_l}. For example, in our $1s^2$ He example, we found that the effective atomic number was less than 2, so the wavefunctions would be more diffuse than the $1s$ function obtained from Table 3.2. We can reproduce that effect with the wavefunction φ by starting from a pure $1s$ orbital, setting c_{100} to one and all other c's to zero, and then trading some of the $1s$ character for $2s$ character by reducing c_{100} and raising c_{200}. The major effect of the added $2s$ character is to increase the average distance of the electron from the nucleus, which is the effect of the electron shielding. The one-electron wavefunctions form a complete basis set, so in principle any wavefunction of the electron can be formed by adding these together with the right set of coefficients.

In practice, the ψ_{nlm_l}'s in Eq. 4.24 are usually not the pure one-electron wavefunctions. When these methods are extended to molecules, the computational problem becomes very difficult because the Schrödinger equation has to be integrated over orbital wavefunctions centered on different atoms. There is a manipulation that makes it *much* easier to integrate **Gaussian functions** e^{-r^2} centered in different places, and therefore the ψ_{nlm_l}'s are almost always replaced by sums of Gaussians made to resemble the one-electron orbitals:

$$\varphi_i(i) = \left(a_i e^{-\alpha_i r^2_i} + b_i e^{-\beta_i r^2_i} + c_i e^{-\gamma_i r^2_i} + \ldots\right) Y(x_i, y_i, z_i),$$

where $Y(x,y,z)$ is a function used to generate the right symmetry for $l \neq 0$ orbitals: for example, $Y = x$ for a p_x orbital, and $Y = z^2$ for a d_{z^2} orbital. The combining of Gaussians to resemble the e^{-r/a_0} function of the one-electron radial wavefunctions is shown in Fig. 4.14.

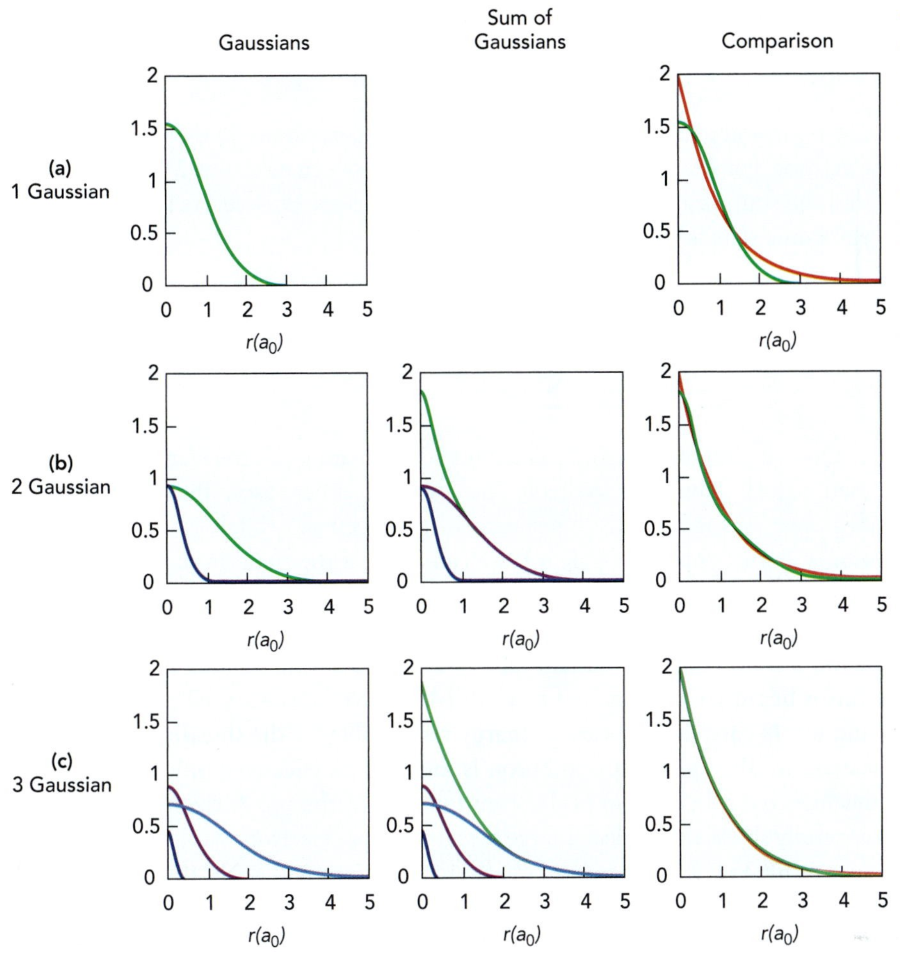

◀ **FIGURE 4.14 Gaussian basis functions.** The 1*s* radial wavefunction $2e^{-r/a_0}$ compared to **(a)** one Gaussian function—showing the difference at the top and bottom of both curves—and compared to sums of **(b)** two and **(c)** three Gaussian functions. With three Gaussian functions, it is possible to get a close match to the 1*s* function even at small values or *r*.

In general, each electron i may occupy its own unique orbital $\varphi_i(i)$. Whatever the specific basis set is, the φ_i's can then be optimized by computing the repulsion term from the average field of all the other electrons. In other words, each $\varphi_i(i)$ becomes a solution of

$$\hat{H}(i)\varphi_i(i) = \left[-\frac{\hbar^2}{2m_e}\nabla(i)^2 - \frac{Ze^2}{4\pi\varepsilon_0 r_i} + \sum_{j\neq i}\left(\int |\varphi_j(j)|^2 \frac{e^2}{4\pi\varepsilon_0 r_{ij}}\, d\tau_j\right)\right]\varphi_i(i) = \varepsilon_i\varphi_i(i), \quad (4.26)$$

with an orbital energy ε_i for each electron.[6] The integral with volume element $d\tau_j$ is a triple integral over the three coordinates of electron j. This treatment of the electron–electron repulsion energy is called the **self-consistent field approximation** (SCF). With additional terms to account for the effects of electron spin, the one-electron wavefunctions that result from optimizing all the c_{nlm_l}'s are called

[6]The index in parentheses represents a coordinate dependence. For example, $\hat{H}(i)$ indicates that the Hamiltonian is a function of the coordinates of electron i, whereas the energy ε_i for electron i is a numerical value that changes (in general) for each value of i.

Hartree-Fock wavefunctions. The orbital energies ε_i each represent contributions from the kinetic, nuclear attraction, and electron repulsion terms:

$$\varepsilon_i = \varepsilon_i(\text{kinetic}) + \varepsilon_i(\text{attraction}) + \varepsilon_i(\text{repulsion}). \tag{4.27}$$

The repulsion energy for each pair of electrons contributes to the energy of both electrons. For example, the repulsion energy between electrons 2 and 3 is added into the sum for both ε_2 and ε_3. Therefore, the total Hartree-Fock energy of the electronic state is computed by this method as

$$\begin{aligned} E_{\text{HF}} &= \sum_{i=1}^{N}\left[\varepsilon_i(\text{kinetic}) + \varepsilon_i(\text{attraction}) + \frac{1}{2}\varepsilon_i(\text{repulsion})\right] \\ &= \sum_{i=1}^{N}\varepsilon_i - \frac{1}{2}\sum_{i=1}^{N}\varepsilon_i(\text{repulsion}). \end{aligned} \tag{4.28}$$

One-electron energies calculated for selected atoms by the Hartree-Fock SCF method are tabulated in Table 4.1. Among other uses, these calculations demonstrate the qualitative behavior of the electron orbitals in many-electron atoms. For example, in Table 4.1, we can see that the trend in the energy of the highest-energy electron is consistent with the trend in effective atomic number. As we move from Li to Ne, the nuclear charge increases by one unit at each new atom, but the shielding remains about the same for the highest-energy orbitals. This is because electrons are being added only to the $n = 2$ shell across this row, and the $1s$ core electrons are primarily responsible for the shielding of the $n = 2$ electrons. The $2p$ electron in boron is much more effectively shielded from the nuclear charge by the two $1s$ electrons than it is by the two $2s$ electrons. Electrons generally shield the nuclear charge best from those electrons with larger n values. As a result, the effective nuclear charge increases from Li to Ne; the gain in actual

TABLE 4.1 Hartree-Fock orbital energies. The energy in E_h is given for each electron, as well as the total computed and experimental energies, in selected ground state atoms.

H	He	Li	Be	B	C	Ne	Na	orbital
−0.500	−0.917	−2.487	−4.733	−7.702	−11.348	−32.763	−40.488	$1s$
	−0.917	−2.469	−4.733	−7.687	−11.302	−32.763	−40.485	$1s$
		−0.196	−0.309	−0.545	−0.830	−1.919	−2.797	$2s$
			−0.309	−0.446	−0.584	−1.919	−2.797	$2s$
				−0.318	−0.439	−0.840	−1.520	$2p$
					−0.439	−0.840	−1.520	$2p$
						−0.840	−1.520	$2p$
						−0.840	−1.518	$2p$
						−0.840	−1.518	$2p$
						−0.840	−1.518	$2p$
							−0.182	$3s$
−0.500	−2.862	−7.433	−14.573	−24.533	−37.694	−128.547	−161.859	E_{HF}
−0.500	−2.903	−7.460	−14.646	−24.630	−37.820	−128.900	−161.960	E_{CI}^{a}
−0.500	−2.903	−7.478	−14.668	−24.658	−37.855	−129.05	−162.43	E_{expt}

[a]Computed energy using a method of configuration interaction.

nuclear charge is greater than the contribution to the shielding from each additional electron. We see this reflected in the Hartree-Fock energies of Table 4.1, where the highest orbital energy progresses steadily from $-0.196\,E_h$ (Li) to $-0.840\,E_h$ (Ne).

Once we fill the $n = 2$ shell, however, the next electron is added to the 3*s* orbital, which we expect to have a higher energy than the $n = 2$ electrons by virtue of the Bohr model, $E_n = -Z^2 E_h/(2n^2)\,E_h$, if nothing else. The energy shift is even higher than we might have guessed, however, because the 3*s* electron is subject to considerable shielding from the nuclear charge by all the $n = 2$ electrons as well as the $n = 1$ electrons.

EXAMPLE 4.3 Hartree-Fock Energies and Effective Atomic Number

CONTEXT One common technique for analyzing the composition of materials or surfaces is **Auger spectroscopy.** The process is illustrated for neon in Fig. 4.15. When an electron is kicked out of a core orbital by a high-energy electron or photon, the ion that remains behind is missing one of its lowest-energy electrons, meaning that the ion is in a highly excited electronic state. When one of the remaining electrons loses energy to fill the core orbital, the energy released is sufficient to eject a second, outer electron. This second electron is detected, and the energy of the electron can be used to determine the energy of the orbitals involved, and from that the elemental composition.

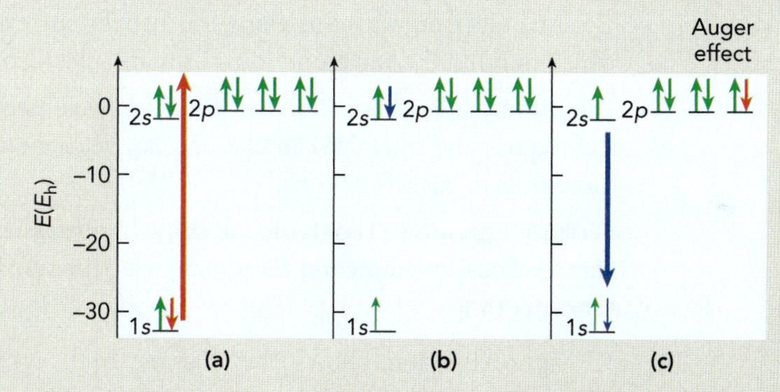

▲ FIGURE 4.15 **Auger spectroscopy.** **(a)** A high-energy photon or electron hitting a neon atom can ionize one of the 1*s* core electrons, leaving **(b)** an ion in an excited state. The ion achieves lower energy by allowing one of the outer electrons to fill the vacancy, but this releases enough energy to **(c)** ionize a 2*p* electron. The kinetic energy of the emitted 2*p* electron is a constant for the element and does not depend on knowing the energy of the first ionization.

Whereas other studies of electronic states usually probe only the valence electrons, Auger spectroscopy relies on the orbital energies of inner electrons as well. The differences among these orbital energies can be interpreted on the basis of the effective atomic number.

PROBLEM Estimate Z_{eff} for the 1*s*, 2*s*, and 2*p* electrons in Ne using only the Hartree-Fock SCF energies in Table 4.1.

SOLUTION Each subshell in Ne experiences a different amount of shielding and therefore has a different characteristic Z_{eff}. The effective atomic number represents the apparent shift in the nuclear charge felt by one electron because of the shielding of the other electrons. Therefore, the effective nuclear charge is essentially the value of Z that would appear in the Bohr model energy equation if we wanted to evaluate the energy ε of a single electron in our many-electron atom:

$$\varepsilon_i = -\frac{Z_{\text{eff}}(i)^2}{2n^2} E_\text{h}.$$

Solving for Z_{eff}, we have the equation

$$Z_{\text{eff}}(i) = \left(-\frac{2\varepsilon_i n^2}{E_\text{h}}\right)^{1/2}. \tag{4.29}$$

The results:

- $Z_{\text{eff}}(1s) = [-2(-32.763)(1^2)]^{1/2} = 8.09$
- $Z_{\text{eff}}(2s) = [-2(-1.919)(2^2)]^{1/2} = 3.92$
- $Z_{\text{eff}}(2p) = [-2(-0.840)(2^2)]^{1/2} = 2.59$

The actual atomic number is 10. The $1s$ electrons are only slightly shielded from the nuclear charge, whereas the $n=2$ electrons experience substantial shielding. These values *underestimate* the effective atomic number because we have over-represented the electron–electron repulsion.

The Hartree-Fock wavefunctions possess three great strengths:

- The one-electron wavefunctions are usually an excellent basis set from which to build the wavefunctions for many-electron atoms and molecules.
- The variational optimization lends itself to well-established computational techniques, and tractable calculations can be carried out on problems with thousands of basis functions.
- Because these are still composed of distinct, one-electron wavefunctions, the Hartree-Fock wavefunction for a many-electron atom remains conceptually very appealing.

The SCF approximation allows the Hartree-Fock orbital wavefunctions to include a substantial portion of the electron–electron interaction energy. What it cannot account for is the ability of electron j to dynamically respond to the distribution of electron i. The density of electron j will tend to be higher when the density of electron i is somewhere else. In other words, the wavefunctions for electrons i and j should be directly coupled. By forcing $\varphi_i(i)$ to be completely separable from $\varphi_j(j)$, the Hartree-Fock wavefunction forces the electrons to be closer together on average than they actually are, and the energies are typically in error by about $0.05\,E_\text{h}$ or 1 eV per pair of electrons.

There are approximation methods for dealing with the remaining **correlation energy,** E_{corr}, where

$$E_{\text{corr}} \equiv E_{\text{experimental}} - E_{\text{SCF}}, \tag{4.30}$$

but precise values require a significantly more complicated calculation. One popular extension is **configuration interaction** (CI), which writes the N-electron wavefunction as a linear combination of different electron configurations, usually using the Hartree-Fock wavefunctions for each configuration.

BIOSKETCH Sylvia Ceyer

Sylvia Ceyer is the J. C. Sheehan Professor of Chemistry at MIT, where she and her research group investigate how molecules interact with solid surfaces. One of her goals has been a better understanding of the pressure-dependence of chemical reactions that occur on a surface. Surface chemistry is normally investigated under ultra-high vacuum conditions, at pressures of 10^{-13} bar or less, in order to allow methods like Auger spectroscopy (Example 4.3) and electron diffraction (Section 1.3) to characterize the reaction. These conditions make it difficult to study how pressure affects the reaction, however. The Ceyer group developed one technique that they christened "Chemistry with a Hammer." In this method, the reactant—methane, for example—is gently laid on the solid surface with too little energy to react. A high-speed beam of non-reactive noble gas atoms then strikes the surface, raising the effective temperature and pressure at the surface—simulating the reaction conditions the group wants to study, but only at the point at which the beam hits the surface. Overall, the pressure is still low enough to allow diagnostic tools like Auger spectroscopy and electron diffraction to function.

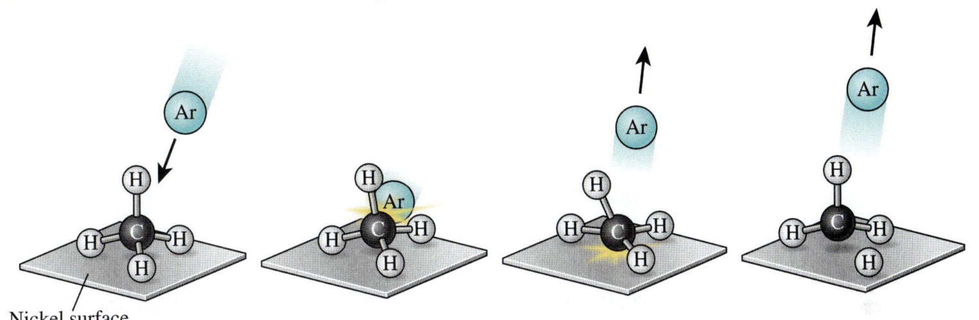

▲ Chemistry with a hammer. (After an image by Tom Dunne, *American Scientist* **87**, p. 21, 1999).

4.3 Spin Wavefunctions and Symmetrization

All of the preceding work describes the *spatial* wavefunctions for many-electron atoms. The complete wavefunction should also describe the spin of the electrons, and so we look at that next and find some remarkable results.

Indistinguishability and the Pauli Exclusion Principle

Let's simplify the notation by using the atomic orbital designations. The notation $1s(2)$, for example, represents the wavefunction for electron 2 in the $1s$ orbital:

$$1s(2) = \left(\frac{\sqrt{Z^3}}{\pi a_0^3}\right)e^{-Zr_2/a_0}. \tag{4.31}$$

Now we add the spin wavefunctions to our picture of the $1s^2$ He atom. We know the spatial part of the wavefunction must be of the form $1s(1)1s(2)$, but there are different possible combinations for the spin orbitals. To find the available choices for the spin wavefunctions, assume that: (*i*) each electron is in either an α spin orbital ($m_s = 1/2$) or a β spin orbital ($m_s = -1/2$), and (*ii*) that the two electrons are **indistinguishable.** The first rule is based on quantum mechanical results discussed earlier, but the second premise introduces a new concept.

It wouldn't be easy, but let's say that we had an experiment that would prepare He^+ ions with the single electron (call it electron 1) in the state $m_s = 1/2$. We then add electron 2 to the He^+ to form $1s^2$ He, so now there is one electron with $m_s = 1/2$ and another with $m_s = -1/2$. If we now pull the $m_s = 1/2$ electron off the atom, it turns out that no experimental measurement can determine whether this particle is electron 1 or electron 2. Electrons 1 and 2 have been in the same atom, with overlapping de Broglie wavelengths. When raindrops falling on a puddle make waves that intersect one another, the waves we see are not carrying water from one place to another. Instead, the peak of each wave is simply a place where the volume of water has temporarily increased as the water shifts back and forth. There is no physical parameter that tells us whether the intersecting waves *pass through* one another or *bounce off* one another, or some combination. In the same way, when two electrons overlap, we cannot separate the charge and mass of one from the other. As far as we can tell, every electron has exactly the same mass and same charge and same spin as any other electron. Therefore, we have no way of knowing if electron 1 and electron 2 switched m_s values. A *rigorously correct* wavefunction must treat the electrons as indistinguishable particles. Every electron in our wavefunction, however we have labeled it, must have an equal chance of being in any of the available orbitals and of having any of the corresponding m_s values.

Experiments detect either the probability density $|\psi|^2$ or an expectation value $\langle A \rangle = \int \psi^* \hat{A} \psi \, d\tau$. The requirement that all the electrons are indistinguishable in a two-electron atom can therefore be written as

$$|\Psi(1,2)|^2 = |\Psi(2,1)|^2, \tag{4.32}$$

(where the upper case Ψ indicates the combined spin-spatial wavefunction) and for a three-electron atom,

$$|\Psi(1,2,3)|^2 = |\Psi(1,3,2)|^2 = |\Psi(2,1,3)|^2 = |\Psi(2,3,1)|^2 = |\Psi(3,1,2)|^2 = |\Psi(3,2,1)|^2. \tag{4.33}$$

The properties of the atom (or molecule) never change as a result of what number we assign the electron, because all electrons are identical.

Our rule (*i*) above limits us to three options: both electrons in α orbitals, both in β orbitals, or one in α while the other is in β. If we write these two-electron spin wavefunctions

$$\alpha(1)\alpha(2) \quad \alpha(1)\beta(2) \quad \beta(1)\alpha(2) \quad \beta(1)\beta(2),$$

we see that $\alpha(1)\alpha(2)$ and $\beta(1)\beta(2)$ respect the indistinguishability of the electrons. For example, $\beta(2)\beta(1)$ and $\beta(1)\beta(2)$ are mathematically identical because multiplication is commutative. We can switch the labels on the electrons, and the wavefunction remains unchanged. However, $\alpha(1)\beta(2)$ and $\beta(1)\alpha(2)$ are not identical expressions, in precisely the same way that xy^2 is not the same as yx^2; the spin of electron 1 is a different coordinate from the spin of electron 2.

The electrons in these spin wavefunctions are distinguishable, because we can identify which electron in the function $\alpha(1)\beta(2)$ is electron 1 by measuring the direction of the spin magnetic moment on each electron.

We can **symmetrize** the $\alpha(1)\beta(2)$ and $\beta(1)\alpha(2)$ wavefunctions to obtain two new functions that reflect electron indistinguishability. We do this by using the two as a basis set from which we construct linear combinations, in the same way that we combined the $2s$ and $2p$ orbitals to create two sp hybrid orbitals in Eqs. 2.5 and 2.6:

$$[\alpha(1)\beta(2)+\beta(1)\alpha(2)]/\sqrt{2}, \qquad [\alpha(1)\beta(2)-\beta(1)\alpha(2)]/\sqrt{2}.$$

The factor of $1/\sqrt{2}$ is a normalization constant. The labels for electrons 1 and 2 are still a mathematical convenience, but they have no experimental meaning because the probability density of each function is unchanged when the labels are reversed:

$$\begin{aligned}\{[\alpha(1)\beta(2)+\beta(1)\alpha(2)]/\sqrt{2}\}^2 &= [\alpha(1)^2\beta(2)^2 \\ &\quad +2\alpha(1)\beta(1)\alpha(2)\beta(2)+\beta(1)^2\alpha(2)^2]/2 \\ &= \{[\alpha(2)\beta(1)+\beta(2)\alpha(1)]/\sqrt{2}\}^2\end{aligned} \tag{4.34}$$

$$\begin{aligned}\{[\alpha(1)\beta(2)-\beta(1)\alpha(2)]/\sqrt{2}\}^2 &= [\alpha(1)^2\beta(2)^2 \\ &\quad -2\alpha(1)\beta(1)\alpha(2)\beta(2)+\beta(1)^2\alpha(2)^2]/2 \\ &= \{[\alpha(2)\beta(1)-\beta(2)\alpha(1)]/\sqrt{2}\}^2\end{aligned} \tag{4.35}$$

We have introduced a symmetry that was not present in the original $\alpha(1)\beta(2)$ and $\beta(1)\alpha(2)$ functions; hence the term *symmetrized* wavefunctions.

The ground $1s^2$ state of helium now appears to have the following four possible spin-spatial wavefunctions:

$$\begin{aligned}\Psi_a &= 1s(1)1s(2)\alpha(1)\alpha(2) \qquad & \Psi_b &= 1s(1)1s(2)\chi_{\text{sym}} \\ \Psi_c &= 1s(1)1s(2)\chi_{\text{anti}} & \Psi_d &= 1s(1)1s(2)\beta(1)\beta(2),\end{aligned} \tag{4.36}$$

where

$$\chi_{\text{sym}} = \frac{1}{\sqrt{2}}[\alpha(1)\beta(2) + \beta(1)\alpha(2)] \qquad \chi_{\text{anti}} = \frac{1}{\sqrt{2}}[\alpha(1)\beta(2) - \beta(1)\alpha(2)]. \tag{4.37}$$

Only one of these results is valid, however.

Let $\hat{P}_{21}$ represent a **permutation operator** that exchanges the labels on electrons 1 and 2,

$$\hat{P}_{21}\psi(1,2) = \psi(2,1). \tag{4.38}$$

Wavefunctions may be either **symmetric,**

$$\psi_{\text{sym}}(2,1) = \psi_{\text{sym}}(1,2), \tag{4.39}$$

or **antisymmetric,**

$$\psi_{\text{anti}}(2,1) = -\psi_{\text{anti}}(1,2), \tag{4.40}$$

with respect to the exchange of those electrons. In other words, when the labels are switched, the wavefunction may remain the same, or the phase (but the phase only) may change. Because the physical properties of the atom always vary as the square modulus of the wavefunction, these properties do not vary with electron exchange, even if the phase changes. In other words, $|\psi|^2$ has to be constant under exchange of electrons, but ψ itself does not have to be. If the permutation results in a new wavefunction ψ' such that $|\psi'|^2 \neq |\psi|^2$, we will call the wavefunction **asymmetric** or **non-symmetric** with respect to that permutation. For permutation of the electron labels, a valid wavefunction *cannot* be asymmetric.

For indistinguishable particles, a valid quantum mechanical wavefunction must be either symmetric or antisymmetric with respect to permutation of any two labels for those particles. We break fundamental particles into two classes: those with integer spin are **bosons** and those with half-integer spin are **fermions.** Paul Dirac's analysis of the spin angular momenta in electrons (Section 3.4), by allowing the construction of complete spin-spatial wavefunctions, revealed the following:

> **A valid wavefunction must be antisymmetric with respect to the exchange of any two indistinguishable fermions.**

If we switch the labels on any two electrons in the same atom or molecule, the wavefunction *must change sign.* In contrast, Dirac's solution for integer spin particles predicts that a wavefunction must be *symmetric* if we exchange the labels on any two indistinguishable bosons. Let's call this the **symmetrization principle.** The proton, neutron, and electron each has a spin quantum number s of $\frac{1}{2}$, and thus they are fermions.

The symmetrization principle applies only to the many-particle spin-spatial wavefunction. We can factor $\Psi(1,2)$ into the spin and spatial parts:

$$\hat{P}_{21}\Psi(1,2) = \hat{P}_{21}[\psi(1,2)\chi(1,2)] = \psi(2,1)\chi(2,1). \tag{4.41}$$

If both parts are symmetric, the overall wavefunction will be symmetric:

$$\begin{aligned}\hat{P}_{21}\Psi(1,2) &= \psi_{\text{sym}}(2,1)\chi_{\text{sym}}(2,1) = \psi_{\text{sym}}(1,2)\chi_{\text{sym}}(1,2) \\ &= \Psi(1,2).\end{aligned} \tag{4.42}$$

And if both parts are antisymmetric, the overall wavefunction will be symmetric:

$$\begin{aligned}\hat{P}_{21}\Psi(1,2) &= \psi_{\text{anti}}(2,1)\chi_{\text{anti}}(2,1) = [-\psi_{\text{anti}}(1,2)][-\chi_{\text{anti}}(1,2)] \\ &= \Psi(1,2).\end{aligned} \tag{4.43}$$

The overall wavefunction is valid only if the spatial and spin parts of the wavefunction have *different* symmetry with respect to permutation of two electrons:

$$\begin{aligned}\hat{P}_{21}\Psi(1,2) &= \psi_{\text{anti}}(2,1)\chi_{\text{sym}}(2,1) = [-\psi_{\text{anti}}(1,2)][\chi_{\text{sym}}(1,2)] \\ &= -\Psi(1,2),\end{aligned} \tag{4.44}$$

$$\begin{aligned}\hat{P}_{21}\Psi(1,2) &= \psi_{\text{sym}}(2,1)\chi_{\text{anti}}(2,1) = [\psi_{\text{sym}}(1,2)][-\chi_{\text{anti}}(1,2)] \\ &= -\Psi(1,2).\end{aligned} \tag{4.45}$$

For the electron configuration $1s^2$, the spatial part of the wavefunction ψ is symmetric because $1s(1)1s(2) = 1s(2)1s(1)$. Therefore, the spin part must be antisymmetric. Of our four spin wavefunctions in Eq. 4.36, only one is antisymmetric: χ_{anti}. The ground state wavefunction for the He atom is therefore

$$\frac{1}{\sqrt{2}}1s(1)1s(2)[\alpha(1)\beta(2) - \beta(1)\alpha(2)]. \tag{4.46}$$

Notice that the two terms of the spin wavefunction require that electrons 1 and 2 have opposite spin: when electron 1 is α, electron 2 is β, and vice versa. So if nothing else, we've come up with a fancy way of arriving at a result you already know as the **Pauli exclusion principle:** electrons in the same orbital (the same spatial wavefunction) must have different values of m_s. The spatial wavefunction that puts two electrons in the same orbital ψ_i is symmetric: $\psi_i(1)\psi_i(2)$. The spin wavefunction in which both electrons have the same m_s is also symmetric: $\alpha(1)\alpha(2)$ or $\beta(1)\beta(2)$.

Since neither wavefunction changes sign when the labels are switched, the product of these spatial and spin functions is a symmetric overall wavefunction, and is therefore invalid. This is true for every pair of electrons, no matter how many there are in the wavefunction, and it applies to molecules as well as to atoms. The Pauli exclusion principle is one direct result of the more general symmetrization principle, but it has a particularly deep meaning for chemistry. Because we cannot put more than two electrons in each orbital, the ground states of the atoms vary greatly from one element to the next. The electrons become distributed among orbitals with different angular momenta, allowing the complex bonding arrangements that we find in molecular structure. The shell structure of the atom, with higher energy electrons averaging greater distances from the nucleus, justifies much of the chemical reactivity that formed the original basis for the periodic table. We will revisit this at the end of the chapter.

Spin Multiplicity and Excited States

At this point, let's adopt an even more compact shorthand for these many-electron product wavefunctions, dropping the labels (1) and (2) and always writing the terms in order by electron number. For example, $1s(1)1s(2)[\alpha(1)\beta(2) - \beta(1)\alpha(2)]$ becomes $1s1s(\alpha\beta - \beta\alpha)$.

Our unused, symmetric two-electron spin wavefunctions are then $\alpha\alpha$, χ_{sym}, and $\beta\beta$. If these are ever to come into play, we need an antisymmetric spatial wavefunction. We've just found that we can't obtain that for both electrons in the same orbital, but what if we form an excited state of helium, with the electrons in different orbitals? Let's take the simplest case, the lowest excited electron configuration $1s^1 2s^1$. Now there is a new two-electron spatial wavefunction, and we have the symmetrization problem again. The symmetrized spatial wavefunctions are

$$\psi_{\text{sym}} = \frac{1}{\sqrt{2}}(1s2s + 2s1s), \qquad \psi_{\text{anti}} = \frac{1}{\sqrt{2}}(1s2s - 2s1s).$$

The spin wavefunction that goes with the symmetric spatial wavefunction ψ_{sym} must again be χ_{anti}, as for the spatially symmetric ground state. But for ψ_{anti} the spin wavefunction may be any of the three symmetric spin functions shown in Eq. 4.36. The number of spin wavefunctions that accompany a particular spatial state is called the **spin multiplicity** of the state. A state with only one spin wavefunction is called **singlet,** one with three spin wavefunctions is called **triplet,** and so on.

We can now use all four of our two-electron spin wavefunctions, obtaining four distinct quantum states within the $1s2s$ excited configuration of He. One of these, with ψ_{sym}, comprises a singlet spin state; the others make up a triplet spin state:

$$\psi_{\text{sym}}[\alpha\beta - \beta\alpha]/\sqrt{2} \qquad \psi_{\text{anti}}\alpha\alpha \qquad \psi_{\text{anti}}[\alpha\beta + \beta\alpha]/\sqrt{2} \qquad \psi_{\text{anti}}\beta\beta. \tag{4.47}$$

Let's pause a moment to point out two features of these symmetrized wavefunctions. First, if the electrons in a wavefunction are indistinguishable, then we can start to see how many terms must appear in the wavefunction. For example, the $1s^2$ ground state of He is constructed from two distinct one-electron states: $1s\alpha$ and $1s\beta$. The electrons must both be in the $1s$, according to the electron configuration, and one must be spin up while the other is spin down, according to the Pauli exclusion principle. There are two ways to put the two electrons into those two states, and *both* ways must appear in the complete wavefunction for the electrons to be indistinguishable. Sure enough, multiplying out the wavefunction in Eq. 4.46 we find that the $1s^2$ ground state of He has two terms: $1s1s\alpha\beta - 1s1s\beta\alpha$. Similarly, for the cases in $1s^1 2s^1$ He in which the two electrons have opposite spins α and β, there are *four* distinct possible states for each electron: $1s\alpha$, $1s\beta$, $2s\alpha$, and $2s\beta$. Each possibility has to be represented in the wavefunction, and the complete wavefunctions each have four terms: $1s2s\alpha\beta - 1s2s\beta\alpha + 2s1s\alpha\beta - 2s1s\beta\alpha$ for the singlet and $1s2s\alpha\beta + 1s2s\beta\alpha - 2s1s\alpha\beta - 2s1s\beta\alpha$ for the triplet.

The second feature of these wavefunctions to notice results from requiring the function to be antisymmetric. The wavefunction changes sign when the labels switch because there are equal numbers of terms with and without minus signs, and these signs get reversed by the permutation of labels. Using the examples we've just looked at, the $1s^2$ wavefunction has two terms, one with a minus sign and one without, while the two $1s^1 2s^1$ wavefunctions above each have four terms, two with minus signs and two without. So, on a brief examination, the general form of these (admittedly complicated) wavefunctions does make some sense.

The energies we calculated for helium by perturbation and variational methods depended only on the spatial wavefunction, but we can now show that the symmetry of the spin-spatial wavefunctions influences the energy as well. First, note that for the $1s^2$ and $1s^1 2s^1$ configurations of helium, there is no orbital angular momentum, so we have no spin-orbit energy to consider. The wavefunctions just given are zero-order approximations. The corresponding zero-order energy can be calculated using Eq. 1.15, $E_n(E_h) = -Z^2/(2n^2)$ for each electron because the electron–electron repulsions are turned off in zero order:

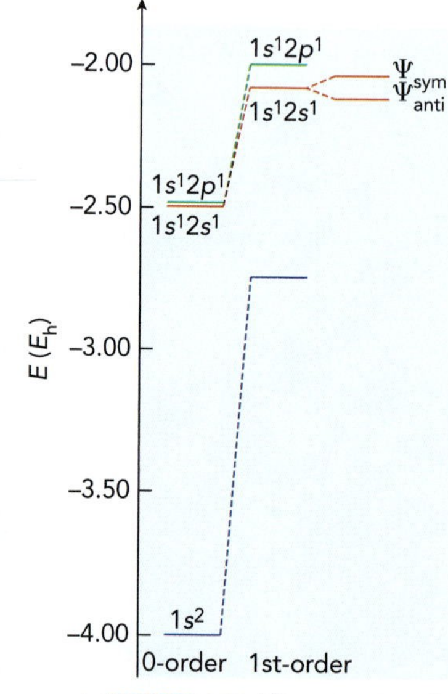

▲ **FIGURE 4.16 The energies of $1s^2$, $1s^1 2s^1$, and $1s^1 2p^1$ atomic He at zero and first order in perturbation theory.** In zero order, the $1s^1 2s^1$ and $1s^1 2p^1$ excited states are degenerate, but adding the electron–electron repulsion to first order breaks that degeneracy. The exchange energy splits the ψ_{sym} and ψ_{anti} energies in the $1s^1 2s^1$ first-order energy.

$$E_0 = E(1s) + E(2s) = \left(-\frac{Z^2}{2(1^2)} - \frac{Z^2}{2(2^2)}\right)E_h = -\left(\frac{4}{2} + \frac{4}{8}\right)E_h = -2.5\,E_h.$$

The first-order energy correction depends on the electron–electron repulsion term, and if we expand that contribution to the energy, we find

$$E_1' = \int\int \psi_{sym}^*\left(\frac{e^2}{4\pi\varepsilon_0 r_{12}}\right)\psi_{sym}\,d\tau_1 d\tau_2$$

$$= \underbrace{\frac{1}{2}\int\int\left(\frac{e^2}{4\pi\varepsilon_0 r_{12}}\right)1s(1)^2\,2s(2)^2\,d\tau_1 d\tau_2 + \frac{1}{2}\int\int\left(\frac{e^2}{4\pi\varepsilon_0 r_{12}}\right)1s(2)^2 2s(1)^2\,d\tau_1 d\tau_2}_{\text{Coulomb integral}}$$

$$+ \underbrace{\int\int\left(\frac{e^2}{4\pi\varepsilon_0 r_{12}}\right)1s(1)1s(2)2s(1)2s(2)\,d\tau_1 d\tau_2}_{\text{exchange integral}}, \tag{4.48}$$

$$\text{and } E_1' = \int\int \psi_{anti}^*\left(\frac{e^2}{4\pi\varepsilon_0 r_{12}}\right)\psi_{anti}\,d\tau_1 d\tau_2$$

$$= \underbrace{\frac{1}{2}\int\int\left(\frac{e^2}{4\pi\varepsilon_0 r_{12}}\right)1s(1)^2\,2s(2)^2\,d\tau_1 d\tau_2 + \frac{1}{2}\int\int\left(\frac{e^2}{4\pi\varepsilon_0 r_{12}}\right)1s(2)^2\,2s(1)^2\,d\tau_1 d\tau_2}_{\text{Coulomb integral}}$$

$$- \underbrace{\int\int\left(\frac{e^2}{4\pi\varepsilon_0 r_{12}}\right)1s(1)1s(2)2s(1)2s(2)\,d\tau_1 d\tau_2}_{\text{exchange integral}}, \tag{4.49}$$

The first two integrals in Eqs. 4.48 and 4.49, together called the **Coulomb integral,** are the same for ψ_{sym} and ψ_{anti}, and are equal to $0.420\,E_h$. The cross term, called the **exchange integral,** changes sign between the singlet and triplet states and has a value $0.044\,E_h$.[7] The difference in energies is therefore $0.088\,E_h$, with the triplet state lower in energy than the singlet state. We are still considering only the electric field contributions to the energy, neglecting any magnetic contributions.

CHECKPOINT Experiments find that the triplet $1s^1 2s^1$ state of helium has lower energy than the singlet $1s^1 2s^1$ state. The symmetrization principle predicts this, because, by allowing only certain combinations of spin and spatial wavefunctions to exist, the principle makes it impossible for two electrons with the same spin (in the triplet) to have as high a repulsion energy as two electrons with opposite spin (in the singlet).

Why is the triplet state lower in energy? A physical justification requires only that we remember that the n, l, and m_l quantum numbers specify the *spatial* part of the electron's wavefunction. According to the symmetrization principle, two electrons with the same value of m_s (our triplet state) must not have the same spatial wavefunction, but two electrons with different values of m_s (our singlet state) may. In the singlet state of $1s^1 2s^1$ He, the two electrons are permitted to have overlapping wavefunctions, because m_s is different. This increases the contribution of the electron–electron repulsion term to the energy, making the singlet state less stable. The symmetrization principle effectively keeps the triplet state electrons at greater average distance, with correspondingly lower electron–electron repulsion. This effective force separating electrons with the same value of m_s is called the **exchange force.**

Slater Determinants

For atoms with more than two electrons, the wavefunctions must become more elaborate to satisfy the symmetrization principle. However, John Slater developed a general method for reliably generating many-electron spin-spatial wavefunctions, antisymmetric with respect to $\hat{P}_{21}$ (exchange of the electron labels 1 and 2), for *any number* of electrons. We call these wavefunctions **Slater determinants,** because they are obtained by taking the determinant of a matrix of possible one-electron wavefunctions. For example, for ground state He, there are two possible one-electron spin-spatial wavefunctions for each electron: $1s\alpha$ and $1s\beta$. We set up a 2×2 matrix in which each row corresponds to a different electron and each column to a different wavefunction:

$$\begin{pmatrix} 1s(1)\alpha(1) & 1s(1)\beta(1) \\ 1s(2)\alpha(2) & 1s(2)\beta(2) \end{pmatrix}.$$

Next we take the determinant of this matrix and multiply by a normalization factor of $1/\sqrt{2}$:

$$\frac{1}{\sqrt{2}}\begin{vmatrix} 1s(1)\alpha(1) & 1s(1)\beta(1) \\ 1s(2)\alpha(2) & 1s(2)\beta(2) \end{vmatrix} = \frac{1}{\sqrt{2}}\left[1s(1)\alpha(1)1s(2)\beta(2) - 1s(1)\beta(1)1s(2)\alpha(2)\right].$$

This is the same wavefunction obtained in Eq. 4.46, which we showed to be antisymmetric with respect to $\hat{P}_{21}$.

For any electron configuration of N electrons, the Slater determinant is

$$\Psi_{det} = \frac{1}{\sqrt{N!}}\begin{vmatrix} \psi_1(1)\chi_1(1) & \psi_2(1)\chi_2(1) & \cdots \\ \psi_1(2)\chi_1(2) & \psi_2(2)\chi_2(2) & \cdots \\ \vdots & & \end{vmatrix},$$

[7] It is not obvious that the exchange integral is positive, because the function $1s2s$ is positive in some regions and negative in others. It turns out that the factor of $1/r_{12}$ ensures that the positive contribution is always larger in the exchange integrals.

where each $\psi_i\chi_i$ is a different one-electron spin-spatial wavefunction. The Slater determinant always delivers an antisymmetric wavefunction, and for most energy calculations this is sufficient. However, the Slater determinant has one major failing: it does not accurately predict all the wavefunctions when two different spin states (such as singlet and triplet) are near in energy.

To show this, let's apply this to the $1s^1 2s^1$ excited electron configuration of He. We need to solve a different determinant for each wavefunction. For example, the state with both electrons having α spin gives the following Slater determinant:

$$\frac{1}{\sqrt{2}}\begin{vmatrix} 1s(1)\alpha(1) & 2s(1)\alpha(1) \\ 1s(2)\alpha(2) & 2s(2)\alpha(2) \end{vmatrix} = \frac{1}{\sqrt{2}}[1s(1)\alpha(1)2s(2)\alpha(2) - 2s(1)\alpha(1)1s(2)\alpha(2)],$$

which is the same as the $\psi_{\text{anti}}\alpha\alpha$ wavefunction in Eq. 4.47. However, if we look at the case when the two electrons have opposite spins, we get new results. For example, one Slater determinant gives

$$\frac{1}{\sqrt{2}}\begin{vmatrix} 1s(1)\alpha(1) & 2s(1)\alpha\beta(1) \\ 1s(2)\alpha(2) & 2s(2)\beta(2) \end{vmatrix} = \frac{1}{\sqrt{2}}[1s(1)\alpha(1)2s(2)\beta(2) - 2s(1)\beta(1)1s(2)\alpha(2)],$$

which is indeed antisymmetric under exchange of the labels 1 and 2, but is different from the wavefunctions $\psi_{\text{sym}}[\alpha\beta - \beta\alpha]/\sqrt{2}$ and $\psi_{\text{anti}}[\alpha\beta + \beta\alpha]/\sqrt{2}$ in Eq. 4.47. What is the origin of this difference, and which function is more accurate?

The problem is with the Slater determinant. When we build the matrix, we are sometimes forced to make an arbitrary choice which spin function to associate with which spatial function. In the example above, the $1s$ electron is always the α electron. But in the same way that the $1s$ electron has an equal chance of being electron 1 or electron 2, the $1s$ electron also has an equal chance of being the α or β spin electron. The Slater determinant above is an artificial mixture of the singlet and triplet states, and does not correctly predict the energy of the excited state. However, the Slater determinants for the ground state and the $\alpha\alpha$ and $\beta\beta$ spin excited states are accurate. Most of the time, this method for generating antisymmetric many-electron wavefunctions works well.

4.4 Vector Model of the Many-Electron Atom

Precise evaluations of the atomic energy for many-electron systems require the sort of explicit analysis we were just looking at, in which the wavefunctions are written out and symmetrized. In a properly symmetrized many-electron wavefunction, reversing the labels of *any* two electrons in the function must change the sign of the function. This leads to complicated wavefunctions for many-electron atoms, particularly because in most cases the spin and spatial parts of the wavefunction can no longer be separated. Commonly, matrix algebra is used to determine these wavefunctions in a task we usually leave to computers.

Fortunately, we can construct an accurate, qualitative picture of the energy level structure of many-electron configurations by means of the **vector model of the atom.** This is essentially the vector coupling picture we used earlier for the spin-orbit coupling in the hydrogen atom, but we extend it by allowing each electron to contribute to the overall orbital and spin angular momenta. For example, in the He atom, the spin angular momenta of the two electrons,

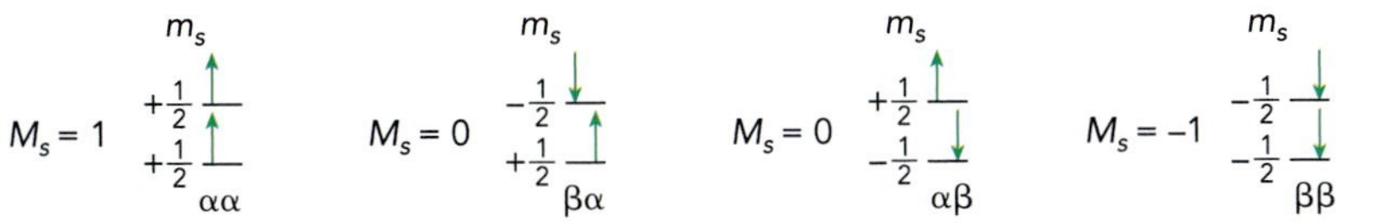

m_s m_s m_s m_s

$M_s = 1$ $+\frac{1}{2}$ $+\frac{1}{2}$ $\alpha\alpha$ $M_s = 0$ $-\frac{1}{2}$ $+\frac{1}{2}$ $\beta\alpha$ $M_s = 0$ $+\frac{1}{2}$ $-\frac{1}{2}$ $\alpha\beta$ $M_s = -1$ $-\frac{1}{2}$ $-\frac{1}{2}$ $\beta\beta$

◀ FIGURE 4.17 **Vector model illustration of spin-spin couplings between the two electrons in the $1s^1 2s^1$ state of helium.** The $M_S = \pm 1$ states result from vector sum of the two $\vec{s}$ vectors to form an $S = 1$ triplet state.

each with $s = \frac{1}{2}$, sum according to $\vec{s_1} + \vec{s_2} = \vec{S}$. This gives a total spin magnitude S of either 0 or 1, as drawn in Fig. 4.17.

These give rise to the singlet and triplet states, respectively. The antisymmetric $(\alpha\beta - \beta\alpha)$ spin state yields a value for the spin projection quantum number $M_S \equiv m_{s_1} + m_{s_2} = 0$, while the triplet state spin wavefunctions correspond to $M_S = -1\,(\beta\beta)$, $0\,(\alpha\beta + \beta\alpha)$, and $1(\alpha\alpha)$. We could have predicted the multiplicity of the $1s^1 2s^1$ He atom without ever considering the explicit wavefunctions or their symmetry at all, although we still need the Pauli exclusion principle to exclude cases where both electrons have the same spin and spatial wavefunction.

In the same way, we can add the one-electron orbital angular momenta l_i to obtain the total orbital angular momentum L. With both electrons in the He atom in s orbitals, $l_1 + l_2 = 0$, so $L = 0$. Once we start exciting the electrons into p orbitals, however, L can take on other values. In the $2p^2$ state, S can again take on only the values 0 and 1, but L can take on values 0, 1, and 2. The quantum numbers L and S represent only the magnitude of the vectors $\vec{L}$ and $\vec{S}$, and are therefore never negative. The z axis projection of the vector is again given by a magnetic quantum number M_L or M_S, where $M_L = -L, -L+1, -L+2, \ldots, L$, and $M_S = -S, -S+1, -S+2, \ldots, S$.

One might think that many-electron atoms must have huge angular momentum quantum numbers, but that's not generally the case. Atoms at all but extremely high excitation energies (which are short-lived states and not of general interest) have a core of electrons occupying all the orbitals with low quantum number n. When a subshell is filled, all the available values of m_l and m_s are taken, and L and S sum to zero for those electrons. It is usually the valence electrons that contribute to L and S. In the ground state Na atom, all the $1s$, $2s$, and $2p$ subshells are filled, with resulting total spin and orbital angular momenta equal to zero from their contributions. Adding the lone $3s$ electron gives $S = \frac{1}{2}$ and $L = 0$.

There is a third quantum number used to specify the state of a many-electron atom: J, analogous to the quantum number j introduced in Section 3.4 for the one-electron spin-orbit effect. This is the magnitude of the total angular momentum vector $\vec{J} = \vec{L} + \vec{S}$, therefore taking any value in the range $|L-S|, |L-S|+1, \ldots, L+S$. The direction of the total angular momentum vector is quantized by the magnetic quantum number $M_J = M_L + M_S$, which lies in the range $-J, -J+1, -J+2, \ldots, J$. The electronic state of an atom specified by L and S is called a **term** and represented by its **term symbol,** $^{(2S+1)}L$, where the value of L is indicated by the S, P, D, etc., labeling scheme. Similarly, the energy of the state is called its **term value.** This is old terminology, predating the interpretation of the quantum spin states, and the multiplicity (which was a direct experimental observable) is given instead of the quantum number S. A given J value within the term is specified by writing J as a subscript to the term symbol. The ground state term symbol of sodium would therefore be written $^2S_{1/2}$, as would the term symbol for ground state hydrogen. The term symbol only describes the relative orientations of the angular momenta that arise from a particular electron configuration. It gives additional information, but doesn't replace the electron configuration.

This means of describing the angular momentum contributions is called **LS-coupling** or **Russell-Saunders coupling.** This is only one way to describe the coupling of angular momenta in many-electron atoms, most appropriate to the low-lying electronic states of elements in the first three periods of the periodic table but useful for other cases as well. Another method, called **jj-coupling,** involves the coupling of the one-electron angular momenta first, $\vec{l} + \vec{s} = \vec{j}$, followed by coupling of the j's to get the total angular momenta. This coupling and similar schemes are used for heavy atoms and for highly excited states with weak electron–electron interactions.

Several terms may be available for a particular electronic configuration. To find the term symbols, the following recipe may be useful.[8]

1. For each subshell that is partly occupied in the electron configuration, list in a row the available values of m_l. Each of these now represents an orbital into which we can put up to two electrons.
2. The electron configuration states the number of electrons that go into each subshell. Your job is to list every possible way of putting those electrons into the available orbitals, labeling the spins with $\uparrow$ or $\downarrow$ to indicate m_s values of $\frac{1}{2}$ and $-\frac{1}{2}$, respectively. As examples, reversing the spin of a single electron in an orbital adds a new entry to this list, because the set of m_s values changes, but reversing the spins of two electrons in the same orbital is not a new entry, because the list of quantum numbers is left unchanged. The complete set of these combinations of m_l and m_s values is our basis set, from which we will construct the term states.
3. For each of these basis states, write down the M_L and M_S values by adding up the m_l and m_s values.
4. Find the largest value of M_S and set this equal to $S_{\max}$.
5. Among all the states with that value of M_S, find the largest value of M_L and set this equal to $L_{\max}$.
6. Now find $(2S_{\max} + 1)(2L_{\max} + 1)$ states such that $-L_{\max} \le M_L \le L_{\max}$ and $-S_{\max} \le M_S \le S_{\max}$, *without duplicating any combination of values* M_L, M_S. Label each of these states by those values of $L_{\max}$ and $S_{\max}$. These basis states are assigned to that term.
7. Ignoring the assigned basis states, repeat the process from (4) until all the basis states are assigned to a term.

CHECKPOINT The goal of the vector model is to determine all the different ways that the magnetic fields of the electrons can combine to give distinct energy levels. The relative orientations of the orbital and spin magnetic fields (which depend on the magnetic quantum numbers) determine these energies. The electron configuration only gives the n and l values of the electrons, so one electron configuration can result in many different term states.

The ordering of term energy levels within a given configuration can generally be found for LS-coupled atoms by using **Hund's rules.** They are these, in order of priority:

1. The energy of the state increases as S decreases.
2. The energy of the state increases as L decreases.
3. The energy of the state increases as J increases if the valence electron subshell is half filled or less, but the energy of the state increases as J decreases if the valence electron subshell is more than half filled.

[8]This version is more instructive but also more tedious than related schemes used in inorganic chemistry, where the couplings can be quite complicated.

EXAMPLE 4.4 The Vector Model of Atomic Oxygen

CONTEXT In the uppermost region of the earth's atmosphere, ultraviolet radiation from the sun ionizes atoms, and the ions and electrons may become trapped by the magnetic field near Earth's poles. Occasionally, the ions and electrons recombine to form neutral atoms in excited states, often excited term states of the lowest-energy electron configuration. The excited state atoms emit radiation as they relax back down to the ground state, and these emissions occur at distinct wavelengths characteristic of the atom. Oxygen and nitrogen, being most abundant in the atmosphere, are responsible for the strongest of these emissions, which give the colors to the aurora borealis seen in the northern skies.

PROBLEM What are the terms of the ground electronic configuration oxygen atom?

SOLUTION The ground state oxygen atom has six valence electrons in the configuration $2s^22p^4$. The $2s$ electrons fill their subshell and contribute no net angular momentum. Recalling the Pauli exclusion principle, the four remaining $2p$ electrons can be arranged in 15 non-identical ways (see Fig. 4.18) in the three orbitals for given m_l and m_s, giving 15 basis states of M_L and M_S. Some of these have duplicate values—(M_L, M_S) = (1,0), (0,0), and (−1,0)—but none comes from identical electron placement. Now we identify the term states:

	$m_l = 0$			M_L	M_S	
$2s^2$	↑↓			0	0	

	$m_l = -1$	0	1	M_L	M_S	
$2p^4$	↑↓	↑	↑	−1	+1	^3P
	↑↓	↑	↓	−1	0	^3P
	↑↓	↓	↑	−1	0	^1D
	↑↓	↓	↓	−1	−1	^3P
	↑↓	↑↓	—	−2	0	^1D
	↑↓	—	↑↓	0	0	^1S
	↑	↑↓	↑	0	+1	^3P
	↑	↑↓	↓	0	0	^1D
	↓	↑↓	↑	0	0	^3P
	↓	↑↓	↓	0	−1	^3P
	—	↑↓	↑↓	+2	0	^1D
	↑	↑	↑↓	+1	+1	^3P
	↑	↓	↑↓	+1	0	^1D
	↓	↑	↑↓	+1	0	^3P
	↓	↓	↑↓	+1	−1	^3P

▲ **FIGURE 4.18 Vector model of the ground state oxygen atom.**

1. Starting with the largest magnitude of M_S ($M_S = 1$), we choose the largest value of M_L for which $M_S = 1$. This is the basis state $(M_L, M_S) = (1, 1)$. We set $L = M_L = 1$, and $S = M_S = 1$, which predicts that our first (and lowest-energy) term state is a ^3P state with J values of 0, 1, and 2. For $L = 1$, the allowed values of M_L are −1, 0, and 1. Likewise, for $S = 1$, the allowed values of M_S are −1, 0, and 1. The ^3P term therefore accounts for nine of the basis states: $(M_L, M_S) = (-1, -1), (-1, 0), (-1, 1), (0, -1), (0, 0), (0, 1), (1, -1), (1, 0), (1, 1)$. They are labeled accordingly. Notice that we do not label *all three* (0, 0) basis states as arising from the ^3P term; a particular combination of values of (M_L, M_S) cannot appear twice in one term state.
2. Ignoring all the basis states we have assigned to the ^3P term, we take the largest value of M_S remaining, which is $M_S = 0$. In fact, all six of the remaining basis states have this value of M_S. We next choose the largest value of M_L among these states and find that $M_L = 2$ is the largest. Setting $L = M_L = 2$ and $S = M_S = 0$, we determine that our next lowest-energy term is s ^1D term, having only one J value, $J = 2$, but five basis states: (−2, 0), (−1, 0), (0, 0), (1, 0), and (2, 0). We assign five basis states to the ^1D term.
3. Finally, only one basis state remains unassigned; it is one of the three $(M_L, M_S) = (0,0)$ states. It doesn't matter which one, because the drawings with the electron spins pointing in specific directions are just a basis set—most of them do not correspond to a single quantum state of the atom. This remaining state must have $L = 0$ and $S = 0$, since it has no value of M_L or M_S greater than 0. It is therefore a ^1S state with $J = 0$.

The terms are $^3P_{2,1,0}$, $^1D_2$, and $^1S_0$.

So by these criteria, for the oxygen atom the ordering of the energy levels would be $^3P_2$, $^3P_1$, $^3P_0$, $^1D_2$, $^1S_0$, from lowest to highest. The first term state identified by the recipe just given should always be the ground state term because it has the largest value of S and largest value of L for that S. The known ground term states for the atoms are shown in the periodic table at the back of this text. It is possible to determine the term states of a given electron configuration by a more qualitative application of the vector model, and for simple problems this may be much faster than the recipe used previously. For example, if we wanted to know only the ground state term for the oxygen atom, we could have invoked Hund's first rule, that the lowest-energy term has the greatest value of S. With a configuration $1s^2 2s^2 2p^4$, we can have at most two unpaired electrons, and they must be in the $2p$ orbital. Therefore our greatest value of S (when these two unpaired electron spins are co-aligned) is $\frac{1}{2} + \frac{1}{2} = 1$, so our lowest-energy state must be a state with $S = 1$. Hund's second rule tells us that among states of a given S, the lowest-energy state will have the largest value of L. With four electrons in the $2p$ subshell, two of which must be in different m_l orbitals in order to maintain $S = 1$, the largest value of M_L that will be possible is 1. Therefore our ground state term is ^3P.

In the low-lying excited state configurations, the energy differences among the term states are often small compared to the excitation energy to a new configuration, but this is by no means a safe assumption. Spectroscopists working with highly excited states, or excited states of the transition metals, often face a significant challenge just in determining what quantum state they are observing from among the possibilities.

The splitting of the term into different J values is called the fine structure splitting, and it is the result of the same spin-orbit interactions that led to fine structure in the one-electron atom. For light atoms, the energy differences between J states of the same term are normally much smaller than the energy differences between the terms—generally only 10–1000 cm^{-1}, or less than 0.2 eV. Transitions between different fine structure levels of the same term, and between different terms of the same configuration, are quite weak compared to allowed transitions between different electron configurations, because these are magnetic dipole transitions, not electric dipole transitions. Yet some term transitions lie in the visible, including the $^1D \rightarrow {}^3P$ or $^1S \rightarrow {}^1D$ oxygen atom transitions seen in the aurora borealis. Fine structure transitions turn out to be especially important in interstellar clouds of atomic gas because they provide a means of radiative cooling and keep the clouds from overheating.

4.5 Periodicity of the Elements

Before leaving atoms, it is worthwhile to examine the periodicity of the elements—arguably the cornerstone of chemistry—in view of the preceding chapters. Consider the periodic table at the front of the text. Every element in column one (Group 1) has a ground state electron configuration, all subshells filled save for a single unpaired electron in an s orbital. As a result, all seven of the known elements in this group have the same ground state term symbol: $^2S_{1/2}$. They are also all metals (even hydrogen, although only at very high pressures).

Similarly, the ground states of the elements in Groups 2, 3, 11, 12, 13, 14, 15, 16, 17, and 18 are all constant over the group. The other groups contain either transition metals, with many electron configurations of different angular momenta and comparable energies, or transuranium elements for which some ground electron configurations are tentative. Keep in mind that even if tungsten and molybdenum, both in Group 6, do not share similar ground state electron configuration, there *is* a similar configuration nearby. For example, molybdenum, with the configuration $[Kr]5s^1 4d^5$ requires less than 1.5 eV to reach a configuration $[Kr]5s^2 4d^4$, chemically and physically similar to the $[Xe]6s^2 4f^{14} 5d^4$ configuration of tungsten in the same group.

Table 4.2 lists the lowest-energy term states of several atoms and illustrates quantitatively some of the periodic trends we have been discussing, showing a sampling of the electronic properties that group these elements in the modern periodic table. The alkali metals (Group 1, excepting H) and halogens (Group 17) follow very consistent patterns in their electronic states. The lowest term states for the halogens are ${}^2P$, ${}^4P$, ${}^2P$, and ${}^4P$, with a jump of 7–13 eV between the ground and first excited state, and the lowest three excited states are all within 2 eV of each other. Within the Group 10 transition metals, the term symbols are much less consistent, because there are so many atomic orbitals with nearly equal energies, and their relative energies change when we go from one row of the periodic table to the next. The particularly stable electron configurations of the noble gases (Group 18) leads those atoms to have among the largest gaps between ground and first excited electronic states.

Table 4.2 is also a useful indicator of the range of energies one might expect for low-lying electronic states in atoms: very small for the transition metals, and up to 20 eV for He. With the exception of the metals and semiconductors (C, Si, and Ge), these excitation energies are quite high; 8 eV corresponds to over 90,000 K. Therefore, atoms are usually found in these excited term states during very energetic processes, such as combustion or laser excitation. Chemical behavior is largely dictated by the electronic states of the atoms, the same states that we now describe in detail using quantum mechanics. Many features of electronic energy levels in atoms will still be present in molecules, but other degrees of freedom become available also, as we see in the next chapter.

4.6 Atomic Structure: The Key to Chemistry

Finally, now that you're wondering why anyone would want to study anything but atoms, we're ready to move on to molecules. Our introduction to atomic physics lays the foundation for the molecular physics of Part II. Two features of our work up to this point merit special attention.

Classical and Quantum Mechanics

Classical mechanics and quantum mechanics are *not* incompatible. Classical mechanics describes macroscopic systems to high precision. Quantum mechanics describes microscopic systems to high precision, and (with greater effort) agrees with classical mechanics in the limit of large, massive, and energetic matter. In Part I, we have described models for the atom and for electromagnetic radiation which are based on both classical and quantum mechanics, combining a predominantly wave-like electron with a classical, point-mass and point-charge nucleus, for example.

TABLE 4.2 Electron configurations, term symbols, energies (*E*, in eV relative to the ground state) of the lowest term states for selected atoms. Ground state spin-orbit splittings (ΔE_{SO}) between highest and lowest J values are also given when relevant. Excited state configurations are indicated by listing only the highest-energy subshell(s); other subshells are unchanged from the ground state configuration. Configurations and term symbols are approximate for many of the larger atoms, and fine structure energies for excited state terms have been averaged.

Atom	Ground state		1st Excited State		2nd Excited State		3rd Excited State	
	state	ΔE_{SO} (cm^{-1})	state	E (eV)	state	E (eV)	state	E (eV)
Group 1								
H	$1s^1\,^2S$		$2s^1\,^2S$	10.2	$2p^1\,^2P$	10.2	$3p^1\,^2P$	12.1
Li	$[He]2s^1\,^2S$		$2p^1\,^2P$	1.8	$3s^1\,^2S$	3.4	$3p^1\,^2P$	3.8
Na	$[Ne]3s^1\,^2S$		$3p^1\,^2P$	2.1	$4s^1\,^2S$	3.2	$3d^1\,^2D$	3.6
K	$[Ar]4s^1\,^2S$		$4p^1\,^2P$	1.6	$5s^1\,^2S$	2.6	$3d^1\,^2D$	2.7
Rb	$[Kr]5s^1\,^2S$		$5p^1\,^2P$	1.6	$4d^1\,^2D$	2.4	$6s^1\,^2S$	2.5
Group 10								
Ni	$[Ar]4s^23d^8\,^3F$	2217	$3d^94s^1\,^3D$	0.1	$3d^94s^1\,^1D$	0.4	$3d^84s^2\,^1D$	1.7
Pd	$[Kr]4d^{10}\,^1S$		$4d^95s^1\,^3D$	1.0	$4d^95s^1\,^1D$	1.5	$4d^85s^2\,^3F$	3.4
Pt	$[Xe]4f^{14}5d^96s^1\,^3D$	10132	$5d^86s^2\,^3F$	0.1	$5d^{10}\,^1S$	0.8	$5d^86s^2\,^3P$	0.8
Group 14								
C	$[He]2s^22p^2\,^3P$	44	$2s^22p^2\,^1D$	1.3	$2s^22p^2\,^1S$	2.7	$2s^12p^3\,^5S$	4.2
Si	$[Ne]3s^23p^2\,^3P$	223	$3s^23p^2\,^1D$	0.8	$3s^23p^2\,^1S$	1.9	$3p^14s^1\,^3P$	4.9
Ge	$[Ar]4s^23d^{10}4p^2\,^3P$	1410	$4s^24p^2\,^1D$	0.9	$4s^24p^2\,^1S$	2.0	$4s^24p^15s^1\,^3P$	4.7
Sn	$[Kr]5s^24d^{10}5p^2\,^3P$	3428	$5s^25p^2\,^1D$	1.1	$5s^25p^2\,^1S$	2.1	$5s^25p^16s^1\,^3P$	4.5
Group 17								
F	$[He]2s^22p^5\,^2P$	404	$2p^43s^1\,^4P$	12.7	$2p^43s^1\,^2P$	13.0	$2p^43p^1\,^4P$	14.4
Cl	$[Ne]3s^23p^5\,^2P$	882	$3p^44s^1\,^4P$	9.0	$3p^44s^1\,^2P$	9.2	$3p^44p^1\,^4P$	10.3
Br	$[Ar]4s^23d^{10}4p^5\,^2P$	3685	$4p^45s^1\,^4P$	8.1	$4p^45s^1\,^2P$	8.4	$4p^45p^1\,^4P$	9.3
I	$[Kr]5s^24d^{10}5p^5\,^2P$	7603	$5p^46s^1\,^4P$	7.3	$5p^46s^1\,^2P$	7.4	$5p^46p^1\,^4P$	8.2
Group 18								
He	$1s^2\,^1S$		$1s^12s^1\,^3S$	19.8	$1s^12s^1\,^1S$	20.6	$1s^12p^1\,^3P$	21.0
Ne	$[He]2s^22p^6\,^1S$		$2p^53s^1\,^3P$	16.6	$2p^53s^1\,^1P$	16.8	$2p^53p^1\,^3D$	18.5
Ar	$[Ne]3s^23p^6\,^1S$		$3p^54s^1\,^3P$	11.6	$3p^54s^1\,^1P$	11.8	$3p^54p^1\,^3D$	13.0
Kr	$[Ar]4s^23d^{10}4p^6\,^1S$		$4p^55s^1\,^3P$	10.2	$4p^55s^1\,^1P$	10.6	$4p^55p^1\,^3D$	11.4
Xe	$[Kr]5s^24d^{10}5p^6\,^1S$		$5p^56s^1\,^3P$	8.4	$5p^56s^1\,^1P$	9.5	$5p^56p^1\,^3D$	9.8

Although classical mechanics alone fails to accurately describe the characteristics of individual photons, atoms, and molecules, it is common to use the language and concepts of the more familiar classical world as much as possible to describe the microscopic world. This is fairly safe, as long as one's classical picture is informed by the non-classical results from quantum mechanics. The major non-classical results we have encountered so far are the following.

1. Electromagnetic radiation interacts with matter only in integer numbers of photons, each with energy $E_{\text{photon}} = h\nu$, where ν is the radiation frequency.
2. Any system (other than the ideal free particle) has only quantized energy levels; there will normally be a set of energy levels that are the *only* energies allowed to the system. Other parameters, such as the angular momentum and velocity, are therefore also quantized.

3. At sufficiently low momentum, particles take on wave properties such as constructive and destructive interference; electrons *can* occupy the same space at the same time if they have different m_s values, but the overall energy, mass, momentum, and charge will still be conserved.
4. No finite potential barrier is insurmountable; a particle can tunnel through a potential barrier, existing partially in a region where the total energy is less than the potential energy.

Four Parameters to Watch: E, U, N_{dof}, g

From beginning to end of this text, the masses and volumes of our systems grow by some 20 orders of magnitude. Yet in our model of chemical behavior, these extremes of scale are joined by a single set of physical laws, and we find that properties of particular importance early on remain in focus throughout our study.

The typical course in general chemistry revolves around the central role of energy. The study of quantized energy levels in atoms and molecules has led directly to our present understanding of chemical structure, and relative stability—a measure of energy—largely determines the motion and reactivity of molecules. Much of chemistry is concerned with predicting and exploiting the ways that the energy function varies from one system to another. That change is controlled principally by two contributions: the potential energy function, U, and the kinetic energy. From our study of the Schrödinger equation, we find that the kinetic energy operator always has the same expression, $-\hbar^2/(2m)\frac{\partial^2}{\partial q^2}$ for each coordinate q. The big difference from one system to the next is just how many coordinates there are. Any coordinate that our system is allowed to move along becomes one of its N_{dof} degrees of freedom.

As N_{dof} increases, not only do we see more terms in the kinetic energy expression, we also see many more possible states available to the system. The degeneracy, g, increases rapidly with N_{dof}, and this parameter will evolve into our principal tool for predicting how chemical systems behave.

By starting with the smallest systems in chemistry, we allow ourselves to see how these four parameters behave in the simplest limit. We've taken N_{dof} from 1 coordinate (x in the one-dimensional box) to $3N$ (in the N-electron atom), seeing in some detail just how this complicates solving the energy. The degeneracy as well has gone from $g = 1$ for all energy levels in the one-dimensional box to roughly $2(2l + 1)$ for each subshell in the many-electron atom. The variation of the energy equation from the one-dimensional box (E proportional to n^2) to the one-electron atom (E proportional to $-1/n^2$) is determined by the changes in U and N_{dof}, and will eventually inform our qualitative predictions about the results on much more complex quantum mechanical, and later classical, systems.

The challenge ahead, understanding the structures and energetics of small molecules, remains at the microscopic scale, only slightly larger than the distance scales of individual atoms. But it involves very significant increases in the available degrees of freedom and the complexity of the potential energy function, and the results can look quite different from what we have done so far. Seeing that there are common threads, including an abiding interest in these four fundamental parameters, reminds us of the relatively simple conceptual basis behind the chemical model: molecular structure dictates chemical behavior.

CONTEXT ***Where Do We Go From Here?***

We've worked out a model of the interactions among electrons and atomic nuclei that allows us to predict the properties of single atoms. The next step is to add more atomic nuclei in order to make molecules. The Coulomb force will continue to govern the interactions of all the particles we are working with, so up to a point we will see that the physics is only an extension of the principles we've just used to describe atoms. However, the addition of more atomic nuclei fundamentally changes the nature of the degrees of freedom that our particles can use to store and transfer energy. The principles of atomic structure will get us started, but then we will soon encounter the rich complexity of function and application that makes molecular structure the foundation of chemistry.

KEY CONCEPTS AND EQUATIONS

4.1 Many-electron spatial wavefunctions.

a. The Hamiltonian for a two-electron atom includes an electron–electron repulsion potential energy term, as well as kinetic and nuclear-electronic potential terms for each electron:

$$\hat{H} = -\frac{\hbar^2}{2m_e}\nabla_1^2 - \frac{\hbar^2}{2m_e}\nabla_2^2 - \frac{Ze^2}{4\pi\varepsilon_0 r_1} - \frac{Ze^2}{4\pi\varepsilon_0 r_2} + \frac{e^2}{4\pi\varepsilon_0 r_{12}}. \quad (4.1)$$

We can extend this to any N-electron atom:

$$\hat{H} = \sum_{i=1}^{N}\left[-\frac{\hbar^2}{2m_e}\nabla(i)^2 - \frac{Ze^2}{4\pi\varepsilon_0 r_i} + \sum_{j=1}^{i-1}\frac{e^2}{4\pi\varepsilon_0 r_{ij}}\right]. \quad (4.2)$$

b. An approximate solution to the Schrödinger equation that uses this Hamiltonian is provided by multiplying together the one-electron wavefunctions to obtain an electron configuration.

4.2 Computational methods. The electron configuration is only a first approximation to the many-electron wavefunction, however, because it doesn't allow the distribution of one electron to change in response to the position of another electron. This many-body problem can be solved by different approximations.

a. In perturbation theory, we split the Hamiltonian into a zero-order $\hat{H}_0$, which contains all the one-electron kinetic energy and potential energy terms, and a perturbation Hamiltonian $\hat{H}'$, which contains all the electron–electron repulsion terms. Then we use the Taylor series to expand the Schrödinger equation in $\hat{H}'$, solving the wavefunctions and energies for each term independently. For a two-electron atom, the first-order correction to the energy is

$$E_1' = \int \psi_0^* \frac{e^2}{4\pi\varepsilon_0 r_{12}} \psi_0 \, d\tau, \quad (4.16)$$

where ψ_0 is the zero-order wavefunction (the eigenstate of $\hat{H}_0$). The first-order wavefunction is

$$\psi_1 = \psi_0 + \sum_{k=1}^{\infty} \frac{\int \psi_0^* \hat{H}' \psi_{0,k} \, d\tau}{E_0 - E_{0,k}} \psi_{0,k} \quad (4.20)$$

and the second-order correction to the energy is

$$E_2' = \sum_{k=1}^{\infty} \frac{\left[\int \psi_0^* \hat{H}' \psi_{0,k} \, d\tau\right]^2}{E_0 - E_{0,k}}. \quad (4.21)$$

b. Although perturbation theory works well for a relatively rapid improvement in the accuracy of the energies and wavefunctions predicted for the many-electron atom, the variational method is more often used for problems demanding accurate results because it is relatively simple to code into software and works to very high accuracy (if everything else is in good shape) without changing the math. Variational calculations adjust the wavefunction to find the minimum value for the energy of the ground state. The idea is that the electrons do this naturally, finding the distribution with the lowest possible potential energy, so the lower the energy of our adjustable wavefunction, the closer it is to the right answer.

c. Most variational wavefunctions for atomic and molecular systems are built initially as a product of one-electron wavefunctions $\varphi(i)$, where each $\varphi(i)$ is a linear combination of several atomic orbital wavefunctions:

$$\varphi_i(i) = c_{1s}\psi_{1s}(i) + c_{2s}\psi_{2s}(i) + c_{2p_x}\psi_{2p_x}(i) + \ldots$$

$$= \sum_{n=1}^{n_{max}} \sum_{l=0}^{n-1} \sum_{m_l=-l}^{l} c_{nlm_l} \psi_{nlm_l}(i). \quad (4.25)$$

The N-electron wavefunction is then $\varphi_1(1)\varphi_2(2) \ldots \varphi_N(N)$. The values of the coefficients c_{nlm_l} for the different electrons are the adjustable parameters in the wavefunction.

d. To find the expectation value for the energy, we often start from a Hartree-Fock calculation, which approximates the electron–electron repulsion for electron i by averaging over all the positions of the other electrons, allowing us to solve the many-electron Schrödinger equation one electron at a time.

4.3 Spin wavefunctions and symmetrization.

a. Spin-spatial wavefunctions of electrons in the same atom must change sign when we reverse the labels on any two of the electrons:

$$\Psi(1,2) = -\Psi(2,1).$$

As an example, the spin-spatial wavefunction of the $1s^2$ configuration of He may be written

$$\frac{1}{\sqrt{2}} 1s(1)1s(2)[\alpha(1)\beta(2) - \beta(1)\alpha(2)], \quad (4.46)$$

where α and β are the two possible one-electron spin states $m_s = -1/2$ and $m_s = +1/2$, respectively.

4.4 Vector model of the many-electron atom. The vector model allows us to find all the possible ways that the angular momenta of the electrons in an atom may add together into net orbital and spin angular momenta L and S, and how those in turn may be combined into total angular momentum values J.

4.5 Periodicity of the elements. Atoms are grouped into columns in the periodic table according to similarity of their chemical properties, which we can see largely correspond to similarities in electron configuration.

4.6 Atomic structure. Throughout physical chemistry, our attention is drawn to four parameters of the system: the total energy, the potential energy, the number of coordinates, and the degeneracy.

KEY TERMS

- **Valence electrons** are the electrons likeliest to participate in chemical bonding, because they are the most weakly bound to the nucleus of their original atom. As a general rule, these are the electrons having the highest value of the principal quantum number n, plus any electrons in partially filled subshells.
- **Core electrons** are the non-valence electrons, found in lower energy orbitals and bound strongly enough to the nucleus that they are little affected by other atoms.
- **Fermions** are particles with half-integer spin, such as the electron or proton (both with spin of 1/2) or ^{35}Cl nucleus (with spin of 3/2).
- **Bosons** are particles with integer spin, such as a ^4He nucleus (with spin of 0) or a ^2H nucleus (with spin of 1).
- The **symmetrization principle** requires that changing the labels on two indistinguishable fermions in a spin-spatial wavefunction will invert the sign on the wavefunction, while changing the labels on two indistinguishable bosons leaves the wavefunction unaffected.
- An **antisymmetric** wavefunction is one that changes sign when an operation is performed, such as switching the labels on two indistinguishable fermions.
- The **Pauli exclusion principle** forbids any two electrons in the same atom from sharing the same set of values for the quantum numbers n, l, m_l, m_s, and arises as a result of applying the symmetrization principle to the electrons.
- The **term symbol** is a designation of the values of total orbital and spin angular momenta L and S for the quantum state of an atom. One electron configuration may have several term states with different energies.
- **Hund's rules** give a recipe for determining the relative energies of a set of term states that come from the same electron configuration. The most important of these is that the term states with greatest spin S are the lowest-energy of the set.

OBJECTIVES REVIEW

1. *Write the Hamiltonian for any atom.*
   Write the complete Hamiltonian for Be^+.
2. *Find a zero-order wavefunction (the electron configuration) and zero-order energy of any atom.*
   Write the electron configuration and calculate the zero-order energy (in E_h) for Be^+.
3. *Use the Hartree-Fock orbital energies to estimate the effective atomic number of the orbital.*

Use Table 4.1 to estimate Z_{eff} for the valence electron in Na.

4. *Determine the permutation symmetry of a many-particle function.*
Determine whether the function $\cos(-x_1)\cos(y_2) - \cos(-x_2)\cos(y_1)$ is symmetric or antisymmetric under exchange of the labels 1 and 2.

5. *Determine the term symbols and energy ordering of the term states resulting from any electron configuration.*
Find the term symbols and energy ordering (from lowest-energy to highest-energy) of the states arising from the electron configuration $1s^1 2p^1$.

PROBLEMS

Discussion Problems

4.1 The two-electron Hamiltonian (Eq. 4.1)

$$\hat{H} = \underbrace{-\frac{\hbar^2}{2m_e}\nabla_1^2}_{A} \underbrace{-\frac{\hbar^2}{2m_e}\nabla_2^2}_{B} \underbrace{-\frac{Ze^2}{r_1}}_{C} \underbrace{-\frac{Ze^2}{r_2}}_{D} \underbrace{+\frac{e^2}{r_{12}}}_{E}$$

contains five terms: A, B, C, D, and E. Write the correct letter or letters in response to the questions below.

a. Which term(s) make positive contributions to the energy?
b. Which term(s) make an exact solution to the Schrödinger equation impossible?
c. Which term(s) have the greatest magnitude?
d. Which term(s) appear in the Hartree-Fock hamiltonian for calculating orbital energy ε_1?

4.2 Is the effective atomic number for the valence electron higher for a neutral sodium atom or a neutral potassium atom?

4.3 One way to look at variational theory is to think of it as a sort of quantum mechanical version of Le Châtelier's principle. We start from an ideal system, where we know the solutions to the Schrödinger equation exactly. Now we stress that system by applying some kind of force; in other words, we change the potential energy curve a little. In response to the stress, our ground state wavefunction borrows characteristics from excited states to restore itself to the most stable distribution possible. Well, if the ground state has to borrow from the excited state to do this, doesn't that generally leave the excited states *less* stable in response to the stress? For example, the $1s$ ground state of hydrogen will deform in an electric field, borrowing character from the $2p_0$ wavefunction to create a more stable charge distribution where there is more negative charge (higher electron density) near the positive end of the electric field. In doing so, the excited state becomes the *opposite* mix of the original $2p_0$ and the $1s$ states, creating a state that would be distorted to *increase* in energy as the external field is turned on. If we start with an atom that's already in the $2p_0$ excited state, does it actually respond to the electric field by being polarized in the reverse direction, with more negative charge at the negative end of the field? Does the quantum mechanics allow the atom in this state to respond in exactly the "wrong" direction to an applied force?

4.4 Explain briefly why the energy gap between the $2s$ and $2p$ orbitals should be greater than the gap between the $3s$ and $3p$ orbitals in the many-electron atom.

4.5 Explain briefly why the values of $\langle e^2/(4\pi\varepsilon_0 r_{12})\rangle$ in Eqs. 4.48 and 4.49, roughly $0.42\,E_h$, are so small compared to the value of $1.25\,E_h$ for $\langle e^2/(4\pi\varepsilon_0 r_{12})\rangle$ from Eq. 4.17.

4.6 In the Hamiltonian for Be,

a. how many electron kinetic energy terms appear?
b. how many electron–electron repulsion terms appear?

4.7 What are the values of L and S for the ground electron configurations of any of the noble gases?

4.8 For $2s$ electrons in the ground state atoms Be, B^+, and Li^-,

a. which has the greatest amount of shielding?
b. which has the lowest-energy $2s$ electrons (having the highest ionization energy)?

Many-Electron Hamiltonians and Wavefunctions

4.9 Write the electron configuration for the lowest-energy excited state of the Na^+ ion.

4.10

a. In accordance with our version of perturbation theory, calculate the zero-order energy in E_h of the $1s^2 3s^1$ excited state Li atom.
b. Explain whether the first-order correction raises or lowers this energy.

4.11 The following is a proposed wavefunction for ground state Be:

$$\Psi = A\left[1s1s2s2s + 1s2s2s1s + 1s2s1s2s + 2s2s1s1s\right] \times \left[\alpha(1)\beta(2) - \beta(1)\alpha(2)\right]\left[\alpha(3)\beta(4) - \beta(3)\alpha(4)\right].$$

a. Find the normalization constant A.
b. What is wrong with this wavefunction?

4.12

a. Write a Hamiltonian to describe three electrons in a one-dimensional box of length a along the x axis.
b. Write the terms from your Hamiltonian you would treat as the perturbation if you had to solve the Schrödinger equation for this system using perturbation theory.

c. Write the equation for the **zero-order** energy E_0 in the ground state of this system.

4.13 Write the Hamiltonian for the electrons in the C^{3+} ion.

4.14 A few terms from the Hamiltonian for a many-electron atom or atomic ion are given next, but some are missing. Add the *minimum* number of additional terms necessary to complete the Hamiltonian, and identify the atom or ion.

$$\hat{H} = -\frac{5e^2}{(4\pi\varepsilon_0)r_3} - \frac{\hbar^2}{2m_e}\nabla_1^2 - \frac{\hbar^2}{2m_e}\nabla_2^2$$

4.15 Write the Hamiltonian for a *three*-electron atom, including the kinetic energy and Coulomb potential terms. Briefly define all the symbols you use.

4.16 The positron is a particle identical in all respects to the electron except that its charge is $+e$, exactly opposite to that of the electron. Write the Hamiltonian for the atom constructed from one nucleus with atomic number Z, two electrons, and one positron.

4.17 The following four wavefunctions for an electron are normalized.

a. $\psi_a(r < a_0) = \sqrt{3/a_0^3}\, Y_0^0(\theta, \phi),\ \psi_a(r \geq a_0) = 0$

b. $\psi_b(r) = \sqrt{\frac{4}{a_0^3}}\, e^{-r/a_0} Y_0^0(\theta, \phi)$

c. $\psi_c(r < a_0) = \sqrt{\frac{30}{a_0^5}}\,(a_0 - r) Y_0^0(\theta, \phi),$
$\psi_c(r \geq a_0) = 0$

d. $\psi_d(r) = \sqrt{\frac{8}{45a_0^5}}\, r^2 e^{-r/a_0} Y_0^0(\theta, \phi)$

We begin with a helium atom that already has one $1s$ electron. Pretend that we could add to that atom one more electron in a state given by any of these four wavefunctions. For which of these wavefunctions would that second electron experience the *least* shielding? The *most* shielding?

4.18 The hyperfine effect is the interaction between the nuclear spin and electron spin. Is the electron spin magnetic field greatest at the nucleus for in the ground state neutral atoms ^{1}H, ^{3}He, or ^{7}Li?

4.19 Estimate a reasonable value for Z_{eff} of an electron in the Na atom $2s$ orbital.

4.20 Start from an atom with electron configuration $1s^2 2s^2 3s^1$. Indicate for each parameter whether an increase in its value will cause Z_{eff} of the $3s$ electron to increase ("+"), decrease ("−"), or stay the same ("0").

Increase	Effect
Z	
number of $2p$ electrons	
number of $3p$ electrons	
nuclear spin I	
l for the $3s$ electron from 0 to 1	

4.21 Indicate whether the parameters listed should increase ("+"), decrease ("−"), or stay the same ("0") as we go from ground state atomic S ($Z = 16$) to ground state Cl^+ ion ($Z = 17$).

Increase	Effect from S to Cl^+
Z_{eff} for the $2s$ electrons	
total electron repulsion energy for the $2s$ electrons	
ε_i for the $2s$ electrons	
total zero-order energy E_0	
L for the ground term state	

4.22

a. Write the ground state electron configuration of sodium (Na).

b. Which subshell ($1s$, $2p$, $3d$, etc.) in this configuration has the electrons that are most strongly shielded?

c. Which subshell in this configuration has the smallest average distance $\langle r \rangle$ from the nucleus?

d. Which subshell in this configuration has the electrons with the greatest orbital angular momentum?

4.23 The following equation is proposed as a (relatively) simple way to estimate the effective atomic number Z_{eff} for electron j in a many-electron atom. It's not going to be exact, but assuming that the basic idea is sound, correct any qualitative errors you find so that the answer will at least be reasonable.

$$Z_{\text{eff}} \approx \left\{ \int_0^{r_j} \sum_{i=1}^{j-1} \left[Z - \int_0^{r_i} R^2_{n_i,l_i}(r_i) r_i^2 dr_i \right] R^2_{n_j,l_j}(r_j) \left(\frac{e^2}{4\pi\varepsilon_0 r_j} \right) r_j^2 dr_j \right\} \Big/ \left[\int_0^{r_j} R^2_{n_j,l_j}(r_j) r_j^2 dr_j \right]$$

4.24 According to the correspondence principle, which member of the following pairs will behave more like a classical particle?

a. core or valence electrons in the same atom

b. core electrons in lithium or iron atoms

c. ground state nitrogen atoms with the same *total* energy at a distance of 10 Å or at a distance of 2 Å (where attractive forces pull the atoms together)

4.25 Write an explicit, many-electron wavefunction ψ based on the electron configuration (what we're now calling the "zero-order" wavefunction) for the lowest excited state of Be^+.

4.26 Consider the suggestion that the $1s^1 2s^1$ excited state of helium is lower in energy than the $1s^1 2p^1$ excited state because the $2s$ orbital penetrates the $1s$ core electron density more effectively, thereby seeing greater stabilization nearer the nucleus. Let's estimate that stabilization by calculating an r-dependent effective atomic number $Z_{\text{eff}}(r)$ equal to the actual atomic number minus the fraction of $1s$ electron charge found within the distance r from the nucleus:

$$Z_{\text{eff}}(r) \equiv Z - \int_0^{r_1} R_{1,0}(r_1)^2\, r_1^2\, dr_1.$$

(We need to consider only the radial parts of the wavefunctions in this problem.) Use the average value theorem to find $\langle Z_{\text{eff}} \rangle$ for electron 2 in the $2s$ and in the $2p$ orbitals. Based on this argument, the higher value of $\langle Z_{\text{eff}} \rangle$ should be found for the less shielded, lower energy orbital.

4.27 How many distinct electronic energies, including hyperfine, would be obtained from a sample of ground state ($1s^2 2s^2 2p^1$) boron atoms, given that there are two abundant isotopes of boron, mass 11 ($I = 3/2$) and mass 10 ($I = 3$). Each isotope has a unique value of the nuclear magnetic moment μ_I.

4.28

a. In an atom with N electrons, how many distinct pairs of electrons are there (i.e., how many individual electron–electron repulsion terms need to be included in the Hamiltonian)?
b. Given this, use Table 4.1 to calculate the average repulsion energy *per pair of electrons* in He, Li, Be, B, and C.

4.29 Each of the following atoms, has two electrons in a $2s$ orbital:

$$\text{C, N, N}^+\text{, N}^*.$$

These are ground state atoms, except the N* atom, which has an electron configuration $1s^2 2s^2 2p^2 3s^1$. List these atoms from left to right in order of increasing energy of the $2s$ electrons.

4.30 Write the explicit integral necessary to find the electron–electron repulsion energy in the triplet $1s^1 2s^1$ excited state of atomic helium to first order in perturbation theory.

Spin and Symmetrization

4.31 A proposed spin-spatial wavefunction for the lithium atom $1s^2 2s^1$ ground state is

$$[1s1s2s + 2s1s1s + 1s2s1s]\,\alpha\alpha\alpha.$$

Is this a correct wavefunction?

4.32 One of the symmetrized wavefunctions for the Li atom ground state ($1s^2 2s^1$) is

$$\frac{1}{\sqrt{6}}[1s1s2s\alpha\beta\alpha + 2s1s1s\alpha\alpha\beta + 1s2s1s\beta\alpha\alpha$$
$$- 1s2s1s\alpha\alpha\beta - 1s1s2s\beta\alpha\alpha - 2s1s1s\alpha\beta\alpha].$$

a. Is this symmetric, antisymmetric, or non-symmetric (asymmetric) with respect to reversing electron labels 1 and 2?
b. What are the values of L, M_L, S, and M_S for this lithium atom wavefunction?
c. How many distinct, symmetrized wavefunctions are there for this system? Write one of these other wavefunctions.
d. What is the symmetry of this wavefunction with respect to cyclic permutation of the electrons: $1 \rightarrow 2$, $2 \rightarrow 3$, $3 \rightarrow 1$?

4.33 In $1s^1 2s^1$ excited-state helium, the Coulomb integral is equal to $0.420\,E_{\text{h}}$ and the exchange integral is equal to $0.044\,E_{\text{h}}$. Estimate the average distance $\langle r_{12} \rangle$ in Å between the two electrons (a) in the singlet state and (b) in the triplet states, using these numbers and assuming that $\langle r \rangle \approx \langle 1/r \rangle^{-1}$.

4.34 Show that the symmetric and antisymmetric spin wavefunctions χ_{sym} and χ_{anti} are normalized and orthogonal if α and β are normalized and orthogonal.

4.35 Indicate whether each of the following functions is symmetric, antisymmetric, or neither with respect to operation by the label exchange operator $\hat{P}_{21}$:

a. $e^{-(r_1^2 - r_2^2)/a_0^2}\left(\frac{r_1^2}{a_0^2} - \frac{r_2^2}{a_0^2}\right)$

b. $(\cos\theta_1 + \cos\theta_2)\,e^{-r_1/a_0}e^{-r_2/a_0}$

4.36 Show whether the following three-electron wavefunction is symmetric, antisymmetric, or asymmetric with respect to exchange of electrons 2 and 3:

$$[1s2s2p\alpha\beta\alpha + 2p1s2s\alpha\alpha\beta + 2s2p1s\beta\alpha\alpha],$$

where $1s2s2p\alpha\beta\alpha = 1s(1)2s(2)2p(3)\alpha(1)\beta(2)\alpha(3)$, etc.

4.37 We put three electrons into a three-dimensional box and label the states by the n_x, n_y, n_z quantum numbers of each electron as follows: $(n_{x1}, n_{y1}, n_{z1})(n_{x2}, n_{y2}, n_{z2})(n_{x3}, n_{y3}, n_{z3})$. Because electrons are identical particles and we have put these three into the same enclosure, we can no longer distinguish between the electrons. This means, for example, that the three states

$$(2,1,1)(1,1,1)(1,1,1)\ (1,1,1)(2,1,1)(1,1,1)\ (1,1,1)(1,1,1)(2,1,1)$$

are all *the same state* as far as we can tell, and they add into the degeneracy only *once*, not three times. We can still tell the difference between excitation along x, y, or z, however. Add this constraint to the three-electron, cubical box problem, require that no more than two electrons can occupy the same state, and enter the degeneracies in the following table:

E	g	states
12		(2,1,1)(1,1,1)(1,1,1), (1,2,1)(1,1,1)(1,1,1), . . .
15		(2,2,1)(1,1,1)(1,1,1), (2,1,2)(1,1,1)(1,1,1), . . .
17		(3,1,1)(1,1,1)(1,1,1), (1,3,1)(1,1,1)(1,1,1), . . .
18		(2,2,2)(1,1,1)(1,1,1), (2,2,1)(2,1,1)(1,1,1), . . .

4.38 Three ^{1}H atoms are prepared in a three-dimensional box in the states $a = (1,1,2)$, $b = (1,2,1)$, and $c = (2,1,1)$. Each atom has an electron and proton spin that happens to combine to give a total atomic spin $F = |\vec{I}+\vec{S}| = 1$. We label the M_F states as follows: α ($M_F = 1$), β ($M_F = 0$), γ ($M_F = -1$). Find whether or not the following is a valid spin-spatial wavefunction for these three **atoms:**

$$[abc + cab + bca - cba - acb - bac]$$
$$[\alpha\beta\gamma + \gamma\alpha\beta + \beta\gamma\alpha - \gamma\beta\alpha - \alpha\gamma\beta - \beta\alpha\gamma].$$

4.39 In the circles in the following wavefunction, fill in the correct signs (+ or −) that make the overall wavefunction antisymmetric with respect to exchange of the labels on electrons 1 and 2.

$$\Psi(1,2) = exp(-r_1/a_0)exp(-2r_2/a_0)\alpha(1)\beta(2)$$
$$\bigcirc\, exp(-2r_1/a_0)exp(-r_2/a_0)\beta(1)\alpha(2)$$
$$\bigcirc\, exp(-r_1/a_0)exp(-2r_2/a_0)\beta(1)\alpha(2)$$
$$\bigcirc\, exp(-2r_1/a_0)exp(-r_2/a_0)\alpha(1)\beta(2)$$

4.40 What is the one term missing from the following symmetrized wavefunction for the $1s^2 2s^1 3s^1$ excited state of Be?

$$\begin{aligned}\Psi = &1s1s2s3s\alpha\beta\alpha\alpha - 1s1s3s2s\alpha\beta\alpha\alpha - 1s2s1s3s\alpha\alpha\beta\alpha\\ &+ 1s2s3s1s\alpha\alpha\alpha\beta + 1s3s1s2s\alpha\alpha\beta\alpha - 1s3s2s1s\alpha\alpha\alpha\beta\\ &+ 1s1s2s3s\beta\alpha\alpha\alpha - 1s1s3s2s\beta\alpha\alpha\alpha - 1s2s1s3s\beta\alpha\alpha\alpha\\ &+ 1s2s3s1s\beta\alpha\alpha\alpha + 1s3s1s2s\beta\alpha\alpha\alpha - 1s3s2s1s\beta\alpha\alpha\alpha\\ &+ 2s1s1s3s\alpha\alpha\beta\alpha - 2s1s3s1s\alpha\alpha\alpha\beta - 2s1s1s3s\alpha\beta\alpha\alpha\\ &+ 2s1s3s1s\alpha\beta\alpha\alpha + 2s3s1s1s\alpha\alpha\alpha\beta - 3s1s1s2s\alpha\alpha\beta\alpha\\ &+ 3s1s2s1s\alpha\alpha\alpha\beta + 3s1s1s2s\alpha\beta\alpha\alpha - 3s1s2s1s\alpha\beta\alpha\alpha\\ &- 3s2s1s1s\alpha\alpha\alpha\beta + 3s2s1s1s\alpha\alpha\beta\alpha\end{aligned}$$

4.41 If we expand the products for our triplet state He $1s2s$ spin-spatial wavefunctions, we find that there must be at least two terms present (ψ_1 and ψ_3), although there can be more (ψ_2):

$$\psi_1 = \frac{1}{\sqrt{2}}(1s(1)2s(2)\alpha(1)\alpha(2) - 2s(1)1s(2)\alpha(1)\alpha(2))$$

$$\psi_2 = \frac{1}{2}(1s(1)2s(2)\alpha(1)\beta(2) - 2s(1)1s(2)\alpha(1)\beta(2) + 1s(1)2s(2)\beta(1)\alpha(2) - 2s(1)1s(2)\beta(1)\alpha(2))$$

$$\psi_3 = \frac{1}{\sqrt{2}}(1s(1)2s(2)\beta(1)\beta(2) - 2s(1)1s(2)\beta(1)\beta(2))$$

What is the *minimum* number of terms possible for a properly symmetrized, spin-spatial wavefunction for ground state atomic beryllium?

4.42 The permutation operator $\hat{P}_{ij}$ exchanges the labels i and j on the electrons. For the following wavefunction, $\Psi = [1s2s3s + 2s3s1s + 3s1s2s - 1s3s2s - 2s1s3s - 3s2s1s]\alpha\alpha\alpha$, give the eigenvalue for the following permutations:

a. $\hat{P}_{12}$ b. $\hat{P}_{23}$ c. $\hat{P}_{12}\hat{P}_{23}$

Is the wavefunction a valid, symmetrized wavefunction?

4.43 State why the following is not a valid spin wavefunction for the ground state of Be:

$$\chi = [\alpha(1)\beta(2) - \beta(1)\alpha(2)][\alpha(3)\beta(4) - \beta(3)\alpha(4)].$$

4.44 Explain whether the following could be a valid symmetrized spin wavefunction for the $1s^2 2s^2$ ground state of beryllium:

$$[\alpha\alpha\beta\beta + \alpha\beta\alpha\beta + \alpha\beta\beta\alpha + \beta\alpha\alpha\beta + \beta\alpha\beta\alpha + \beta\beta\alpha\alpha]/\sqrt{6}.$$

4.45 The deuteron is a single particle (the combination of one proton and one neutron), having a charge of +1 and a total spin of 1. There are three spin wavefunctions: $\alpha(m_s = +1)$, $\beta(m_s = 0)$, and $\gamma(m_s = -1)$. Two deuterons become bound to a large, negatively charged particle the same way electrons become bound to an atomic nucleus. Using the same orbital labels ($1s$, $2s$, etc.) that we use for electrons, write *two* normalized, symmetrized, spin-spatial two-deuteron wavefunctions for this system. Use a symmetric spatial wavefunction for the first case and an antisymmetric spatial wavefunction in the second case.

4.46 One of the symmetric spatial wavefunctions for the $1s^2 2s^1$ lithium atom has the form $[1s1s2s + 1s2s1s + 2s1s1s]$. Normalize this wavefunction, and write the exchange integral for this wavefunction using the same form as Eqs. 4.48 and 4.49. (It is not necessary to write out the atomic orbitals in r, θ, and ϕ.)

4.47 Imagine a particle identical to the electron but with spin $s = 3/2$ instead of 1/2. Give the ground state electron configuration for neon with the electrons replaced by such particles.

4.48 Which of the following are possible electronic states of the helium atom?

a. $^2P_{1/2}$ b. $^3D_3$
c. $^4S_0$ d. $^1S_1$

4.49 The following integrals appear in the calculation of the valence electron–electron repulsion term in $1s^2 2s^2 2p^4$ ^3P atomic oxygen. Identify any integrals that vanish. (Here x means $2p_x$, etc.)

a. $\int\int\int\int \hat{H}'(xxyz\alpha\beta\alpha\alpha)(xxyz\alpha\beta\alpha\alpha)\, d\tau_1 d\tau_2 d\tau_3 d\tau_4$

b. $\int\int\int\int \hat{H}'(xxyz\alpha\beta\alpha\alpha)(xxyz\beta\alpha\alpha\alpha)\, d\tau_1 d\tau_2 d\tau_3 d\tau_4$

c. $\int\int\int\int \hat{H}'(xxyz\alpha\beta\alpha\alpha)(zxxy\alpha\beta\alpha\alpha)\, d\tau_1 d\tau_2 d\tau_3 d\tau_4$

d. $\int\int\int\int \hat{H}'(xxyz\alpha\beta\alpha\alpha)(yxxz\alpha\beta\alpha\alpha)\, d\tau_1 d\tau_2 d\tau_3 d\tau_4$

e. $\int\int\int\int \hat{H}'(xxyz\alpha\beta\alpha\alpha)(xyxz\alpha\alpha\beta\alpha)\, d\tau_1 d\tau_2 d\tau_3 d\tau_4$

4.50 The $1s^1 2p^1$ excited electron configuration of atomic helium has two terms: ^1P and ^3P. Using the values for the integrals listed next, estimate the following energy differences in E_h:

a. $\Delta E_s = E(1s^1 2p^1\, ^1P) - E(1s^2\, ^1S)$
b. $\Delta E_t = E(1s^1 2p^1\, ^3P) - E(1s^2\, ^1S)$

$$\int\int\left(\frac{e^2}{4\pi\varepsilon_0 r_{12}}\right)1s(1)1s(1)1s(2)1s(2)\,d\tau_1\,d\tau_2 = 1.250$$
$$\int\int\left(\frac{e^2}{4\pi\varepsilon_0 r_{12}}\right)1s(1)1s(1)2p(2)2p(2)\,d\tau_1\,d\tau_2 = 0.485$$
$$\int\int\left(\frac{e^2}{4\pi\varepsilon_0 r_{12}}\right)1s(1)2p(1)1s(2)2p(2)\,d\tau_1\,d\tau_2 = 0.034$$
$$\int\int\left(\frac{e^2}{4\pi\varepsilon_0 r_{12}}\right)2p(1)2p(1)2p(2)2p(2)\,d\tau_1\,d\tau_2 = 0.390$$

Perturbation and Variation Theory

4.51 Write a normalized and symmetrized zero-order spin-spatial wavefunction for the $1s^1 3s^1$ He ^1S term in terms of the coordinates of electrons 1 and 2.

4.52 Write the *explicit* zero-order wavefunction in terms of all relevant coordinates for the ground state boron atom. Do not normalize or symmetrize the wavefunction, but include the one-electron spin wavefunctions α and β.

4.53 Write, but do not evaluate, the integral in terms of all the relevant coordinates for first-order correction to the energy of the ground state lithium atom, $\int \psi_0^* \hat{H}' \psi_0 d\tau$. Do not normalize or symmetrize the wavefunction, do not include the spin terms, but do include the limits of integration.

4.54 For an atom in a ^2S term with spin-spatial wavefunction Ψ, the Zeeman contribution $\hat{H}_Z$ to the Hamiltonian operates only on the spin part of the wavefunction, χ (because for an S term no magnetic field arises from the orbital motion). The remaining term in the Hamiltonian, $\hat{H}_0$, operates only on the spatial part ψ. If the Zeeman energy is treated (using perturbation theory) as a small correction to the electronic energy, show the integrals, in terms of $\hat{H}_0$, $\hat{H}_Z$, ψ, χ, and $d\tau$ necessary to obtain the zero-order and first-order energies for this case. Use $d\tau_s$ to represent the spin volume element.

4.55 Figure 3.18 illustrates the energies of the hydrogen atom in a magnetic field. The magnetic field energies are usually included in spectroscopic analysis as perturbations to the electronic Hamiltonian. In one formulation, the first-order energy correction is given by Eq. 3.59,

$$E'_1 = g_j \mu_B m_j B_Z,$$

and the second-order correction by the integral (labeling the wavefunctions by ψ_{j,m_j})

$$\int \psi^*_{3/2,m_j} \hat{H}' \psi_{1/2,m_j}\, d\tau = \frac{1}{3}\left[\left(\frac{3}{2}\right)^2 - m_j^2\right]^{1/2} (g_s - g_l)\,\mu_B B_Z.$$

(Integrals between states with different m_j values are all zero.) What are these corrections for the $j,m_j = \frac{1}{2},\frac{1}{2}$ state at a field B_Z of 0.20 T? Compute these energies in units of cm^{-1}, and use these values for the constants: $g_{j=1/2} = 2/3$, $g_s = 2.00$, $g_l = 1.00$, $\mu_B = 9.27 \cdot 10^{-24}\ \text{J T}^{-1}$, and $E_0(j = 1/2) - E_0(j = 3/2) = -A_{so} = -0.365\ \text{cm}^{-1}$.

4.56 In our discussion of perturbation theory, Fig. 4.10 graphs contributions to the first-order wavefunction of ground state helium, showing that the major share comes from the $1s^2$ and $1s^1 2s^1$ zero-order wavefunctions. (a) Show that the contribution from $1s^1 2p^1$ is exactly zero by symmetry, which is why it does not appear in the figure (even though in zero order it is as close in energy as the $1s^1 2s^1$). (b) What rule must be satisfied for these contributions to be non-zero? Would a $2p^2$ wavefunction or a $2s^2$ configuration contribute to the first-order ground state wavefunction?

4.57 We take the particle in a one-dimensional box and add a "barrier" of height ε in the middle of the box to obtain the following potential energy curve. The potential energy barrier is treated as a perturbation.

a. Write the zero-order Hamiltonian $\hat{H}_0$ for this system for the region inside the box ($0 \le x < a$).
b. Write the perturbation Hamiltonian $\hat{H}'$ for this system.
c. Write the integral that must be solved to find the first-order correction to the energy E'_1 for the ground state wavefunction.

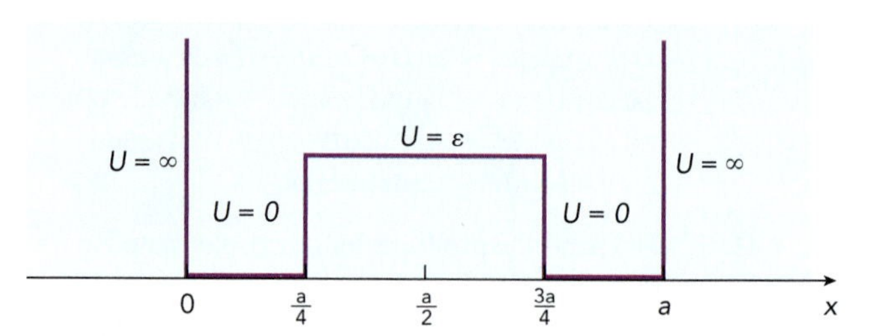

4.58 We want to use perturbation theory to estimate the energy of a box with a slope. The zero-order system is the ordinary particle of mass m in a one-dimensional box that runs from $x = 0$ to a. The perturbation is an added potential energy term, $U_a(x) = E_1\left(1-\frac{x}{a}\right)$, where E_1 is the zero-order ground state energy and is equal to $0.100\,E_h$. The following integral may be useful:

$$\int_0^a \sin^2\left(\frac{n\pi x}{a}\right)\left(1 - \frac{x}{a}\right)dx = \frac{a}{4}$$

a. Make a graph of the potential energy function.
b. On your graph, draw a line for the zero-order energy.
c. Calculate the ground state energy to first order.
d. On your graph, draw a line for the first-order energy.
e. Finally, on your line for the first-order energy, sketch (approximately) what you think the first-order wavefunction will look like.

4.59 Using perturbation theory, we found the Schrödinger equation to second order in λ for ψ and E.

a. If the first-order correction to the wavefunction is $\left(\frac{\partial\psi}{\partial\lambda}\right)\big|_{\lambda=0}$, then what is the *similar* expression for the *third-order* correction to the wavefunction, in terms of λ? (This is *not* looking for a derivation in terms of the energies.)

b. The part of the Schrödinger equation that depends on λ to first order reads $\hat{H}_0\psi_1 + \hat{H}'\psi_0 = E_0\psi_1 + E_1\psi_0$. Write the part of the Schrödinger equation that is proportional to λ^3.

4.60 For the electrons in the lowest excited state of Li^+, write the explicit zero-order wavefunction and find the zero-order energy in E_h.

4.61 A proposed solution for the $1s^12s^1$ excited state wavefunction of helium is

$$\psi_a = 1s2s\alpha\beta - 1s2s\beta\alpha.$$

a. Which property of electrons does this wavefunction violate?

b. Write a short equation for the energy of this state E_a in terms of the energies E_s and E_t of the singlet and triplet states, respectively.

4.62 Construct a wavefunction in terms of the electron coordinates for a variational calculation of the ground state energy of the lithium atom. Include four parameters that can be adjusted to minimize the energy and define them briefly. Do not normalize the function.

4.63 A possible linear variational wavefunction $\psi(r)$ for the hydrogen atom is the following. Try to find the values of c_1 and c_2 that minimize the energy and calculate the average energy. The function is already normalized.

$$\psi(r) = \left(\frac{30c_2^3}{c_1^5}\right)^{1/2}(c_1 - c_2r) \ if\ 0 \leq r < c_1/c_2$$

$$\psi(r) = 0 \ if\ r \geq c_1/c_2$$

$$\langle E \rangle = \int_0^{c_1/c_2} \psi(r)\hat{H}\psi(r)r^2\, dr$$

$$\hat{H} = -\frac{1}{2r^2}\frac{\partial}{\partial r}r^2\frac{\partial}{\partial r} - \frac{1}{r}$$

4.64 We change the particle in a one-dimensional box to add a well in the middle, as shown. Our variational wavefunction is $\psi(x)^{\text{var}} = \sum_{n=1}^{3} c_n \sin n\pi/a$ where we vary the three parameters c_1, c_2, and c_3. After the variational problem is solved, complete the following:

a. Which of the three parameters will have the greatest magnitude?

b. Which of the three parameters will be equal to zero?

c. Sketch what you think the wavefunction will look like.

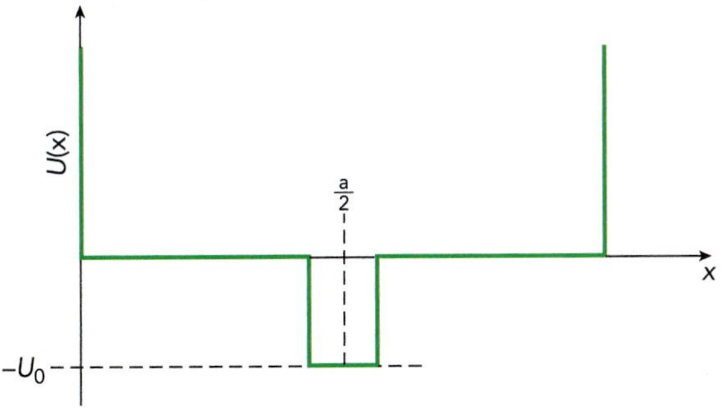

4.65 The variational method is used to solve the Schrödinger equation for a particle of mass m with potential energy $U(r) = kr$, starting with a normalized variational wavefunction $\psi(r) = (2a^{3/2})e^{-ar}$. The variational parameter is a. The expectation value of the energy obeys the equation

$$\langle E \rangle = \frac{\hbar^2 a^2}{2m} + \frac{3k}{2a}.$$

Find the optimized wavefunction and its energy.

4.66 We defined the polarizability, α, by Eq. 3.36:

$$\mu_{\text{induced}} = \alpha\mathcal{E}.$$

We can roughly estimate the value of α for an electron with charge $-e$ in a one-dimensional box using perturbation theory. The zero-order Hamiltonian $\hat{H}_0$ is the ordinary one-dimensional box potential, of length d and *centered* at $x = 0$, and the perturbation $\hat{H}'$ is the electric potential $-e\,\mathcal{E}x$. (The box is centered at zero in this problem to make the symmetry more obvious.) The dipole moment of the box is calculated from the integral

$$\mu = e\int_{-d/2}^{d/2} \psi^*(x)\, x\,\psi(x)\, dx = e\int_{-d/2}^{d/2} |\psi(x)|^2\, x\, dx.$$

The dipole moment of the box is zero if no field is applied, so we set $\mu = \mu_{\text{ind}} = \alpha\mathcal{E}$. Find an equation for α in terms of d and m_e by calculating μ for the $n = 1$ state at first order in order perturbation theory, including the interaction with $n = 2$ (assume all interactions with higher n are negligible). The zero-order wavefunctions you need are then $\psi_1(x) = \sqrt{2/d}\cos(\pi x/d)$ and $\psi_2(x) = \sqrt{2/d}\sin(2\pi x/d)$. You can simplify the final equation by requiring that the energy due to $\hat{H}'$ is small compared to the zero-order energy gap $E_2^{(0)} - E_1^{(0)}$.

4.67 One can estimate the polarizability of the hydrogen atom using perturbation theory (this is a slightly more advanced version of the previous problem). Assume that only the $n = 2$ shell needs to be included in the treatment, and calculate the polarizability of the ground state hydrogen atom to first order in perturbation theory. (As rewarding as this calculation is, it should be mentioned at the outset that the result is good only to within a factor of three.)

4.68 In this problem we solve the one-electron part of the variational problem for the helium atom in the $1s^2$ ground state, where the sole adjustable parameter is the effective atomic number Z_{eff}, as shown in Eq. 4.23:

$$\psi_1 = 4\left(\frac{Z_{\text{eff}}}{a_0}\right)^3 e^{-Z_{\text{eff}}r_1/a_0}e^{-Z_{\text{eff}}r_2/a_0}.$$

The one-electron part of the Hamiltonian has two terms: the kinetic energy and the electron-nuclear potential energy. We can treat these two terms for a single electron at a time and then deal separately with the electron–electron repulsion.

a. Find the kinetic energy of our variational wavefunction by solving the expectation value of the one-electron radial kinetic energy operator,

$$\hat{K}_r = -\frac{\hbar^2}{2m_e}\frac{1}{r^2}\frac{\partial}{\partial r}r^2\frac{\partial}{\partial r}.$$

There is no angular kinetic energy in this problem because we are using only s orbitals.

b. Find the expectation value of the electron-nuclear potential energy of a single electron in helium,

$$U = -\sum_{i=1}^{2}\frac{Ze^2}{4\pi\varepsilon_0 r}.$$

c. Finally, because there are two electrons in helium, double the sum of these two contributions.

This last result, combined with the electron–electron repulsion energy, gives the variational solution to the energy and wavefunction of the ground state helium atom, as shown in Section 4.2. THINKING AHEAD ▶ [What sign should each result have? What should the result be if we set $Z_{\text{eff}} = Z$?]

Hartree-Fock Energies and Orbitals

4.69 The hydride ion, H^-, has the same number of electrons as atomic helium, He. In the following table, the values for certain parameters are given for He. In the next column, write the value of the same parameter for H^- if you can calculate it using algebra or arithmetic. If you cannot solve for the value exactly, then write "$>$" if the value for H^- is *greater* than the value for He, and write "$<$" if the value for H^- is *less than* the value for He.

	He	H^-
Z	2	
Z_{eff}	1.69	
ε_1 (E_h)	−0.917	
E_{HF} (E_h)	−2.862	
zero-order energy (E_h)	−4.00	
first-order correction (E_h)	+1.25	

4.70 Use the corresponding letters (a–e) to rank the following atomic orbitals in order of *increasing Hartree-Fock energy* (from *most negative* to *least negative*).

a. Cl 3s b. F 2s c. Na 3s

d. K 4s e. F^+ 2s

4.71 Referring to the Hartree-Fock orbital energies for ground state atomic lithium,

a. what is the ionization energy of the 2s electron?

b. what is the energy required to *completely* ionize the atom to form Li^{3+}?

c. write the electron configuration for the lowest excited state of Li.

d. is the total Hartree-Fock energy *higher* (less negative) or *lower* (more negative) in the excited state than in the ground state?

e. is the 1s orbital energy *higher* or *lower* in the lowest excited state than in the ground state?

4.72 For a particular atom, count all of the degenerate orbitals as *one* energy level. How many *distinct* Hartree-Fock energy levels are found for the electrons in each of the following atomic states?

a. ground state Ca

b. ground state K

c. lowest excited state Mg

4.73

a. Rank the following in order of increasing orbital energy (less negative = higher energy) of the *valence* electrons: He, Ar, K.

b. Rank the following in order of increasing orbital energy (less negative = higher energy) of the 1s *core* electrons: He, Ar, K.

4.74 Rank the following letters (a–d) in order of *increasing* Hartree-Fock energy for the corresponding atomic subshells (where increasing energy means *decreasing stability*):

a 2s S b. 1s Al

c. 2p O d. 2p P

4.75 Table 4.1 lists orbital energies based on Hartree-Fock calculations but does not include values for the N ($Z = 7$) and O ($Z = 8$) atoms. When changing from N to O, put a "+" sign in the right-hand column of the following table if the energy in the left-hand column increases, a "−" sign if the energy decreases, and "0" if it stays exactly the same.

Property	From N to O
1s orbital energy	
2s orbital energy	
total repulsion energy	
total kinetic energy	
total energy	

4.76 In Table 4.1, we see that there are three different orbital energy values given for the Li atom, whereas only two distinct energies are listed for the orbitals of Be. How many *distinct* Hartree-Fock orbital energies will there be for the electrons of atomic calcium in its *lowest excited* electronic state?

4.77

a. A Hartree-Fock self-consistent field (SCF) calculation of the carbon atom ground state electronic wavefunction yields the following energy values in E_h for each of the six electrons: 1*s*: −11.3475, 1*s*: −11.3021, 2*s*: −0.8296, 2*s*: −0.5840, 2*p*: −0.4388, 2*p*: −0.4388. What is the shift in the energies of these electrons from the values they would have in the same orbitals but in the absence of the other electrons? In other words, what is the difference between these energies and the corresponding one-electron orbital energies?

b. In part (a), we have three subshells (1*s*, 2*s*, and 2*p*), each of which have two electrons. Note the difference in energy between the electrons in each of these pairs: 0.0454 E_h for the 1*s* electrons, 0.2456 E_h for the two 2*s* electrons, and 0.0000 E_h for the 2*p* electron energies. The ground state term for carbon is ^3P. Briefly explain why these energy differences are so small for the 1*s*, so large for the 2*s*, and zero for the 2*p*.

c. The total energy for the state in part (a) is calculated to be −37.6906 E_h. (It is not the same as the sum of all the ε_i's, which counts the electron–electron terms twice.) If the calculation in part (a) is repeated for the carbon atom in the $1s^2 2s^1 2p^3\ ^5S$ state, the total energy for the electronic state is computed to be −37.5957 E_h. This is one of the states atomic carbon can enter in order to free up four electrons for bonding (for example, to form the sp^3 hybrid orbitals). Calculate the amount of energy necessary to reach this state from the ^3P ground state in problem 1, and give the answer in kJ mol^{-1}. A typical chemical bond energy is about 400 kJ mol^{-1}.

4.78 Based on the energies in Table 4.1, estimate the energies of the occupied atomic orbitals in the ground state F atom.

The Vector Model

4.79 Find the term states of the ground electron configuration for the C^+ atomic ion.

4.80 Find the symbol for the neutral atom with *lowest atomic number* that has a *G* term (meaning $L = 4$) among the terms states in its *ground state* electron configuration. Then determine the spin multiplicity $2S + 1$ that goes with this *G* term and the value of *J* for the lowest-energy state of this term. (The *G* term itself need not be the lowest-energy term state.)

4.81 The valence electron configuration for atomic niobium reads $5s^2 4d^3$. Determine how many individual microstates are generated by this electron configuration.

4.82 Pretend you want to solve the vector model for ground state protactinium with electron configuration $[\text{Rn}]7s^2 6d^1 5f^2$. Draw any two valid microstates (the diagrams with the little arrows) for this configuration, and write the M_L and M_S values.

4.83 Find the term states, including *J* values, of neutral platinum atom in its ground electron configuration, $[\text{Xe}]6s^1 4f^{14} 5d^9$. Rank these from left to right in order of increasing energy.

4.84 Find all the term states, including *J* values, resulting from the electron configuration $[\text{Kr}]5s^1 4d^1$ of strontium, and list the term symbols in order of increasing energy.

4.85 The $4s^1 3d^4$ excited electron configuration of atomic vanadium leads to 24 distinct term states. Several proposed term states are listed next, including the *J* values. Only three are correct. Eliminate any term states that cannot possibly arise from this electron configuration, based on the *L*, *S*, and/or *J* values, and write the remaining states, in order of *increasing energy.*

$^4F_{1/2}$ $^1S_0$ $^2P_3$ $^6D_{1/2}$

$^2F_{7/2}$ $^3F_1$ $^1D_2$ $^3S_{3/2}$

$^4P_2$ $^2S_{1/2}$ $^8D_6$ $^7F_3$

4.86 A professor solving the vector model for $1s^2 2s^1 3p^2$ excited state boron obtained the M_L and M_S values given next, but mistakenly skipped one combination. Identify the term states (but not *J* values) resulting from this configuration and identify M_L and M_S for the missing combination.

M_L	M_S		M_L	M_S		M_L	M_S	
−1	$\frac{3}{2}$	^4P	−1	$+\frac{1}{2}$	^2D	−1	$+\frac{1}{2}$	^2P
0	$\frac{3}{2}$	^4P	0	$+\frac{1}{2}$	^2D	−1	$-\frac{1}{2}$	^4P
+1	$\frac{3}{2}$	^4P	+1	$+\frac{1}{2}$	^2D	−1	$-\frac{1}{2}$	^2P
−1	$-\frac{3}{2}$	^4P	−2	$+\frac{1}{2}$	^2D	0	$+\frac{1}{2}$	^4P
0	$-\frac{3}{2}$	^4P	+2	$+\frac{1}{2}$	^2D	0	$+\frac{1}{2}$	^2P
+1	$-\frac{3}{2}$	^4P	−2	$-\frac{1}{2}$	^2D	0	$-\frac{1}{2}$	^2P
−1	$-\frac{1}{2}$	^2D	+2	$-\frac{1}{2}$	^2D	0	$-\frac{1}{2}$	^2P
0	$-\frac{1}{2}$	^2D	0	$-\frac{1}{2}$	^4P	+1	$+\frac{1}{2}$	^2P
+1	$-\frac{1}{2}$	^2D	−1	$+\frac{1}{2}$	^4P	+1	$-\frac{1}{2}$	^4P
			+1	$+\frac{1}{2}$	^4P	+1	$-\frac{1}{2}$	^2P

4.87 All of the *non-negative values* of M_L and M_S from the vector model of palladium in the excited electron configuration $[\text{Kr}]5s^2 4d^8$ are given next. The term state with $L = 4$ is ^1G, and it has already been assigned to the correct M_L,M_S combinations in the table. Find the term symbols for the remaining states, and write them in order of increasing energy. You do not need to provide *J* values this time.

M_L	M_S	term	M_L	M_S	term	M_L	M_S	term
3	1		3	0	^1G	1	0	
2	1		3	0		1	0	
1	1		2	0	^1G	0	0	^1G
1	1		2	0		0	0	

(continued)

M_L	M_S	term	M_L	M_S	term	M_L	M_S	term
0	1		2	0		0	0	
0	1		1	0	^1G	0	0	
4	0	^1G	1	0		0	0	

4.88 List the two lowest-energy electron configurations of atomic beryllium, and give the term states for each, including J values.

4.89 Find an electron configuration for the Be atom that results in a $^3S_1$ state.

4.90 Find the ground state term, including J value, of the $1s^13d^1$ excited state He atom.

4.91 Using the vector model, find and give a term symbol for each fine structure state of the $1s^22p^1$ excited state lithium atom.

4.92 In trying to obtain the term symbols for the $[Ar]4s^14p^3$ excited state configuration of V^+, a professor obtains the $^4F_{7/2}$ term symbol, among others. State two reasons why this is clearly incorrect, without attempting to solve the entire vector model.

4.93 Write the lowest-energy excited electron configuration for the Mg atom and find the term symbols (including J values) for all the term states.

4.94 Ground state atomic zirconium has an electron configuration $[Kr]5s^24d^2$ and a lowest-energy term state $^3F_2$. Rearranging only the electrons in the 5*s* and 4*d* valence subshells, write the electron configuration and term symbol (including J value) for the *lowest-energy* state that has the *highest possible* total electron spin.

4.95 Find the term symbols, including J value, for the excited electron configuration of fermium, $[Rn]7s^15f^{13}$, and place them in order of increasing energy. You can solve this by working through the complete vector model, but it's not necessary.

4.96 What low-energy excited electron configuration of titanium has the following characteristics?

a. The lowest-energy term state is $^5D_0$.

b. The term state with highest L is $^1I_6$.

c. There is only one partially filled subshell.

4.97 The periodic table in the front of the book tells us that the lowest-energy term state of atomic titanium is ^3F. What is the lowest excited term state and its lowest-energy J value?

4.98 The vector model predicts that the $[Ar]4s^23d^2$ electron configuration of titanium generates five term states: ^3F, ^3P, ^1G, ^1D, ^1S. Based on this (i.e., without working through the whole vector model), predict all the term states of scandium with the excited electron configuration $[Ar]4s3d^2$. List them, with J values, in order of increasing energy.

4.99 Find all the term states, including J values, for the ground electronic configuration of hafnium (element 72). The periodic table in the front of the text gives the configuration and lowest-energy term state.

4.100 Using the vector model, find term symbols for the states within the $1s^22s^22p^5$ ground state of fluorine, and give their degeneracies.

4.101 Find the term symbol for only the lowest-energy term (including J value) of the $[Ar]4s^13d^2$ excited state of scandium. Do not use the full vector model for this problem (there are 90 individual M_L, M_S states).

4.102 The Zeeman contribution to the energy of the ground state F^+ ion is approximately $g_s\mu_B J_Z B_Z$. Calculate these contributions in MHz for all the M_J states of the lowest-energy J state of F^+ at a magnetic field along the Z axis of 1.0 T.

4.103 The vector model is applicable to nuclear spin states as well as electronic states. In the abundant isotopic form of ethene, for example, there are four ^1H nuclei with spin 1/2 and two ^{12}C nuclei with zero spin. In the same way that electron spins add together to form a resultant spin angular momentum quantum number S, the individual nuclear spin states of the protons can be added to obtain an overall nuclear spin quantum number I. Also as for electronic states, it is possible to have multiple states of the same spin I arising from one nuclear configuration. For ethene, identify all the possible nuclear spin states: the possible values of I and the number of distinct states for each value.

PART I

ATOMIC STRUCTURE

PART II

MOLECULAR STRUCTURE

PART III

MOLECULAR INTERACTIONS

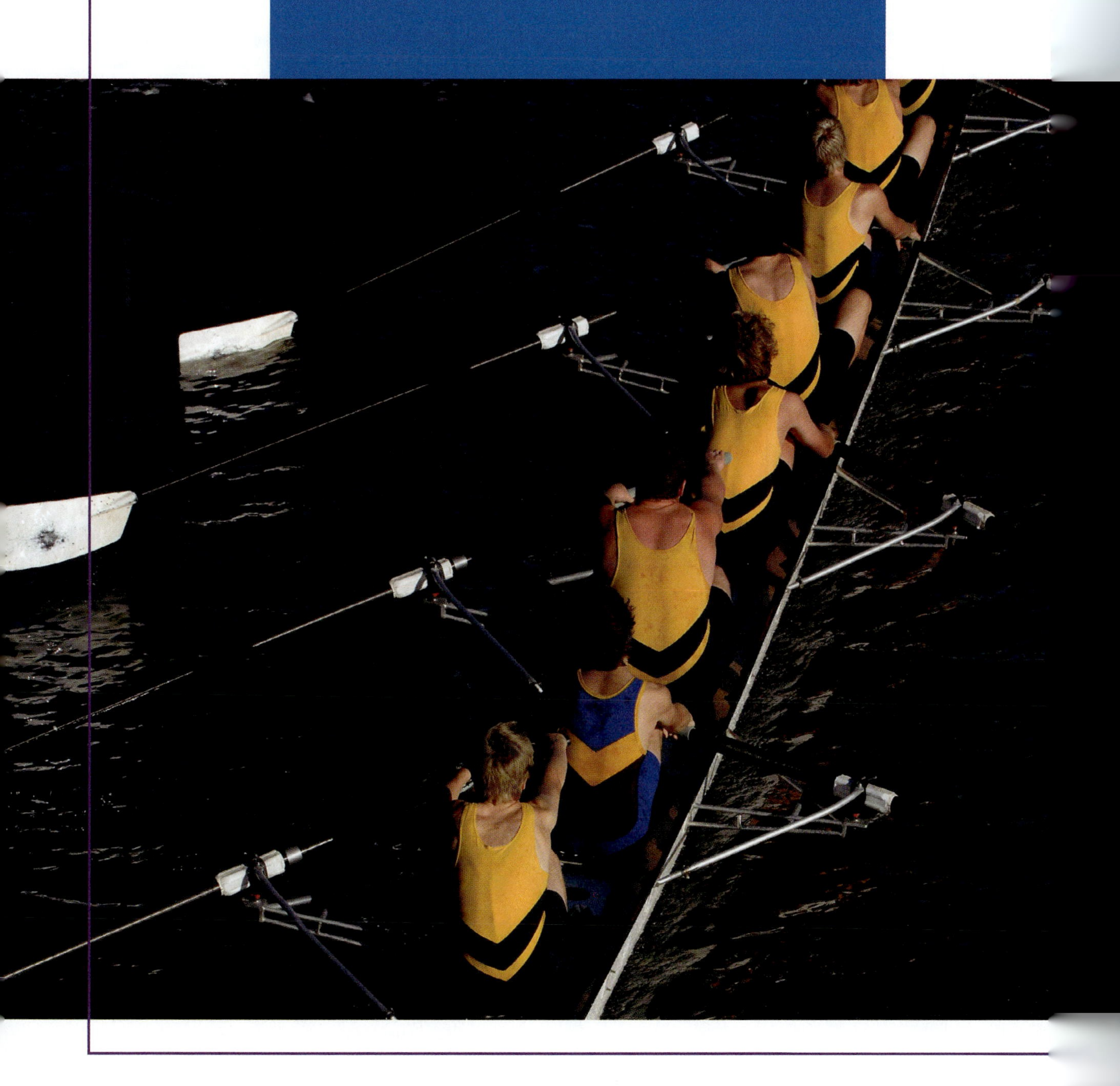

5 Chemical Bonds and Nuclear Magnetic Resonance

LEARNING OBJECTIVES

After reading this chapter, you will be able to do the following:

❶ Write the Hamiltonian for any molecule.

❷ Write and sketch zero-order molecular orbital wavefunctions by combining atomic orbital wavefunctions.

❸ Draw approximate potential energy curves for the motion of two nuclei in a chemical bond.

❹ Determine the shape and other properties of hybrid orbitals.

❺ Predict and interpret basic proton and ^{13}C NMR spectra.

GOAL *Why Are We Here?*

The goal of this chapter is to present the basic physics underlying the formation of chemical bonds.

CONTEXT *Where Are We Now?*

Our model of the individual atom has advanced to the point that we can introduce it to another atom and see what happens. We've been building our theoretical framework for chemistry by first examining the one-electron atom, then adding more electrons. Next, we're going to add one more positively charged nucleus. Coulomb's law tells us what to expect: although the two nuclei repel each other, the added positive charge will immediately pull electrons into the gap between the two nuclei. The shielding of the nuclear charge by core electrons, a destabilizing effect when we looked at the valence electrons of the single atom, becomes a *stabilizing* effect when we combine atoms, because it reduces the nuclear–nuclear repulsion.

The basic principles of atomic structure form the foundation for our investigation into the nature of molecules. We shall define a molecule as *any group of atoms that is held together by strong interatomic forces*—strong enough, that is, to withstand the pummeling a molecule gets in a typical terrestrial environment. This joining of the atoms constitutes a **chemical bond.**

SUPPORTING TEXT *How Did We Get Here?*

The main *qualitative* concept we draw on for this chapter is that the electrons in a molecule have quantized energies and other properties that we can predict by solving the Schrödinger equation with a

Hamiltonian that takes into account all the particles and forces present. To support this contention and more, we will draw on the following equations and sections of text in this chapter:

- In three dimensions, the kinetic energy operator is (Eq. 2.36):

$$\hat{K} = -\frac{\hbar^2}{2m}\nabla^2.$$

  The kinetic energy operator for an electron in polar coordinates may then be written (Eqs. 3.4, 3.5, leaving in the center of mass term)

$$\hat{K} = -\frac{\hbar^2}{2\mu r^2}\frac{\partial}{\partial r}r^2\frac{\partial}{\partial r} + \frac{1}{2\mu r^2}\hat{L}^2(\theta,\phi) - \frac{\hbar^2}{2(M_{\text{nuc}} + m_e)}\nabla^2_{\text{COM}},$$

  where

$$\hat{L}^2(\theta,\phi) = -\hbar^2\left[\frac{1}{\sin\theta}\frac{\partial}{\partial\theta}\sin\theta\frac{\partial}{\partial\theta} + \frac{1}{\sin^2\theta}\frac{\partial^2}{\partial\phi^2}\right].$$

- We often choose a basis set of the one-electron orbitals formulated in Tables 3.1 and 3.2, and then add these together to form a linear combination of atomic orbitals (Eq. 4.25):

$$\varphi(i) = c_{1s}\psi_{1s}(i) + c_{2s}\psi_{2s}(i) + c_{2p_x}\psi_{2p_x}(i) + \cdots = \sum_{n=1}^{n_{\max}}\sum_{l=0}^{n-1}\sum_{m_l=-l}^{l} c_{nlm_l}\psi_{nlm_l}(i),$$

  where the $\psi_{nlm_l}(i)$'s are the pure one-electron orbitals.

- Section 4.2 showed how the Hartree-Fock method estimates electron–electron repulsion energies—the tough nut to crack in solving the many-electron Schrödinger equation—by computing the repulsion experienced by one electron when we average over the positions of all the others. Then the method uses the variational principle to adjust the coefficients c_{nlm_l} in the basis set until the lowest possible energy is found. Table 4.1 gives several examples of the resulting energies. Configuration interaction (CI) extends the Hartree-Fock results to calculate more accurate energies.
- The angular momentum $\vec{L}$ of an object with linear momentum $\vec{p} = m\vec{v}$ and position vector $\vec{r}$ relative to the system's center of mass is defined to be the vector cross-product (Eq. A.46)

$$\vec{L} = \vec{r} \times \vec{p} = m\vec{r} \times \vec{v}.$$

- In Section 3.4, we learned that there are magnetic moments associated with the orbital motion of an electron (Eq. 3.41)

$$\vec{\mu}_l = -g_l\mu_B\,\vec{l}/\hbar$$

  and with the spin of the electron (Eq. 3.46)

$$\vec{\mu}_s = -g_s\mu_B\,\vec{s}/\hbar.$$

  Each of these magnetic moments can interact with an external magnetic field B_0 along the z axis to shift the energy of the quantum state. For example, for the interaction of the spin with the external field we obtain (Eq. 3.47)

$$E_{\text{mag},s} = g_s\mu_B m_s B_0.$$

  This dependence of the energy on the strength of an applied field is called the Zeeman effect.

5.1 The Molecular Hamiltonian

We want to solve the molecular Schrödinger equation. No, really, we do, because the resulting energies and wavefunctions will predict the behavior of molecules in situations that we care about in all branches of chemistry: why the molecule has the spectrum it has, whether it forms a liquid or solid or gas, why it reacts the way it does. The molecular Hamiltonian unlocks the whole world of chemistry. But it's a big world, and we're going to travel it in steps, beginning with how the Coulomb force dictates the shape of the potential energy term.

Chemical bonds exhibit a wide range of features, leading to bonds being classified according to various criteria: by the amount of charge transfer between the atoms, for example, or by the number of participating electrons. However, all chemical bonds are governed by the Coulomb force, an extension of the same interaction that binds the electron to the nucleus in a single atom.

The **covalent bond** gets its name from the sharing of valence electrons by neighboring atoms. If we start with two atoms, each with electrons centered on the atomic nucleus, and bring them together, the electron density may shift to occupy the space *between* the two atoms. Effectively, those electrons may then belong to the valence of both atoms simultaneously: the atoms become *co*-valent. But how does this cause a bond to form? A convincing way to see this is by forgetting about the energy for a while and examining the Coulomb force directly, starting with the simplest possible molecule, H_2^+.

Electrostatics of Covalent Bonding

The simplest molecule, the molecular ion H_2^+, is not as familiar in chemistry as common aqueous ions such as NH_4^+, because it is much more reactive, capable of pulling electrons off of water and most other neutral molecules. However, the chemical bond between the two hydrogen nuclei is strong enough that H_2^+ can survive indefinitely in the gas phase at typical energies, provided it doesn't bump into anything more reactive than, say, helium or neon atoms.

Having two nuclei and one electron, H_2^+ is a three-body system like the helium atom. As was the case for helium, the Schrödinger equation for H_2^+ cannot be exactly solved by analytical integration, but again we can find tractable and accurate approximations. The coordinates we will use for diatomics are drawn in Fig. 5.1. The two nuclei are labeled A and B and lie on the z axis. The position of the electron is given in spherical coordinates, with the origin at either nucleus A or nucleus B.

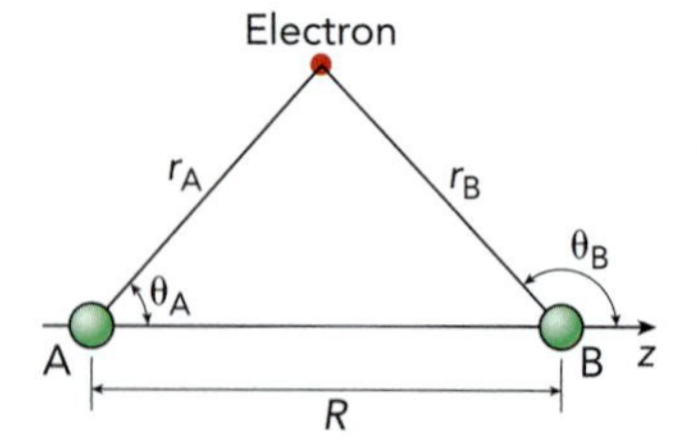

▲ **FIGURE 5.1 The coordinate system used for the electrostatics treatment of H_2^+ and other diatomics.**

But if our interest is in the chemical bond, then the critical coordinate is the distance R between the two nuclei, because changing this distance corresponds to breaking and forming the chemical bond. As long as the nuclei can be held together—as long as R stays small—we have a chemical bond. The two nuclei repel each other, so the electron must be what holds the molecule together. We can show how this works classically by considering the forces at work when the electron and nuclei are at rest. This kind of analysis, using stationary charges, is called **electrostatics.**

Parameters key: the binding force

symbol	parameter	SI units
e	fundamental charge	C
F_{binding}	binding force	N
r_i	distance of electron from nucleus i	m
R	distance between nuclei A and B	m
Z_i	atomic number of nucleus i	unitless
θ_i	angle of vector $\vec{r_i}$ from internuclear axis	unitless

The forces exerted by nuclei A and B on the electron have magnitude $Ze^2/(4\pi\varepsilon_0 r_A^2)$ and $Ze^2/(4\pi\varepsilon_0 r_B^2)$, respectively, and directions that depend on the angles θ_A and θ_B. We're going to distinguish *binding* (the net force that draws nuclei together) from *bonding* (the tendency of the nuclei to *stay* together). The **binding force** is the sum of the Coulomb forces between each pair of particles, projected onto the z axis because only the force along that axis determines whether the nuclei will move toward or away from each other:

$$F_{\text{binding}} = \frac{Z_A e^2}{4\pi\varepsilon_0 r_A^2}\cos\theta_A - \frac{Z_B e^2}{4\pi\varepsilon_0 r_B^2}\cos\theta_B - \frac{Z_A Z_B e^2}{4\pi\varepsilon_0 R^2}. \tag{5.1}$$

The first two terms are the attractions of the nuclei for the electron, projected onto the z axis. The minus sign before the second term, for nucleus B, appears because nucleus B must move to the left in Fig. 5.1 if it is attracted toward A, while A must move toward the right if it is attracted toward B. The last term is the nucleus–nucleus repulsion term, which the binding force of the electron must overcome. Bonding occurs when the attractive contribution to the binding force outweighs the nucleus–nucleus repulsion. If F_{binding} is positive, the two nuclei are pulled toward each other; if negative, the nuclei will be pushed apart.

Consider some simple cases for H_2^+, for which $Z_A = Z_B = 1$. If the electron is on the internuclear axis between the two nuclei (Fig. 5.2a), $\theta_A = 0, \theta_B = \pi$; thus,

$$\cos\theta_A = 1,\ \cos\theta_B = -1,\quad \text{and } R = r_A + r_B,$$

so

$$F_{\text{binding}}(\theta_A = 0, \theta_B = \pi) = \frac{e^2}{4\pi\varepsilon_0}\left[\frac{1}{r_A^2} + \frac{1}{r_B^2} - \frac{1}{(r_A + r_B)^2}\right] > 0. \tag{5.2}$$

Since $(r_A + r_B)^2$ is larger than both r_A^2 and r_B^2, the third term in brackets is always smaller magnitude than the other two, and therefore F_{binding} is positive anywhere on this axis between the nuclei. However, if we move the electron to the right side of the molecule (changing θ_B to 0, Fig. 5.2b), the electron pulls less on the far nucleus than on the near nucleus, which tends toward *breaking* the bond. Combined with the nucleus–nucleus repulsion, we have:

$$\cos\theta_A = 1,\ \cos\theta_B = 1,\ \text{and } R = r_A - r_B,$$

$$F_{\text{binding}}(\theta_A = 0,\ \theta_B = 0) = \frac{e^2}{4\pi\varepsilon_0}\left[\frac{1}{r_A^2} - \frac{1}{r_B^2} - \frac{1}{(r_A - r_B)^2}\right] < 0. \tag{5.3}$$

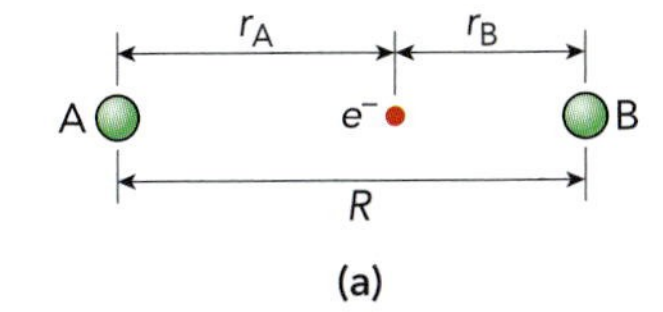

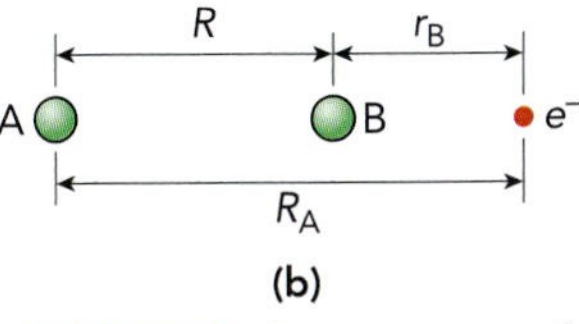

▲ **FIGURE 5.2 Arrangement of particles in H_2^+.** Shown for **(a)** $\theta_A = 0, \theta_B = \pi$ ($\cos\theta_A = 1$, $\cos\theta_B = -1$), and **(b)** $\theta_A = 0, \theta_B = 0$ ($\cos\theta_A = \cos\theta_B = 1$).

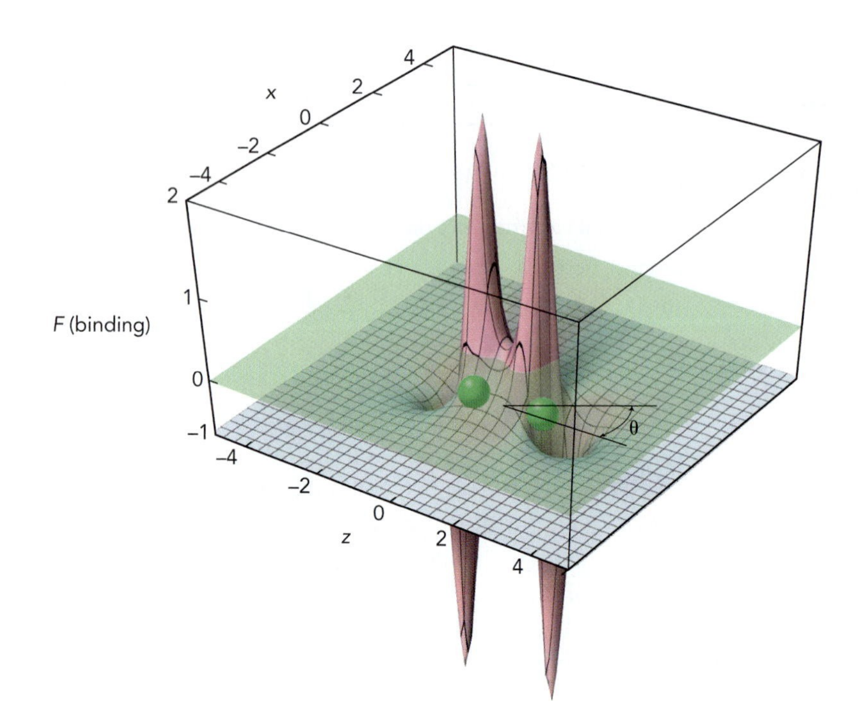

▶ FIGURE 5.3 **The binding force in Cartesian coordinates.** The function F_{binding}, defined in Eq. 5.1, is shown for H_2^+ with x and z measured in units of R. The angle θ in Eq. 5.1 is the angle measured in any direction (such as in the xz plane) from the z axis. The binding force here is given in units of $e^2/(4\pi\varepsilon_0 R^2)$, approaching a constant value of $-e^2/(4\pi\varepsilon_0 R^2)$ at distances far from both nuclei. The positive peaks indicate the regions of greatest binding force, and the negative spikes are the regions where the electron most effectively pulls the nuclei apart. The transparent floor at $F_{\text{binding}} = 0$ divides the surface into binding and antibinding regions.

In this case, r_B^2 or $(r_A - r_B)^2$ or both must be less than r_A^2, so for these values of θ the binding force is negative and the molecule begins to fall apart.

Using this electrostatic analysis, the space surrounding the molecule can be separated into **binding** and **antibinding** regions, as drawn in Fig. 5.3. Valence electrons in molecules will generally be smeared across both binding and antibinding regions, and it is largely some average of the electron density in each region that determines the stability of the chemical bond. Roughly speaking, in order for the bond to be stable, the electron density should be greater in the binding region than in the antibinding region.

The General Form of the Hamiltonian

The electrostatic treatment just discussed can lend an intuitive, classical perspective on the interactions between the particles in molecules, but we need to describe molecules more precisely. We again turn to quantum mechanics, starting from the Hamiltonian. For any molecule, the Hamiltonian may be written as the sum of several contributions: the kinetic energies of the electrons and nuclei and the potential energies derived from the forces that appear in Eq. 5.1 for the binding force,

$$\hat{H} = \hat{K}_{\text{elec}} + U_{\text{nuc-elec}} + U_{\text{elec-elec}} + U_{\text{nuc-nuc}} + \hat{K}_{\text{nuc}}, \tag{5.4}$$

where $\hat{K}_{\text{elec}}$ and $\hat{K}_{\text{nuc}}$ are the electron and nucleus kinetic energy operators, respectively, and where the U's are the Coulomb potential terms for the nucleus–nucleus, nucleus–electron, and electron–electron interactions. The kinetic energy operator is one that we have used before (Eq. 2.36):

$$\hat{K} = -\frac{\hbar^2}{2m}\nabla^2,$$

and there is one such operator for each nucleus and one for each electron. The potential energy terms are all examples of the same, classical Coulomb potential energy function, $q_1 q_2/(4\pi\varepsilon_0 r)$, that we used for the atom in Chapters 3 and 4. We will neglect the electron spin at first, as we did for the atom, and return to it later.

For H_2^+, which has two nuclei and one electron, the Hamiltonian is

$$\hat{H} = -\frac{\hbar^2}{2m_e}\nabla(1)^2 + \frac{e^2}{4\pi\varepsilon_0}\left[-\frac{Z_A}{r_{A1}} - \frac{Z_B}{r_{B1}} + \frac{Z_A Z_B}{R_{AB}}\right] - \frac{\hbar^2}{2m_A}\nabla(A)^2 - \frac{\hbar^2}{2m_B}\nabla(B)^2, \quad (5.5)$$

with the electron labeled 1 and the nuclei labeled A and B. For a molecule with two electrons and two nuclei, such as H_2, the following terms would appear in the Hamiltonian:

$$\hat{H} = -\frac{\hbar^2}{2m_e}\nabla(1)^2 - \frac{\hbar^2}{2m_e}\nabla(2)^2 + \frac{e^2}{4\pi\varepsilon_0}\left[-\frac{Z_A}{r_{A1}} - \frac{Z_B}{r_{B1}} - \frac{Z_A}{r_{A2}} - \frac{Z_B}{r_{B2}} + \frac{1}{r_{12}} + \frac{Z_A Z_B}{R_{AB}}\right] - \frac{\hbar^2}{2m_A}\nabla(A)^2 - \frac{\hbar^2}{2m_B}\nabla(B)^2. \quad (5.6)$$

To make this completely general, if we have a molecule of N electrons and M nuclei, then the molecular Hamiltonian (neglecting spin-orbit and magnetic interactions as usual) may be written as

$$\hat{H} = -\frac{\hbar^2}{2m_e}\sum_{i=1}^{N}\nabla(i)^2 - \sum_{i=1}^{N}\sum_{k=1}^{M}\frac{Z_k e^2}{(4\pi\varepsilon_0)r_{ik}} + \sum_{i=2}^{N}\sum_{j=1}^{i-1}\frac{e^2}{(4\pi\varepsilon_0)r_{ij}} + \sum_{k=2}^{M}\sum_{l=1}^{k-1}\frac{Z_k Z_l e^2}{(4\pi\varepsilon_0)R_{kl}} - \frac{\hbar^2}{2}\sum_{k=1}^{M}\frac{1}{m_k}\nabla(k)^2. \quad (5.7)$$

Separating the Variables

Let's return to the Hamiltonian for H_2^+ in Eq. 5.5. In general, there must be nine coordinates to specify the locations of the three particles of H_2^+, three coordinates (such as x, y, and z) for each nucleus and for the electron. All nine coordinates can be important in chemistry, but we'd like to break the problem into pieces so that we deal with only one group of coordinates at a time.

The **Born-Oppenheimer approximation** takes us a long way in that direction by assuming that the rapid motion of the electrons is separable from the nuclear motion. The Coulomb forces acting on the electrons and nuclei in a molecule are of comparable magnitude, but force is mass times acceleration, and each nucleus has at least 10^3 times more mass than each electron. Therefore, the forces accelerate the electrons to much higher speeds than the nuclei. The Born-Oppenheimer approximation extends this difference in speed to the limit in which we treat the electrons as traveling around *stationary* nuclei. This means we don't have to solve for the motions of the electrons and nuclei at the same time: the electronic and nuclear coordinates are *separable,* in a way similar to the separation of angular and radial coordinates for the one-electron atom (Section 3.1). In the one-electron atom, we solve the angular part first in order to get the angular momentum term we need to solve the radial part. Here, we need to solve the electronic part first, to get the electric field in which the nuclei move:

- To obtain the wavefunctions and energies for the electrons, we first neglect motion of the nuclei, because they move slowly by comparison, and solve the electronic Schrödinger equation at different, fixed positions of the nuclei.
- To obtain the wavefunctions and energies of the nuclei, we solve a nuclear Schrödinger equation in which the potential energy term depends on the electron distributions determined in the first step.

Using this logic, we set the nuclear motion in H_2^+ to zero for a moment and treat the electron in terms of an **effective Hamiltonian,** $\hat{H}_{\text{eff}}$, which includes all the terms of Eq. 5.5 except the nuclear kinetic energy operator $\hat{K}_{\text{nuc}}$. Then we solve the Schrödinger equation for this effective Hamiltonian at a single value of R to get the electronic wavefunctions and energies at that bond length:

$$\hat{H}_{\text{eff}}(R)\psi_{\text{elec}}(R) = E_{\text{elec}}(R)\psi_{\text{elec}}(R), \tag{5.8}$$

where

$$\hat{H}_{\text{eff}}(R) = \hat{K}_{\text{elec}} + U_{\text{nuc-nuc}} + U_{\text{nuc-elec}} + U_{\text{elec-elec}}. \tag{5.9}$$

For H_2^+ and more complicated molecules, the electronic energy $E_{\text{elec}}(R)$ can be obtained using the same Hartree-Fock method that we used to obtain atomic orbital and overall atomic energies in Section 4.2. However, the resulting electronic wavefunctions and energies depend on the geometry of the nuclei. Every time the nuclei move a little bit, the potential energies in the Hamiltonian change, and we have to solve the Schrödinger equation again at this new geometry (Fig. 5.4).

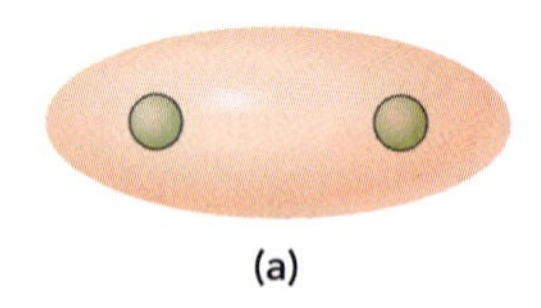

(a)

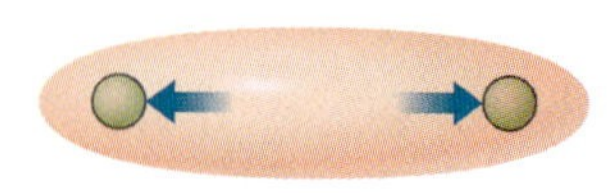

(b)

▲ FIGURE 5.4 **The Born-Oppenheimer approximation.** The approximation assumes that we can solve the Schrödinger equation for the electrons independently of the nuclear motion. **(a)** At the initial positions of the nuclei, we can find the distribution of the electrons by solving the Schrödinger equation. **(b)** If the nuclei move to another geometry, the electron distribution adjusts on a timescale much faster than the nuclear motion, and the electronic Schrödinger equation needs to be solved again for this new geometry.

The Born-Oppenheimer approximation separates the electronic and nuclear coordinates. For H_2^+, we figured out there are nine coordinates overall. The three coordinates of the electron are separated into the effective Hamiltonian. Once we have obtained the electronic energy as a function of R, we can add $\hat{K}_{\text{nuc}}$ and solve the Schrödinger equation along the remaining six coordinates for motion of the nuclei:

$$\hat{H}\psi_{\text{nuc}} = [E_{\text{elec}}(R) + \hat{K}_{\text{nuc}}]\psi_{\text{nuc}} = E_{\text{nuc}}\psi_{\text{nuc}}. \tag{5.10}$$

Even though we could write $\hat{K}_{\text{nuc}}$ in terms of the Cartesian coordinates, we find again (as we did for atoms) that it's much easier to solve the Schrödinger equation if we use coordinates that line up with the forces working on the particles. In this case, the force we expect to find is a binding force that pulls the two nuclei together. So we want a set of coordinates that includes the bond axis R connecting the two nuclei.

For the electronic wavefunctions, we defined the location of the electron using the **molecule-fixed coordinates** r, θ, and ϕ, which are defined using only the atomic nuclei as reference points (Fig. 5.5). If the molecule is rotated or translated, the molecule-fixed coordinates move with it. This is a sensible coordinate system for K_{elec} and the electronic wavefunction, because only the positions of the nuclei determine the electric field that controls the electron's motion.

For motion of the whole molecule, however, it is the molecule's orientation and location in the laboratory that matters, and so we use the **lab-fixed** or **space-fixed coordinates** Θ and Φ to describe the orientation of the nuclei in the lab, and the coordinates X, Y, and Z, which pinpoint the center of mass of the molecule.

When writing the Hamiltonian for one electron in an atom, we incorporated a kinetic energy operator formulated to use the spherical coordinates r, θ, ϕ instead of the Cartesian coordinates x, y, z (Eq. 3.4 plus the center of mass term),

$$\hat{K} = -\frac{\hbar^2}{2\mu r^2}\frac{\partial}{\partial r}r^2\frac{\partial}{\partial r} + \frac{1}{2\mu r^2}\hat{L}(\theta,\phi)^2 - \frac{\hbar^2}{2(M_{\text{nuc}}+m_e)}\nabla^2_{\text{COM}},$$

because the force in that case depends only on the distance r between the electron and the nucleus. We face a similar situation now, because the forces in the molecule move with the nuclei—for example, the binding force acts along the internuclear axis z—but the location and orientation of the nuclei may be changing constantly as the molecule travels and rotates. Therefore, we express the kinetic term for the nuclei in terms of the coordinates R, Θ, and Φ. These are not spherical coordinates, because R is a bond length and not the distance from the origin, but the mathematics is similar. We will also need a new angular momentum symbol J, which represents the angular motion of the nuclei, to replace the orbital angular momentum of the electron L. Sparing ourselves the trigonometric details, as we did in obtaining Eq. 3.4, we arrive at the kinetic term in this form:

$$\hat{K}_{\text{nuc}} = -\frac{\hbar^2}{2\mu R^2}\frac{\partial}{\partial R}R^2\frac{\partial}{\partial R} + \frac{1}{2\mu R^2}\hat{J}(\Theta,\Phi)^2 - \frac{\hbar^2}{2M_{\text{nuc}}}\nabla^2_{\text{COM}}, \quad (5.11)$$

where M_{nuc} is the combined mass of all the nuclei. The eigenvalue of the new operator $\hat{J}(\Theta,\Phi)^2$ is the square of the nuclear angular momentum, and the operator conveniently absorbs all the angular dependence of the nuclear kinetic energy.

We are left with four kinetic energy contributions to the Hamiltonian: **electronic** from $\hat{K}_{\text{elec}}$, **vibrational** from the $\partial/\partial R$ term, **rotational** from the $\hat{J}^2$ term, and translational from the center of mass (COM) term. These correspond to four different ways of storing energy in the molecule through motions of the particles (see Fig. 5.6), and to four different sets of quantum levels. We shall again neglect the translational term, the motion of the molecule's center of mass, and confine ourselves to the energy in the center of mass reference frame.

One principle that guides the variable separations we've just carried out is this: *contributions to the Hamiltonian that operate at very different energy scales can be separated,* with little cost to the accuracy. Unless we confine our molecule

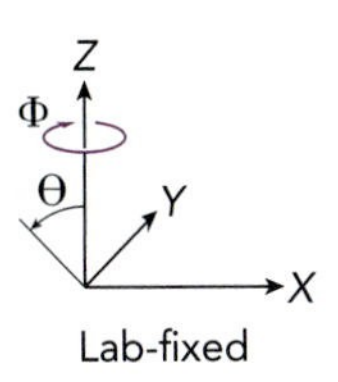

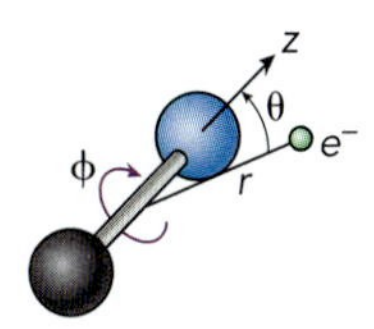

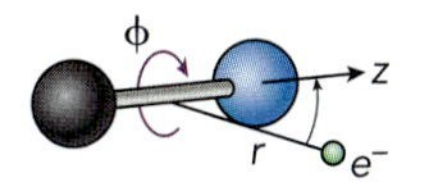

▲ **FIGURE 5.5 The distinct lab-fixed and molecule-fixed coordinate systems.** As the molecule rotates, the lab-fixed axes (X, Y, Z) remain stationary while the molecule-fixed axes (x, y, z) rotate with the molecule. The angles Θ and Φ measure rotations of the molecule relative to the space-fixed axes, while the coordinates r, θ, ϕ give the position of the electron relative to the molecule-fixed axes.

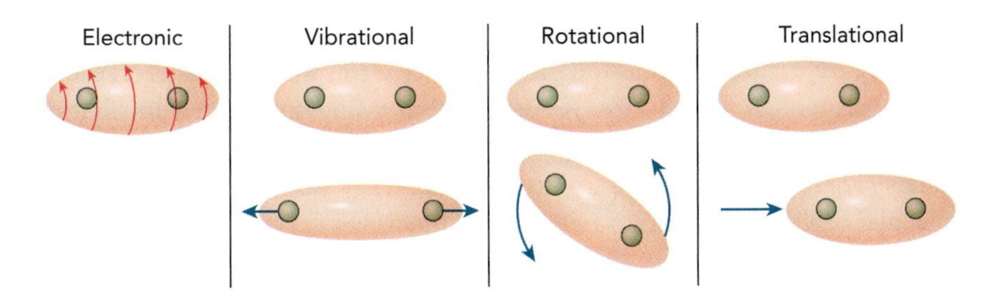

◀ **FIGURE 5.6 The four major molecular degrees of freedom.** The principal ways a molecule stores energy are electronic motion, which establishes the distribution of the electrons around the nuclei; vibrational motion, for motions that change the bond lengths and bond angles; rotational motion, in which the orientation of the molecule changes; and translational motion of the center of mass.

to a very small box, the translational motion is akin to the motion of a free particle, with a continuous distribution of energies. The electrons, having much less mass, are much more quantum-mechanical in behavior and have very large energy gaps. The vibrational and rotational motions of the nuclei lie between these two extremes. We have separated the nine-coordinate Schrödinger equation into these three distinct regimes—translational (low energy gaps), rotational and vibrational (intermediate), and electronic (high energy gaps)—each with a fairly manageable number of coordinates.

5.2 The Molecular Wavefunction

Using the Born-Oppenheimer approximation, we describe the electronic wavefunctions of the diatomic molecule by first assuming the nuclei to be separated at some constant distance R on the potential surface. **Molecular orbital (MO)** theory is the most widely used method for writing those electronic wavefunctions. Each molecular orbital wavefunction defines the distribution of a *single* electron over the entire molecule, just as an atomic orbital is a one-electron spatial wavefunction for an atom.

The simplest MOs are constructed using the LCAO (linear combination of atomic orbitals) method introduced by Eq. 4.25, adding atomic orbital wavefunctions centered on different nuclei in different proportions to obtain one MO wavefunction, often with the electron density **delocalized** across several nuclei. Extended to several nuclei, Eq. 4.25 becomes

$$\begin{aligned}\varphi(i) &= c_{1s_A}\psi_{1s_A}(i) + c_{2s_A}\psi_{2s_A}(i) + c_{2p_{x,A}}\psi_{2p_{x,A}}(i) + \cdots \\ &\quad + c_{1s_B}\psi_{1s_B}(i) + c_{2s_B}\psi_{2s_B}(i) + c_{2p_{x,B}}\psi_{2p_{x,B}}(i) + \cdots \\ &= \sum_{j=A}^{M}\sum_{n=1}^{n_{max}}\sum_{l=0}^{n-1}\sum_{m_l=-l}^{l} c_{nlm_l,j}\psi_{nlm_l,j}(i), \end{aligned} \tag{5.12}$$

where $j = A, B, \ldots, M$ indicates which nucleus in the molecule is used as the center of the atomic orbital $\psi_{nlm_l,j}$. Let's apply this generic expression to the simplest specific case: the single electron that holds two protons together to make the molecular ion H_2^+.

The One-Electron Covalent Bond: H_2^+

Since we're going to spend a short while focusing on a molecule you probably didn't encounter in general chemistry or organic chemistry, it's worth pointing out that H_2^+ is formed in interstellar space and by electrical discharge through hydrogen gas. Although the charge makes H_2^+ a reactive molecule, when left alone it provides the simplest example of a chemical bond, and is stable indefinitely.

For H_2^+, we write the MO as the sum of one-electron wavefunctions on each of the two protons A and B. The ground state is going to reflect contributions mainly from the ground state atomic wavefunctions, the hydrogen $1s_A$ and $1s_B$ orbitals, where $1s_A$ is centered on nucleus A and $1s_B$ on nucleus B, so our first molecular orbitals will be sums of these two functions.

But we have a choice: when we add two wavefunctions to get a new wavefunction, we can vary their relative phase. The two functions can be given the same phase ($1s_A + 1s_B$) or opposite phases ($1s_A - 1s_B$). To write these two options formally, we have

$$\psi_+(r,\theta,R) = C_+(R)(1s_A + 1s_B) \quad \psi_-(r,\theta,R) = C_-(R)(1s_A - 1s_B), \quad (5.13)$$

where the $C(R)$'s are R-dependent normalization constants. When the two functions have the same phase, we get constructive interference in the middle of the molecule. When they have opposite phase, the two wavefunctions cancel in the middle (Fig. 5.7). The two nuclei in the H_2^+ molecule are indistinguishable, which means that the *probability density* must be the same at each nucleus. It's okay in ψ_- for the phase of the *wavefunction* to be different at nucleus B than at nucleus A, because what we measure is $|\psi_-|^2$, not ψ_-.

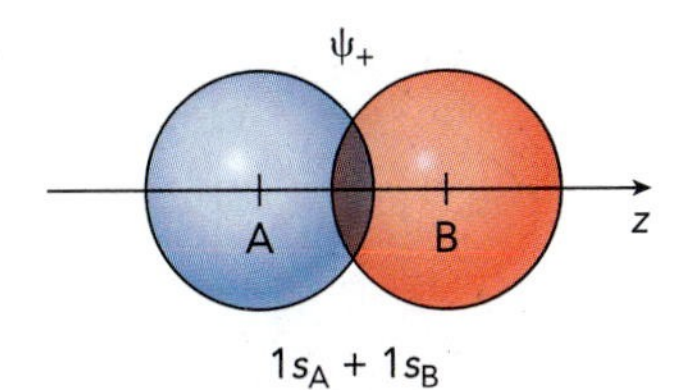

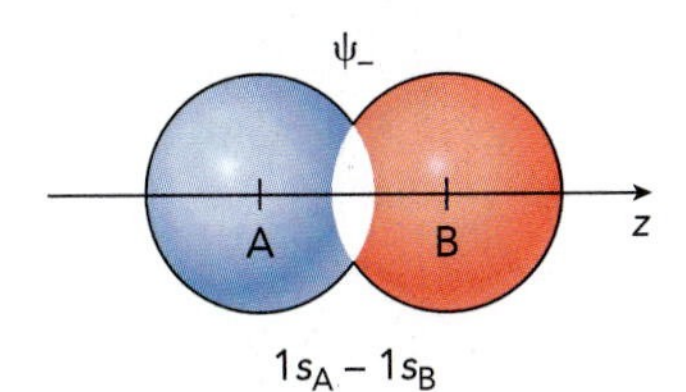

▲ **FIGURE 5.7 The ψ_+ and ψ_- MOs.** We combine the two spherically symmetric $1s$ orbitals on nuclei A and B to get the simplest molecular orbitals.

We are combining two atomic orbital wavefunctions, $1s_A$ and $1s_B$, and getting two MO wavefunctions, ψ_+ and ψ_-. Just as we found for atomic orbitals, each MO can be occupied by one electron or by two electrons with opposite values of m_s.

The wavefunctions along the z axis (setting $\theta = 0$ or $\theta = \pi$) are sketched in Fig. 5.8 as functions of z for a particular value of R. This lets us know how the electron density varies along the bond axis, but the binding region extends for some distance away from that axis (Fig. 5.3). Figure 5.9 shows the same wavefunctions and probability densities, graphed as functions of x and z, so that we can see how much electron charge can be found as we move away from the bond axis.

What about the third coordinate, y? We don't need to worry about graphing that separately in this case. When the nuclei lie in a line (as must be the case for any diatomic molecule), the electric field formed by the nuclei has cylindrical symmetry: it does not depend on the angle of rotation around the z axis. Consequently, the electron density is also cylindrically symmetric. Figure 5.9 would look the same if we used the y axis instead of the x axis.

CHECKPOINT The probability densities $|\psi|^2$ in Figs. 5.8 and 5.9 show us how the charge and mass of the electron would be distributed around the two protons in H_2^+ if the electron were in either the ψ_+ or the higher energy ψ_- molecular orbital. The key feature for us here is that in the ψ_- orbital, the density goes to zero at the midpoint between the two nuclei. This is expected for a wavelike electron: with enough energy, it should be able to oscillate with a node in the middle. But this node reduces the electron charge in the binding region, causing the orbital to have the net impact of weakening the bond between the two nuclei.

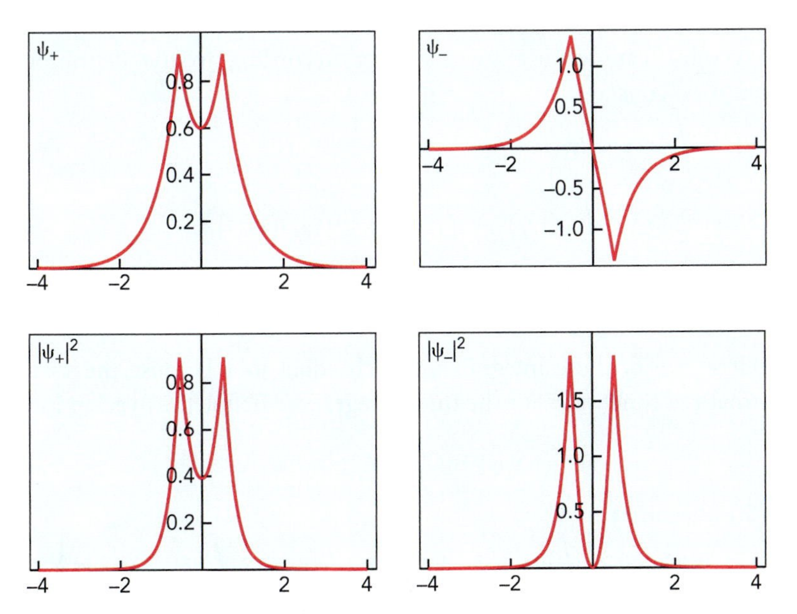

◀ **FIGURE 5.8 The ψ_+ and ψ_- wavefunctions and probability densities, graphed as functions of z.**

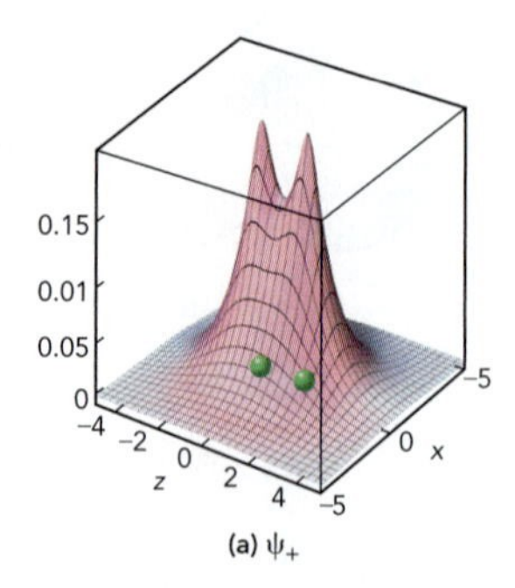

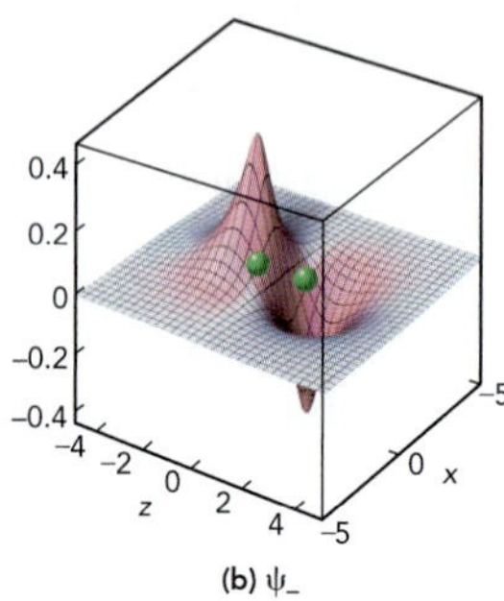

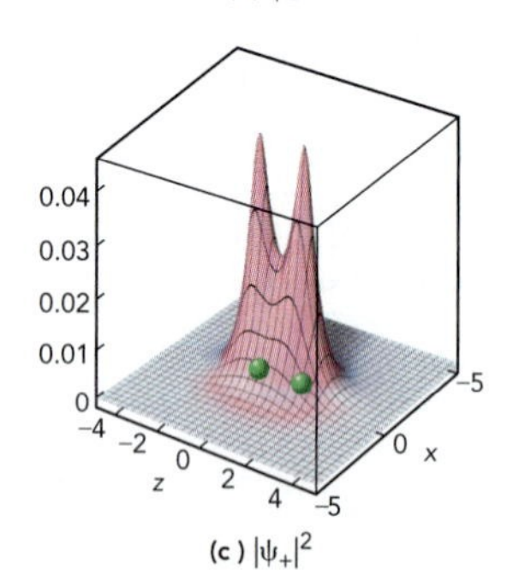

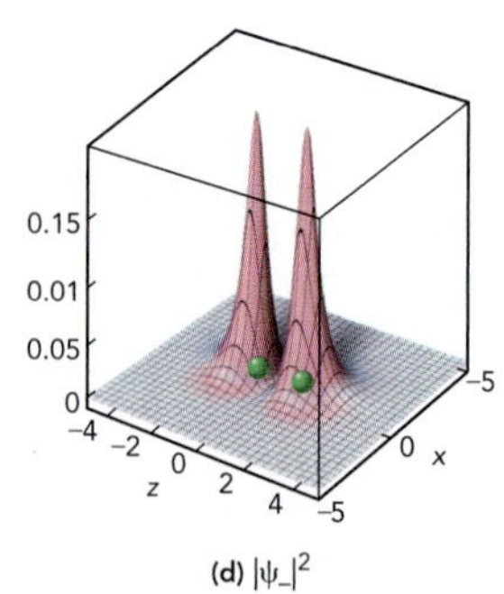

▲ FIGURE 5.9 **The normalized wavefunctions and probability densities of H_2^+. (a) ψ_+ and (b) ψ_-, (c) $|\psi_+|^2$ and (d) $|\psi_-|^2$,** plotted in the xz plane, with z, the internuclear axis, at the equilibrium bond length R_e of 1.06 Å. All distance units are a_0 (including normalization factors).

Where the wavefunction has a node, the probability density goes to zero. From the electrostatics analysis in Section 5.1 we know that for strong binding the electron needs to be between the nuclei. The electron density is concentrated between the nuclei in the ψ_+ state, but it is largely excluded from that region in ψ_-. So we can guess that ψ_+ is a **bonding orbital,** which strengthens the chemical bond, and ψ_- is an **antibonding orbital,** which weakens the bond.

We can test that guess by checking how the energies of these states vary as functions of R. Formation of a chemical bond requires that the molecule be more stable (lower energy) than the separated atoms. If we write the two wavefunctions as $\psi_\pm = C_\pm(R)[1s_A \pm 1s_B]$, then the average value theorem allows us to estimate the electronic energy:

$$E_\pm(R) = \langle H_{\text{eff}}(R)\rangle = \int \psi_\pm^* \hat{H}_{\text{eff}}\psi_\pm d\tau$$

$$= C_\pm(R)^2 \int [1s_A \pm 1s_B]\hat{H}_{\text{eff}}[1s_A \pm 1s_B]d\tau$$

$$= C_\pm(R)^2\left\{\int 1s_A\hat{H}_{\text{eff}}1s_A d\tau + \int 1s_B\hat{H}_{\text{eff}}1s_B d\tau \pm \int 1s_A\hat{H}_{\text{eff}}1s_B d\tau \pm \int 1s_B\hat{H}_{\text{eff}}1s_A d\tau\right\}.$$

We have chosen the origin of our energy scale so that $E_\pm(R)$ is zero at infinite R, when we have split the H_2^+ into one H^+ ion and one H atom, both not moving and too far apart to interact with each other. Since the $1s_A$ and $1s_B$ functions are identical except for location in space, we may merge the notation for some of the terms:

$$\int 1s_A\hat{H}_{\text{eff}}1s_A d\tau = \int 1s_B\hat{H}_{\text{eff}}1s_B d\tau \equiv H_{AA}(R)$$

$$\int 1s_A\hat{H}_{\text{eff}}1s_B d\tau = \int 1s_B\hat{H}_{\text{eff}}1s_A d\tau \equiv H_{AB}(R),$$

so our energy simplifies to

$$E_\pm(R) = C_\pm(R)^2\{2H_{AA}(R) \pm 2H_{AB}(R)\}. \tag{5.14}$$

What value can we use for $C_\pm(R)^2$? According to the definition of a normalization constant,

$$1 = \int |\psi_\pm|^2 d\tau$$

$$= C_\pm(R)^2 \int [(1s_A)^2 + (1s_B)^2 \pm 2(1s_A)(1s_B)]d\tau$$

$$= C_\pm(R)^2\left\{\int (1s_A)^2 d\tau + \int (1s_B)^2 d\tau \pm 2\int (1s_A)(1s_B)d\tau\right\}. \tag{5.15}$$

The first two of these integrals are each equal to 1 because the 1s orbitals are normalized functions. For the third integral, we define the **overlap integral,**

$$S(R) \equiv \int (1s_A)(1s_B)d\tau, \tag{5.16}$$

so Eq. 5.15 becomes

$$1 = C_\pm(R)^2\left\{\int (1s_A)^2 d\tau + \int (1s_B)^2 d\tau \pm 2\int (1s_A)(1s_B)d\tau\right\}$$

$$= C_\pm(R)^2\{1 + 1 \pm 2S(R)\}.$$

Solving for $C_\pm(R)$, we find the normalization constants are given by

$$C_\pm(R)^2 = \frac{1}{2 \pm 2S(R)}. \tag{5.17}$$

Combining this equation for $C_\pm(R)$ into Eq. 5.14, we get the energies

$$E_+(R) = \frac{H_{AA}(R) + H_{AB}(R)}{1 + S(R)} \quad E_-(R) = \frac{H_{AA}(R) - H_{AB}(R)}{1 - S(R)}. \tag{5.18}$$

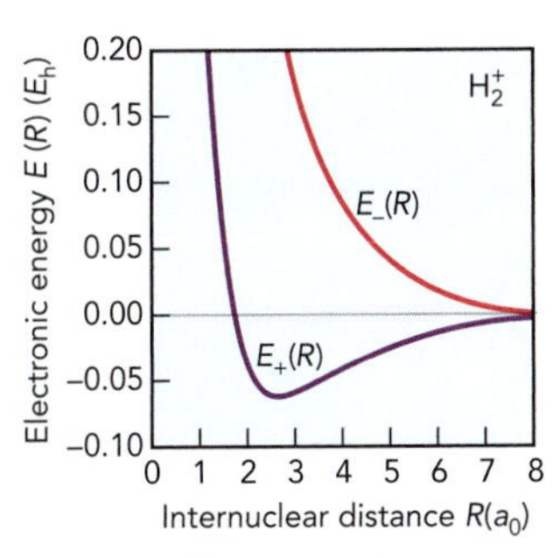

▲ **FIGURE 5.10 The electronic energy functions $E_+(R)$ and $E_-(R)$ for H_2^+.**

Plotting these two energies as functions of R in Fig. 5.10, we find that indeed ψ_+ is bonding and ψ_- is antibonding.

One way to read these potential energy curves is to recall that the slope of the potential energy gives the force acting on the system. The curve for $E_+(R)$ has a minimum potential energy, at roughly $R = 2.5a_0$. If we push the nuclei closer together than $2.5a_0$, then the force pushes them apart again. If the bond stretches to greater than $2.5a_0$, the force works in the opposite direction, pushing the nuclei together again. A minimum in a potential energy function allows a **bound state** to exist, such that the forces keep the interacting particles—in this case our two nuclei—stuck together at some intermediate distance.

The curve $E_-(R)$ has no minimum—it just glides toward zero as R approaches infinity—so the force is *always* working to push the two nuclei apart. This means that ψ_- is an **unbound state,** in which the particles spontaneously separate.

Both of these are valid electronic states for the H_2^+ molecule, even though the ψ_- excited state is predicted to be unstable.[1] In many-electron molecules, the overall molecule may be stable even if several of the electrons are in antibonding orbitals, so long as there is enough electron density overall in the binding region to cancel the contributions from the antibinding regions.

Before we move on to the two-electron chemical bond, we should take note that our treatment of the H_2^+ bond was a simple one with limited accuracy. Requiring the electron in H_2^+ to use only the $1s$ orbitals on the nuclei to form the bond is rather like requiring the electrons in He to keep the same orbital shape as they would in He^+. The actual wavefunctions for the H_2^+ electron borrow characteristics from atomic orbitals other than the $1s$, resulting in a more stable bond than the bond we form using the $1s$ orbitals alone. Symmetry requires that to preserve the cylindrical symmetry around the z axis, only atomic orbitals with $m_l = 0$ contribute to the lowest energy molecular orbital. A Hartree-Fock calculation estimates that 98.2% of the variationally optimized wavefunction comes from s atomic orbitals, 1.8% from p_z orbitals, and less than 0.05% from d_{z^2} and other orbitals. Contributions from the p_z orbitals can improve the molecular orbital by shifting some of the electron density from the antibinding to the binding region. Figure 5.11 shows that the actual minimum of the ground state potential energy is about 30% deeper than predicted by the $E_+(R)$ curve. As the variational principle predicts, the actual energy of the H_2^+ ground state is lower than our approximate $E_+(R)$ function at every value of R.

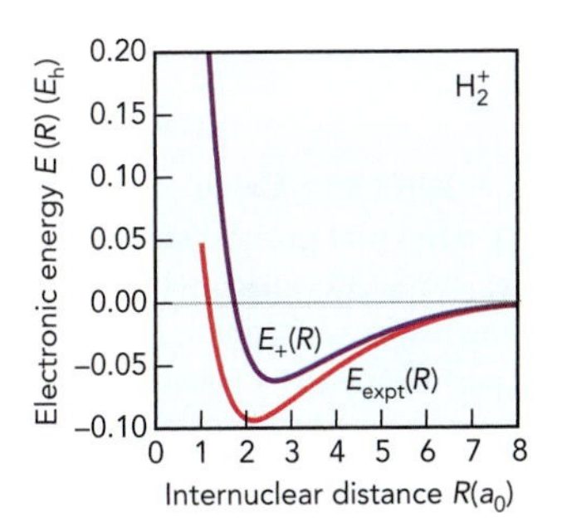

▲ **FIGURE 5.11 Calculated and experimental ground state electronic energies of H_2^+.** Shown are the approximate $E_+(R)$ function and the electronic energy determined by experiment.

[1] In experiments, this state turns out to be barely stable enough to be observed. A very shallow minimum exists, only 13 cm^{-1} ($6 \cdot 10^{-4}\, E_h$) deep and at a separation of roughly $12a_0$, arising from the long-range attraction between the H^+ ion and neutral H atom.

Many-Electron Molecular Orbital Wavefunctions

What happens when we try to extend this method beyond the one-electron molecule to the next simplest case, the H_2 molecule? The pattern may be familiar from our treatment of the atom: we add another particle, the math gets harder, and we lower our expectations a bit, favoring a qualitatively simple picture over the most precise results.

For many-electron molecules, we start from an electron configuration, the same initial approximation that we used to approach many-electron atoms. We obtain the ground state **molecular orbital configuration** by following the same *Aufbau* principle that we use for atoms: fill the available orbitals starting from the lowest energy and on up, until we run out of electrons. Two electrons completely fill any MO for the same reason that two electrons fill any atomic orbital: the Pauli exclusion principle forbids any two electrons in the same molecule from sharing the same spin and spatial wavefunctions. In H_2, we have only two electrons, so we can put them both in the ψ_+ ground state MO. But when dealing with the two-electron atom helium in Section 4.3, we found that if two electrons had the same spatial wavefunction, then the only available two-electron spin wavefunction was $[\alpha(1)\beta(2) - \beta(1)\alpha(2)]/\sqrt{2}$, where α represents the $m_s = +\frac{1}{2}$ spin wavefunction and β the $m_s = -\frac{1}{2}$ wavefunction. Combining this spin wavefunction with the ground state MO for H_2, we have

$$\begin{aligned}\Psi_{\text{MO}}(\text{ground}) &= C_+(R)^2\psi_+(1)\psi_+(2)[\alpha(1)\beta(2) - \beta(1)\alpha(2)]/\sqrt{2} \\ &= C_+(R)^2[1s_A(1) + 1s_B(1)][1s_A(2) + 1s_B(2)][\alpha(1)\beta(2) - \beta(1)\alpha(2)]/\sqrt{2},\end{aligned} \tag{5.19}$$

where $1s_A(1)$ indicates a $1s$ orbital for electron 1 around nucleus A, and so on. This molecular orbital wavefunction is a simple (and approximate) formula for the wavefunction ψ_{elec} that appears in the electronic Schrödinger equation Eq. 5.8. Once we have this guess wavefunction, we can estimate the energy function $E_{\text{elec}}(R)$ of the state at different bond lengths R, using the average value theorem.

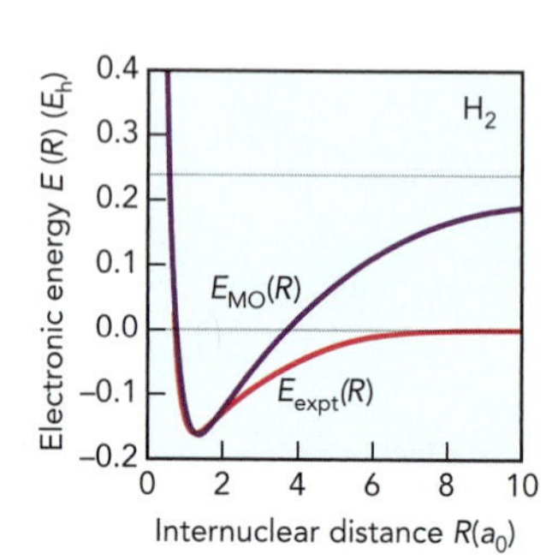

▲ **FIGURE 5.12 Calculated and experimental ground state electronic energies of H_2.** Shown are results obtained by experiment and by Hartree-Fock calculations using MO theory.

The results, plotted in Fig. 5.12, turn out to be good up to a point, and then awful after that. The agreement with experiment at low R is not so bad—better than it was for H_2^+—but at large R the error is egregious. The trouble occurs because we are combining the failings of LCAO with those of a many-electron wavefunction where the electrons are uncorrelated. Looking at the differences between $E_0(\text{HF})$ and $E_0(\text{corr})$ in Table 4.1, we don't see huge errors when correlation in the atoms is neglected. However, for molecular potential energy curves, that correction becomes critical.

We now show why this basic MO wavefunction fails at large R, the **separated atom limit.** The spatial part of our ground state MO wavefunction can be expanded to

$$\begin{aligned}\psi_{\text{MO}}(\text{ground}) &= C_+(R)^2[1s_A(1) + 1s_B(1)][1s_A(2) + 1s_B(2)] \\ &= C_+(R)^2[1s_A(1)1s_A(2) + 1s_A(1)1s_B(2) + 1s_B(1)1s_A(2) + 1s_B(1)1s_B(2)].\end{aligned} \tag{5.20}$$

In the limit of very large R, we now square this to obtain the probability density:

$$\begin{aligned}\lim_{R\to\infty}|\psi_{\text{MO}}(\text{ground})|^2 = C_+(\infty)^4\{&[1s_A(1)1s_A(2)]^2 + [1s_A(1)1s_B(2)]^2 \\ &+ [1s_B(1)1s_A(2)]^2 + [1s_B(1)1s_B(2)]^2\end{aligned} \tag{5.21}$$

$$+ 2[1s_A(1)1s_A(2)1s_A(1)1s_B(2) + 1s_A(1)1s_A(2)1s_B(1)1s_A(2)$$
$$+ 1s_A(1)1s_A(2)1s_B(1)1s_B(2) + 1s_A(1)1s_B(2)1s_B(1)1s_A(2)$$
$$+ 1s_A(1)1s_B(2)1s_B(1)1s_B(2) + 1s_B(1)1s_A(2)1s_B(1)1s_B(2)]\}.$$

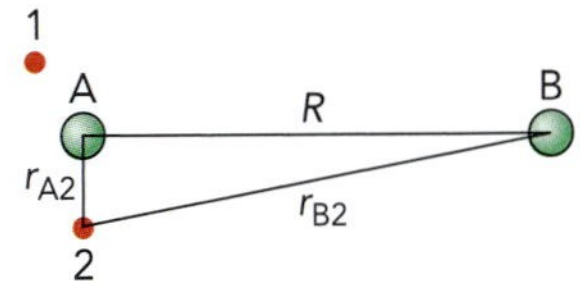

▲ **FIGURE 5.13 One arrangement of two electrons and two hydrogen nuclei in the limit of large *R*.** In this case, both electrons 1 and 2 are drawn near nucleus A, with the result that r_{A1} and r_{A2} are small numbers while r_{B1} and r_{B2} are large numbers. The cross terms are therefore all zero when R is large. This drawing represents the case in which the first term in Eq. 5.22 is significant, the term corresponding to $H_A^- + H_B^+$.

The last six of these terms are cross terms, each of which contains at least one product of the type $1s_A(1)1s_B(1)$ or $1s_A(2)1s_B(2)$. From Table 3.2, we know that each $1s$ orbital has a radial function of the form e^{-r/a_0}, and therefore each of these cross terms has a product $e^{-r_{A1}/a_0}e^{-r_{B1}/a_0}$ or $e^{-r_{A2}/a_0}e^{-r_{B2}/a_0}$. These products vanish in the limit of large R. Figure 5.13 shows why for the product $e^{-r_{A2}/a_0}e^{-r_{B2}/a_0}$: at large separation R, the electron cannot be close to both nuclei simultaneously, so either r_{A2} or r_{B2} is large. That means in turn that either e^{-r_{A2}/a_0} or e^{-r_{B2}/a_0} approaches zero, so the whole cross term vanishes.

We can also find the value of the normalization constant $C_+(\infty)$ in this limit where the cross terms go to zero:

$$\begin{aligned} 1 &= \int \{C_+(\infty)[1s_A(1) + 1s_B(1)]\}^2 d\tau_1 \\ &= C_+(\infty)^2 \int \{[1s_A(1)]^2 + 2[1s_A(1)1s_B(1)] + [1s_B(1)]^2\} d\tau_1 \\ &= C_+(\infty)^2 \left\{ \int [1s_A(1)]^2 d\tau_1 + \int [1s_B(1)]^2 d\tau_1 \right\} \\ &= 2C_+(\infty)^2 \end{aligned}$$

$$C_+(\infty) = \frac{1}{\sqrt{2}}.$$

Returning to Eq. 5.21, we can now show that the probability density at large R is given by:

$$\lim_{R\to\infty} |\psi_{MO}(\text{ground})|^2 = \left(\frac{1}{\sqrt{2}}\right)^4 [1s_A(1)1s_A(2)]^2 + [1s_A(1)1s_B(2)]^2 + [1s_B(1)1s_A(2)]^2 + [1s_B(1)1s_B(2)]^2. \quad (5.22)$$

What is wrong with this? The first of the four terms is significant only when both electrons are near nucleus A and none is near nucleus B. The first term therefore corresponds to the system $H_A^- + H_B^+$. The second and third terms give the probability distributions for nuclei A and B each holding one $1s$ electron apiece, the system $H_A + H_B$. Finally, the fourth term corresponds to the system $H_A^+ + H_B^-$. So Eq. 5.22 can be written this way:

$$\lim_{R\to\infty} |\psi_{MO}(\text{ground})|^2 = \frac{1}{4}[(H_A^- + H_B^+) + (H_A + H_B) + (H_A + H_B) + (H_A^+ + H_B^-)]. \quad (5.23)$$

Half of the terms predict ions, half predict neutrals. Our MO wavefunction breaks down at large distances because it becomes a mixture of two completely distinct experimental results. We would like the potential energy curve in Fig. 5.12 to connect states of the molecule that might actually exist, and instead the curve predicts that when we break one H_2 molecule we get a combination of two states—$H^+ + H^-$ or $H + H$—that can't both exist at the same time.

CHECKPOINT The MO wavefunction is an approximation to the distribution of the electrons in H_2, and like any approximation, it can fail. Here, we're finding that the MO wavefunction can work well to describe the bound H_2 molecule, but the same characteristic that makes it an appealing approximation—its separation of the electrons—also makes it incapable of correctly predicting the energy of this bond-breaking reaction. To predict the energy of this process successfully, we need to allow the position of one electron to affect the position of the other.

In terms of an experimental observable, the MO wavefunction predicts much too high an energy necessary to break the bond. The ionization energy of H atoms is $0.500\,E_h$ (13.6 eV),[2] whereas the electron affinity is only $0.028\,E_h$ (0.75 eV). The $H^+ + H^-$ pair is less stable than the H+H pair by $0.472\,E_h$ (12.9 eV), and so the $E_{MO}(R)$ curve in Fig. 5.12 climbs toward $0.472\,E_h/2$ at large R because half of the energy is from the ions at $0.472\,E_h$ and half is from the neutral atoms at 0.

Valence Bond Wavefunctions

One solution to this problem is available using the **valence bond representation** of the molecular wavefunction.[3] The primary assumption is that two-electron wavefunctions from the atomic valence orbitals are all that contribute to the formation of chemical bonds. We begin with two H atoms at the separated atom limit, the limit where R is so large that there is no interaction between the atoms. Given the indistinguishability of the electrons, the ground state wavefunction may be written as

$$\psi_{VB}(\text{ground}) = [1s_A(1)1s_B(2) + 1s_B(1)1s_A(2)]. \tag{5.24}$$

Assuming, as before, that the singlet state (symmetric spatial wavefunction, antisymmetric spin wavefunction) is the ground state, the spatial probability density at large R is given by

$$\lim_{R\to\infty} |\psi_{VB}(\text{ground})|^2 = [1s_A(1)1s_B(2)]^2 + [1s_B(1)1s_A(2)]^2, \tag{5.25}$$

where the cross terms have been dropped. This gives us the right behavior at large R, two separated neutral hydrogen atoms, and turns out to be slightly superior to the MO representation everywhere for H_2 (although the difference is negligible at low R).

Why then do we even consider using the MO theory? The valence bond (VB) representation is a correlated picture in which the separation of the electronic wavefunctions in the MO theory is abandoned. This makes the functions much more complicated to manipulate mathematically, particularly as more and more electrons become involved. Also, the MO result is not so badly wrong near the minimum of the potential energy curve, where the molecule is most stable. If we want to solve the electronic wavefunction and energy only near the minimum, then the simplicity of the MO wavefunction may be worth the relative error of 10% or so.

In practice, chemists who need to compute the more accurate behavior of a valence bond function usually achieve this by mixing together different MO configurations. In the case just discussed, for example, we can obtain the VB

[2]The electron volt (eV) remains a unit in common usage for electronic energies, because a straightforward way to measure and control the kinetic energies of free electrons has been to accelerate or decelerate the electrons using a known voltage, where voltage is the relative electrostatic potential energy between two points.

[3]Valence bond theory predates molecular orbital theory, being an extension of G. N. Lewis's theory of bonding. It had the early advantage of a conceptually simple, yet accurate depiction of electron distributions in chemical bonds. It has subsequently been eclipsed by MO theory, which shares its conceptual simplicity but is much easier to formulate mathematically.

representation by combining the MOs obtained from in-phase $1s$ atomic orbitals (the ground state MO) and out-of-phase $1s$ atomic orbitals (an excited state MO):

$$\psi_{\mathrm{VB}}(\text{ground}) = \frac{1}{2}[\psi_{\mathrm{MO}}(\text{ground}) - \psi_{\mathrm{MO}}(\text{excited})], \tag{5.26}$$

where

$$\psi_{\mathrm{MO}}(\text{excited}) = [1s_{\mathrm{A}}(1) - 1s_{\mathrm{B}}(1)][1s_{\mathrm{A}}(2) - 1s_{\mathrm{B}}(2)].$$

This combines two different MO configurations: one in which both electrons are in the ground MO, and one in which both electrons are in an excited MO.

This general method of writing the wavefunction as a linear combination of MO wavefunctions is known as **configuration interaction** (CI). Agreement can be improved over that from either the MO or VB approaches by letting the contribution from the excited state MO vary until the lowest energy is found, taking advantage of the variational principle. Even better agreement is found by varying the coefficients for a large number of higher MO configurations. For example, we could augment our earlier wavefunction with configurations in which each electron is in a $2s$ orbital:

$$\begin{aligned}\psi = {} & c_1[1s_{\mathrm{A}}(1) + 1s_{\mathrm{B}}(1)][1s_{\mathrm{A}}(2) + 1s_{\mathrm{B}}(2)] \\ & + c_2[1s_{\mathrm{A}}(1) - 1s_{\mathrm{B}}(1)][1s_{\mathrm{A}}(2) - 1s_{\mathrm{B}}(2)] \\ & + c_3[2s_{\mathrm{A}}(1) + 2s_{\mathrm{B}}(1)][2s_{\mathrm{A}}(2) + 2s_{\mathrm{B}}(2)].\end{aligned}$$

If we vary the coefficients c_1, c_2, and c_3 to obtain the lowest energy at a given value of R, we can obtain an improved wavefunction and more accurate prediction of the electronic energy. By including configurations from MOs constructed from the $2s$, $2p$, and higher energy atomic orbitals, the electronic energy of H_2 can be predicted to within experimental precision.

Effective Potential Energy Functions in Diatomics

For the atom, the potential energy is a one-dimensional surface; it varies only along the coordinate r. For molecules, the surface is always much more complicated. Even for a one-electron diatomic molecule, the potential energy is a function of R, r_{A}, and r_{B} (or R, θ_{A}, and θ_{B}). However, if our primary interest lies solely in the chemical bond, then we must focus on the distance between the nuclei, R, because we define the chemical bond by what the *nuclei* are doing, regardless of the specifics of the electronic motions.[4] But this nuclear degree of freedom cannot be factored apart from the electrons entirely: the electrons are bound to the nuclei, and it is the mutual attraction of two nuclei for their neighboring electrons that forms a chemical bond in the first place.

[4]Given our strict attention to electrons in the preceding chapters, and the limelight devoted to electrons in introductory chemistry courses, this may seem counter-intuitive. Recall, however, that we ascribe each atom in chemistry to a particular element of the periodic table based only on the number of protons present in the nucleus. In chemistry, the nucleus is the conserved particle of the atom. We routinely redistribute electrons among atoms, but we don't change the identity of any of the atomic nuclei. That process usually occurs at energies higher than we associate with chemistry, so we entrust it to our friends in physics.

This is where the Born-Oppenheimer approximation is useful. Because the electron motions are much faster than the nuclear motions, the electron distribution responds instantly to compensate for any changes we make to the relative positions of the nuclei. The eigenvalue of $\hat{H}_{\text{eff}}$ in Eq. 5.10 is the only contribution to the energy other than the nuclear kinetic energy, and in that sense it resembles the nuclear potential energy. Because it also contains kinetic energy contributions from the electrons, however, it is not strictly a potential energy, so instead it is commonly called the **effective potential energy** function $U_{\text{eff}}(R)$ of the nuclei.

The energy curves graphed in Figs. 5.11 and 5.12 are the effective potential functions for the H_2^+ and H_2 molecules. *These curves encapsulate the principles of chemistry.* In the shape of these energy functions, we see the competing forces that keep the chemical bond stable, how these factors influence molecular geometry, and the energy scales involved in chemical reaction. The effort in modern chemical research to predict molecular geometries and reactivities is in large part based on these potential energy functions. We will see later what complications arise in polyatomic molecules, but the basic principles underlying the effective potential energy functions of diatomics apply widely to more complex systems. The effective potential energy functions for diatomic chemical bonds share the following features, illustrated in Fig. 5.14:

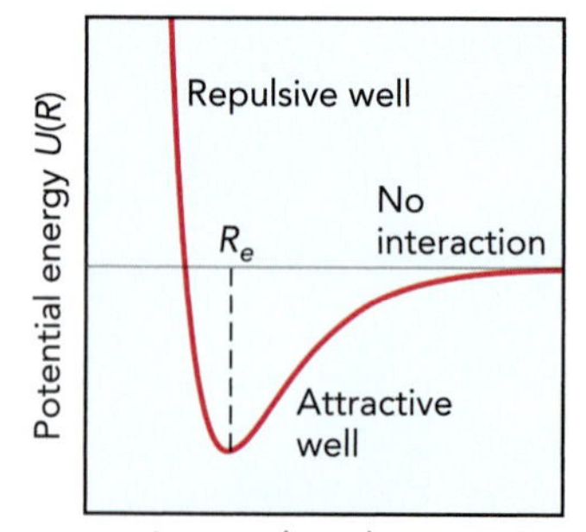

▲ FIGURE 5.14 **The three regions of any typical chemical bond potential energy curve.** You'll see a lot of this curve.

1. At very small values of R, core electrons of different atoms and the atomic nuclei themselves are pushed close together. The repulsive forces dominate, causing the potential energy to become positive and climb rapidly as R approaches zero.
2. At fairly small values of R (1–5 Å), there is an attractive force (if a chemical bond can form), mediated by the electrons that hold the two nuclei together, resulting in a negative potential energy or **potential well,** with minimum energy at the **equilibrium bond length,** abbreviated R_e. The energy required to break a chemical bond is called the **dissociation energy.** This is the energy required to climb out of the well and reach the zero energy plateau where the nuclei are no longer bound.
3. At very large R, the atoms interact less and less, and the potential energy becomes zero. This is the lowest energy at which the bond is broken, and the two atoms are free to separate.

If we start from a bound molecule with little kinetic energy near the bottom of the potential energy well, there are two competing forces that keep the chemical bond stable. As R decreases and the nuclei approach each other, the distances between like charges shrink, and repulsions push the nuclei apart again. As R increases, the nuclei pull against the electron charge in the binding region and eventually turn back. Any molecular bonding has these features, but they are easiest to illustrate for the diatomic molecule because there is only one coordinate, R, needed to define the geometry of the molecule. We shall see the same principles in molecules of more than two atoms, **polyatomic molecules,** which we describe next.

5.3 Covalent Bonds in Polyatomic Molecules

Now at last we extend our description of molecular structure to several atoms at a time, all bound together by the same charge–charge interactions that hold the diatomic molecule together, indeed the same interaction that holds the atom itself together. We'll have to keep track of more particles, but the basic physics changes little. Still, we don't easily visualize all these particles at once, and we will consider different ways of looking at these molecules that help us to make sense out of the increasingly complex interplay among the electrons and nuclei.

Wavefunctions for Polyatomics and Local Bonding

For H_2^+ and H_2, we used quantum mechanics to get some idea where the electrons could be found, but most of our discussion was based just on the classical mechanics of the system, specifically the potential energy function. Let's write the Hamiltonian for a triatomic molecule, say BeH_2, just to see how many terms there are and what the potential energy terms look like. We label the nuclei Be_A, H_B, and H_C, and start again from Eq. 5.4; there are no new terms in the Hamiltonian, only more of them:

$$\begin{aligned}\hat{H} &= \hat{K}_{\text{elec}} + U_{\text{nuc-elec}} + U_{\text{elec-elec}} + U_{\text{nuc-nuc}} + \hat{K}_{\text{nuc}} \\ &= -\frac{\hbar^2}{2m_e}\sum_{i=1}^{6}\nabla_i^2 - \sum_{i=1}^{6}\sum_{k=\text{A,B,C}}\frac{Z_k e^2}{r_{ik}} + \sum_{i=2}^{6}\sum_{j=1}^{i-1}\frac{e^2}{r_{ij}} \\ &\quad + \left(\frac{4e^2}{R_{\text{AB}}} + \frac{4e^2}{R_{\text{AC}}} + \frac{e^2}{R_{\text{BC}}}\right) - \frac{\hbar^2}{2}\sum_{k=\text{A,B,C}}\frac{1}{m_k}\nabla_k^2. \end{aligned} \tag{5.27}$$

Let's very quickly consider what the electronic wavefunctions will be like. Notice that, as required, the Hamiltonian treats the electrons as indistinguishable particles. The electrons cannot be labeled as Be atom electrons and H atom electrons. We're not even allowed to say which electrons are in the BeH bonds. On solving the wavefunctions, we would find that all electrons participate equally in the bonding. We could write one of the MO wavefunctions by combining the lowest-energy atomic orbitals on all three atoms:

$$\psi_a = c_{\text{A1}}\phi_{1s_\text{A}} + c_{\text{B1}}\phi_{1s_\text{B}} + c_{\text{C1}}\phi_{1s_\text{C}}. \tag{5.28}$$

Other MO wavefunctions ψ_b and ψ_c could combine these with different phase, or combine the $1s$ hydrogen and $2s$ beryllium atomic orbitals. The trouble is that when we combine these MO wavefunctions to write a many-electron wavefunction, enforcement of the symmetrization principle—that all six electrons be indistinguishable—leads to a long function of the form:

$$\begin{aligned}\Psi &= \psi_a\psi_a\psi_b\psi_b\psi_c\psi_c\alpha\beta\alpha\beta\alpha\beta - \psi_a\psi_a\psi_b\psi_b\psi_c\psi_c\beta\alpha\alpha\beta\alpha\beta \\ &\quad + \psi_a\psi_b\psi_a\psi_b\psi_c\psi_c\beta\alpha\alpha\beta\alpha\beta - \psi_a\psi_b\psi_a\psi_b\psi_c\psi_c\alpha\alpha\beta\beta\alpha\beta + \cdots \end{aligned} \tag{5.29}$$

This approach works well for the precise prediction of molecular properties. However, this formula is clearly too cumbersome for routine thinking about a simple polyatomic, and a simpler, more qualitative approach is a desirable option.

It is a challenge just to describe the potential energy. Even as simple a polyatomic as BeH_2 has an effective potential energy that depends on three coordinates: the two bond lengths and the HBeH bond angle. The potential energy function for the nuclei is no longer a curve, but a **potential energy surface.** The determination of these surfaces, and their use in predicting chemical behavior, keeps the lights burning late in many a laboratory.

To get around this, chemists tend to think of chemical bonds one pair of atoms at a time. However complex a multi-step chemical synthesis is, the chemist will commonly draw arrows illustrating an imagined sequence of electron migrations from one bond to another. Each bond is dealt with in turn. This is a simplification: instead of treating the full three-dimensional potential energy surface of BeH_2, we reduce the molecule to its two bonds, each of which we can describe individually...up to a point.

This way of looking at the molecule appeals to us for the same reason that we use the one-electron atomic orbitals to write electron configurations that approximate the much more complicated many-electron atomic wavefunctions. Reducing the number of coordinates, the *dimensionality* of the problem, makes the problem more tractable and can lead to a useful intuition about how these systems behave.

Lewis dot structures exemplify this approach, and they comprised the first broadly successful description of chemical bonding. In a Lewis structure, the electrons are restricted to atoms or pairs of atoms, rather than being smeared all over the molecule. Taking Lewis's assumptions to a greater level of detail leads to the **local bond model** of molecules. Each chemical bond may be represented by a potential energy curve similar to that used for the diatomics. These potential curves each represent the interaction of two parts of the molecule, independent of all other interactions. As one of these chemical bonds is shortened, repulsive forces between the core electrons of the two nuclei cause the potential to climb rapidly. As the bond is stretched to large distances, the attractive forces diminish and the potential energy converges to that of the separated fragments.

The simplicity of this picture is extremely attractive. It also addresses a weakness of the MO picture that we will want to keep in mind, namely that electron–nucleus interactions are often a great deal more localized than the general equation for the MOs may imply, particularly in the case of the core electron orbitals. The erroneous prediction that H_2 dissociates into an equal mixture of neutral atoms and ion pairs in Section 5.2 is an example of where the delocalization of the electrons in simple MO theory goes too far.

However, the chemical bonds in a polyatomic molecule are not truly independent, and the local bonding picture will always break down in some limit. As we break one of the $C{=}O$ bonds in CO_2, for example, the other $C{=}O$ bond shortens, approaching the triple bond in $C{\equiv}O$. Moreover, the length and the orientation of a chemical bond at some atom depend on the other bonds present. We can't deduce the geometry of a large polyatomic molecule if we look at the interactions only two atoms at a time. However, the solution we often employ is not far away.

BIOSKETCH **Douglas Grotjahn**

Douglas Grotjahn is a professor of chemistry at San Diego State University who designs catalysts for organic reactions. One of his achievements is the development of organometallic compounds that use hydrogen bonding to stabilize reaction intermediates and transition states. Grotjahn's research group connects transition metals such as ruthenium or iridium to ligand molecules that contain nitrogen atoms. Organic molecules such as alkenes attach easily to these metals, becoming primed for reaction. The accompanying figure shows a vinylidene group, H_2CC, attached to the ruthenium atom of one of these catalysts (a). As a water molecule approaches, it forms a hydrogen bond to one of the catalyst's nitrogen atoms (b). As the nitrogen provides a temporary host for the hydrogen atom, the OH group from the water attaches to the vinylidene group (c), transforming an *sp* hybridized carbon to sp^2. Professor Grotjahn has been able to show the importance of hydrogen bonding by ^{15}N nuclear magnetic resonance (NMR) spectroscopy of the catalyst at an intermediate stage of the reaction. (NMR is discussed in Section 5.5.) When the nitrogen is weakly bonded to a hydrogen atom, it becomes possible to see a splitting in the spectrum from the interaction between the nitrogen nucleus and the proton. The bonding in this case is weak compared to the chemical bonding of the reactants and catalyst, and a variety of spectroscopic methods are used to confirm that hydrogen bonding is taking place.

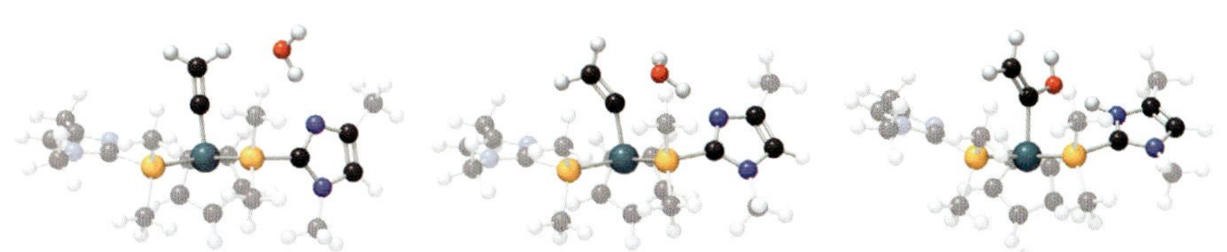

▲ **Chemical bonding in catalysis.** A nitrogen (dark blue) accepts a hydrogen (white) from water during the hydrogenation of an organic compound bound to the catalyst.

Hybrid Orbitals

Instead of restricting ourselves to only two atoms at a time, let's instead look at all the bonds that share a common *central atom.* We can then justify why the atomic orbitals of the central atom must often be reconstructed into **hybrid molecular orbitals,** molecular orbitals formed by linear combination of atomic orbitals *centered on the same atom.* In particular, by mixing *s* and *p* atomic orbitals we are able to distribute the electrons in directions that minimize the interaction between adjacent chemical bonds while maximizing the electron density in binding regions. Because many atoms participate in three or more chemical bonds simultaneously, the resulting geometry is often cumbersome to calculate by MO theory, but easily predicted (at least qualitatively) by orbital hybridization. Hybrid orbital theory, like the MO wavefunction we constructed for H_2, is a *model* for the properties of electrons in a molecule: helpful for visualizing how electrons distribute themselves in molecules, but not infallible. We take advantage of the model only as long as it accurately predicts experimental results.

The hybrid orbital model is widely used to explain molecular geometries in organic chemistry, and for those molecules its predictions are generally consistent with calculations based on molecular orbital theory. The most common application is the construction of sp, sp^2, and sp^3 hybrid orbitals at carbon, nitrogen, and oxygen atoms, and we will confine the discussion here to those cases.[5]

What do these hybrid orbitals look like? In the case of the H_2^+ and H_2 bonds, we formed MOs by combining the atomic orbitals on different atoms, but it is not necessary that we combine orbitals only from different atoms. We are using the one-electron orbitals only because they are simple starting points for describing the electron distributions in many-electron systems. When we form a chemical bond, we expect the electron wavefunction, according to the variational principle, to take whatever shape it needs to find its lowest energy. In the case of methane, for example, the $2s$ orbital is completely non-directional, whereas for a deep potential well and lower average energy, we would like a wavefunction that pushes the electron into the bonding region between the two nuclei. Even the $2p$ orbitals, although not spherically symmetric like the s orbital, make terrible orbitals for a chemical bond because they put electron density equally on opposite sides of the same atom.

This is all corrected by allowing the four orbitals on the carbon atom to share character, becoming mixed sp^3 hybrid orbitals in the same way that we combined s and p orbitals in Eqs. 2.5 and 2.6 to get the directional sp orbitals, labeled ψ_a and ψ_b:

$$\psi_a = \frac{1}{\sqrt{2}}[(2s) + (2p_z)]$$

$$\psi_b = \frac{1}{\sqrt{2}}[(2s) - (2p_z)].$$

In both cases, the new orbitals are hybrids of the original s and p atomic orbitals.

In combining the s and p orbitals, the distribution of the chemical bonds is also determined. Using the simpler example of the sp hybrids, an explicit expression is obtained if we rewrite the atomic orbitals in terms of their radial components $R_{n,l}(r)$ and angular components $Y_l^{m_l}(\theta,\phi)$:

$$\psi_a = \frac{1}{\sqrt{2}}[R_{2,0}Y_0^0 + R_{2,1}Y_1^0]$$

$$\psi_b = \frac{1}{\sqrt{2}}[R_{2,0}Y_0^0 - R_{2,1}Y_1^0].$$

These may be written in detail as follows, using Tables 3.1 and 3.2:

$$\psi_{a/b} = \frac{1}{\sqrt{2}}\left[\left(1-\frac{Zr}{2a_0}\right)e^{-Zr/(2a_0)}\left(\sqrt{\frac{1}{4\pi}}\right) \pm \frac{Zr}{a_0}e^{-Zr/(2a_0)}\left(\sqrt{\frac{3}{4\pi}}\cos\theta\right)\right], \quad (5.30)$$

where the plus sign gives ψ_a and the minus sign gives ψ_b.

[5]More elaborate geometries, such as the trigonal bipyramidal structure of PF_4Cl, can be explained as a hybrid of s, p, and d orbitals, mixed together to minimize the interaction among the bonded atoms. However, it has long been known that d orbitals rarely participate in bonding through orbital hybridization. Systems of more than four covalent bonds to the same central atom can often be explained instead by a combination of two effects: (i) partial electron transfer from the outer atom to unhybridized d orbitals on the central atom, and (ii) distribution of the traditional valence electron octet among all the bonds, assigning fewer than two electrons per bond on average (in effect, an average of distinct resonance structures).

EXAMPLE 5.1 Hybrid Orbital Bond Angle

CONTEXT The *sp* hybrid orbitals appear in organic chemistry to justify the linear bonding found in alkynes, allenes, and nitriles—all cases involving multiple π bonds. The *sp* hybrids direct electron density into two different bonding regions while leaving two *p* orbitals unhybridized and available for π bond formation. These unsaturated carbon chains burn at high temperatures because the π bonds have a higher potential energy than the corresponding σ bonds, and more energy is released when these electrons are used for formation of the very stable chemical bonds in water and carbon dioxide.

PROBLEM Show that the two *sp* hybrid orbitals ψ_a and ψ_b will form bonds with a bond angle of 180° by finding the values of θ with the greatest electron density.

SOLUTION These functions depend only on r and θ, because they are cylindrically symmetric about the bond axis. To find the optimum bond angle for atoms using these hybrid orbitals, we don't need to consider the r-dependence of the wavefunction, and that's why we only need to look for the maximum electron density as a function of θ.

The maximum electron density occurs where the magnitude of the wavefunction is greatest, that is, where it reaches a maximum (if $\psi > 0$) or minimum (if $\psi < 0$). So let's fix r to some constant value r_0 and find the derivative of the *sp* orbital with respect to θ to find the minima and maxima:

$$\frac{d\psi_{a/b}}{d\theta} = \frac{1}{\sqrt{2}}\left[\pm\frac{Zr_0}{a_0}e^{-Zr_0/(2a_0)}\left(\sqrt{\frac{3}{4\pi}}\sin\theta\right)\right]$$

$$= 0. \qquad \text{for minimum or maximum}$$

This condition is met only if $\sin\theta = 0$, which requires the angle θ to be either 0 or π; the values are 180° apart. Checking the actual values of $\psi_{a/b}$, we find that $\theta = 0$ is indeed the angle of maximum magnitude for ψ_a, and that $\theta = \pi$ is the angle of maximum magnitude for ψ_b.

We don't have to make the two *sp* hybrid orbitals equivalent to one another, however. By shifting the amount of *s* and *p* character in each of the two hybrids, we can obtain two orbitals—still normalized and mutually orthogonal—that have different fractions of electron density in the bonding region. Figure 5.15 shows that by varying the fractions of *s* and *p* character, we can control how much of the electron density lies along the *z* axis toward a neighboring atom. This in turn affects the chemical bond strength. In acetylene, HCCH, the *sp* hybrids on each carbon are predicted to be roughly 52:48 mixtures of *s* and *p*, with slightly more *p* character in the CH bond than in the CC.

If there are **lone pair** electrons (valence shell electrons that are paired but bound to a single atom instead of a chemical bond), they will also favor a molecular geometry that minimizes their overlap with other electron clouds.

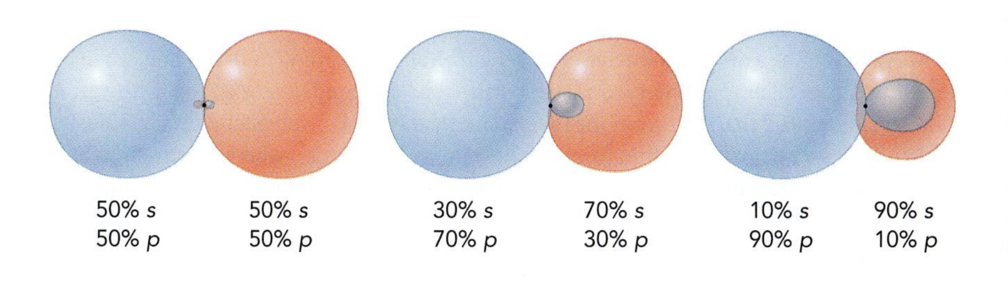

◀ FIGURE 5.15 **Angular components of *sp* hybrids formed from different amounts of *s* and *p* character.** The greater the *p* character, the more strongly the electron density is focused along the *z* axis, but also the more electron density is shifted to the opposite side of the nucleus. Most *sp* hybrids will be close to 50:50 mixtures, and the other cases shown here are extreme in order to make the effect more visible.

The resulting variety of geometries is accounted for largely by **valence shell electron pair repulsion** (VSEPR) theory, discussed in many general chemistry texts. In NCO^-, for example, where carbon bonds to atoms supporting lone pairs, the *sp* hybrids on the carbon atom are 55% *p*, 45% *s* in the CO bond hybrid and 45% *p*, 55% *s* in the CN bond hybrid. The hybrids that form σ-bonds toward more electronegative atoms tend to have more *p* character, as in this case in which O is more electronegative than N. This principle is **Bent's rule.** We justify it by recalling that the 2*s* atomic orbital is lower in energy than the 2*p*, and therefore more *p* character creates an orbital higher in energy, which more readily donates electron density toward the electronegative atom at the other end of the bond (Fig. 5.16).

The sp^2 and sp^3 hybrid orbitals similarly lead to molecular bonds that are as widely separated as possible. In the case of the sp^3 hybrids, the bond angles are 109.5° compared to 90°, and the carbon is bound to four atoms, which offsets the promotion energy of the 2*s* electron to 2*p*.

Like the molecular orbitals, the hybrid orbitals should comprise a mutually orthogonal set. For example, the two *sp* hybrid orbitals just used are orthogonal:

$$\begin{aligned}\int (sp_a)^*(sp_b)d\tau &= \frac{1}{2}\int [(2s)+(2p_z)]^*[(2s)-(2p_z)]\,d\tau \\ &= \frac{1}{2}\left[\int (2s)^*(2s)d\tau - \int (2s)^*(2p_z)d\tau + \int (2p_z)^*(2s)d\tau - \int (2p_z)^*(2p_z)d\tau\right] \\ &= \frac{1}{2}[1+0-0-1] \\ &= 0.\end{aligned}$$

This proof relied on the atomic orbitals used to create the hybrid orbitals being themselves normalized and orthogonal. Other combinations, such as the three sp^2 or the four sp^3 hybrid orbitals, are also orthogonal sets.

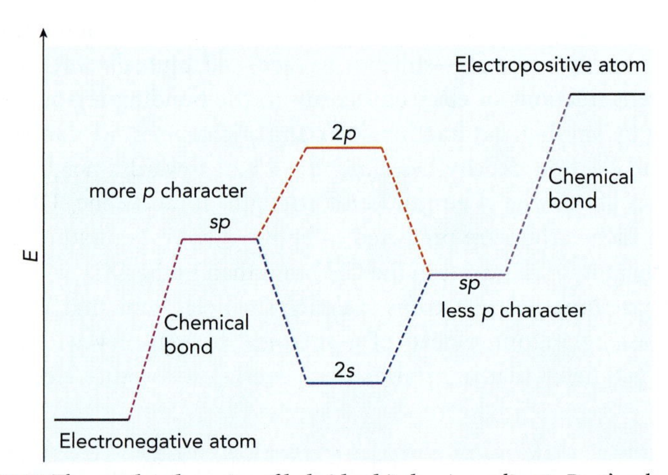

▲ **FIGURE 5.16 The *s* and *p* character of hybrid orbitals.** According to Bent's rule, when mixing *s* and *p* orbitals to make σ-bonding hybrid orbitals, a bond to an electronegative atom will tend to use more character from the higher energy *p* orbital. That encourages the transfer of electron density out of the hybrid toward the electronegative atom, where some of that electron charge will be more stable. Similarly, bonds to electropositive atoms tend to employ more *s* character, because that provides a lower energy hybrid that better stabilizes electron charge being transferred *away* from the electropositive atom.

EXAMPLE 5.2 Hybrid Orbitals

CONTEXT The carbonate ion, CO_3^{2-}, is a resonance-stabilized ion used to regulate the pH of blood as one of the chemical species in the equilibrium

$$CO_3^{2-}(aq) + H^+(aq) \rightleftharpoons HCO_3^-(aq)$$
$$HCO_3^-(aq) + H^+(aq) \rightleftharpoons H_2CO_3(aq)$$
$$H_2CO_3(aq) \rightleftharpoons CO_2(aq) + H_2O(liq)$$
$$CO_2(aq) \rightleftharpoons CO_2(gas).$$

If the pH climbs too high, a condition known as *alkalemia*, the kidneys remove HCO_3^- ion, driving the equilibrium between HCO_3^- and H_2CO_3 to the reactant side. More H^+ is liberated and the pH is reduced. If the blood pH is too low (leading to *acidemia*), Le Châtelier's principle naturally leads the excess H^+ to drive the equilibrium to the products, toward production of CO_2 gas, which the body removes by exhalation. The pK_a of HCO_3^- is only 10.3, rather low, given that its conjugate base is the doubly charged CO_3^{2-}. The low pK_a is partly explained by resonance stabilization in CO_3^{2-}, which distributes the two excess electrons over the three equivalent oxygen atoms. The bonding in carbonate ion does not favor one oxygen atom over another, relying instead on the formation of three equivalent sp^2 hybrids centered at the carbon atom.

PROBLEM Write the expressions for an orthonormal set of the three sp^2 hybrid orbitals in carbonate ion, all exactly equivalent, lying in the xy plane with one orbital pointing along the x axis. Use the $2s$ and $2p$ subshells.

SOLUTION To get orbitals that lie in the xy plane, we will use only the $2s$, $2p_x$, and $2p_y$ atomic orbitals. Any contribution from $2p_z$ will cause the orbital to have different densities above and below the xy plane. In order for all three orbitals to be equivalent, they must each have the same ratio of s character to p character, because that determines the shape of the orbital (more s character makes the orbital more spherical, more p character makes it more elongated along the p orbital axis). To completely use up the $2s$ orbital, therefore, we put 1/3 of its *density* into each sp^2 hybrid, corresponding to a coefficient in the *wavefunction* of $\sqrt{1/3}$. For the orbital pointing along the x axis, the rest of the hybrid will be formed from the $2p_x$ orbital:

$$(sp_a^2) = \sqrt{\frac{1}{3}}(2s) + \sqrt{\frac{2}{3}}(2p_x).$$

The coefficient for the $(2p_x)$ is determined by knowing that the sum of the two squares must be 1.
The remaining two sp^2 orbitals must use up the rest of the $(2p_x)$ and must each take half of the $(2p_y)$ orbital:

$$(sp_b^2) = \sqrt{\frac{1}{3}}(2s) - \sqrt{\frac{1}{6}}(2p_x) + \sqrt{\frac{1}{2}}(2p_y)$$
$$(sp_c^2) = \sqrt{\frac{1}{3}}(2s) - \sqrt{\frac{1}{6}}(2p_x) - \sqrt{\frac{1}{2}}(2p_y).$$

The phases for each term are determined by arbitrarily selecting the coefficient of the $(3s)$ term to be positive, and then recognizing that (sp_b^2) and (sp_c^2) must point in the opposite direction along the x axis from (sp_a^2) (and therefore must have a coefficient of opposite sign). Finally, the phases for the $(2p_y)$ terms must be opposite so that (sp_b^2) and (sp_c^2) point in opposite directions along the y axis. The resulting orbitals are shown in Fig. 5.17.

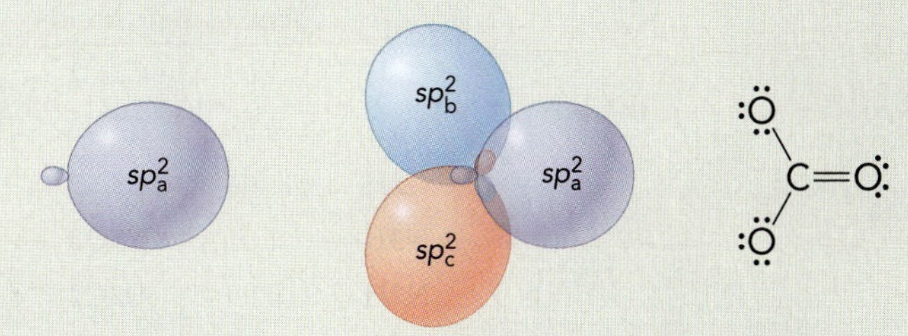

▲ **FIGURE 5.17** **The sp_a^2 hybrid and three equivalent sp^2 hybrid orbitals.**

EXTEND The three equivalent hybrid orbitals have maxima separated by 120°, so that no two are closer together than any other pair. This is ideal if the three orbitals contribute to equivalent chemical bonds, as for example in BH_3. However, if the bonds are not equivalent, then the optimal geometry will be for angles at least slightly different from 120°. For example, in F_2CO, Bent's rule predicts that the σ-bonds to the highly electronegative F atoms should have greater *p* character than the bond to the O atom. The molecule constructs sp^2 hybrids on the carbon atom such that each of the orbitals pointing toward a fluorine has 29% *s* character and 71% *p* character. Because both orbitals have more p_z character than equivalent sp^2 hybrids do, the two orbitals lie closer together and give an optimal bond angle of only 108°(Fig. 5.18). One additional note: the three equivalent sp^2 hybrids shown for CO_3^{2-} can be seen to overlap, and in fact they form a disk of equal electron density *at all angles.* There is not more electron density at 0° than at 10°, for example. To completely explain the triangular arrangement of the bonds, we need to include the H atoms at the other end of the bond, and *their* electronic wavefunctions.

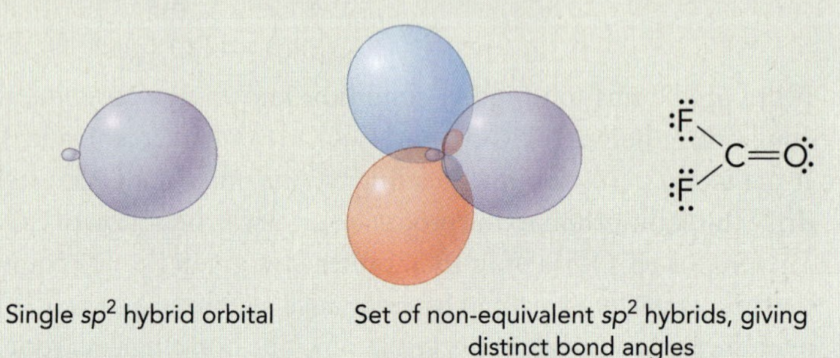

▲ **FIGURE 5.18 Hybrid orbitals predicted for the carbon atom in F_2CO.** The sp_a^2 hybrid (41% *s*, 59% p_z) and the three inequivalent sp^2 hybrids are shown.

The local bond model allows us to successfully predict many aspects of molecular structure without reliance on any sophisticated calculations. For most molecules, the distribution of valence electrons near any group of three or four atoms is sensitive primarily to the bonding requirements of *those* atoms. For example, it is not necessary to consider the entire $(CH_3CH_2)_3CCCH$ molecule to predict that the C≡C bond is roughly 1.20 Å long. Typical bond lengths for H, C, N, and O in covalently bound molecules are given in Table 5.1. Even in molecules for which the local bond model is a poor approximation, the bond lengths in Table 5.1 are useful guides for determining the range of likely values.

TABLE 5.1 Approximate covalent bond lengths for H, C, N, and O (Å). The number of atoms connected to each C, N, and O atom is indicated by the number following the atomic symbol; for example, C4C4 designates the bond between two carbon atoms, both of which are bound to four atoms, an example being the CC bond in CH_3CH_3.

Formal single bonds			
C4H	1.09	C3C2	1.45
C3H	1.08	C3N3	1.40
C2H	1.06	C3N2	1.40
N3H	1.01	C3O2	1.36
N2H	0.99	C2C2	1.38
O2H	0.96	C2N3	1.33
C4C4	1.54	C2N2	1.33
C4C3	1.52	C2O2	1.36
C4C2	1.46	N3N3	1.45

Formal double bonds			
C3C3	1.34	C2O1	1.16
C3C2	1.31	N3O1	1.24
C3N2	1.32	N2N2	1.25
C3O1	1.22	N2O1	1.22
C2C2	1.28	O1O1	1.21
C2N2	1.32		

Formal triple bonds			
C2C2	1.20	N1N1	1.10
C2N1	1.16		

(continued)

TABLE 5.1 continued Approximate covalent bond lengths for H, C, N, and O (Å).

Formal single bonds			
C4N3	1.47	N3N2	1.45
C4N2	1.47	N3O2	1.36
C4O2	1.43	N2N2	1.45
C3C3	1.46	N2O2	1.41

Aromatic bonds			
C3C3	1.40	N2N2	1.35
C3N2	1.34		

EXAMPLE 5.3 Covalent Bond Lengths

CONTEXT Acetamide is used as a **plasticizer,** a chemical agent that softens materials to make them more pliable in manufacturing. One substance in which acetamide serves this purpose well is *soy protein isolate,* a highly purified form of soy proteins that can be used to make strong, biodegradable plastics. The acetamide structure, which we examine in this example, enables the formation of hydrogen bonds to the soy proteins, providing an environment that binds the proteins together with more open space than when the proteins bind directly one to the other. The extra open space makes the plastic more flexible. Acetamide is itself a biological product, secreted by mammals, and so provides an environmentally friendly component in the treatment of soy protein isolate plastics.

PROBLEM From Table 5.1, estimate the bond lengths of all the chemical bonds in acetamide, CH_3CONH_2.

SOLUTION The methyl carbon is bonded to four atoms: three hydrogens and another carbon. Therefore, we estimate the C—H bond lengths from the C4-H value, 1.09 Å. The carbonyl carbon is bound to three atoms: carbon, oxygen, and nitrogen. For the C—C bond length, we take the value for C4-C3, 1.52 Å. The nitrogen is bound to three atoms: two hydrogens and the carbonyl carbon, indicating that the C—N bond length should be obtained from the value 1.40 Å for C3-N3. The predicted and actual values are shown next:

bond	form	predicted	actual
C—H	C4-H	1.09	1.124
C—C	C4-C3	1.52	1.519
C—O	C3-O1	1.22	1.220
C—N	C3-N3	1.40	1.380
N—H	N3-H	1.01	1.022

Some of the discrepancies can be attributed to the electron-withdrawing nature of the carbonyl group. In general, the predictions are quite good.

5.4 Non-Covalent Bonds

All of the bonds between atoms that we will consider can be modeled using the molecular orbital method described earlier, and this is the most common way to approach quantitative studies of chemical bonding at present. However, the intuitive models that we apply to chemical bonds vary depending on the system, and in some cases the relationship between the MO picture and the qualitative bonding mechanism is not obvious.

We commonly divide chemical bonds into covalent (shared electrons in localized bonds), **ionic** (electrons transferred from one atom to another, rather than shared), and **metallic** (shared, delocalized electrons stabilizing metal atom cations). Let's quickly examine these last two types of bonding and see how the MO description deals with their characteristics.

Ionic Bonds

Coulomb's law explains the stability of an ionic bond more easily than any other chemical bond, once we have formed the ions. If we take for granted that an electron from neutral atom A can be relocated to neutral atom B, forming an ionic pair A^+B^-, then the attractive term in the potential energy surface for the system can be approximated by treating the ions as two point charges:

$$U_{\text{ionic attr}} = \frac{q_A q_B}{4\pi\varepsilon_0 R}, \tag{5.31}$$

where q_A and q_B are the charges on the two ions and R is the distance between them. As long as q_A and q_B have opposite signs, $U_{\text{ionic attr}}$ is negative at all R. For the alkali halides (such as NaCl) and alkali hydrides (such as LiH), $q_A = e$ and $q_B = -e$.

The potential energy in Eq. 5.31 reaches zero as R approaches an infinite separation between the two ions. We saw, however, that this was a major problem in the simplest MO treatment of H_2. Equation 5.22 incorrectly predicted that H_2 could just as easily dissociate into ionic $H^+ + H^-$ as into neutral H + H. The trouble in Eq. 5.31 looks similar, but the problem this time is that the ionic potential assumes that the charges on the atoms are fixed, that the charge transfer to form the pair of ions is a permanent arrangement. In fact, if we select *any two* elements from the periodic table and separate them, the two neutral atoms are always more stable than the combination of ions they may form. It always takes more energy to remove an electron to form the cation than can be retrieved by adding that electron to the other atom. Therefore, for the ground electronic state of any ionic molecule, an accurate wavefunction would show a charge separation near the equilibrium geometry, but it would allow the electron density to even out smoothly as the bond breaks. This is shown in Fig. 5.19 for the HF molecule, where the positive charge on the H atom near the equilibrium geometry (R_e = 0.92 Å) fades away as we pull the two atoms apart.

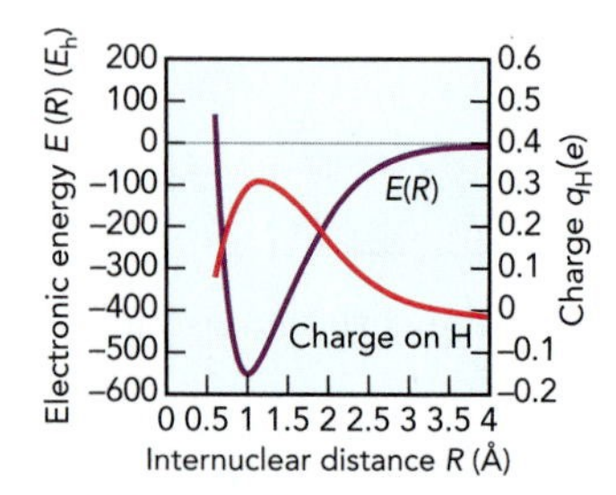

▲ FIGURE 5.19 **Computed potential energy curve for dissociation of HF into neutral atoms.** The second curve indicates the charge on the H atom as a function of *R*, showing that the bond is strongest (the well is deepest) where the bond is most ionic, and that the charge approaches zero for the neutral atom as the bond breaks.

Metallic Bonds

Metallic bonds pose a greater challenge than ionic bonds to the standard MO theory approach because the LCAO method we normally use to build the molecular orbitals from atomic orbitals turns out to be an inefficient way of treating the valence electrons in metal–metal bonds. The LCAO method relies on basis functions that position the electrons near the nuclei, with exponentially decreasing density as the distance from the nuclei increases. This contrasts with the qualitative picture of bonding in a network of metal atoms.

If we start with a single metal cation, the electric field of the positive charge strongly localizes an electron. If we start instead with two metal cations, a couple of angstroms apart, the additional positive charge makes the neighborhood of the nuclei even more attractive to an electron, but the electric field *between* the nuclei does not vary as dramatically. The more cations we contribute to the volume, the more evenly distributed the electric field lines become, and the less the tendency for electron density to bunch up in the neighborhood of the nuclei. The long-range distribution of positive charges effectively reduces the attraction of the electrons for the nuclei. As a result, the LCAO approach converges slowly on accurate wavefunctions for the metals.

Figure 5.20 shows a crude representation of the potential energy curve for a valence electron in the vicinity of several positive charges. The electrons are bound to the cations in the low depression that appears near the spikes; the spikes indicate the Coulomb and exchange repulsions that resist the valence electrons trying to penetrate the core of each atom. This illustrates that the bound electrons are not stuck near a single atom or pair of atoms. The region of stability is a continuous, if twisty, valley in the potential surface of the electron. Given a little kick, an electron can sail, with a few twists, from one side of the molecule to the other. This is the property that separates the metals from other elements in many material applications, a topic we return to in Chapter 13.

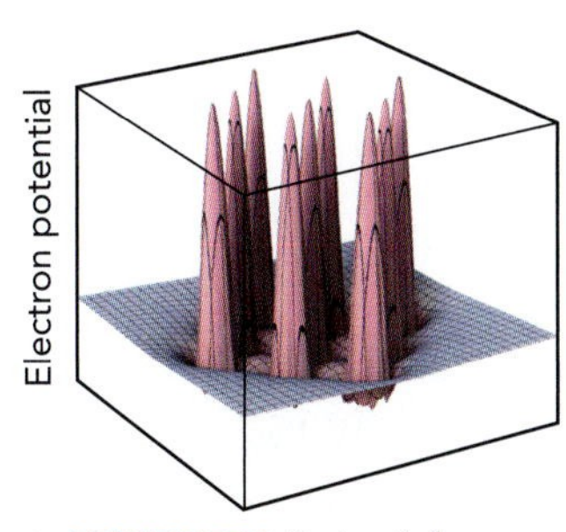

▲ FIGURE 5.20 **Potential energy of an electron in a metal.** An approximate potential energy function is plotted for a valence electron in the neighborhood of nine metal cations arranged in a planar 3×3 grid. The potential energy spikes near the core electrons on the cations, but a shallow pool of negative potential energy surrounds and connects all of those spikes, allowing valence electrons to be distributed evenly throughout the structure.

5.5 Nuclear Magnetic Resonance Spectroscopy

One of the key methods we use for analyzing chemical structures is based on the local bonding model described in Section 5.3. The C—C bond in an alkane is longer and weaker than the C≡C bond length in an alkyne because the alkyne has a greater average electron density in the binding region than the alkane, regardless of what else may be attached to the carbon atoms at each end of the bond. Therefore, we could distinguish between molecules with only carbon–carbon single bonds and molecules with carbon–carbon triple bonds if we had a technique that somehow probed this average electron density with high precision.

By far the most common way we accomplish this is by **nuclear magnetic resonance** (NMR) spectroscopy. At its heart, all this method does is measure how strongly the electron cloud around a nucleus shields the nucleus from an applied magnetic field. The amount of shielding depends on the distribution of the electrons near that nucleus. Let's note here that this shielding (which is shielding by the surrounding electrons of the external *magnetic* field B_0 experienced by the *nucleus*) differs from the shielding in Section 4.1 (which is shielding by inner electrons of the *electric field of the nucleus,* as experienced by *outer electrons*). The principle here is analogous to that for most other forms of spectroscopy:

1. The molecule has several energy levels and we introduce radiation to excite a change from one energy level to another.
2. By determining the precise photon energy that causes this transition, we learn the energy gap between the two states in the transition.
3. From this, we can work back to determine fundamental properties of the molecule that interests us.

The energy levels in this case are the result of the interactions between the nuclear spin magnetic fields in the molecule and an external magnetic field. We begin by way of a glance at nuclear spins.

TOOLS OF THE TRADE Nuclear Magnetic Resonance Spectroscopy

In 1924, Wolfgang Pauli first suggested that nuclear angular momentum might be detected spectroscopically, and in 1927 S. A. Goudsmit and Ernst Back (though unaware of Pauli's paper) published experimental data on bismuth in a magnetic field, in which they correctly justified one observation by proposing that the major isotope, ^{209}Bi, possessed a nuclear spin of 9/2. This work opened the door to the experimental study of laboratory magnetic fields interacting with nuclear magnetic fields in atoms, but the development of a reliable spectroscopic technique was slow. Working with a beam of gas-phase LiCl molecules at low pressure in 1938, Isador Rabi (1944 Nobel Prize in Physics) observed the first NMR spectrum as the absorption of radio waves by a transition between ^{7}Li nuclear spin states in a magnetic field. In 1945, Felix Bloch and Edward M. Purcell independently observed the first NMR transitions in the condensed phase—Bloch in liquid water spiked with manganese sulfate, Purcell in solid paraffin—for which they shared the 1952 Nobel Prize in Physics.

Applying NMR to chemical analysis posed two major instrumental challenges: (*i*) the spectroscopic signals are *very* weak compared with most other methods, and (*ii*) the magnetic field must be of exceptionally high quality (and ideally high strength) where it interacts with the sample. The first challenge was largely addressed by the intense development of radiofrequency (rf) electronics during World War II, which led to powerful and precise radiation sources at long wavelengths.

The second challenge arises from the natural tendency of magnetic fields to weaken with distance from the source, remaining constant only inside a perfect solenoid. If the magnetic field in an NMR spectrometer has different values at different locations within the sample, then the nuclear spin transitions occur at different frequencies of the radiation, and the signals in the spectrum become broadened and weaker. Stronger magnetic fields can help resolve separate transitions by increasing the energies of the nuclear spin transitions, and the introduction of superconducting magnets into NMR spectrometers through the 1960s and 1970s increased the available field strengths from under 4 T to over 20 T. However, while magnet design is vastly improved over the earliest days of NMR, maintaining the magnetic field homogeneity—the precisely consistent field strength over the entire sample volume—continues to be a challenge.

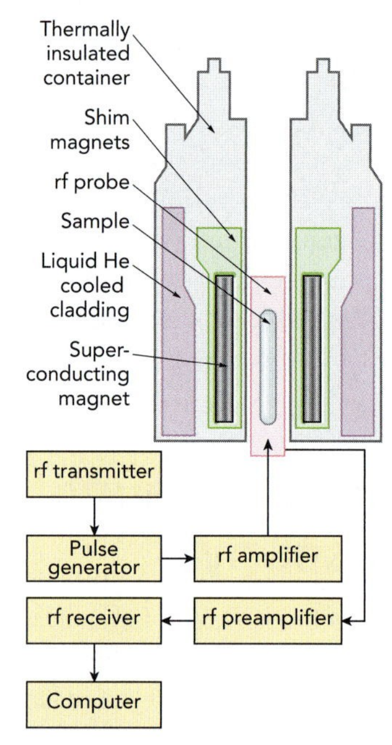

Schematic of an NMR spectrometer.

How does it work? A typical NMR spectrometer combines the following subsystems: (1) a superconducting magnet, (2) rf generation and detection electronics, and (3) a sample probe.

The magnet. The magnet is responsible for splitting the energies of the nuclear spin quantum states so that a transition may occur from a lower energy state to a higher energy state. Without the magnetic field, these spin states would have the same (or nearly the same) energy. The magnets used today consist of a coil of liquid helium-cooled, superconducting $(NbTaTi)_3$ Sn wire, protected by a shell of copper. The wire coil forms a solenoid magnet and, because a perfect superconductor has zero resistance, can support electrical currents of over 100 A, which enables magnetic fields of over 10 T to be generated over a volume of several cubic centimeters. Partnered with the main magnet are several shim magnets, which help to correct inhomogeneities, variations in the magnetic field.

Radiofrequency electronics. The radiation needed to induce the nuclear spin transition is typically generated by a frequency synthesizer—a flexible, low-power rf source—that is processed by a pulse programmer and amplified to be released in high-power bursts of up to 100 W. The radiation is delivered to the sample from another set of wire coils, called the probe. Following the burst, which is only a few microseconds long, the energy stored in the excited spin states is gradually released over the next several milliseconds as new rf photons. These photons are collected by the same probe that delivered the pulse and form the free induction decay signal used to generate the spectrum.

Sample probe. The sample probe includes the rf coils, a spinner for the sample (to further reduce the impact of field inhomogeneity), and a temperature control system. The rf coils are part of the sample probe because they must be optimized for radiation at a particular frequency, in order to handle the high powers involved. The temperature control provides a valuable tool for assessing the barriers of low-energy processes such as conformational changes.

The NMR spectrometer has become the primary means for determining structures of compounds in solution, making it *the* key tool for qualitative analysis in chemical synthesis. Its ability to characterize a molecule in solution in a fraction of a second has also led to the extensive use of NMR in studies of reaction rates and mechanisms.

Nuclear Spins

Atomic nuclei are composed of protons and neutrons, each of which, like electrons, has a spin angular momentum of magnitude $\hbar/2$. This means that any proton or neutron generates a tiny magnetic field, with a magnitude measured in units of tesla (T). As with electrons in Section 3.3, we calculate the strength of the magnetic field interactions using the magnetic moment vector $\vec{\mu}$. In the case of one proton, the magnitude of the magnetic moment is

$$\mu_p = 2.793\mu_N = 1.411 \cdot 10^{-26}\,\mathrm{J\,T^{-1}}, \tag{5.32}$$

where the nuclear magneton,

$$\mu_N = \frac{e\hbar}{2m_p} = 5.051 \cdot 10^{-27}\,\mathrm{J\,T^{-1}}, \tag{5.33}$$

is the conventional measuring unit. The greater mass of the proton leads to a smaller magnetic moment than the electron's,

$$\mu_e = -1.001\mu_B = -9.285 \cdot 10^{-24}\,\mathrm{J\,T^{-2}},$$

which is measured in units of the Bohr magneton μ_B. Both μ_N and μ_B serve to convert magnetic field strengths into energies, at the scale appropriate to the strength of the interaction, so we can expect the nuclear magnetic field to be roughly 600 times weaker than the electron spin magnetic field.

When the protons and neutrons ("nucleons") are combined, chemists treat the resulting atomic nucleus as a single particle. As with electrons, the spins of the nucleons may add together or may cancel, and this affects the strength of the magnetic moment. The total spin of the nucleus is labeled by the nuclear spin quantum number I, and the nuclear magnetic quantum number m_I gives the orientation of the magnetic moment relative to a lab-fixed Z axis. I is always either an integer (for an even number of nucleons) or a half-integer (for an odd number of nucleons), and its magnetic quantum number m_I can have values of $-I, -I + 1, -I + 2, \ldots, I - 1$, and I. Values of I range from 0 to at least 16 (for a metastable isotope of polonium). For example, ^{1}H has a nucleus with one proton, the proton has a spin of 1/2, so the total nuclear spin for ^{1}H is $I = 1/2$.

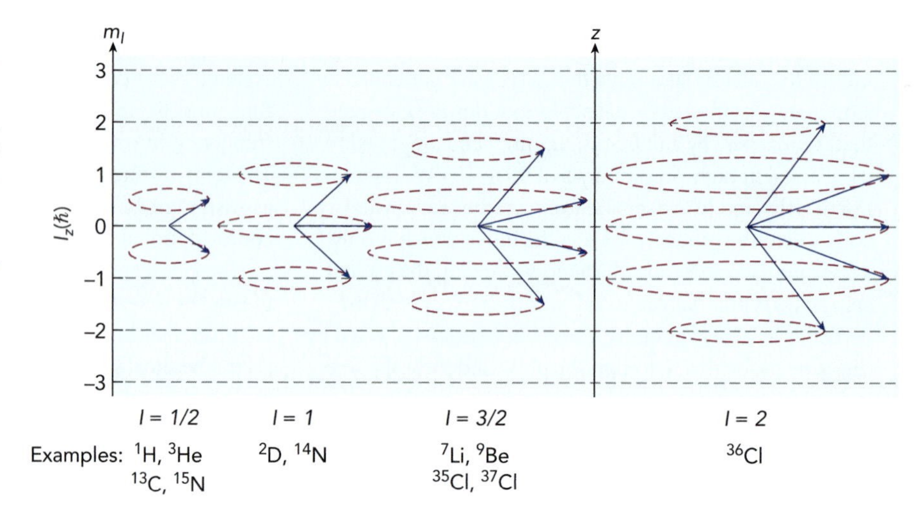

FIGURE 5.21 Nuclear spin and nuclear magnetic quantum numbers. A nucleus with spin I has a spin angular momentum vector of magnitude $\hbar\sqrt{I(I+1)}$ and possible projections onto the Z axis of $\hbar m_I$, where m_I can vary from $-I$ to I in increments of one. (Interestingly, all the nuclear isotopes with $I = 2$, such as ^{36}Cl, are radioactive.)

There are only two possible values of m_I: $-1/2$ and $1/2$. This and other examples are shown in Fig. 5.21.

For other nuclei, the I value is the result of angular momentum couplings similar to those covered in Section 4.4, and is not necessarily easy to predict from the numbers of protons and neutrons. If the mass number of the nucleus is a multiple of 4, the spin is usually 0. So ^{4}He, ^{12}C, ^{16}O, and ^{32}S all have $I=0$, whereas ^{2}H and ^{14}N have $I = 1$. The values of I for all the discovered isotopes are known, and in chemistry we treat those values as fundamental parameters, meaning we look them up instead of calculating them.

Aside from ^{1}H, other common half-integer spin nuclei are ^{13}C, ^{15}N, ^{31}P, and ^{33}S, all having $I = 1/2$. The two most abundant isotopes of oxygen, ^{16}O and ^{18}O, have zero nuclear spin, so $m_I = 0$ always. However, a rare (but naturally occurring) isotope, ^{17}O, has a spin $I = 5/2$, and therefore the allowed m_I values are $-5/2, -3/2, -1/2, 1/2, 3/2$, and $5/2$.

Nuclear Spin Transitions

If there is no magnetic field other than that generated by the nuclear spin, then all the different m_I states of a nucleus have the same energy. As soon as we introduce any other magnetic field, the energies of the different m_I states begin to split, because different orientations of the magnetic moment in that external field will have different energies. In the simplest NMR spectroscopy, we use a powerful laboratory magnet to introduce a strong external magnetic field lined up along the Z axis. We will represent the strength of the applied magnetic field in the lab by B_0, and, because the electrons in an atom affect the magnetic field at the nucleus, B_{local} will be the strength of this field as seen by the nucleus.

If $I = 0$, then there is only one nuclear spin quantum state: $I, m_I = 0,0$. In that case, no transition between different magnetic spin states is available, and NMR spectroscopy does not probe that nucleus. There is no such thing as ^{12}C NMR or ^{16}O NMR. However, if there is more than one m_I state, then those

Parameters key: nuclear magnetic resonance

symbol	parameter	SI units
B_0	applied magnetic field	T
B_{local}	applied magnetic field at the nucleus	T
e	fundamental charge	C
g_I	nuclear gyromagnetic factor	unitless
$\hbar$	(Planck's constant)/(2π)	J s
$\vec{I}$	nuclear spin vector	J s
m_I	nuclear spin magnetic quantum number	unitless
m_p	proton mass	kg
μ	magnetic moment	J T^{-1}
μ_B	Bohr magneton	J T^{-1}
μ_N	nuclear magneton	J T^{-1}
$\vec{\mu}_I$	nuclear magnetic moment	J T^{-1}
σ	shielding factor	unitless
σ_{iso}	isotropic shielding contribution	unitless

states have different energies in the external field, and spectroscopy can measure the energy gaps between them.

The transitions we excite in an NMR experiment change the value of the magnetic nuclear spin quantum number, m_I, by one unit, so $\Delta m_I = \pm 1$. The energy of the interaction between two magnetic fields is equal to the dot product of the magnetic moment $\vec{\mu}$ for one with the field strength $\vec{B}$ of the other. If we place one of these atomic nuclei in an external magnetic field B lined up along the Z axis of our lab, then the energy of the nuclear spin state becomes the nuclear equivalent of our Zeeman energy in Eq. 3.58:

$$\begin{aligned} E_{\text{mag},I} &= -\vec{\mu}_I \cdot \vec{B} \\ &= -g_I\mu_N \vec{I} \cdot \vec{B}/\hbar && \vec{\mu}_I = g_I\mu_N \vec{I}/\hbar \\ &= -g_I\mu_N m_I B_{\text{local}}, && \vec{B} = (0,0,B_{\text{local}}), I_Z = \hbar m_I \end{aligned} \tag{5.34}$$

where g_I is the unitless nuclear gyromagnetic factor, similar to the electron gyromagnetic factors introduced in Section 3.4.[6] In nuclear magnetic fields, the gyromagnetic factor accounts for the variation in magnetic moment from one nucleus to the next. The value of g_I varies among different elements and different isotopes, with magnitudes ranging from 0 (for nuclei with no spin, $I = 0$) to 11.06 (for a metastable isotope of indium). The value of g_I, and therefore the

[6]NMR spectroscopy is often explained instead in terms of a classical angular momentum analysis, using as a central parameter the *Larmor frequency*, which is equal to $\hbar g_I \mu_N B_{\text{local}}$. The classical approach doesn't extend as easily to nuclei with spins greater than ½, however.

TABLE 5.2 Nuclear spin properties of selected nuclei. The predicted frequency ν for the lowest energy nuclear spin state transition is included.

nucleus	I	abundance	g_I	ν @9.4 T (MHz)
^1H	1/2	99.98%	5.586	400 MHz
^2H	1	0.02%	0.857	61 MHz
^{13}C	1/2	1.10%	1.405	100 MHz
^{14}N	1	99.63%	0.404	29 MHz
^{15}N	1/2	0.37%	−0.283	21 MHz
^{31}P	1/2	100.00%	2.263	163 MHz
^{33}S	3/2	0.75%	0.429	31 MHz

CHECKPOINT Recall that the photon acts like a particle with one unit of angular momentum. Although our photon energies are much too low to change the absolute spin I of the nucleus, we can change the orientation of the nuclear spin, using the photon's angular momentum to shift the m_I value up or down by one unit.

magnetic moment μ_I, can also be negative. Selected values are given in Table 5.2. The transition energy is therefore

$$\Delta E_{\mathrm{mag},I} = g_I \mu_N B_{\mathrm{local}} \Delta m_I = g_I \mu_N B_{\mathrm{local}}. \tag{5.35}$$

SAMPLE CALCULATION **The Frequency of an NMR Spectrometer.** For the proton, there are two nuclear spin states ($m_I = -1/2$ and $m_I = 1/2$) and the value of g_I is 5.586. If we use a bare proton, then $B_{\mathrm{local}} = B_0$, and at a magnetic field of 9.4 T, there is a transition energy between the two nuclear spin states of

$$(5.586)(5.05 \cdot 10^{-27})(9.4) = 2.65 \cdot 10^{-25}\,\mathrm{J} = 400\mathrm{MHz}.$$

For this reason, an NMR spectrometer with a 9.4 T magnet is called a 400 MHz NMR.

NMR transitions occur at low photon energies, in the radiofrequency (rf) range of the spectrum. For example, the $2.82 \cdot 10^{-25}$ J NMR transition of atomic hydrogen in the Sample Calculation just shown lies in the rf range of the electromagnetic spectrum, whereas the $2.18 \cdot 10^{-18}$ J for the $1s \rightarrow 2p$ electronic transition in the same atom is in the UV. With more than a century of technology behind us, we have learned to generate and manipulate radiofrequency radiation with high power and high precision. Our ability to control the radiation has been crucial to the success of NMR, because these nuclear spin transitions are *very* weak, meaning the probability of the nucleus absorbing the radiation to flip its m_I value is quite small (for example) in comparison to the $1s \rightarrow 2p$ absorption. The low transition energy is one reason the transition is weak. These are wavelengths of 1 m and longer, which have a hard time exchanging energy with atoms less than 10^{-9} m in diameter.

Another critical problem with exciting these transitions is that they are driven not by the electric field, but by the *magnetic* field of the radiation. There is no significant change in the electric field of the atom when we flip the nuclear spin, so the electric field of the photon cannot cause the transition to occur. The photon's magnetic field can cause the flip, but magnetic field interactions are typically much weaker than electric field interactions. We make up for this partly by using powerful rf sources.

Our ability to manage static magnetic fields does not nearly match our ability to control the rf radiation. For these experiments to work well, the magnetic field strength needs to be the same throughout the sample to less than 0.1 ppm.

While we can accomplish this, it is very hard to do reliably with the magnetic field strength changing.

Table 5.2 shows that, among the most common nuclei we encounter in chemistry, the ^{1}H nucleus has the greatest sensitivity to an external magnetic field. Many other nuclei can be probed by NMR as well, although they are not equally easy to use. The precision and resolution of NMR tends to be limited by the transition energies, which in turn depend on the g_I of the nucleus being studied and on the strength of the magnetic field. At higher fields, it is possible to get better measurements on nuclei such as ^{15}N, which has a transition frequency 20 times smaller than ^{1}H.

Nuclear Shielding

If that were all there were to it, this would not be interesting to chemists, because Eq. 5.35 has no dependence on the molecular structure, only on the particular atomic nuclei. However, the electrons in the molecule affect these energy levels by responding to the external magnetic field in ways that change the field strength at the nucleus. The energy shift $\Delta E_{\text{shielding}}$ at any particular nucleus depends on the relative orientation of the applied magnetic field and the molecule's magnetic moment at that nucleus:

$$\Delta E_{\text{shielding}} = \mu_\alpha \sigma_{\alpha\beta} B_\beta, \tag{5.36}$$

where α and β are any combination of the lab-fixed Cartesian axes X, Y, and Z, and where $\sigma_{\alpha\beta}$ are the elements of the **nuclear magnetic shielding tensor.** For our purposes, a tensor is a 3×3 matrix of parameter values. In this case, the tensor tells us how the magnetic moment of the molecule along an axis α is affected by an external magnetic applied along an axis β, where α and β can be but don't have to be the same.

The values of $\sigma_{\alpha\beta}$ may be found by taking the second derivative of this expression with respect to μ_α and B_β:

$$\sigma_{\alpha\beta} = \frac{\partial^2 E_{\text{mag},I}}{\partial \mu_\alpha \partial B_\beta}. \tag{5.37}$$

In general, all nine elements of the tensor are different. We simplify matters by dividing the shielding tensor into two components:

1. The **isotropic shielding** is the shielding measured if we average the effect over all relative orientations of the molecule in the field. This is what we expect to observe if we have a sample where the molecule is found at many random orientations, as in the gas or liquid phases. The isotropic shielding is equal to

$$\sigma_{\text{iso}} = \frac{\sigma_{XX} + \sigma_{YY} + \sigma_{ZZ}}{3}. \tag{5.38}$$

2. The **anisotropic shielding tensor** is what's left after we subtract the isotropic part. We use the anisotropic tensor to interpret the NMR spectra of molecules that we can align in particular orientations with the magnetic field. Because this gives us additional orientation-dependent information, it can be helpful in determining the molecular structure, but it complicates the analysis by replacing the single number σ_{iso} by a 3×3 matrix.

Isotropic Diamagnetic Shielding

We begin with the isotropic term, which is the largest contribution to the *shielding* in closed shell molecules (although we shall see that it may not be considered the largest contributor to the NMR *spectrum* we obtain). We define the σ tensor based on the energy shift $\Delta E_{\text{shielding}}$, as in Eq. 5.36, but to describe physically how shielding affects the energies, we often look instead at how the shielding shifts the magnetic field B_{local} seen at the nucleus. Assuming the molecules are randomly oriented, we effectively average over the components of the shielding tensor, and the local field depends only on the isotropic component:

$$B_{\text{local}} = B_0(1 - \sigma_{\text{iso}}). \quad (5.39)$$

The shift in the local field is proportional to the strength of the applied field B_0, and to the value of σ_{iso}.

How does the shielding σ_{iso} come to vary so specifically and reliably with the chemical environment of the nuclei?

All ordinary matter, when placed in an external magnetic field, generates a weak magnetic field of its own that *repels* the applied field, a phenomenon labeled **diamagnetism** by Michael Faraday.[7] Magnetic fields generated in matter through net electron spin or orbital motion are much stronger than this opposing field, so the effects of diamagnetism may not be observable, but they are always present. For example, ground state O atom and H_2^+ molecule are both systems with unpaired electrons and are therefore **paramagnetic,** meaning they generate their own magnetic fields. In the presence of an applied magnetic field, O atom and H_2^+ will respond by trying to line up their own magnetic moments with the external field—an attractive interaction (lower potential energy)—but there will also be a weak diamagnetic response that diminishes the attraction. (You may notice that these definitions differ slightly from the way you may have used these terms before—perhaps in general chemistry, where we commonly call *paramagnetic* any atom with unpaired electrons and *diamagnetic* any atom with all its electrons paired. That usage recognizes that the spin of unpaired electrons typically generates the strongest intrinsic magnetic field in an atom—hence unpaired electrons cause paramagnetism—and that any atom that is not paramagnetic still has diamagnetism. The definitions we use here are generally consistent with that usage, but are based purely on the response of the material to an external field, whatever the origin of the effect.)

DERIVATION SUMMARY Diamagnetism. Diamagnetism is central to how NMR works, but it takes some careful thinking about how charges move in a magnetic field for it to make sense. Briefly, it works as follows:

1. Any external magnetic field induces a circular current flow in conductors. The electron cloud in the molecule we want to study is a conducting substance.
2. Therefore, the external magnetic field introduces a new component of circular motion in the electrons.

[7]This principle has been used to levitate low-mass objects (and even live frogs and other organisms) in very strong fields, paving the way for future studies that require an effectively weightless environment.

3. This induced circular motion of the electrons contributes an additional magnetic field, B_{induced}, which is proportional to the applied field B_0 but aligned in the opposite direction:

$$B_{\text{induced}} = -\sigma_{\text{dia}} B_0, \tag{5.40}$$

where σ_{dia} is the **diamagnetic nuclear shielding** parameter. There are other sources of nuclear shielding, but there will be time for those later.

Now we will step through the derivation in some detail. To see better how diamagnetism arises, we apply some fundamental definitions and results from physics to a general system of charged particles having randomly distributed angular momentum. This is not a rigorous derivation of the diamagnetic shielding constant, but it shows why the effect counters an applied field and why it is a universal property of ordinary matter.

We choose a particular particle with angular momentum $\vec{L}$ and charge q; it could be a single electron or a group of electrons in an atom, and we will show that it doesn't matter if the charges or the angular momenta sum to zero for the overall system. The equations we need are

$$\vec{L} = m\vec{r} \times \vec{v} \qquad \text{angular momentum, Eq. A.46} \tag{5.41}$$

$$\vec{\mu} = \frac{q}{2m}\vec{L} \qquad \text{magnetic dipole movement} \tag{5.42}$$

$$\vec{F}_{\text{mag}} = q\vec{v} \times \vec{B}_0. \qquad \text{magnetic force} \tag{5.43}$$

Equation 5.42 is a different form of the relationship between angular momentum and magnetic moment seen in Eqs. 3.41 and 3.46, specific to a charge q in a circular orbit.

Figure 5.22 shows two possible orientations of the external field $\vec{B}_0$, relative to the trajectory of the charged particle. We have chosen the $\vec{r}$ and $\vec{v}$ vectors to lie in the XY plane such that the angular momentum vector $\vec{L}$ is perpendicular to the page in the $+Z$ direction. The magnetic field is chosen in either the $-Z$ or $+Z$ direction, and the magnetic moment μ is in the $+Z$ direction if $q > 0$ and in the $-Z$ direction if $q < 0$. There are compromises in this model between convenience and making sure we get the answer we want: (1) restricting the particle to the XZ plane (so that $\vec{L}$ and $\vec{\mu}$ are easy to calculate) and (2) drawing the particle as a point charge (to simplify our use of the classical magnetic force law in Eq. 5.43). However, the results we get still apply when averaged over all the position and velocity vectors of the system, including averages over the *distributed* charge of electrons in molecules.

We wish to find the effect of the external field $\vec{B}_0$ on the motion (and therefore the magnetic moment) of the charge q. The external field exerts a force on the particle which changes the velocity vector by an amount $d\vec{v}_{\text{mag}}$:[8]

$$\vec{F}_{\text{mag}} = m\vec{a} = m\frac{d\vec{v}_{\text{mag}}}{dt} \qquad \text{Newton's second law}$$

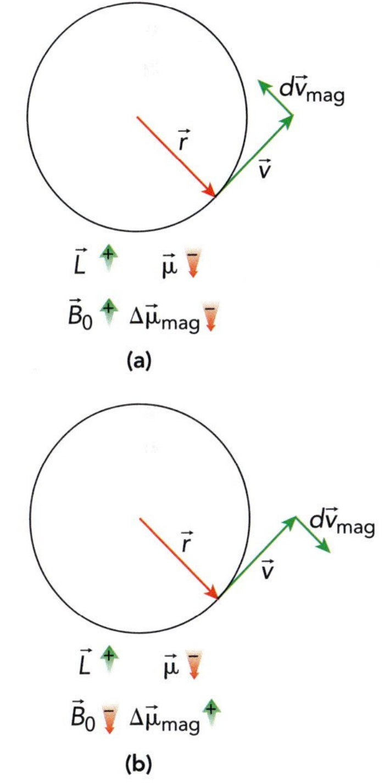

▲ **FIGURE 5.22 The principle of diamagnetism.** Two arrangements of vectors for the interaction of a magnetic field $\vec{B}_0$ with an orbiting electron: **(a)** with $\vec{B}_0$ in the same direction as $\vec{L}$ and **(b)** with $\vec{B}_0$ opposite to $\vec{L}$. The change in velocity $d\vec{v}_{\text{mag}}$ is in the same direction as $\vec{F}_{\text{mag}}$. Vectors out of the page are labeled "+"; vectors into the page are labeled "−."

[8]The cumbersome subscript "mag" is so we don't forget that $\vec{v}$ must be changing all the time even before we apply the external field; otherwise, the particle would have no angular momentum. So we are disregarding here, in particular, the $d\vec{v}$ due to the Coulomb force that binds the electrons to the nuclei.

CHECKPOINT Our goal with this bit of math is to show that the electrons in a molecule will always reduce the effect of an external magnetic field on the nuclei. In the same way that sending current through a coil of wire around a nail turns the nail into an electromagnet, applying a magnetic field to the molecule causes small currents that generate their own magnetic fields. It turns out that those induced magnetic fields always partly cancel the applied field, weakening its effect.

$$d\vec{v}_{\text{mag}} = \frac{\vec{F}_{\text{mag}}dt}{m} = \frac{qdt}{m}\vec{v} \times \vec{B}_0. \qquad \text{by Eq. 5.43} \qquad (5.44)$$

Parameters key: diamagnetism

symbol	parameter	SI units
$\vec{a}$	acceleration	m s^{-2}
$\vec{B}_0$	applied magnetic field	T
$\vec{F}_{\text{mag}}$	magnetic force	N
$\vec{L}$	orbital angular momentum	J s
m	mass	kg
q	charge	C
$\vec{r}$	orbital position vector	m
$\vec{v}$	orbital velocity	m s^{-1}
$\vec{\mu}$	magnetic moment	J T^{-1}
$\Delta\vec{\mu}_{\text{mag}}$	induced magnetic moment	J T^{-1}
σ	overall shielding parameter	unitless
σ_{dia}	diamagnetic nuclear shielding	unitless
σ_0	shielding of reference substance	unitless

This $d\vec{v}_{\text{mag}}$ is a change in velocity parallel to the position vector $\vec{r}$, along the axis between our particle and the center of its orbit. In circular motion, the derivative of the velocity is always toward the center of the orbit, so the particle has a derivative $d\vec{v}$ toward the center before we add the external field. The effect of diamagnetism is to either increase or decrease this derivative, which in turn speeds up or slows down the particle.

For the specific case of an electron (with negative charge, $q < 0$) orbiting a nucleus, there are two possible orientations to consider:

1. If $\vec{B}_0$ is in the same direction as $\vec{L}$, then the cross products lead to a $d\vec{v}_{\text{mag}}$ that points toward the nucleus, increasing the rate of rotation and increasing the magnetic moment μ (Fig. 5.22a).
2. If $\vec{B}_0$ lies in the opposite direction to $\vec{L}$, then the cross products lead to a $d\vec{v}_{\text{mag}}$ that points away from the nucleus, decreasing the rate of rotation and decreasing the magnetic moment μ (Fig. 5.22b).

However, in both cases the magnetic moment is aligned *opposite* to the angular momentum vector $\vec{L}$ (because the electron has a negative charge). Therefore, increasing the magnetic moment in case (1) ($\vec{B}_0$ and $\vec{L}$ in the same direction) is going to add to a field from the electron that *opposes* the applied magnetic field, and decreasing the magnetic moment in case (2) ($\vec{B}_0$ and $\vec{L}$ in opposite directions) will decrease a magnetic field from the electron that is aligned with the applied field. In both cases, the net result is to *oppose the applied field.*

The value of $d\vec{v}_{\text{mag}}$ in Eq. 5.44 is proportional to q and to B_0. The change in magnetic moment is proportional to $qd\vec{v}_{\text{mag}}$, or q^2B_0. Therefore, the strength of the opposing magnetic field is proportional to the strength of the applied field.

To calculate the value of the isotropic diamagnetic shielding constant, σ_{iso}, we integrate the change in μ over the entire three-dimensional distribution of *all* the electrons in the atom. Willis Lamb did this and obtained the **Lamb formula** for diamagnetism, showing that in the simplest case, the shielding is proportional to the expectation value of $1/r$, averaged over all the electrons in the atom:

$$\sigma_{\text{iso}} = \frac{4\pi e^2}{3m_e c^2}\left\langle\frac{1}{r}\right\rangle. \tag{5.45}$$

The factor of 4π comes from integrating over all the angles, so this result is isotropic. Also notice that as we increase atomic number, the atomic nucleus attracts the electrons more strongly, so the average r drops and $\langle 1/r\rangle$ and σ_{iso} *increase* with atomic number. This trend is clear among the values of σ_{iso} for single atoms given in Table 5.3.

Chemical Shift

The shielding shifts the magnetic field seen by the nucleus, so it shifts the frequency of the nuclear spin transition:

$$\begin{aligned} \nu &= \Delta E_{\text{mag},I} = -g_I\mu_N B_{\text{local}}\Delta m_I && \text{by Eq. 5.34} \\ &= -g_I\mu_N B_0(1-\sigma)\Delta m_I && \text{by Eq. 5.39} \\ &= |g_I\mu_N B_0(1-\sigma)|. && \Delta m_I = \pm 1,\ \text{choose so } \nu > 0 \quad (5.46) \end{aligned}$$

This shows that the transition frequency depends on things that we know already (the g_I value for the nucleus, the applied magnetic field B_0) and the σ value.

TABLE 5.3 Nuclear shielding values for selected neutral atoms. Also listed are the shielding for the same atoms in corresponding reference compounds and a representative alkane, in parts per million of the applied magnetic field. The estimated chemical shift δ is the difference between the σ values for the reference and the alkane.[a]

Z	atom	σ_{iso} in atom (ppm)	reference molecule	σ_{iso} (ppm) in reference molecule	alkane molecule	σ_{iso}(ppm) in alkane	δ(ppm) in alkane
1	H	18	$(CH_3)_4Si$	31	C_2H_6	30	1
6	C	260	$(CH_3)_4Si$	192	C_2H_6	182	10
7	N	336	NH_3	269	$C_2H_5NH_2$	240	59
8	O	388	D_2O	338	C_2H_5OH	299	39
9	F	479	$CFCl_3$	184	C_2H_5F	403	−219
16	S	1050	CS_2	527	C_2H_5SH	611	−84
17	Cl	1150	NaCl(aq)	980	C_2H_5Cl	834	146

[a]Values of σ in many cases are estimated from calculations.

However, we don't want to measure absolute values of σ every time we want to calibrate the experiment, so instead we compare this transition frequency to the transition frequency ν_0 of some reference compound. The difference, scaled by ν_0, provides the **chemical shift,** δ:

$$\begin{aligned}\delta &\equiv \frac{\nu - \nu_0}{\nu_0} \\ &= \frac{[-g_I\mu_N B_0(1-\sigma)\Delta m_I] - [-g_I\mu_N B_0(1-\sigma_0)\Delta m_I]}{[-g_I\mu_N B_0(1-\sigma_0)]\Delta m_I} && \text{by Eq. 5.46} \\ &= \frac{(1-\sigma) - (1-\sigma_0)}{(1-\sigma_0)} && \text{cancel like factors} \\ &= \frac{\sigma_0 - \sigma}{1 - \sigma_0}. && (5.47)\end{aligned}$$

Like the shielding σ, the chemical shift δ is a unitless number often expressed in units of parts per million (ppm).

The chemical shift is a convenient measure for several reasons. It doesn't depend on the value of B_0, so experiments with slightly different magnetic field strengths should still measure the same value of δ for any given sample. Furthermore, δ doesn't even depend on which isotope of a particular atom we happen to be probing (as long as the isotope has $I \neq 0$). There are several elements where two isotopes can be probed by NMR (such as ^{10}B and ^{11}B, ^{14}N and ^{15}N, ^{35}Cl and ^{37}Cl). There are reasons why one isotope or the other might be preferable, but both isotopes will give essentially the same values of δ, because δ depends only on the σ values, and σ is independent of the mass of the nucleus (to a good approximation).[9]

The reference compounds used to set the value of ν_0 are usually chosen to have σ_0 values higher (more shielding) than that element will have in most other compounds. This helps to avoid having the reference signal overlap with signals from the molecule being studied. For proton and ^{13}C NMR spectra, we usually use the NMR frequencies of $(CH_3)_4Si$ (tetramethylsilane, or TMS) as the reference values. Silicon has both a greater electron density and a lower electronegativity than either carbon or hydrogen, so it tends to donate electrons to the rest of the molecule, giving high shielding values to the C and H atoms. The reference standards used for S, P, and Cl are actually ionic solutions where the nucleus being probed is part of an anion (SO_4^{2-}, PO_4^{3-}, Cl^-), because adding electrons will increase the shielding around the nucleus and give a high σ_0.

A typical NMR spectrum plots the intensity of the signal versus the chemical shift δ, with δ decreasing toward zero as the plot reads from left to right as shown in Fig. 5.23. Because δ varies as $-\sigma$ in Eq. 5.47, the left end of the spectrum ("downfield") represents lower shielding, and the right end of the spectrum ("upfield," often closer to the reference) represents higher shielding.

---

[9]This does not mean that the NMR spectra of two different isotopes of the same element look exactly the same. The magnetic moments are never the same for two different isotopes, and this affects splittings in the lines that we discuss soon and the spins may differ as well.

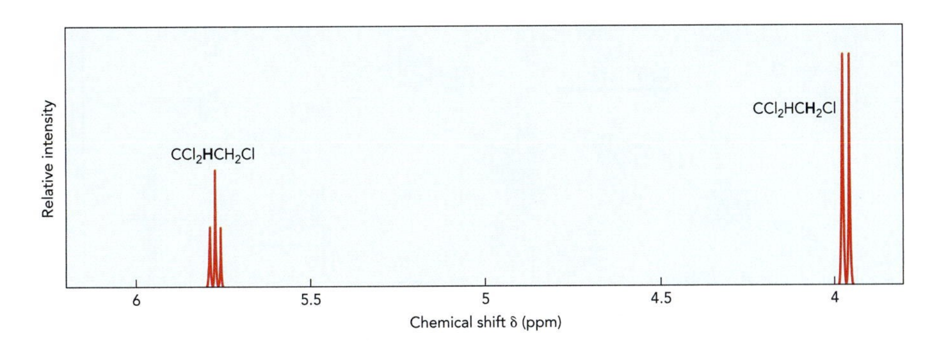

◀ **FIGURE 5.23 The proton NMR spectrum of 1,1,2-trichloroethane.** The spectrum plots signal versus δ, with δ decreasing toward the reference signal at $\delta = 0$.

Once we put the atom in a chemical bond, the electron density near the nucleus changes, and so does the nuclear shielding. In particular, the following general conclusions usually hold up:

1. The electron density at the hydrogen nucleus is the smallest of any atom, and putting the hydrogen in a chemical bond tends to increase the shielding. Notice in Table 5.3 that the value of σ for H increases from 18 ppm to roughly 30 ppm when we form a CH bond.
2. Covalent bonds tend to draw electron density away from the other nuclei, reducing the value of σ_{iso} when compared to the bare atom. Except for hydrogen, all the σ values in Table 5.3 decrease when we move the nucleus from a single atom into a chemical bond.
3. From Eq. 5.45, one can predict that nuclei near more electronegative atoms will lose electron density, have lower shielding, and appear downfield in the NMR spectrum. The more electronegative element tends to pull electron density toward itself, **deshielding** the other nucleus (reducing the value of σ). We see this trend in the ^{13}C NMR chemical shifts listed in Table 5.4.
4. Resonance structures that convey anionic character to an atom will increase the electron density and the σ_{iso} at that nucleus. Cationic character will reduce σ_{iso}.

TABLE 5.4 Representative ^{13}C chemical shifts for selected compounds. The table shows the increase in δ with increasing electronegativity of adjacent atoms, and with increasing ionic character through resonance. The nuclei being probed are in boldface.

molecule	$\delta(^{13}C)$ (ppm)	$\delta(^{1}H)$ (ppm)	electronegativity	
F**CH$_2$**CH$_2$CH$_3$	85.2	4.2	4.0	F
OH**CH$_2$**CH$_2$CH$_3$	63.6	3.5	3.5	O
Cl**CH$_2$**CH$_2$CH$_3$	46.2	3.4	3.0	Cl
NH$_2$**CH$_2$**CH$_2$CH$_3$	44.6	2.5	3.0	N
Br**CH$_2$**CH$_2$CH$_3$	34.6	3.4	2.8	Br
CH$_3$**CH$_2$**CH$_2$CH$_3$	25.1	1.1	2.5	C
I**CH$_2$**CH$_2$CH$_3$	9.2	3.2	2.5	I

molecule	$\delta(^{13}C)$ (ppm)	resonance form
$H_2\mathbf{C}{=}C(CH_3)OCH_3$	80	$H_2C^- - C(CH_3) = O^+CH_3$
$H_2\mathbf{C}{=}C(CH_3)_2$	110	none
$H_2\mathbf{C}{=}C(CH_3)C{\equiv}N$	131	$H_2C^+ - C(CH_3) = C{=}N^-$

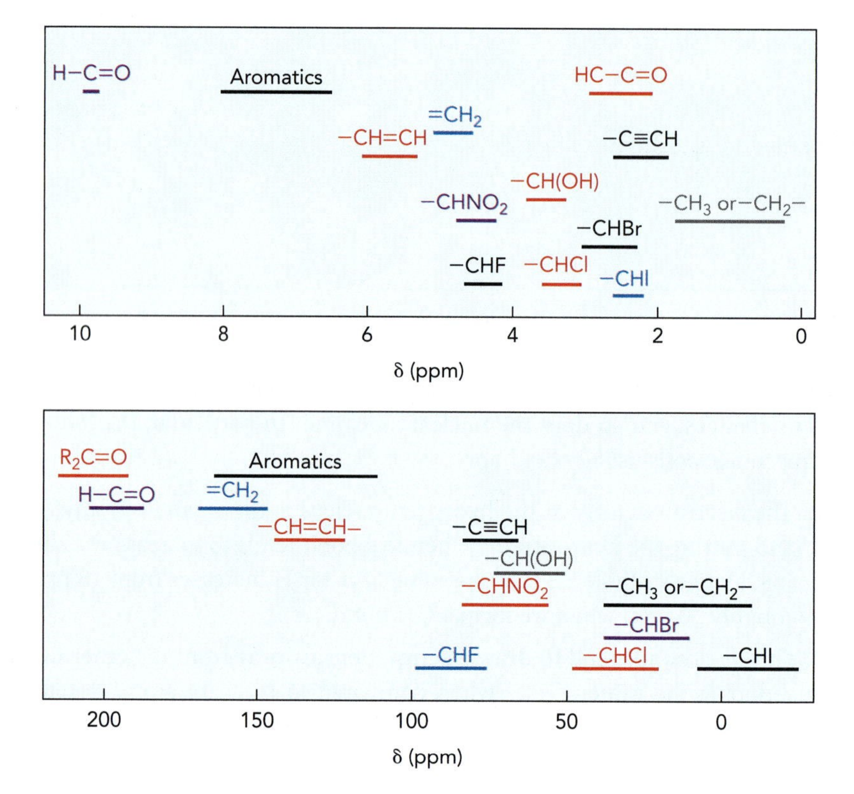

▶ **FIGURE 5.24 Approximate chemical shift ranges for H nuclei in various functional groups.**

▶ **FIGURE 5.25 Approximate chemical shift ranges for C nuclei in various functional groups.**

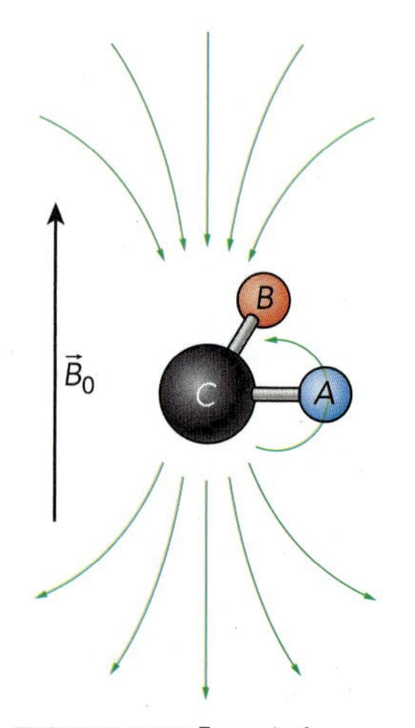

▲ **FIGURE 5.26 Isotropic diamagnetism.** Although the isotropic diamagnetic shielding of the electrons at one nucleus always opposes the applied field, the effect of that field on surrounding nuclei will average to zero if the orientations in space are random. In this example, the isotropic response of atom C leads to a deshielding effect (adding to the applied field $\vec{B}_0$) at nucleus A and a shielding effect (opposing $\vec{B}_0$) at nucleus B.

Because the chemical shifts depend primarily on the local structure of the molecule, it is possible to make the maps, familiar to every organic chemistry student, that depict the range within which the chemical shift will appear for a particular environment of the nucleus. Examples of these maps for hydrogen and carbon NMR are shown in Figs. 5.24 and 5.25. The same dependence of δ on the electronegativity of a neighboring atom (such as F, Cl, Br, and I) that appears in Table 5.4 is evident in these charts.

It is less obvious in these charts why the alkenes, alkynes, aldehydes and ketones, and aromatic compounds show up where they do. Although the isotropic diamagnetic term is usually the major contributor to the *overall shielding* σ_0 of the nucleus, the reference compound shares much of that same contribution, which is subtracted out when we calculate δ. Therefore, other terms may be more significant when it comes to the *chemical shift*, which is a smaller number than σ_0.

Anisotropic Diamagnetic Chemical Shift

The contributions to the chemical shift that we have discussed so far are all *local*, based on the shielding of electrons surrounding the nucleus being probed. What the Lamb formula, Eq. 5.45, fails to take directly into account is that the magnetic field at the nucleus B_{local} is also affected by magnetic fields that are induced in neighboring atoms and functional groups, which we call *non-local* effects.

Non-local contributions to δ are observed only if the origin is anisotropic, because isotropic diamagnetism generates a magnetic field whose *non-local* effect averages to zero, as shown in Fig. 5.26. Only a few arrangements of attached atoms give an isotropic distribution, so anisotropic effects are common, but they range widely in strength.

One of the strongest anisotropic effects in NMR spectroscopy is the deshielding of protons in benzene or connected to a phenyl group. The effect of the applied magnetic field is to induce an added circular motion to the movements of the electrons. Conjugated ring systems are made to order for this, providing a racetrack in which electrons can circulate easily over distances much greater than electrons that are confined to individual atoms (Fig. 5.27). In aromatic molecules, the protons outside the ring have high δ values, and those (if any) pointing toward the middle of the ring have low, even negative, chemical shifts. A number of large, conjugated hydrocarbon rings called **annulenes** were synthesized specifically to verify this effect.

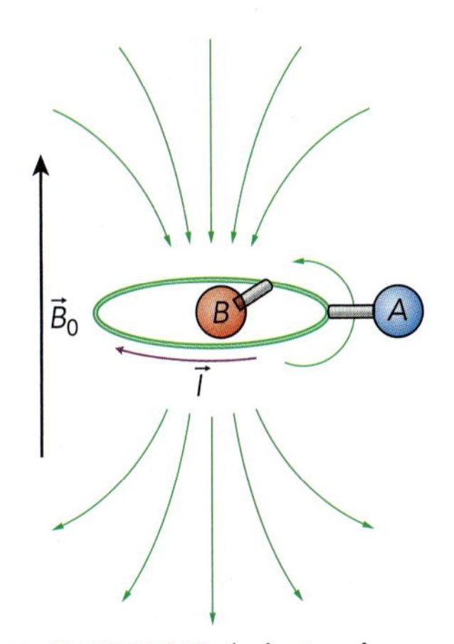

▲ FIGURE 5.27 **Anisotropic diamagnetism in aromatic rings.** Conjugated rings allow large-scale circulation of electrons (current vector $\vec{I}$), such that aromatic molecules have a magnified diamagnetic response to an applied magnetic field $\vec{B}_0$ when the plane of the ring is perpendicular to the applied field. This creates a large opposing field in the center of the ring and a similarly large field *parallel* to the external field outside the ring. The result is that nuclei outside the aromatic ring **(A)** become strongly deshielded, while those near the center **(B)** are strongly shielded. This effect diminishes when the ring is parallel to the applied field, so this is a strong *anisotropic* effect.

A similar effect contributes to the chemical shifts in alkenes (C=C) and alkynes (C≡C), but the effects are in opposite directions. The double bond, in the local bonding model, is formed from a combination of overlapping sp^2 hybrid orbitals (along the axis between the two carbon nuclei) and overlapping unhybridized p orbitals (with a planar node that contains the sp^2 hybrids). The p orbital electrons circulate more easily than the sp^2 electrons but are not permitted to circulate around the bond axis because the electron density must go to zero at the planar node (Fig. 5.28a). Therefore, the anisotropic diamagnetism in alkenes generates a magnetic moment that opposes the field through the middle of the double bond, but *deshields* the atoms adjacent to the double bond.

The triple bond in alkynes is similar but now consists of overlapping sp hybrids along the internuclear axis and two pairs of overlapping p orbitals. One pair of p orbitals lies in the nodal plane of the other pair, with the result that there is now some density from these electrons at every angle around the internuclear axis. Now the electrons can circulate freely around that axis, and the diamagnetic response is strongest along the bond axis, which enhances the *shielding* of the nuclei at either end of the triple bond (Fig. 5.28b). As a result, chemical shifts for H and C nuclei tend to be higher for alkenes than for alkanes, whereas the chemical shifts for alkynes tend to be similar to those of the single-bonded alkanes.

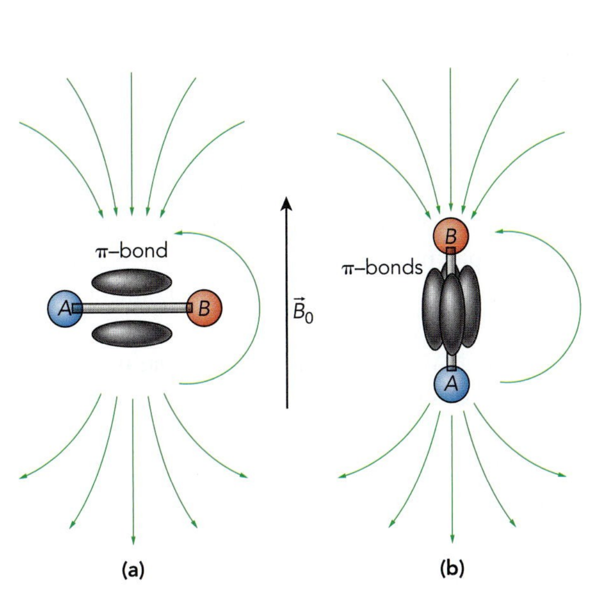

◀ FIGURE 5.28 **Anisotropic diamagnetism in alkenes and alkynes.** The strongest anisotropic diamagnetic response of **(a)** a double bond tends to be when the π bond lies perpendicular to the applied field $\vec{B}_0$, *deshielding* the connected nuclei, whereas the strongest response of **(b)** a triple bond tends to be when the π-bonds are parallel to $\vec{B}_0$, *shielding* the connected nuclei. This is because the induced electron current can circulate about the nuclear axis in (b), but not in (a).

These are among the strongest anisotropic effects, but most functional groups will contribute some anisotropic diamagnetism to the chemical shifts of nearby nuclei, and a qualitative understanding of the δ ranges given in Figs. 5.24 and 5.25 is usually as far as most of us get. For greater detail, we rely on splittings in the lines, and our increasingly reliable computational predictions of NMR spectra from molecular electronic wavefunctions.

Paramagnetic Chemical Shift

Diamagnetism is universal, but it is not the only possible response that a material has to an external magnetic field. All molecules could be also said to be at least weakly *paramagnetic,* but often the effect is so weak that, unlike diamagnetism, it may not be measurable at all in a typical NMR experiment. On the other hand, strong paramagnetic responses lead to huge chemical shifts that exceed the detection limits of any garden-variety NMR spectrometer.

This strong paramagnetism is the simplest, so let's look at that first. Molecules with partly filled orbitals—molecular free radicals—have a net electron spin or orbital magnetic field, and their response to an external field is to line up $\vec{\mu}$ with $\vec{B}_0$, the opposite of the effect of diamagnetism. Furthermore, the permanent magnetic field of a paramagnetic molecule is usually much stronger than the induced field that appears through diamagnetism, so paramagnetism, when it's present, tends to dominate the NMR response. The problem with this in NMR spectroscopy is that the spin of a single unpaired electron is thousands of times more sensitive to the external field than the diamagnetic response the NMR was built to detect. The transition energy for an *electron* spin flip, $m_s = -\frac{1}{2} \rightarrow \frac{1}{2}$, is approximately

$$
\begin{aligned}
E_{\text{mag},s} &= g_s \mu_B m_s B_0 && \text{by Eq. 3.47} \\
\Delta E &= E_{\text{mag},s}(m_s = \tfrac{1}{2}) - E_{\text{mag},s}(m_s = -\tfrac{1}{2}) && (5.48) \\
&= g_s \mu_B B_0.
\end{aligned}
$$

In a 9.4 T field, this transition energy would be 263,000 MHz, no longer in the rf part of the spectrum but in the microwave, and requiring different radiation-handling techniques.

However, the unpaired electrons of a free radical may be localized to a specific region of the molecule, and the magnetic field of these electrons diminishes in strength as we get farther away from the source, even if we stay within the same molecule. So paramagnetic effects in the NMR chemical shift may be seen, for example, in protons that are several angstroms away from an open-shell transition metal atom. The proton chemical shift in the triplet molecule chromacene ($C_6H_6CrC_6H_6$) is observed as a broad peak at 324 ppm, where a typical proton NMR spectrum of a non-paramagnetic molecule will only run out to δ of 25 or less. Although the electron density at the protons is similar to what it would be in benzene, the protons respond as though they were strongly *deshielded,* because the magnetic field of the unpaired electrons *adds to the external field.*

However, the most common appearance of paramagnetic shifts in NMR spectra is less spectacular: for spectra of molecules that have a low-lying, electronic state with at least one unpaired electron. The applied magnetic field encourages mixing of the closed-shell wavefunction with the open-shell (paramagnetic), excited state wavefunction. This happens dramatically in the case of antiaromatic compounds, such as the [12]annulene drawn in Fig. 5.29, where the lone interior proton has a chemical shift of 16.4 ppm.

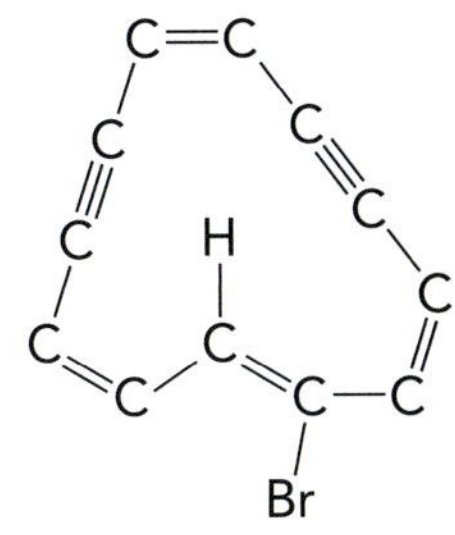

▲ FIGURE 5.29 **Paramagentism in NMR.** This [12]annulene is an antiaromatic system, having a roughly planar structure and 12 electrons in the conjugated bonds. In molecules with this planar, cyclic structure, when the number of conjugated electrons is a multiple of 4, there tends to be a low-lying electronic state, allowing the paramagnetic contribution to the chemical shift to dominate. The chemical shift on the H nucleus shown is 16.4 ppm.

Spin-Spin Coupling

A typical molecule in synthetic organic chemistry today may have more than 20 distinct carbon and proton signals, and the chemical shifts alone are not sufficient to determine the structure of such a complicated system. But after one has looked at the chemical shifts to get a rough idea of what environments exist in the molecule, there is often much more information available from the interactions between neighboring nuclear spins. Protons commonly have several neighbors with non-zero nuclear spin (such as other H atoms, N, ^{13}C), so we will examine splittings in the proton NMR spectrum.

Each magnetic nucleus ($I \neq 0$) generates a magnetic field that interacts with the other nuclear magnetic fields. This is a small effect compared to the interaction with the external magnetic field, and it is strongest between nuclei that are very close together in the molecule. The **spin-spin coupling term** is proportional to an interaction constant J that may be positive or negative (depending on details of the electron distribution) and proportional to the product of m_I for the interacting nuclei. Overall, the effect also diminishes rapidly with distance. The strength of a magnetic dipole field drops as $1/r^3$, where r is the distance from the nucleus, and the spin-spin effect depends on the product of the two interacting dipoles, so its strength drops as $1/r^6$. From a distance of 2 Å to 4 Å, the spin-spin coupling loses a factor of 64 in strength.

Protons have spin of 1/2, and so each proton has two spin states, which we label α (m_s=$1/2$) and β (m_s= $-1/2$). Two protons A and B next to each other therefore have four possible spin states: $\alpha_A\alpha_B$, $\alpha_A\beta_B$, $\beta_A\alpha_B$, $\beta_A\beta_B$. All four of these states will have different energy levels when the external magnetic field is turned on, but how this affects the NMR spectrum depends on whether or not the nuclei are chemically equivalent.

Chemically equivalent nuclei are those that have exactly identical environments in the molecule. Another way of saying this is that if we list all the distances between every pair of atomic nuclei in the molecule, two equivalent nuclei will have the same set of distances to nuclei of the same elements. So, for example, all six H atoms in ethane (CH_3CH_3) are equivalent, but propane ($CH_3CH_2CH_3$) has two groups: the six H atoms at the ends are all equivalent, and the two in the middle are equivalent. Equivalent nuclei experience the same chemical shifts because the symmetry of the molecule requires the electron density at each nucleus to be the same.

If protons A and B are near each other but *not equivalent*, then the four nuclear spin states result in energy levels

$$E = g_I \mu_N B_0 [\, m_{s,\mathrm{A}}(1 - \sigma_\mathrm{A}) + m_{s,\mathrm{B}}(1 - \sigma_\mathrm{B})] + m_{s,\mathrm{A}} m_{s,\mathrm{B}} J$$

$$\equiv E^0_\mathrm{A} m_{s,\mathrm{A}} + E^0_\mathrm{B} m_{s,\mathrm{B}} + m_{s,\mathrm{A}} m_{s,\mathrm{B}} J,$$

where J is the **spin-spin coupling constant.** Let's make A the nucleus with the smaller shielding and larger transition energy. Indicating the sign of the two m_I values by subscripts, the energy levels (in order of decreasing energy) are

$$E_{++} = E^0_\mathrm{A}/2 + E^0_\mathrm{B}/2 + J/4$$

$$E_{+-} = E^0_\mathrm{A}/2 - E^0_\mathrm{B}/2 - J/4$$

$$E_{-+} = -E^0_\mathrm{A}/2 + E^0_\mathrm{B}/2 - J/4$$

$$E_{--} = -E^0_\mathrm{A}/2 - E^0_\mathrm{B}/2 + J/4,$$

as shown in Fig. 5.30. The allowed transitions are those that change one nuclear spin at a time, so the transition energies are

$$E_{++} - E_{+-} = E^0_\mathrm{B} + J/2$$

$$E_{++} - E_{-+} = E^0_\mathrm{A} + J/2$$

$$E_{+-} - E_{--} = E^0_\mathrm{A} - J/2$$

$$E_{-+} - E_{--} = E^0_\mathrm{B} - J/2.$$

Therefore, each proton has its transition at E^0_A or E^0_B split into doublets.

If the two protons are chemically equivalent, there is an important difference: there is an overall nuclear spin $I_t = 1$ if the two nuclear spins are roughly parallel and $I_t = 0$ if they cancel. Each of these states has possible M_I states that obey the same rule as for single nuclei: $M_I = -1,0,1$ for $I_t = 1$ and $M_I = 0$ for $I_t = 0$. The selection rules allow us to change M_I by one, but not I_t. Therefore, only two transitions are possible: $I_t = 1, M_I = -1 \rightarrow 0$ and $I_t = 1, M_I = 0 \rightarrow 1$. Because the spins are roughly parallel in all of these states, the magnitude of the spin-spin coupling is the same $(+J/4)$, and it does not affect the transition energy.

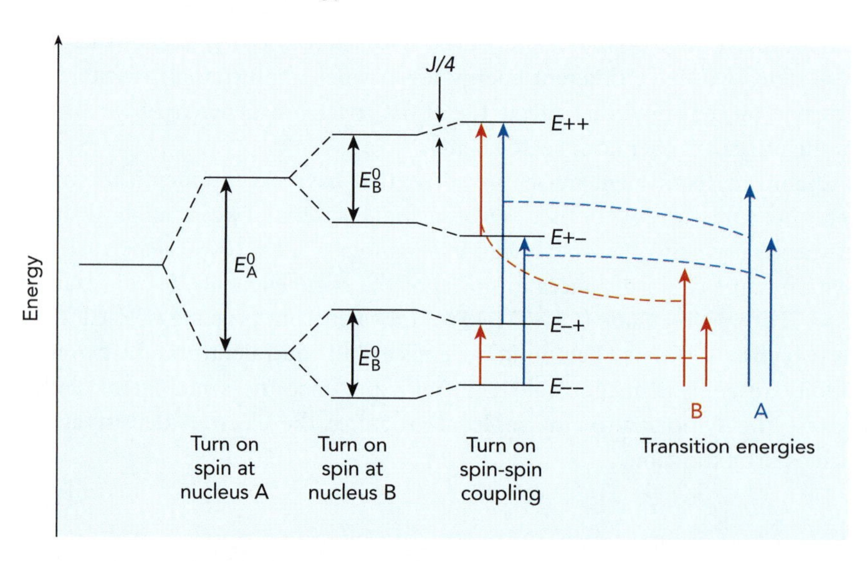

▶ **FIGURE 5.30 Two non-equivalent protons A and B result in four nuclear spin states.** If there were no spin-spin interaction between them, there would be only two transition energies: one for flipping the spin on nucleus A, and a different transition energy for flipping the spin on nucleus B. Adding the spin-spin interaction shifts the energy levels to spit those two transition energies into four distinct energies.

So equivalent nuclei do not cause NMR fine structure in each other, but they will cause splittings in signals from other non-equivalent nuclei. The $(n + 1)$th row of Pascal's triangle describes how the signal from one nucleus will be affected by n neighboring equivalent nuclei. Two equivalent nuclei split a non-equivalent third into a 1:2:1 pattern, three split a fourth into a 1:3:3:1 pattern, etc.

CHECKPOINT Pascal's triangle is formed by adding two adjacent numbers in a row to obtain the value between those numbers on the next row, counting the blank area outside the triangle as zero. Beginning with 1, we add 1 to zero on each side to get two 1's below. Now we add those two 1's to get the value 2 in the middle of the third row, and 1's on the outside:

1
1 1
1 2 1
1 3 3 1
1 4 6 4 1
etc.

CONTEXT *Where Do We Go From Here?*

This chapter justifies the formation of chemical bonds from the charge–charge interactions of the atomic particles. With that understanding in place, we can approach more sophisticated ways of analyzing molecular structure. In Chapter 6, we take chemical bonding for granted as we examine the principles of molecular symmetry, which provide our labeling system for the quantum states of the molecule, as well as a basis for understanding much of the remaining molecular physics. Chapters 7, 8, and 9 revisit the molecular Hamiltonian that we introduced in this chapter, one piece at a time, to complete our picture of how energy can be distributed within molecules.

KEY CONCEPTS AND EQUATIONS

5.1 The molecular Hamiltonian.

a. Around any pair of nuclei A and B we can calculate a binding force,

$$F_{\text{binding}} = \frac{Z_A e^2}{4\pi\varepsilon_0 r_A^2}\cos\theta_A - \frac{Z_B e^2}{4\pi\varepsilon_0 r_B^2}\cos\theta_B - \frac{Z_A Z_B e^2}{4\pi\varepsilon_0 R^2}, \tag{5.1}$$

such that in regions where $F_{\text{binding}} > 0$, an electron will contribute attractive pulls on both the nuclei that overcome the repulsion between the nuclei. This is the essence of the covalent chemical bond.

b. The Hamiltonian for a molecule combines kinetic energy terms for each electron i, kinetic energy terms for each nucleus k, and potential energy terms for the interaction between each pair of particles

$$\hat{H} = \hat{K}_{\text{elec}} + U_{\text{nuc-elec}} + U_{\text{elec-elec}} + U_{\text{nuc-nuc}} + \hat{K}_{\text{nuc}}, \tag{5.4}$$

or to write it out more explicitly,

$$\hat{H} = -\frac{\hbar^2}{2m_e}\sum_{i=1}^{N}\nabla_i^2 - \sum_{i=1}^{N}\sum_{k=1}^{M}\frac{Z_k e^2}{(4\pi\varepsilon_0)r_{ik}} + \sum_{i=2}^{N}\sum_{j=1}^{i-1}\frac{e^2}{(4\pi\varepsilon_0)r_{ij}} + \sum_{k=2}^{M}\sum_{l=1}^{k-1}\frac{Z_k Z_l e^2}{(4\pi\varepsilon_0)R_{kl}} - \frac{\hbar^2}{2}\sum_{k=1}^{M}\frac{1}{m_k}\nabla_k^2. \tag{5.7}$$

c. In order to solve the Schrödinger equation, we successively separate the variables into several groups. The first step in this process is to separate the electronic and nuclear coordinates, applying the Born-Oppenheimer approximation:

$$\hat{H}_{\text{eff}}(R)\psi_{\text{elec}}(R) = E_{\text{elec}}(R)\psi_{\text{elec}}(R), \tag{5.8}$$

where

$$\hat{H}_{\text{eff}}(R) = \hat{K}_{\text{elec}} + U_{\text{nuc-nuc}} + U_{\text{nuc-elec}} + U_{\text{elec-elec}}. \tag{5.9}$$

d. The nuclear kinetic energy term we further break up into vibrational, rotational, and translational terms:

$$\hat{K}_{\text{nuc}} = -\frac{\hbar^2}{2\mu R^2}\frac{\partial}{\partial R}R^2\frac{\partial}{\partial R} + \frac{1}{2\mu R^2}\hat{J}(\Theta,\Phi)^2 - \frac{\hbar^2}{2M_{\text{nuc}}}\nabla_{\text{COM}}^2. \tag{5.11}$$

5.2 The molecular wavefunction.

a. We write the wavefunction for an electron in a molecule by adding (with different possible relative phases) the wavefunctions of atomic orbitals on the nuclei. For example, for H_2^+ we have two possible wavefunctions made by combining $1s$ atomic orbital wavefunctions centered on each of the two nuclei A and B:

$$\psi_+(r,\theta,R) = C_+(R)(1s_A + 1s_B)$$
$$\psi_-(r,\theta,R) = C_-(R)(1s_A - 1s_B). \tag{5.13}$$

b. In a molecular orbital wavefunction, we multiply these one-electron wavefunctions together.
For example, the simplest spin-spatial wavefunction for H_2 puts both electrons in the ψ_+ MO:

$$\Psi_{\text{MO}}(\text{ground}) = C_+(R)^2\psi_+(1)\psi_+(2)\left[\alpha(1)\beta(2) - \beta(1)\alpha(2)\right]/\sqrt{2}. \quad (5.19)$$

5.3 Covalent bonds in polyatomic molecules. We can form hybrid molecular orbitals by forming linear combinations of atomic orbitals centered on the same nucleus. The number of hybrid orbitals formed is equal to the number of atomic orbitals combined.

5.4 Non-covalent bonds. In addition to the covalent bonding method, we define ionic bonds (where one or more electrons are effectively transferred from one nucleus to another) and metallic bonding (where metal cations are bound by a delocalized electron).

5.5 Nuclear magnetic resonance spectroscopy. In NMR spectroscopy, an external magnetic field B_0 splits the magnetic spin states of certain atomic nuclei into different energies, allowing us to measure transition frequencies

$$\nu = \left|g_I\mu_N B_0(1 - \sigma)\right|, \quad (5.46)$$

which depend on the shielding σ experienced by the nucleus. This in turn is a sensitive probe of the electron distribution near the nucleus. The NMR spectrum is usually a function of the chemical shift δ,

$$\delta = \frac{\sigma_0 - \sigma}{1 - \sigma_0}, \quad (5.47)$$

which is measured relative to the position of some reference compound or solution.

KEY TERMS

- A **chemical bond** is the strong binding of two neighboring atoms, such that energies of roughly 50 kJ mol^{-1} or more are required to break the bond.
- In a **covalent bond,** the two atoms are held together by the mutual attraction of the two nuclei for electrons in a **binding region** that lies between the two.
- The **Born-Oppenheimer approximation** is the assumption that we can accurately solve the electronic part of the Schrödinger equation with the nuclei held in fixed positions.
- A **molecular orbital (MO)** is a one-electron wavefunction representing the spatial distribution of the electron in a molecule. In general, the MO is distributed across two or more atoms in the molecule.
- The **overlap integral** $S(R)$ measures how much two different atomic orbitals in an MO intersect in space. If the orbitals are the same, then $S(R) = 1$, and if they are orthogonal then $S(R) = 0$.
- A **molecular orbital configuration** is a many-electron spatial wavefunction written as a product of individual MOs.
- The **effective potential energy** U_{eff} combines the electronic energy (obtained by solving the electronic Schrödinger equation under the Born-Oppenheimer approximation) with the nuclear–nuclear potential energy. If we consider only the nuclei, with the electron motions averaged to give a net electric field for interaction with the nuclei, then this function describes the potential energy of the vibrational motions of the nuclei.
- The **equilibrium geometry** is the set of bond lengths and bond angles that give the minimum value of U_{eff}.
- The **dissociation energy** is the minimum energy necessary to break a chemical.
- **Nuclear magnetic resonance (NMR)** is a widely used tool for structural analysis in synthetic chemistry, as well as many other applications.
- **Diamagnetism** is the universal response of a material with moving charged particles to generate a magnetic field that weakly opposes an applied magnetic field.
- A **paramagnetic** molecule has a net magnetic field of its own, resulting from unpaired electrons in the ground or low-lying excited electronic states.
- The **nuclear magnetic shielding tensor** is a matrix showing how much an applied magnetic field along each of the molecule-fixed axes x, y, or z affects the diamagnetic response along each of the axes. The observed shielding of the molecule in solution may be separated into the angle-independent **isotropic shielding** contribution and the angle-dependent **anisotropic shielding tensor.**
- The **chemical shift** is the horizontal axis of a typical NMR spectrum, roughly proportional to the shielding strength at a nucleus relative to some reference substance.
- The **spin-spin coupling term** in NMR is the shift in the nuclear spin energy levels of one nucleus from its interaction with another nucleus. It is a highly distance-dependent effect and is used to map out which features in the NMR spectrum arise from adjacent nuclei.

OBJECTIVES REVIEW

1. *Write the Hamiltonian for any molecule.*
   Write the Hamiltonian for LiH^+.
2. *Write and sketch zero-order molecular orbital wavefunctions by combining atomic orbital wavefunctions.*
   Sketch the appearance of a bonding molecular orbital formed by combining in equal proportions a $1s$ orbital on nucleus A with a $2p_z$ orbital on nucleus B. Show nodes if there are any.
3. *Draw approximate potential energy curves for the motion of two nuclei in a chemical bond.*
   Sketch an approximate graph of $U_{\text{eff}}(R)$ versus R for a diatomic molecule with an equilibrium bond length of 1.5 Å and a bond energy of roughly 200 kJ mol^{-1}.
4. *Determine the shape and other properties of hybrid orbitals.*
   If three equivalent sp^2 hybrid orbitals are formed in the xy plane, with orbital 1 pointing in the $+x$ direction, what is the effect on the bond angles of increasing the proportion of p_x in orbital 1?
5. *Predict and interpret basic H and C NMR spectra.*
   Sketch the appearance of the proton NMR and ^{13}C spectra for 1-bromo,2-chloroethane.

PROBLEMS

Discussion Problems

5.1 The binding force in the H_2^+ molecule is positive in binding regions and negative in antibinding regions. State whether the potential energy of the electron is positive or negative in (a) the binding region; (b) the antibinding region.

5.2 The binding force for a one-electron diatomic molecule is independent of one of our electronic coordinates. Which coordinate does not affect F_{binding}, and what is the form of the electronic wavefunction in this coordinate?

5.3 For the H_2^+ MO ψ_+ wavefunction in Eq. 5.13,

$$\psi_+(r,\theta,R) = C_+(R)(1s_A + 1s_B),$$

do you expect the normalization constant $C_+(R)$ to increase or decrease as R increases?

5.4 We interpreted Eq. 5.22 to predict (incorrectly) that the dissociation of a sample of H_2 molecules leads to a 50:50 mixture of neutral and ionic hydrogen atoms. All of a sudden we were talking about a large group of H_2 molecules, rather than what happens to a single H_2 molecule on dissociation. This is a very different use of the electron probability density than we saw for atoms, where we interpreted the electron probability density as measuring how much of the total electron character (charge, mass, spin, etc.) will be found in any given region of space around a single atom. Why in this case do we not describe the electron distribution in similar terms, using the fractional distribution to show (correctly) that each electron has a 50% chance of being found near either nucleus? Is it even *possible* for Eq. 5.22 to describe the outcome of the dissociating a single H_2 molecule?

The Molecular Hamiltonian

5.5 For a diatomic molecule with two electrons, list the coordinates of which the wavefunction and Hamiltonian are functions (more than one answer is acceptable). What two sets of these coordinates are made separable by the Born-Oppenheimer approximation?

5.6 Which terms ($\hat{K}_{\text{elec}}$, $U_{\text{elec-elec}}$, etc.) in $\hat{H}_{\text{eff}}$ (Eq. 5.9) dominate the energy at (a) $R < 0.1$ Å, (b) $R = R_e$, and (c) $R > 10$ Å?

5.7 After each of the following Hamiltonian terms contributing to the overall molecular energy, write "+" if that contribution is always positive, "−" if it is always negative, "±" if it may be either positive or negative, and "0" if it is always zero:

a. $\hat{K}_{\text{elec}}$ b. $\hat{K}_{\text{nuc}}$ c. $\hat{U}_{\text{elec-elec}}$
d. $\hat{U}_{\text{nuc-elec}}$ e. $\hat{U}_{\text{nuc-nuc}}$

5.8 The electronic Schrödinger equation of H_2^+ looks like it could be straightforward to solve, because under the Born-Oppenheimer approximation we can freeze the two nuclei, leaving only the dynamics of one electron to solve. We were able to simplify the three-dimensional one-electron atom by first finding a set of coordinates (r,θ,ϕ) such that the problem was separable: we could solve the Schrödinger equation one coordinate at a time.

a. Is the H_2^+ Schrödinger equation separable in cylindrical coordinates, (s,z,ϕ), where z is the coordinate along the internuclear axis, s is the distance measured perpendicular to z, and ϕ is the angle of rotation about z?
b. Is it possible for any other set of coordinates to exist such that this Schrödinger equation is separable?

5.9 For molecules with many electrons, the Hamiltonian can be automatically generated by computer. Next are written the beginning and ending terms for each contribution to a particular molecular Hamiltonian. (You could fill in all the missing terms if you had to, but it's not necessary.) Give the molecular formula and ionic charge, if any, for the molecule with this Hamiltonian:

$$\hat{H} = -\frac{\hbar^2}{2m_e}(\nabla_1^2 + \nabla_2^2 + \cdots + \nabla_{14}^2) - \frac{\hbar^2}{2m_A}\nabla_A^2 - \frac{\hbar^2}{2m_B}\nabla_B^2$$
$$-\frac{e^2}{4\pi\varepsilon_0}\left(\frac{6}{r_{A1}} + \frac{6}{r_{A2}} + \cdots + \frac{6}{r_{A14}} + \frac{7}{r_{B1}} + \frac{7}{r_{A2}} + \cdots + \frac{7}{r_{B14}}\right)$$
$$+\frac{e^2}{4\pi\varepsilon_0}\left(\frac{42}{R} + \frac{1}{r_{1,2}} + \frac{1}{r_{1,3}} + \cdots + \frac{1}{r_{13,14}}\right).$$

Binding Forces

5.10 It's time to check those unit analysis skills. Let the forces in Fig. 5.3 be in units of $e^2/(4\pi\varepsilon_0 a_0^2)$. Find the conversion factor to change these units to newtons.

5.11 Calculate the binding force F_{binding} in units of $e^2/(4\pi\varepsilon_0 R^2)$ if a single electron is placed on the z axis halfway between two fluorine nuclei.

5.12 Prove that for the electron between the nuclei in H_2^+,

$$F_{\text{binding}}(\theta_A = 0, \theta_B = \pi) = \frac{e^2}{4\pi\varepsilon_0}\left[r_A^{-2} + r_B^{-2} - (r_A + r_B)^{-2}\right],$$

and that this is positive for all r_A.

5.13 In the molecule HeH^{2+}, show whether a point at $r_H = 1$ Å, $\theta_H = \pi/2$ is in the binding region or antibinding region, if $R = 1.0$.

5.14 For the molecule HeH^{+2} with bond length R, find the minimum value of the binding force [in units of $e^2/(4\pi\varepsilon_0 R^2)$] if the electron lies along the bond and between the two nuclei. Also find the largest binding force.

5.15 Use F_{binding} values to predict whether or not HeH^{+2} is a stable molecule.

5.16 Calculate the binding force at the midpoint of the chemical bond in (a) H_2^+ 2p and (b) He_2^{3+} for $R = 1.00$.

5.17 Calculate the binding force at $r_A = R, \theta_A = \pi/2$ for H_2^+ 2p. Leave the answer in terms of R.

5.18 For H_2^+ and HeH^{2+}, the region between the nuclei on the z axis always gives a positive binding force. Find the smallest values of the atomic number Z_X and the charge n for the one-electron carbide XC^{n+} such that there is an antibinding region on the z axis exactly halfway between the nuclei.

5.19 Find an equation for F_{binding} when the electron is in the xy-plane, through the middle of the bond. Is this region binding or antibinding?

Molecular Orbitals

5.20 The properties of H_2 are predicted by integrating the probability density of the MO wavefunction. Draw a line through any of the integrals in that calculation below that approach zero when $R \rightarrow \infty$.

a. $\iint 1s_A(1)1s_A(1)1s_A(2)1s_A(2)d\tau_1 d\tau_2$

b. $\iint 1s_A(1)1s_A(1)1s_A(2)1s_B(2)d\tau_1 d\tau_2$

c. $\iint 1s_A(1)1s_A(1)1s_B(2)1s_B(2)d\tau_1 d\tau_2$

d. $\iint 1s_A(1)1s_B(1)1s_A(2)1s_B(2)d\tau_1 d\tau_2$

5.21 Write formulas for the *lowest* energy and *highest* energy molecular orbitals of the linear molecule LiH_2^+ that can be formed from sums of the 1s orbitals. Use $1s_A$ and $1s_B$ to represent the orbitals on the two hydrogens, and $1s_{Li}$ for the 1s on the lithium. The lithium is between the H atoms. Don't worry about normalization.

5.22 Identify the molecule represented by the following MO wavefunction for all the electrons, where r_{A1} is the distance of electron 1 from nucleus A, and so on:

$$\psi_{MO} = \left[\sqrt{\frac{64}{\pi a_0^3}}e^{-4r_{A1}/a_0}\right]\left[\sqrt{\frac{64}{\pi a_0^3}}e^{-4r_{A2}/a_0}\right]$$
$$\left[\sqrt{\frac{8}{\pi a_0^3}}\left(1 - \frac{2r_{A3}}{a_0}\right)e^{-2r_{A3}/a_0} + \sqrt{\frac{1}{\pi a_0^3}}e^{-r_{B3}/a_0}\right]$$
$$\times\left[\sqrt{\frac{8}{\pi a_0^3}}\left(1 - \frac{2r_{A4}}{a_0}\right)e^{-2r_{A4}/a_0} + \sqrt{\frac{1}{\pi a_0^3}}e^{-r_{B4}/a_0}\right].$$

5.23 Our lowest energy molecular orbital for H_2 is written $A(R)[1s_A(1) + 1s_B(1)]$. Write a similar equation for the lowest energy molecular orbital of C_3H_8 with the atoms labeled as drawn in the accompanying figure. Include only the most important contributions.

H$_D$ H$_E$ H$_C$ H$_H$ C$_J$ H$_B$ C$_I$ C$_K$ H$_G$ H$_A$ H$_F$

5.24 Write an expression for the lowest excited VB wavefunction for H_2 in terms of the 1s atomic orbitals.

5.25 Find the numerical values of E_+ and E_- for H_2^+ as R becomes infinite.

5.26 What does the two-electron H_2 wavefunction

$$\psi_s = \frac{1}{2}[\psi_{MO}(\text{ground}) + \psi_{MO}(\text{excited})]$$

predict for the limit of very large R, the dissociated atom limit?

5.27 Prove that the two H_2^+ MOs shown are orthogonal: $\psi_+(r,\theta,R) = A_+(R)(1s_A+1s_B)\, \psi_-(r,\theta,R) = A_-(R)(1s_A-1s_B)$.

5.28 The orbitals $1s_A$ and $1s_B$ are not orthogonal to each other at finite R (otherwise the overlap integral $S(R)$ would always be zero). State whether or not the following pairs of orbitals are orthogonal and whether or not your answer depends on R:

a. $1s_A$ and $2p_{zB}$

b. $1s_A$ and $2p_{xB}$

5.29 Graph the H_2^+ excited state molecular orbitals $\psi_{\pm}^{2p_z} = (2p_{zA} \pm 2p_{zB})$ along the internuclear axis (the z axis) for $R = 2a_0$. Don't worry about the normalization constant.

5.30 Show that the following is an invalid wavefunction for H_2:

$$C(R)[1s_A(1) - 2p_{zB}(1)][2p_{zA}(2) - 1s_B(2)][\alpha(1)\beta(2) + \beta(1)\alpha(2)].$$

5.31 Write the integral (including limits and volume element) necessary to normalize the $\psi_+(R) = (1s_A + 1s_B)$ wavefunction for H_2^+ when the A and B atoms are separated by $2a_0$.

5.32 In terms of the ψ_+ and/or ψ_- MO wavefunctions and α and β spin functions, write equations for the four lowest energy excited state spin-spatial wavefunctions of the H_2 molecule. Do not normalize the functions, but they should have proper symmetry under the operator $\hat{P}_{21}$. Put a star next to the wavefunction that has the *highest* energy of these four. For example, the ground state wavefunction is

$$\Psi_{\text{ground}} = \psi_+(1)\psi_+(2)[\alpha(1)\beta(2) - \beta(1)\alpha(2)] \text{ or } \psi_+\psi_+[\alpha\beta - \beta\alpha].$$

5.33 We saw that ψ_{MO} for H_2 predicts that pulling the atoms apart yields neutral atoms half the time and ions half the time, for a final mixture of 50% H atoms, 25% H^- ions, and 25% H^+ ions. Using our simple MO wavefunction of H_3^+, we could write the two-electron H_3^+ wavefunction this way:

$$\psi(1,2) = A[1s_A(1) + 1s_B(1) + 1s_C(1)][1s_A(2) + 1s_B(2) + 1s_C(2)].$$

What does this predict are the products, and in what percentages, when all three nuclei are pulled apart ($R_{AB} = R_{BC} = R_{AC} \rightarrow \infty$)? It is not necessary to work through all the terms to solve this.

5.34 Give a rough sketch of the $\psi_+ = C_+(1s_{He} + 1s_H)$ wavefunction for HeH^{2+}, illustrating qualitatively the effect of having different nuclear charges.

5.35 A suitable spatial wavefunction for Li_2 using the valence bond method can be written

$$\psi_{VB} = [1s_A(1)1s_B(2) + 1s_B(1)1s_A(2)][1s_A(3)1s_B(4) - 1s_B(3)1s_A(4)][2s_A(5)2s_B(6) + 2s_B(5)2s_A(6)].$$

Determine whether this can be written as a sum of MO wavefunctions, and if so, give the sum.

5.36 Write any reasonable equation for two molecular orbitals ψ_1 and ψ_2 that represent the two π bond orbitals in acetylene, HCCH, as a sum of appropriate atomic orbitals (i.e., in a form similar to the $\psi_{\pm}$ orbitals of H_2^+).

5.37 Determine whether or not the following molecular orbitals for a triatomic molecule are orthogonal.

a. $C_1[(2s_A) + (2s_B) + \sqrt{2}(2s_C)]$

b. $C_2[(2s_A) + (2s_B) - \sqrt{2}(2s_C)]$

5.38 The following table lists orbital energies in E_h from a Hartree-Fock calculation on the N_2 molecule. The total Hartree-Fock energy of the molecule is $-108.983\,E_h$.

a. Assign each energy to the correct molecular orbital in the MO configuration.

−15.682	−0.632
−15.678	−0.612
−1.470	−0.612
−0.777	

b. Calculate the total electron–electron *repulsion* energy in E_h.

c. To estimate the strength of the electron–nuclear interaction, calculate the effective atomic number of an electron in the highest energy orbital.

Hybrid Orbitals

5.39 The CCC bond angle in cyclopropane is $\pi/3$. Given the coordinate axes drawn in the accompanying figure, write equations for the two hybrid orbitals on carbon atom A that would form CC bonds with the correct bond angle.

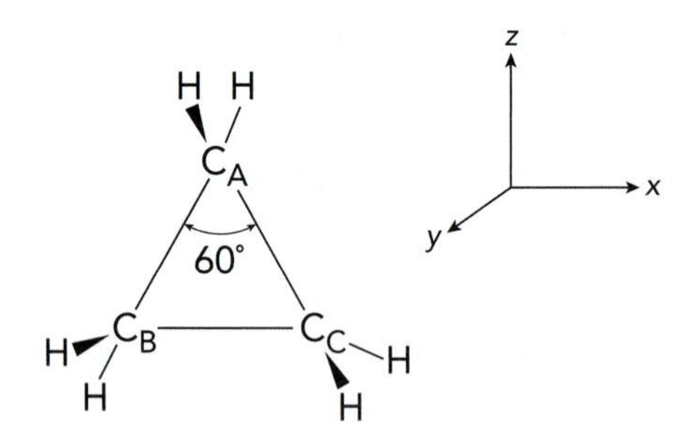

5.40 The sp^2d hybrid orbital configuration continues to be invoked to explain some binding geometries with significant covalent character. Write any two normalized and mutually orthogonal sp^2d hybrid molecular orbitals in terms of the atomic orbitals $4s$, $4p_x$, $4p_y$, and $3d_{xy}$.

5.41 On the coordinate system provided, draw a simple sketch to show the direction and phase of greatest electron density for the sp^2d hybrid molecular orbital represented by

$$-\frac{1}{2}(4s) + \frac{1}{2}(4p_x) + \frac{1}{2}(4p_y) - \frac{1}{2}(3d_{xy}).$$

The $3d_{xy}$ orbital has the shape shown in the accompanying figure.

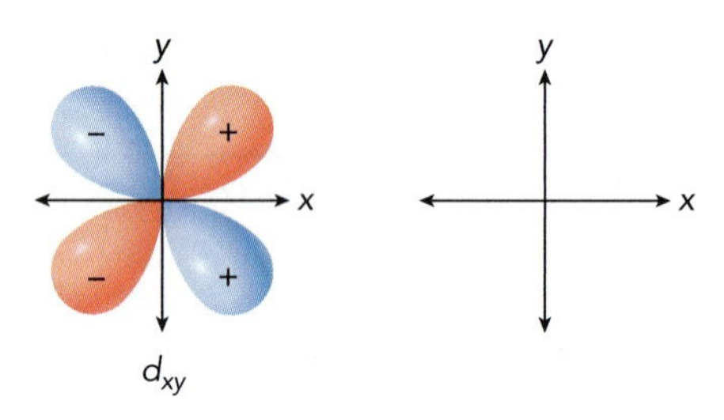

5.42 A $2s$ orbital and a $2p_z$ orbital are combined to make a pair of non-equivalent sp hybrids, sp_a and sp_b. If the sp_a hybrid has the formula given next, give the formula for the normalized function sp_b:

$$(sp_a) = \sqrt{\frac{3}{5}}(2s) - \sqrt{\frac{2}{5}}(2p_z).$$

5.43 Show that the maxima of the sp_b^2 and sp_c^2 hybrid wavefunctions in the xy plane are at $2\pi/3$ and $4\pi/3$ from the x axis, where $sp_b^2 = [\sqrt{2}(2s) - (2p_x) + \sqrt{3}(2p_y)]/\sqrt{6}$, $sp_c^2 = [\sqrt{2}(2s) - (2p_x) - \sqrt{3}(2p_y)]/\sqrt{6}$, $(2p_x) = [(2p_1) + (2p_{-1})]/\sqrt{2}$, and $(2p_y) = i[(2p_{-1}) - (2p_1)]/\sqrt{2}$.

5.44 Prove that the angle between the bonds formed from these two sp^3 hybrid orbitals is 109.47°:

$$(sp_a^3) = \frac{1}{2}[(2s) + (2p_x) + (2p_y) + (2p_z)]$$

$$(sp_d^3) = \frac{1}{2}[(2s) - (2p_x) - (2p_y) + (2p_z)].$$

5.45 Construct any two normalized and orthogonal sp^3d orbitals using the set of atomic functions $3s$, $3p_x$, $3p_y$, $3p_z$, $3d_{z^2}$.

5.46 The following are a group of three valid, orthonormal sp^3 hybrid orbitals. Find the fourth orbital, (sp_a^3).

$$(sp_b^3) = \sqrt{\frac{1}{6}}(2s) - \sqrt{\frac{2}{3}}(2p_x) + \sqrt{\frac{1}{6}}(2p_z)$$

$$(sp_c^3) = \sqrt{\frac{1}{6}}(2s) + \sqrt{\frac{1}{6}}(2p_x) + \sqrt{\frac{1}{2}}(2p_y) + \sqrt{\frac{1}{6}}(2p_z)$$

$$(sp_d^3) = \sqrt{\frac{1}{6}}(2s) + \sqrt{\frac{1}{6}}(2p_x) - \sqrt{\frac{1}{2}}(2p_y) + \sqrt{\frac{1}{6}}(2p_z)$$

5.47 Written next are two sets of sp^3 hybrid orbitals, one for methane with all bond angles equal to 109.5°, and one for chloroform ($CHCl_3$) where the Cl—C—Cl bond angle has climbed to 111.3°. Correct equations are given for the four CH_4 hybrid orbitals. Write "+," "−," or "0" next to each of the $CHCl_3$ orbitals that follow to show whether the coefficients for $CHCl_3$ are larger (+), smaller (−), or the same (0) as the corresponding coefficients for methane. (If a negative coefficient gets more negative, write "−.")

$$(sp^3)_a = \frac{1}{2}(2s) + (0)(2p_x) + (0)(2p_y) + \sqrt{\frac{3}{4}}(2p_z)$$

$$(sp^3)_b = \frac{1}{2}(2s) + (0)(2p_x) + \sqrt{\frac{2}{3}}(2p_y) - \sqrt{\frac{1}{12}}(2p_z)$$

$$(sp^3)_c = \frac{1}{2}(2s) + \sqrt{\frac{1}{2}}(2p_x) - \sqrt{\frac{1}{6}}(2p_y) - \sqrt{\frac{1}{12}}(2p_z)$$

$$(sp^3)_d = \frac{1}{2}(2s) - \sqrt{\frac{1}{2}}(2p_x) - \sqrt{\frac{1}{6}}(2p_y) - \sqrt{\frac{1}{12}}(2p_z)$$

$CHCl_3$

$$(sp^3)_a = \quad (2s) + \quad (2p_x) + \quad (2p_y) + \quad (2p_z)$$

$$(sp^3)_b = \quad (2s) + \quad (2p_x) + \quad (2p_y) + \quad (2p_z)$$

$$(sp^3)_c = \quad (2s) + \quad (2p_x) + \quad (2p_y) + \quad (2p_z)$$

$$(sp^3)_d = \quad (2s) + \quad (2p_x) + \quad (2p_y) + \quad (2p_z)$$

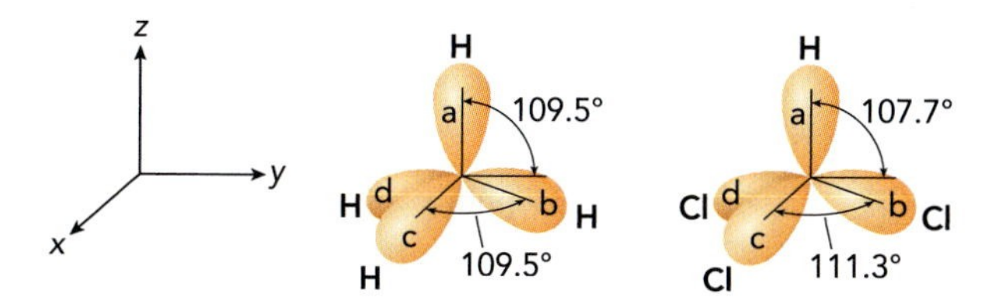

5.48 The bonds along the molecular axis of carbon dioxide can be thought of as forming from combinations of hybrid orbitals on each of the three atoms. Assume the molecule lies on the z axis in the arrangement O_ACO_B.

a. According to VSEPR theory, what is the hybridization at each oxygen atom?
b. According to VSEPR theory, what is the hybridization at the carbon atom?
c. Write a formula in terms of the *atomic orbitals* of these atoms for the wavefunction that forms the CO_B bonding orbital. Assume the hybrid orbitals at each atom are all equivalent, but don't worry about the overall normalization.
d. Sketch the amplitude of that wavefunction along the z axis, showing the approximate location of any nodes. (For simplicity, assume that a $2s$ orbital has the same shape as a $1s$.)

5.49 Construct a set of sp^3 hybrid orbitals from s, p_x, p_y, and p_z atomic orbitals such that one orbital has exactly 40% s character and the other three orbitals are all equivalent.

5.50 For cyclobutanone (shown in the accompanying figure), the following are the equations for the normalized sp^2 molecular orbital hybrids that correspond to the observed bond angle α of 93.1° at the carbonyl carbon.

a. Identify the *total* percentage of contribution from p orbitals in each MO.

$$\psi_a = (0.9470)(2s) + (0.3214)(2p_x)$$

$$\psi_b = (0.2272)(2s) - (0.6696)(2p_x) + \frac{1}{\sqrt{2}}(2p_y)$$

$$\psi_c = (0.2272)(2s) - (0.6696)(2p_x) - \frac{1}{\sqrt{2}}(2p_y)$$

b. On the accompanying figure, draw the coordinate axes consistent with the orbitals defined.

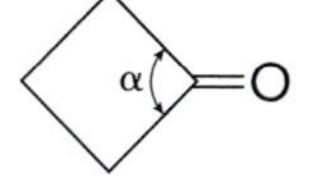

5.51 Show that the sp hybrid function

$$\psi_a = \frac{1}{\sqrt{2}}[(2s) + (2p_z)]$$

is an eigenfunction of $\hat{H}$ for the H atom, and find the eigenvalue.

5.52 Find the coefficients for a normalized sp_a hybrid orbital at which the angular node exactly disappears, assuming an r value such that $R_s(r) = R_p(r)$ (in other words, just look at the angular part of the wavefunction and don't worry about the radial part). Does the companion orbital sp_b also lose its angular node?

Empirical Chemical Structure and NMR

5.53 From Table 5.1, determine how closely (i.e., to within what percentage) the predicted aromatic CC bond length in benzene agrees with the average of the CC bond lengths predicted from benzene's two resonance structures.

5.54

a. Using Table 5.1, predict the bond lengths in formaldehyde, H_2CO.
b. Within each box of pictures below, circle the picture that puts the electron (e^-) where the binding force is strongest:

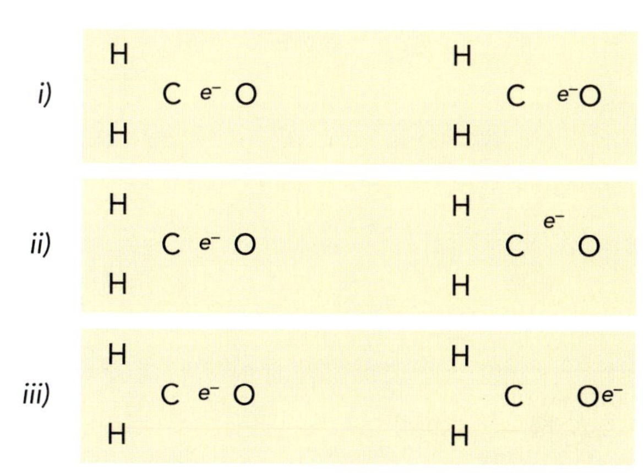

5.55 Find the chemical shift of the carbon atoms in the ^{13}C NMR spectrum of benzene if the shielding constant is 65.2 ppm, while the shielding constant for the carbons in TMS is 192.3 ppm.

5.56 The following table is started for data from nitrogen NMR spectra of the atoms at each end of the linear azide anion, NNN^-. Separate entries are given for two common reference substances, NH_3 ($\sigma_0 = 269$ ppm) and CH_3NO_2 ($\sigma_0 = -112$ ppm).

a. Fill out the remaining entries in the table.

nucleus	^{15}N	^{15}N	^{15}N
external field B_0 (T)	9.4	9.4	14.1
reference	NH_3	CH_3NO_2	NH_3
δ (ppm)	99		
$B_0 - B_{local}$ (T)			

b. Which of the following describes the δ value of nitrogen atom N_a in neutral hydrazoic acid, $HNN\,^{15}N_a$, at $B_0 = 9.4$ T, using NH_3 as a reference?

< 99 ppm   99 ppm > 99 ppm

6 Molecular Symmetry

LEARNING OBJECTIVES

After reading this chapter, you will be able to do the following:

1. Carry out algebraic manipulations of point group symmetry operators.
2. Determine the point group and symmetry elements of a molecule.
3. Find the symmetry selection rules for electric dipole and Raman transitions in a molecule.
4. Assign a molecular orbital to its symmetry representation.

GOAL *Why Are We Here?*

Our goal in this chapter is to learn to recognize and take advantage of symmetry in molecules. So crucial a consideration is symmetry that it forms the basis of the notation we use for molecular wavefunctions and energy levels. Initially we will develop our skills in identifying molecular symmetry, so that we can put those skills to use in solving the molecular Schrödinger equation and—more importantly—in *understanding* those solutions. The laws of symmetry that we will develop in this chapter can be applied to questions of molecular structure and reactivity in organic chemistry (e.g., chirality, aromaticity, Woodward-Hoffmann rules) and inorganic chemistry (e.g., crystal structure, bonding to transition metals), and these laws have broad implications for all forms of spectroscopy.

CONTEXT *Where Are We Now?*

Our look at chemical bonding in Chapter 5 led us to a general molecular Hamiltonian (Eq. 5.7), which can be divided in different ways (as in Eqs. 5.9 and 5.10, for example) according to how we wish to classify the different contributions to the energy. Each term in our Hamiltonian now corresponds to electronic, vibrational, or rotational motion, and we're anxious to examine what implications each of these motions has for the material's structural and chemical properties. Chapters 7 and 8 make extensive use of the symmetry analysis developed here, and Chapter 13 extends some of these methods to the study of crystals.

But before we move forward to tackle each component of the Schrödinger equation individually, we pause to consider a crucial aspect of molecular structure that may simplify the problem in practical terms, and that may also lead us towards a more intuitive understanding of molecular structure.

SUPPORTING TEXT **How Did We Get Here?**

The main *qualitative* concept we draw on for this chapter is that the structure and properties of a molecule depend largely on its symmetry. We will use the material in this chapter to better understand some aspects of that dependence. We will also draw on the following equations and sections of text:

- Section 2.1 described the rules governing operators. We will meet a new set of operators in this chapter, ones that manipulate the spatial coordinates of the molecule, but they follow the same rules of algebra as the Hamiltonian and other operators. For example, one rule is that in a sequence of operations $\hat{A}\hat{B}\psi$, we carry out the operations from right to left. In other words, $\hat{A}\hat{B}\psi = \hat{A}(\hat{B}\psi)$.
- In Section 2.2, under *Degeneracy and the Three-Dimensional Box,* the degeneracy of an energy level is defined as the number of distinct quantum states that share that energy. When the degeneracy arises from a symmetry in the system, as is the normal case, the analysis that we cover in this chapter will help us determine the value and origin of the degeneracy.
- If we have a molecule of N electrons and M nuclei, then the molecular Hamiltonian can be written

$$\hat{H} = -\frac{\hbar^2}{2m_e}\sum_{i=1}^{N}\nabla^2(i) - \sum_{i=1}^{N}\sum_{k=1}^{M}\frac{Z_k e^2}{(4\pi\varepsilon_0)r_{ik}} + \sum_{i=2}^{N}\sum_{j=1}^{i-1}\frac{e^2}{(4\pi\varepsilon_0)r_{ij}} + \sum_{k=2}^{M}\sum_{l=1}^{k-1}\frac{Z_k Z_l e^2}{(4\pi\varepsilon_0)R_{kl}} - \frac{\hbar^2}{2}\sum_{k=1}^{M}\frac{1}{m_k}\nabla^2(k). \quad (5.7)$$

  Applying the Born-Oppenheimer approximation groups the electron-dependent terms into an effective Hamiltonian:

$$\hat{H}_{\text{eff}}(R)\psi_{\text{elec}}(R) = E_{\text{elec}}(R)\psi_{\text{elec}}(R), \quad (5.8)$$

  where

$$\hat{H}_{\text{eff}}(R) = \hat{K}_{\text{elec}} + U_{\text{nuc-nuc}} + U_{\text{nuc-elec}} + U_{\text{elec-elec}}. \quad (5.9)$$

  Once we have obtained the electronic energy as a function of R, we can add $\hat{K}_{\text{nuc}}$ and solve the Schrödinger equation along the remaining coordinates for motion of the nuclei:

$$\hat{H}\psi_{\text{nuc}} = [E_{\text{elec}}(R) + \hat{K}_{\text{nuc}}]\psi_{\text{nuc}} = E_{\text{nuc}}\psi_{\text{nuc}}. \quad (5.10)$$

- Section 5.2 showed how we can combine atomic orbitals to get molecular orbital wavefunctions:

$$\psi_+(r,\theta,R) = C_+(R)(1s_A + 1s_B), \quad \psi_-(r,\theta,R) = C_-(R)(1s_A - 1s_B). \quad (5.13)$$

  In this chapter we will analyze the symmetry properties of those molecular orbitals.

6.1 Group Theory

In this chapter, we examine the **symmetry** of a molecule: the characteristic of certain structures that lets us divide them into *separate but identical* pieces. A mathematical function is symmetric if there is an operation that alters its coordinates in some manner but leaves the graph of the function unchanged.

For example, the graph of $\cos x$ is identical to the graph of $\cos(x+2\pi)$, and also to the graph of $\cos(-x)$. Shifting the x value by 2π and multiplying x by -1 are, in that example, symmetry operators. For any given function, there may be many symmetry operators, and the symmetry does not have to be restricted to the spatial symmetry of Cartesian coordinates like x, y, z.

A molecule is much more than an algebraic expression, but we can define symmetry operators in the same way that rearrange or compare similar regions of the molecule. The study of these sets of symmetry operators is a branch of **group theory,** and tools from this area of mathematics can greatly strengthen our understanding of molecular properties.

Symmetry Operators

For a molecule that has a symmetric arrangement of the atomic nuclei, we label all of its electronic and vibrational states according to that symmetry. The labels may help us determine how many electrons are bonding or antibonding, how much electron density lies along the internuclear axis, and what excited states are accessible by electric dipole selection rules.

Symmetry serves another purpose: it may simplify the mathematics when we predict the properties of a molecule. Solving the Schrödinger equation for molecules normally requires a lot of numerical integration, and symmetry makes numerical integrals easier to evaluate. For example, when we evaluate the integral of $\cos^4\theta \sin^4\theta$ from $\theta = 0$ to 4π (Fig. 6.1), we don't have to evaluate it everywhere. By inspection we can see that the function repeats itself every period of $\pi/2$. We need only evaluate the function from 0 to $\pi/2$ and then multiply by eight to get the integral. In the same way, symmetric molecules may have several equivalent contributions to the kinetic and potential terms of the Hamiltonian. Solving the Schrödinger equation to predict the energy levels (which means integrating a differential equation) may be simplified if we take that symmetry into account.

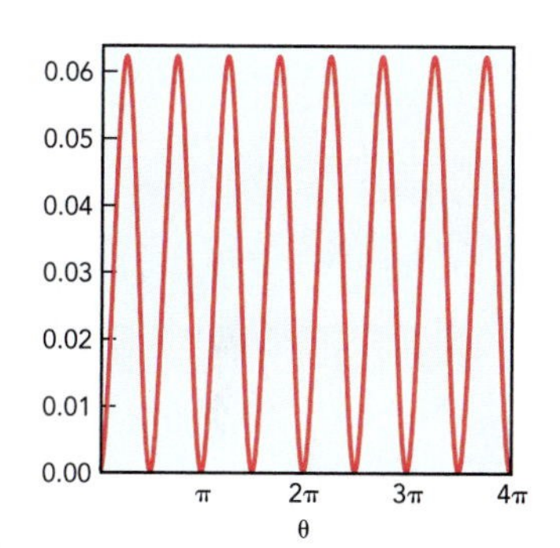

▲ **FIGURE 6.1 Symmetry in mathematical functions.** The function $\cos^4\theta \sin^4\theta$, showing that the function is periodic with period $\pi/2$.

In Section 4.3, we considered one kind of symmetry: the symmetry with respect to permutation of the electrons in an atom, changing the coordinates that label one electron with those that label another. We now consider a second kind of symmetry—*spatial* symmetry. What is the effect on the wavefunction, imagining that we could always write it as some algebraic formula, of changing the spatial coordinates in any of several ways? We are interested only in symmetry operations that leave the *geometry* of the nuclei unchanged, although equivalent nuclei may exchange positions. If we pick a specific molecule and find all of the distinct symmetry operations that leave its nuclear geometry unchanged, those operations are the **symmetry elements.** Here's one way to think of it:

1. You have in front of you the arrangement of atomic nuclei for some fascinating molecule.
2. You close your eyes.
3. A friend carries out operation $\hat{A}$ on the structure.
4. You open your eyes.
5. If you cannot tell that anything was done to the molecule—i.e., if the arrangement is the same except for *exactly* equivalent (indistinguishable) nuclei changing places—then $\hat{A}$ is a symmetry element of the molecule.

We are still considering only one molecule at a time, and therefore our symmetry elements cannot move the molecule's center of mass. If $\hat{A}$ moves the center of mass, you'll notice that the whole molecule has shifted when you open your eyes.

Any such symmetry operation $\hat{A}$ can be classified as one of the following five types, with examples shown in Fig. 6.2.

1. $\hat{E}$ is the **identity operator.** It leaves the function completely unaltered. It is a convenient symbol for any combination of operations that together have no effect on the coordinates of the function, such as rotation by a full 2π.
2. $\hat{C}_n$ is the **proper rotation operator.** It rotates the system by an angle $2\pi/n$ about a particular axis. For example, $\hat{C}_2$ is rotation by $2\pi/2 = \pi$ (or 180°), and $\hat{C}_4(x)$ is rotation by $2\pi/4$ (or 90°, as shown in Fig. 6.2b). If the rotation is a symmetry element of the molecule, the axis is called a **symmetry axis.** The symmetry axis with the greatest value of n for a given molecule is assigned to z and is the **principal rotation axis** (for the rest of this chapter, we shall simply call it the "principal axis"). In a few cases, the results are easily expressed in Cartesian coordinates. For example, proper rotation by π about the z axis transforms a function as follows:

$$\hat{C}_2(z)\psi(x,y,z) = \psi(-x,-y,z). \tag{6.1}$$

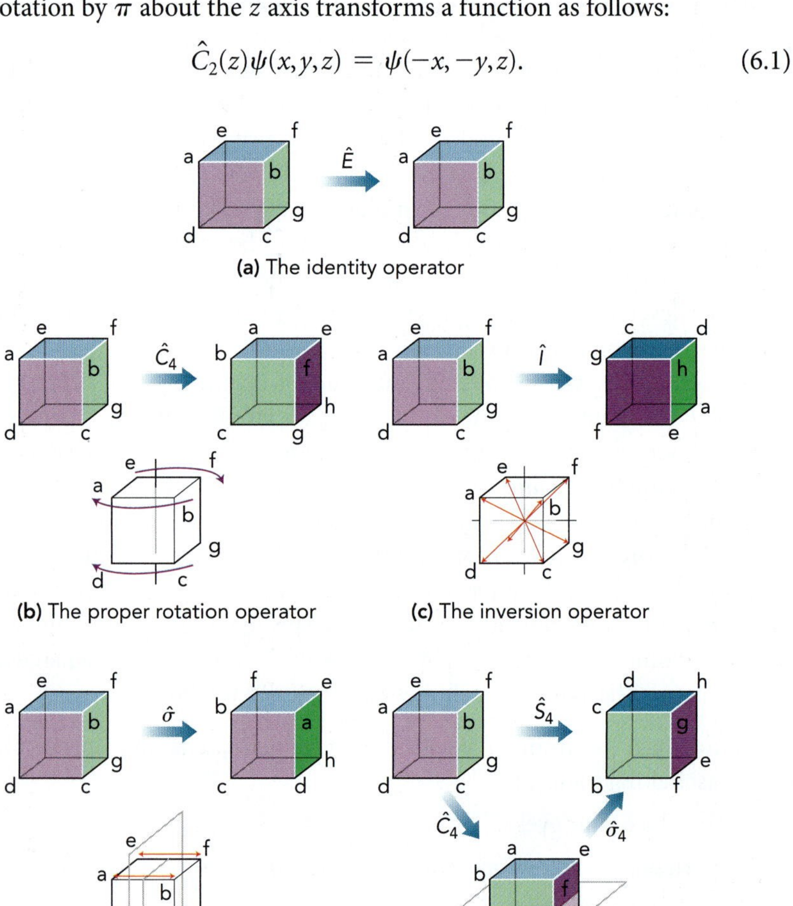

(a) The identity operator

(b) The proper rotation operator

(c) The inversion operator

(d) The reflection operator

(e) The improper rotation operator

▶ FIGURE 6.2 **The five types of point group symmetry operations.** The operations are performed on a cube with vertices labeled a–h. Each face of the cube is a different color. The changing positions of the vertices and faces show the results of the various symmetry operations. Inversion, reflection, and improper rotation each cause a change in the relative positions of the corners and sides that cannot be reproduced by proper rotations through space.

Proper rotation leaves all the points in the rotation axis unchanged.

For a diatomic or any other linear molecule, we define the z axis to be the axis containing all the nuclei, because this is the principal axis. Rotation of a linear molecule about the z axis by *any* angle leaves the nuclear arrangement unchanged. We index the operator according to the smallest rotation $2\pi/n$ that leaves the geometry unchanged, and in the case of proper rotation about the z axis of a linear molecule that angle is infinitesimal, effectively $2\pi/\infty$. Therefore, the symmetry operator for this infinitesimal rotation is labeled $\hat{C}_\infty$.

3. $\hat{I}$ is the **inversion operator,** and it changes the sign of every x and y and z value in the system:

$$\hat{I}\psi(x,y,z) = \psi(-x,-y,-z), \tag{6.2}$$

where the origin is at the molecule's center of mass. Molecules that are symmetric with respect to inversion possess an **inversion center** or **center of symmetry** at the center of mass. To test for inversion symmetry, draw a line segment that runs from any nucleus in the molecule through the center of mass and extends an equal distance on the opposite side. If $\hat{I}$ is a symmetry element, then the final point must lie at a nucleus exactly equivalent to the nucleus at the starting point. Inversion leaves only the center of mass unchanged.

4. $\hat{\sigma}$ is the **reflection operator,** moving every point to its mirror image on the opposite side of a specified mirror plane. When proper rotation is a symmetry element, then two kinds of mirror plane reflections are defined: reflection $\hat{\sigma}_h$ through the **horizontal mirror plane,** which is perpendicular to the principal axis, and reflection $\hat{\sigma}_v$ through a **vertical mirror plane,** which contains the principal axis. Since we choose the principal axis to be the z axis, the only possible horizontal mirror plane is the xy plane. In contrast, there may be many vertical mirror planes. If these planes lie along the Cartesian coordinate axes, then the effect of these reflections can again be written in algebraic form. For example, the horizontal mirror reflection will be

$$\hat{\sigma}_h\psi(x,y,z) = \hat{\sigma}_{xy}\psi(x,y,z) = \psi(x,y,-z), \tag{6.3}$$

where $\hat{\sigma}_{xy}$ denotes reflection through the xy plane. Similarly,

$$\hat{\sigma}_{yz}\psi(x,y,z) = \psi(-x,y,z) \quad \hat{\sigma}_{xz}\psi(x,y,z) = \psi(x,-y,z) \tag{6.4}$$

are two possible vertical plane reflections. In some circumstances, the vertical mirror reflection operators may be labeled $\hat{\sigma}_d$ (**dihedral** mirror reflections), but for our purposes this is only to distinguish between two different sets of $\hat{\sigma}_v$ planes.

All planar molecules have a mirror plane reflection as one of their symmetry elements, because atoms lying in the mirror plane are not affected by the reflection.

5. $\hat{S}_n$ is the **improper rotation operator,** equal to $\hat{\sigma}_h\hat{C}_n$, proper rotation followed by reflection, as shown in Fig. 6.2e. Like inversion, improper rotation leaves only the center of mass unchanged.

CHECKPOINT The five basic symmetry operations we've just seen account for every possible way that equivalent atoms in a molecule can be exchanged, without altering which atoms are connected to which. We need these symmetry operations in order to determine how the shape of the molecule may limit the possible molecular orbitals and vibrational motions, and also how the molecule interacts with radiation.

The symmetry operators obey our other rules for operators from Section 2.1. Operations $\hat{A}$ and $\hat{B}$ are equal if for every function $\psi(x,y,z)$,

$$\hat{A}\psi(x,y,z) = \hat{B}\psi(x,y,z).$$

For example, if we carry out two successive rotations by $\pi/2$ or 90°, that would be $\hat{C}_4(z)$ twice in a row, which we write as $\hat{C}_4(z)^2$. The result of the combined operation is the same as rotation by π, or the result of applying the operator $\hat{C}_2(z)$. Therefore, we may set $\hat{C}_4^{\,2}(z) = \hat{C}_2(z)$. They do not count as separate operators.

An important glitch in the notation occurs for the improper rotations. Normally if we write $\hat{C}_3^2$, for example, we mean the same operation as $\hat{C}_3\hat{C}_3$—in other words, two successive proper rotations by 120° or a single proper rotation by 240°. However, the rule for improper rotations is $\hat{S}_n^m = \hat{\sigma}_h\,\hat{C}_n^m$. In other words, we carry out the proper rotation m times, but the horizontal reflection just once. For example, $\hat{S}_3^2$ means $\hat{\sigma}_h\hat{C}_3\hat{C}_3$, not $\hat{S}_3\hat{S}_3$. If we carried out the reflection twice, then the two reflections would cancel and $\hat{S}_3^2$ would be the same as $\hat{C}_3^2$. Because we need a symbol for the operator that means "rotate by 240° and then reflect," we define $\hat{S}_n$ raised to the power m to carry out the reflection just once. (Not all references adopt this convention, so read your sources carefully.)

EXAMPLE 6.1 Symmetry Operator Equivalence

CONTEXT Physicists often define a vector more narrowly than we have, requiring a *vector* to behave the same under proper and improper rotations, whereas a **pseudovector** has an additional sign change under improper rotation. A vector cross-product is one example of a pseudovector, because if $\vec{a}$ and $\vec{b}$ both change sign, the cross product vector $(-\vec{a}) \times (-\vec{b})$ remains unchanged. The distinction matters in programming the lighting for video game computer graphics (and other essential technologies). Each normal vector of a drawn surface (say, a castle wall) may be conveniently calculated by taking a cross product of the vectors that form the surface. But there's a problem then, because the cross product may change sign unintentionally during transformations that allow the program to render, for example, the reflection of a castle in its moat. The reflection of light is calculated by determining angles of incidence and reflection relative to the normal vectors of the reflective surface. Fundamental computer game programming therefore demands an understanding of the nature of improper rotations and related symmetry operators.

PROBLEM Prove that $\hat{S}_2 = \hat{I}$. In other words, show that inversion is a particular example of improper rotation.

SOLUTION For operators involving reflection through one of the Cartesian planes or $\hat{C}_2$ rotation about a Cartesian axis, it is often simplest to write what is happening in terms of the Cartesian coordinates. In this example, if we assign the proper rotation axis to the z axis, then the horizontal mirror plane (defined to be perpendicular to the principal axis) is the xy plane. Then from Eqs. 6.1 and 6.3 we have $\hat{C}_2\psi(x,y,x) = \psi(-x,-y,z)$ and $\hat{\sigma}_h\psi(x,y,z) = \psi(x,y,-z)$. Combining these with the definition of the improper rotation, we find

$$\begin{aligned}\hat{S}_2\psi(x,y,z) &= \hat{\sigma}_h\hat{C}_2\psi(x,y,z)\\ &= \hat{\sigma}_h\psi(-x,-y,z)\\ &= \psi(-x,-y,-z) = \hat{I}\psi(x,y,z).\end{aligned}$$

Therefore, $\hat{S}_2 = \hat{\sigma}_h\hat{C}_2 = \hat{I}$.

Although it's easy to find molecules without symmetry (where $\hat{E}$ is the only symmetry element), many of the simplest molecules have a great deal of symmetry, and results from symmetric molecules often give us a convenient reference point for more complicated, asymmetric molecules.

Our first job is to learn to find, given the geometry of some molecule, which symmetry operators are symmetry elements—operators that appear to leave the distribution of the atoms unchanged. There may be lots. For example, we can count how many distinct symmetry elements a cube has as follows. The cube drawn in Fig. 6.2 has 8 corners, which we assume are all equivalent (except for the useful labels). Consider corner *a*. After we operate with some symmetry element, *a* can end up in any of the 8 locations previously occupied by a corner. Corner *a* is connected to edges—*ab*, *ad*, and *ae*—and there are 6 ways to rearrange those 3 edges. Altogether, there must be

$$(8 \text{ locations of corner}) \times (6 \text{ edge arrangements}) = 48 \tag{6.5}$$

different possible configurations of the cube after some symmetry operation. There must be one symmetry element for each of those possible outcomes, so there must be 48 *distinct* symmetry elements for the cube.

Most nonlinear molecules will have fewer symmetry elements, but they will all be drawn from the five classes of point group symmetry operations listed previously.

As a first example, let's find the symmetry elements of the linear molecule HCN. In addition to the identity operation, there are two sets of symmetry elements: rotation about the z axis, which is an infinite-fold $\hat{C}_\infty$ axis, and reflection through any of an infinite number of $\hat{\sigma}_v$ mirror planes (Fig. 6.3).

H—C≡N— z

$\hat{C}_\infty$

(a)

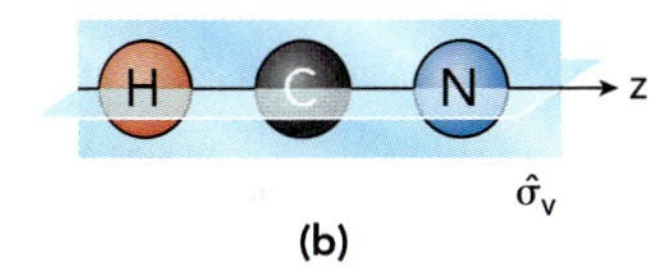

(b)

▲ **FIGURE 6.3 Symmetry and linear molecules.** The $\hat{C}_\infty$ and $\hat{\sigma}_v$ operations on HCN have no effect on the locations of the nuclei and are therefore symmetry elements.

EXAMPLE 6.2 Molecular Symmetry Elements and Pseudorotation

CONTEXT Phosphorus pentachloride, PF_5, is a popular molecule in general chemistry classes for exemplifying the *trigonal bipyramidal* shape in VSEPR theory. The structure is highly symmetric, having two equivalent axial fluorines (so-called because they lie on the principal rotational axis of the molecule) and three equivalent equatorial fluorines. However, studies of the molecular symmetry led to the discovery of an unexpected type of molecular motion.

Herbert Gutowsky and coworkers carried out some of the first spectroscopic studies on PF_5 in the 1950s, and obtained surprising results. The infrared spectrum showed vibrational motions consistent with a trigonal bipyramid, but in the ^{19}F NMR spectrum, all five fluorines appeared to be equivalent! Gutowsky recognized the implications immediately: the fluorines were effectively exchanging positions on a timescale that was too rapid for NMR to tell the difference between axial and equatorial, but too slow for the IR spectrum to see the exchange at all. The difference in timescales for the two experiments does indeed provide a large window. The time required to determine a transition frequency with a precision of $\delta\nu$ is roughly $1/\delta\nu$,and NMR transitions are much more precisely determined ($\delta\nu \approx 100\,\text{s}^{-1}$) than infrared transitions ($\delta\nu \approx 10^8\,\text{s}^{-1}$). The question was, what motion could exchange equatorial and axial positions on a timescale between 10^{-2}s and 10^{-8}s?

R. Stephen Berry answered this question seven years later when he recognized that this exchange was an example of **pseudorotation,** when atoms that lie in a circle oscillate perpendicular to that circle, mimicking rotation. (Fans doing "the wave" in a baseball stadium are pseudorotating: all the motion is actually just up and down, but it conveys an impression of rotation around the stadium.) The motion involved in PF_5 is illustrated in Fig. 6.4. This was the first chemical example of pseudorotation discovered, but the motion is associated with many intramolecular rearrangements and even plays a part in chemical reactions such as the hydrolysis of phosphate esters.

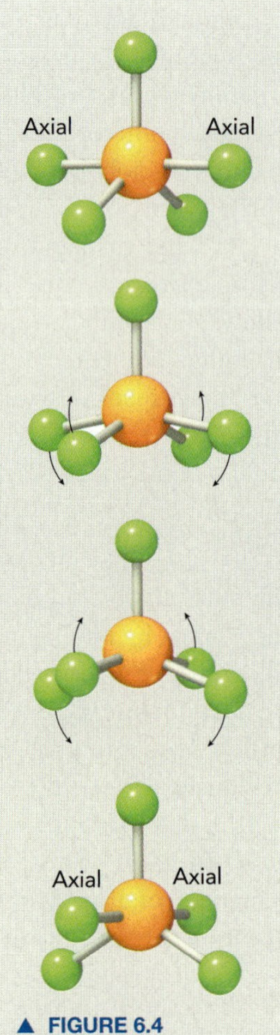

▲ **FIGURE 6.4** **Pseudorotation of PF_5.** Equatorial and axial positions of the F atoms are exchanged by a motion where two equatorial atoms bend in one direction while the axial atoms bend in the opposite direction.

PROBLEM Find the symmetry elements of PF_5 in its equilibrium geometry, as drawn in Fig. 6.5, with two sets of equivalent (indistinguishable) fluorine atoms: the three equatorial atoms and the two axial F atoms.

SOLUTION

1. There is always the identity operation $\hat{E}$.
2. There is a $\hat{C}_3$ axis that contains the axial F atoms. This is the principal axis. There are *two* operations associated with this axis, however: $\hat{C}_3$ (rotation by $2\pi/3$ or 120°) and $\hat{C}_3^2$ (rotation by $4\pi/3$ or 240°). The next rotation, $\hat{C}_3^3$, would be rotation by a full circle and is equal to $\hat{E}$.
3. There are three distinct $\hat{C}_2$ proper rotation operators, one for each of the axes that contains the P atom and one of the three equatorial F atoms. A single molecular geometry may have several different proper rotation axes.
4. Inversion is not a symmetry element, because inversion flips each equatorial atom 180° to the opposite side of the nucleus. There are no atoms at those locations before the operation, so the distribution of atoms changes under inversion.
5. There is a horizontal plane of reflection that contains the three equatorial atoms.
6. There are three vertical mirror planes that contain the axial atoms and one of the equatorial atoms.
7. The $\hat{C}_3$ axis coincides with an improper rotation axis $\hat{S}_3$. If we carry out a rotation by 120°, all the equatorial atoms shift by one position, but the distribution of equivalent atoms is left unchanged. If we then carry out the horizontal reflection (to complete the improper rotation), all we do is switch the locations of the two axial atoms, and again the distribution remains the same.

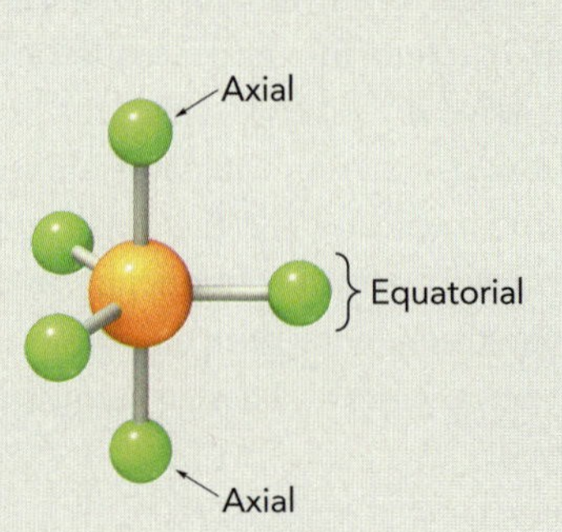

◀ **FIGURE 6.5** **The equilibrium geometry of PF_5.**

Point Groups

A mathematical **group** is a complete set of operators $\hat{A}_i$ such that any combination of those operators $\hat{A}_i\hat{A}_j$ is another member of the set. A group of symmetry operators that leave a single point unchanged, the way our operators leave the center of mass unchanged, is called a **point group.** If all of those operators are symmetry elements of a particular molecule, then the molecule is said to "belong" to that point group. The point group of the molecule determines the nomenclature, degeneracies, and selection rules for all the electronic and vibrational states of the molecule.

When dealing with single molecules, the point groups are commonly labeled by their **Schönflies symbols,** which hint at the symmetry elements contained in the group. The groups we will examine are listed in Table 6.1. In these designations, n can be as large an integer as you want, and so there are an infinite number of point groups. Happily, chemical structures tend to restrict n to values of 6 or less.

EXAMPLE 6.3 Point Group Operations

CONTEXT Problems in quantum mechanics can often be approached from different perspectives, and it becomes important to see when two processes, although described differently, are actually the same. For example, quantum mechanical tunneling has a dramatic impact on many chemical reactions that involve hydrogen transfer, because hydrogens are relatively light (which increases their tunneling probability). When there are several equivalent hydrogens in the same molecule, tunneling can also allow them to exchange places. In the 2-butyne molecule shown, it is possible for tunneling to exchange H atoms 1 and 2. A second tunneling exchange can then reverse the positions of atoms 2 and 3. Each of these exchanges is similar to a reflection of the methyl group through a mirror plane.

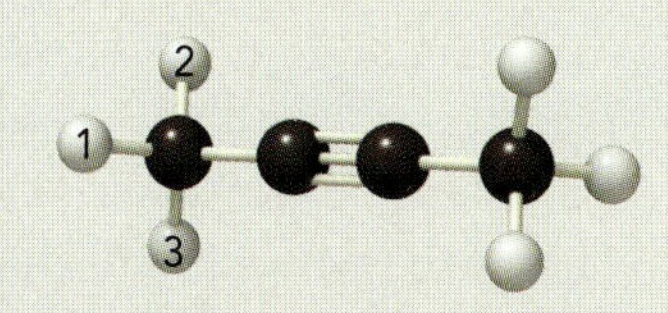

The combination of those two exchanges (switching atoms 1 and 2, and then switching 2 and 3) results in exactly the same arrangement as if the methyl group were rotated by 120°. The combination of two reflections in this case is equivalent to a single rotation by a third of a turn. This example illustrates in another way how we can determine that one combination of operations has the same result as another, single operation.

PROBLEM If $\hat{C}_2(z)$ indicates rotation by π about the z axis, $\hat{\sigma}_{xy}$ indicates reflection through the xy plane, and so on, then find the single operation that is identical to

$$\hat{\sigma}_{xy}\hat{C}_2(z)\hat{\sigma}_{yz}.$$

SOLUTION Remembering to carry out the operations from right to left, we have

$$\begin{aligned}\hat{\sigma}_{xy}\hat{C}_2(z)\hat{\sigma}_{yz}\psi(x,y,z) &= \hat{\sigma}_{xy}\hat{C}_2(z)\psi(-x,y,z)\\ &= \hat{\sigma}_{xy}\psi(x,-y,z) = \psi(x,-y,-z)\\ &= \hat{C}_2(x)\psi(x,y,z)\\ \hat{\sigma}_{xy}\hat{C}_2(z)\hat{\sigma}_{yz} &= \hat{C}_2(x).\end{aligned}$$

Therefore, a point group that contains $\hat{\sigma}_{xy}$, $\hat{C}_2(z)$, and $\hat{\sigma}_{yz}$, must also contain $\hat{C}_2(x)$.

TABLE 6.1 Schönflies labeling scheme for common point groups. In this list n must be an integer greater than 1. The operations listed are not necessarily all the symmetry elements of the group, but they are sufficient to distinguish it from other groups.

symbol	contains
C_1	$\hat{E}$ only
C_s	$\hat{E}$ and $\hat{\sigma}$ only
C_i	$\hat{E}$ and $\hat{I}$ only
C_n	$\hat{E}$ and one $\hat{C}_n$ axis only
C_{nv}	$\hat{E}$, one $\hat{C}_n$ axis, and n $\hat{\sigma}_v$ planes only
C_{nh}	$\hat{E}$, one $\hat{C}_n$ axis, and one $\hat{\sigma}_h$ plane only
D_n	$\hat{E}$, one $\hat{C}_n$ axis with n $\hat{C}_2$ axes perpendicular to it, but no $\hat{\sigma}_h$ or $\hat{\sigma}_v$
D_{nh}	$\hat{E}$, one $\hat{C}_n$ axis, one $\hat{\sigma}_h$ plane, and n $\hat{C}_2$ axes that lie in the $\hat{\sigma}_h$ plane
D_{nd}	$\hat{E}$, one $\hat{C}_n$ axis, n $\hat{\sigma}_v$ planes, and n $\hat{C}_2$ axes perpendicular to the $\hat{C}_n$ axis, but no $\hat{\sigma}_h$
S_{2n}	$\hat{E}$, one $\hat{S}_{2n}$ axis
T	$\hat{E}$, three perpendicular $\hat{C}_2$ axes, and four $\hat{C}_3$ axes
T_d	$\hat{E}$, three perpendicular $\hat{C}_2$ axes, four $\hat{C}_3$ axes, and six mirror planes
O	$\hat{E}$, two sets of three perpendicular $\hat{C}_4$ axes, and eight $\hat{C}_3$ axes
O_h	$\hat{E}$, two sets of three perpendicular $\hat{C}_4$ axes, eight $\hat{C}_3$ axes, and $\hat{I}$
I	$\hat{E}$, four sets of three perpendicular $\hat{C}_5$ axes, and twenty $\hat{C}_3$ axes
I_h	$\hat{E}$, four sets of three perpendicular $\hat{C}_5$ axes, twenty $\hat{C}_3$ axes, and $\hat{I}$

TABLE 6.2 Multiplication table for the group C_{2v}.

	$\hat{E}$	$\hat{C}_2$	$\hat{\sigma}_V$	$\hat{\sigma}_V'$
$\hat{E}$	$\hat{E}$	$\hat{C}_2$	$\hat{\sigma}_V$	$\hat{\sigma}_V'$
$\hat{C}_2$	$\hat{C}_2$	$\hat{E}$	$\hat{\sigma}_V'$	$\hat{\sigma}_V$
$\hat{\sigma}_V$	$\hat{\sigma}_V$	$\hat{\sigma}_V'$	$\hat{E}$	$\hat{C}_2$
$\hat{\sigma}_V'$	$\hat{\sigma}_V'$	$\hat{\sigma}_V$	$\hat{C}_2$	$\hat{E}$

If we examine the group C_{2v}, which is the point group for H_2O and H_2S, its symmetry elements are $\hat{E}$, $\hat{C}_2$, $\hat{\sigma}_v$, and $\hat{\sigma}_v'$. We can make a multiplication table, Table 6.2, giving the operation obtained when any two of the elements are combined. As required, every combination of operations is another operation in the group.

The groups at the bottom of Table 6.1, labeled T for tetrahedral, O for octahedral, and I for icosahedral, are collectively known as the **cubic point groups.**[1] The cubic groups have high symmetry (meaning they contain many symmetry elements) and are distinct from the other point groups in that they have several *different* proper rotation axes $\hat{C}_n$ where n is greater than 2. For example, the tetrahedral groups have

[1]You may notice in these the names of three of the five *Platonic solids,* the only three-dimensional solids for which each face and each vertex is exactly identical to any other. In addition to the tetrahedron (4 triangular faces), octahedron (8 triangular faces), and icosahedron (20 triangular faces), there are the cube (6 square faces) and dodecahedron (12 pentagonal faces). The cube and the dodecahedron do not get their own point groups because they are *isomorphs* of the octahedron and the icosahedron, respectively, meaning that they have the same symmetry elements. If you place a point in the middle of each face of the octahedron and then connect those 8 dots, you find that they are the 8 corners of a cube. All the rotations, reflections, and other symmetry properties of the octahedron are shared by the cube, and vice versa. Similarly, the centers of the 20 faces of an icosahedron work out to be the 20 corners of a dodecahedron. The tetrahedron is by itself in this list, because if we connect the centers of its 4 faces, we just get the 4 corners of another tetrahedron.

four different $\hat{C}_3$ axes. Samples of the rotational symmetries from these groups are illustrated in Fig. 6.6.

For molecules with a lot of symmetry, it usually isn't necessary to find *all* of the symmetry elements in order to determine the point group. With practice, you will be able to identify the point groups for molecules of ten or fewer atoms

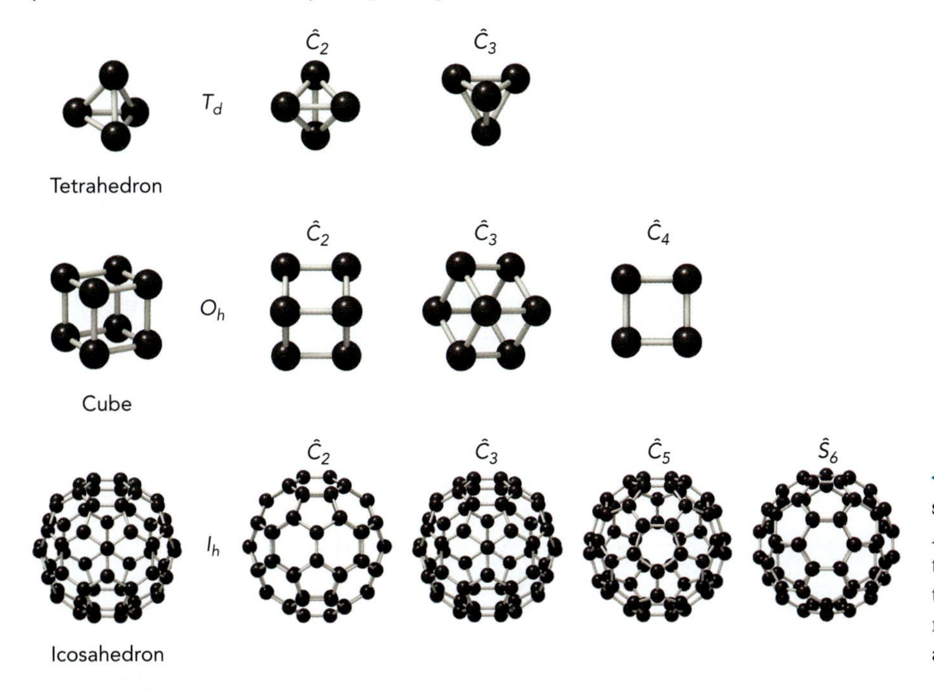

◀ **FIGURE 6.6 Rotational symmetry of the cubic groups.** A tetrahedron, a cube, and a truncated icosahedron are used to illustrate the surprising number of different rotational axes in three of the cubic groups.

EXAMPLE 6.4 Assigning Point Groups

H—F
(a)

O=N—O ⟷ O—N=O
(b)

$F_2C=CH_2$
(c)

HFC=CHF
(d)

CONTEXT Physical chemistry is usually easiest to digest when the examples are small molecules, and that is most of what you will find in this chapter. However, molecular symmetry is relevant to much larger structures than these, extending well into the realm of biomedicine.

Following their Nobel Prize-winning work on the structure of DNA, James Watson and Francis Crick made a signal contribution to the microbiology when they proposed a theory correctly predicting the high degree of symmetry in the structures of viruses. Viruses are composed primarily of a nucleic acid core protected by a protein shell. Watson and Crick realized that the nucleic acid at the core of a typical virus was too short to code for any very complicated protein. Instead, they hypothesized, a virus makes a lot of one very simple protein, and the structure of that one protein allows each molecule to easily interlock with identical neighbors, forming the shell that will contain the nucleic acid. If every protein is the same, they will most easily fit together if the *environment* of each protein is also the same—in other words, if after the shell is formed, each protein is still equivalent to every other. That means the shell itself must be a highly symmetric structure, such as an icosahedron. Watson and Crick were proved right with the first electron micrographs that could resolve virus structure, and the subsequent decades have revealed viruses to rank among the most beautiful natural structures uncovered by science, however terrible their consequences may be for human health.

There are other biochemical applications of symmetry that arise from **multimers**—structures assembled from several identical protein subunits. The acetylcholine receptor, which has approximate C_5 symmetry, forms a guarded doorway into the cell for ion transport that regulates, for example, muscle control. Cobra venom includes a protein that blocks that receptor, thereby preventing breathing. This exercise tests our ability to find the same kind symmetry but in smaller and more innocuous molecules.

PROBLEM Assign the point group for each of the molecules shown.

SOLUTION Going by the flow chart, Fig. 6.7:

- HF: is linear $\rightarrow$ but has no $\hat{\sigma}_h$. Thus we can assign it to $C_{\infty v}$.
- NO_2: is not linear $\rightarrow$ and has a $\hat{C}_2$ axis in the NO_2 plane, bisecting the NO bonds (recall that in resonance theory the two NO bonds are equivalent) $\rightarrow$ no $\hat{C}_n$ with $n > 2$ $\rightarrow$ no $\hat{\sigma}_h$ $\rightarrow$ 2 $\hat{\sigma}_v$ mirror planes (the NO_2 plane and the plane perpendicular to the NO_2 plane between the two NO bonds) $\rightarrow$ no perpendicular $\hat{C}_2$ axes. It is therefore C_{2v}.
- F_2CCH_2: is not linear $\rightarrow$ and has a $\hat{C}_2$ axis along the C=C bond $\rightarrow$ no $\hat{C}_n$ with $n > 2$ $\rightarrow$ no $\hat{\sigma}_h$ $\rightarrow$ 2 $\hat{\sigma}_v$ mirror planes (the molecular plane and the plane perpendicular to the molecular plane that also contains the C=C bond) $\rightarrow$ no perpendicular $\hat{C}_2$ axes. It is thus C_{2v}.
- C_2H_3F: is not linear $\rightarrow$ has no $\hat{C}_n$ axes $\rightarrow$ but has $\hat{\sigma}$ in the plane of the molecule. Therefore it is C_s.

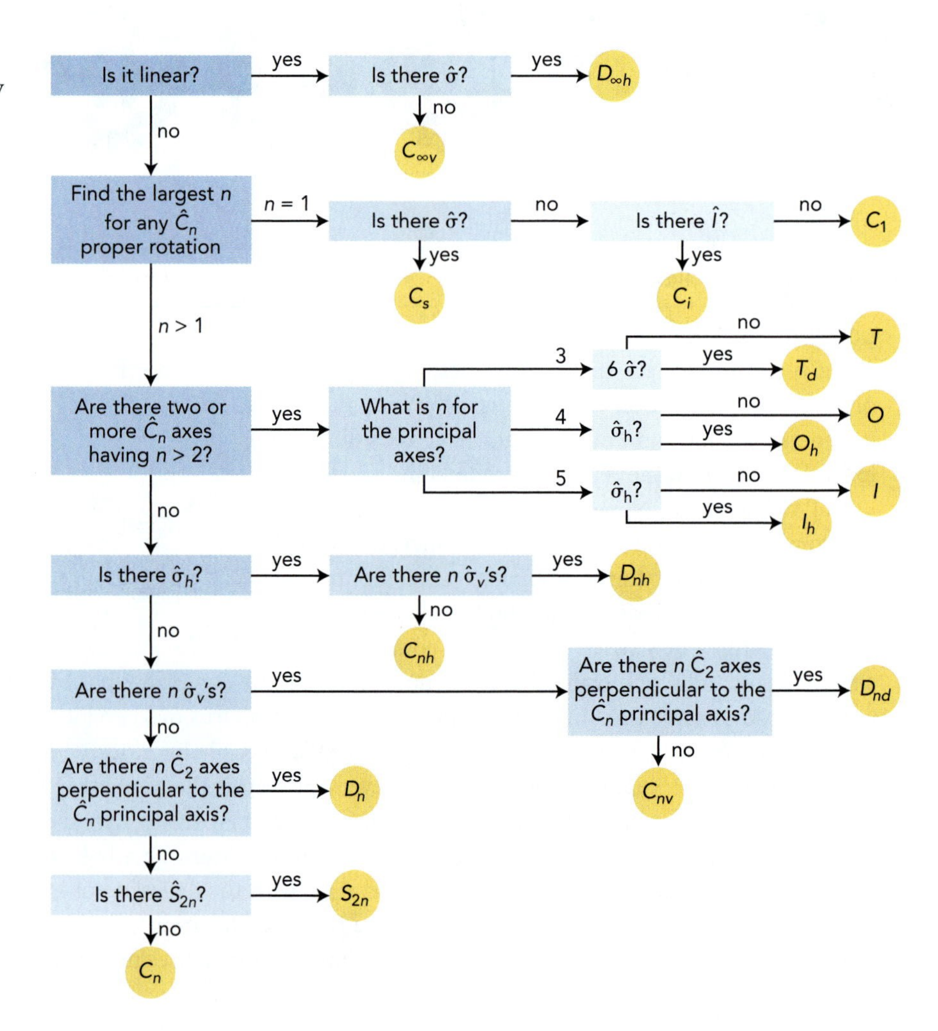

▶ FIGURE 6.7 **Flow chart for identifying point groups.** Only the most common cubic point groups are listed.

by inspection. In the meantime, a flow chart such as the one in Fig. 6.7 may be helpful for determining the point group of a molecule in just a few steps. From the flow chart, for example, we would observe that PF_5 in Example 6.2 (*i*) is not linear, (*ii*) has a principal $\hat{C}_3$ axis (through the axial PF bonds), (*iii*) which is the only $\hat{C}_n$ axis with $n > 2$, (*iv*) has a $\hat{\sigma}_h$ horizontal mirror plane and (*v*) three $\hat{\sigma}_v$ mirror planes, and therefore PF_5 belongs to the point group D_{3h}. In the same manner, we find that HCN in Fig. 6.3 (*i*) is linear but (*ii*) has no $\hat{\sigma}_h$ mirror plane and therefore belongs to the point group $C_{\infty v}$.

One immediate application of group theory to chemistry is this: once we know a molecule's point group, we can tell whether or not the molecule is polar. The overall dipole moment can be written as the vector sum of individual bond dipole moments in the molecule. However, individual bond dipole moments in a molecule will cancel if they

1. lie perpendicular to the rotation axis of a $\hat{C}_n$ symmetry element;
2. lie perpendicular to the mirror plane of a $\hat{\sigma}$ symmetry element;
3. lie anywhere in a molecule with a $\hat{S}_{2n}$ symmetry element (including inversion, because $\hat{I} = \hat{S}_2$).

Our reliable PF_5 molecule provides examples of the first two cases, as shown in Fig. 6.8.

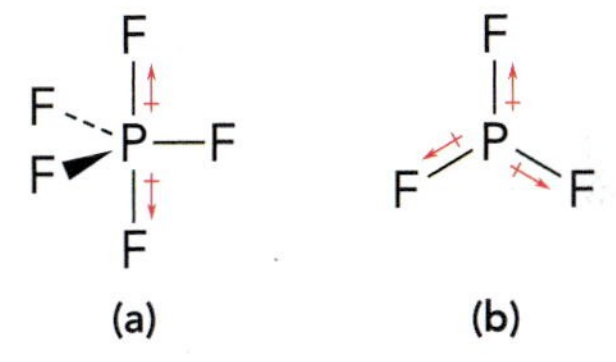

▲ FIGURE 6.8 **Symmetry and molecular dipole moments.** Individual bond dipole moments must cancel (the vector sum is zero) if they lie **(a)** perpendicular to a mirror plane, such as the $\hat{\sigma}_h$ plane in PF_5, or **(b)** perpendicular to a proper rotation axis, such as the $\hat{C}_3$ axis in PF_5.

Molecules are nonpolar if they belong to any point group that combines these symmetry elements so as to cancel the bond dipoles in all directions. For example, molecules in any of the dihedral groups D_n, D_{nh}, and D_{nd} are nonpolar, because the principal rotation axis will cancel any bond dipoles perpendicular to the axis, and the perpendicular $\hat{C}_2$ rotation axes require any remaining dipole components to cancel. The cubic groups, with multiple proper rotation axes, lead to nonpolar molecules by the same reasoning. Molecules in the C_{nh} point groups have bond dipole components that lie either perpendicular to the rotation axis or perpendicular to the $\hat{\sigma}_h$ plane, so they all cancel and the molecules are nonpolar. And molecules in the C_i or S_{2n} groups are nonpolar by the last criterion (they contain $\hat{I}$ or $\hat{S}_{2n}$). That leaves only the following:

All polar molecules belong to C_n (including C_1), C_s, or C_{nv} point groups.

Because PF_5 is in the point group D_{3h}, it must be nonpolar. Knowing that HCN is in the point group $C_{\infty v}$ tells us that it should be polar, with the dipole moment along the internuclear axis—the $\hat{C}_\infty$ axis—because there can be no net dipole moment perpendicular to that axis. This is just a qualitative test, however. Knowing that NO_2 belongs to the point group C_{2v} tells us that the molecule is polar, but not how large the dipole moment is.

Point Group Representations

Assigning the molecule to a point group establishes all the symmetry elements for that distribution of atomic nuclei. Carrying out any of those operations on the nuclei leaves their distribution effectively unchanged. The probability densities $|\psi|^2$ for the electronic and vibrational wavefunctions must also be symmetric under these operations, because identical regions of the molecule should yield identical contributions to our experimental observables. However, the *wavefunction* ψ itself is *not* an experimental observable and is not necessarily symmetric with respect to each of the molecule's symmetry elements.

For example, the point group C_s contains only the identity operator and a reflection operator $\hat{\sigma}$. Let's confine ourselves for now to the molecular orbital (MO) wavefunctions, and let's set the mirror plane to the xy plane. Molecules assigned to the C_s point group have some MOs that are symmetric under $\hat{\sigma}_{xy}$ and some that are antisymmetric—that change sign—when reflected through the mirror plane. To be antisymmetric, the MO must change sign from one side of the xy plane to the other, and must therefore have a nodal plane that coincides with the xy plane.

The point groups help us organize the symmetry operators into sets. But now we see that the point group alone does not specify the symmetry properties of the functions we can associate with a given point group. Is there a way to organize the functions—such as the MO wavefunctions—that these operators operate *on*?

There is: we can associate any function with a *representation* in the point group. Any two functions that share the same representation behave the same way under the operations of the point group. For example, in the point group C_s, the functions z^2 and xy are unaffected by the $\hat{\sigma}_{xy}$ operator (which changes the sign of the z coordinate):

$$\hat{\sigma}_{xy}(z^2) = (-z)^2 = z^2$$

$$\hat{\sigma}_{xy}(xy) = xy$$

and both functions (by definition) are unchanged by the identity operator $\hat{E}$. Because z^2 and xy are both unchanged by the symmetry operators in C_s, they must belong to the same representation. The point group C_s has only two possible representations: A' (which is symmetric with respect to $\hat{\sigma}_{xy}$) and A'' (which is antisymmetric with respect to $\hat{\sigma}_{xy}$).

But the functions we want to work with can be extremely complicated. How can we be sure that *any* function can be classified according to the representations of the point group? In each point group, it is possible to express the symmetry properties of any function in terms of a finite set of **irreducible representations** of the group. In the same way that any whole number can be broken down into a unique product of prime numbers, we can break down the symmetry properties of any function into a unique product of irreducible representations. Each irreducible representation Γ_i stands for a particular combination of symmetry properties that make it distinct from every other irreducible representation Γ_j, such that functions given by any two different irreducible representations are always orthogonal:

$$\int_{\text{all space}} \psi(\Gamma_i)\psi(\Gamma_j)d\tau = 0 \quad \text{if } \Gamma_i \neq \Gamma_j. \tag{6.6}$$

CHECKPOINT The point group describes the symmetry of the nuclei in the molecule. The representations tell us the symmetry properties of the quantum states that will be based on that arrangement of nuclei. It's as though the point group establishes the frame, and the representations describe different ways you can build on that frame. For example, a square hole in a wall for a window means the window will be square, but the window frame can have different designs in it—vertical and horizontal bars dividing it into smaller squares or rectangles, or diagonal bars that form diamond shapes and triangles. The overall structure is still square, but the designs within that structure (like the representations) can have more intricate symmetries.

Character Tables

The irreducible representations and their symmetry characteristics are given in the **character table** for the point group, several of which are in Tables 6.3 and 6.4. (Additional character tables make up the Appendix.) The symmetry characteristics of a representation in the left-hand column are identified by the representation's eigenvalue when it is operated on by the operator in the top row. The irreducible representation is given by a lower case letter (a_1, π_u, etc.) when applied to a single component of the overall wavefunction, such as a single MO or a single vibrational motion. The capital letter (A_1, Π_u, etc.) is used to identify the symmetry of the wavefunction for the overall electronic or vibrational state.

Let's look at three examples of irreducible representations:

1. The A'' representation of the C_s point group has character -1 under the operator $\hat{\sigma}$ in Table 6.3. This means that A'' is the representation for any of those MOs that are antisymmetric with respect to reflection through the mirror plane (Fig. 6.9).
2. The character table for C_{2v} in Table 6.3 has a representation A_1, which is symmetric with respect to all four symmetry operations of the point group:

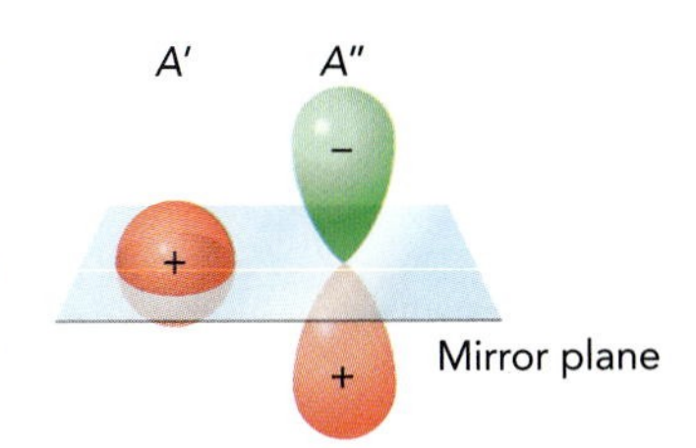

▲ **FIGURE 6.9 Symmetry representations of point group C_s.** A function with A' symmetry has the same sign and magnitude on both sides of the mirror plane, whereas a function with A'' symmetry has the same magnitude on both sides but opposite sign.

TABLE 6.3 Selected character tables for point groups of non-linear molecules. More tables appear in the Appendix.

C_s	$\hat{E}$	$\hat{\sigma}$	Functions
$A'(a')$	1	1	$x, y; x^2, y^2, z^2, xy; R_z$
$A''(a'')$	1	-1	$z; yz, xz; R_x, R_y$

C_{2h}	$\hat{E}$	$\hat{C}_2$	$\hat{I}$	$\hat{\sigma}_h$	Functions
$A_g(a_g)$	1	1	1	1	$x^2, y^2, z^2, xy; R_z$
$B_g(b_g)$	1	-1	1	-1	$xz, yz; R_x, R_y$
$A_u(a_u)$	1	1	-1	-1	z
$B_u(b_u)$	1	-1	-1	1	x, y

C_{2v}	$\hat{E}$	$\hat{C}_2(z)$	$\hat{\sigma}_v(xz)$	$\hat{\sigma}_v'(yz)$	Functions
$A_1(a_1)$	1	1	1	1	$z; x^2, y^2, z^2$
$A_2(a_2)$	1	1	-1	-1	$xy; R_z$
$B_1(b_1)$	1	-1	1	-1	$x; xz; R_y$
$B_2(b_2)$	1	-1	-1	1	$y; yz; R_x$

D_{2h}	$\hat{E}$	$\hat{C}_2(z)$	$\hat{C}_2(y)$	$\hat{C}_2(x)$	$\hat{I}$	$\hat{\sigma}(xy)$	$\hat{\sigma}(xz)$	$\hat{\sigma}(yz)$	Functions
$A_g(a_g)$	1	1	1	1	1	1	1	1	x^2, y^2, z^2
$B_{1g}(b_{1g})$	1	1	-1	-1	1	1	-1	-1	$xy; R_z$
$B_{2g}(b_{2g})$	1	-1	1	-1	1	-1	1	-1	$xz; R_y$
$B_{3g}(b_{3g})$	1	-1	-1	1	1	-1	-1	1	$yz; R_x$
$A_u(a_u)$	1	1	1	1	-1	-1	-1	-1	
$B_{1u}(b_{1u})$	1	1	-1	-1	-1	-1	1	1	z
$B_{2u}(b_{2u})$	1	-1	1	-1	-1	1	-1	1	y
$B_{3u}(b_{3u})$	1	-1	-1	1	-1	1	1	-1	x

(continued)

TABLE 6.3 continued Selected character tables for point groups of non-linear molecules.

D_{6h}	$\hat{E}$	$2\hat{C}_6$	$2\hat{C}_3$	$\hat{C}_2$	$3\hat{C}_2'$	$3\hat{C}_2''$	$\hat{i}$	$2\hat{S}_3$	$2\hat{S}_6$	$\hat{\sigma}_h$	$3\hat{\sigma}_v$	$3\hat{\sigma}_d$	Functions
$A_{1g}(a_{1g})$	1	1	1	1	1	1	1	1	1	1	1	1	$x^2 + y^2, z^2$
$A_{2g}(a_{2g})$	1	1	1	1	−1	−1	1	1	1	1	−1	−1	R_z
$B_{1g}(b_{1g})$	1	−1	1	−1	1	−1	1	−1	1	−1	1	−1	
$B_{2g}(b_{2g})$	1	−1	1	−1	−1	1	1	−1	1	−1	−1	1	
$E_{1g}(e_{1g})$	2	1	−1	−2	0	0	2	1	−1	−2	0	0	$(xz, yz); (R_x, R_y)$
$E_{2g}(e_{2g})$	2	−1	−1	2	0	0	2	−1	−1	2	0	0	$(x^2 - y^2, xy)$
$A_{1u}(a_{1u})$	1	1	1	1	1	1	−1	−1	−1	−1	−1	−1	
$A_{2u}(a_{2u})$	1	1	1	1	−1	−1	−1	−1	−1	−1	1	1	z
$B_{1u}(b_{1u})$	1	−1	1	−1	1	−1	−1	1	−1	1	−1	1	
$B_{2u}(b_{2u})$	1	−1	1	−1	−1	1	−1	1	−1	1	1	−1	
$E_{1u}(e_{1u})$	2	1	−1	−2	0	0	−2	−1	1	2	0	0	(x, y)
$E_{2u}(e_{2u})$	2	−1	−1	2	0	0	−2	1	1	−2	0	0	

$\hat{E}$, $\hat{C}_2$, $\hat{\sigma}_v$, and $\hat{\sigma}_v{}'$. An a_1 MO of a C_{2v} molecule (note the lower case "a_1" used for a single orbital) such as the HOMO−2 MO shown in Fig. 6.10 is also symmetric under all four operations. There are three other possible representations, each of which is antisymmetric with respect to two of the operations.

3. Degenerate representations pose a more complex problem. Consider the Π representation listed in Table 6.4 for the point group $C_{\infty v}$. The character under $\hat{E}$ is 2 rather than 1. This means we need to think of Π as representing two orbitals simultaneously.

Degenerate representations work like the subshell label for atomic orbitals. There is only one 1*s* orbital in an atom, because the 1*s* subshell is non-degenerate, but the 2*p* subshell stands for three different orbitals ($m_l = -1, 0, 1$) at the same time. In the same way, all the states within a degenerate representation are treated simultaneously. The characters for that representation may therefore appear with coefficients other than just 1 and −1. The results when operating with one of the group's symmetry elements may be difficult to visualize because it operates simultaneously on two or more states. Degeneracy appears in the irreducible

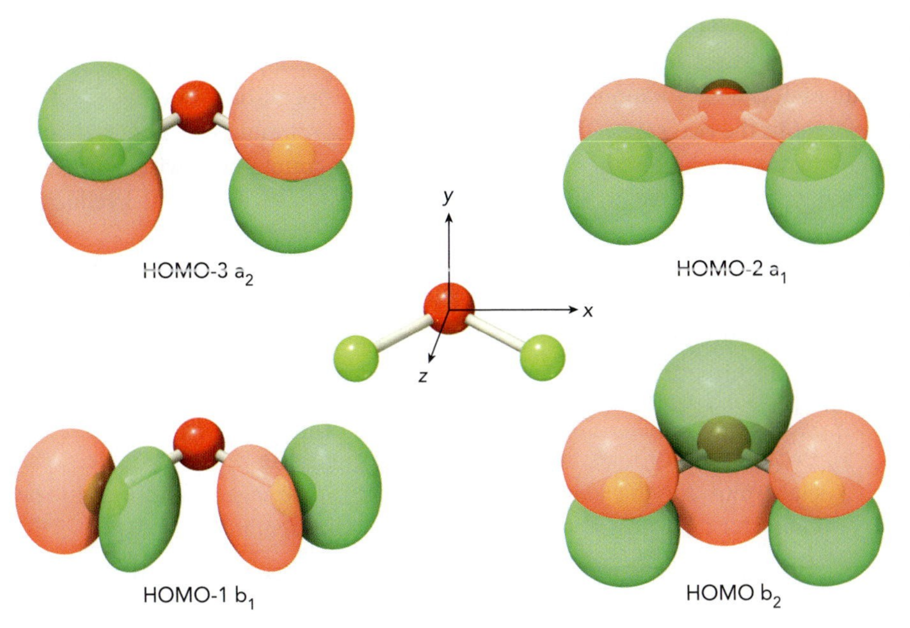

◀ **FIGURE 6.10 F_2O molecular orbital wavefunctions.** The four highest-energy occupied MOs illustrate each of the four representations of the C_{2v} point group. The HOMO is the highest energy occupied MO and the HOMO−1 is the next highest. Different colors indicate regions of opposite phase.

TABLE 6.4 Character tables for the linear molecule point groups $C_{\infty v}$ and $D_{\infty h}$. Under $\hat{C}_\infty$, ϕ is the angle of rotation about the z axis.

$C_{\infty v}$	$\hat{E}$	$\infty \hat{C}_\infty$	$\infty \hat{\sigma}_v$	Functions
$\Sigma^+(\sigma)$	1	1	1	$z; x^2 + y^2, z^2$
Σ^-	1	1	−1	R_z
$\Pi(\pi)$	2	$2\cos\phi$	0	$x, y; xz, yz; R_x, R_y$
$\Delta(\delta)$	2	$2\cos 2\phi$	0	$xy, x^2 - y^2$
$\Phi(\varphi)$	2	$2\cos 3\phi$	0	
⋮	⋮	⋮	⋮	

C≡O $C_{\infty v}$ H—H $D_{\infty h}$

$D_{\infty h}$	$\hat{E}$	$\infty \hat{C}_\infty$	$\infty \hat{\sigma}_v$	$\hat{I}$	$2\hat{S}_\infty$	$\infty \hat{C}_2$	Functions
$\Sigma_g^+(\sigma_g)$	1	1	1	1	1	1	$x^2 + y^2, z^2$
Σ_g^-	1	1	−1	1	1	−1	R_z
$\Sigma_u^+(\sigma_u)$	1	1	1	−1	−1	−1	z
Σ_u^-	1	1	−1	−1	−1	1	
$\Pi_g(\pi_g)$	1	$2\cos\phi$	0	2	$-2\cos\phi$	0	$xz, yz; R_x, R_y$
$\Pi_u(\pi_u)$	2	$2\cos\phi$	0	−2	$2\cos\phi$	0	x, y
$\Delta_g(\delta_g)$	2	$2\cos 2\phi$	0	2	$2\cos 2\phi$	0	xy
$\Delta_u(\delta_u)$	2	$2\cos 2\phi$	0	−2	$-2\cos 2\phi$	0	
⋮	⋮	⋮	⋮	⋮	⋮	⋮	

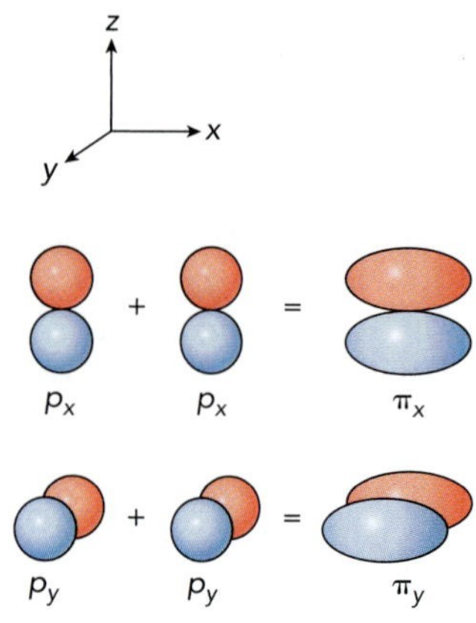

▲ FIGURE 6.11 **Combining p_x and p_y atomic orbitals to get the degenerate π MOs.**

representations whenever the group contains an operator that can be carried out in opposite directions with different results. For example, the clockwise and counterclockwise rotations of 120° are both $\hat{C}_3$ operations, but the result is different in the two cases. This is not true of the $\hat{C}_2$ rotation, which always gives the same result whether the rotation is clockwise or counterclockwise. Therefore, all the groups that contain any $\hat{C}_n$ or $\hat{S}_n$ with $n \geq 3$ have degenerate irreducible representations.

Let's return to the doubly degenerate Π representation in the point group $C_{\infty v}$. What are the two MOs that the Π symbol stands for? There is more than one way to think of it, but for now let's break the π MOs into a π_x MO formed from p_x atomic orbitals, say, and a π_y formed from p_y orbitals, as shown in Fig. 6.11.

1. Because two orbitals are represented by the same set of characters, the eigenvalue for the identity operator $\hat{E}$ is 2. The eigenvalue of $\hat{E}$ in the character table gives the degeneracy of the representation.
2. The eigenvalue for rotation about the z axis depends on the angle of rotation. If we rotate the function through an angle $\phi = \pi$, the π_x and π_y orbitals are both inverted, and the eigenvalue is
$$2\cos\phi = 2\cos\pi = -2.$$
If we rotate by a full 2π, then we should get the same two functions back, and sure enough the eigenvalue is
$$2\cos\phi = 2\cos(2\pi) = 2.$$
3. The result of vertical plane reflection on the π_x and π_y functions depends on the choice of vertical mirror plane. If we choose the $\hat{\sigma}_{xz}$ reflection, for example, the eigenvalue for operation on π_x is 1, but the eigenvalue for operation on π_y is -1, because this reflection will invert the π_y but not the π_x. The reverse is true if we choose the yz plane instead. In either case, one function changes while the other remains the same. Consequently, the character in the character table is 0.

CHECKPOINT The characters in the character table are not exactly eigenvalues. When the degeneracy of the representation is 2 or more, the character represents a sum of the possible eigenvalues, and is determined by a matrix operation. Our simplification will suffice for the work in this text, however.

6.2 Symmetry Representations for Wavefunctions

What good is it to assign a molecule to its point group? Section 6.1 points out that a major benefit of recognizing symmetry in molecules is that it reduces the work involved in integrating properties around the molecular structure—for example, integrating the Schrödinger equation to find the electronic wavefunctions. In addition to this quantitative advantage, there are important qualitative conclusions that can be drawn by exploiting molecular symmetry. We introduce here two major applications of group theory to molecules: determining the symmetry of overall wavefunctions from the symmetry representations of their components and using those representations to determine spectroscopic selection rules. These methods are used extensively in Chapters 7 and 8.

Molecular orbitals are a model, a simplified picture that we use as a first, big step toward understanding a complicated system. For any many-electron molecule, there is no one true set of equations for its MOs. Instead, we are free to decide how we want to build them, because they are only an approximation of the many-electron wavefunction, and even the wavefunction itself is just a mathematical construct. With a few exceptions, for the rest of the text we choose to write our MOs as **symmetry orbitals,** one-electron wavefunctions that share the symmetry properties of an irreducible representation in the molecule's point group. For example, F_2O in Fig. 6.10 has C_{2v} symmetry, and so we label all of its MOs by the representations a_1, a_2, b_1, and b_2, which are the four representations of C_{2v} appearing in Table 6.3. This bears closer examination, however, so let us return to the simplest molecules for a start.

Molecular Orbital Symmetry in Diatomics

You may be wondering at this point why so much attention is given to symmetry in molecules when, after all, most molecules are not symmetric. Let's begin with perhaps the most fundamental reason: chemists use the symmetry properties of diatomic molecules to describe individual chemical bonds in *any* molecule. In essence, we apply a local bond model to the chemical bond, allowing us to consider—just as a first approximation—that the bond depends only on the two atoms that it connects. In this section we see how to label the MOs of diatomic molecules, and this will inform many of our conclusions about bonding in polyatomic molecules.

The diatomic molecules, being linear, belong either to the point group $C_{\infty v}$ (heteronuclear diatomics, such as CO) or to the group $D_{\infty h}$ (homonuclear diatomics, such H_2). These two groups have an infinite number of symmetry elements and irreducible representations, and therefore they have some properties that distinguish them from the lower symmetry point groups. Let's review the symmetry elements appropriate to $C_{\infty v}$ and $D_{\infty h}$, and consider any aspects that are specific to the diatomic molecules.

1. $\hat{C}_n(z)$, which rotates the wavefunction by $2\pi/n$ about the z axis or internuclear axis, and is the same as rotation through the angle ϕ. The angular momentum projection quantum number m_l that we used for atoms is still a good quantum number for linear molecules, as long as we project the angular momentum onto the internuclear axis rather than an arbitrary z axis. The quantum number for this projection is labeled λ rather than m_l. As in atoms, this leads to a symmetry for rotation about ϕ such that $\psi(\phi + \frac{\pi}{\lambda}) = -\psi(\phi)$ (for $\lambda \neq 0$) and $\psi(\phi + \frac{2\pi}{\lambda}) = \psi(\phi)$. Therefore the symmetry with respect to this operation also tells us about the orbital angular momentum of the wavefunction.[2]

[2]Because spherical symmetry in the potential is lost, there is no good quantum number in molecules equivalent to the atomic l. In nonlinear molecules, λ also ceases to be a good quantum number.

2. $\hat{\sigma}_v$. The vertical mirror planes are the infinite number of planes that contain the z axis, including the xz and yz planes.
3. $\hat{\sigma}_h$. The horizontal mirror plane is the xy plane, perpendicular to the internuclear axis and passing through the center of the bond. For the homonuclear diatomic, reflection through the xy plane is a symmetry element; it is not for the heteronuclear diatomic.
4. $\hat{I}$. For diatomics, this is again a symmetry element for the homonuclear case (such as H_2) but not the heteronuclear (such as CO).

Now let's go back through those four operations and consider in some detail the implications for the MOs of the diatomic molecules and see how this determines the notation:

1. $\hat{C}_n(z)$. $|\lambda| = 0, 1, 2, 3, 4, \ldots$ correspond to $\sigma, \pi, \delta, \phi, \gamma, \ldots$ orbitals (these are the Greek equivalents of the atomic state designations $s, p, d, f, g, \ldots$). All of these except for σ are doubly degenerate representations, because the same symbol applies to both $+|\lambda|$ and $-|\lambda|$ (Fig. 6.12).

   Fig. 6.11 shows one way to think of the formation of a pair of π MOs: from the p_x and p_y atomic orbitals. But there is another way. We saw in Section 3.2 (under *Cartesian orbitals*) that the p_x and p_y atomic orbitals are themselves formed by combining the $m_l = -1$ and $m_l = +1$ p orbitals. By starting from those p orbitals, where the projection of the angular momentum is well-defined, we again obtain two π orbitals, but this time one with $\lambda = +1$ and one with $\lambda = -1$, as drawn in Fig. 6.13. The symbol π stands for a *subshell* of two degenerate orbitals having $\lambda = +1$ and $\lambda = -1$, in the same way that the symbol $2p$ stands for an atomic subshell with degenerate orbitals having $m_l = -1, 0, +1$.
2. $\hat{\sigma}_v$. The degenerate orbitals, with $\lambda \neq 0$, can be broken into components that are symmetric and antisymmetric with respect to $\hat{\sigma}_v$, and in the character tables for $C_{\infty v}$ and $D_{\infty h}$, these characters are zero. The σ MOs must always be symmetric with respect to $\hat{\sigma}_v$, and those characters are always 1. Therefore, no additional symbol is used to denote this reflection symmetry. However, we shall see that it is possible to build overall Σ^- wavefunctions—combining more than one orbital—that do not change under rotation about z but nonetheless change sign under $\hat{\sigma}_v$ reflection.
3. $\hat{\sigma}_h$. If the wavefunction has a node in the middle of the chemical bond, then the electron density must go to zero in the middle of the bonding region, and hence the orbital is antibonding. In $D_{\infty h}$ diatomics, this means the orbital wavefunction is antisymmetric with respect to $\hat{\sigma}_h$, whereas a wavefunction that is symmetric under $\hat{\sigma}_h$ is bonding. That is

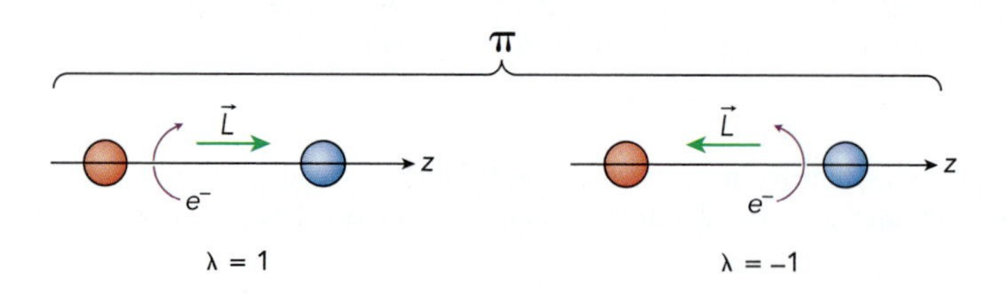

▶ FIGURE 6.12 **Degenerate representations in linear molecules.** Molecular orbital subshells with orbital angular momentum are doubly degenerate, with the two states corresponding to angular motion in opposite directions and opposite signs of the quantum number λ.

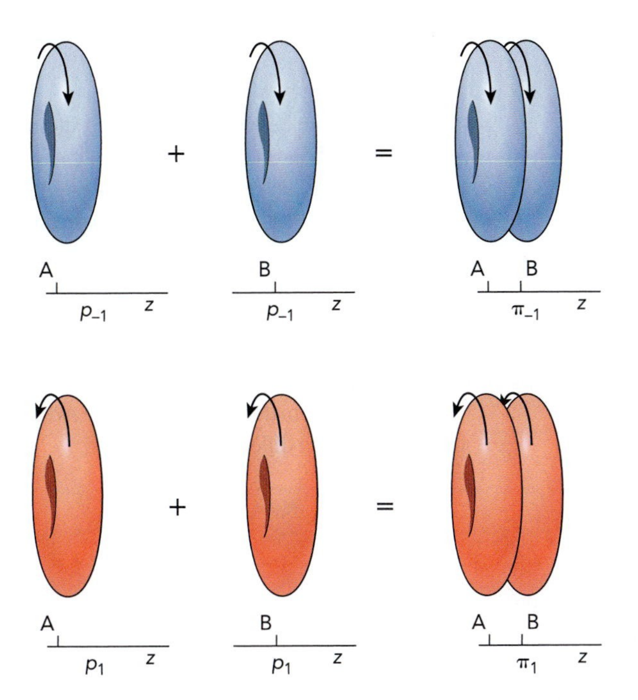

FIGURE 6.13 Combining $m_l = \pm 1$ p atomic orbitals to get the degenerate π MOs. In a linear molecule, we can choose λ to be a good quantum number instead of choosing the orientation of the orbital to be fixed. In that case, we can combine p atomic orbitals with the same value of m_l but centered on different atoms to form a π MO with λ equal to the m_l values of the original p orbitals. The arrows convey the direction of angular motion in each orbital.

not necessarily the case for *polyatomic* linear molecules of the point group $D_{\infty h}$. In any case, the symmetry of the orbital with respect to $\hat{\sigma}_h$ is known once λ and the inversion symmetry are determined, and the reflection symmetry is not included in the symbol for the irreducible representation. Antibonding orbitals are commonly indicated by a superscript "*," as in "σ*," but this is *not* part of the symbol for the symmetry representation.

4. $\hat{I}$. For a molecule with an inversion center, the wavefunction is labeled g (*gerade*) if it is symmetric with respect to inversion and u (*ungerade*) if it is antisymmetric.[3] This appears in the orbital designation as a subscript, for example, σ_u or π_g.

All of this specifies only the spatial distribution of the electron. There is still an electron spin, which associates each electron with an m_s value of either $-1/2$ or $+1/2$, and Pauli exclusion still requires that electrons in the same orbital have different m_s values. Therefore, σ orbitals are like atomic s orbitals and can hold up to two electrons, whereas the π orbitals come in pairs, which together can hold up to four electrons. A consequence of all this is that the σ orbitals, because they have no angular momentum for rotation about the z axis, are the only orbitals with electron density between the two atoms. All the other orbitals have $|\lambda|$ nodal planes through the internuclear axis. This recalls the feature that atomic s orbitals, having no angular momentum, are the only orbitals with density at the nucleus, while the others have l angular nodes.

[3] The terms are German for "even" (*gerade*) and "odd" (*ungerade*), consistent with common mathematical nomenclature in English: a symmetric function is often called "even," an antisymmetric function is "odd."

BIOSKETCH **James Rondinelli**

James Rondinelli, a professor of materials science and engineering at Drexel University, uses group theory as an aid in designing new materials. The symmetry of a crystal plays a role in its properties, but the nature of the relationship is not always obvious. Rondinelli solves the electronic Schrödinger equation of various crystalline materials and uses the results to uncover connections between the molecular structure and the material properties. For example, perovskite, a mineral composed of calcium titanium oxide, has several possible crystal structures and can be forced from one to another. These transitions from one crystal phase to another normally change the symmetry of the crystal. However, Rondinelli, working with Sinisa Coh of Rutgers, has predicted that when stress is applied along a plane in one particular form of perovskite, individual cells in the crystal structure rotate, changing the overall material properties but preserving the same symmetry. If that rotation can be connected to the electronic structure of the crystal, it might pave the way for the design of new materials that could be switched back and forth between different properties.

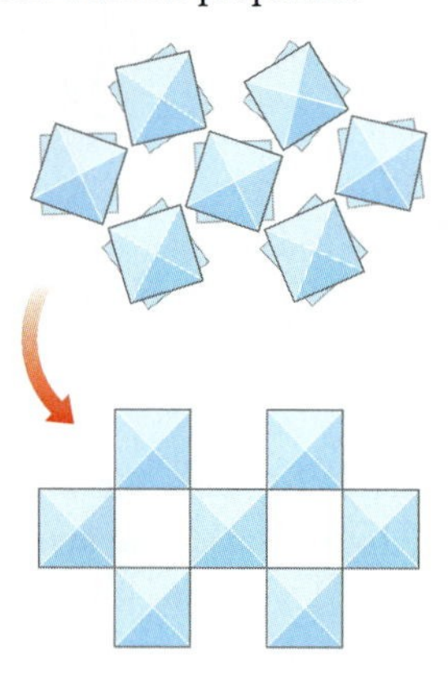

In Section 5.2, we combined two 1*s* orbitals with the same phase to make the ψ_+ MO of H_2^+ (Eq. 5.13):

$$\psi_+(r,\theta,R) = C_+(R)(1s_A + 1s_B), \quad \psi_-(r,\theta,R) = C_-(R)(1s_A - 1s_B).$$

The H_2^+ ion is a homonuclear diatomic and belongs to the point group $D_{\infty h}$. The ψ_+ MO has electron density on the internuclear axis, so it must be a σ orbital. We also know that this orbital must be a σ orbital ($\lambda = 0$, no orbital angular momentum) because it was created from two *s* orbitals, which have no angular momentum to begin with.

If we invert ψ_+, we get exactly the same function. In fact, because the ψ_+ orbital is positive everywhere, *it must be symmetric under all the symmetry elements* of the point group $D_{\infty h}$. In order to be antisymmetric under some operation, a wavefunction must have equal regions of positive and negative phase that could change places. Therefore, the ψ_+ orbital has the characters of the σ_g representation in the table for $D_{\infty h}$, and we now relabel that orbital the "$1\sigma_g$." The quantum number 1 at the beginning of "$1\sigma_g$" is similar to the principal quantum number *n* in atoms, but we don't need to assign any fancy meaning to it. The lowest energy σ_g orbital is the $1\sigma_g$, the next lowest will be the $2\sigma_g$, and so on.[4]

[4]There are alternatives to our labeling scheme for distinguishing orbitals of one symmetry (say, σ_u) from others of the same symmetry. The *n* label is the simplest and is applied to polyatomic as well as diatomic molecules.

The ψ_- orbital behaves similarly: symmetric with respect to reflection in the xz plane and zero orbital angular momentum, but antisymmetric with respect to inversion and reflection in the xy plane. We combine the symmetry under rotation about z (which makes this a σ orbital) and antisymmetry under inversion (which makes this a u orbital) to relabel this antibonding orbital the $1\sigma_u$.

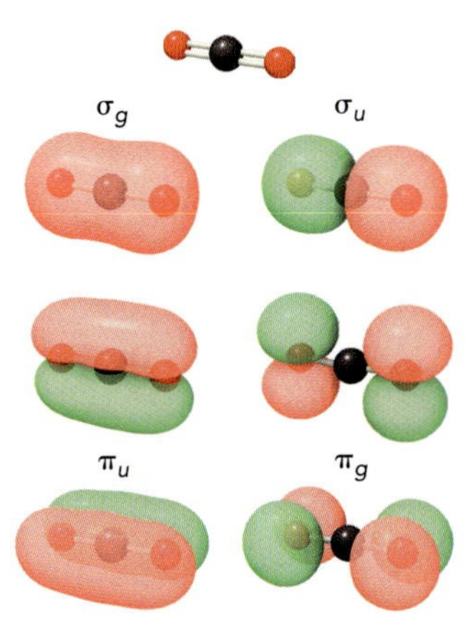

▲ FIGURE 6.14 **The CO_2 bonding orbitals.** In CO_2, the bonding orbitals represent four different symmetries. The σ_u and π_g MOs have nodes, but they are bonding orbitals because the nodes do not pass through the middle of the bond.

Molecular Orbital Symmetry in Polyatomics

The irreducible representations of the groups $D_{\infty h}$ and $C_{\infty v}$ label all the MOs of any linear molecule, whether diatomic or polyatomic. For example, CO_2, OC_3O, HCN, N_3^-, and HC_9N are all linear molecules and they each have σ and π orbitals. However, an orbital with a nodal plane through the molecular axis is not necessarily antibonding. For example, CO_2 has a σ_u orbital that is bonding for each of the two C—O bonds (Fig. 6.14). The nodal plane cuts through the carbon nucleus in this case, not through the middle of the bond.

The representations of other point groups label the MOs of all nonlinear molecules. For linear molecule wavefunctions, we considered only four symmetry operations: rotation about the z axis, inversion, reflection in the xy horizontal plane, and reflection in the vertical planes. Remember that when these operators act on the wavefunction, they may change ψ but not $|\psi|^2$. The same principle remains true when we move on to polyatomic molecules, now with other possible symmetry elements. The symmetry properties of the orbital are denoted by the representation used to label the orbital. The a_2 MOs of the C_{2v} molecule F_2O, for example, have electronic wavefunctions that are antisymmetric with respect to reflection in either of the two vertical mirror planes (Fig. 6.10).

Degenerate states exist for any molecule belonging to a point group with a proper or improper rotation axis where $n \geq 3$ (including the linear molecules, which have $n = \infty$). Just as the Π representations in the linear molecule groups are doubly degenerate, doubly degenerate E representations appear in the character tables for C_{3v} and D_{6h}, for example. The tetrahedral group T_d has two non-degenerate A representations, a doubly degenerate E representation, and two triply degenerate T representations.

The rules for labeling these irreducible representations are summarized in Table 6.5. The labels Σ, Π, Δ, and so on are used *only* for the linear molecule groups $C_{\infty v}$ and $D_{\infty h}$. For the nonlinear molecules, A and B states are non-degenerate (with A states symmetric with respect to rotation about the principal axis), E states are doubly degenerate, T or F states are triply degenerate, and G states are fourfold degenerate. Subscripts and superscripts then provide keys for other symmetry elements (if any), and the u or g subscript *always* appears if inversion is a member of the group.

Beware, though: in many point groups, the choice of representation depends on how the coordinate system is defined, and there's no universally recognized standard. For example, the representations for C_{2v} molecules are not always assigned the same way, because reversing the definition of the x and y axes interchanges the B_1 and B_2 representations. When analyzing the orbitals of FO_2, different researchers may refer to the same orbital as either b_1 or b_2. This cannot affect the outcome

TABLE 6.5 Summary of the representation symbols in nonlinear molecule point groups. Representations are of the general form x'_{nq} or x''_{nq}, where x is determined by the degeneracy as shown below. The prime or double prime indicates the symmetry with respect to a horizontal mirror plane ($\hat{\sigma}_h$), q indicates symmetry with respect to inversion ($\hat{I}$), and n is a number used to distinguish among representations when these other indicators alone are insufficient.

representation	degeneracy	symmetry property
$A(a)$	1	symmetric with respect to $\hat{C}_n(z)$ (or $\hat{S}_{2n}$ in D_{nd} and S_{2n})
$B(b)$	1	antisymmetric with respect to $\hat{C}_n(z)$
$E(e)$	2	
$T(t)$	3	
$G(g)$	4	
$H(h)$	5	
x_g		symmetric with respect to $\hat{I}$
x_u		antisymmetric with respect to $\hat{I}$
x'		symmetric with respect to $\hat{\sigma}_h$
x''		antisymmetric with respect to $\hat{\sigma}_h$

of an experiment, but it certainly can lead to annoying misunderstandings. Once you have decided on your coordinate system, just stick with it.

Every irreducible representation is orthogonal to each of the other irreducible representations in the group. For example, any function with π_g symmetry must be orthogonal to a function with π_u or σ_g symmetry. The complete set of MOs for a molecule is an orthogonal set anyway, but when two orbitals have different representations we say that they are orthogonal *by symmetry*. Different orbitals of the same symmetry will also be orthogonal, but the group theory does not prove that they are.

EXAMPLE 6.5 Symmetry and Choice of Coordinate System

CONTEXT The allyl radical delocalizes three electrons over three carbon atoms, resulting in a relatively stable intermediate in many chemical reactions. For example, it takes about 44 kJ mol^{-1} less energy to break the C—H bond in propene (which leaves an allyl radical) than in propane. This allows techniques in organic synthesis that take advantage of radical intermediates, but the allyl radical also appears in biochemical systems. Lipoxygenases and cyclooxygenases form peroxides from fatty acids, with the remarkable control over rate and stereochemistry that makes enzymes the most highly skilled chemists on Earth. The mechanism for these reactions is not firmly established, but an allyl radical has been proposed as one of the key intermediates.

PROBLEM Identify the point group representation for the π-bond orbital constructed from the $2p$ atomic orbitals in the C_{2v} allyl radical, as drawn next, using the coordinate system illustrated and Table 6.3.

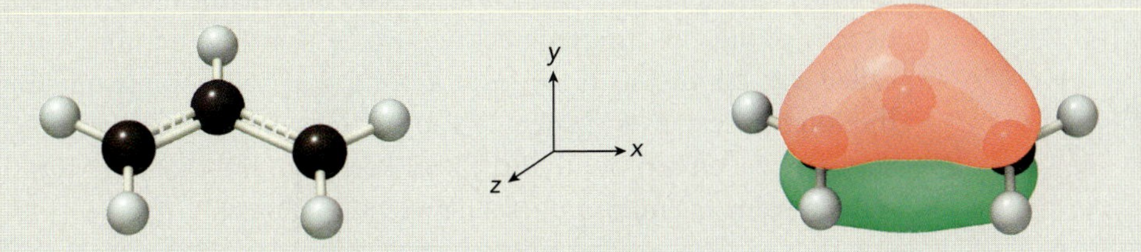

SOLUTION As drawn, the π orbital is composed from $2p_y$ atomic orbitals, and will be antisymmetric with respect to both the $\hat{C}_2$ rotation about the z axis and the $\hat{\sigma}_v$ reflection through the xz plane. According to the designations in Table 6.3, that means the representation for this orbital must be b_2. However, if we had reversed the labels on the x and y axes, the same orbital would be represented by b_1.

Local Bond Models and Symmetry Molecular Orbitals

How are these symmetry MOs related to the Lewis structures that we enjoyed so much in general chemistry? The Lewis structure is a local bond model of the molecule that accounts for all the electrons as follows: core electrons are associated with individual nuclei; each valence electron either contributes to a chemical bond between two atoms or else occupies a lone pair or radical center on a single atom.

We can also account for these electrons using MO theory. Both the local bond picture and the MO picture are models of the molecular electronic structure; neither is rigorously correct. The local bond model provides a fairly accurate and intuitive picture of the strongly localized core electrons, whereas the MO theory presents a far more accurate image of the strongly delocalized bonding electrons in π-conjugated molecules such as benzene and can be advanced to provide a precise, quantitative description of the molecule. The local bonding model also tends to be less useful in describing excited electronic states. With symmetry molecular orbitals, we can characterize all electronic states using common principles of group theory.

The equivalence of the local bond and MO models is an example of how a single system can be expressed in terms of different basis sets. For each electron in the local bond model there is a counterpart in the MO model, but it will not usually be possible to assign each electron in the local bonding model to a specific symmetry MO. Instead, a group of MOs (such as all of the lowest energy σ orbitals in a linear molecule) can be associated with a group of local bonding orbitals—perhaps the core $1s$ electrons of the atoms.

The MO model and local bonding model should each predict the same qualitative ordering of the energy levels. In the local bonding model, orbital energy increases as the electron interacts less with the nuclei: the core electrons interact most strongly and have the lowest energy, followed by the electrons in the relatively strong σ bonds. The higher energy orbitals include π bond, lone pair, and antibonding electrons, but the ordering varies from molecule to molecule.

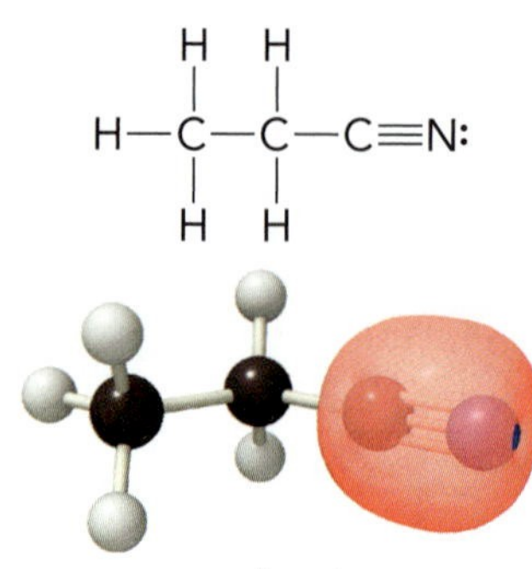

σ-bond
(but not a σ MO)

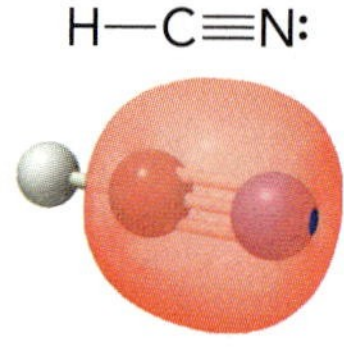

σ-bond
(and a σ MO)

▲ FIGURE 6.15 **When is a σ-bond not a σ MO?** Ethyl nitrile has C_s symmetry and several σ bonds, including the CN bonding MO shown. However, the MO for that bond has a' symmetry. Hydrogen cyanide also has a CN σ-bond, and because HCN is a linear molecule, the MO for that bond is also a σ MO.

Even though MOs of nonlinear molecules are not assigned to the σ and π representations, this notation is used anyway when describing *individual chemical bonds.* An MO with electron density along the axis between the atoms in a diatomic must have σ symmetry, and the chemical bond formed by electron density along the axis connecting two nuclei in a nonlinear molecule is still called a σ bond (Fig. 6.15). And a bond that has a single nodal plane through the internuclear axis (generally the same as the plane formed by neighboring atoms) is called a π bond. But please keep in mind that the MOs themselves are written using σ and π *only* if the molecule is linear.

Overall Symmetry

We now know how to assign an MO to a particular symmetry representation, but MOs describe the wavefunctions of only one electron at a time. For example, in the H_2 molecule, we can put two electrons in the $1\sigma_g$ MO. Does the resulting wavefunction for *both* electrons still have the symmetry of the σ_g representation? How about if we put one electron in the $1\sigma_g$ and one in the $1\sigma_u$—what is the representation for the resulting two-electron wavefunction then?

When a wavefunction is created by multiplying together a bunch of component wavefunctions, it seems sensible that the symmetry of the total wavefunction is determined somehow by the symmetries of its components. This is indeed the case: the symmetry of a product wavefunction is given by the **direct product** of the representations for the wavefunction's components. The direct product is a representation itself, with its characters obtained by multiplying together—one symmetry element at a time—the characters of the original representations.

EXAMPLE 6.6 Local Bond and MO Models

CONTEXT Acetylene is the simplest alkyne, having the same number of electrons and a similar bonding structure to molecular nitrogen but spectacularly different chemical properties. Whereas the *formation* of N_2 gas is the driving force that makes hydrazine a useful rocket fuel—until recently used in auxiliary rockets for the space shuttle—it is the *reaction* of acetylene with O_2 that powers oxy-acetylene welding torches. Chemical changes can often be explained in terms of motion of the electrons, but don't forget that the atomic nuclei are what define the identity of the elements. Electronic structure is only one aspect crucial to understanding chemistry.

PROBLEM Give an MO description of all the electrons in acetylene, H—C≡C—H, in terms of the atomic core and local bonding orbitals. What would you change to obtain the equivalent MO description of nitrogen gas?

SOLUTION The lowest energy electrons will be the four in the carbon 1*s* core orbitals. The molecular orbitals formed from these two 1*s* orbitals are like our ψ_+ and ψ_- orbitals for H_2: one will have σ_g symmetry and the other will have σ_u symmetry. That accounts for 4 of the 14 electrons.

The C—C and C—H σ bonds are likely to be next in energy because they occupy the binding region between the nuclei more fully than the π bonds. The *sp* hybrid orbitals are useful in describing these bonds because they allow the carbons each to bond to two other atoms, keeping the bonds 180° apart. The C—C bond is formed from two overlapping *sp* hybrids, one from each carbon. In order for it to be bonding, the *sp* hybrids must have the same phase; the resulting MO is σ_g and contains two electrons. Each carbon atom has one *sp* hybrid remaining, used to form the C—H bonds. These can be written as MOs in which the *sp* hybrids have the same phase (σ_g) or opposite phase (σ_u). The two possibilities account for four more electrons.

We've used four electrons from each carbon: two for the 1*s* core, one for the C—C bond, and one for the C—H bond (the other C—H bond electron is provided by the hydrogen). This leaves two electrons in 2*p* orbitals on each carbon. These form two C—C π bonds, which with the C—C σ bond yield a triple bond.

number of electrons	local bond model	MO model
4	1*s* C core	$1\sigma_g$, $1\sigma_u$
2	C—C σ bond	$2\sigma_g$
4	C—H σ bonds	$3\sigma_g$, $2\sigma_u$
4	C—C π bonds	$1\pi_u$

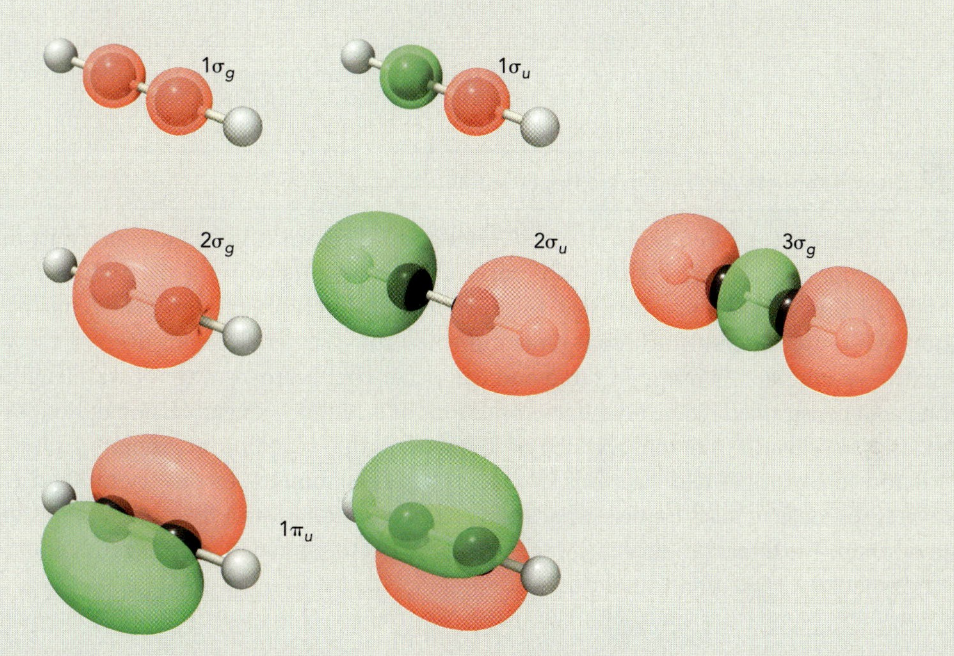

The only difference between this result and the result we would obtain for N_2 is that the $3\sigma_g$ and $2\sigma_u$ MOs would represent the two nitrogen atom lone pairs, rather than the C—H bonds.

For example, to obtain the direct product $B_1 \otimes B_2$ (the symbol "$\otimes$" is used for the direct product) in the group C_{2v}, we multiply the characters of B_1 and B_2 for each symmetry element:

	$\hat{E}$	$\hat{C}_2$	$\hat{\sigma}_v$	$\hat{\sigma}_v'$	
B_1	1	−1	1	−1	
B_2	1	−1	−1	1	
$B_1 \otimes B_2$	1	1	−1	−1	$= A_2$

The representation for the direct product has characters 1,1,−1,−1, respectively, for the four symmetry elements of C_{2v} as listed previously. The −1 for $\hat{\sigma}_v{}'$, for example, comes from multiplying the B_1 character for $\hat{\sigma}_v{}'$, −1, by the B_2 character, +1.

This is the set of characters for the representation A_2, so we write

$$B_1 \otimes B_2 = A_2.$$

The direct product in $B_1 \otimes B_2$ yields another irreducible representation. Often the result is not so simple, however. When we multiply two functions, the symmetry of the result can be different from those of the original functions, and in some cases the new function can be rewritten as the sum of two functions with different symmetries. We express this using the tools of group theory as follows. If the point groups have degenerate representations, then the direct product of irreducible representations may yield a **reducible representation.** You would recognize that the new representation is reducible because its characters would not appear in the character table for the point group. A reducible representation can always be written as a *direct sum* of the irreducible representations in the point group, where the direct sum is represented by the symbol $\oplus$. We call this process **decomposing** the reducible representation

EXAMPLE 6.7 Direct Products and Reducible Representations

CONTEXT Color centers (or F-centers, from the German *farbe,* meaning color) are locations in an ionic crystal where an anion is missing, but the location is occupied by one or more electrons to maintain the neutral charge of the crystal. These turn out to provide excellent laser sources, providing the basis for **color-center** or **F-center lasers,** which emit radiation when color center electrons drop to a lower energy quantum state. The emission is in the infrared and can occur over a broad range of wavelengths because the electron is not constrained by the boundary conditions of an atom. Using precision optics, it is possible to tune the laser wavelength over roughly 25% of its wavelength—a significant amount. Color centers in BeO were found to consist of two electrons in regions of C_{3v} symmetry in the crystal, and the excited electronic states responsible for the emission arise from a configuration with the two electrons in a doubly degenerate e orbital. The direct product $e \otimes e$ then tells us what the symmetries of the individual quantum states are and what laser transitions can occur.

PROBLEM Find the direct product of $e \otimes e$ in the point group C_{3v} and decompose it to its irreducible representations. The character table is as follows.

C_{3v}	$\hat{E}$	$2\hat{C}_3$	$3\hat{\sigma}_v$	Functions
$A_1(a_1)$	1	1	1	$z; x^2 + y^2, z^2$
$A_2(a_2)$	1	1	−1	R_z
$E(e)$	2	−1	0	$i(x, y); (x^2 - y^2, xy), (xz, yz); (R_x, R_y)$

SOLUTION The direct product $e \otimes e$ has the characters 4, 1, 0. The only way to get the 4 for the $\hat{E}$ character is by either adding four A representations $(1 + 1 + 1 + 1 = 4)$, two A's and the E $(1 + 1 + 2 = 4)$, or $E \oplus E$ $(2 + 2 = 4)$. The four A's added together will give 4 for the $\hat{C}_3$ character as well, so that's not right, and $E \oplus E$ will give −2 for the $\hat{C}_3$ character, so that's also not right. Therefore, we add two A's and the E, and in order to get the final character to be 0, the reducible representation must be equal to $A_1 \oplus A_2 \oplus E$.

into its component irreducible representations. *When decomposing the direct product into its components, there will always be only one valid reduction into irreducible representations.* For cases where the degeneracy is not very high, it may be possible to deduce which irreducible representations contribute to the direct product by inspection. See Example 6.7 for one way to do this.

Decomposing Reducible Representations

You can see that for direct products with a high degeneracy, the reasoning in Example 6.7 becomes too inefficient. For any direct product, there is a systematic and fail-safe method of arriving at the reduction.

We'll use the following symbols:

Γ = reducible representation	$\chi[\hat{A}_j]$ = character of Γ for operator $\hat{A}_j$
Γ_i = irreducible representation	$\chi_i[\hat{A}_j]$ = character of Γ_i for operator $\hat{A}_j$
P = number of irreducible representations	$\Gamma = \sum_{i=1}^{P} a_i\Gamma_i$
Q = order of the group	

The order of the point group Q can be found either by counting the *total* number of symmetry elements in the group or by adding together the squares of the degeneracies of all the irreducible representations in the group. For example, the point group D_{3d} (character table in Table 6.6) is of order (1+2+3+1+2+3) = 12, because there is one $\hat{E}$, two $\hat{C}_3$'s, three $\hat{C}_2$'s, one $\hat{I}$, two $\hat{S}_6$'s, and three $\hat{\sigma}_d$'s. Or we could obtain the same number by adding the squares of the numbers in the column under $\hat{E}$:

$$1^2 + 1^2 + 2^2 + 1^2 + 1^2 + 2^2 = 12.$$

The number of representations P is equal to the number of columns in the character table, which is always less than or equal to Q. For D_{3d}, the value of P is 6.

The problem of decomposing the representation is solved by finding the values of the a_i's in the expression

$$\Gamma = a_1\Gamma_1 \oplus a_1\Gamma_2 \oplus \ldots \oplus a_P\Gamma_P.$$

Each of those a_i's can be determined from this equation:

$$a_i = \frac{1}{Q}\sum_{j=1}^{Q}\chi[\hat{A}_j]\chi_i[\hat{A}_j]^*. \tag{6.7}$$

TABLE 6.6 The character table for D_{3d}.

D_{3d}	$\hat{E}$	$2\hat{C}_3$	$3\hat{C}_2$	$\hat{I}$	$2\hat{S}_6$	$3\hat{\sigma}_d$	Functions
$A_{1g}(a_{1g})$	1	1	1	1	1	1	$x^2 + y^2, z^2$
$A_{2g}(a_{2g})$	1	1	−1	1	1	−1	R_z
$E_g(e_g)$	2	−1	0	2	−1	0	$(x^2 - y^2, xy), (xz, yz); (R_x, R_y)$
$A_{1u}(a_{1u})$	1	1	1	−1	−1	−1	
$A_{2u}(a_{2u})$	1	1	−1	−1	−1	1	z
$E_u(e_u)$	2	−1	0	−2	1	0	(x, y)

There are some groups having complex characters, in which case it is important to use the complex conjugate $\chi_i[\hat{A}_j]^*$ in this equation. Note that the sum is over *all* the operators in the group, even those that have been combined into a single

EXAMPLE 6.8 Methodical Decomposition of a Reducible Representation

CONTEXT **Jahn-Teller distortion** is a coupling between degenerate electronic and degenerate vibrational states of a molecule, resulting in the degeneracies being partly broken. This can be seen as the direct product of the symmetry representations for the electronic and vibrational states giving rise to a reducible representation. For example, the staggered conformation of ethane has D_{3d} symmetry. The ethane radical cation is formed during photoionization of ethane and has low-lying degenerate electronic states of E_g and E_u symmetry. There are also vibrational motions with e_g symmetry. By taking the direct product of E_u and e_g, we obtain a reducible representation that can be written as a sum of different symmetries, and each symmetry corresponds to a different quantum state. The degeneracy of the electronic state is broken: the states no longer have quite the same energy. In this example, we factor in an additional E_g symmetry for the rotational motion as well, to see how many different symmetries may result from one combination of states.

PROBLEM Decompose the direct product

$$\Gamma = E_g \otimes E_g \otimes E_u$$

for the point group D_{3d} with the character table in Table 6.6.

SOLUTION The characters for the reducible representation are

$\hat{E}$	$2\hat{C}_3$	$3\hat{C}_2$	$\hat{i}$	$2\hat{S}_6$	$3\hat{\sigma}_d$
8	−1	0	−8	1	0

To decompose this reducible representation, we need the coefficients a_1 through a_6 in the expression

$$\Gamma = a_1 A_{1g} \oplus a_2 A_{2g} \oplus a_3 E_g \oplus a_4 A_{1u} \oplus a_5 A_{2u} \oplus a_6 E_u.$$

Here are the results from applying Equation 6.7:

$$a_1 = \frac{1}{12}[(8)(1) + 2(-1)(1) + 3(0)(1) + (-8)(1) + 2(1)(1) + 3(0)(1)] = 0$$

$$a_2 = \frac{1}{12}[(8)(1) + 2(-1)(1) + 3(0)(-1) + (-8)(1) + 2(1)(1) + 3(0)(-1)] = 0$$

$$a_3 = \frac{1}{12}[(8)(2) + 2(-1)(-1) + 3(0)(0) + (-8)(2) + 2(1)(-1) + 3(0)(0)] = 0$$

$$a_4 = \frac{1}{12}[(8)(1) + 2(-1)(1) + 3(0)(1) + (-8)(-1) + 2(1)(-1) + 3(0)(-1)] = 1$$

$$a_5 = \frac{1}{12}[(8)(1) + 2(-1)(1) + 3(0)(-1) + (-8)(-1) + 2(1)(-1) + 3(0)(1)] = 1$$

$$a_6 = \frac{1}{12}[(8)(2) + 2(-1)(-1) + 3(0)(0) + (-8)(-2) + 2(1)(1) + 3(0)(0)] = 3$$

So the answer is

$$E_g \otimes E_g \otimes E_u = A_{1u} \oplus A_{2u} \oplus 3E_u.$$

This can be verified by checking the character sums. We could have saved half the work by realizing that $g \otimes g \otimes u$ must be composed of only u representations.

column for the character table. For example, in D_{3d} there are *three* contributions to the sum that come from the three $\hat{C}_2$ operations.

Every group has a **totally symmetric representation,** whose characters are all ones. The direct product of any representation Γ with the totally symmetric representation Γ_s gives the original representation Γ back again. Furthermore, the direct product of any representation with itself yields the totally symmetric representation, although if the original representation is degenerate there will be other components as well. This last rule results from our requirement that any symmetry wavefunction ψ must give rise to a probability density $|\psi|^2$ that is symmetric under all the operations. Because ψ may correspond to any of the representations, the product $\psi^*\psi \equiv |\psi|^2$ corresponds to the direct product of any representation with itself, and must also be totally symmetric.[5]

Using the direct product, we can determine the symmetry of a many-electron wavefunction from the symmetries of the individual MOs. As an exercise, let's return to the ground state two-electron MO wavefunctions of H_2 (Eq. 5.19):

$$\Psi_{\text{MO}}(\text{ground}) = C_+(R)^2\psi_+(1)\psi_+(2)[\alpha(1)\beta(2) - \beta(1)\alpha(2)]/\sqrt{2}.$$

The lowest singlet excited state has a similar form, but moves one electron from the ψ_+ to the higher energy ψ_- orbital:

$$\Psi_{\text{MO}}(\text{excited}) = C_+(R)C_-(R)[\psi_+(1)\psi_-(2)+\psi_-(1)\psi_+(2)][\alpha(1)\beta(2)-\beta(1)\alpha(2)]/\sqrt{2}. \quad (6.8)$$

The symmetry analysis that we're doing now pertains only to the spatial part of these two wavefunctions, so let's pull those out and relabel the ψ_+ and ψ_- labels according to the symmetry representations σ_g and σ_u that we determined earlier:

$$\psi_{\text{MO}}(\text{ground}) = C_+(R)^2\psi_+(1)\psi_+(2) = C_+(R)^2 1\sigma_g(1)1\sigma_g(2) \quad (6.9)$$
$$\psi_{\text{MO}}(\text{excited}) = C_+(R)C_-(R)[\psi_+(1)\psi_-(2) + \psi_-(1)\psi_+(2)]$$
$$= C_+(R)C_-(R)[1\sigma_g(1)1\sigma_u(2) + 1\sigma_u(1)1\sigma_g(2)].$$

CHECKPOINT The character tables provide a way for learning the symmetry of the product of two things when we know the symmetry of each thing individually. No matter how complicated the functions are (and the molecular wavefunctions can have terribly complicated formulas), the symmetry of the product of those functions is given by the direct product of their representations. We don't have to know the mathematical formula for either function, as long as we have some way of determining their symmetries.

Now we can establish the symmetries of these *two*-electron wavefunctions of H_2. The ground state MO wavefunction, with two σ_g electrons, has an overall symmetry

$$\sigma_g \otimes \sigma_g = \Sigma_g^+.$$

(Remember that we use the capital letter for the representation of an overall wavefunction). On the other hand, the excited state with one σ_g electron and one σ_u electron has an overall symmetry

$$\sigma_g \otimes \sigma_u = \Sigma_u^+.$$

The two-electron wavefunctions, like the MOs they are made from, have symmetry properties described by the point group representations. The next section explains one significance of these overall wavefunction symmetries.

[5] If the characters of the representation include any complex values, notice that the same representation is used for both the value and its complex conjugate, so any imaginary component to the phase can be eliminated in this direct product.

6.3 Selection Rules

Section 4.3 explains that a transition between two atomic energy levels occurs readily only if certain selection rules are satisfied, particularly the spin selection rule $\Delta S = 0$ and the conservation of photon orbital angular momentum $\Delta l = \pm 1$. The $\Delta S = 0$ selection rule applies to molecules as well, but the spatial part of the wavefunction is more complicated now, and the corresponding selection rules, which are called **symmetry selection rules,** depend entirely on each molecule's point group.

Electric Dipole Selection Rules

The good news is that group theory gives us a simple, general method for finding these selection rules. The transition strength for a transition from an initial state with wavefunction ψ_i to a final state with wavefunction ψ_f is given by the integral

$$\mathrm{I} \propto \left| \int \psi_f^* \hat{\mu}_t \psi_i \, d\tau \right|^2, \tag{6.10}$$

where $\hat{\mu}_t$ is the transition moment operator, and where the two states are assumed to have the same spin. For electric dipole transitions, $\hat{\mu}_t$ is ex or ey or ez, where the coordinate x, y, or z gives the polarization of the radiation's electric field along the axes of the molecule and e is the fundamental charge. Because e is a constant, the electric dipole selection rule[6] can be restated as follows: the transition is allowed if any one of the integrals $\int \psi_f^* x \psi_i \, d\tau$, $\int \psi_f^* y \psi_i \, d\tau$, or $\int \psi_f^* z \psi_i \, d\tau$ is non-zero.

In our character tables, the representation that gives the behavior of x, y, and z under each of the relevant symmetry operations is identified on the far right-hand side, under the heading "functions." For example, in the character table for the point group C_s, the symmetry plane is chosen to be the xy plane. Therefore, reflection through the symmetry plane leaves the x and y axes unchanged, and the functions x and y are A' functions, symmetric with respect to $\hat{\sigma}$. On the other hand, reflection through the mirror plane switches the sign of any z coordinates, and the function z itself therefore has A'' symmetry (Table 6.7).

TABLE 6.7 The character table for C_s.

C_s	$\hat{E}$	$\hat{\sigma}$	Functions
$A'(a')$	1	1	x, y; x^2, y^2, z^2, xy; R_z
$A''(a'')$	1	−1	z; yz, xz; R_x, R_y

For some cases, as in the group C_3, neither x nor y alone act as irreducible representations of the point group. Instead, combinations of x and y can be found that have the proper symmetry. In these cases, the functions x and y are grouped together in the character table: (x,y). For our purposes, this is still sufficient to indicate which representations can be used for the transition moment operator. In the point group C_3, the representations A (for z) and E (for x and y) can both be used to represent the transition moment. We would use one or the other if the radiation were somehow polarized along the same axis for every molecule, for example, if we could force all of the molecules to line up in the same direction.

[6]By convention, these are often called "infrared" selection rules, to distinguish between the selection rules for vibrational spectra obtained by infrared absorption and Raman spectroscopy (described later in this section). We will stick to the more cumbersome "electric dipole" to avoid the suggestion that the rules are wavelength-dependent.

As an example, let's identify the representations of the point group D_2 that correspond to the functions xy, xz, yz, and z^2. Recall that a $\hat{C}_2$ rotation about any Cartesian axis leaves that coordinate unchanged and changes the sign of the other two (Eq. 6.1). The function xy is therefore antisymmetric under $\hat{C}_2(x)$ and $\hat{C}_2(y)$ but not $\hat{C}_2(z)$, which means it has the symmetry of the B_1 representation. Similarly, xz and yz have the symmetries of B_2 and B_3, respectively. Because z^2 does not change sign under any of the rotations, it has A symmetry.

D_2	$\hat{E}$	$\hat{C}_2(z)$	$\hat{C}_2(y)$	$\hat{C}_2(x)$	functions
A	1	1	1	1	z^2
B_1	1	1	−1	1	z, xy
B_2	1	−1	1	−1	y, xz
B_3	1	−1	−1	1	x, yz

To satisfy our selection rule, we need $|\int \psi_f^* \mu_t \psi_i d\tau|^2$ to be non-zero. Here's where the group theory helps. The symmetry of the integrand $\psi_f^* \hat{\mu}_t \psi_i$ is given by the direct product of the symmetry representations for ψ_f^*, $\hat{\mu}_t$, and ψ_i. If that direct product does not contain the totally symmetric representation Γ_s, then the integral will be exactly zero, and the transition from state i to state f is forbidden. In all the other representations of the point group, there are equal volumes of the function that have negative and positive phase, and these will cancel out when the integral is carried out over all space.

As an example, the σ MOs are the totally symmetric orbitals for molecules in the point group $C_{\infty v}$. Integrals carried out over any other orbital will sample equally the positive and negative phases of the lobes symmetrically distributed about the z axis and cancel out, as shown in Fig. 6.16.

Now the electric dipole selection rule can be stated as follows: for an allowed transition,

$$\Gamma_f \otimes \Gamma_\mu \otimes \Gamma_i = \Gamma_s \oplus \ldots \qquad (6.11)$$

where Γ_f and Γ_i are the representations for the final and initial states of the transition, Γ_μ is the representation for the function x, y, or z, and Γ_s is the totally symmetric representation. For the transition to be allowed, the direct product must either be exactly equal to Γ_s or must be a reducible representation that contains Γ_s. There's a simpler way to write that:

$$\Gamma_f \otimes \Gamma_\mu \otimes \Gamma_i \ni \Gamma_s, \qquad (6.12)$$

where the " $\ni$ " symbol means that Γ_s must be *contained* within the right-hand expression.

This can be further simplified by exploiting the fact that the direct product of any representation with itself yields the totally symmetric representation. We take the direct product of both sides with Γ_μ:

$$\Gamma_\mu \otimes [\Gamma_f \otimes \Gamma_\mu \otimes \Gamma_i] \ni \Gamma_\mu \otimes [\Gamma_s].$$

Then we drop $\Gamma_\mu \otimes \Gamma_\mu = \Gamma_s$ on the left, and replace $\Gamma_\mu \otimes \Gamma_s$ by Γ_μ on the right, obtaining

$$\Gamma_f \otimes \Gamma_i \ni \Gamma_\mu. \qquad (6.13)$$

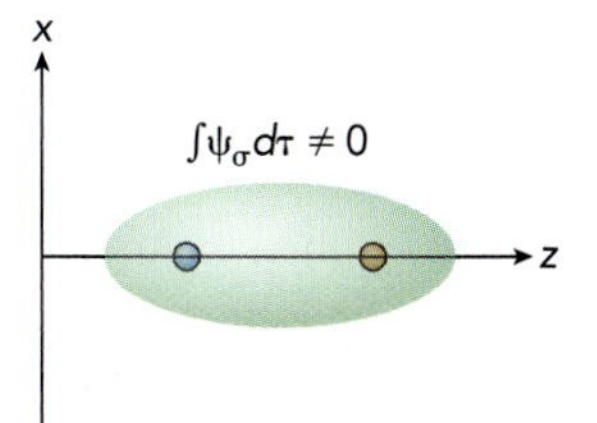

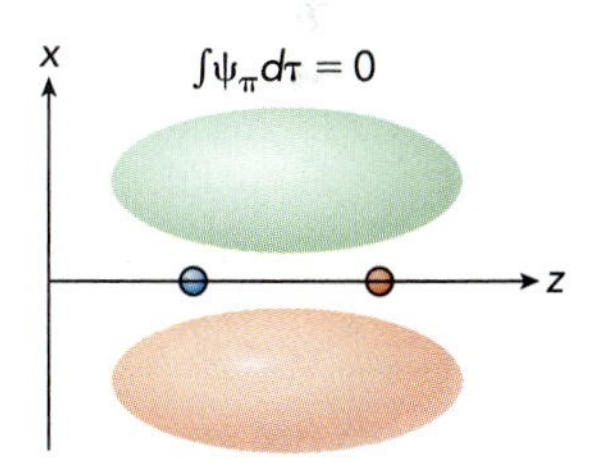

▲ **FIGURE 6.16 Symmetry representations and vanishing integrals.** Only the totally symmetric representation of a group has a non-zero integral. The σ representation of the $C_{\infty v}$ group may be positive everywhere, and therefore yield a positive volume integral. The π representation has equal volumes of opposite phase, and therefore the volume integral can be broken into two equal and opposite contributions, giving a total of zero.

The direct product of the representations of two states gives the representations for the corresponding transition moments. If $\Gamma_f \otimes \Gamma_i$ does not correspond to the representation for any transition moment, the transition $i \rightarrow f$ is forbidden.

By the same argument, the symmetry of one quantum state in a transition can often be determined if the other state and the transition moment symmetry are known:

$$\Gamma_\mu \otimes \Gamma_i \ni \Gamma_f. \tag{6.14}$$

If the direct product yields a reducible representation, then more than one final state symmetry is possible.

EXAMPLE 6.9 Selection Rules

CONTEXT Most of our atmosphere is composed of N_2 and O_2, diatomic molecules belonging to the highest symmetry point group: $D_{\infty h}$. High symmetry creates more possibilities for intensity integrals to vanish, meaning that fewer spectroscopic transitions are allowed than for similar, lower symmetry molecules. For example, the one vibrational motion in N_2 and O_2—stretching of the bond—gives a series of vibrational quantum states that all have *g* inversion symmetry. No transitions among these states are allowed by electric dipole selection rules, so N_2 and O_2 cannot strongly absorb light at those photon energies. Carbon dioxide, a major component of the atmosphere on Venus, also belongs to $D_{\infty h}$, but having three atoms allows CO_2 to have vibrational states with both *g* and *u* symmetry. Many transitions between states of *g* and *u* symmetry are allowed. Hence, our atmosphere absorbs little sunlight compared to the atmosphere of Venus. On the other hand, N_2 and O_2 can scatter light efficiently, because both Rayleigh and Raman scattering leave the inversion symmetry of the quantum states unchanged.

PROBLEM For a homonuclear diatomic, prove that the excitation of a single electron from a σ_g to a π_u MO is allowed by electric dipole selection rules, whereas σ_g to π_g is not.

SOLUTION The functions *x* and *y* are given by the representation Π_u, while *z* is given by Σ_u^+. The MOs σ_g, π_u, and π_g correspond to the representations Σ_g^+, Π_u, and Π_g, respectively. The direct products

$$\Sigma_g^+ \otimes \Pi_u = \Pi_u$$
$$\Sigma_g^+ \otimes \Sigma_u^+ = \Sigma_u^+$$

show that the only possible upper state MOs for an electric dipole transition for σ_g are π_u and σ_u.

Raman Spectroscopy

When a photon passes through matter, the possible outcomes break down into three categories:

1. **Transmission,** in which there is no significant interaction between radiation and matter, and the photon continues moving with its original energy and trajectory
2. **Absorption,** in which the energy and angular momentum of the photon are absorbed by the matter, usually governed by the electric dipole selection rules we've just described

3. **Scattering,** in which the photon traverses the matter but with its energy or trajectory altered by the interaction

While we learn much about atoms and molecules from the wavelengths at which they absorb radiation, often we can measure completely different molecular properties by studying the scattering. In some cases, the chemical sample we want to study cannot be probed by absorption spectroscopy, and scattering provides the only avenue to probe the quantum states of the sample.

According to one description, scattering occurs when the radiation's electric field deforms the distribution of the electrons to a form that does *not* satisfy our time-independent Schrödinger equation. The molecule cannot remain in this **virtual state,** and therefore it cannot completely absorb the photon. So the molecule returns to a valid quantum state, and the photon leaves with its direction of motion altered by the interaction with the molecule (hence the name "scattering").

If the scattered photon has the same energy as the incident photon, then we learn little or nothing about the molecule's properties. This we will call **Rayleigh scattering.** However, if the energy of the scattered photon differs from the incident photon energy, then the molecule ends up in a different state, and the difference in photon energies gives the separation between the initial and final states. Figure 6.17 illustrates this process, known as **Raman scattering.** If the final state is lower than the initial state, the scattered light is higher in energy than the incident light, and the transition is labeled an **anti-Stokes transition.** If the final state is higher energy than the initial state, the transition is a **Stokes transition.**

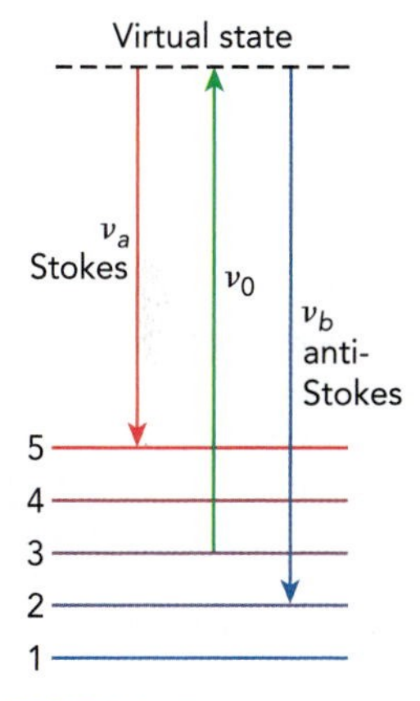

▲ FIGURE 6.17 **Raman transitions.** An incident photon with frequency ν_0 may result in a Raman transition to a higher energy state (Stokes transition) or a lower energy state (anti-Stokes transition) by way of an intermediate "virtual" (non-stationary) state. The difference in energy between states 3 and 5 shown would be obtained by measuring ν_0 and the frequency ν_a of the emitted radiation and calculating $\Delta E_{5,3} = h(\nu_0 - \nu_a)$.

The formation of the virtual state relies on the distortion of the electron wavefunction by the incident radiation, rather than on the strength of a molecular dipole moment. Therefore, the likelihood of this process depends on the polarizability α of the molecule—the ability of the electric field of the radiation to momentarily deform the electron cloud around the molecule and *induce* a dipole moment. For a spherical electron distribution, this induced dipole moment has a magnitude given in Eq. 3.36:[7]

$$\mu_{\text{induced}} = \alpha \mathcal{E}.$$

In SI units, the dipole moment would be expressed in C m and the electric field in V m^{-1}, which means that the polarizability must have SI units of C m^2 V^{-1}, which nearly everyone finds unappealing. The long history of CGS units of electromagnetism has led to the polarizability being expressed more commonly

[7]Molecules in the tetrahedral, octahedral, and icosahedral point groups behave like spherical electron distributions in this respect; the induced dipole moment is independent of the molecule's orientation in the electric field. Most molecules, however, are more easily polarized along one axis than another. The polarizability in this case is actually represented by a matrix called the *polarizability tensor*, with elements that describe the polarizability along the molecule's principal inertial axes.

in units of volume,[8] either cm^3 or $Å^3$. To use α in any of our equations for calculating energies, we will have to first convert to SI units by

$$\alpha(\mathrm{C m^2 V^{-1}}) = (10^{-30}) 4\pi\varepsilon_0\, \alpha(Å^3). \tag{6.15}$$

Incidentally, the probability of radiation scattering off small molecules is proportional to λ^{-4} at wavelengths longer than x-ray. Infrared wavelengths are a micron or more long, much larger than a small molecule. As we decrease the wavelength, approaching the size of the molecule, the scattering efficiency grows rapidly. For example, blue light undergoes Rayleigh scattering at a rate of about $(700\,\mathrm{nm}/400\,\mathrm{nm})^4 = 10$ times more often than red light does. Light of all visible wavelengths, emanated by the sun, passes through our atmosphere, encountering kilometers of N_2 and O_2 molecules as well as dust and water droplets and other particles. These encounters have an order of magnitude stronger preference for scattering the blue light relative to the red. This is why the sky is blue: blue light from the sun is scattered all over the sky and appears to be coming from all directions, whereas the longer wavelength yellows and reds are less scattered and travel more nearly in straight lines (Fig. 6.18). At sunrise and sunset, the light must penetrate much more of the atmosphere to reach the observer, and only the longest wavelength reds travel directly to the observer; the sun appears red and the surrounding sky may be brilliantly colored.

The Raman transition selection rules are available the same way as the electric dipole selection rules, but the transition moment operator has the symmetry of the second order functions x^2, y^2, z^2, xy, yz, and xz. If we think of the Raman transition represented in Fig. 6.17 as a dual process—absorption and then emission—then this makes sense: the probability of the Raman transition depends on the transition moment for reaching the virtual state (when the incident photon hits the molecule)

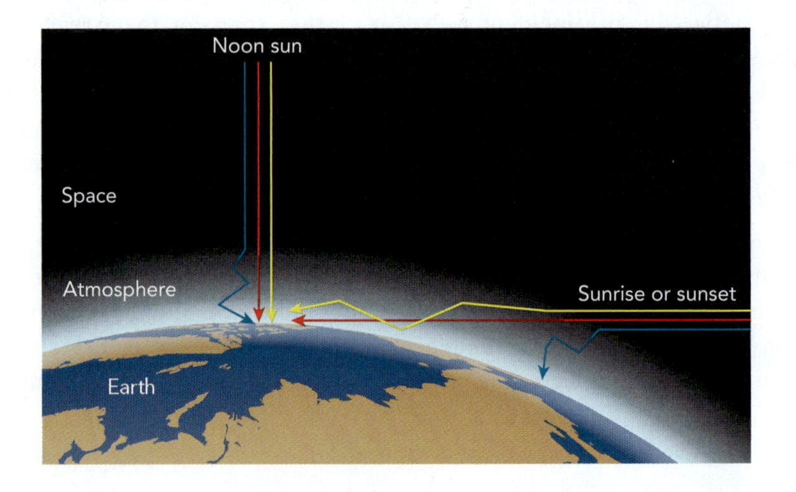

▶ FIGURE 6.18 **The effect of atmospheric scattering on the color of the sky.** The sun emits radiation at all colors of the visible spectrum, but the shorter wavelength blue light is ten times more likely to be scattered than the red. Much of this scattered light eventually reaches the earth's surface, but is no longer in line with the sun; the sky therefore appears blue.

[8] In CGS units, the dipole moment would be written in esu · cm (the basis for the debye, which is 10^{-18} esu cm), and the electric field in statvolt/cm = dyn/esu. Taking advantage of the convenient conversion for esu in CGS units, that would put α in units of

$$\frac{\mathrm{esu^2\,cm}}{\mathrm{dyn}} = \frac{(\mathrm{g\,cm^3\,s^{-2}})\,\mathrm{cm}}{\mathrm{g\,cm\,s^{-2}}} = \mathrm{cm}^3. \tag{6.16}$$

In these units, the molecular polarizability is typically of magnitude 10^{-26}, and so α is often given instead in $Å^3$.

times the transition moment for going from the virtual state down to the final state. Therefore, we need functions that go as x or y or z (for the incident photon) times x or y or z (for the scattered photon). In some cases we will see linear combinations of these functions, such as $x^2 + y^2$ and $x^2 - y^2$, but there will still be some *reducible* representation that has the same symmetry as x^2 or y^2. The irreducible representations for the functions like $x^2 + y^2$ are still valid representations of the Raman transition moment operator and can be used to find the selection rules.

For degenerate representations, you may also see functions in the character table grouped together in parentheses. For example, the C_3 point group has a doubly degenerate representation E, and its functions column contains (x,y), $(x^2 - y^2, xy)$, and (yz,xz). Because the x and y axes are exactly equivalent in this point group, it is not possible for any of the irreducible representations to behave differently for x than for y. The E representation allows us to consider the two functions x and y simultaneously, and therefore the two are grouped together inside parentheses. For the purposes of finding selection rules, the transition moment operator may be represented by E for electric dipole transitions (because it characterizes the x and y functions) or for Raman transitions (because E also characterizes the $x^2 - y^2$, xy, yz, and xz functions).

TOOLS OF THE TRADE | Raman Spectrometers

In 1928, Chandrasekhara Venkata Raman, working with others in the physics department of Calcutta University, discovered that he could send light of one wavelength through a sample of benzene and obtain a *different* wavelength on the other side. This discovery, which earned Raman the Nobel Prize and his name on a broad field of spectroscopy, was extended by the progress of theory (particularly the use of group theory in chemistry by Weyl and Wigner in the late 1920s) and the progress of technology (especially the development of the visible laser by Theodore Maiman in 1960) to form the basis of a laboratory instrument in common use today.

What is Raman spectroscopy? Most light scattering off of matter changes only its direction of travel. This is Rayleigh scattering. However, there is a small chance (usually less than 0.1%) that the photon energy changes, with the energy difference balanced by a change in the quantum state of the matter. This is Raman scattering. The difference between the photon energies of the incoming and the Raman scattered radiation gives the energy gap ΔE between two quantum states of the sample. If ΔE is positive (the sample absorbs energy from the radiation), then the scattering is called Stokes scattering. If ΔE is negative, then the sample loses energy to the radiation, and this is anti-Stokes scattering.

Raman spectroscopy typically probes vibrational quantum states. Electronic transitions usually have large values of ΔE, which weakens the Raman scattering because the probability of scattering is actually proportional to $(h\nu_0 - \Delta E)^4$, where ν_0 is the photon frequency of the incident radiation. Rotational transitions are rarely studied by Raman for two reasons: (1) the ΔE's of rotational transitions are normally much too small to easily separate the scattered light from the incident light, and (2) rotations only occur in gases, and the relatively high density of liquids and solids helps to strengthen the intrinsically weak Raman scattering.

Why do we use Raman spectroscopy? Competing methods for studying vibrational states of molecules include infrared (IR) spectroscopy (see Chapter 8) and fluorescence spectroscopy, both of which are guided by the electric dipole selection rules described in this chapter. While Raman selection rules do offer access to many quantum states that are forbidden to IR spectrometers, the ongoing development of Raman-based technology is actually driven by other needs.

Unlike IR spectroscopy, which uses radiation 2.5μ to 25μ in wavelength, Raman spectroscopy usually involves wavelengths less than 0.6 nm. Light cannot be focused to a spot much smaller than its wavelength, so an IR spectrometer cannot image a sample with resolution better than a few microns. The shorter wavelength light in a Raman experiment, on the other hand, can produce a map of the sample with sub-micron resolution. In other words, it is possible to build microscopes in which the image depends on the vibrational energies of different parts of the sample, allowing us to see where the chemical composition changes. For example, Raman microscope images have been used to determine the distribution of minerals in a meteorite and the pattern of a microchip.

How does it work? A typical Raman spectrometer schematic is sketched in the following figure. A laser of known wavelength provides the incident radiation. Ideally, the user can select from several wavelengths because shorter wavelengths increase the intrinsic scattering probability, but too short a wavelength may cause the sample to fluoresce (producing radiation much stronger than the desired Raman signal). Furthermore, if the vibrational energy gap is small, then it is more difficult to separate the Rayleigh and Raman scattering at short wavelengths.[9] Biological samples tend to be especially sensitive to radiation at short wavelengths, and for those experiments the incident wavelength may be 800 nm or more.

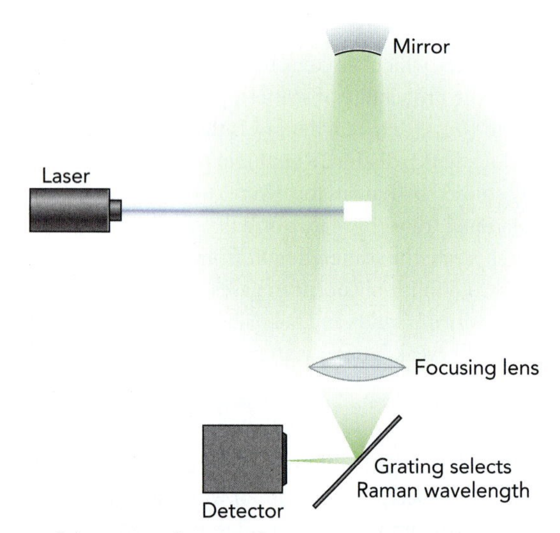

▲ **Schematic of a typical Raman spectrometer.**

The incident beam hits the sample, and light is scattered in all directions. Some designs detect the light at an angle 90° from the incident radiation to avoid interference from the laser light. However, for imaging systems the resolution is normally better if the detector measures the radiation scattered back toward the incident beam, using a beam splitter to separate the two. Furthermore, detecting the light that is scattered back along the axis of the incident beam can be used to generate the Raman spectrum by measuring the interference pattern of the two wavelengths. Because the interference pattern depends *only on the difference in wavelengths,* it is not necessary to measure the wavelengths of the incident and scattered radiation. Otherwise, the wavelength of the scattered light is usually determined by means of a diffraction grating.

There are many variations on the basic methods described here. Two of the most commonly used in the research community are briefly described here.

- **Surface-Enhanced Raman Spectroscopy** (SERS). When a molecule is stuck to a rough metal surface, its Raman scattering often becomes orders of magnitude stronger than scattering from the same molecule free in the gas phase or in solution. The effect can be largely attributed to the highly polarizable electrons that characterize bulk metals. The electric field of the incident radiation drives large oscillations in the motion of these electrons, which effectively increases the incident electric field driving the Raman scattering. The surface of the metal must be rough rather than smooth and flat, because scattering off the surface is enhanced only by the electric field oscillations perpendicular to the surface.
- **Coherent Anti-Stokes Raman Spectroscopy** (CARS). CARS probes the same transitions as traditional Raman spectroscopy, but by much more elaborate and sensitive means based on the interaction of three incident laser beams. The combination of two of these laser beams drives the Raman transition resulting from the third beam, producing a (usually) much stronger response than the Raman signal of the one laser by itself. Efficient mixing of the beams requires significant laser power, and this is achieved by pulsing the lasers, so CARS is normally a technique that probes the sample for very short windows of time.

[9]Isolation of the Raman scattering from the Rayleigh scattering is usually with a resolution proportional to the difference in *wavelengths* rather than the photon energies. Assume the vibrational ΔE is 500 cm^{-1}. For an incident laser wavelength of 400 nm, the difference in Rayleigh and Raman wavelengths is 8 nm. If instead the incident wavelength is 600 nm, then the difference between Raman and Rayleigh more than doubles, to 19 nm. Depending on how broad the Rayleigh signal is, this can be the difference between seeing the Raman line and having it swamped by the stronger Rayleigh scattering.

EXAMPLE 6.10 Raman Selection Rules

CONTEXT Even when a transition can be observed by both IR and Raman spectroscopy, it can be useful to have both techniques available. Titanium (IV) oxide, TiO_2, has attracted enormous interest recently as a *photocatalyst*, using absorbed visible and UV radiation to enhance the oxidation rates of a variety of compounds and forming the structural and electronic framework for dye-sensitized solar cells. Experiments in the 1990s investigated the mechanism for this activity by measuring, among other parameters, the *surface acidity* of TiO_2. The surface acidity was determined partly by the tendency of the surface to bond to the nitrogen lone pair in ammonia. The metal–nitrogen bonding strength could be assessed by IR spectroscopy of the ammonia. Initial IR spectra revealed what appeared to be two different signals for the same N—H stretching transition, suggesting that the ammonia bonded to TiO_2 in two different places. Chemical engineers Gregory Went and Alexis Bell at the Lawrence Berkeley Laboratory used Raman spectroscopy to probe the system, discovering that the extra N—H stretching signal arose from a higher energy transition ($v = 0 \rightarrow 2$) of the degenerate bending motion, rather than the N—H stretch.

PROBLEM Given the following character table for C_{3v}, could the transition from the ground A_1 state of ammonia to an excited E state be allowed by Raman selection rules? By electric dipole selection rules?

C_{3v}	$\hat{E}$	$2\hat{C}_3$	$3\hat{\sigma}_v$	Functions
$A_1(a_1)$	1	1	1	$z; x^2 + y^2, z^2$
$A_1(a_2)$	1	1	−1	R_z
$E(e)$	2	−1	0	$(x, y); (x^2 - y^2, xy), (xz, yz); (R_x, R_y)$

SOLUTION We take the direct product of the representations for the initial and final states, and then look to see if the result contains a representation for any of the functions that correspond to electric dipole (x, y, or z) or Raman (x^2, xy, . . .) transition moments. Because A_1 is the totally symmetric representation, we know right away that $A_1 \otimes E = E$, so we can go straight to the character table for the point group of NH_3, C_{3v}. We see there that the functions column for E includes (x,y), which is sufficient for electric dipole transitions, and (xy,yz), which means Raman transitions will also be allowed.

6.4 Selected Applications

In addition to labeling states and determining selection rules, group theory appears in chemistry in many forms. Here we briefly touch on some examples, showing how group theory has its uses outside perfectly symmetric molecules, and even outside physical chemistry.

Asymmetric Molecules

While we examine these applications of group theory to some molecular problems, a cynical observer could point out that group theory is less helpful to the many molecules in the C_1 point group, which has only one representation: A. If all the states have the same symmetry, what can be simplified or classified by the group theory? Group theory does tell us that, for molecules in the C_1 point group, transitions between *any* two electronic states of the same spin satisfy both Raman and electric dipole selection rules, and that no degenerate electronic states exist. However, those results do little to simplify our analysis of C_1 systems.

Molecules adore symmetry, and when exact symmetry isn't available they'll often settle for something close. In that case, a qualitative picture of the bonding or selection rules in a low-symmetry molecule may be available when we consider a more symmetric cousin. Isotopic substitution provides one simple example. Strictly speaking, monodeuterated water (HDO) is a C_s molecule; it has only a mirror plane and the identity operator as its symmetry elements. However, the deuterium substitution has very little effect on the electron distribution in the molecule, and the electronic states of HDO are essentially identical to those of C_{2v} H_2O. There will even be MOs comparable to the a_2 H_2O orbitals that have a nodal plane through the oxygen atom perpendicular to the molecular plane, even though this plane is no longer a plane of symmetry for HDO. Beware of carrying the analogy too far, however. In this example, H_2O is a very poor model of the vibrational states of HDO, because the vibrational part of the Hamiltonian *is* very closely tied to the mass of the nuclei, and doubling the mass of the hydrogen isotope has a profound effect on those states.

Many asymmetric molecules can be constructed from highly symmetric components, such as phenyl rings based on D_{6h} benzene, D_{5h} cyclopentadienyl ligands, or C_{3v} methyl groups. The electron distributions and vibrational motions of these groups usually retain much of the symmetry that they have in the isolated molecule. How much of this effective symmetry is lost depends on the environment into which the group is substituted. For example, the electronic wavefunctions of the phenyl group may be dramatically affected if the conjugated π system of the phenyl group can be extended beyond the phenyl carbon atoms, as in naphthalene and other polycyclic aromatic compounds.

When we replace one of the H atoms in benzene with a methyl group to form toluene, the molecular point group becomes C_s, which has no degenerate representations. Therefore, toluene has no degenerate orbitals. But we've done relatively little to change the original electronic structure of the benzene, so we can expect two of the MOs to be similar in energy and to have electron distributions similar to those of benzene's e orbitals, as shown in Table 6.8. Similar predictions hold for the vibrational energy levels, but the significant change in nuclear masses causes a proportionately larger effect on the vibrations than on the MO energies.

TABLE 6.8 Effect of symmetry-breaking substitutions X on degenerate vibrational and MO energies of benzene. The HOMO and LUMO MOs in benzene are both doubly degenerate and so is the e_g ring squeeze vibration, so each pair of energies is equal for benzene. The weakly-interacting methyl group breaks the degeneracy, so the energies split into two different values for toluene. The electron-withdrawing chlorine atom splits and also shifts the energies.

X		vibrational excitation energy (cm^{-1})		E (HOMO) (E_h)		E (LUMO) (E_h)	
H	benzene	663	663	−0.336	−0.336	0.136	0.136
CH_3	toluene	562	680	−0.334	−0.324	0.136	0.140
Cl	chlorobenzene	672	760	−0.353	−0.339	0.116	0.117

In the C_{2v} derivative chlorobenzene, the strongly electronegative Cl atom exerts a stronger influence on the electronic and vibrational structure, shifting both the highest occupied (HOMO) and lowest unoccupied molecular orbital (**LUMO**) energies down, but maintaining an approximation of the original degeneracy.

Hückel's Rule

Before we leave this chapter, let's look at one important application of group theory to chemistry. The π-electron systems of aromatic compounds provide an excellent example of the application of group theory to molecules of low symmetry. Hückel's $4n + 2$ rule predicts that a symmetric, monocyclic compound with $4n + 2$ π-electrons, where n is any whole number, is substantially more stable than a corresponding compound with $4n$ π-electrons. Benzene, for example, has six π-electrons ($n = 1$) and is much less reactive than the $4n$ π-electron compounds cyclobutadiene ($n = 1$) and cyclooctatetraene ($n = 2$). This seems at odds with the fact that all three of these compounds—cyclobutadiene, benzene, and cyclooctatetraene—may be drawn with two equivalent resonance structures and therefore may be presumed to have similar resonance stabilization.

The relative stability of the $4n + 2$ compounds can be justified by the symmetry and ordering of the MOs. The $4n + 2$ electrons donated from $2p$ orbitals construct $4n + 2$ MOs in the π system, and likewise $4n$ electrons generate $4n$ MOs. These MOs increase in energy with the number of vertical nodal planes they contain: the lowest energy π orbital has no vertical nodes, the next highest has one, and so on. The lowest and highest energy orbitals in the π system are always non-degenerate, while the orbitals in between are doubly degenerate.[10]

This arrangement of orbitals, pictured for benzene and for cyclobutadiene in Figure 6.19, means that $4n + 2$ compounds have just enough electrons to fill all the orbitals with net bonding character. The highest occupied molecular orbital, or **HOMO,** in benzene is a doubly degenerate e_{1g} orbital. Depending on how the nodes are drawn, there are either (a) four bonds and two nodes through bonds (which contribute antibonding character) or (b) two bonds and nodes through two atoms (which don't affect the bonding character). Either way, there is more bonding character than antibonding character. The $4n$ compounds like cyclobutadiene, on the other hand, unload their highest-energy electrons by half-filling a doubly degenerate orbital that has *as much antibonding character as bonding character.* The e_g HOMO in cyclobutadiene has two bonds, but these are counterbalanced by nodes through the other two bonds. Unlike the HOMO in $4n + 2$ molecules, the HOMO in the $4n$ molecule does *not* stabilize the molecule. Consequently, the $4n + 2$ molecules are more stable by comparison.

A further analysis of these systems results in another valid prediction. Cyclobutadiene has only two electrons in the doubly degenerate HOMO. By stretching two opposite bonds and tightening the other two, the molecule can reduce the symmetry of the molecule to D_{2h} symmetry. This splits the e_g orbital

[10]If a non-degenerate orbital has any nodes, there must be enough nodes to divide every bond in the ring. Otherwise the orbital cannot have a unique eigenvalue for the operation $\hat{C}_n$, and it must then have some degeneracy. If there are six atoms in the ring, but only two vertical nodes, the $\hat{C}_6$ rotation will have a different effect on the sign of the wavefunction at different atoms.

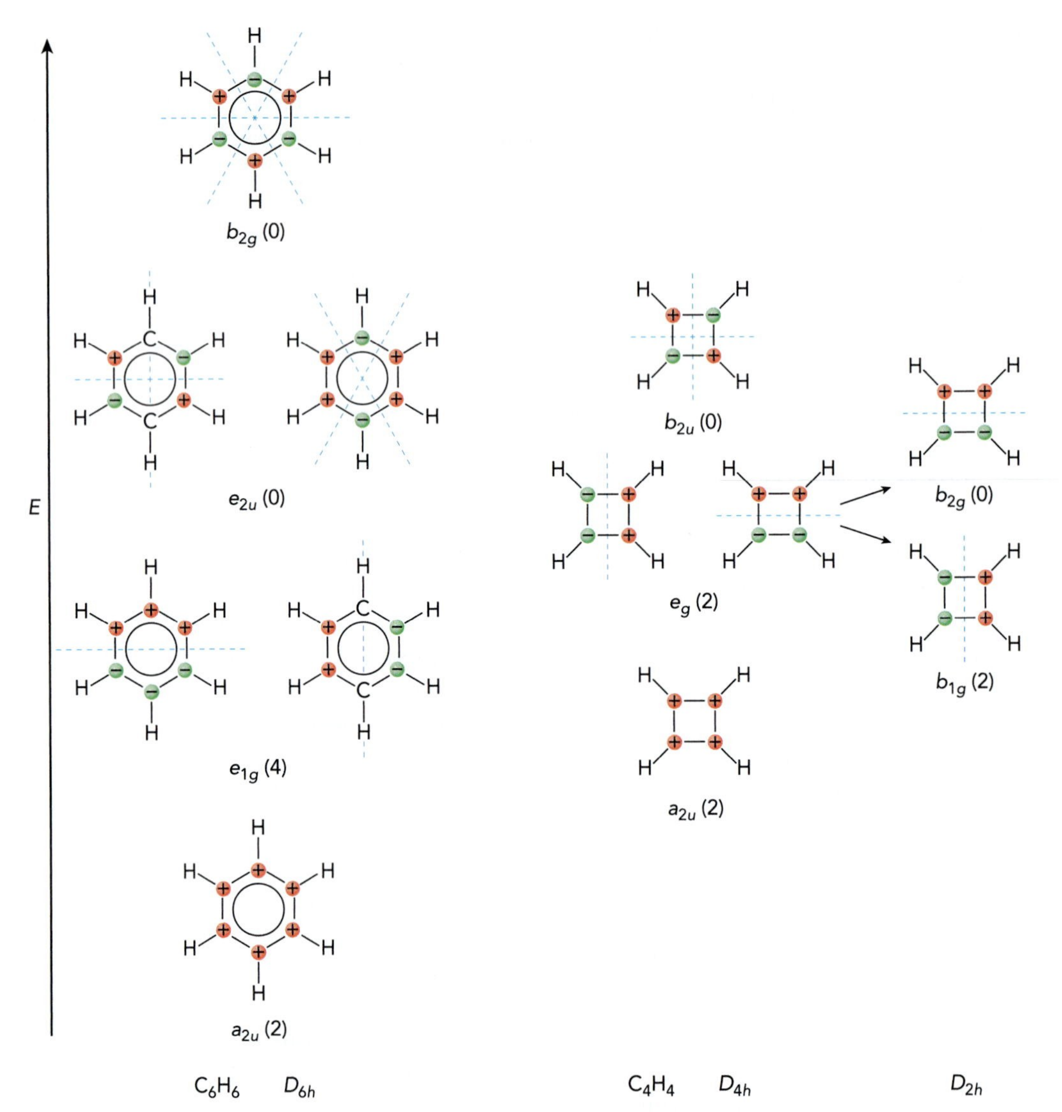

▲ **FIGURE 6.19 The π system MOs of cyclic C_6H_6 and C_4H_4.** The signs indicate the phase of the top part of the $2p$ orbital at a given carbon atom; those carbon atoms with no sign do not contribute to that MO. The number of electrons that go into each orbital in the ground state is given in parentheses. The C_4H_4 e_g orbital would therefore be only half filled with two electrons in the D_{4h} geometry, so the molecule deforms to a D_{2h} geometry that stabilizes those two electrons.

into two non-degenerate orbitals: one at lower energy (b_{1g}) and one at higher energy (b_{2g}). Since there are only two electrons in the e_g to begin with, the more stable b_{1g} orbital (which puts the nodes through the longer, weaker bonds) can be filled, making the D_{2h} form more stable than the D_{4h} form. Anti-aromatic compounds (those with $4n$ electrons in the cyclic π system) tend to reject highly symmetric geometries in order to split a degenerate HOMO into distinct bonding and antibonding orbitals.

Incidentally, the degeneracies in polyatomic molecules share the same origin as the degeneracies of diatomic MOs and atomic orbitals: angular momentum. When the molecule has a proper rotation axis with $n > 2$, the axes perpendicular to the $\hat{C}_n$ axis are equivalent. This is not an obvious point, but for now we can see this from the way x and y are always paired together in the character tables for those point groups that have a $\hat{C}_{n>2}$ axis. This equivalence allows the projection of the angular momentum onto the $\hat{C}_n$ axis to be well-defined. (Whereas if x and y were not equivalent, the value of L_z would change as the angular motion took place, and we would not be able to assign a fixed value of L_z to any wavefunctions.) But because the energy of the state cannot depend on the direction of motion around the axis, we will find two states with the same energy, represented in the character tables by a degenerate representation. The degeneracy may be higher in the cubic groups, where multiple $\hat{C}_{n>2}$ axes exist, and so angular momentum around different axes may yield several distinct states with the same energy and symmetry properties.

As one example, the lowest energy a_{1g} orbital of the π-electron system in benzene is analogous to a linear molecule's σ MO, having no angular momentum; the doubly degenerate e orbital of benzene is similar to the π orbital of a linear molecule and has angular momentum. Classically, this angular momentum arises from circular motion of the electrons around the ring; the motion can be either clockwise or counterclockwise, resulting in two equal but opposite vectors for the angular momentum.

Molecular symmetry offered a relevant diversion from our major task at hand, but the task hasn't disappeared. We still need to solve, to within reasonable approximations, the molecular Hamiltonian. Now we return to that challenge, but this chapter gives away much of the philosophy behind our approach. We will simplify the electronic and vibrational and rotational degrees of freedom by first examining in each case the highest symmetry molecules—homonuclear diatomics—and will only then (and rather reluctantly) move on to the more ornery cases of asymmetric systems. Aside from its utilitarian value, group theory should be appreciated for its ability to elicit some of the native beauty of molecular structure, which is reason enough to study chemistry.

CONTEXT *Where Do We Go From Here?*

The principles of symmetry summarized in this chapter provide a powerful tool for predicting and understanding the behavior of molecules at every level. With the mathematics available, we can return to the molecular Schrödinger equation first presented in Chapter 5 and begin to deal with the results at each level, beginning with the most quantum-mechanical (the electronic states, Chapter 7) and proceeding through increasingly classical degrees of freedom (vibrations and rotations, Chapters 8 and 9) until at last we arrive at the translational motions that allow molecular interactions, the topic of Part III.

KEY CONCEPTS AND EQUATIONS

6.1 Group theory.

a. Symmetry elements are operators that manipulate the coordinate labels of a molecule but leave the distribution of indistinguishable atoms the same. A point group is a set of symmetry elements for a particular structure, in which the symmetry elements are chosen from the identity $\hat{E}$, inversion $\hat{I}$, reflections $\hat{\sigma}$, proper rotations $\hat{C}_n$, and improper rotations $\hat{S}_n$.

b. Wavefunctions built on a molecule that belongs to a particular point group may have nodes that cause the wavefunction to change sign or change in other ways when operated on by the symmetry elements. Symmetry elements operate on a particular distribution in an eigenvalue equation. The eigenvalues are collected into a character table for the point group, and each distinct set of eigenvalues corresponds to a particular irreducible representation.

6.2 Symmetry representations for wavefunctions.

a. We may have functions that are composed of several different irreducible representations. To break a function down (decompose the function) into its component representations, we find the values of the a_i's in the expression

$$\Gamma = a_1\Gamma_1 \oplus a_1\Gamma_2 \oplus \ldots \oplus a_P\Gamma_P.$$

Each of those a_i's can be determined from the equation

$$a_i = \frac{1}{Q}\sum_{j=1}^{Q}\chi[\hat{A}_j]\chi_i[\hat{A}_j]^*. \tag{6.7}$$

6.3 Selection rules.

a. The transition strength for a transition from an initial state with wavefunction ψ_i to a final state with wavefunction ψ_f is given by the integral

$$\mathrm{I} \propto \left|\int \psi_f^* \hat{\mu}_t \psi_i d\tau\right|^2, \tag{6.10}$$

where $\hat{\mu}_t$ is the transition moment operator. The transition moments can be identified in the character tables by the Cartesian coordinates in the *functions* column: x, y, or z signify electric dipole transitions, whereas x^2, xy, or other quadratic terms signify Raman transitions.

b. The transition is allowed by symmetry selection rules if the symmetry representation Γ_i for the initial state and Γ_f for the final state obey the relations

$$\Gamma_\mu \otimes \Gamma_i \ni \Gamma_f \tag{6.13}$$

$$\Gamma_f \otimes \Gamma_i \ni \Gamma_\mu \tag{6.14}$$

where Γ_μ is the representation of the transition moment operator.

KEY TERMS

- A **group** is a collection of operators such that the product of any two is the same as one of the operators already in the group.
- **Group theory** is the branch of mathematics that studies the properties of groups.
- A **point group** is a group of symmetry operators that leave the center of mass unchanged.
- A **symmetry element** is a symmetry operator that leaves a molecular geometry unchanged, except perhaps for the exchange of equivalent nuclei.
- The **identity operator** leaves a structure completely unchanged.
- The **inversion operator** moves every point (x,y,z) to the location $(-x,-y,-z)$.
- A **proper rotation operator** turns the molecule by a fixed angle about a particular axis.
- If proper rotation is a symmetry element of a molecule, the axis of that proper rotation is a **symmetry axis** of the molecule.
- The **principal rotation axis** is the symmetry axis having the smallest possible proper rotation as a symmetry element.
- A **reflection operator** generates a mirror image of the molecule.
- A **horizontal mirror plane** is a mirror reflection plane that lies perpendicular to the principal rotation axis, whereas a **vertical mirror plane** is any reflection plane that contains the principal rotation axis.
- An **improper rotation operator** is a proper rotation followed by a horizontal mirror reflection.
- The **character table** for a point group lists all the simplest sets of eigenvalues for operations by the symmetry elements of a point group.
- An **irreducible representation** is any of the simplest possible sets of responses to the symmetry elements in a group.
- The **totally symmetric representation** is the representation of any function that is completely

unchanged by any of the symmetry elements in a group. It is the only representation that appears in every character table.

- A **reducible representation** is composed of two or more irreducible representations summed together.
- **Symmetry orbitals** are molecular orbitals that have been constructed specifically to correspond to the irreducible representations of the molecule's point group.
- The **cubic point groups** are point groups containing symmetry elements for proper rotation by less than 180° around more than one axis.
- In **Rayleigh scattering,** a photon interacts with matter to alter its direction of motion, but the photon energy is unchanged.
- In **Raman scattering,** photons striking a sample are redirected with energies either greater **(anti-Stokes transition)** or less **(Stokes transition)** than the original photon energy.
- The **HOMO** is the highest occupied molecular orbital in a molecule, and is often related to the **LUMO,** the lowest unoccupied molecular orbital.

OBJECTIVES REVIEW

1. *Carry out algebraic manipulations of point group symmetry operators.*
   Find the single operator equal to $\hat{I}\,\hat{\sigma}_{xy}$.
2. *Determine the point group and symmetry elements of a molecule.*
   Find the point group and list all the symmetry elements of 1,1-dibromoethene.
3. *Find the symmetry selection rules for electric dipole and Raman transitions in a molecule.*
   What are the possible states in symmetry-allowed Raman and electric dipole transitions from the $^1A_1$ state of 1,1-dibromoethene?
4. *Assign a molecular orbital to its symmetry representation.*
   What is the symmetry representation of the π bond in *trans*-1,2-dibromoethene?

PROBLEMS

Discussion Problems

6.1 What are all the possible point groups of the non-linear, planar molecules?

6.2 Identify the point group of each of the molecules shown in the box on top of the next page.

6.3 Why is there no point group S_m where m is an odd number?

6.4 Among all the point groups, D_{2d} is the only one that has degenerate representations without having a single $\hat{C}_n$ symmetry element where $n > 2$. Why does D_{2d} have a degenerate representation when C_2, C_{2v}, C_{2h}, D_2, and D_{2h} have only non-degenerate representations?

Symmetry Elements

6.5 Consider the following functions:

a. $f(x,y,z) = e^{-r^2/a^2}$

b. $f(x,y,z) = \left[e^{-(z-R/2)^2/a^2} + e^{-(z+R/2)^2/a^2}\right]\dfrac{x}{\sqrt{x^2+y^2}}$

c. $f(x,y,z) = \left[e^{-(z-R/2)^2/a^2} + e^{-(z+R/2)^2/b^2}\right](x^2+y^2)$

d. $f(r,\theta,\phi) = re^{-r^2/a^2}\cos\theta e^{i\phi}$

where $r = \sqrt{x^2+y^2+z^2}$, $\cos\theta = \dfrac{z}{\sqrt{x^2+y^2}}$, and $\cos\phi = \dfrac{x}{\sqrt{x^2+y^2}}$, and a and b are constants. Find the symmetry of each of the functions with respect to (*i*) inversion, (*ii*) reflection in the *xy* plane and reflection in the *xz* plane, and (*iii*) rotation by π around the z axis.

6.6 Prove that

a. $(\hat{S}_4)(\hat{S}_4) = \hat{C}_2$. b. $\hat{I} = \hat{\sigma}_{xy}\hat{\sigma}_{yz}\hat{\sigma}_{xz}$.

6.7 Solve for $\hat{A}$ in the following equation for symmetry operators:

$$\hat{A}\hat{S}_4(z) = \hat{I}.$$

6.8 Obtain the multiplication table for the point group C_{2h}. (A sample is seen in Table 6.2 for C_{2v}.)

6.9 One corner of a multiplication table for the group D_{4h} is left blank. For convenience, we have labeled the four dihedral $\hat{C}_2$ rotation axes by v, w, x, and y, as shown. The v and w axes (like x and y) do not move when we carry out an operation. Fill in the missing entries. (You may choose

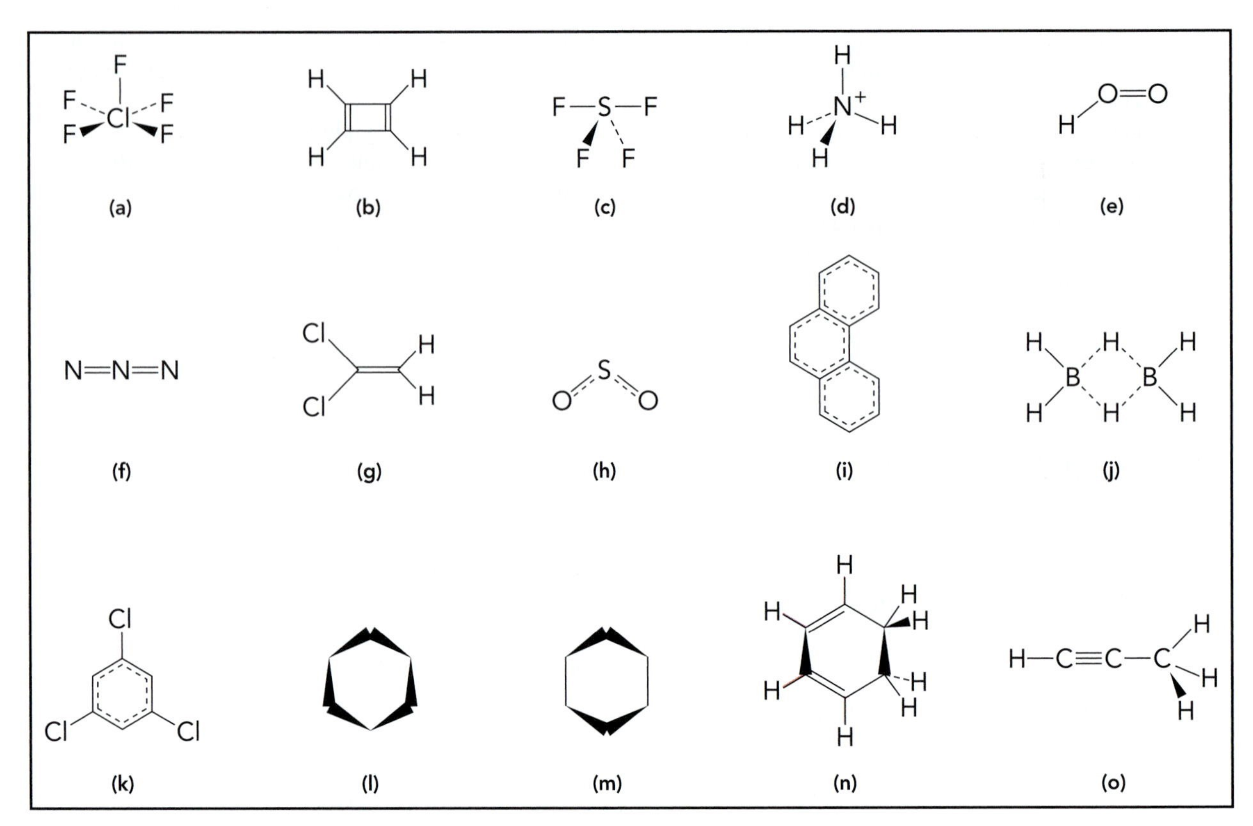

any direction of rotation, and you may choose which operation is carried out first, but be consistent.)

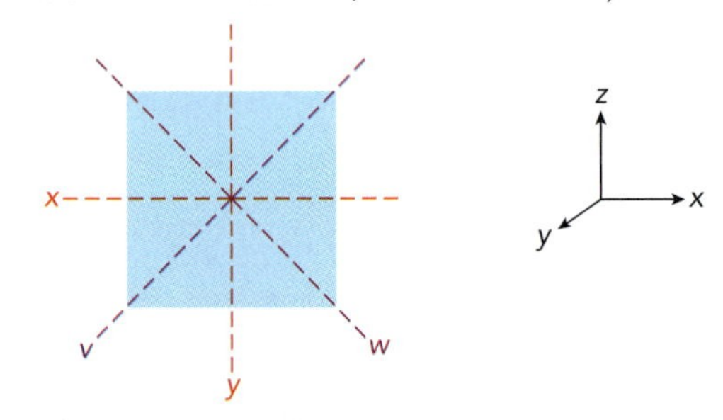

D_{4h}	$\hat{E}$	$\hat{C}_4(z)$	$\hat{C}_2(w)$	$\hat{I}$
$\hat{E}$				
$\hat{C}_4(z)$				
$\hat{C}_2(w)$				
$\hat{I}$				

6.10 Any combination of successive symmetry operations from one group must be identical to a single operation by some element of the same group. Find the single point group operation that is equal to the following,

a. $\hat{S}_4\hat{C}_4\hat{I}$ b. $\hat{I}\,\hat{C}_3$

c. $\hat{\sigma}_{xy}\hat{E}\hat{I}\,\hat{C}_2(z)$ d. $\hat{I}\,\hat{C}_4\hat{\sigma}_h$,

where the $\hat{C}_4$ axis is the principal axis.

6.11 In the group D_{3h}, find a combination of any two symmetry elements *except* $\hat{E}$ that performs the same operation as $\hat{\sigma}_v$.

6.12 In the expression $\hat{C}_3\hat{\sigma}_v = \hat{A}\hat{\sigma}_v\hat{C}_3$, solve for the symmetry operation $\hat{A}$, using the labels for the individual operations given here.

6.13 Using the accompanying figure to identify specific rotation axes and mirror planes that don't line up with the Cartesian axes, find the single symmetry operator that is equal to $\hat{C}_6(z)\hat{C}_2(x)$. So, for example, $\hat{\sigma}_r$ is reflection through the plane perpendicular to the page through the r axis, and $\hat{C}_2(r)$ is a 180° rotation about the r axis. Note that the r, s, t and u axes do *not move* under any operation; they are fixed axes in the coordinate system, like x.

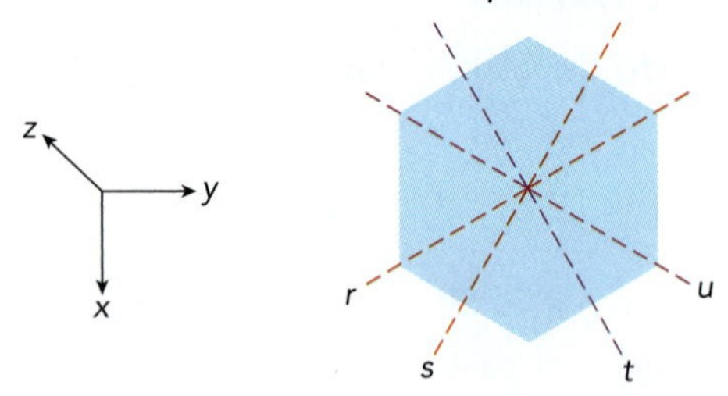

6.14 The symmetry elements of the point group D_{6h} can be identified by the labels for the axes and planes shown. For example, $\hat{C}_2(4)$ corresponds to a $\hat{C}_2$ rotation around axis 4 (the x axis), while $\hat{\sigma}_1$ is reflection through the vertical mirror plane that contains the z axis and axis 1 (the yz plane). Find the single symmetry operator that is equivalent to

$$\hat{C}_6(z)\hat{C}_2(2)\hat{C}_2(3).$$

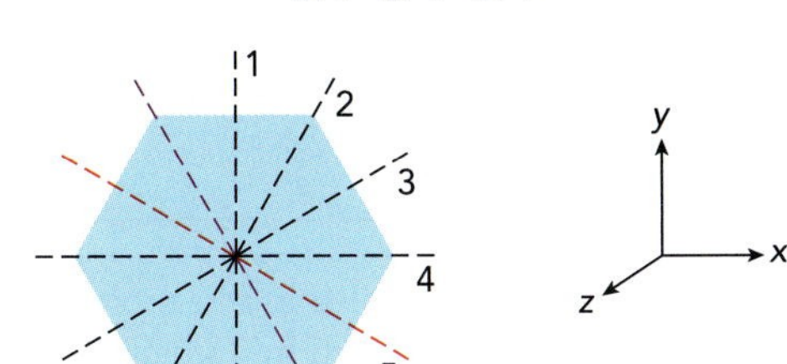

6.15 The following triangles are labeled **A** on the top and **B** (not shown) on the bottom. Draw the result of carrying out each specified operation. (Direction of rotation is up to you, but be consistent.)

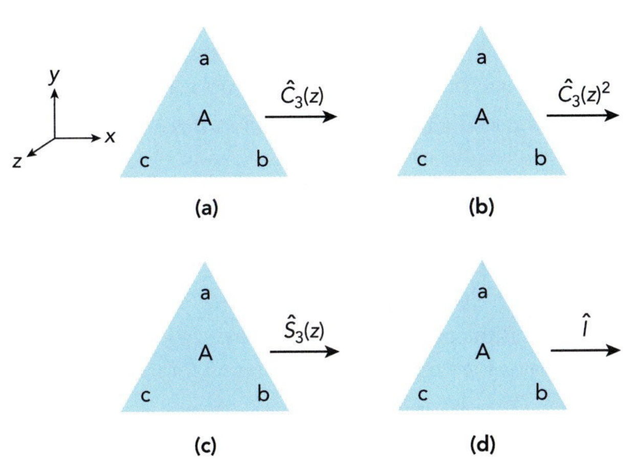

6.16 What two *single* operations can carry out the transformations shown on the following cube? If you need to identify an axis or plane, use the coordinate system provided.

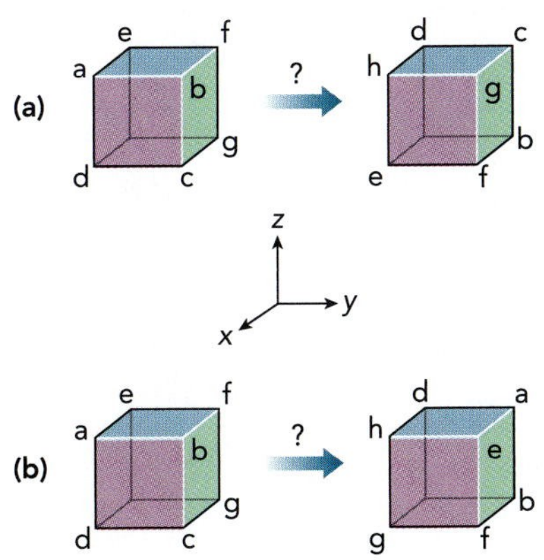

6.17 If two symmetry operators $\hat{A}$ and $\hat{B}$ commute, then $\hat{A}\hat{B}\psi$ is equal to $\hat{B}\hat{A}\psi$. For each pair $\hat{A}$ and $\hat{B}$ in the following table, put "Y" in the last column if the operators commute. If the operators do not commute, instead write the operator $\hat{C}$ such that

$$\hat{A}\hat{B}\psi = \hat{C}\hat{B}\hat{A}\psi.$$

$\hat{A}$	$\hat{B}$	
$\hat{\sigma}_{xz}$	$\hat{C}_3(z)$	
$\hat{E}$	$\hat{S}_4(z)$	
$\hat{I}$	$\hat{C}_2(z)$	

6.18 Two operators $\hat{A}$ and $\hat{B}$ commute if $\hat{A}\hat{B} = \hat{B}\hat{A}$. The symmetry elements of the point group D_{2h} are $\hat{E}$, $\hat{C}_2(z)$, $\hat{C}_2(y)$, $\hat{C}_2(x)$, $\hat{\sigma}_{xy}$, $\hat{\sigma}_{xz}$, $\hat{\sigma}_{yz}$, and $\hat{I}$. List all the operators from this set that commute with $\hat{C}_3(y)$.

6.19 Point group operators are not always commutative, meaning that for two operators $\hat{A}$ and $\hat{B}$ it may be that $\hat{A}\hat{B}\psi \neq \hat{B}\hat{A}\psi$. But are the point group operators always associative? In other words, is it necessarily the case that $(\hat{A}\hat{B})\hat{C}\psi = \hat{A}(\hat{B}\hat{C})\psi$?

6.20 For every operator $\hat{A}$, there is also an *inverse* operator $\hat{A}^{-1}$ in the point group that undoes that operation, such that

$$\hat{A}^{-1}\hat{A}\psi = \psi.$$

For example, each of the operators $\hat{C}_2$, $\hat{\sigma}$, and $\hat{I}$ is its own inverse, because $\hat{A}^2\psi = \psi$ (or, to put it another way, $\hat{A}^2 = \hat{E}$). What operator is the inverse of $\hat{S}_3$?

6.21 Find the eigenvalue when the angular momentum vector $\vec{L}$ is acted on by the symmetry operator $\hat{S}_2$.

Point Groups

6.22 Molecules are never rigid, as we learn in Chapter 8. For this reason, determining the point group of a molecule is not always as straightforward as we have assumed in this chapter. For example, the methyl group in toluene rotates about the CC single bond, and the apparent point group for the molecule depends on its orientation. A rule of thumb is that one should assign the molecule to the point group that corresponds to the *average* structure of the molecule, where the averaging is done over the time scale of the experiment. (For example, a UV/vis spectrum of toluene would generally be fast compared to the methyl rotation, but an NMR spectrum would be slow.) Identify the point group for toluene (a) in the conformation shown in the accompanying figure *and* (b) assuming that the methyl group rotates fast enough to appear like a perfect cone attached to the ring.

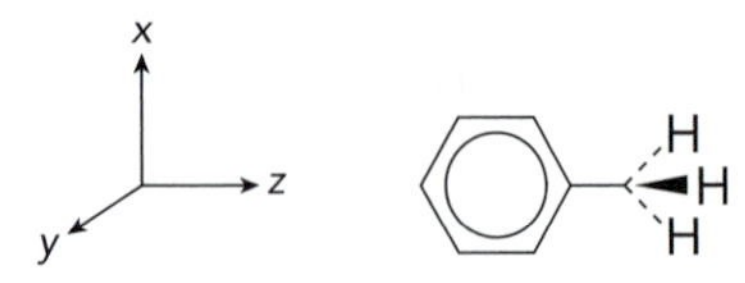

6.23 Find the point group for each the following household furnishings, as shown: (a) a rectangular table, (b) a simple chair.

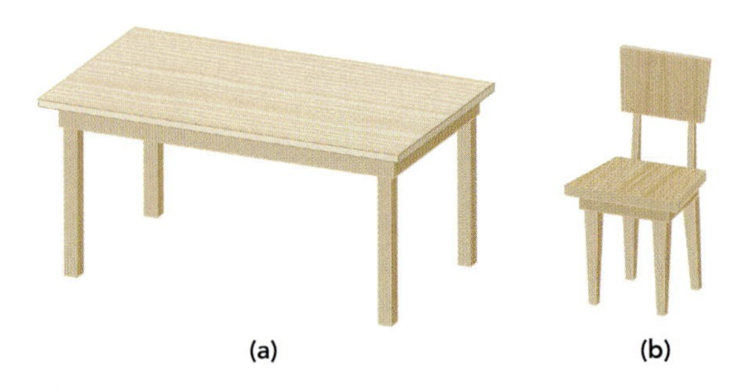

6.24 Group theory can be applied not only to small molecules but also to large ones. The structure shown at the bottom of this page is that of the protein Wza viewed from three angles, as given in the literature (*J. Biol. Chem.* **279** 28227-28232, 2004). Identify its point group.

6.25 List all the point groups possible for the fluorobenzenes, C_6H_5F, $C_6H_4F_2$, $C_6H_3F_3$, $C_6H_2F_4$, C_6HF_5, and C_6F_6, with any arrangement of the F atoms.

6.26 Find the point group of the isotopically substituted *closo*-$B_6H_4D_2$ dianion drawn next. The B_6H_6 dianion is octahedral, but the isotopic substitution spoils the octahedral symmetry in vibrational spectra.

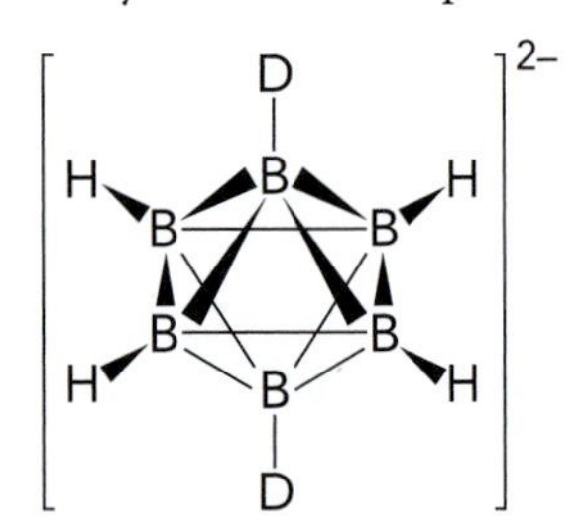

6.27 Provide *short* answers to these questions about some features of the point group C_{4v}.

C_{4v}	$\hat{E}$	$2\hat{C}_4$	$\hat{C}_2$	$2\hat{\sigma}_v$	$2\hat{\sigma}_v$	Functions
$A_1(a_1)$	1	1	1	1	1	$z; x^2 + y^2, z^2$
$A_2(a_2)$	1	1	1	−1	−1	R_z
$B_1(b_1)$	1	−1	1	1	−1	$x^2 - y^2$
$B_2(b_2)$	1	−1	1	−1	1	xy
$E(e)$	2	0	−2	0	0	$(x, y); (xz, yz); (R_x, R_y)$

a. Is the transition from a b_2 orbital to an e orbital allowed by electric dipole selection rules? by Raman?

b. In some point groups, your choice of the x and y axes can affect which representation label you assign to a particular function. Is this one of them?

c. What is the *minimum* number of atoms possible in a molecule that belongs to this point group?

6.28 If $\hat{C}_4(z)$ and $\hat{\sigma}(xy)$ and $\hat{I}$ are members of a point group, what is the *minimum* set of other operations that must also belong to that group? THINKING AHEAD ▶ [What property makes a set of operators a group?]

6.29 Consider an electron trapped in a three-dimensional box with lengths a = 1.0 Å, b = 2.0 Å, and c = 4.0 Å along x, y, and z, respectively. The wavefunctions are given by the equation

$$\psi(x,y,z) = \left(\frac{8}{abc}\right)^{1/2} \sin\left(\frac{n_x \pi x}{a}\right) \sin\left(\frac{n_y \pi y}{b}\right) \sin\left(\frac{n_z \pi z}{c}\right).$$

a. What is the point group for the box?

b. What is the symmetry representation for the ground state wavefunction?

c. What is the symmetry representation for the lowest excited state $(n_x, n_y, n_z) = (1,1,2)$ wavefunction?

d. Write the selection rule for Δn_z for electric dipole-allowed transitions in this system; in other words, what are the allowed values for Δn_z, assuming $\Delta n_x = \Delta n_y = 0$?

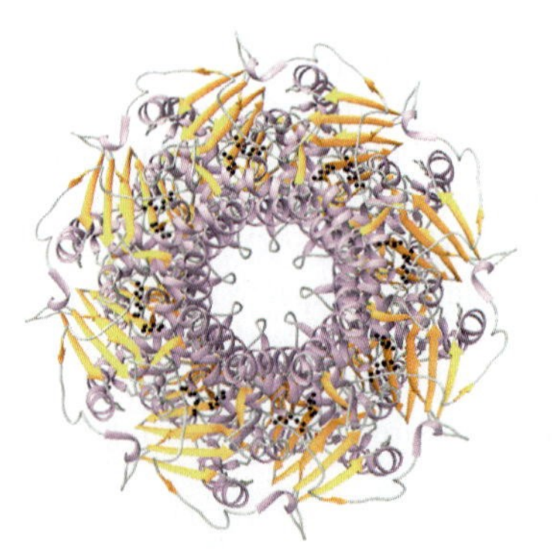

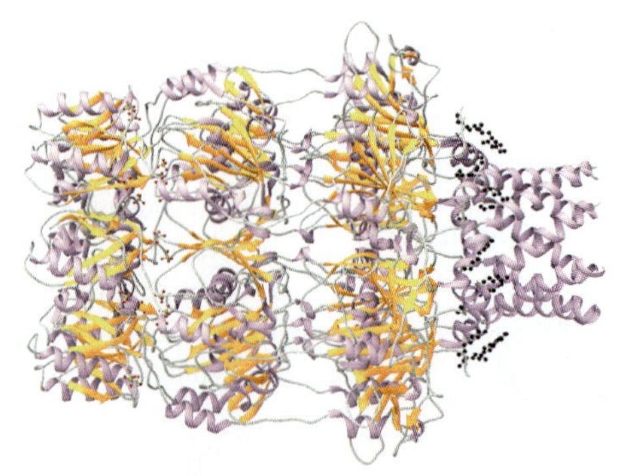

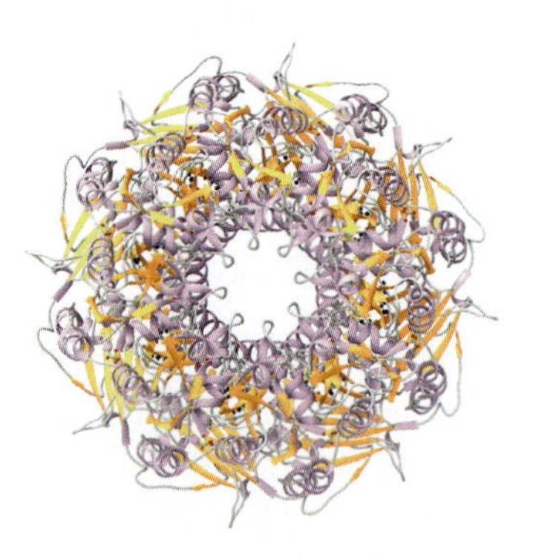

6.30 A chiral molecule is one that cannot be superimposed on its mirror image. State how one would recognize a chiral molecule by its point group, and give examples of three such point groups.

6.31 The point group I is listed in Table 6.1, but its character table is not among those given in this chapter. Write the character table for the point group I, given the essential symmetry elements listed in Table 6.1 and the character table for the similar point group I_h. Keep in mind that the order of the group must be reflected in the number of symmetry elements and the degeneracies of the irreducible representations.

6.32 In many applications of quantum mechanics, the **parity** of the wavefunctions—their eigenvalue under operation by the inversion operator $\hat{I}$—is a conserved quantity. However, we may label some quantum states by assigning a group of quantum numbers that does not preserve the parity. For example, in linear molecules with unpaired electrons, there may be angular momenta arising simultaneously from the net electronic orbital motion $\overrightarrow{L}$ and from the molecular rotation. (We will neglect the electron spin in this problem, although it would normally be here as well.) It is conventional to label the sum of the rotational and orbital angular momenta by the vector $\overrightarrow{N}$, with projection quantum number Λ onto the molecular z axis. We can write the combined electronic and nuclear wavefunction then in the following general form:

$$\psi_{\Lambda,N,M_N} = \psi_0\, Y_N^{M_N}\, e^{i\Lambda\phi}.$$

In this equation, Y is the spherical harmonic rotational wavefunction for rotational quantum number N and projection M_N, $e^{i\Lambda\phi}$ is a phase factor that arises from the orbital motion of the electron about the internuclear axis, analogous to the $e^{im_l\phi}$ term in the one-electron orbitals, and ψ_0 contains all the other details of the electron distribution. The wavefunction in this problem does not conserve parity, however. Prove that the set of wavefunctions shown does have a fixed parity $p = \pm 1$, by operating on this formula with $\hat{I}$:

$$\psi_{\Lambda,N,M_N,p} = \psi_0\, Y_N^{M_N}\left[e^{i\Lambda\phi} + p(-1)^N e^{-i\Lambda\phi}\right].$$

THINKING AHEAD ▶ [Why does the original function *not* conserve parity?]

6.33 The solution to Problem 4.103 finds several nuclear spin states for $^{12}C_2{}^1H_4$ ethene. Because these differ only in relative orientations of the nuclear magnetic fields, conversion from one nuclear spin state of the molecule to another is slow, but it can occur. However, unless the inversion symmetry of the molecule is broken by a collision or other interaction, the u or g symmetry of the nuclear spin state is forbidden to change. For this reason, nuclear spin states of molecules with equivalent hydrogens are used in astronomy as an indicator of the collision conditions for the gas in a particular region of space. Each of the nuclear spin states in ethene may be assigned to an irreducible representation of the molecule's point group. Lay the molecule in the xy plane with the C—C bond along the x axis, to remove any ambiguity about how to label the representations. Next, label the hydrogen nuclei α (spin up) or β (spin down) in every way possible that corresponds to either an overall M_I value of 1 or an overall M_I value of 2. Now combine these as necessary to identify the irreducible representations Γ_{ns} that correspond to each of the distinct $M_I = 1$ and $M_I = 2$ quantum states of the molecule. (Our spatial symmetry operators can exchange equivalent nuclei by moving them through space, but they do *not* operate on the spin coordinate ω, so the m_I value of each nucleus is unchanged as it is rotated, reflected, or inverted.) You should then be able to associate each of these M_I components with an overall value of I for the molecule, assuming that the symmetry is the same for all the M_I components of a given nuclear spin state I. Finally, determine which pairs of these M_I states can interconvert in the absence of collisions—that is, while preserving parity.

Character Tables

6.34 Prove that any two MOs in a C_{2v} molecule having different representations (a_1, a_2, b_1, b_2) are orthogonal.

6.35 Find any algebraic combination of x, y, and z that has symmetry representation A_u in the point group D_{2h}.

6.36 Find the direct product of the following representations in the point group D_{3h}:

$$a'_1 \otimes a''_2 \otimes e'' =$$

6.37 Reduce the direct product $A_{2g} \otimes T_{2g} \otimes E_u$ in the point group O_h.

6.38

a. Evaluate and reduce the direct product $E' \otimes E''$ for the point group D_{3h}.

b. For the point group D_{2d}, evaluate and reduce the direct product $a_1 \otimes b_2 \otimes e \otimes e$.

6.39 Reduce the direct product $E'' \otimes E''$ in the point group D_{3h}.

6.40 Using the character table for D_{6h}, find the representation given by (a) $A_{1g} \otimes E_{1g}$; (b) $A_{1u} \otimes B_{1u}$; (c) $B_{2u} \otimes E_{2g}$; and (d) $E_{2u} \otimes E_{1g} \otimes B_{2g}$.

6.41 Prove that for the point group S_4 the function $x^2 - y^2$ belongs to the representation B.

6.42 The symbols R_x, R_y, and R_z in the functions column of the character tables correspond to rotation about the Cartesian axes. Draw a molecule with C_{2v} symmetry, draw the Cartesian coordinate axes, and then draw arrows to show how the atoms would move if you rotated the molecule in the plane of your drawing, about the axis perpendicular to the drawing. Write whether your rotation is R_x, R_y, or R_z, and then show that the symmetry of those arrows is given by the representation for the corresponding R function.

Wavefunction Symmetry

6.43 What is the symmetry representation for the CC σ-bonding orbital in ethane?

6.44 Write the correct MO symmetry labels for the following combinations of atomic orbitals, and indicate whether they are bonding or antibonding:

a. $1s_A + 1s_B$

b. $1s_A + 2p_{zB}$

c. $2p_{xA} - 2p_{xB}$

6.45 Find the correct symmetry (δ_u, π^*, etc.) label for the diatomic MO wavefunction given by

$$\psi(x,y,z) = 5z(x^2 + y^2)e^{-\sqrt{x^2+y^2+z^2}}.$$

Is it bonding or antibonding?

6.46 Write one properly symmetrized wavefunction in terms of r_A and r_B for each electron, including the spin wavefunctions α and β, for the $1\sigma_g 2\sigma_g$ MO configuration in H_2.

6.47 This problem concerns H_3^+.

a. If H_3^+ has the geometry of an equilateral triangle, what is its point group?

b. Write an equation for the lowest-energy molecular orbital wavefunction of H_3^+, labeling the nuclei A, B, and C and the electrons 1 and 2. Include a possible spin wavefunction, but do not symmetrize or normalize the wavefunction.

c. Write a properly symmetrized version of the same wavefunction.

d. Finally, what is the symmetry representation of this wavefunction?

6.48 Identify the symmetry representations of all the σ bond orbitals for the following molecules. If there is any ambiguity in the labels, show any coordinate choices that you make to resolve the ambiguity (for example, if you have decided which axis is the x axis, draw your coordinate system).

a. XeF_4 (square planar)

b. CH_2Cl_2 (tetrahedral)

6.49 The possible molecular orbital designations for a *planar* molecule in the point group C_{2h} are given in the character table. Consider the *lowest energy* MO for each of these representations and put these MOs in order of increasing energy.

6.50 For the molecule biphenylene shown, do the following:

a. Find the point group.

b. Find the irreducible representation of the lowest energy molecular orbital in the π system (having only one nodal plane, which contains all the nuclei).

Label the coordinate axes in the figure to show your choice of coordinates.

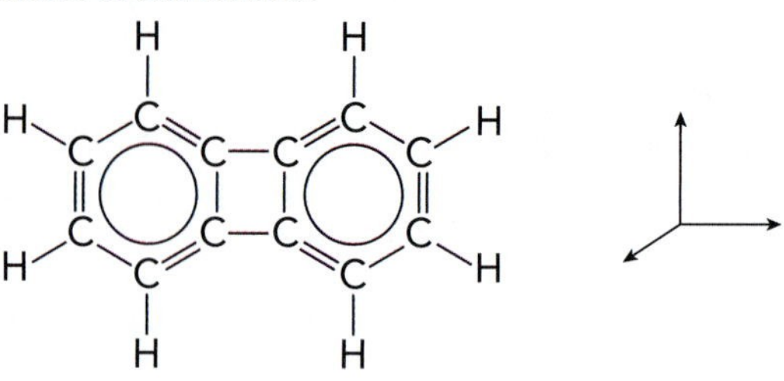

6.51 Identify the point group representations of the MOs associated with the CCl bond electrons in *trans*-dichloroethylene.

6.52 For each of the MOs for hexatriene drawn at the top of the next page, give the point group symmetry representation.

6.53 Drawn next are two of the MOs for Br_2. Assign their representations.

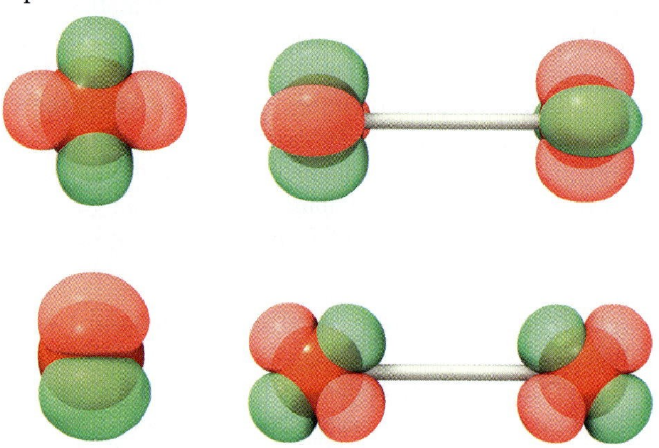

6.54 One possible set of angular wavefunctions for the four σ bond orbitals of square planar XeF_4 is as follows :

$$\psi_1 = A\sin\theta$$
$$\psi_2 = A\sin\theta\cos\phi$$
$$\psi_3 = A\sin\theta\sin\phi$$
$$\psi_4 = A\sin\theta\cos 2\phi.$$

Recall that θ is the angle measured from the z axis (the principal axis) and ϕ is the angle measured in the xy plane from the x axis.

a. Next to each of the equations, write the irreducible representation to which the orbital belongs. (One case depends on how the axes are drawn, and you may pick any of the valid representations.)

b. Write the irreducible representation for each of the corresponding probability densities:

i. $|\psi_1|^2$

ii. $|\psi_2|^2 + |\psi_3|^2$

iii. $|\psi_4|^2$

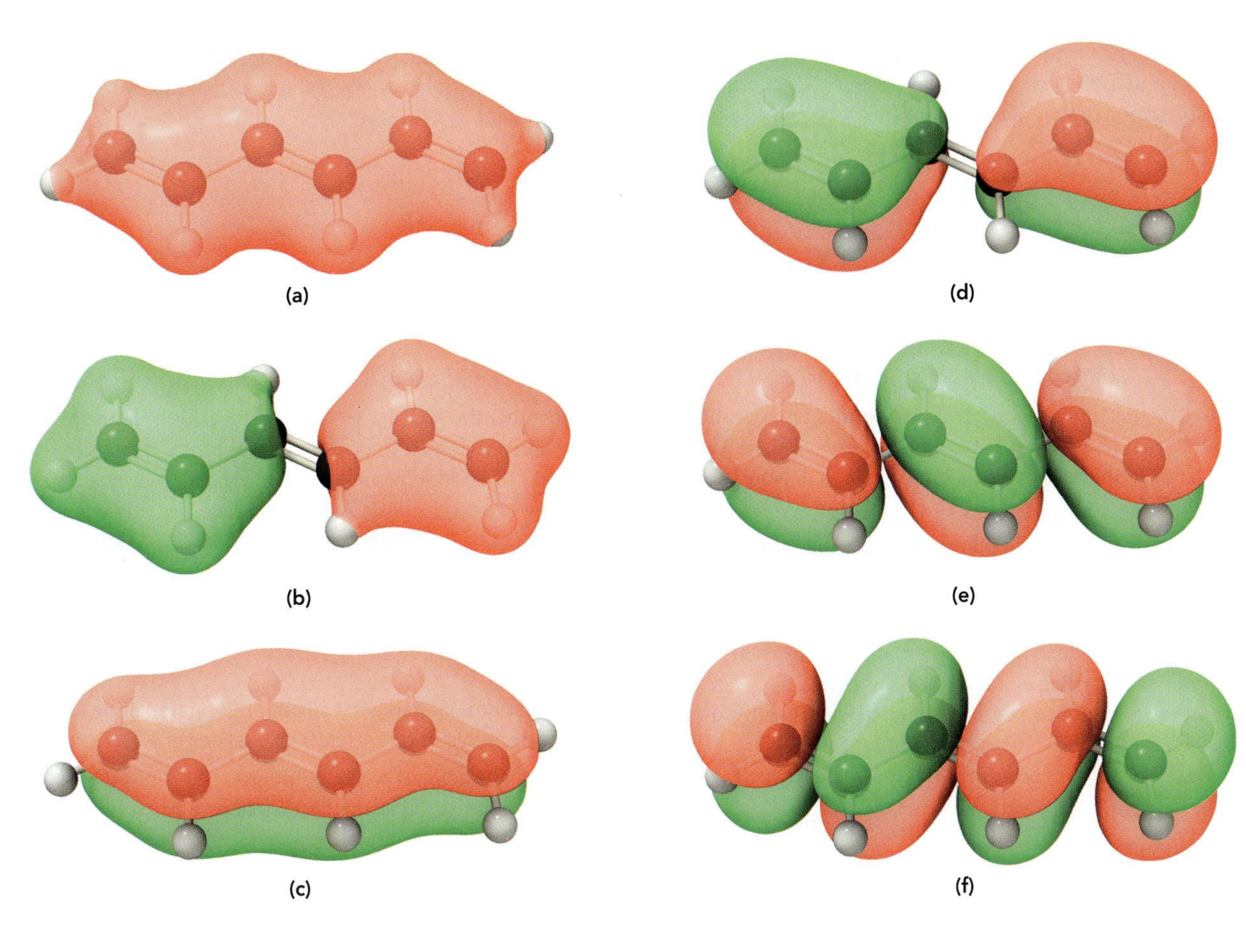

6.55 Cyanogen is a linear molecule with the structure NCCN. List the number of electrons in MOs of each symmetry representation in the ground state of this molecule. For example, for H_2O, there are six a_1 electrons, two b_1 electrons, and two b_2 electrons. Don't worry about the order of the energies.

6.56

a. Find whether the wavefunction ψ is symmetric, antisymmetric, or non-symmetric with respect to the operators $\hat{C}_2$ and $\hat{C}_4$, where

$$\psi(r,\theta,\phi) = (e^{-r_A/a_0} + e^{-r_B/a_0})\cos\theta\, e^{i\phi},$$

and where r_A and r_B are distances from identical nuclei A and B separated by a distance R, and where θ is the angle measured from the internuclear axis and ϕ the angle measured around the internuclear axis. Use the molecular center of mass as the origin.

b. Find the symmetry of the same wavefunction with respect to inversion.

c. Write the correct symmetry label for this wavefunction. Is it bonding or antibonding?

6.57 The lowest energy π bonding orbital of the planar polyphenyl shown at left in the accompanying figure has no nodes between the atomic p orbitals. An edge-on view of this orbital is shown at right. What is the symmetry representation of this orbital?

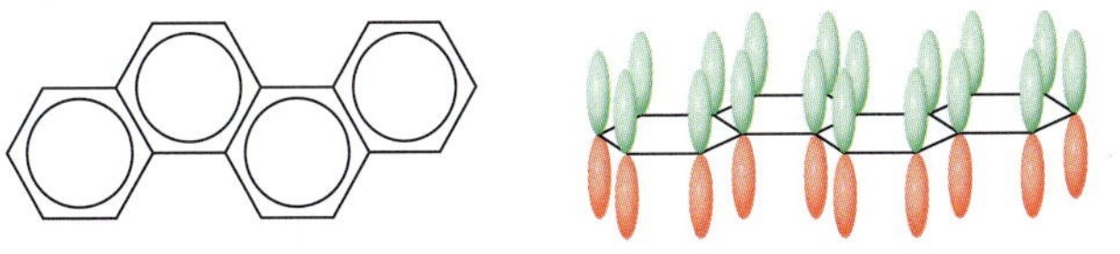

6.58 Find the MO representations for *all* the CC bond electrons in the planar molecule shown.

6.59 Correlate the electrons in the local bonding model of ground state NH_3 molecule to those in its MO configuration $1a_1^2 2a_1^2 3a_1^2 1e^4$.

6.60 Solve for the direct product $\Pi \otimes \Delta$ in the group $C_{\infty v}$. You can do this without using Eq. 6.7 if you use a little trigonometry.

6.61 A molecule with D_{4h} symmetry has an MO configuration $a_{1g}^2 e_u^2$. Find all the representations of the electronic states resulting from this configuration.

6.62 Give the symmetry representation for the MOs that correspond to the two π bonds of 1,3-butadiene in the conformation drawn next.

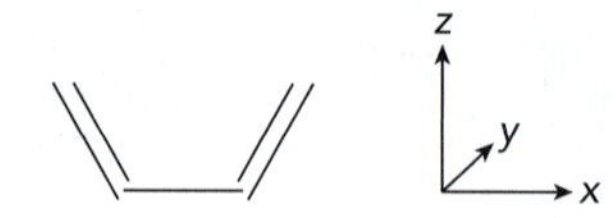

6.63 The molecular orbital configuration for ground state CO_2 is

$$1\sigma_g^2 1\sigma_u^2 2\sigma_g^2 2\sigma_u^2 3\sigma_g^2 1\pi_u^4 3\sigma_u^2 4\sigma_g^2 1\pi_g^4.$$

Describe this molecule using a local bonding model such as that used to describe C_2H_2, and then match the MO electrons to those in the local bonding model. Note that, unlike C_2H_2, there are lone pair electrons in CO_2.

6.64 The C_{2v} molecule cyclopropene has the MO configuration

$$(1a_1)^2(2a_1)^2(1b_1)^2(3a_1)^2(4a_1)^2(2b_1)^2(5a_1)^2(3b_1)^2(6a_2)^2(1b_2)^2(2b_2)^2,$$

where the nuclei lie in the xz plane and the representations are defined according to Table 6.3. Group these electrons according to their place in the local bond model of this molecule. For example, the first six electrons $((1a_1)^2(2a_1)^2(1b_1)^2)$ correspond to the carbon atom $1s^2$ cores.

Selection Rules

6.65 In the point group D_{3h}, is the transition that promotes an electron from an e' orbital to an a''_2 allowed by (*i*) electric dipole or (*ii*) Raman selection rules?

6.66 Boron trihydride, BH_3, has an MO configuration $1a'^2_1 2a'^2_1 1e'^4$, and the lowest unoccupied MO is the $1a''_2$. Is the transition from the ground electronic state of BH_3 to the lowest excited state of BH_3 allowed by (*i*) electric dipole or (*ii*) Raman selection rules?

6.67 Is the transition between the ground and lowest excited singlet states of cyclobutadiene allowed by electric dipole selection rules, Raman selection rules, both, or neither?

6.68 Give the possible upper state symmetries for transitions from the A_{1g} ground state of benzene according to (a) electric dipole selection rules and (b) Raman selection rules.

6.69 In the tetrahedral group T_d, determine whether the electronic transition ${}^1T_1 \rightarrow {}^1T_1$ is allowed by electric dipole selection and/or Raman rules.

6.70 The cyclopentadienyl radical has the structure drawn next. All the carbons are equivalent, and all the hydrogens are equivalent.

a. Find the point group for this molecule.

b. Show that the HOMO $\rightarrow$ LUMO transition (highest occupied to lowest unoccupied MO) in the π-bonding system of cyclopentadienyl must be allowed by electric dipole selection rules. (You do not need the character table for this point group to answer the question.)

6.71 Make a chart of allowed and forbidden transitions for the point group C_{2h} labeling the electric dipole-allowed transitions "ED," the Raman-allowed transitions "R," and the forbidden transitions "X."

6.72 What Raman and what electric dipole transitions are allowed from one of the A_2 electronic states of NH_3?

6.73 Ethylene belongs to the point group D_{2h}. List all representations for the states accessible by Raman transitions from the A_u-symmetry excited electronic state.

6.74 Determine whether the transition $E' \rightarrow E'$ is allowed (a) by electric dipole or (b) by Raman selection rules for the molecule BH_3.

6.75 We were able to write the angular momentum selection rules ($\Delta l = \pm 1, \Delta m_l = 0, \pm 1$) for atoms based on the idea that the photon has one unit of angular momentum. Using the same idea, what are the selection rules on $\Delta\lambda$ in linear molecules

a. for electric-dipole transitions?

b. for Raman transitions?

THINKING AHEAD ▶ [How does λ correspond to the atomic quantum numbers?]

6.76 Find one example of a transition forbidden by both electric dipole and Raman selection rules in the group D_{2h}.

7 Electronic States and Spectroscopy

LEARNING OBJECTIVES

After reading this chapter, you will be able to do the following:

1. Construct and interpret correlation diagrams for certain highly symmetric molecules.
2. Predict molecular orbital configurations for homonuclear diatomics and other simple molecules.
3. Determine selection rules for electronic transitions in molecules.
4. Use potential energy diagrams and selection rules to predict the processes resulting from electronic excitation of a molecule.

GOAL *Why Are We Here?*

The goal of this chapter is to describe the quantum states and energies that result from solving the electronic part of the molecular Schrödinger equation.

CONTEXT *Where Are We Now?*

Molecular symmetry provides tools that make it easier for us to describe the electronic and nuclear wavefunctions of the molecule. Now we return to the molecular Hamiltonian, and for this chapter we just look at the contributions to the electronic energy term E_{elec}, which determines the effective potential energy for our nuclei in Eq. 5.10:

$$\hat{H}\psi_{\text{nuc}} = [E_{\text{elec}}(R) + \hat{K}_{\text{nuc}}]\psi_{\text{nuc}} = E_{\text{nuc}}\psi_{\text{nuc}}.$$

SUPPORTING TEXT *How Did We Get Here?*

The main *qualitative* concept we draw on for this chapter is that we can solve the Schrödinger equation for the electrons in a molecule while we hold the atomic nuclei at fixed positions. We can do this at a number of positions to generate the potential energy curves seen in Chapter 5. We will also take advantage of the following equations and sections of text:

- Section 4.2 describes the variational principle that we use for approximate solutions to the molecular Schrödinger equation.
- Eq. 4.26 gives the key equation of Hartree's method,

$$\hat{H}(i)\varphi(i) = \left[-\frac{\hbar^2}{2m_e}\nabla(i)^2 - \frac{Ze^2}{4\pi\varepsilon_0 r_i} + \sum_{j\neq i}\left(\int|\varphi(j)|^2\frac{e^2}{4\pi\varepsilon_0 r_{ij}}d\tau_j\right)\right]\varphi(i) = \varepsilon_i\varphi(i),$$

which lets us solve for an orbital energy ε_i for each electron.

- Section 4.4 describes how the vector model of the atom can predict overall symmetries of many-electron atoms. A similar approach can be used to determine overall symmetries of molecular electronic states.
- Section 5.2 gives us (Eq. 5.13)

  $$\psi_+(r,\theta,R) = C_+(R)(1s_A + 1s_B) \quad \psi_-(r,\theta,R) = C_-(R)(1s_A - 1s_B)$$

  and

  $$\varphi(i) = \sum_{j=A}^{M} \sum_{n=1}^{n_{max}} \sum_{l=0}^{n-1} \sum_{m_l=-l}^{l} c_{nlm_l,j}\psi_{nlm_l,j}(i), \qquad 5.12$$

  $j = A, B, \ldots, M$ indicates which nucleus in the molecule is used as the center of the atomic orbital $\psi_{nlm_l,j}$.
- Sections 6.1 and 6.2 introduce the tools of symmetry analysis, which we use here for labeling MOs and electronic states, and also for building correlation diagrams to determine the energy ordering of the MOs.
- Example 6.6 shows how we can use our understanding of symmetry and molecular orbitals to deduce the MO configuration of a simple molecule.

7.1 Molecular Orbital Configurations

In Chapter 5, we examined chemical bonding from the standpoint of MO theory, valence bond theory, and local bonding, always focusing on the electrons involved in the bond formation: the valence electrons. To accurately predict the geometries and properties of molecules with many electrons, however, we have to take into account *all* of the electrons, including the core electrons.

The most common approach describes all the electrons using MO theory, which conveniently allows us to apply any point group symmetry of the molecule to the orbitals. Molecular orbitals offer the additional advantage of any orbital model; namely, the many-electron wavefunction can be expressed as a product of simpler one-electron orbital wavefunctions. Slater determinants (Section 4.3) then use combinations of these MO wavefunctions to assemble a properly antisymmetric many-electron wavefunction, but each term in the Slater determinant puts the electrons one at a time into a given set of orbitals. In other words, we can express the contributions of the many electrons in terms of a **molecular orbital configuration,** similar to the atomic orbital configurations used in Chapter 4. For example, our lowest energy state in H_2 has two electrons in a σ_g MO. Therefore, we write the MO configuration σ_g^2. The C_2H_2 system described in Example 6.6 has a molecular orbital configuration

$$1\sigma_g^2 1\sigma_u^2 2\sigma_g^2 2\sigma_u^2 3\sigma_g^2 1\pi_u^4.$$

(Recall that the number before the MO symbol "σ" or "π" only indicates relative energy, and is not necessarily related to the "n" value of the atomic orbitals involved.) One difficulty in writing electron configurations for atoms is finding the energy ordering: whether to fill the 4f before or after the 5d orbital, for example. The same difficulty exists in many-electron molecules. To write the MO configurations, we need to have some way of deducing the energy ordering of the MOs.

Correlation Diagrams for Diatomics

One useful tool for approaching this problem (though not necessarily solving it) is the correlation diagram. The correlation diagram takes a system such as our molecule and extrapolates it toward simplifying cases in two opposite directions, so that the system of interest lies in between. In the case of the MO configuration, we will use the limit of atomic orbitals—for which we do know the ordering of the energies—to provide clues to the energy ordering of our *molecular* orbitals.

Homonuclear Diatomics

We will first consider homonuclear diatomics, which belong to the point group $D_{\infty h}$. The molecular orbitals are the one-electron wavefunctions that result when we place the two atoms near one another. The distance R between the two nuclei lies between two extreme limits: the **united-atom limit,** when R is reduced to zero and the two nuclei merge, and the **separated-atom limit,** when R is infinite and the two atoms and their orbitals are independent of one another.

In Section 5.2, we introduced the idea of combining atomic orbitals to make an MO. The decision to use only $1s$ orbitals was easy; we wanted the lowest energy MOs so we used the lowest energy atomic orbitals available. But with larger molecules, we'll need to combine all sorts of atomic orbitals—how do we know which ones we can use?

In this chapter, we will require that the orbitals we form are symmetry orbitals, so that each orbital wavefunction corresponds to an irreducible representation in the molecule's point group. For the homonuclear diatomic MOs, we combine only orbitals of the same n, l and $|m_l|$ values on each of the two atoms, because the electron distributions around identical nuclei in the same molecule must be identical if the symmetry is to be preserved. We will not mix a $1s$ orbital on nucleus A with a $2p_x$ orbital on nucleus B, for example.

When we do that, we can take advantage of the symmetry principles and require that *the symmetry representation of the orbital stays unchanged as we vary all the distances between the atoms by the same factor*. For example, the H_2^+ ψ_+ orbital in Eq. 5.13 is formed from two $1s$ orbitals with the same phase. This pair of functions is symmetric with respect to inversion and reflection in either mirror plane. The angular momenta of both orbitals projected onto the internuclear axis are zero. All of these features remain if we multiply the distance between the two nuclei by a thousand, or divide it by ten. We can imagine, therefore, that we prepare a symmetry MO by starting with the atoms at some huge distance from one another. In this separated atom limit, we put atomic orbitals on each nucleus that correspond to some representation, say σ_g, and then we make the MO just by bringing the two atoms closer together.

In a sense, this is an application of our variational principle from Chapter 4: the lowest energy state of a particular *symmetry* can be found by the variational method. Say that we start with a set of separated atomic orbitals that has π_u symmetry. Then, as we change the distance between the atoms and try to adjust that wavefunction to minimize the energy, the variational method will still only give us a function of π_u symmetry, as long as we don't change the point group of the molecule.

The p orbitals have an angle-dependence that the s orbitals do not have, so there are different possible relative orientations, as well as different relative phases. The p_z atomic orbitals ($l, m_l = 1, 0$) have electron density along the z axis (the internuclear axis), and therefore combine to form two σ orbitals, as shown in Fig. 7.1. The p_x and p_y atomic orbitals (constructed from the $l, m_l = 1, \pm 1$ functions) have no electron density along the z axis, and they combine to form π_x and π_y orbitals. The s atomic orbitals, on the other hand, combine to form only σ orbitals.

Now let's consider the united atom limit. Setting the bond length R to zero makes the two nuclei overlap, forming a single atom.[1] What then is the single atomic orbital formed from the two original separated atomic orbitals? This atomic orbital must have the same $D_{\infty h}$ symmetry characteristics as the original MO, because changing R does not affect the symmetry with respect to the rotation, reflection, and inversion operators. The appearances of selected MOs in the united atom, molecular orbital, and separated atom cases are shown in Fig. 7.1. It often will not be obvious that a particular united atom orbital (the $4p_z$, for example) is formed from a particular pair of separated atom orbitals (in this case, the $2p_z$ orbitals). We are effectively using the

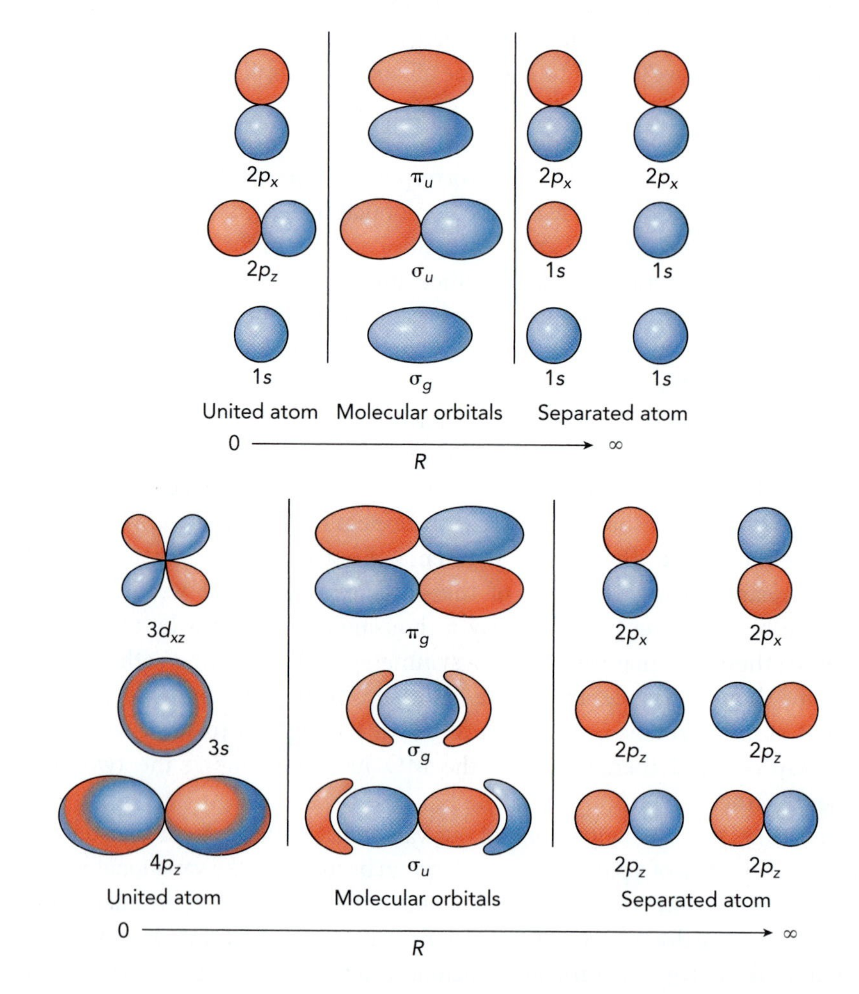

▶ FIGURE 7.1 **The principle of MO correlation, applied to homonuclear diatomics.** When separated atoms (on the right) are combined to form MOs and ultimately united atom orbitals (on the left), the symmetry of the set of orbitals remains constant with respect to inversion, rotation about the internuclear axis, or reflection through the mirror planes. Only the x component of the doubly degenerate π orbitals is shown here.

[1]We can forget about the nuclear–nuclear repulsion; we're only concerned with the electronic wavefunctions here.

variational principle (Section 4.2) to correlate the united and separated atom orbitals, and this means we mix orbitals to obtain the lowest energy wavefunction for each irreducible representation. For each line drawn on the correlation diagram, the symmetry characteristics shared by the united and separated atom limits are preserved everywhere in between. Our $4p_z$ united atom orbital isn't formed 100% from the $2p_z$ separated atom orbitals, but by a mixture of $2p_z$, $3p_z$, $1s$, and other orbitals, because all of these share the same symmetry properties: symmetric under $\hat{C}_\infty$, antisymmetric under $\hat{I}$ and $\hat{\sigma}_h$. This principle of preserving the symmetry from one end of the diagram to the other provides a first guess of the energy ordering of the MOs, because the united atom and separated atom relative energies are known, and the MO energies generally lie between these two extremes.

The chart that shows these energies as a function of R is one example of a **correlation diagram.** A schematic correlation diagram for homonuclear diatomics is shown in Fig. 7.2. A correlation diagram is a common tool in physics to help describe some complicated system by positioning it between two relatively simple limits.

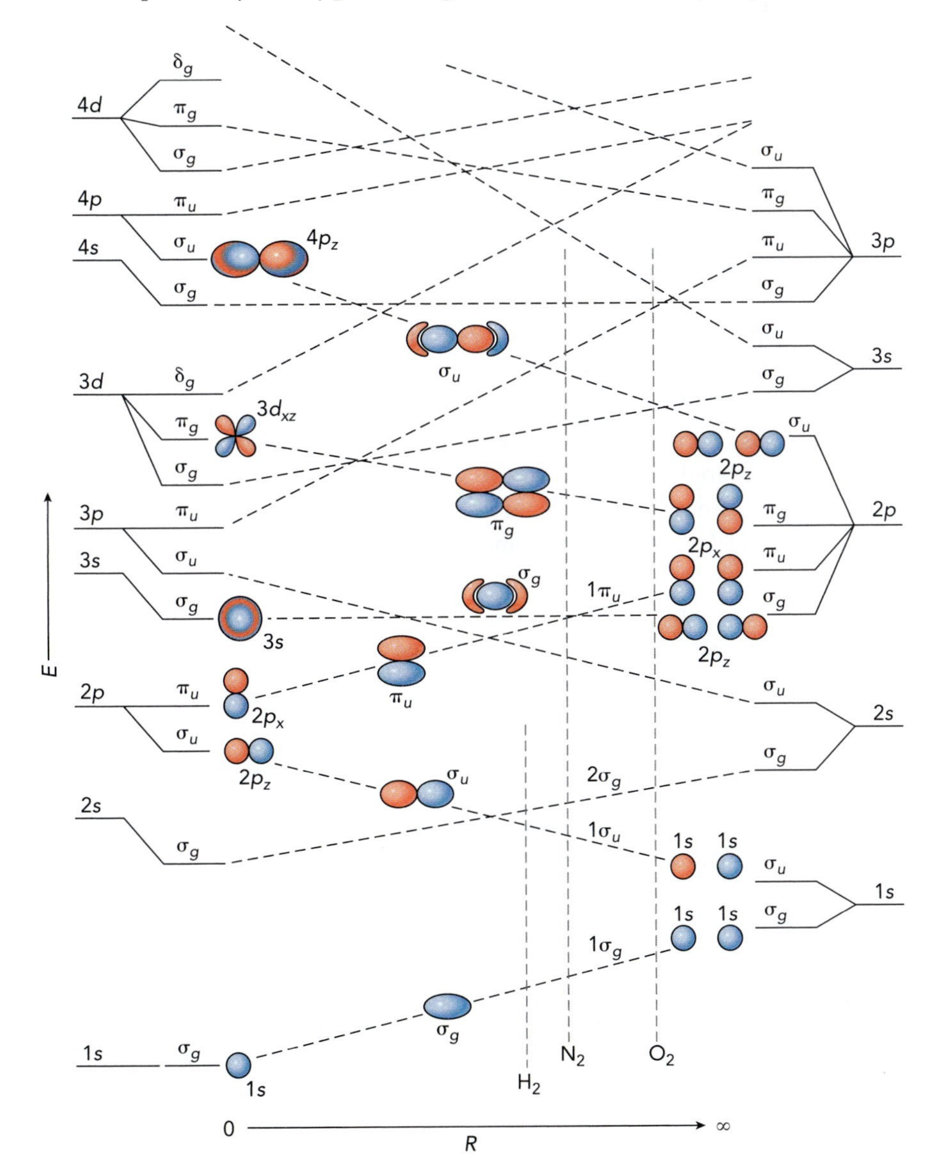

◀ **FIGURE 7.2 Schematic correlation diagram for the homonuclear diatomics.** Energies are not given to scale.

CHECKPOINT The correlation diagram is carrying out wholesale the same process used in Chapter 5 to combine the 1s orbitals on atoms A and B to make the ψ_+ and ψ_- MOs. On the right-hand side of the correlation diagram are always two separated atomic orbitals, and we bring them together with different relative phases (adding them or subtracting one from the other) to form all the molecular orbitals.

The left-hand side of Fig. 7.2 gives the united atom limit, with the atomic orbitals labeled. Also given is the λ symmetry designation, determined by the symmetry of the orbital with respect to rotation about the z axis. The value of λ is constant along any of the lines that connects the united atom limit to the separated atom limit at the right-hand side of the diagram. In the separated atom limit, we list each of the atomic orbitals of one of the individual atoms. Each of these orbitals is combined with another atomic orbital of the same n, l, and m_l values, but not necessarily the same phase. As shown in Fig. 7.1, combining two 1s orbitals can lead to either a σ_g or a σ_u MO, depending on the relative phase. We carry out the same process when we combine 1s orbitals to generate the ψ_+ and ψ_- MO functions of H_2^+ in Section 5.2. The relative phase determines the inversion symmetry, g or u. Similarly, the m_l values of the atomic orbitals determine the λ value of the MO. The p_z orbitals have $m_l = 0$, and combine to form $\lambda = 0$ orbitals (σ MOs). The p_x and p_y orbitals, which are obtained from the $m_l = \pm 1$ states, yield $\lambda = \pm 1$ orbitals (π MOs).

There's a basic recipe for drawing the lines that connect the states on each side of the diagram:

1. Draw a line from the lowest state on the left to the *lowest energy state of the same symmetry* representation on the right.
2. Continue the process, always moving to the very next higher energy state on the left, and never drawing a line to the same state more than once.

One test that this has worked: lines for states of the same symmetry representation (for example, σ_g) should never cross in the diagram.

On the other hand, lines for states of different symmetry often do cross. The energy ordering in the separated atom limit is different from that in the united atom limit, and the MO energy ordering lies in between. With the exception of H_2 (which has a very short bond length), the homonuclear diatomic orbital energies are ordered roughly as in the separated atom limit, as shown by the dashed vertical lines in Fig. 7.2, molecules with long bond lengths tending toward the right (separated atom) and those with short bonds toward the left (united atom). Accurate relative energies require solving the Schrödinger equation, but this schematic correlation diagram gives an idea of how levels are grouped, and the order in which MOs are filled as electrons are added to form a many-electron molecule.

EXAMPLE 7.1 Finding the MO Configuration

CONTEXT Molecular oxygen, O_2, poses an awkward problem in general chemistry courses. Although it's an abundant gas of obvious importance to everyone who enjoys breathing air or starting fires, O_2 does not fit comfortably into the Lewis model of chemical bonding. This problem is compounded by the fact that in the periodic table, oxygen is sandwiched between two other elements—nitrogen and fluorine—which provide elegant examples of diatomic gases with triple and single bonds, respectively. Surely, O_2 must therefore have a double bond. Yes it does, which makes it all the more frustrating that you cannot adequately demonstrate that with Lewis structures. Lewis structures that are drawn for the molecule either show the double bond by depositing the non-bonding electrons into lone pairs, or show (correctly) that there are two unpaired electrons, but at the expense of drawing a single bond between the two atoms. The MO model of electrons in molecules replaces the simpler, more restrictive electron distributions in the Lewis structure with a more flexible model, in this case allowing non-bonding electrons to be distributed equally on both atoms rather than fixed at one atom or the other.

PROBLEM What is the MO configuration for O_2?

SOLUTION The O_2 molecule has 16 electrons. We fill the MOs in the correlation diagram Fig. 7.2 from the bottom up (remembering that the π representations stand for two MOs each, and can hold up to 4 electrons), and we get

$$1\sigma_g^2 1\sigma_u^2 2\sigma_g^2 2\sigma_u^2 3\sigma_g^2 1\pi_u^4 1\pi_g^2,$$

using the ordering near the separated atom limit.

You may already be familiar with slices of the diatomic correlation diagram called **molecular orbital diagrams.** These are correlation diagrams drawn for a single set of separated atomic subshells, showing the relative energies of the resulting MOs at a particular internuclear distance R, with the atomic orbital energies of the two atoms drawn on either side. For example, when combining two nitrogen atoms to form N_2, the $2p$ atomic subshells combine to yield the $1\pi_u$, $3\sigma_g$, $1\pi_g$, and $3\sigma_u$ MOs. The MO diagram (Fig. 7.3) shows that the antibonding orbitals are less stable than the bonding orbitals. Fig. 7.3 also demonstrates that only the $1\pi_u$, $3\sigma_g$ orbitals should be occupied in the ground state, because the separated atoms each have three electrons in the $2p$ subshell, and therefore only six electrons are available to the MOs.

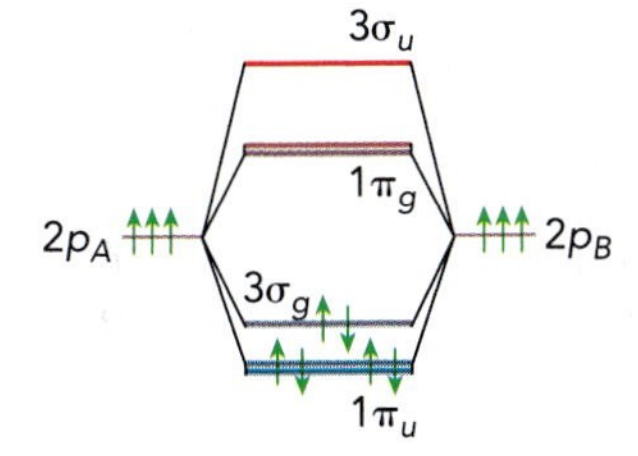

▲ **FIGURE 7.3 MO diagram for the N_2 orbitals correlating to the atomic $2p$ orbitals.**

Fig. 7.2 is a crude approximation. The orbital energies do not vary in straight lines, and they are functions of the nuclear charges and of the number of electrons in other orbitals, as well as functions of R. Furthermore, while the symmetry of the orbitals in the point group $D_{\infty h}$ is preserved, that does not prevent us from combining, for example, an *sp* hybrid orbital on nucleus A with an equivalent *sp* hybrid on nucleus B. In forming the chemical bond, we expect the *molecular* orbitals to borrow character from several *atomic* orbitals to obtain the lowest possible energy (Section 5.3). Even the concept of individual orbitals for the electrons is a simplifying approximation. One example of a more accurate correlation diagram is drawn in Fig. 7.4, which graphs the orbital energies for two nitrogen atoms as functions of R. Now you see why we'll use Fig. 7.2 instead.

But before we leave Fig. 7.4, notice one important discrepancy. Unlike our simpler version in Fig. 7.2, this diagram places the $1\pi_u$ orbital above the $3\sigma_g$. It is commonly asserted that the $1\pi_u$ is lower in energy than the $3\sigma_g$ because the $1\pi_u^4 3\sigma_g^1\ {}^2\Sigma_g^+$ state of the N_2^+ ion is about 1 eV more stable than the $1\pi_u^3 3\sigma_g^2\ {}^2\Pi_u$ state. In other words, it takes less energy to remove an electron from $3\sigma_g$ than from $1\pi_u$, so the $3\sigma_g$ must be higher in energy. But this is a deceptively arbitrary criterion, because the bond lengths and orbital energies all shift with any change in the electron configuration. The ordering given in Fig. 7.4 reflects the fact that the $1\pi_u^3 3\sigma_g^2 1\pi_g^1\ A^3\Sigma_u^+$ excited state of N_2 is about 1 eV more stable than the $1\pi_u^4 3\sigma_g^1 1\pi_g^1\ B^3\Pi_g$ state. In other words, it takes less energy to shift a $1\pi_u$ electron into $1\pi_g$ than to shift a $3\sigma_g$ electron into $1\pi_g$, suggesting that

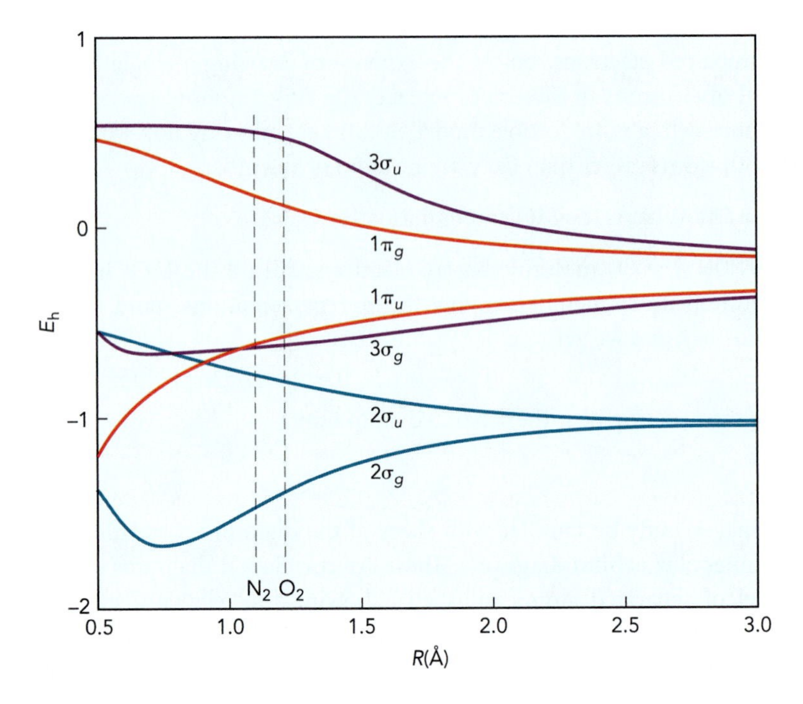

FIGURE 7.4 Correlation diagram determined from Hartree-Fock N_2 orbital energies. Dashed lines indicate the equilibrium bond lengths of ground state N_2 and O_2. The HF energies predict the switch in the $1\pi_u$ and $3\sigma_g$ orbital energies, but at a shorter distance than the actual N_2 bond length. The energy is measured relative to the ionization limit, defined as $E = 0$.

the $1\pi_u$ orbital should indeed be above the $3\sigma_g$ in the correlation diagram. Let this be a reminder that the MO picture is only an approximation, convenient but also potentially misleading in its simplicity.

That discrepancy appears more for N_2 than for most other molecules (such as O_2 in the previous example) because the $1\pi_u$ and $3\sigma_g$ MOs are so close in energy.

In the correlation diagram, every combination of atomic orbitals in the separated atom limit gives rise to an equal number of bonding and antibonding MOs, but *the bonding orbital is always lower in energy than the corresponding antibonding orbital.* As a result, when filling the orbitals to obtain the lowest energy MO configuration for a given molecule, the filled antibonding orbitals never outnumber the filled bonding orbitals. Chemical bonds form easily, although the strength of these bonds varies considerably.

Any occupied antibonding orbitals will partly cancel the binding force of the bonding orbitals, and the **bond order,** which is a crude measure of the bond strength, is calculated as one-half the number of bonding electrons minus the number of antibonding electrons:

$$b = [\,\textit{number of bonding electrons} - \textit{number of antibonding electrons}\,]/2. \quad (7.1)$$

The homonuclear diatomics whose ground state MO configurations predict zero bond order include all the noble gas diatomics, such as He_2 and Ne_2. If we apply the *Aufbau* principle to He_2, we obtain the ground electronic configuration $1\sigma_g^2 1\sigma_u^2$. Since σ_g is bonding and σ_u antibonding, and since each of these has an equal occupation (number of electrons) in He_2, we expect the bond strength to be much weaker in this molecule than in H_2. The prediction is accurate; He_2 has such a weak bond that its first detection was reported in 1993. This is one perspective from which to explain why the noble gas elements are monatomic gases under typical laboratory conditions, whereas many other elements that are easily found in the gas phase—particularly H, O, N, F, Cl—all form diatomic molecules.

BIOSKETCH **Arthur Suits**

Arthur Suits is a professor of chemistry at Wayne State University, where he and his group study the many roles of *free radicals*—atoms or molecules with unpaired electrons—in chemical processes. One of the techniques developed in his lab, DC Slice Velocity Map Imaging, uses a video camera to record images of ions formed by photoionization after (for one example) a dissociation reaction occurs. For instance, Suits has used laser photodissociation (Section 7.4) to break deuterium bromide (^{2}HBr or DBr) into deuterium and bromine radicals. The photodissociation occurs at a fixed photon energy $h\nu_{\text{laser}}$ determined by the laser, but the DBr molecules start off with different rovibrational rotational and vibrational energies E_{exc}. After the dissociation, any excess energy from the laser photon is converted into kinetic energy $K(\text{frag})$ of the D and Br fragments. Molecules that start with lower vibrational energy use more of the laser energy to dissociate, so their fragments end up with less kinetic energy. Therefore, the fragments from molecules with more vibrational energy travel farther from where the molecule was split.

To map their energy distribution, the fragments are photoionized and the resulting ions are extracted by electric fields and imaged onto a detector. The distance of an ion signal from the center tells how much kinetic energy that group of ions possessed. If we label the original molecule A, then conservation of the energy before and after the interaction of A with the laser gives the relation

$$h\nu_{\text{laser}} + E_{\text{exc}}(\text{A}) = D_0 + K(\text{frag}) + E_{\text{exc}}(\text{frag}),$$

where D_0 is the dissociation energy of the molecule's ground state. The values of $h\nu_{\text{laser}}$ and $K(\text{frag})$ are determined by experiment, allowing the excitation energies of the original molecule *and* the fragments to be calculated. These fragment kinetic energy images give an exquisitely detailed map of the way energy is distributed in the initial and final states of the dissociation.

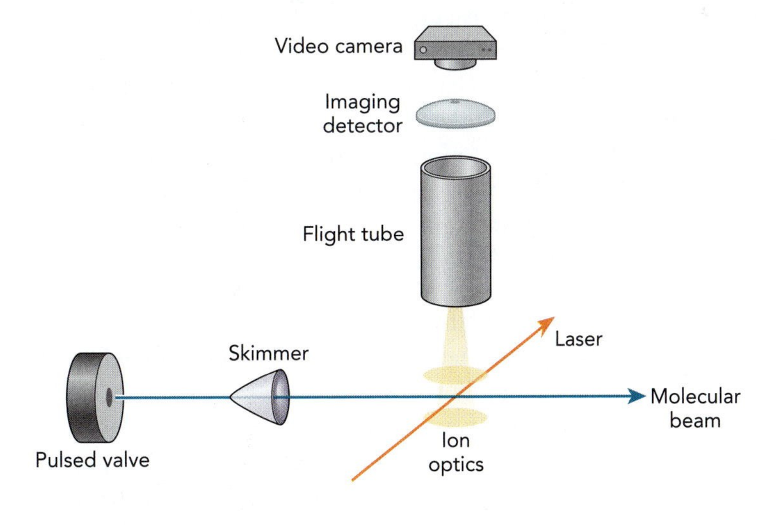

The bond order rarely exceeds three. Realistically, even though the core electrons are represented as filling delocalized MOs, they will be closely confined near the nuclei and contribute little to the overall bonding or antibonding character of the state. In Eq. 7.1, the contribution to b from electrons in filled cores is always zero.

EXAMPLE 7.2 Bond Order

CONTEXT The formation of the very stable oxidized forms of hydrogen and carbon (H_2O and CO_2) releases the huge amounts of energy we associate with the combustion of hydrocarbons. But the complete combustion of nitrogen-containing compounds doesn't yield oxidized nitrogen; it yields nitrogen gas, N_2. Diatomic nitrogen is more stable than any of its oxidized forms in the gas phase and at the same temperature. We ascribe that great stability in part to the fact that it has the highest bond order of any of the homonuclear diatomics. There are plenty of electrons in bonding orbitals and few in antibonding orbitals. Moving from nitrogen to the first of the period 2 elements, we find that elemental lithium, in contrast, forms oxides so easily that it will even pull the oxygen off of water, releasing enough heat to ignite the hydrogen gas that is produced simultaneously. The lithium dimer Li_2 is found in high temperature lithium vapor, and we can see to what extent the MO picture is able to characterize these two very different homonuclear diatomics.

PROBLEM Find the bond orders for Li_2 and N_2.

SOLUTION Li_2 has an MO configuration $1\sigma_g^2 1\sigma_u^2 2\sigma_g^2$, netting two electrons in a bonding orbital. This configuration predicts $b = 2/2 = 1$, a single bond. Nitrogen, N_2, on the other hand has 14 electrons and occupies its MOs as follows: $1\sigma_g^2 1\sigma_u^2 2\sigma_g^2 2\sigma_u^2 1\pi_u^4 3\sigma_g^2$. The first 4 electrons are core electrons and contribute little to the bonding one way or the other. The next four are divided between $2\sigma_g$ bonding and $2\sigma_u$ antibonding orbitals, and together make no net contribution to the bond order. The last six electrons all go into bonding orbitals, resulting in a bond order $b = \frac{6}{2} = 3$, a triple bond that happens to be nine times stronger than the Li—Li bond. (Beware of assuming that bond strength is proportional to bond order.)

EXTEND Although the MO model gives us a qualitative picture of the distinction between the bonding in these two compounds, keep in mind that the MO picture itself is an approximation and subject to errors. In the example just given, if we remove an electron from the highest occupied orbital, which is a bonding σ_g MO, we form Li_2^+, which should then have a weaker bond with a bond order of only 1/2. However, the Li_2^+ bond is stronger than the Li_2 bond. A qualitative explanation abandons our version of MO theory, pointing out instead that electrons in metals become easier to ionize as the number of metal atoms increases. That means, for example, that it takes less energy to ionize Li_2 to Li_2^+ than to ionize Li atom to Li^+. Breaking the Li_2 bond forms two Li atoms, whereas breaking the Li_2^+ bond forms one Li atom and one Li^+ ion. Since the atomic ion takes more energy to form than the molecular ion, we have to put more energy into Li_2^+ to break its bond than into Li_2, so the Li_2^+ bond must be stronger. Although MO theory can predict this successfully, it takes more than the linear combination of atomic orbitals and correlation diagram to get the right answer.

It is common but not necessary to distinguish the antibonding orbitals with *'s. Unfortunately, it is also common to confuse the g and u labels with bonding and antibonding character. An orbital should not be called "antibonding" just because it has u symmetry. In particular, for our homonuclear diatomics, the π_uMO is bonding, the π_g is antibonding. For polyatomics, most MOs have some combination of bonding and antibonding character anyway. The locations of the nodes determine the net bonding/antibonding character of an orbital; the symmetry alone is almost never sufficient.

Heteronuclear Diatomics

If we keep to a diatomic molecule but allow the two nuclei to be different, inversion and reflection through the xy plane are no longer symmetry operations of the nuclei, and therefore cannot be symmetry operations of the electronic orbitals.

This loss of symmetry makes the correlation diagram for the MOs considerably more complicated by reducing the limits on which atomic orbitals can be combined with which. However, the symmetry of the $\hat{C}_\infty$ rotation remains, and therefore only atomic orbitals of the same l and m_l are combined (which also amounts to conserving the angular momentum). It also becomes more difficult to see the relative energies of the MOs, since the relative energies of the separated atoms are now distinct (e.g., $1s_A + 2s_B \neq 2s_A + 1s_B$). The ordering of the resulting MOs is therefore strongly dependent on the particular atoms involved and needs to be determined case-by-case; it is no longer possible to draw a single correlation diagram to cover the entire class of molecules. However, atoms of similar atomic number can usually be treated approximately using the simpler homonuclear correlation diagram, keeping in mind the loss of symmetry. For example, NF has an MO configuration similar to that of O_2:

$$1\sigma^2 2\sigma^{*2} 3\sigma^2 4\sigma^{*2} 1\pi^4 5\sigma^2 2\pi^{*2},$$

where the antibonding σ orbitals are now labeled σ^* instead of σ_u, and so on, since there is now no inversion symmetry. Note that we renumber the σ orbitals because they now all have the same symmetry representation, whether bonding or antibonding. Both Figs. 7.2 and 7.3 show the $3\sigma_g$ and $1\pi_u$ orbitals as quite close in energy, and one simply can't be sure *a priori* which is lower in energy. In the case of NF, the nitrogen inverts the relative energy of the lowest π and highest σ orbitals, relative to the ordering in O_2.

EXAMPLE 7.3 Heteronuclear Diatomic Electron Configurations

CONTEXT Molecular oxygen poses one challenge to Lewis structure analysis in general chemistry, but we can blame that on its having two unpaired electrons, an unusual characteristic among the ground electronic states of abundant molecules. Carbon monoxide has no such excuse. The most accurate Lewis structure is the bizarre $^-$:C $\equiv$ O:$^+$, which depicts the carbon as a carbanion (an *extremely* reactive reagent in organic chemistry) and the highly electronegative oxygen as donating an electron to the carbon atom. This structure does correctly predict the triple bond, and, despite what the electronegativities suggest, the direction of the dipole moment that this picture predicts is one of the triumphs of the Lewis structure model: the carbon atom *does* have a net negative charge. However, the actual electron density in the triple bond lies near the oxygen, leading to a dipole moment that is nearly zero, much less than we would expect from the Lewis structure. General chemistry students, in an effort to suppress the formal charges in the Lewis structure, will often violate the octet rule, predicting a double bond for the molecule. Our correlation diagram can be used to show, by a different line of reasoning, that CO should indeed have a triple bond.

PROBLEM Predict the electron configuration for CO, based on the correlation diagram Fig. 7.2.

SOLUTION CO has the same number of electrons as N_2, and the atomic numbers are not very different, so we predict that both molecules have essentially the same MO configuration, taking into account the lack of inversion symmetry in CO. For N_2 the MO configuration is

$$1\sigma_g^2 1\sigma_u^2 2\sigma_g^2 2\sigma_u^2 1\pi_u^4 3\sigma_g^2.$$

Replacing σ_g with σ, σ_u with σ^*, and π_u with π, we obtain the following configuration for CO:

$$1\sigma^2 2\sigma^{*2} 3\sigma^2 4\sigma^{*2} 1\pi^4 5\sigma^2,$$

which is indeed correct.

Correlation Diagrams for Polyatomics

Correlation diagrams can also be helpful for analyzing the molecular orbitals of polyatomic molecules; these go by the name **Walsh diagrams.** We may draw a correlation diagram for a specific class of compound, XH_3 molecules, for example, in order to estimate which symmetry the molecule will have for a particular element X. Fig. 7.5 shows a correlation diagram for the XH_3 molecules between a tetrahedral C_{3v} limit and the planar D_{3h} limit. From the correlation diagram, it can be seen that an eight-electron system such as BH_3 will be more stable in the planar D_{3h} form, whereas the ten-electron NH_3 is more stable with a C_{3v} geometry. During the transformation from D_{3h} to C_{3v}, there is always a $\hat{C}_3$ proper rotation axis and three $\hat{\sigma}_v$ mirror planes. States are correlated that have the same symmetry with respect to those operations: e' in D_{3h} with e in C_{3v}, and a_1' or a_2'' with a_1. The $\hat{\sigma}_h$ mirror plane of D_{3h} is lost when the molecule becomes non-planar, however, and therefore both the a_1' and a_2'' representations (which have opposite symmetry with respect to $\hat{\sigma}_h$) can be correlated to the same representation, a_1 in the C_{3v} limit.

One would also expect from Fig. 7.5 that the nine-electron molecules CH_3 and NH_3^+ would be closer to planar than NH_3, because they have only one electron in the highest orbital. This prediction is accurate; in fact, CH_3 and NH_3^+ are planar on average, but can bend easily with little increase in the effective potential energy. Some general qualitative rules governing the relative energies of the MOs help to determine the energy ordering of polyatomic MOs, even without a correlation diagram.

1. In a double or triple bond, the σ bond is stronger (lower potential energy, greater dissociation energy) than the π bonds because it occupies the region where F_{binding} is greatest.
2. In any molecule, the more nodes there are in the orbital wavefunction along a given coordinate, the higher the orbital energy. Among the valence electron orbitals of benzene shown in Fig. 7.6, the e_{1g} orbital (a π system

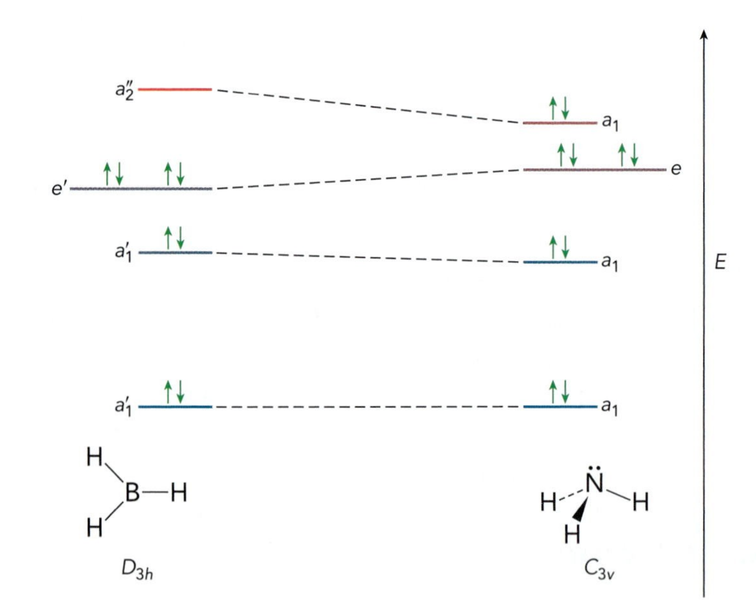

▶ **FIGURE 7.5 Correlation diagram for the XH_3 molecules between C_{3v} and D_{3h} geometries.** Arrows indicate electrons occupying the ground state MO configurations of BH_3 (D_{3h}) and NH_3 (C_{3v}).

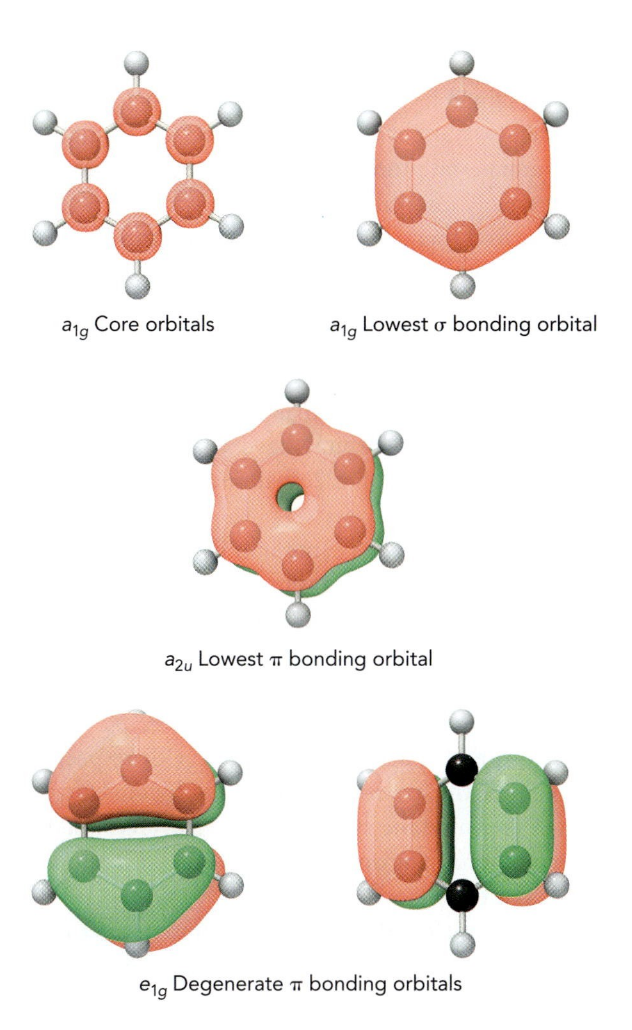

FIGURE 7.6 Selected MO wavefunctions of benzene, C_6H_6.

orbital with one node perpendicular to the molecular plane and one in the molecular plane) is the highest energy of the occupied orbitals, and a_{1g} (a σ bonding orbital with no nodes) is the lowest energy.

7.2 Electronic States

Once we have an idea how the electrons will group themselves into the molecular orbitals, we can back off a step and examine the entire many-electron quantum state of the molecule. Symmetry guided our construction of the individual MOs, and so it continues to play a dominant role in our consideration of the resulting electronic states.

Electronic State Symmetry

The notation for the electronic states and energies of many-electron molecules is based on the symmetry MOs. We use the *Aufbau* principle to fill the individual electronic orbitals to obtain a ground state MO configuration for the molecule.

The ground electronic state is labeled by the representation for the configuration's overall symmetry, obtained by taking the direct product of the representations for any unfilled orbitals. We may form excited electron configurations in molecules as we do in atoms, by promoting an electron from one of the occupied orbitals to a higher energy unoccupied orbital.

In Section 4.4, we find that a single electron configuration in atoms may be shared by several distinct *term states* with different energies. The term states differ in how the angular momenta of the individual electrons combine to give different values of the *total* spin and orbital angular momentum quantum numbers S and L. The change from atoms to molecules has little effect on the electron spin, and the total spin S is obtained in the same way: paired electrons that share the same orbital have exactly canceling spins, while unpaired electrons have spins of 1/2 that may add to or cancel those of other unpaired electrons.

But whereas we treat S about the same in molecules as in atoms, the orbital angular momentum L is not a good quantum number in molecules, so we cannot do the same kind of vector addition of individual electron l values to find a total L. Instead, we note that the l quantum number in atomic orbitals determines the rotation and inversion symmetry of the orbital. For example, an s orbital is totally symmetric, but a p orbital changes sign under 180° rotation about any axis in its nodal plane, and is also antisymmetric with respect to inversion. In molecules, we combine the *symmetries* of the individual orbitals—using the direct product—to determine the symmetry of the overall electronic wavefunction. As with atoms, more than one term state exists if there are two or more unpaired electrons in the MO configuration.

The labeling scheme for electronic term states of molecules is as follows:

1. Take the direct product of representations for any electron that is in a partially-filled orbital. If all the orbitals are filled, the result is the totally symmetric representation. The term state is labeled by the capital letter for the result of this direct product. For example, in a C_{2v} molecule with all its orbitals filled, such as water, the term is labeled A_1. For ground state H_2, with MO configuration $1\sigma_g^2$, the electronic state must also be labeled by the totally symmetric representation, in this case Σ_g^+.
2. The spin multiplicity $2S + 1$ is still indicated by a prefixed superscript to the orbital angular momentum designation. The H_2 ground state has both electrons in the same orbital with opposite spins, and consequently is a singlet state, ${}^1\Sigma_g^+$.

In the case of linear molecules, the direct product of the orbital representations gives the projection of the overall orbital angular momentum, Λ:

$$\Lambda = \sum_i \lambda_i, \tag{7.2}$$

where the sum is over all the electrons. This sum is equivalent to adding all the m_l values for the valence electrons in an atom to get the total M_L. As in atoms, all the filled core orbitals contribute nothing to this sum. In the state symbol, the value of $|\Lambda|$ (which must be an integer) is given by the capital *Greek* letter corresponding to the atomic orbital designation for the same value of l, that is, $0 \rightarrow \Sigma$, $1 \rightarrow \Pi$, $2 \rightarrow \Delta$, $3 \rightarrow \Phi$. For example, the ground state of the H_2 molecule is a Σ state, meaning $\Lambda = 0$, because no angular momentum is available from the σ MOs.

As another example, let's find the electronic term symbol for H_2^-, the simplest (and perhaps the most elusive) molecular anion, with the MO configuration $\sigma_g^2\sigma_u$. The point group is $D_{\infty h}$ which is the only one with σ_g representations. The σ_g orbital is filled, and any filled orbital representations contribute only the totally symmetric representation under $D_{\infty h}$, Σ_g^+, to the overall symmetry. We then take the direct product of any remaining partly filled orbitals to find the possible term symbol symmetries. In this case, because there is one unpaired electron in the $1\sigma_u$ MO, the overall symmetry is $\sigma_g \otimes \sigma_u = \Sigma_u^+$. Also, because exactly one of the electrons is unpaired, the total spin must be the spin of that electron, $S = 1/2$, so $2S + 1 = 2$. The state is a ${}^2\Sigma_u^+$ state.

In the same way that filled atomic subshells contributed nothing to the angular momentum of the atom's term symbol, the angular momenta cancel among the electrons in a fully occupied set of degenerate MOs. Electronic Σ states in which the π, δ, and other $\lambda \neq 0$ MOs are all filled are *always* Σ^+ states.

Some of the symmetry properties of the direct product can be predicted by inspection. Inversion and horizontal reflection symmetry are binary properties: when a point group has either of these symmetry elements, every representation in that group is either symmetric or antisymmetric with respect to those operations. Products of like symmetry functions give symmetric functions, products of opposite symmetry functions give antisymmetric functions. For inversion, this means that $g \otimes g$ and $u \otimes u$ give g functions, but $g \otimes u$ gives a u function. For horizontal reflection (indicated by primes and double-primes), $x' \otimes y'$ and $x'' \otimes y''$ give z', but $x' \otimes y''$ gives z''. The ground state of H_2, with the configuration $1\sigma_g^2$, is therefore ${}^1\Sigma_g^+$, whereas the first excited states, within the $1\sigma_g 1\sigma_u$ configuration, are ${}^1\Sigma_u^+$ and ${}^3\Sigma_u^+$.

Notice how similar H_2 is to the case of atomic helium, with a ground state electron configuration $1s^2$ that results in a single term state ${}^1S_0$, and with a lowest excited electron configuration $1s^1 2s^2$ that results in both singlet (${}^1S_0$) and triplet (${}^3S_1$) terms. For helium, we could use Hund's rules to hypothesize the energy ordering of the states, but which of the H_2 excited states is lower in energy? Hund's first rule (from Section 4.4), that higher electron spin quantum numbers S imply lower energy, still holds quite well. In particular, triplet spin states tend to be lower in energy than singlet states that arise from the same electron configuration. However, the much more complicated spatial distribution of MOs and their dependence on the bond length prevent the formulation of any reliable and straightforward rules that further dictate the energy ordering for molecular electronic states.

Beyond the fairly well-defined labeling conventions for the electronic states of diatomics given earlier, each electronic state of the molecule is distinguished from the others by a letter prefix. The distinction is necessary because different states may share the same symmetry characteristics, just as the $1s$ and $2s$ atomic orbitals are both spherically symmetric but have very different distributions and energies. The convention for molecules is that the ground electronic state is always the X state. Beyond that, each excited state with the same spin multiplicity as the ground state is usually designated by an upper case letter (A, B, C, etc.), and each state with spin multiplicity different from the X state is assigned to a lower case letter (a, b, c, etc.). The ordering of these excited state labels is not obvious; excited states are often labeled not in order of increasing energy but in order of discovery, and it is often very difficult to determine precise relative energies for excited states anyway. The two lowest energy states of H_2 discussed previously are labeled $X^1\Sigma_g^+$ (the ground state) and $b^3\Sigma_u^+$ (the lowest triplet excited state).

When looking at atoms, we discovered that more than one term symbol could result from a single electron configuration. The same holds for molecules: the MO configuration is sometimes not enough to specify the electronic state. The angular momenta due to individual electrons within an orbital may sum together or cancel, resulting in different total angular momenta. In such a case, the direct product of the unfilled orbital representations yields a reducible representation, a direct sum of the representations for the possible electronic states.

Diatomic oxygen, O_2, has a ground MO configuration $[Be_2]\pi_u^4\sigma_g^2\pi_g^2$. Consider just the π_g^2 part of the configuration. Again, the point group is $D_{\infty h}$. To obtain the term states resulting from this configuration we could work through a vector model of the angular momentum. According to this group theory algebra, however, we only need to look at the square of the characters of the π_g representation (Table 6.4), obtaining

$$4, 4\cos^2\phi, 0, 4 \;=\; 4, 2 \;+\; 2\cos 2\phi, 0, 4,$$

where we have used the trigonometric identity

$$\cos^2\phi = \frac{1}{2}(\cos 2\phi \;+\; 1).$$

This set of characters does not match any of the available representations in our character table, but that's because more than one term is possible from this configuration. The result we have obtained is the sum of the characters for those terms; that is, this is a reducible representation. To find the individual states, we need to decompose this into its irreducible representations. The methods were covered in Section 6.2, and we would find in this case that the configuration leads to Δ_g, Σ_g^+, and Σ_g^- term states. (See also Example 7.4.)

This vector model tells us what spatial symmetries to expect, but not what the spins are that go with the electronic term symbol. For that, we may once again rely on the vector model, which can be applied with some success to molecules with degenerate electronic states.

Several term energies T_e for excited electronic states of diatomic molecules are given in Table 7.1. For small molecules, these energies are similar to those seen for individual atoms in Table 4.2. Notice also how the energy gaps between the electronic states tend to be smaller as the electrons approach a more classical limit. For example, the gap between the higher energy B and W states in N_2 is 0.1 eV, compared to the 6.2 eV gap between the X and A states. Similarly, I_2 and AuPb, where the valence electrons begin with high values of the n quantum number, require relatively little energy (1.5 eV and 2.0 eV) to reach the lowest excited state.

EXAMPLE 7.4 O_2 Vector Model

CONTEXT The electronic structure of O_2 is remarkably complex, even in its ground state. The lowest energy electron configuration of O_2 gives rise to several electronic states, with different spins and reactivities. Triplet oxygen is reactive enough to be the key reactant in flames, but the excited singlet state can also be destructive, attacking organic molecules to form alcohols, and participating in the damage of living tissue as one of several **reactive oxygen species** found in cells.

PROBLEM Construct a vector model diagram of the two electrons in the unfilled π_g orbital of ground state O_2, and find the three resulting term symbols, including the spin multiplicities.

SOLUTION Because the π orbital implies $\lambda = \pm 1$, we use those two values in place of the m_l values we would use for an atom, but the m_s quantum number is no different in molecules than in atoms:

λ		Λ	M_S
+1	−1		
↑↓		2	0
	↑↓	−2	0
↑	↑	0	1
↑	↓	0	0
↓	↑	0	0
↓	↓	0	−1

The two configurations with $\Lambda = \pm 2$ comprise a Δ electronic state; the rest, with $\Lambda = 0$, are associated Σ states. Because $M_S = 0$ for both components of the Δ state, we know that its spin is zero, so one of the electronic term symbols of molecular oxygen is $^1\Delta$. Of the Σ states, we can associate three (with $M_S = -1, 0, 1$) with a triplet state ($S = 1$), leaving one last configuration with $\Lambda = 0, S = 0$, which leads to a singlet Σ state. Therefore, the three term states arising from the ground MO configuration of O_2 are $^1\Delta$, $^1\Sigma$, and $^3\Sigma$. We need the group theory to further show that the three term symbols must be symmetric with respect to inversion, and still more detailed analysis to show that the triplet state must be the term with Σ_g^- symmetry (Problem 7.27). This MO configuration gives rise to three electronic states, and the ground state of O_2 is one in which the electron spins are unpaired, the triplet state $X^3\Sigma_g^-$.

TABLE 7.1 Term symbols and term energies of selected diatomic molecules. The term energy is T_e, in eV (and E_h) relative to the ground state.

molecule	ground state	excited states					
H_2	$X^1\Sigma_g^+$	$B^1\Sigma_u^+$	11.4 (0.42 E_h)				
I_2	$X^1\Sigma_g^+$	$A^3\Pi_u$	1.5 (0.06 E_h)	$B^3\Pi_u$	2.0 (0.07 E_h)		
N_2	$X^1\Sigma_g^+$	$A^3\Sigma_u^+$	6.2 (0.23 E_h)	$B^3\Pi_g$	7.3 (0.27 E_h)	$W^3\Delta_u$	7.4 (0.27 E_h)
LiH	$X^1\Sigma^+$	$A^1\Sigma^+$	3.3 (0.12 E_h)	$B^1\Pi$	4.3 (0.16 E_h)		
NO	$X^2\Pi$	$A^2\Sigma^+$	5.4 (0.20 E_h)	$B^2\Pi$	5.7 (0.21 E_h)	$C^2\Pi$	6.5 (0.24 E_h)
AuPb	$X^2\Pi$	$^2\Sigma^+$	2.0 (0.07 E_h)				

Once the overall electronic state symmetry of the molecule is determined, it is subject to the same rules that governed the characteristics of individual MOs. In particular, the selection rules for transitions between electronic states are dictated by the symmetry representations of the transition moment operators.

TOOLS OF THE TRADE UV-vis Spectroscopy

Driven partly by the need for a fast and easy method to test foods for its soldiers as World War II loomed ahead, the U.S. government invested in the development of a factory-manufactured instrument to put spectroscopy to work as an analytical tool. That investment led to the release, in 1941, of the Beckman DU ultraviolet and visible (UV-vis) spectrophotometer, the first commercially available spectrometer at any wavelength. For more than forty years, the basic design of the UV-vis spectrometer changed little from that original model. In the 1980s, the first diode array detectors were introduced, allowing the simultaneous measurement of the entire UV-visible spectrum. Charge-coupled device (CCD) detectors now offer the same advantages of array detection, and with higher resolution than diode arrays. Today, UV-vis spectroscopy is about as painless as an analytical tool can be, operating with push-button efficiency and delivering a spectrum in a matter of seconds.

What is UV-vis spectroscopy?

A UV-vis spectrometer measures the absorption of ultraviolet and visible light by the substance—a pure element, compound, or mixture—being analyzed, usually a liquid sample in a transparent rectangular container called a **cuvette.** The radiation sources typically generate light with wavelengths between 200 and 1000 nm. In this region of the electromagnetic spectrum, most molecules containing heavy elements (below period 2) or conjugated π electron systems will undergo electronic transitions. The common components of air, on the other hand, and many organic solvents—the typical environments for a compound being studied—do not strongly absorb light in this region, so the method usually enjoys relatively low interference compared to that encountered in other spectroscopic methods. (Proton NMR spectroscopy, for example, is more widely used to determine structures of organic compounds but must rely

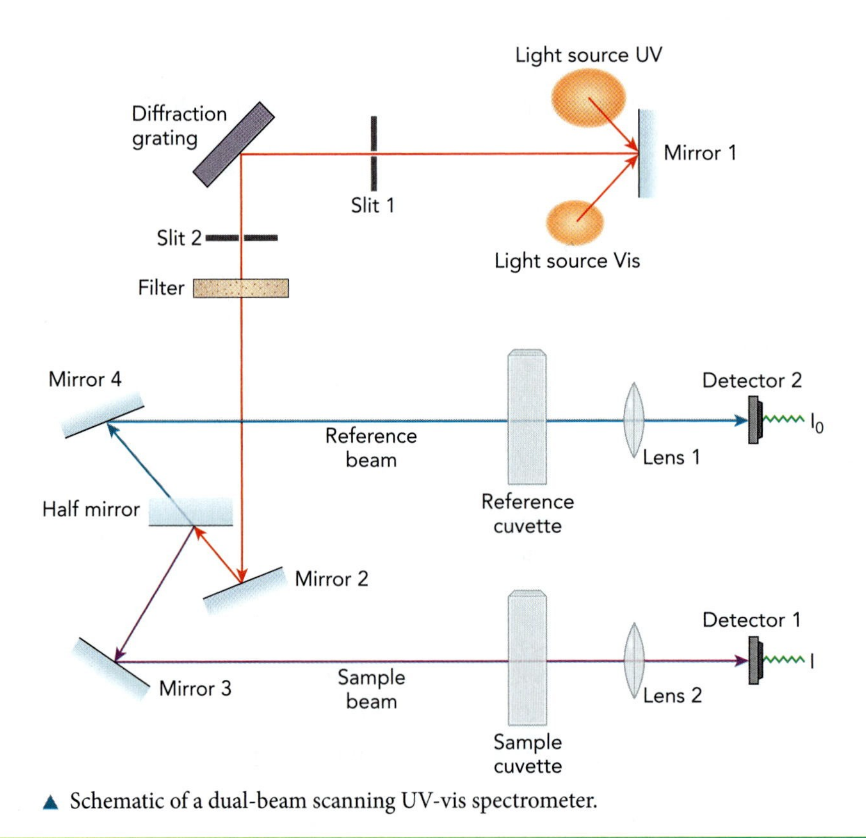

▲ Schematic of a dual-beam scanning UV-vis spectrometer.

on deuterated solvents to prevent the signals of solvent protons from overwhelming those of the target molecules. Similarly, strong absorption by atmospheric water and CO_2 can obscure infrared spectra.)

Why do we use UV-vis spectroscopy?

Transition metals and extended π electron systems have low-lying electronic states, accessible from the ground electronic states by excitation with photons of visible and low-energy ultraviolet light. The resulting spectrum is then particularly sensitive to processes that affect the electron distribution, such as changes in the symmetry of a central metal atom, electron transfer to a ligand molecule, or protonation of a double bond. For example, the working part of many acid–base indicators is a series of conjugated double bonds that acts like a one-dimensional box potential: π electrons experience a fairly constant potential energy along the length of the double bond chain, and the ends of the chain are marked by a rapid increase in potential energy—the walls of the box. When the chain is protonated at one end, one link in the conjugated chain is lost, reducing the length of the box and increasing all the electronic transition energies. Electronic transitions in the deprotonated indicator that absorbed, say, red light may absorb blue light when the dye is protonated. The color of a solution that contains the dye is determined by the light that is transmitted, so typically the color of a dye shifts from a long wavelength red or yellow at low pH (where the dye's high transition energies absorb short wavelengths) to a short wavelength blue or green or purple at high pH (where the dye's low transition energies absorb long wavelengths). This dramatic change in the wavelength of the absorbed light allows UV-vis spectroscopy to monitor acid–base reactions, obtaining quantitative and time-resolved measurements of the reaction's progress.

How does it work?

Any ordinary spectrometer consists primarily of a light source, a wavelength separation method (which may be intrinsic to the light source), a sample holder, and a detector. In UV-vis spectrometers, often two sources supply the radiation in order to provide a fairly even distribution of power from long to short wavelengths. A quartz tungsten halogen lamp emits radiation in the visible and near-ultraviolet, down to wavelengths of about 350 nm. These lamps consist of a quartz bulb housing a tungsten filament and filled with a mixture of iodine or bromine vapor in some noble gas. A high electrical current runs through the filament to achieve high emission powers, particularly at short wavelengths. The presence of halogen in the gas increases the lifetime of the filament, and the quartz is necessary to withstand the very high heat dissipated by the filament. To get further into the ultraviolet, a second lamp is used, consisting of two electrodes that generate an electrical arc discharge through deuterium (D_2) gas. The discharge excites D_2 into an excited state that quickly relaxes to a dissociating state, producing atomic deuterium that rapidly recombines into D_2. These dissociations produce a continuous band of emission wavelengths from roughly 180 nm to 380 nm.

There are two common methods for separating the wavelengths in UV-vis spectrometers. In the older method, the light is dispersed by a diffraction grating so that only light of a particular wavelength passes through the sample at any one time. The wavelength is then scanned by changing the angle of the grating. With the advent of array detectors, it has become possible to simultaneously detect the light at different points along a strip, so it is increasingly common for the spectrometer to send all the light through the sample, and *then* disperse it with a grating so that the different wavelengths strike different parts of the array detector.

The most common UV-vis spectrometers are designed to accept samples in glass or quartz cuvettes, 1 cm on a side. Fiber optics now allow remarkable versatility in how the radiation is transmitted to the sample and returned to the spectrometer. Dip probes combine two optical fibers for the delivery and return of the radiation, with a tiny sample compartment and mirror arrangement at the probe tip, allowing the probe to sample analytes in a wide range of environments.

Detection schemes in UV-vis spectrometers vary. The dual-beam spectrometer design shown in the previous figure is one option that allows the spectrometer to obtain simultaneously the signal from the sample and the signal from a reference cuvette containing pure solvent. The instrument calculates and displays the ratio of the two signals, so that in effect only the absorptions by the analyte are observed. Scanning spectrometers generally use a photomultiplier tube (PMT), an extremely sensitive device that detects the radiation by converting each photon to a tiny electrical current that it then amplifies by orders of magnitude. A CCD detector is far less sensitive, using tiny capacitors to detect the photons, but the low cost and small size of each capacitor allows the CCD to measure the signal from all the wavelengths at once, whereas the PMT detects only one wavelength at a time.

UV-vis spectroscopy does not give the detailed structural information offered by NMR spectroscopy. Instead it provides a direct probe of the larger-scale electronic structure of the molecule with an ease of use that few analytical techniques can match.

EXAMPLE 7.5 Selection Rules

CONTEXT In an effort to better understand the quantum mechanics of metals, spectroscopists have long studied the electronic states of small groups of metal atoms, exploring the gradual build-up to the properties of bulk metals as the number of atoms increases. Among the earliest of these studies were measurements of the spectra of the homonuclear alkali metal diatomics. Alkali metals are as simple as metals get, because each atom contributes only one electron into the sea of electrons, and *d* orbitals play little or no role in the quantum states of the conduction electrons. Sodium dimer Na_2 has been studied more extensively than the simpler lithium dimer partly because sodium has a higher vapor pressure, so it is easier to get into the gas phase where these measurements are made. For years, spectra of sodium dimer included the transitions from the $X^1\Sigma_g^+$ ground state to the $A^1\Sigma_u^+$ and $B^1\Pi_u$ states but no data on low-lying g-symmetry states.

PROBLEM For a homonuclear diatomic, prove that a transition from Σ_g^+ to Π_u is electric-dipole allowed, while Σ_g^+ to Π_g is not.

SOLUTION For a transition to be electric-dipole allowed, we need the direct product $\Gamma_i \otimes \Gamma_f$ to yield one of the repesentations for x, y, or z (Section 6.3). The transition $\Sigma_g^+ \rightarrow \Pi_u$ is allowed while $\Sigma_g^+ \rightarrow \Pi_g$ is forbidden. The functions x and y are given by the representation Π_u, while z is given by the representation Σ_u^+. We evaluate the direct products

$$\Sigma_g^+ \otimes \Pi_u = \Pi_u$$
$$\Sigma_g^+ \otimes \Pi_g = \Pi_g$$

One of the transition moment representations is available only from the first of these direct products, and indeed $\Sigma_g^+ \rightarrow \Pi_u$ is the only allowed transition of those considered. In fact, it also tells us that the transition is only allowed if the radiation is polarized along the x or y axes of the molecule; it will not occur if the radiation is polarized exclusively along the z axis.

The extension of overall electronic state symmetry to polyatomic molecules is straightforward. Individual MOs are labeled by the lowercase letters for the representations of the molecule's point group; overall electronic state symmetries are labeled by the uppercase letter. Transitions between electronic states of a polyatomic molecule must obey the selection rules for the molecule's point group.

SAMPLE CALCULATION **Obtaining the Electronic State Term Symbol from an MO Configuration.**

To find the lowest energy electronic state arising from the MO configuration for the valence orbitals of $XeCl_4^+$ (in the point group D_{4h})

$$1a_{1g}^2 1a_{1u}^2 2a_{1g}^2 2a_{2u}^2 1b_{1u}^2 1a_{2u}^1,$$

we reason as follows. The filled MOs all give the totally symmetric representation, so we evaluate only the direct product $a_{1g} \otimes a_{2u} = A_{2u}$. One unpaired electron means $S = \frac{1}{2}$. Therefore the state is $^2A_{2u}$.

Potential Energy Surfaces of Electronic States

The lowest excited states of H_2 were discussed earlier in this section, and they are representative of diatomic electronic states. Building on our introduction to MO theory in Chapter 5, with the advantage of the molecular symmetry properties discussed in Chapter 6, we now examine the common characteristics of these states. The effective potential energy curves for selected electronic states of H_2 are sketched in Fig. 7.7.

Note the energy scale. The equilibrium **well depth,** abbreviated D_e, the energy difference between the potential at the equilibrium R value and at $R = \infty$, is about 4.75 eV for $X^1\Sigma_g^+$ H_2. The well depth gives the approximate dissociation energy or **bond energy** for the molecule, and is typically a few times smaller than the ionization energy. This potential energy curve resembles the potential function for our one-dimensional particle in a box in Section 2.2, and as with that example there is a zero-point energy. Once we have worked out the nature of the electronic energy in this chapter, we will be able to estimate the zero-point energy in Chapter 8 and obtain a more accurate value for the bond energy.

As one example, the ionization energy of H_2 is 16 eV (the energy gap between the $X^1\Sigma_g^+$ H_2 and $X^2\Sigma_g^+$ H_2^+ curves), about three times the H_2 bond energy. Bond energies for excited states are usually less than for the ground state, and

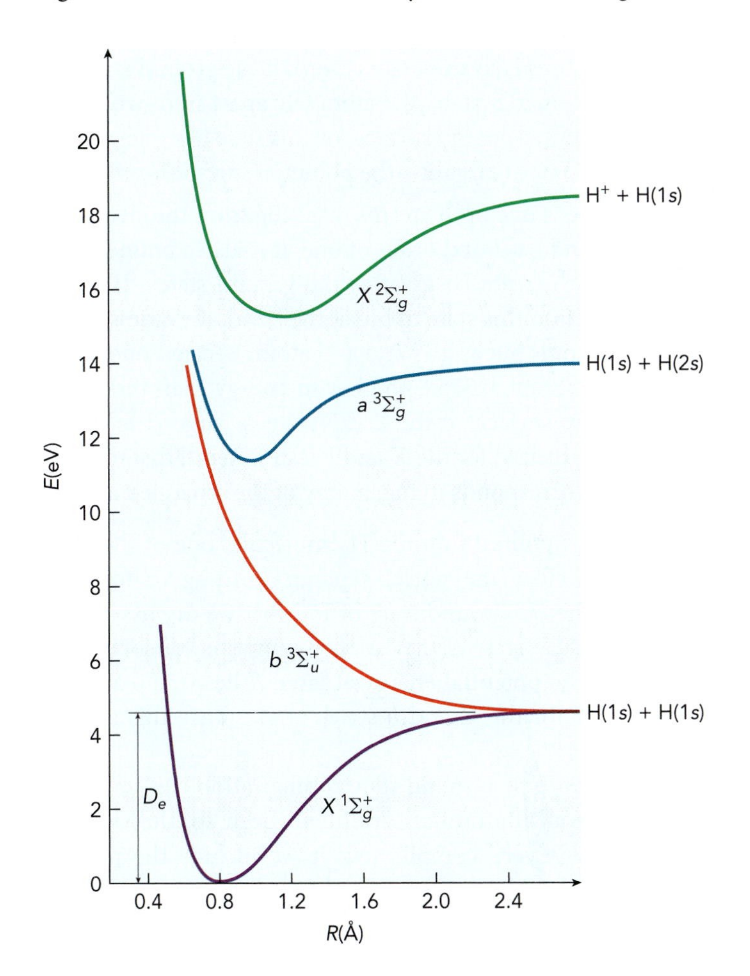

◀ **FIGURE 7.7 Potential energy curves for selected states of the H_2 molecule.** Energies are measured relative to the ground state equilibrium potential energy.

often a *lot* less. Bond strengths for diatomics vary from small values for noble gas diatomics (0.05 eV for ArAr) to about 10 eV for several triply bonded nitrogen- and oxygen-containing diatomics.

From Fig. 7.7 it can be seen that the equilibrium bond length R_e—the R value at the minimum of the potential energy curve—is 0.742 Å in the ground state of H_2. For diatomics involving larger atoms, this value increases, but it rarely exceeds 2.5 Å. The principal exceptions are weak bonds involving noble gas atoms, for which the equilibrium bond lengths are typically between 3 and 5 Å.

Let's take a closer look at the three electronic states of H_2 shown in Fig. 7.7:

1. $X^1\Sigma_g^+$. This is the $1\sigma_g^2$ ground electronic state of H_2. The $X^1\Sigma_g^+$ is a **bound** state, meaning the atoms form a bond that is stable with respect to motion of the nuclei. The bond is apparent from the deep minimum in the potential energy curve for this state. For energy values between 0 and 4.75 eV in Fig. 7.7, the two H atoms are trapped between the walls of the potential well. They are bound together until sufficient energy is added to break the bond.
2. $b^3\Sigma_u^+$. For the $1\sigma_g^1 1\sigma_u^1$ configuration of H_2, the triplet state $b^3\Sigma_u^+$ is **dissociative,** meaning that it is repulsive at all values of R; the atoms are not bound. We could predict that this state dissociates by noting that the bond order for H_2 in this state is zero. If the ground state H_2 molecule is excited into this $b^3\Sigma$ state, the atoms fly apart into two 1*s* H atoms. At large R, the potential energy surface for this state becomes the same as that for the ground state at large R—the potential surface for two 1*s* H atoms.
3. $a^3\Sigma_g^+$. This state is one of the terms resulting from the electron configuration $1\sigma_g^1 2\sigma_g^1$, and has bond order of one. It is also a bound state, although its bond strength is not so great as the ground state's. If enough energy were deposited into this state to break the bond, the atoms would separate not as two 1*s* atoms but as a 1*s* and a 2*s* atom. Because one of the resulting atoms is in a 2*s* state, $0.375\,E_h$ higher in energy than the 1*s* H atom, the potential energy surface of the $a^3\Sigma_g^+$ state at large R is drawn $0.375\,E_h$ higher than the surface for the X and b states. Recall that the potential surface at large R corresponds to the energy of the separated atoms at rest.

If enough energy is poured into the H_2 molecule, one of the electrons may leave the molecule before the nuclei separate, leading to formation of the molecular ion H_2^+. The ψ_+ ground state of H_2^+ that we discussed earlier, which corresponds to the $X^2\Sigma_g^+$ state of the $1\sigma_g$ configuration, dissociates into a 1*s* H atom and a proton. The potential energy at large R lies $0.5\,E_h$ or 13.6 eV above the surface at large R for the X and b states of H_2, since this is the ionization energy for $H(1s) \rightarrow H^+ + e^-$.

The N_2 molecule, with the ground state configuration $[Be_2]\sigma_g^2 1\pi_u^4$, has a $^1\Sigma_g^+$ ground state because all the orbitals are filled. Even so, the electronic energy level structure of N_2 is very complicated. If we include the potential energy curves, as in Fig. 7.8, the problem of identifying excited electronic states becomes apparent.

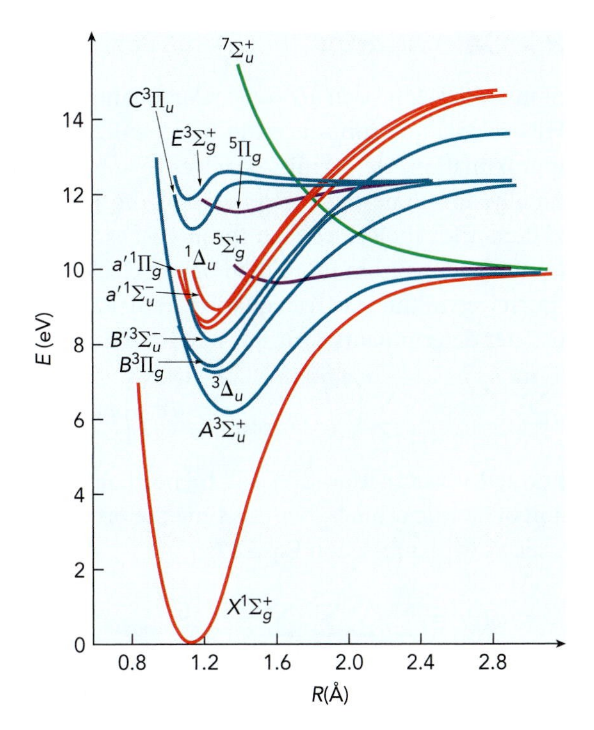

FIGURE 7.8 Potential energy surfaces for selected states of the N_2 molecule. The curves are drawn in red for singlet states, blue for triplet, purple for quintet, and green for septet.

7.3 Computational Methods for Molecules

Since shortly after the formulation of quantum mechanics, chemists have sought reliable and efficient methods to compute the properties of individual molecules, particularly the molecular geometries, the relative energies of different species in a chemical reaction, and the electronic spectra. All three of these properties may be determined by the solutions to the electronic Schrödinger equation at different nuclear geometries; in other words, by using the Born-Oppenheimer approximation to find the effective potential energy surface of the system. In research, the computed properties of a molecule may be used to explain experimental observations or to predict the results of experiments that cannot yet be carried out (perhaps for reasons of expense or limitations of available technology).

For small molecules, chemists believe that we understand the quantum mechanics well enough that given enough resources and care, we could solve the Schrödinger equation with sufficient accuracy to predict the majority of measurable properties reasonably well. The danger is that we rarely have enough resources. As we've seen, the number of dimensions involved even in the many-electron atom is daunting, and molecules make the problem worse in several ways:

- The potential energy function for the electrons loses much, if not all, of its symmetry.
- The potential energy function you start with may not even be the right one if the nuclei are not in the right geometry from the start.
- The integral needed to compute the electron–electron repulsion is much more difficult to evaluate when the orbitals are centered on different nuclei.

So what do we do?

Hartree-Fock Calculations

The most common initial approach to molecular quantum mechanics is the Hartree-Fock (HF) calculation, introduced in Section 4.2. Applying the Born-Oppenheimer approximation, we permit ourselves to solve only the electronic wavefunction initially, for some fixed geometry of the nuclei. Therefore, the approach is no different for molecules than for atoms, except that the potential energy includes contributions from more than one nucleus. For simplicity here, we will confine ourselves to the Hartree equation, rather than the HF equation that is applied to Slater determinants. Our Eq. 4.26 for the atom becomes

$$\hat{H}(i)\varphi(i) = \left[-\frac{\hbar^2}{2m_e}\nabla(i)^2 - \sum_{k=1}^{M}\frac{Z_k e^2}{4\pi\varepsilon_0 r_{ik}} + \sum_{j\neq i}\left(\int |\varphi_j|^2 \frac{e^2}{4\pi\varepsilon_0 r_{ij}} d\tau\right)\right]\varphi(i) = \varepsilon_i\varphi(i). \tag{7.3}$$

Once again, the orbital wavefunction $\varphi(i)$ will be normally be constructed as a linear combination of atomic orbitals, but this time the orbitals may be centered on any of the nuclei, as we expressed in Eq. 5.12:

$$\varphi(i) = \sum_{j=1}^{M}\sum_{n=1}^{n_{\max}}\sum_{l=0}^{n-1}\sum_{m_l=-l}^{l} c_{nlm_l,j}\psi_{nlm_l,j}(i),$$

where j specifies which nucleus is the center of the basis function $\psi_{nlm_l,j}$.[2]

Equation 5.12 conceals a dreadful computational hazard. We obtain a variational solution to the HF Eq. 7.3 normally by a numerical integration to solve the average value theorem for the energy,

$$\langle E\rangle = \sum_i \int_{\text{all space}} \varphi(i)\,\hat{H}(i)\,\varphi(i)\,d\tau_i + U_{nuc-nuc}, \tag{7.4}$$

and then we vary the LCAO coefficients $c_{nlm_l,j}$ to minimize $\langle E\rangle$. Because the $\hat{H}(i)$ operators contain integrals of $\varphi(j)^2$ in the electron–electron repulsion term (Eq. 4.27), each integral contains four φ orbital functions, each (in general) centered on a different nucleus. If each φ is a sum of many atomic orbitals then we have roughly N^4 of these four-center integrals to calculate, where N is the number of functions in our basis set. These cannot be solved analytically if the orbitals have the shape e^{-ar} and are centered on different nuclei. It would take too long to numerically calculate all of the four-center integrals (and they would have to be recalculated at each new geometry), so a clever substitution is used instead. We replace the familiar e^{-ar} shape of our orbitals with a Gaussian function, $e^{-\alpha r^2}$. The reason for this is that the four-center integrals can then be replaced by two-center integrals, because it turns out that the product of two Gaussian centered in different places can be written as a constant times just one Gaussian:

$$e^{-\alpha|\vec{r}-\vec{R}_A|^2}\, e^{-\beta|\vec{r}-\vec{R}_B|^2} = e^{-\alpha\beta|\vec{R}_A-\vec{R}_B|^2/(\alpha+\beta)}\, e^{-(\alpha+\beta)|\vec{r}-\vec{R}_0|^2} \tag{7.5}$$

[2]In practice, one would not specify the $n_{\max}$ values (since these do not determine the symmetry of the atomic orbital), but would simply choose as many basis functions as practical to adequately model the shape of the radial probability density. The maximum l value, on the other hand, is normally specified for each atom, and does not need to be limited to the atom's ground state l value. For example, the atomic orbital functions on a hydrogen nucleus typically include up to $l = 1$ functions, while a carbon atom in the same calculation might be given basis functions up through $l = 2$.

where

$$\vec{R}_0 \equiv \frac{\alpha\vec{R}_A + \beta\vec{R}_B}{\alpha + \beta}. \tag{7.6}$$

Even if we conserve resources by using Gaussian functions, quantum mechanical calculations that attempt to model large atoms, such as those of Period 4 and beyond in the periodic table, are often too demanding to carry out if the electrons are all included explicitly in the wavefunction. Furthermore, the core electrons of atoms with large nuclear charges have relativistic velocities, and special relativity alters the electron distribution.[3] It has therefore become popular to treat these heavy atoms by separating the electrons into the valence electrons, which are included explicitly as for the other atoms, and the core electrons, which are dropped out of the wavefunction and included instead as an **effective core potential** (ECP), an electric field contribution to the potential energy term in the Hamiltonian, rather than as explicit particles. By replacing the electrons with the electric field they generate, the number of basis functions drops dramatically. In the case of a platinum atom, for example, we would need basis functions for 78 electrons in an all-electron basis set, but only 10 electrons when we use an effective core potential for the electrons in the subshells $1s$ through $5p$. Furthermore, the effect of relativity on the distribution of the core electrons can be taken into account by the shape of the effective core potential, rather than having to solve the relativistic Schrödinger equation. As shown by the terrible example of the simplest H_2 MO wavefunction in Section 5.2, the prediction of basic chemical properties such as the dissociation energy generally requires a more demanding approach than we needed when looking at the orbital energies of atoms. Furthermore, HF calculations give significant errors in other experimental properties, such as relative energies of different isomers of the same compound. For these reasons, HF calculations are often only a starting point in calculations of molecular properties.

Density Functional Calculations

A particularly valuable addition to the arsenal of methods has been **density functional theory** (DFT), which aims to predict molecular properties with greater accuracy than Hartree-Fock calculations. Rather than directly integrating the electronic Schrödinger equation to get the N-electron wavefunction (where N is the number of electrons in the molecule), DFT methods solve instead for the overall electron *density*, ρ_0. The many-electron wavefunction is a function of $3N$ coordinates (x_i, y_i, z_i for each electron i), but ρ_0 depends on just 3 coordinates: x, y, and z. Only this density function is needed to calculate the energy, and many other molecular properties. By skirting the need for the explicit, many-electron wavefunction, DFT methods provide a fast alternative route to predicting properties of molecules. Other methods make use of perturbation theory, variational configuration interaction, and extensions of valence bond theory.

[3] At relativistic speeds, the electron mass is greater, and therefore the second derivative $\nabla_i^2\psi(i)$ needs to be greater to keep the same kinetic energy. A higher second derivative means a more rapidly varying wavefunction, and therefore a smaller average orbital radius. Relativistic electron cores are more compact than the non-relativistic Schrödinger equation predicts.

The many-electron density of the electronic ground state wavefunction ψ may be written by picking any electron, say electron 1, and integrating over the positions of all the other electrons (2 through N) to average out their effects on the position of electron 1. Then we multiply the result by the number of electrons N. We can write the density in this way, picking any one electron that we wish, because in the correct wavefunction all the electrons are indistinguishable. The expression for ρ_0 looks like this:

$$\rho_0(x,y,z) = N\int \ldots \int |\psi(x,y,z,x_2,\ldots,z_N)|^2 dx_2 \ldots dz_N. \qquad (7.7)$$

The name "density functional theory" comes from the description of E_0 and other properties as *scalar values* (numbers) that can be determined from multivariable *functions.* A functional $F[f(q)]$ gives the dependence of the scalar F on the ordinary function $f(q)$. For example, the electronic energy E_0 that we obtain from the Schrödinger equation is a functional of the wavefunction ψ, and we represent this relationship using square brackets, $E_0[\psi]$. In density functional theory, we instead write the energy as a functional that depends on the density function, $E_0[\rho_0]$. Pierre Hohenberg and Walter Kohn proved that ρ_0 can be variationally optimized (the same way we variationally optimize the wavefunction in Hartree-Fock calculations) from this density. And from that density—without ever having to write the many-electron wavefunction—we could (in principle) determine all the electronic properties of the ground state, including the ground state energy E_0.

But there's a catch: *we don't know how to write the function we're optimizing.* For a given arrangement of the nuclei, the energy can be written as a sum over average values of the contributors to the Hamiltonian:

$$E_0[\rho_0] = K[\rho_0] + U_{\text{nuc-elec}}[\rho_0] + U_{\text{elec-elec}}[\rho_0] + U_{\text{nuc-nuc}}. \qquad (7.8)$$

Applying the Born-Oppenheimer approximation, we treat the nuclei as stationary, so $U_{\text{nuc-nuc}}$ is a constant. We can evaluate the nucleus-electron potential energy functional by integrating over the sum of the Coulomb energies for the interaction of the electron cloud (represented by the density ρ_0) with each nucleus k:

$$\langle U_{\text{nuc-elec}}[\rho_0]\rangle = -\int \rho_0(x,y,z)\left[\sum_k \frac{Z_k e^2}{r_k}\right] d\tau, \qquad (7.9)$$

where the distances r_k are measured between each nucleus and the point (x,y,z) where we evaluate the electron density. Then we run into trouble: we don't know any way to derive the functionals for the kinetic and electron–electron potential energies.

Now imagine that we have a reference system s that is a group of electrons—no nuclei—with exactly the same density function, so that $\rho_s = \rho_0$ at every point in space, but these electrons are *not interacting.* There must exist some unique potential energy function $U_s(x,y,z)$ which would push the non-interacting electrons into the correct density ρ_0. We call this contribution to the energy the *external* potential energy, and it is given by the integral of $U_s(x,y,z)$ over the electron density:

$$U_{\text{ext}}[\rho_0] = \int U_s(x,y,z)\, \rho_0(x,y,z)\, d\tau. \qquad (7.10)$$

If ρ_0 is known, then $U_{\text{ext}}[\rho_0]$ can be calculated.

The major part of the electron–electron repulsion can be calculated from the density by integrating over the Coulomb repulsion between the electron density values at any two points—the uncorrelated repulsion energy—which we'll label $U_{\rho\rho}$:

$$U_{\rho\rho}[\rho_0] = \frac{1}{2}\iint \frac{\rho(x_1,y_1,z_1)\rho(x_2,y_2,z_2)}{r_{12}}\, d\tau_1 d\tau_2. \tag{7.11}$$

The density is already required to be correct, so we are not neglecting any correlations between the electrons that affect their positions when we calculate $U_{\rho\rho}[\rho_0]$. However, this classical expression for the repulsion energy neglects the exchange force (Section 4.3).

The kinetic energy $\langle K_s[\rho_0]\rangle$ of these non-interacting electrons can also be calculated from the density, but it is not the same as the kinetic energy $\langle K\rangle$ of the real, mutually interacting electrons because it neglects all of the correlation between the movements of the electrons. We roll these missing terms for the exchange and correlation into a single functional, and rewrite the energy in Eq. 7.8 as follows:

$$E_0[\rho_0] = \langle K_s[\rho_0]\rangle + U_{ext}[\rho_0] + U_{\rho\rho}[\rho_0] + E_{xc}[\rho_0] + U_{\text{nuc-nuc}} \tag{7.12}$$

where the one thing we can't calculate directly from the electron density—because we don't know its form—is the **exchange-correlation energy functional** $E_{xc}[\rho_0]$. Getting $E_{xc}[\rho_0]$ now becomes the focus of the problem.

If the electron density is a slowly varying function, then $E_{xc}[\rho_0]$ can be calculated in the **local density approximation** (LDA), where the density only needs to be known at one point to calculate $\varepsilon_{xc}(\vec{r})$ at that point. That needn't be true, because Hohenberg and Kohn's theorem (that we can determine each observable of the system from the density) holds rigorously only if we know the density function *everywhere.* The **generalized gradient approximation** (GGA) provides a correction to the LDA using the gradient of the electron density at each point. A hybrid DFT method incorporates the HF solution to the exchange integrals as all or part of its functional. As an example, the popular B3LYP DFT method combines several functionals as follows:

$$\begin{aligned} E_{xc} &= E_{xc}(\text{LDA}) + (0.20)[E_x(\text{HF}) - E_x(\text{LDA})] \\ &\quad + (0.72)[E_x(\text{GGA}) - E_x(\text{LDA})] + (0.81)[E_c(\text{GGA}) - E_c(\text{LDA})]. \end{aligned} \tag{7.13}$$

Geometry Optimization

Many properties of molecules may be predicted from these computational approaches, but generally these require first knowing the molecular geometry. Therefore, a typical quantum mechanical calculation will begin by adjusting the positions of the atoms to find the minimum energy on the effective potential energy surface for the nuclei. This step, the **geometry optimization,** involves a series of steps such as the following:

1. Calculate the electronic energy at the current nuclear geometry, and add in the nuclear potential energy to find the effective potential energy U_{eff} at that geometry.

2. Calculate (either analytically or numerically) the derivative of U_{eff} with respect to each geometrical coordinate; these derivatives resemble forces, which are the derivatives of a true potential energy function with respect to one of the spatial coordinates.
3. Use the forces to estimate how the geometry should change in order to reach its minimum value. Different algorithms exist for making this estimate; among the simplest and most commonly used is the **Newton-Raphson method,** described in Problem 7.81.
4. Move the atoms according to your estimate in the previous step to obtain a new geometry. Go back to step 1.

The process ends when the forces are all roughly zero, indicating that the calculation has identified a stationary point on the U_{eff} surface. In cases where the surface is complicated, this series of steps may arrive at a stationary point that is not a minimum, but a saddle point. To test this possibility, one may calculate all the second derivatives of U_{eff}: a minimum on the surface will have only positive second derivatives.

These methods work for the ground electronic state of a molecule. Within limits, these methods may also be used to solve for the optimum geometries and properties of excited states as well, especially the lowest energy states of a particular symmetry Γ and total electron spin S. Although these excited states usually do not last long, they may serve as gateways to new species or analytical techniques. Once we know the U_{eff} functions in the ground and excited electronic states, we can start to predict the behavior of the molecule when it absorbs enough energy to reach some of these excited states.

7.4 Energetic Processes

Table 7.1 shows that in typical small molecules, energies of roughly 1 eV or more are required to reach the excited electronic states from the ground state. This energy corresponds to over 10^4 K, much more energy than the molecule is likely to acquire by collisions with its neighbors. Therefore, under ordinary conditions, nearly all of these molecules will be in their ground electronic state. Exceptions include molecules of the transition metals and molecules with extensive conjugated π-orbitals, because both of these systems tend to have a high degeneracy of states near the ground state.

Let's focus on the simpler small molecules for the moment. A selective method for reaching these excited electronic states is by a direct spectroscopic transition from the ground state, induced by the absorption of visible or ultraviolet radiation. For this to be most effective, the selection rules must favor the transition, although forbidden transitions may be observed by increasing the radiation power or operating with a particularly sensitive experiment.

The general selection rules for electric-dipole allowed transitions in diatomic molecules are $\Delta\Lambda = 0, \pm 1$, $\Delta S = 0$, $g \leftrightarrow u$, and (for Σ states only) $+ \leftrightarrow +$ and $- \leftrightarrow -$. The selection rule on S is particularly limiting, because it implies that the relative energies of, for example, the singlet and triplet states in H_2 cannot be easily determined by measuring transitions between them. The measurement is difficult because the forbidden ($\Delta S \neq 0$) transitions rarely occur and are consequently hard to detect.

When we pour energy into a molecule, the energy has to go somewhere, and it usually will not stay long in excited electronic motion. A variety of processes can occur to redistribute this energy, and the energy may end up in any of several different forms. We now examine some of the processes that result when molecules are energized, dividing them into **relaxation** processes, which leave the molecule intact but redistribute the energy from the electronic motion elsewhere, and **fragmentation** processes, which change the chemical nature of the molecule.

Relaxation Processes

1. *Fluorescence.* Perhaps the simplest way for an energized molecule to redistribute its excess energy is by spitting out a photon. The emission of a photon during an *allowed* electronic transition is called **fluorescence.** Because the transition is allowed, the molecule normally spends very little time in the initial excited state (often only a few nanoseconds or less), so fluorescence is observed very soon after the molecule is excited. The final state after fluorescence does not need to be the ground state, nor does the molecule have to arrive at the lowest energy point of the final electronic state. These processes are often illustrated using diagrams such as Fig. 7.9a, in which the energy transfer is shown as a sequence of arrows on the potential energy curves of the electronic states of the molecule. The horizontal axis in these diagrams corresponds to the bond length of a diatomic molecule, and we expect this value to change as the molecule stretches back and forth. However, nuclei are much slower than electrons, and the nuclei may as well be stationary during the change in electron distribution as radiation is absorbed or emitted. Therefore, radiative processes appear in these diagrams as *vertical arrows,* indicating that the nuclei are not moving as the photon energy is transferred into or out of the molecule. Typically, an excited state molecule has time to vibrate back and forth a bit before fluorescence, so the vertical line down for the emission may end up almost anywhere along the potential energy curve of the lower state.
2. *Phosphorescence.* In this scenario, we begin again in a bound excited electronic state, but this time the transition to the lower state is *forbidden.* This radiation is called **phosphorescence.** Because the forbidden transition slows the rate of photon emission, phosphorescence is generally weaker than fluorescence but takes place over relatively long times (often milliseconds, some cases much longer). If the transition down is forbidden, how do we get to that excited state in the first place? A common mechanism is illustrated in Fig. 7.9b. An allowed transition from the ground state X to an excited state A may be followed by a non-radiative transition of the molecule from state A to state a, called an **intersystem crossing.** The transition between A and a does not have to

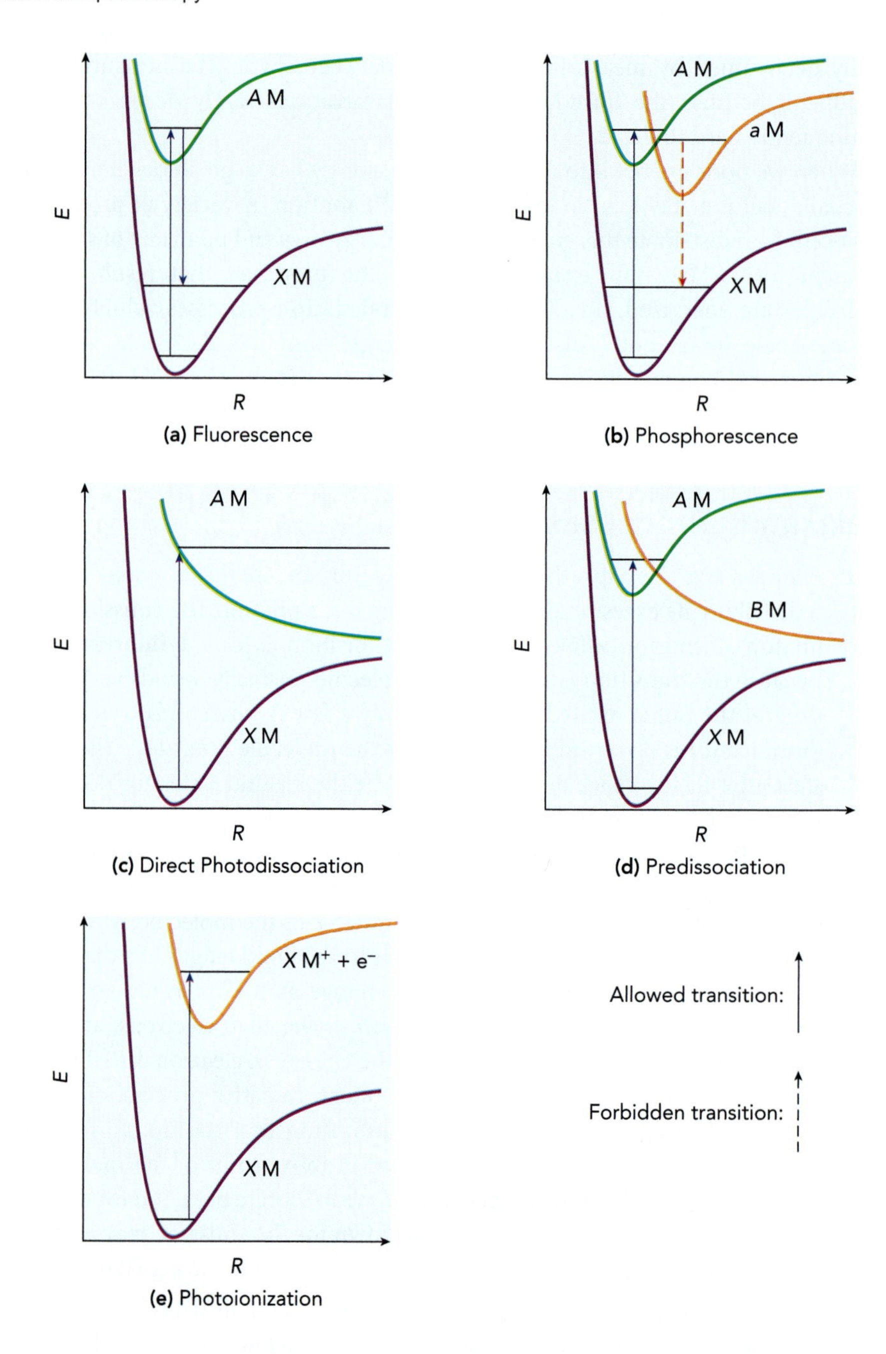

▶ FIGURE 7.9 **Processes that redistribute the energy in energized molecules.**

satisfy selection rules, which apply to radiative processes. It occurs because, in keeping with perturbation theory, when two energy levels are very close to one another, they will mix strongly. For example, if A is a singlet state ($S = 0$) and a is a triplet state ($S = 1$), then at the energies where the two potential energy curves cross, the overall electron spin S becomes a poor quantum number: it is no longer a constant 0 or 1. The A and a states can become so alike in character by this mixing that the molecule smoothly vibrates from the potential energy curve for A onto the curve for a. If state a has a lower energy at equilibrium than A, then small losses of vibrational energy can soon trap the molecule in state a, and there is no longer any possibility of an allowed transition back to the ground state X. The molecule must wait to lose its energy either by phosphorescence or by shifting the energy onto a different molecule.

3. *Vibrational relaxation.* The molecular Hamiltonian contains a vibrational energy contribution among the other kinetic energy terms in Eq. 5.11. Although we analyze the quantum mechanics of vibrations in detail in Chapter 8, the presence of the nuclear kinetic energy terms in the Hamiltonian means that any energy we put into one particular degree of freedom can be redistributed to others. For example, if we excite a π orbital electron in HCN into a π^* antibonding orbital, we expect the equilibrium CN bond length to be affected, and this in turn will usually cause the CN bond to vibrate. If there is enough of that vibrational energy, however, some of it can be channeled out of the CN stretching vibration and into vibration of the HC bond. Lowering the energy of one degree of freedom (the CN vibration in this example) by raising the energy in a vibration somewhere else (the HC vibration) is known as **intramolecular vibrational relaxation** or **internal conversion.** Although, in principle, it is possible for electronic energy, vibrational energy, and rotational energy all to be interconverted, it is usually much more efficient for energy to be transferred between like degrees of freedom. The intersystem crossing previously described as part of the mechanism for phosphorescence involves electronic and vibrational energy being interconverted.
4. *Collisional relaxation.* In fluorescence and phosphorescence, the energy originally put into the molecule is re-expressed in the form of radiation. There are other final forms that the energy can take, however. Excited state molecules may pass their energy onto another molecule during a collision, and in some cases that allows the energy of one chemical component to channel energy selectively into a few quantum states of a second component. This mechanism of energy transfer is exploited in the operation of several types of lasers, including helium-neon lasers (electronically excited He atoms transfer energy into Ne atoms) and CO_2 lasers (vibrationally excited N_2 molecules transfer energy into CO_2). For this to be an efficient path of energy transfer, the colliding molecule should have a quantum state at about the right energy to accept the energy being lost by the first molecule.

Fragmentation Processes

1. *Photodissociation.* If we excite our molecule into a state that no longer has a potential barrier to the separated atoms (an unbound state), then the atoms will fall apart, a process called **direct photodissociation** (Fig. 7.9c). This process redistributes the energy of the incoming photon into the potential energy needed to break the bond, with any excess energy going into kinetic energy of the molecular or atomic fragments. If the transition into the upper state is allowed, dissociation occurs on the timescale of a single vibration, usually less than one picosecond.
2. *Predissociation.* Our next mechanism is related to photodissociation but takes longer. Some excited states are bound states, but at an energy higher

than the ground state dissociation energy. These are states formed in the correlation diagram from excited state atoms. Once we reach one of these states, an intersystem crossing to a dissociative state may occur, at which point the molecule falls apart. Unlike direct photodissociation, the process occurs on a timescale longer than one vibration. The sequence is illustrated in Fig. 7.9d. The photon does not directly break the chemical bond, and this process is called **predissociation.**

3. *Photoionization.* If the energy of the radiation is high enough, it may **photoionize** the atom or molecule, with the excess energy converted into kinetic energy of the electron (Fig. 7.9e). If the incident radiation has energy per photon $h\nu$ and the energy required for ionization is IE, there may be excess energy K_{elec} that goes into the kinetic energy of the electron: $K_{\text{elec}} = h\nu - \text{IE}$. The higher $h\nu$ is relative to IE, the higher K_{elec} and the faster the electron flies away from the molecular ion.

Spectroscopy of Transition Metal Compounds

An appreciation of electronic state symmetry is particularly useful when studying transition metal compounds. The transition metals are defined by a partially filled *d* subshell in the ground electronic state of the atom. The five *d* orbitals are degenerate in the spherically symmetric isolated atom, but that degeneracy is broken when we introduce chemical bonds. The distribution of the orbital energies that results can often be interpreted using the tools of group theory and a little chemical common sense.

For example, we can form ionic complexes of six Cl^- ions attached in an octahedral arrangement to any of several transition metal cations. We should be able to draw a correlation diagram showing how the five *d* orbitals of the metal shift in energy as we slowly bring the atoms closer to the central metal atom. As shown in Fig. 7.10, the five degenerate *d* orbitals separate into two groups: the doubly degenerate e_g and triply degenerate t_{2g} sets of the point group O_h, with the t_{2g} at lower energy. We can explain why this occurs, using the same principles that we used to determine MO symmetries in Chapter 6.

First we need to mix the *d* orbitals to obtain pure real functions, in the same way that we constructed the Cartesian p_x, p_y, and p_z from the three *p* functions defined in terms of m_l. For the *d* orbitals, we do this again by combining orbitals with the same magnitude of m_l so that the imaginary parts cancel (Problem 3.42), and we obtain

$$\begin{aligned} d_{z^2} &= d_0 \\ d_{yz} &= -\frac{i}{\sqrt{2}}(d_1 + d_{-1}) \quad d_{xz} = \frac{1}{\sqrt{2}}(d_1 - d_{-1}) \\ d_{xy} &= -\frac{i}{\sqrt{2}}(d_2 + d_{-2}) \quad d_{x^2-y^2} = \frac{1}{\sqrt{2}}(d_2 - d_{-2}). \end{aligned} \tag{7.14}$$

We define our Cartesian axes to contain the six atoms so that they approach the metal from $+\infty$ and $-\infty$ along *x*, *y*, and *z*. The three axes are equivalent, and the three orbitals d_{yz}, d_{xz}, d_{xy} are therefore still degenerate. All of the *d* orbitals are

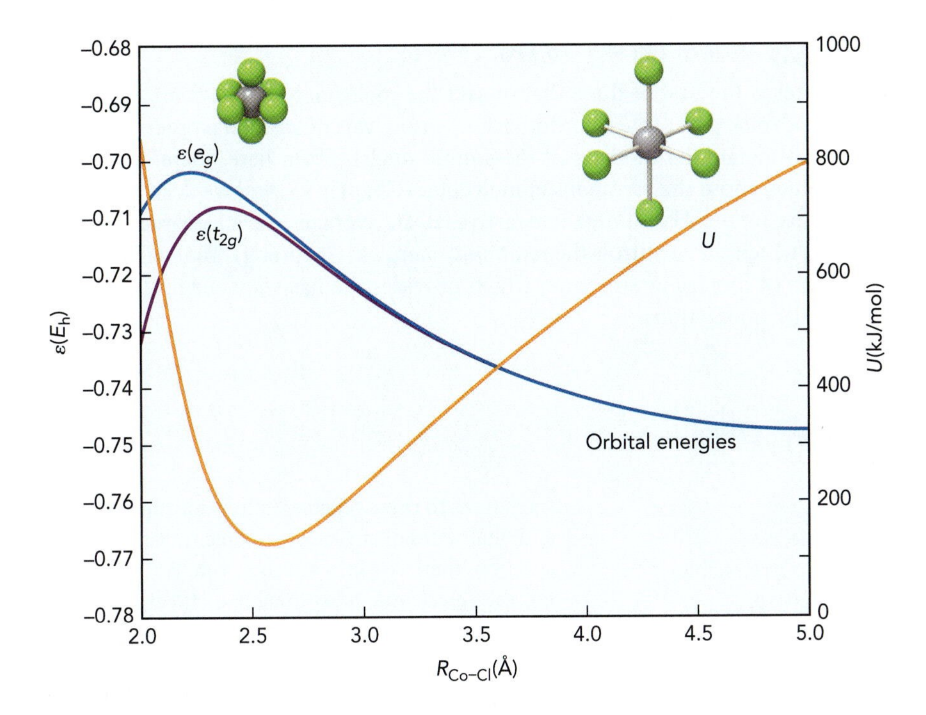

◄ **FIGURE 7.10 Energies of the transition metal atom d orbitals as a function of ligand distance, drawn for $CoCl_6^{3-}$.** The five d orbitals, degenerate at large distance R, split into two groups of orbitals, e_g and t_{2g}, as the Cl^- ions approach the metal and distort the spherical symmetry of the potential energy.

symmetric under inversion, so they will correspond to g representations of the point group O_h. To find whether this triply degenerate set is represented by t_{1g} or t_{2g}, we can try out a $\hat{C}_4(z)$ rotation on all three orbitals: d_{xz} and d_{yz} are converted by this operation into d_{yz} and $-d_{xz}$, respectively, while d_{xy} is rotated into $-d_{xy}$. The net change is $-1+1-1 = -1$, which is the character under $\hat{C}_4$ for t_{2g}.

It is more difficult to show that the two remaining orbitals are still degenerate also, but they cannot belong to the non-degenerate representations. The non-degenerate representations are either symmetric or antisymmetric under 90° rotation about any of the $\hat{C}_4$ axes, and neither the d_{z^2} nor the d_{xy} satisfy this criterion. There is only one doubly degenerate representation with g symmetry: the e_g, and this is the representation for these last two d orbitals.

As non-bonding electrons, the d electrons of the transition metals see less repulsion if they lie away from the bonding electrons localized along the x, y, and z axes. Therefore, the t_{2g} orbitals are lower energy than the e_g orbitals.

This distribution of orbitals is common to a large family of transition metal compounds. The symmetry may be disrupted further if the six atoms coordinated to the metal are not equivalent, but it is still generally possible to group the resulting orbitals into approximate t_{2g} and e_g sets. These symmetry properties have wonderful implications for the spectroscopy of these compounds, because, despite the rich complexity of the electronic states in transition metals, there are some general patterns that emerge for this family of compounds, patterns that we can exploit to analyze the nature of the metal and its bonds.

Our next chapter takes us deeper into these bonds. Although chemists tend to obsess about where electrons are and where they're going, chemical bonds in fact are defined by their ability to stabilize two or more atomic nuclei side by side. It's time, therefore, that we return to the nuclear terms in the molecular Hamiltonian.

CONTEXT *Where Do We Go From Here?*

This chapter is the first of three that dissect the molecular Hamiltonian. By solving just the *electronic* part of the Schrödinger equation, we construct the potential energy curves that dictate the motions of the atomic nuclei. From here, therefore, we can explain much about the *vibrations* of molecules (Chapter 8). And we draw on both of these terms in the Hamiltonian to arrive at an average overall geometry of the molecule, which then controls the *rotational* energies (Chapter 9). That will complete our picture of molecular structure, which provides the basis for our understanding of molecular interactions.

KEY CONCEPTS AND EQUATIONS

7.1 Molecular orbital configurations.

a. If there is any symmetry in the molecule, we normally label each MO according to its irreducible representation in the molecule's point group.

b. To write an MO configuration, we need to know not only the symmetries of the MOs, but also their energy ordering, because we fill the orbitals in order of increasing energy. We can often estimate this ordering using correlation diagrams, which connect states from different geometries that have a shared symmetry.

7.2 Electronic states.

a. The symmetry of the overall electronic state is the direct product of the representations of any partly filled orbitals. We can then apply the symmetry selection rules for electric dipole and Raman transitions to these representations for the initial and final states of any spectroscopic transition. In addition, there is a spin selection rule $\Delta S = 0$.

b. For each electronic state, there is an effective potential energy function that depends on the geometry. The actual energy necessary to move from one state to another will depend not only on what states are involved but also on the geometry of the molecule during the transition.

7.4 Energetic Processes.

Molecules that have been electronically excited may redistribute that energy by shifting the energy to other degrees of freedom (relaxation) or by ionization or chemical bond dissociation (fragmentation).

KEY TERMS

- A **correlation diagram** shows how electronic states shift in energy as the geometry of the molecule changes from one symmetry to another.
- A **molecular orbital diagram** shows schematically how the energies of the MOs formed by combining a set of atomic orbitals lie above and below the energies of the original atomic orbitals.
- **Fluorescence** is the emission of photons soon after excitation via an allowed transition.
- **Phosphorescence** is the emission of photons long after excitation via a forbidden transition.
- **Intramolecular vibrational relaxation** is the redistribution of energy into vibrational motions of a polyatomic molecule.
- **Photodissociation** is the absorption of radiation leading to breaking a chemical bond.
- **Photoionization** is the absorption of radiation leading to the ejection of an electron.
- **Density functional theory** is a semi-empirical method for extracting properties of quantum mechanical systems (such as molecules) from the electron density, rather than from a wavefunction using the Schrödinger equation.
- An **intersystem crossing** is the spontaneous transfer of a molecule from one electronic state to another, usually near the intersection between the potential energy curves of the states.
- **Predissociation** is the breaking of a chemical bond by intersystem crossing to a dissociating electronic state from a non-dissociating state.

OBJECTIVES REVIEW

1. *Construct and interpret correlation diagrams for certain highly symmetric molecules.*
   Use the correlation diagram in Fig. 7.2 to predict the bond order for Be_2^+.
2. *Predict molecular orbital configurations for homonuclear diatomics and other simple molecules.*
   Use the correlation diagram in Fig. 7.2 to write the MO configuration for Be_2^+.
3. *Determine selection rules for electronic transitions in molecules.*
   Determine whether the ${}^1A_g \rightarrow {}^3A_u$ transition in *trans*-1,2-dichloroethene is allowed or forbidden by electric dipole selection rules.
4. *Use potential energy diagrams and selection rules to predict the processes resulting from electronic excitation of a molecule.*
   Based on the potential energy curves in Fig. 7.8, what process could return N_2 in the $A^3\Sigma_u^+$ state to the $X^1\Sigma_g^+$ ground state?

PROBLEMS

Discussion Problems

7.1 Explain why the high Z elements tend to lie toward the separated atom limit of the correlation diagram.

7.2 Transitions between electronic states of the same MO configuration are always forbidden in a homonuclear diatomic molecule. Explain why this must be true.

7.3 According to our correlation diagram for homonuclear diatomics, the $1\pi_u$ MO in N_2 is more stable than the $3\sigma_g$ MO. If this is rigorously true, the dication N_2^{2+} would have two π-bonds, but no σ-bond. Only one of the following statements can be true:

a. This ion does have a σ bond, despite the prediction of the correlation diagram.

b. This ion does not have a σ bond, despite the fact that the π bonds have more nodes and should be higher energy.

Which statement is true, and how do you explain the suggested paradox?

7.4 Explain each of the following briefly:

a. The one-electron atom has only spin-doublet term states.

b. A closed-shell molecule has only spin-singlet states.

c. If a molecule has a triplet-spin electronic state, it will have no doublet-spin states.

7.5 An excimer is a molecule, such as ArF or XeCl, with a potential energy diagram of the sort diagrammed in the following figure. The two lowest-energy electronic states are shown. Explain briefly what happens to the molecule after it is formed in the $A\Pi$ excited state.

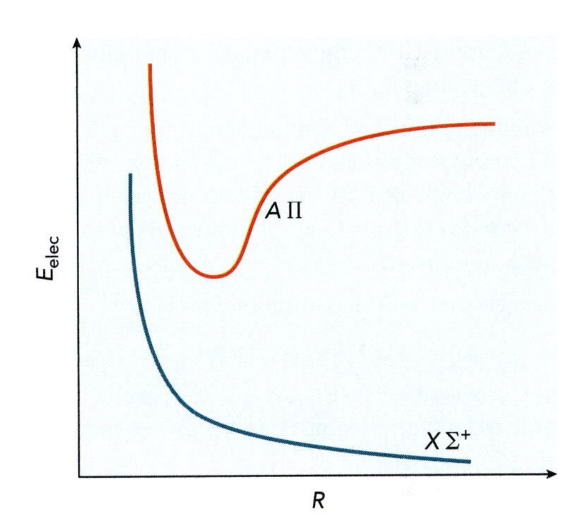

7.6 Complete the following statements:

a. In the point group D_{2h}, $b_{2g} \otimes a_u =$

b. The degeneracy of the $\lambda = \pm 4\ \gamma$ orbital of a linear molecule is ________.

c. The lowest energy molecular orbital in H_2 has symmetry representation ________.

d. The ground electronic state of benzene has symmetry representation ________.

e. In the correlation diagram for homonuclear diatomics, the $1\pi_u$ molecular orbital is formed by combining two atomic $2p$ orbitals with $m_l =$ ________.

f. The complete list of MO representations obtained for a homonuclear diatomic molecule from two $3d$ orbitals in the separated atom limit is ________.

7.7 A mass spectrometer analyzes atomic and molecular ions. Often the ions are created when neutral molecules are struck by electrons accelerated to kinetic energies of 5 to 100 eV, as in the following example for HF:

$$HF + e^- \rightarrow HF^+ + 2e^-.$$

a. Is the HF^+ likely to be formed in the ground or excited *vibrational* state?

b. Is the HF^+ likely to be formed in the ground or excited *rotational* state?

Correlation Diagrams and MO Configurations for Diatomics

7.8 What would be the MO configuration of O_2 in its ground state if it were (a) very near the united atom limit, and (b) very near the separated atom limit.

7.9 For the homonuclear diatomic molecule Be_2, give the MO configuration and bond order for the ground state and first excited state, based on the correlation diagram as the values of R approach (a) the separated atom limit and (b) the united atom limit.

7.10 Following the correlation diagram in Fig. 7.2, write the MO configurations and the term symbols for the *lowest excited state* of B_2 if the orbitals are ordered according to (a) the united atom limit, and (b) the separated atom limit.

7.11 The ground MO configuration for N_2 is

$$1\sigma_g^2 1\sigma_u^2 2\sigma_g^2 2\sigma_u^2 1\pi_u^4 3\sigma_g^2.$$

Find the term symbol(s) (e.g., ${}^3A_2$ or ${}^2\Sigma^-$) for the electronic state(s) originating from the lowest energy excited MO configuration of N_2.

7.12 Based on the correlation diagram drawn in Fig. 7.2 find the electronic states formed when one electron is promoted from the π_u MO of N_2 to the *second* lowest energy unoccupied MO. What is the new bond order?

7.13 Give the MO configurations, bond orders, and term symbols for ground state N_2^+ and N_2^- using the correlation diagram.

7.14 Homonuclear diatomics of heavier elements tend to lie closer to the separated atom limit of the correlation diagram (Fig. 7.2). Based on the dashed line representing O_2 in this figure, what neutral, homonuclear, diatomic molecule would be the first to have a filled δ molecular orbital in its ground state, and how many electrons does it have?

7.15 The correlation diagram in Fig. 7.2 does not go high enough in energy to reach any δ_u molecular orbitals. What are the lowest energy atomic orbitals in (a) the united and (b) in the separated atom limits that would give the correct symmetry for a δ_u molecular orbital?

7.16 The Ne_2 molecule dissociates at room temperature. Using the correlation diagram for homonuclear diatomics as a guide, write the MO configuration and estimate the bond order of NeF in its lowest excited state. (The molecule is $C_{\infty v}$ instead of $D_{\infty h}$, but we can expect the energy ordering of the orbitals to be very similar.)

7.17 Draw only the left-hand side of the correlation diagram for homonuclear diatomics corresponding to the $n = 5$ united atom atomic orbitals, identifying the appropriate MO representations.

7.18 Given that Li_2 and Be_2 will have bond lengths longer than that of H_2, use the correlation diagram for homonuclear diatomics to write the ground state MO configurations for both molecules. Circle the MO configuration that you would expect to form the stronger chemical bond.

7.19 Assume that Li_2^+ follows the line for N_2 in the correlation diagram. Write the MO configuration for the ground state of Li_2^+, and the ground state term symbol.

7.20 Predict the ground state MO configuration and electronic state (${}^2\Pi_g$, etc.) for F_2.

7.21 From the correlation diagram for homonuclear diatomics, determine the MO configuration and bond order for B_2^{2+}, assuming B_2 lies on the same line as N_2.

7.22 The diatomic molecule C_2 lies in the region of the correlation diagram between O_2 and H_2. What are the two possible valence MO configurations for the ground state of C_2 in this region? What are the bond orders for each of these MO configurations?

7.23 Find the MO configuration, bond order, and electronic state symbol for the ground state of FO.

7.24 Find the symbol for the ground electronic state of CF, assuming the MO energy levels are similar to those of O_2.

7.25 What is the MO configuration of the first excited state of N_2?

7.26 Write the complete MO configurations for the ground and first excited states of Ne_2 ($Z = 10$ for Ne) and calculate their bond orders.

7.27 You may recall a strange omission in the character tables of Chapter 6: with one exception, the symbols of the irreducible representations may be written using a capital letter (for overall symmetry) or a lower case letter (for the symmetry of a single-coordinate function, such as an orbital wavefunction). The exception is that the Σ^- representations of the linear molecule point groups are always written using capital letters, because Σ^- symmetry is not possible from a one-coordinate function. For example, σ molecular orbitals are always symmetric upon reflection through any vertical mirror plane, whereas Σ^- electronic states (which are antisymmetric under $\hat{\sigma}_v$)

may be formed when the angular momentum projections from two orbitals cancel, as may happen in the MO configuration π_g^2.

For a homonuclear diatomic molecule with nuclei A and B, consider the two-electron MO wavefunction π_g^2, where the π orbital wavefunctions may be written

$$\pi_+ = (P_{2,1}{}^{A} + P_{2,1}{}^{B})\, e^{i\varphi} \equiv P^{AB}\, e^{i\varphi}$$

$$\pi_- = (P_{2,1}{}^{A} + P_{2,1}{}^{B})\, e^{-i\varphi} \equiv P^{AB}\, e^{-i\varphi}.$$

The functions $P_{2,1}$ stand in for the r- and θ-dependent parts of the $2p$ atomic orbital wavefunctions centered on either nucleus A or nucleus B. For our purposes, only the φ-dependent part of the function really matters. Show that the overall spin-spatial wavefunctions that arise from this MO configuration include one function with Σ^- symmetry.

7.28 The ground state MO configuration of O_2 results in the electronic states $X^3\Sigma_g^-$, $a^1\Delta_g$, and $b^1\Sigma_g^+$. The excited $[N_2]\pi_g\sigma_u$ configuration gives rise to the electronic states $D^3\Pi_u$ and $e^1\Pi_u$. Identify all the allowed transitions between these two MO configurations.

7.29 What are the allowed electric dipole transitions between the states shown below 10 eV for N_2 in Fig. 7.8?

Correlation Diagrams and MO Configurations for Polyatomics

7.30 The following are the MOs for the molecule FO_2 arranged in order of increasing energy. Write the symbol for the lowest energy electronic state of FO_2.

$1b_2\ 1a_1\ 2a_1\ 3a_1\ 2b_2\ 4a_1\ 1b_1\ 5a_1\ 3b_2\ 1a_2\ 4b_2\ 6a_1\ 2b_1\ 7a_1\ 5b_2$

7.31 The correlation diagram for the valence molecular orbitals of XH_2 molecules has been started. Draw the lines that connect orbitals of the proper symmetry on each side of the diagram.

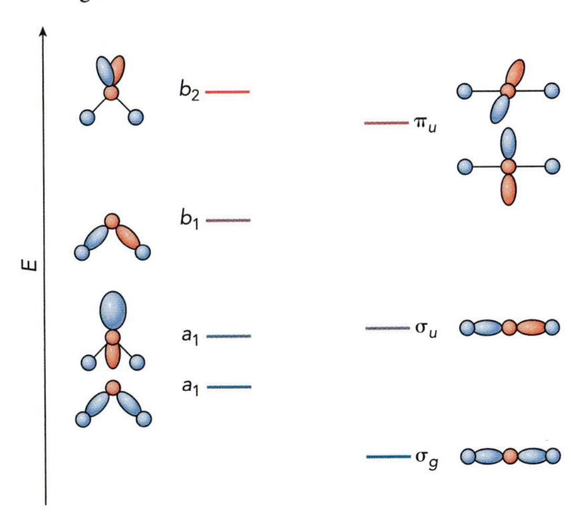

7.32 Here's a different correlation diagram for the dihydrides (compare to Problem 7.31). Finish labeling the MOs in the correlation diagram shown next for the C_{2v} dihydrides, and use this to find the ground state MO configuration and electronic state for BH_2.

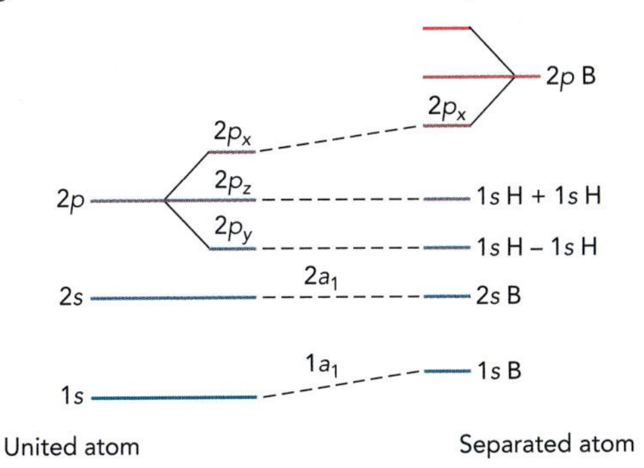

7.33 On the basis of the correlation diagram shown next, predict whether the lowest excited state of H_2CO is planar or not, and give its valence MO configuration. The core electron orbitals are *not* shown on the correlation diagram.

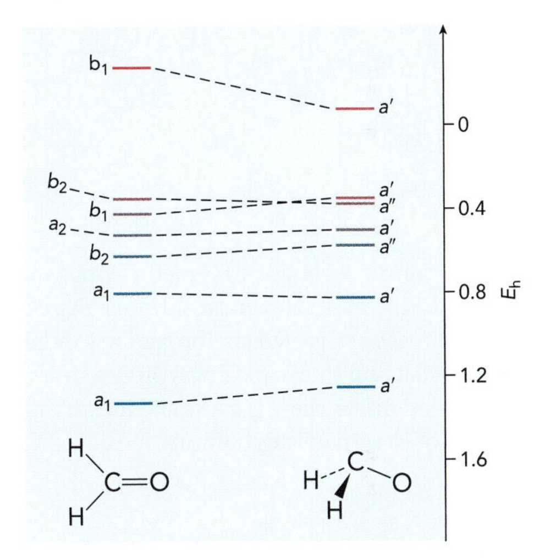

7.34 The most stable geometry of methane is a tetrahedral structure belonging to point group T_d. If we stand methane up on two C—H bonds and squash it until it becomes planar, the molecule goes through a D_{2d} geometry before becoming D_{4h} at the planar form:

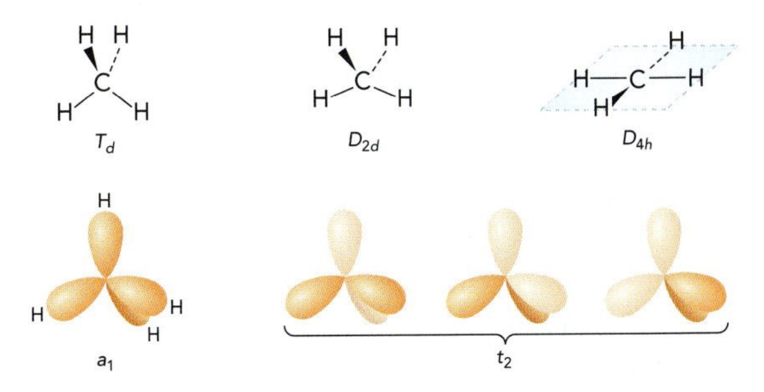

Given that the ground state MO configuration for CH_4 is $1a_1^2\,2a_1^2\,1t_2^6$ (the valence MOs are drawn next), draw the correlation diagram for this transformation. Show your coordinate system and rotational axis labels.

7.35 A correlation diagram for a molecule with three identical atoms is shown next between two limits: the linear $D_{\infty h}$ limit where the three atoms are linear, and the D_{3h} limit where the atoms form an equilateral triangle. Finish filling in the representations for the intermediate C_{2v} case.

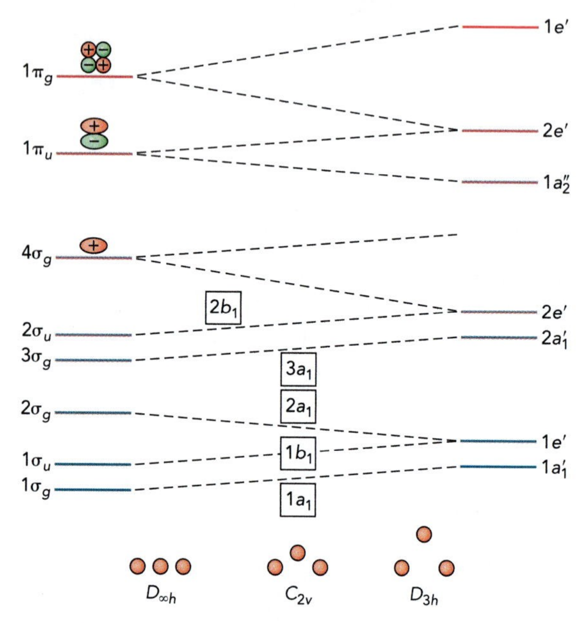

7.36 Ethane is composed of two CH_3 methyl groups joined by a single bond. The molecule can twist from a D_{3h} eclipsed conformation to a D_{3d} staggered conformation. Complete the correlation diagram shown next by writing in the symmetry representations for the D_{3d} states and identifying the point group of the intermediate conformations.

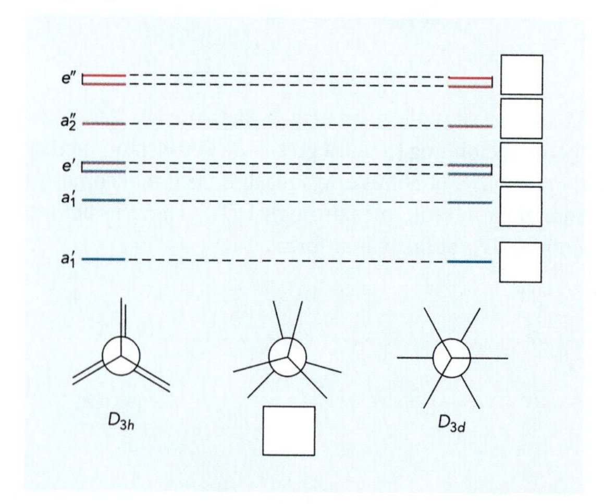

7.37 What is the representation for the lowest energy MO of any molecule in the point group D_{3h}?

7.38 The typical ground state MO configuration of a stable molecule has all its MOs filled. What can one deduce about the spin multiplicities $2S + 1$ of the states of the lowest excited MO configuration? If these are bound states, what processes (photoionization, fluorescence, etc.) can occur when they are excited?

7.39 The ground electronic state of C_2 is the $X\,^1\Sigma_g^+$, but there is a low-lying $^3\Pi_u$ state at nearly the same energy and a $^3\Sigma_g^-$ state roughly 0.7 eV above that. In contrast, Si_2 has a different ordering, with a $^3\Sigma_g^-$ ground state and the $^3\Pi_u$ state 1.5 eV higher in energy. Use these results from spectroscopy to deduce an approximate MO diagram for the Si_2 valence electrons and to explain why acetylene (HCCH) is stable in the gas phase while HSiSiH is not.

7.40 Write the term symbol for the BH_3^+ ion, with MO configuration $1a_1'^2 2a_1'^2 1e'^3$.

7.41 Identify the electronic states arising from the following MO configuration for the molecule *trans*-difluoroethene:

$$(3a_g)^2(4a_g)^2(3b_u)^2(5a_g)^2(4b_u)^2(6a_g)^2$$
$$(5b_u)^2(7a_g)^2(6b_u)^2(1a_u)^2(1b_g)^2(2a_u)^1(2b_g)^1.$$

7.42 If the molecular ion BH_3^+ has the ground MO configuration $(1a')^2(2a')^2(1e')^3$, what is its ground state term symbol? The molecule is planar, and all three hydrogens are equivalent.

7.43 The molecular orbitals for the C_{2v} molecule NH_2 are filled in the order

$$1a_1 2a_1 1b_2 3a_1 1b_1 4a_1.$$

Find the MO configuration and electronic state symbol of the lowest excited *quartet* spin state ($S = 3/2$).

7.44 A correlation diagram is shown for the MOs of the π bond in 1,2-difluoroethene.

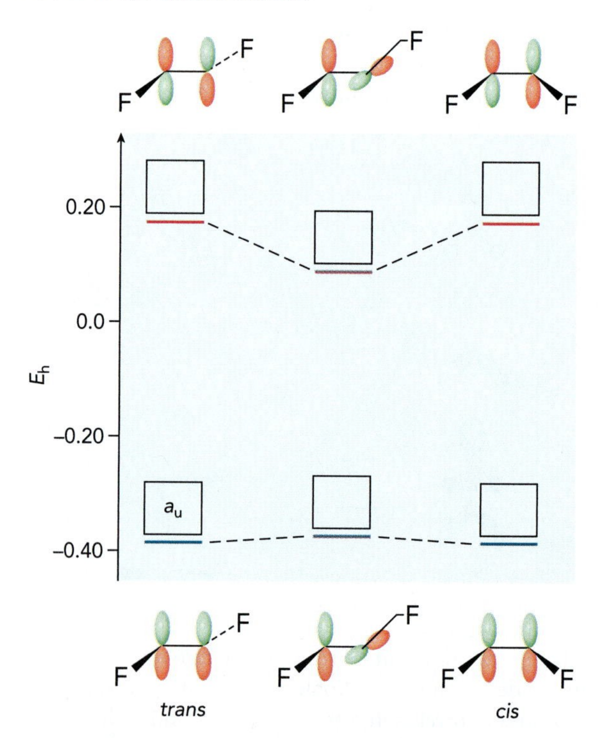

a. Write the five missing MO symmetry representations in the boxes provided.

b. According to this diagram, what is the likely point group for the lowest excited state of the molecule?

7.45 The equilibrium bond length of ground state $X\,^1\Sigma^+$ of CO is 1.128 Å. The equilibrium bond length of the $I\,^1\Sigma^-$ state from the lowest excited MO configuration is 1.391 Å. How could one explain the large change in R_e?

7.46 The ground state of the CN radical has a molecular orbital configuration $1\sigma^2 2\sigma^2 3\sigma^2 4\sigma^2 1\pi^4 5\sigma$. The configuration of the lowest excited state is $1\sigma^2 2\sigma^2 3\sigma^2 4\sigma^2 1\pi^3 5\sigma^2$.

a. What is the term symbol of the ground electronic state (X)?

b. What is the term symbol of the lowest excited electronic state (A)?

c. Is the $X \rightarrow A$ transition allowed by electric dipole selection rules?

d. Is the $X \rightarrow A$ transition allowed by Raman selection rules?

7.47 The *complete* MO configuration of ethene is $1a_g^2 1b_{3u}^2 2a_g^2 3a_g^2 2b_{3u}^2 1b_{1u}^2 1b_{2g}^2 1b_{2u}^2$.

a. Label the coordinate axes x, y, and z in the figure, based on the representations used in this MO configuration.

b. Identify each of the MOs listed here with one of the following groups of electrons:
   i. CH σ-bond electrons
   ii. CC π-bond electrons
   iii. 1s core electrons
   iv. CC σ-bond electrons

$$1a_g^2\ 1b_{3u}^2\ 2a_g^2\ 3a_g^2\ 2b_{3u}^2\ 1b_{1u}^2\ 1b_{2g}^2\ 1b_{2u}^2$$

7.48 Ground state acetylene is linear, but its lowest excited state is a triplet state with C_{2h} symmetry. Give the representations in the C_{2h} limit that correlate to the MOs listed here for the linear molecule.

H—C≡C—H

a. C—H σ_g b. C—H σ_u

c. C≡C σ_g d. C≡C π_u

7.49 Determine the MO configuration for NO_2, a C_{2v} molecule with two equivalent resonance Lewis structures, lying in the xz plane. Try to give the MOs for a given group of electrons (e.g., lone pair electrons) in order of increasing energy, but don't worry about the ordering of different groups. After writing the configuration, identify which orbitals correspond to which electron groups in the Lewis structures.

7.50 Although cyclobutane may appear to be a D_{4h} molecule in chemical structure drawings, it is more stable with the four carbons in a non-planar arrangement, as shown in the following figure.

a. Find the point group of cyclobutane.

b. Write the representation for the lowest energy MO.

c. Identify the ground electronic state term symbol (e.g., $^1\Sigma_g^+$).

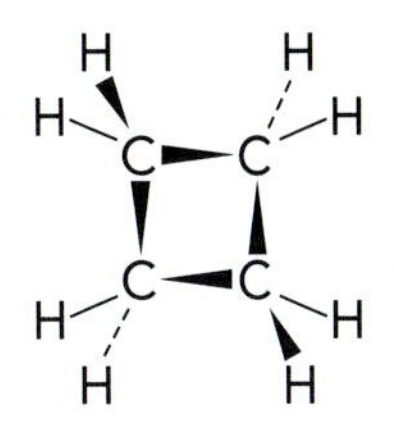

7.51

a. What is the electronic term symbol (e.g., $^1\Sigma_u^+$, $^2A_{1g}$) for the lowest excited state of the chromacene molecule (shown at the left in the following figure), if the HOMO (drawn in the center) has a_{2u} symmetry and the LUMO (shown at the right) has a_{1g} symmetry?

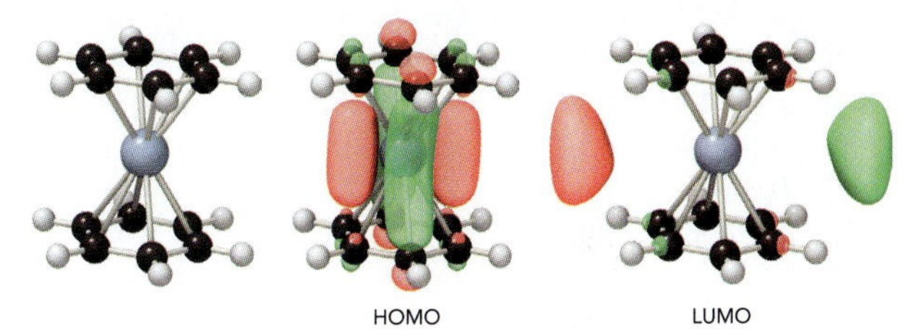

b. Show whether the transition between the two MOs in part (a) is allowed by electric dipole or by Raman selection rules. Is the transition from the electronic ground state of the molecule to the lowest excited state allowed?

7.52 The only electronic terms derived from a particular MO configuration are $^3\Delta_u$, $^1\Delta_u$, $^3\Sigma_u$, and $^1\Sigma_u$. What conclusions could one draw about the MO configuration?

7.53 The only electronic terms resulting from the molecular orbital configuration for the *lowest excited state* of a new molecule are $^2\Delta_g$, $^4\Sigma_g^+$, and $^2\Sigma_g^-$. What conclusions could one draw about the highest occupied molecular orbitals?

7.54 What is the ground electronic state symbol for fluoromethane?

7.55 The molecular orbitals for the C_s molecule HCO are filled in the order

$$1a'2a'3a'4a'5a'6a'1a''7a'2a''8a'.$$

What are the highest occupied molecular orbitals and term symbols (including spin) for the ground states of HCO and HCO^+?

7.56 The planar HC_3O radical is drawn in the following figure. There is one unpaired electron in an sp^2 hybrid MO that lies in the molecular plane.

a. Find the molecule's point group.

b. Determine the term symbol for the ground state of the molecule.

H—C≡C—C=O:

7.57 The $ClOF_4^-$ anion has a square pyramidal geometry, belonging to the point group C_{4v} with the character table shown. The ground state MO configuration is $[\text{core}]\,e^4\,a_1^2$, where the core has no unfilled orbitals, and the lowest excited state MO configuration is $[\text{core}]\,e^4\,a_1\,e^1$.

a. Write the term symbol for the ground electronic state.

b. Write the term symbol(s) for the excited electronic state(s).

c. If any of the excited states is accessible by an allowed transition, write its term symbol and write "ED" if allowed by an electric dipole transition, "R" if allowed by a Raman transition, and "EDR" if allowed by both. If no transitions are allowed, write "none."

C_{4v}	$\hat{E}$	$2\hat{C}_4$	$\hat{C}_2$	$2\hat{\sigma}_v$	$2\hat{\sigma}_d$	Functions
$A_1(a_1)$	1	1	1	1	1	$z; x^2+y^2, z^2$
$A_2(a_2)$	1	1	1	−1	−1	R_z
$B_1(b_1)$	1	−1	1	1	−1	x^2-y^2
$B_2(b_2)$	1	−1	1	−1	1	xy
$E(e)$	2	0	−2	0	0	$x, y; xz, yz; R_x, R_y$

7.58 The molecular orbitals of NH_2^+ in order of increasing energy are $1a_1 2a_1 1b_2 3a_1 1b_1 4a_1$. What are the MO configurations and electronic state symbols for the ground and first excited states of this molecule? Don't forget to label the spin multiplicity.

7.59 Complete the unfinished MO configuration for formaldehyde, H_2CO (assume the molecule is in the xz plane), and give the symbol for its ground electronic state.

MO configuration: $1a_1^2 2a_1^2 3a_1^2 4a_1^2 1b_1^2 5a_1^2 2b_1^2 \ldots$

7.60 Write the ground state MO configuration for the C_{2v} free radical H_2DC, assuming the molecule lies in the xz plane.

7.61 A hypothetical quartet-spin C_9H_9 radical, having C_{3v} symmetry, is drawn in the following figure (the three single dots are unpaired electrons).

a. Write the MO configuration for only the C—C σ-bonding electrons.

b. Write the MO configuration for only the C—C π-bonding electrons.

c. Write the MO configuration for only the unpaired electrons.

Potential Energy Curves and Relaxation Processes

7.62 Based on the H_2 energy level diagram, Fig. 7.7, if we started from ground state H_2 molecule near $E = 0$ eV, what would be the minimum energy necessary to reach (a) the $a^3\Sigma_g^+$ state, (b) the ground state H_2^+ molecule, and (c) the dissociated $H + H^+$ atoms?

7.63 Refer to Fig. 7.7.

a. Sketch on the figure a rough representation of the potential energy curve for H_2^-.

b. Label the electronic state.

c. Show whether R_e and D_e are larger or smaller than the corresponding values for H_2.

d. Finally, give the correct value of the potential energy at large R.

7.64 Unlike many doubly charged ions in the gas phase, the oxygen molecule dication, O_2^{2+}, has been studied experimentally. It is found to have a metastable ground state, in that the molecule is bound by higher energy than the dissociated ground state O^+ ions.

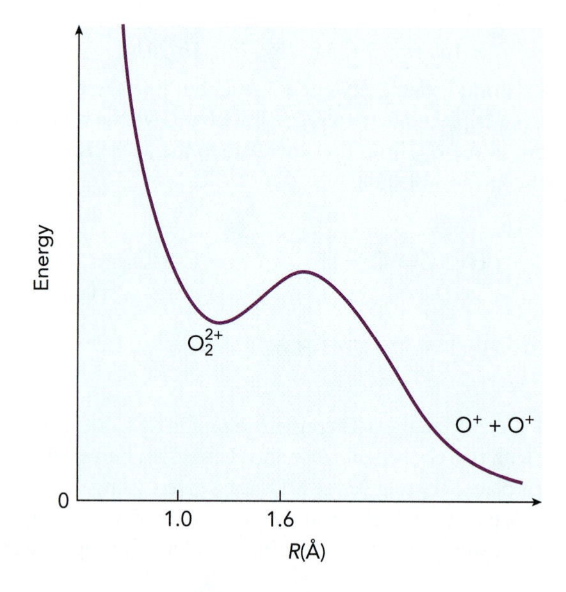

a. Identify the ground state electron configurations and term symbols for the O_2^{2+} molecule and the O^+ atoms. Which *neutral* homonuclear diatomic X_2 has the same configuration as O_2^{2+}?

b. Estimate the energy gap in eV between ground state O_2^{2+} and $2O^+$ by calculating the potential energy for repulsion between two point charge O^+ ions separated by 1.1 Å and adding this to the potential energy of the X_2 molecule at the same separation.

7.65 For the $^7\Sigma_u^+$ state of N_2 in Fig. 7.8, answer the following questions:

a. What is the spin S of this state?

b. What is the minimum number of unpaired electrons in this electronic state?

c. What are the term symbols for the N atoms in the large-R limit?

d. What is the MO configuration for this state?

7.66 It is possible to study some electronic states by going through more than one spectroscopic transition. For example, excited states of acetylene have been studied by inducing a *sequence* of transitions, such as $^1\Sigma_g^+ \rightarrow {}^1\Pi_u$, $^1\Pi_u \rightarrow {}^1\Delta_g$, with $^1\Pi_u$ acting as an intermediate state. For a molecule with point group D_{2h}, a transition directly from a ^1A_g state to ^1A_u is forbidden by both electric dipole and Raman selection rules. Write any sequence of transitions, using as few intermediate states as possible, to go from ^1A_g to ^1A_u for a D_{2h} molecule while obeying

a. electric dipole selection rules only.

b. any combination of Raman and electric dipole selection rules.

7.67 Excitation of H_2 into the lowest excited $^3\Sigma_u^+$ state would probably result in which of the following?

a. photodissociation b. predissociation

c. fluorescence d. phosphorescence

7.68 We can represent the ionization of H_2 as a vertical line between the approximate potential surfaces for ground state H_2 and H_2^+ in Fig. 7.7. Do the same for ionization of the He atom in terms of the potential curves for the remaining electron.

7.69 The ground state electron configuration of F_2 is $1\sigma_g^2 1\sigma_u^2 2\sigma_g^2 2\sigma_u^2 3\sigma_g^2 1\pi_u^4 1\pi_g^4$.

a. Write the term symbol for this electronic state.

b. The lowest excited state of F_2 is a $^3\Pi_u$ term. Call it A. Based on your answer to the previous problem, put an "X" next to *any* of the following that would allow us to observe this state by spectroscopy:

i. by an allowed electric dipole transition from the ground state to A

ii. by an allowed Raman transition from the ground state to A

iii. by fluorescence to the ground state from A

iv. by phosphorescence to the ground state from A

7.70 In a series of experiments in which the starting material is the ground state NO molecule, the following results are obtained:

(i) At 6.0 eV there is a strong absorption of the radiation, followed by phosphorescence and fluorescence emission to the ground state.

(ii) At 7.5 eV, the molecule photodissociates into ground state ^4S N and ^3P O atoms.

(iii) The upper state of the phosphorescence transition is stable long enough to be excited to an even higher state when an additional 4.0 eV are pumped into it. That state dissociates immediately into ground state ^4S N and excited state ^1D O atoms.

Sketch the potential curves for the states involved in these interactions, identifying the relative energies that were determined by these experiments.

7.71 The following graph represents potential energy curves for selected electronic states of the NO radical and its cation. Based on this diagram, do the following:

a. Estimate to within 1 eV the ionization energy of atomic oxygen.

b. Estimate to within 1 eV the minimum photoionization energy of NO from the $X, v = 0$ ground state.

c. Estimate to within 1 eV the minimum photodissociation energy from the $X, v = 0$ ground state.

d. If NO is formed in the $b^4\Sigma^-$ excited state, what are all of the likeliest processes (e.g., photoionization to $X^1\Sigma^+$) by which the molecule can lose its excess electronic energy? Give the term symbols for the final states of any processes you list.

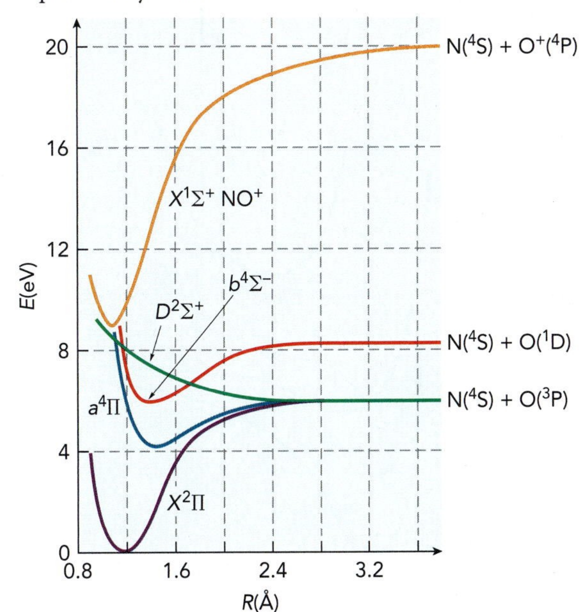

7.72 Imagine that the $X^2\Pi \rightarrow D^2\Sigma^+$ transition is excited in NO by a single intense pulse of 8 eV radiation lasting only $20 \cdot 10^{-9}$ s. The following events occur at the times indicated after the excitation:

(i) From $0.2 - 20 \cdot 10^{-6}$ s, strong emission is observed over the range 6–7.5 eV.

(ii) From $4 - 350 \cdot 10^{-3}$ s, weak emission is observed over the range 5–5.5 eV.

(iv) From 0.30–1.2 s, separated N and O atoms are observed.

Sketch the combination of potential energy curves below which this behavior could result.

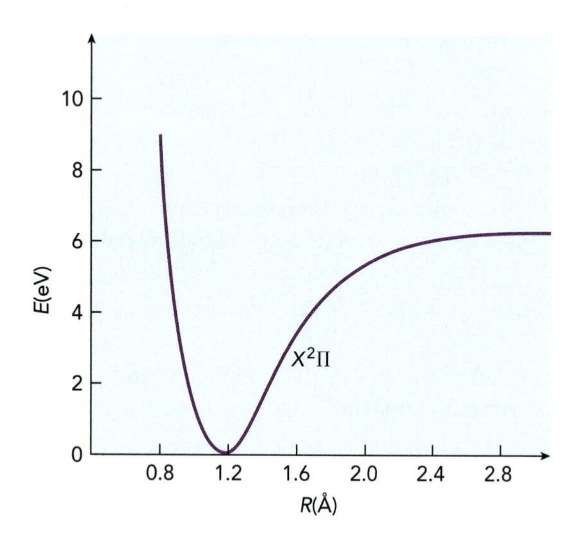

7.73 Given the set of potential energy curves shown in the following figure, what is the likeliest process to be induced by radiation at a photon energy of 8.0 eV, starting from the $v = 0$ state shown? Identify any of the electronic states involved.

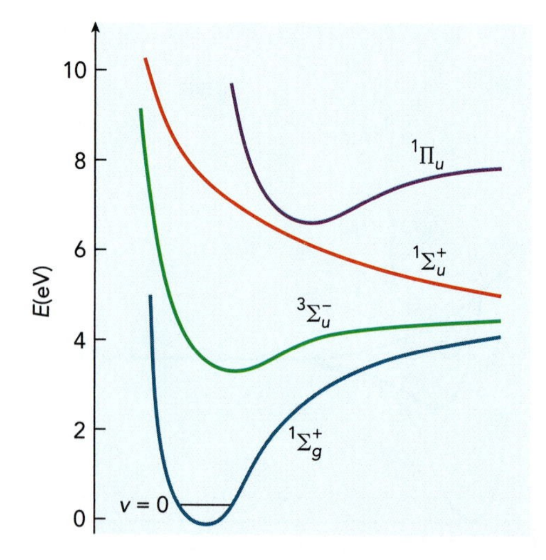

7.74 If we start from the $^5\Pi_u$ excited state (just below 12 eV) of N_2 (Fig. 7.8), what is the likeliest set of processes that will return the system to the ground X state? Give both the transition and the type of process (such as fluorescence, photodissociation, etc.). Assume that forbidden transitions are more likely the fewer selection rules they violate.

7.75 What process or processes are most likely following the excitation illustrated in the following figure? Assume that the $X \rightarrow A$ transition is allowed.

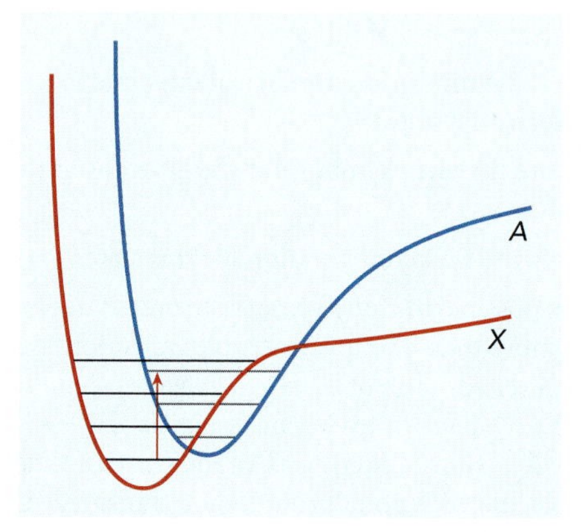

7.76 We ionize the molecule AB in the gas phase to form AB^+ in its lowest excited state. Identify which of the three processes drawn is the most likely from the state given, and identify the type of process it is (e.g., fluorescence, predissociation).

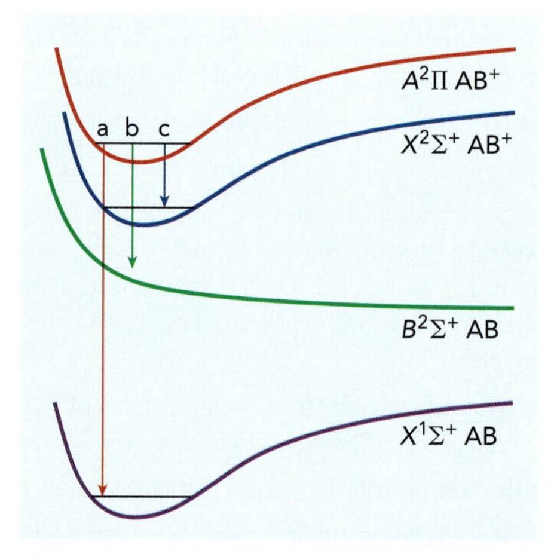

7.77 A pure gas is irradiated with 5 eV radiation. Electrons are released by the gas with velocities of $6.2 \cdot 10^7$ cm s^{-1} or less. What is the ionization energy of the gas?

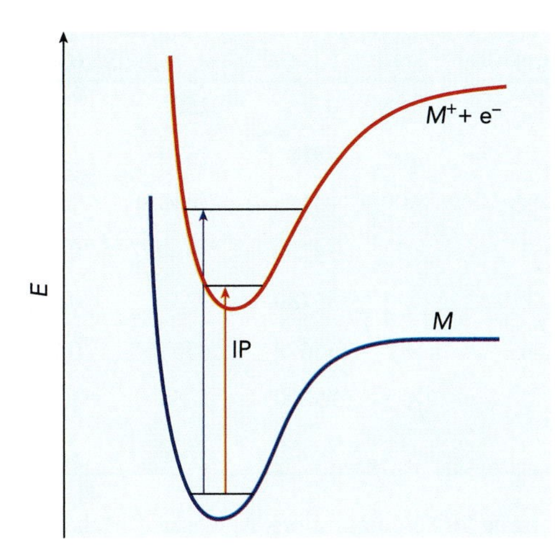

7.78

a. Label the A state symmetry so that the $X \rightarrow A$ transition is allowed for the BeF molecule, with potential curves shown in the following figure.

b. What process or processes occur when a transition is excited to the A state from $v = 0$ in the $X^2\Sigma^+$ state?

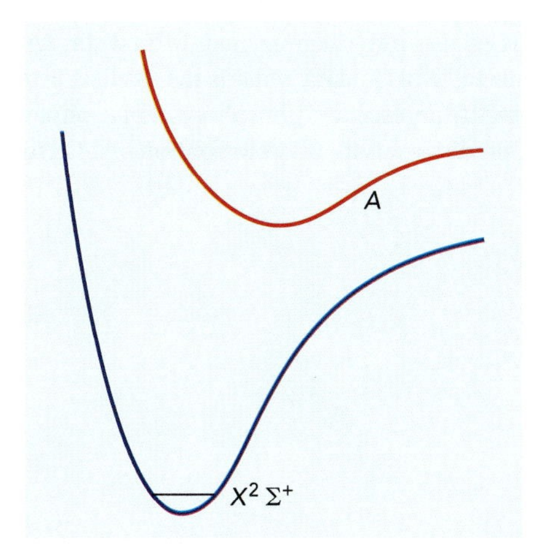

7.79 With high-power lasers, it is sometimes possible to get one molecule to absorb two photons in rapid succession. The following figure is a diagram of the known electronic states of a homonuclear diatomic molecule M and its ion M^+. The X and A states of M are singlet states, the a state is a triplet, and X M^+ is a doublet state. Assume that the transition XM $\rightarrow$ A M is electric dipole allowed. If radiation at frequency ν is transmitted into a sample of molecule M in its X ground state at high power, indicate with a "1" and "2" in the following table which of the listed processes are the likeliest and second likeliest to result.

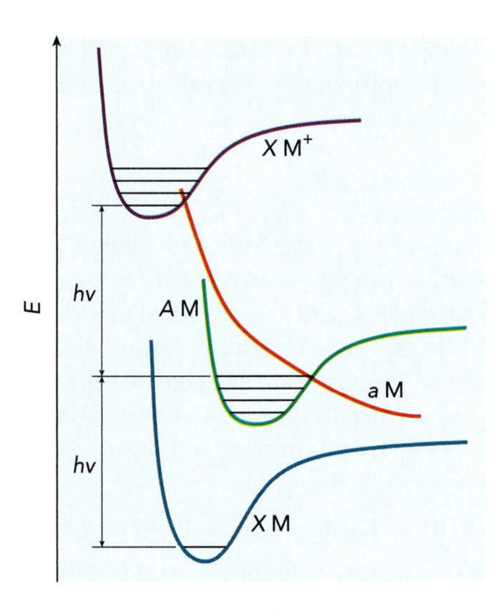

fluorescence from A M to X M
fluorescence from a M to X M
Raman transition from X M to X M via A M
phosphorescence from A M to X M
phosphorescence from a M to X M
photodissociation of A M
photodissociation of a M
predissociation of A M via a M
predissociation of X M^+ via a M
photoionization via A M to X M^+

7.80 For the set of potential energy curves drawn in the following figure, what is the likeliest event following excitation with 10 eV from $v = 0$ of the ground state: fluorescence, phosphorescence, photodissociation, predissociation, photoionization, intersystem crossing, or some combination?

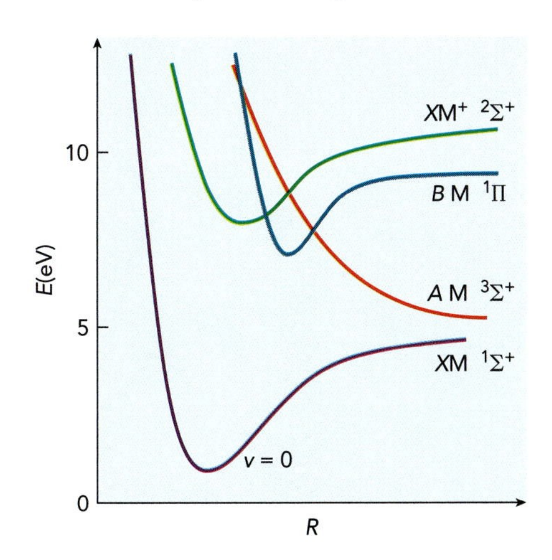

7.81 The Newton-Raphson method approaches the root x_∞ of an expression $f(x_n) = 0$ by calculating the next value x_{n+1} using the equation

$$x_{n+1} = x_n - \frac{f(x_n)}{df(x_n)/dx}.$$

The new value of x is plugged into the righthand side of the equation, and an improved value of x is computed. The process ends when x_n and x_{n+1} are equal to within the desired precision. Computational chemistry uses this method to find the minimum of a potential energy function $U(R)$ by solving for the root of the equation $dU(R)/dR = 0$. Try this out on the relatively simple potential energy function

$$U(R) = 1.5(R - 1.5)^2 + 0.2(R - 1.8)^4.$$

Starting from an initial value of $R = 2.000$, find the equilibrium value of R for this function to three significant digits, using the Newton-Raphson method. (You might wonder why we don't just find the value of R that makes the derivative equal to zero, since we have to find the derivative in order to use the Newton-Raphson method. The reason is that when computing molecular potential energies, we generally don't get an equation for the whole potential energy, so we can't get an equation for its derivative at all R. Instead, we get the value of $U(R)$ only at particular values of R, but we can calculate the first and second derivatives of $U(R)$ *at that point*, and that's all we need to find the next point in the Newton-Raphson method.)

7.82 These four problems use the properties of several electronic states of O_2 listed in the following table. Each question may be answered independently of the others.

configuration	state	T_e (cm^{-1})	ω_e (cm^{-1})	B_e (cm^{-1})
A	$X^3\Sigma_g^-$	0	1580	1.446
A	$a^1\Delta_g$	7918	1509	1.426
A	$b^1\Sigma_g^+$	13195	1433	1.400
B	$C^3\Delta_u$	34200	≈ 700	0.90
B	$A^3\Sigma_u^+$	35780	773	0.895
B	$c^1\Sigma_u^-$	36679	650	0.826
C	$\beta^3\Sigma_u^+$	75500	1850	1.72
C	$\alpha^1\Sigma_u^+$	76350	1820	1.70

a. The three MO configurations A, B, and C each have the form $[Be_2]\ 3\sigma_g^2 1\pi_u^a 1\pi_g^b 3\sigma_u^c 4\sigma_g^d$ (any of a, b, c, or d may be zero). These are not necessarily the lowest energy configurations, but the states listed are the only states arising from configurations A, B, and C. Based on the data in the table, determine the MO configurations A, B, and C.

b. Identify any of the O_2 electronic states listed in the table that are accessible by electric-dipole allowed transitions from the ground electronic state.

c. If O_2 is excited into the molecular orbital configuration B in our table of O_2 data, what is the likeliest process or processes (fluorescence, phosphorescence, photodissociation, predissociation, photoionization) that result?

8 Vibrational States and Spectroscopy

LEARNING OBJECTIVES

After reading this chapter, you will be able to do the following:

1. Approximate vibrational energies and wavefunctions using the simple harmonic oscillator.
2. Sketch approximate solutions to a one-dimensional Schrödinger equation based on the shape of the potential energy curve and the reduced mass.
3. Predict trends in the vibrational constant based on atomic mass and chemical bond strength.
4. Use group theory to predict the symmetries of a molecule's vibrational normal modes and the selection rules for transitions among the vibrational quantum states.

GOAL *Why Are We Here?*

The goal of this chapter is to show how we solve the vibrational part of the molecular Schrödinger equation to obtain wavefunctions and energies that describe the relative motions of atoms in the molecule.

CONTEXT *Where Are We Now?*

After looking into the electronic states of a molecule in Chapter 7, you could reasonably ask why we need to look at the energy of the nuclei at all. Electrons often appear to be our major focus in chemistry because it is the redistribution of bonding electrons that accounts for the transformation of one chemical substance into another. However, we define the identity of a chemical element by its *nucleus*—we use the same atomic symbol for helium no matter how many electrons it has. In chemistry, we can rarely consider either electrons or nuclei in isolation; the two are linked. Electronic energy is easily converted into vibrational energy, and the formation and cleavage of chemical bonds is as much a vibrational effect as an electronic effect. So we move now from considering the electronic degrees of freedom to the movements of the nuclei.

The motions of the nuclei in the molecule can be separated into three terms: translation (motion of the center of mass), rotation (changes in the orientation of the molecule), and vibration (motions of the nuclei relative to one another). The nuclei, being much more massive than the electrons, have energy levels that behave more classically, and the gaps between the energy levels for translation, rotation, and vibration are much smaller than the electronic energy spacings. Among these nuclear motions, the vibrations most resemble the electronic states in energy scale and most directly affect the electron distributions we've just examined. These vibrational motions are next on our survey of molecular degrees of freedom.

SUPPORTING TEXT **How Did We Get Here?**

- Section 3.3 gives a semiclassical picture of the interaction between matter and radiation (Fig. 3.14) which can be helpful for understanding the vibrational selection rules.
- For any molecule, the Hamiltonian may be written as the sum of several contributions: the kinetic energies of the electrons and nuclei and the potential energy derived from the forces that appear in Eq. 5.1 for the binding force (Eq. 5.4),
$$\hat{H} = \hat{K}_{\text{elec}} + U_{\text{nuc-elec}} + U_{\text{elec-elec}} + U_{\text{nuc-nuc}} + \hat{K}_{\text{nuc}}.$$
- The Born-Oppenheimer approximation separates out the terms that depend on the electron coordinates (Eqs. 5.8 and 5.9):
$$\hat{H}_{\text{eff}}\psi_{\text{elec}} = E_{\text{elec}}\psi_{\text{elec}},$$
where
$$\hat{H}_{\text{eff}} = \hat{K}_{\text{elec}} + U_{\text{nuc-nuc}} + U_{\text{nuc-elec}} + U_{\text{elec-elec}}.$$
The Schrödinger equation for this effective Hamiltonian $\hat{H}_{\text{eff}}$ must be solved at each geometry of the nuclei that interests us, so that we obtain the electronic energy as a function of the molecular geometry. Then we can add $\hat{K}_{\text{nuc}}$ and solve the Schrödinger equation along the remaining six coordinates for motion of the *nuclei* (Eq. 5.10):
$$\hat{H}\psi_{\text{nuc}} = [E_{\text{elec}} + \hat{K}_{\text{nuc}}]\psi_{\text{nuc}} = E_{\text{nuc}}\psi_{\text{nuc}}.$$
- Section 5.2 describes the effective potential energy function, which becomes the potential energy term in our vibrational Schrödinger equation in this chapter.
- Section 6.1 describes the point group representations that we use for labeling vibrational states.
- The transition strength for a transition from an initial state with wavefunction ψ_i to a final state with wavefunction ψ_f is given by the integral (Eq. 6.10)
$$\text{I} \propto \left| \int \psi_f^* \hat{\mu}_t \psi_i \, d\tau \right|^2,$$
where $\hat{\mu}_t$ is the transition moment operator and where the two states are assumed to have the same spin.

8.1 The Vibrational Schrödinger Equation

The Born-Oppenheimer approximation separates the electronic and nuclear coordinates in the Schrödinger equation, allowing us to split the one big problem into two smaller problems. But the motions of the nuclei separate still further.

Separation of Variables

Let's examine the nuclear contribution to the Hamiltonian that we dropped in the last chapter. The Hamiltonian is written (Eqs. 5.4 and 5.10)
$$\begin{aligned}\hat{H} &= \hat{K}_{\text{elec}} + U_{\text{nuc-nuc}} + U_{\text{nuc-elec}} + U_{\text{elec-elec}} + \hat{K}_{\text{nuc}} \\ &= \hat{H}_{\text{eff}} + \hat{K}_{\text{nuc}}.\end{aligned}$$

The Born-Oppenheimer approximation allowed us to isolate the electronic energy using an effective Hamiltonian for the molecule with fixed, non-rotating nuclei, solving the Schrödinger equation at each value of R (Eq. 5.9):

$$\hat{H}_{\text{eff}} = \hat{K}_{\text{elec}} + U_{\text{nuc-nuc}} + U_{\text{nuc-elec}} + U_{\text{elec-elec}}.$$

All that we have done is drop out the kinetic term for the nuclei, $\hat{K}_{\text{nuc}}$, which for a diatomic molecule is written (Eq. 5.11)

$$\hat{K}_{\text{nuc}} = -\frac{\hbar^2}{2\mu R^2}\frac{\partial}{\partial R}R^2\frac{\partial}{\partial R} + \frac{1}{2\mu R^2}\hat{J}(\Theta,\Phi)^2 - \frac{\hbar^2}{2M_{\text{nuc}}}\nabla^2_{\text{COM}},$$

where $\hat{J}(\Theta,\Phi)$ is the operator for the rotational angular momentum of the nuclei. By making the molecule rigid, fixing R, we eliminate the first term in $\hat{K}_{\text{nuc}}$. When we prevent rotation, the second term vanishes. Finally, if we fix the location of the center of mass, the last term vanishes. In this way we solved for the energy of the molecule at each value of R, obtaining the electronic energy of the system as a function of R and resulting in the potential energy curves for different electronic states, all as functions of R.

Our Hamiltonians can always be put in the form $\hat{H} = \hat{K} + U$, dividing the energy contributions into kinetic and potential energy. The eigenvalues of our effective Hamiltonian are therefore essentially potential energies for the motion of the nuclei. The Born-Oppenheimer approximation separates two sets of variables: the electronic coordinates r and the nuclear coordinates R, Θ, and Φ. Having solved for the electronic energies and wavefunctions in Chapter 7, we can now solve the quantum mechanics of the nuclei by adding the electronic contribution to the potential energy.

To analyze the nuclear motions, we carry out two further separations of the variables, the same that we also carried out for the atomic orbital wavefunctions in Chapter 2. We first neglect the translational contribution to the kinetic energy operator in Eq. 5.11, the term that depends on ∇^2_{COM}. Moving the center of mass does not directly affect the other degrees of freedom, because all the forces within the molecule that act on the nuclei and electrons move with the center of mass. We next separate the radial and angular coordinates, splitting the nuclear energies and wavefunctions into vibrational and rotational parts. The rotational contribution to the Hamiltonian is straightforward in that there is no potential energy contribution: $\hat{H}_{\text{eff}}$ does not depend on Θ and Φ. The rotational energy is described by the angle-dependent kinetic energy term of the Hamiltonian:

$$\hat{H}_{\text{rot}} = \frac{1}{2\mu R^2}\hat{J}^2(\Theta,\Phi). \tag{8.1}$$

This term is dealt with in the next chapter.

The vibrational term is our present concern, and it is a little more involved because there are both kinetic and potential energy contributions. The nuclear potential energy is not constant as R changes, because the electronic energy described by $\hat{H}_{\text{eff}}$ depends on R. We treat this by taking the eigenvalue of $\hat{H}_{\text{eff}}$ and dividing it into a fixed electronic energy E^0_{elec} and a vibrational potential energy $U_{\text{vib}}(R)$:[1]

$$\hat{H}_{\text{eff}}\psi(r,R) = \left(E^0_{\text{elec}} + U_{\text{vib}}(R)\right)\psi(r,R). \tag{8.2}$$

CHECKPOINT In Eq. 8.2, our color-coding emphasizes that we can now draw on the *solved* energy eigenvalues E^0_{elec} from Chapter 6 to replace the electronic part of the Hamiltonian in this chapter. The Born-Oppenheimer approximation not only makes the electronic part of the Schrödinger equation problem easier to solve, it allows us to address the vibrational motions now without worrying about the motions of all the electrons.

[1]The potential energy term $U_{\text{vib}}(R)$ does also depend on the electron coordinates r, but we use the Born-Oppenheimer approximation to average out that dependence. That is why r seems to disappear from the Hamiltonian.

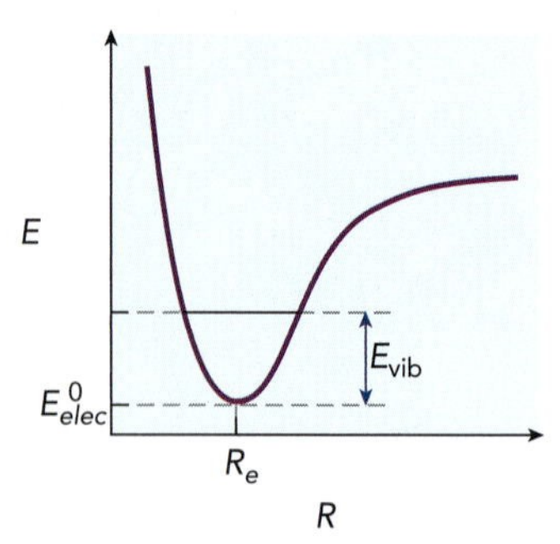

▲ **FIGURE 8.1 The potential energy for vibration.** The lowest energy of the potential well is defined as E^0_{elec} and is located at the equilibrium bond length R_e.

CHECKPOINT Several different expressions are listed here for the molecular Hamiltonian, showing the steps we take to shift from the conceptually straightforward, general form in Eq. 8.4 to the more practical form in Eq. 8.7. Equation. 8.4 is based on dividing the energy into kinetic and potential terms and subdividing each of those by the type of particle involved. But the actual motions of and within the molecule tend to group into types of motion, and some of those types (like vibration) involve moving both electrons and nuclei simultaneously.

The electronic energy E^0_{elec} is the energy of the most stable geometry of the molecule in a given electronic state, the energy at the bottom of the potential well, as shown in Fig. 8.1. The advantage of this separation of U_{eff} is that it defines the nuclear potential energy so that $U_{\text{vib}} = 0$ at the bottom of the well. Now we have kinetic energy and potential energy terms that depend only on R, and these can be combined to yield an R-dependent vibrational Hamiltonian that acts on an R-dependent vibrational wavefunction:

$$\begin{aligned}\hat{H}_{\text{vib}}\psi_{\text{vib}}(R) &= \left[\hat{K}_{\text{vib}} + U_{\text{vib}}(R)\right]\psi_{\text{vib}}(R)\\ &= \left[-\frac{\hbar^2}{2\mu R^2}\frac{\partial}{\partial R}R^2\frac{\partial}{\partial R} + U_{\text{eff}}(R)\right]\psi_{\text{vib}}(R)\\ &= E_{\text{vib}}\psi_{\text{vib}}(R).\end{aligned} \tag{8.3}$$

Our separation of variables into different Hamiltonian contributions can be summarized as follows:

$$\hat{H} = \underbrace{\hat{K}_{\text{elec}} + U_{\text{nuc-elec}} + U_{\text{elec-elec}} + U_{\text{nuc-nuc}}}_{} + \hat{K}_{\text{nuc}} \tag{8.4}$$

$$= \hat{H}_{\text{eff}} + \hat{K}_{\text{nuc}} \tag{8.5}$$

$$= \underbrace{\hat{H}_{\text{eff}} - \frac{\hbar^2}{2\mu R^2}\frac{\partial}{\partial R}R^2\frac{\partial}{\partial R}} + \underbrace{\frac{1}{2\mu R^2}\hat{J}^2(\Theta,\Phi)} - \underbrace{\frac{\hbar^2}{2M_{\text{nuc}}}\nabla^2_{\text{COM}}} \tag{8.6}$$

$$= \quad E^0_{\text{elec}} + \hat{H}_{\text{vib}} \quad + \quad \hat{H}_{\text{rot}} \quad + \quad \hat{H}_{\text{trans}} \tag{8.7}$$

We have separated variables so that the overall Hamiltonian may be written

$$\hat{H}(r,R,\Theta,\Phi) = E^0_{\text{elec}} + \hat{H}_{\text{vib}}(R) + \hat{H}_{\text{rot}}(R,\Theta,\Phi) + \hat{H}_{\text{trans}}(X,Y,Z), \tag{8.8}$$

where the last term is the translational Hamiltonian, which depends on the X, Y, and Z coordinates of the molecule's center of mass. The overall wavefunction will be of the form

$$\psi(r,R,\Theta,\Phi,X,Y,Z) = \psi_{\text{elec}}(r,R)\psi_{\text{vib}}(R)\psi_{\text{rot}}(\Theta,\Phi)\psi_{\text{trans}}(X,Y,Z). \tag{8.9}$$

The energy has likewise been divided among its electronic, vibrational, rotational, and translational contributions:

$$E = E^0_{\text{elec}} + E_{\text{vib}} + E_{\text{rot}} + E_{\text{trans}}. \tag{8.10}$$

We have introduced one more important approximation: that the vibrational wavefunction is only weakly affected by rotation of the molecule, and vice versa. This simplification lets us solve the vibrational term in the absence of rotation and solve the rotational term for the molecule in the absence of vibration (using fixed R). This approximation is called the **rigid rotor** or **rigid rotator,** and is usually good to within 0.1% as long as the rotational energy is small compared to the spacing between vibrational energy levels.

The Harmonic Oscillator

The difficulty with solving the vibrational Schrödinger equation (Eq. 8.3) is the potential energy term, $U_{\text{vib}}(R)$. Because this term is a function of the molecular geometry and the electronic structure of the molecule, there is no single general expression that accurately describes every vibrational potential energy function. There is, however, an easily solved example system that can be applied with surprising success to a wide range of cases. To show that we can use such an approximation, let's try to simplify the vibrational Hamiltonian in a way that does not depend on the particular algebraic form of $U_{\text{vib}}(R)$.

Derivation: The Harmonic Approximation to Vibrational Quantum States

Our goal with this derivation is to find a simple and reasonably accurate solution to the energies and wavefunctions of atoms vibrating in a chemical bond. We will first show that the potential energy curve of a chemical bond can be crudely approximated by the harmonic oscillator potential energy $\frac{1}{2}kx^2$. Then we will turn once more to our friends at the 19th-century French Academy for the solution to the differential equation. The final step is relating the mathematical solution to the physical parameters of the molecule. A **DERIVATION SUMMARY** appears after Eq. 8.24.

Parameters key: the harmonic approximation

symbol	parameter	SI units
U_{vib}	vibrational potential energy	J
R	bond distance	m
R_e	equilibrium bond distance	m
ψ_{vib}	vibrational wavefunction	$\text{m}^{-3/2}$
E_{vib}	vibrational energy	J
$\hbar$	Planck's constant divided by 2π	J s
x	displacement of the harmonic oscillator	m
y	a scaled, unitless version of x	unitless
μ	reduced mass, equal to $\dfrac{m_A m_B}{m_A + m_B}$ for a diatomic	kg
k	harmonic oscillator force constant	N m^{-1}
$\eta_v(x)$ or $\eta_v(y)$	harmonic oscillator wavefunction	$\text{m}^{-1/2}$
ω_e	vibrational constant	J
A_v	normalization constant in η_v	$\text{m}^{-1/2}$
$H_v(y)$	Hermite polynomial in η_v	unitless

We first expand U_{vib} in a Taylor series around the lowest energy geometry, which is at the equilibrium distance R_e:

$$U_{\text{vib}}(R) = U_{\text{vib}}(R_e) + \left(\frac{\partial U_{\text{vib}}}{\partial R}\right)\bigg|_{R_e}(R - R_e) + \frac{1}{2!}\left(\frac{\partial^2 U_{\text{vib}}}{\partial R^2}\right)\bigg|_{R_e}(R - R_e)^2 \quad \text{Taylor series expansion}$$
$$+\frac{1}{3!} + \left(\frac{\partial^3 U_{\text{vib}}}{\partial R^3}\right)\bigg|_{R_e}(R - R_e)^3 + \frac{1}{4!}\left(\frac{\partial^4 U_{\text{vib}}}{\partial R^4}\right)\bigg|_{R_e}(R - R_e)^4 + \ldots \quad (8.11)$$

The first term gives the potential energy at the equilibrium distance R_e, which is a constant that we use to determine the reference point of our potential energy curve. For studies of a single electronic state, and particularly when considering dissociation processes, it is most convenient to set the potential energy to zero at the dissociation limit, which would set $U_{\text{vib}}(R_e) = -D_e$. In this case, the bound states of the molecule have negative energy. However, when we consider more than one electronic state, there may not be a unique dissociation energy, and it becomes more straightforward to set $U_{\text{vib}}(R_e) = 0$.

If the power series in Eq. 8.11 converges rapidly, then only the first few terms will be important. The assumption works best at low vibrational energies, and the higher-order terms become increasingly important as E_{vib} approaches the dissociation energy. At R_e, the potential is at a minimum and its first derivative $(\partial U_{\text{vib}}/\partial R)$ is zero, so the first term in Eq. 8.11 vanishes. Truncating the series after the second term leaves only one potential energy contribution in the vibrational Schrödinger equation:

$$\hat{H}_{\text{vib}}\psi_{\text{vib}}(R) = \left[\hat{K}_{\text{vib}} + \frac{1}{2}\left(\frac{\partial^2 U_{\text{vib}}}{\partial R^2}\right)\bigg|_{Re}(R - R_e)^2\right]\psi_{\text{vib}}(R) \qquad \text{by Eq. 8.11}$$

$$= \left[-\frac{\hbar^2}{2\mu R^2}\frac{\partial}{\partial R}R^2\frac{\partial}{\partial R} + \frac{1}{2}\left(\frac{\partial^2 U_{\text{vib}}}{\partial R^2}\right)\bigg|_{Re}(R-R_e)^2\right]\psi_{\text{vib}}(R) \qquad \text{by Eq. 8.3}$$

$$= E_{\text{vib}}\,\psi_{\text{vib}}(R). \tag{8.12}$$

It turns out that this relationship between energy and position defines an important system in physics, and we will use that system to model our vibrating molecule.

A **harmonic oscillator** is any system in which the force obeys Hooke's law,

$$F = -kx.$$

As we move a displacement x in either direction from the origin, the force pushes us back toward the origin, with a magnitude proportional to the displacement. The Hamiltonian for this system is written in terms of the potential energy, which we obtain by integration:

$$U = -\int F dx = -\int_0^x (-kx')dx' = \frac{1}{2}kx^2, \tag{8.13}$$

where k is called the **force constant.** Finding the resulting equations of motion constitutes a common and ancient problem, because this potential curve closely resembles that for a spring holding two objects together, where the spring's restoring force prevents the objects from getting too close or too far from one another along the coordinate x. Therefore, when we need to consider the internal motions of the nuclei in a molecule, we often think of the chemical bonds as similar to springs that connect the atoms.

The Hamiltonian for the harmonic oscillator is

$$\hat{H} = \hat{K} + U = -\frac{\hbar^2}{2\mu}\frac{d^2}{dx^2} + \frac{1}{2}kx^2, \tag{8.14}$$

and the Schrödinger equation for this system is

$$\left[-\frac{\hbar^2}{2\mu}\frac{d^2}{dx^2} + \frac{1}{2}kx^2\right]\eta(x) = E\eta(x), \tag{8.15}$$

where $\eta(x)$ is our harmonic oscillator wavefunction. The reduced mass μ appears in this kinetic energy operator as a result of factoring out the center-of-mass motion (see Section A.2). As with any of these fundamental differential equations, a French mathematician has already done all the hard work, and we need only look up the solutions. The wavefunctions are written

$$\eta_v(y) = A_v H_v(y)e^{-y^2/2}, \tag{8.16}$$

where v is a quantum number, A_v is a normalization constant, y is a unitless distance equal to $x(k\mu/\hbar^2)^{1/4}$, and $H_v(y)$ is another set of tabulated functions called the **Hermite polynomials,** this time indexed only by the quantum number v. The first several polynomials in this series are given in Table 8.1.

The solution for the energies E_v as a function of the quantum number v is

$$E_v = \left(v + \frac{1}{2}\right)\omega_e, \qquad v = 0,1,2,3,\ldots \tag{8.17}$$

and where

$$\omega_e\,(\mathrm{J}) = \hbar\sqrt{\frac{k}{\mu}}. \tag{8.18}$$

This is called the **harmonic vibrational constant** and has units of energy. However, molecular vibrational constants are usually evaluated in units of cm^{-1} because their determinations have historically been based on wavelength measurements rather than direct energy measurements. In those units,

$$\omega_e(\mathrm{cm}^{-1}) = \frac{1}{2\pi c}\sqrt{\frac{k}{\mu}} = 130.28\sqrt{\frac{k(\mathrm{N\,m}^{-1})}{\mu(\mathrm{amu})}}. \tag{8.19}$$

The energies and wavefunctions of the harmonic oscillator are plotted in Fig. 8.2. A few notes:

1. Unlike some previous examples, the quantum number can equal zero here and still yield a non-trivial solution. For $v = 0$, the wavefunction is a Gaussian function $e^{-y^2/2}$.
2. The energy for $v = 0$ is not zero; it is $\omega_e/2$. This system is bounded and therefore it *is not possible to remove all the kinetic energy.* In other words,

TABLE 8.1 The harmonic oscillator wavefunctions. The Hermite polynomials and normalization constants are given for the first six harmonic oscillator wavefunctions $\eta_v(y) = A_v H_v(y) e^{-y^2/2}$, where the unitless coordinate y is equal to $x(k\mu/\hbar^2)^{1/4}$. Symbolic math programs can compute $H_v(y)$ automatically.

v	A_v	$H_v(y)$	nodes
0	$\left(\frac{k\mu}{\hbar^2}\right)^{1/8}\left(\frac{1}{\sqrt{\pi}}\right)^{1/2}$	1	0
1	$\left(\frac{k\mu}{\hbar^2}\right)^{1/8}\left(\frac{1}{2\sqrt{\pi}}\right)^{1/2}$	$2y$	$1{:}y = 0$
2	$\left(\frac{k\mu}{\hbar^2}\right)^{1/8}\left(\frac{1}{8\sqrt{\pi}}\right)^{1/2}$	$4y^2 - 2$	$2{:}y = \pm 1.414$
3	$\left(\frac{k\mu}{\hbar^2}\right)^{1/8}\left(\frac{1}{48\sqrt{\pi}}\right)^{1/2}$	$8y^3 - 12y$	$3{:}y = 0, \pm 1.225$
4	$\left(\frac{k\mu}{\hbar^2}\right)^{1/8}\left(\frac{1}{384\sqrt{\pi}}\right)^{1/2}$	$16y^4 - 48y^2 + 12$	$4{:}y = \pm 0.525, \pm 1.651$
5	$\left(\frac{k\mu}{\hbar^2}\right)^{1/8}\left(\frac{1}{3840\sqrt{\pi}}\right)^{1/2}$	$32y^5 - 160y^3 + 120y$	$5{:}y = 0, \pm 0.959, \pm 2.020$

$$yH_v(y) = vH_{v-1}(y) + \frac{1}{2}H_{v+1}(y) \qquad y\eta_v(y) = A_v\left[v\frac{\eta_{v-1}(y)}{A_{v-1}} + \frac{1}{2}\frac{\eta_{v+1}(y)}{A_{v+1}}\right]$$

$$\frac{dH_v(y)}{dy} = vH_{v-1}(y) - \frac{1}{2}H_{v+1}(y) \qquad \frac{d\eta_v(y)}{dy} = A_v\left[v\frac{\eta_{v-1}}{A_{v-1}} + \frac{1}{2}\frac{\eta_{v+1}(y)}{A_{v+1}}\right]$$

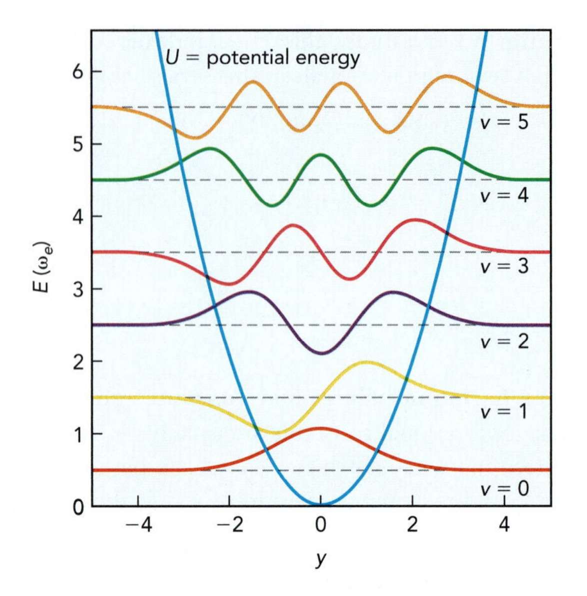

▶ **FIGURE 8.2 The simple harmonic oscillator potential curve** with energy levels and wavefunctions $\eta(y)$ shown to $v = 5$.

there is a zero-point energy, a phenomenon we first discussed for the particle in a one-dimensional box. In this case, the zero-point energy is equal to $\omega_e/2$.

3. As with all the previous examples we have seen, increasing the quantum number increases the number of nodes in the corresponding wavefunction. In this case, the nodes fall at the roots of the Hermite polynomial.
4. Unlike the particle in a box, these wavefunctions can tunnel into the potential walls.

You may have noticed that the Hamiltonian for the harmonic oscillator (Eq. 8.15) does not look quite like our vibrational Hamiltonian. The potential energy terms are actually the same, where the harmonic oscillator force constant k is equal to the second derivative of the potential energy:

$$k = \left(\frac{\partial^2 U_{\text{vib}}}{\partial R^2}\right)\bigg|_{R_e}. \tag{8.20}$$

This substitution is valid as long as our truncation of Eq. 8.11 to obtain Eq. 8.12 is a good approximation. Our vibrational Schrödinger equation is now

$$\hat{H}_{\text{vib}}\psi_{\text{vib}}(R) = \left[\hat{K}_{\text{vib}} + \frac{1}{2}k(R - R_e)^2\right]\psi_{\text{vib}}(R).$$

What about the kinetic energy term? We do something sneaky. Instead of solving for our actual $\psi_{\text{vib}}(R)$, we are going to solve for a different function, $\eta_v(R) = R\psi_{\text{vib}}(R)$. The kinetic term in our Schrödinger equation becomes

$$\frac{1}{R^2}\frac{\partial}{\partial R}R^2\frac{\partial}{\partial R}\left[\frac{\eta_v(R)}{R}\right] = \frac{1}{R}\frac{\partial^2}{\partial R^2}\eta_v(R), \tag{8.21}$$

and the complete Schrödinger equation in terms of η_v is

$$\left[-\frac{\hbar^2}{2\mu R}\frac{\partial^2}{\partial R^2} + \frac{1}{R}\left(\frac{1}{2}\right)k(R - R_e)^2\right]\eta_v(R) = \frac{1}{R}E_{\text{vib}}\eta_v(R). \tag{8.22}$$

We can multiply both sides by R, leaving the harmonic oscillator Schrödinger equation,

$$\left[-\frac{\hbar^2}{2\mu}\frac{\partial^2}{\partial R^2} + \frac{1}{2}k(R - R_e)^2\right]\eta_v(R) = E_{\text{vib}}\eta_v(R).$$

The force constant k is now related to the details of the molecular potential curve.

The harmonic oscillator wavefunctions $\eta_v(R)$ in this equation are identical to the wavefunctions $\eta_v(x)$ described previously, except for a change in coordinate from x to R. Previously, the potential energy was zero at $x = 0$. Now the potential energy is zero at $R = R_e$, so our expression for x is

$$x(R) = (R - R_e), \tag{8.23}$$

and the wavefunctions are then given by our previous expression, Eq. 8.16:

$$\eta_v(R) = A_v H_v(y) e^{-(R-R_e)^2/(2c^2)}, \quad c = \left(\frac{\hbar^2}{k\mu}\right)^{1/4},$$

where y is now $(R - R_e)/c$, a unitless coordinate proportional to R.

Notice that we have not labeled these wavefunctions by $\psi_{\text{vib}}(R)$, for the vibrational wavefunctions that appear in Eq. 8.12. We can restore $\psi_{\text{vib}}(R)$ by dividing η_v by R:

$$\psi_{\text{vib}}(R) = \frac{A_v}{R} H_v(y) e^{-(R-R_e)^2/(2c^2)}. \tag{8.24}$$

However, it is almost always $\eta_v(R)$ that we draw when plotting vibrational wavefunctions. Understanding the reason for this requires some careful thinking. The two wavefunctions $\eta_v(R)$ and $\psi_{\text{vib}}(R)$ result from solving Schrödinger equations that have the same potential energy term but different kinetic energy terms. When we solve the equation for $\eta_v(R)$, we have solved the problem using a one-dimensional kinetic energy operator. This yields wavefunctions with units of $R^{-1/2}$, and the probability density $|\eta_v(R)|^2$ will have units of R^{-1} (because it represents a probability per unit distance). The vibrational wavefunction $\psi_{\text{vib}}(R)$ has been solved using a *three*-dimensional kinetic energy term, reflecting the fact that the molecule is a three-dimensional entity. This gives $\psi_{\text{vib}}(R)$ units of $R^{-3/2}$ and the probability density units of R^{-3}. The distinction is the same as between the square of the one-electron atom radial wavefunction $R_{n,l}(r)$ in Chapter 3, which could give the probability density per unit volume, and the radial probability density $R^2_{n,l}(r)$, which gives the probability per unit distance from the nucleus. The function $\eta_v(R)$ correctly describes the distribution of the nuclei as a function of R alone (which is the way we usually think of it); $\psi_{\text{vib}}(R)$ is appropriate only if the angular wavefunction is simultaneously considered.

DERIVATION SUMMARY The Harmonic Approximation. The harmonic approximation yields a solution to the vibrational energy, Eq. 8.17, which predicts that the vibrational quantum states should be evenly spaced by a vibrational constant $\omega_e = \hbar\sqrt{k/\mu}$, where k is the force constant and μ is the reduced mass. The wavefunctions are Gaussian curves multiplied by polynomials, with the order of the polynomial (and the number of nodes in the wavefunction) equal to the vibrational quantum number v, which can be any whole number.

The energies in Eq. 8.17 and wavefunctions in Eq. 8.24 are valid only within the boundaries of our numerous approximations, including the following:

1. Separability of electronic (r) and nuclear coordinates (R, Θ, Φ) by the Born-Oppenheimer approximation
2. Separability of vibrational (R) and rotational (Θ, Φ) nuclear coordinates by the rigid-rotor approximation
3. Neglect of the higher-order terms—the **anharmonicity**—in the Taylor expansion of $U_{\text{vib}}(R)$ (Eq. 8.11) by the harmonic approximation.

The harmonic oscillator is our first **model potential** of several to come. Determining the exact potential energy functions of real molecules is a lot of work, and instead we normally use relatively simple equations that approximate the molecular potential energy. Using the harmonic oscillator, we solve the Schrödinger equation, within the above approximations. Technically, we end up with a different wavefunction, because the kinetic energy term in Eq. 5.11 is three-dimensional and we replaced it with a kinetic energy term for motion along only one dimension. In working through this, we have glided over some details of changing coordinate systems. Fortunately, the energy solutions are immune to this change of coordinates, and to a first approximation the molecular vibrational energies are given by

$$E_{\text{vib}} = \left(v + \frac{1}{2}\right)\hbar\sqrt{\frac{k}{\mu}} = \left(v + \frac{1}{2}\right)\omega_e. \tag{8.25}$$

These are evenly spaced energies beginning with $\omega_e/2$ for $v = 0$ and continuing in intervals of ω_e.

8.2 Vibrational Energy Levels in Diatomics

Equation 8.25 gives the solution to the energy levels of the simple harmonic oscillator, a model for the vibrational mechanics of a chemical bond. This is only a model, however, and we know that it doesn't succeed under all conditions. Often in vibrational spectroscopy we look no further than the lowest excited state, $v = 1$, and in that case Eq. 8.25 is usually adequate. It predicts rather well, for example, how the transition energy depends on the atomic masses. However, detailed studies of molecular dynamics and interactions demand a more general approach to the vibrational Schrödinger equation. In this section, we look at how the harmonic oscillator model fails and what we can do about it.

Generalizing Potential Surfaces

There is a faster but less precise way to see that the harmonic potential works for molecules near their vibrational ground states. The wavefunctions and energy level structure for real systems are determined by the nature of the Hamiltonian. Every Hamiltonian contains a kinetic energy operator of the form $-(\hbar^2/2m)\nabla^2$. The only differences in kinetic energy operator from one system to another are in the values of the masses m and the number of coordinates.[2] This principle holds true in classical mechanics, where K always looks like $mv^2/2$, as well as in quantum mechanics. The critical thing that makes one mechanical system genuinely different from another is *the shape of the potential energy function.* Two systems with similar potential energy surfaces are going to have similar wavefunctions and energy level structure. We can affect the spacing between energy levels, or between nodes of the wavefunctions, by changing the masses of the particles, but this is similar to enlarging or reducing a photograph without changing its basic image. To get qualitatively different wavefunctions and energies, we need to change the potential energy.

Conveniently, we now have exact solutions to the Schrödinger equation for three ideal cases with starkly different potential energy functions: (*i*) the particle in a box, with infinitely steep walls in the potential; (*ii*) the harmonic oscillator, with walls that get steeper as we get farther from the origin; and (*iii*) the Coulomb potential for the one-electron atom, with walls that get *less* steep as we move away from the origin. Now, compare the energy spacings of the Coulomb $1/r$ potential and the box potential, shown in Fig. 8.3. For the harmonic oscillator the energy levels are evenly spaced. As we make the walls of the potential curve *steeper* than a harmonic oscillator potential (approaching the 1-D box), the energy level spacing begins to diverge at higher E. As the walls become *flatter* (as in the $-1/r$ function in Fig. 8.3c), the spacing collapses at higher E. For any potential energy function we come across from now on, regardless of the system and the source of the potential energy, we can predict the qualitative behavior of the energies and wavefunctions by comparison to these model systems.

For example, the harmonic oscillator potential, $U(x) = \frac{1}{2}kx^2$, leads to evenly spaced energy levels. Some bending vibrations have potential energy functions more similar to $U_4(x) = k_4x^4$. Because that function has walls that are steeper than the harmonic oscillator, approaching the shape of the one-dimensional box potential, we can be sure that the energy levels that solve the Schrödinger equation on $U_4(x)$ will have energy levels that get more widely spaced as the energy climbs.

How about the wavefunctions? Examination of each potential energy function in Fig. 8.3 shows that as the kinetic energy $(E - U)$ gets larger, the wavefunction amplitude becomes weaker near the middle of the potential curve, and its amplitude changes more quickly (the spacing between nodes gets smaller). This trend makes sense in light of the correspondence principle: classically the particles spend the least time (and should have the smallest wavefunction amplitude) in the region where they move the fastest (where the kinetic energy is largest). In this region their de Broglie wavelength is small, and wave characteristics are harder to observe.

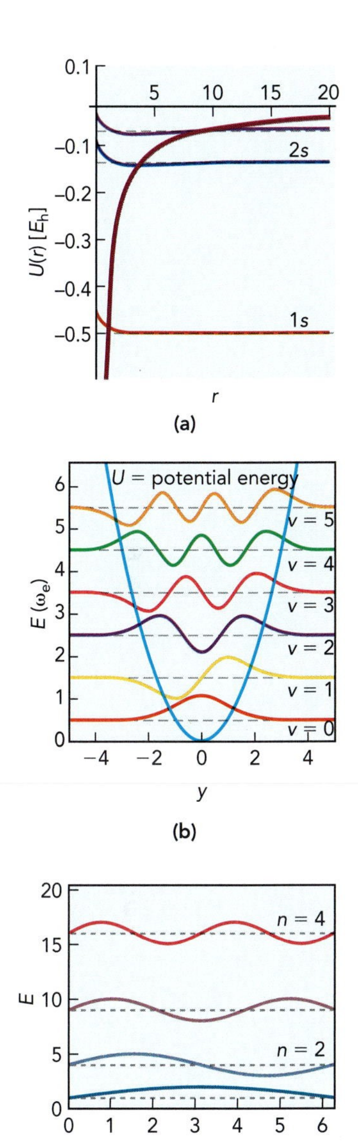

▲ **FIGURE 8.3 Comparison of the energy spacing and wavefunctions of the lowest energy levels:** **(a)** the Coulomb potential with decreasing spacing between levels at higher energy; **(b)** the harmonic oscillator potential, with constant level spacing; and **(c)** the one-dimensional box potential, with increasing level spacing.

[2]The number of coordinates is affected both by the dimensionality of the space and the number of particles. For example, the one-electron atomic system dealt with one particle moving in three dimensions, and so $\hat{K}$ had three terms (derivatives along x, y, and z that we transformed to derivatives along r, θ, and ϕ), whereas we used only the x coordinate to define motion in the one-dimensional box, so there was only one term in $\hat{K}$, proportional to $\frac{\partial^2}{\partial x^2}$. If the system contains two independent particles, then we have different coordinates for each particle. The Hamiltonian for the two-electron atom therefore has six terms in $\hat{K}$, and two particles in a one-dimensional box would need a $\hat{K}$ operator with two terms: one varying as $\frac{\partial^2}{\partial x_1^2}$ and one as $\frac{\partial^2}{\partial x_2^2}$.

There is a fair match between the harmonic oscillator potential and the real molecular potential energy curve we saw in Chapter 5 at low potential energy (this is our requirement that R should be near R_e in the Taylor series). The match becomes worse as we go up in energy and the potential becomes anharmonic. At low vibrational energy, the harmonic oscillator is a good approximation to most molecular vibrations.

Vibrational Energy Levels and Anharmonicity

At high v or for more precise measurements, the energy expression needs the anharmonic terms from the Taylor series expansion of $U_{\text{vib}}(R)$. Traditionally, the anharmonicities are incorporated by means of a power series in $v + 1/2$:

$$E_{\text{vib}} = \left(v + \frac{1}{2}\right)\omega_e - \left(v + \frac{1}{2}\right)^2 \omega_e x_e + \left(v + \frac{1}{2}\right)^3 \omega_e y_e + \ldots, \quad (8.26)$$

which includes the anharmonic constants $\omega_e x_e$ and $\omega_e y_e$ as well as the harmonic vibrational constant ω_e. This kind of power series expansion is common in spectroscopy, but it suffers from the fact that the values for constants such as ω_e depend on how many other terms in the series are included. Each term in the vibrational expansion has a contribution to the energy even at $v = 0$. Examples of these vibrational constants for selected diatomics and the corresponding force constants can be found in Table 8.2.

Several common-sense points can be noted from Table 8.2. The first three molecules in the table are simply the H_2 molecule with different isotopes. The force constants are nearly identical, which is not surprising because the force constant is determined by the electronic wavefunction and, within the Born-Oppenheimer approximation, should be independent of nuclear mass. The vibrational constants ω_e are not equal, however, because they are determined by the kinetic energy of the nuclei on the effective potential and decrease roughly as the square root of the reduced mass μ ($4401\,\text{cm}^{-1}/\sqrt{2} = 3112\,\text{cm}^{-1}$, compared to $3118\ \text{cm}^{-1}$ observed for ω_e of $^2H_2$).

The series HF, HCl, HBr, and HI generally shows the same result: when increasing the mass of the atoms, the vibrational constant decreases. However, the reduced mass rapidly approaches the mass of the H atom. Why does ω_e continue to decrease significantly from HBr to HI, when both have reduced masses of 1.00 amu? There is an added effect: the hydrogen halide bond strength decreases as the halogen becomes larger (having less tightly bound valence electrons) and less electronegative. This effect reduces the force constant k, and so ω_e decreases as well. The same effect appears in the series Cl_2, Br_2, IBr, I_2.

The relationship between force constant and bond order can also be seen in the series of homonuclear diatomics N_2, O_2, and F_2. For this series, the bond orders (approximate indicators of the bond strength) are 3, 2, and 1, respectively. The corresponding force constants are 2290, 1180, and $440\,\text{N}\,\text{m}^{-1}$. CO and NO^+, which also have bond orders of 3, have force constants of $1900\,\text{N}\,\text{m}^{-1}$ and $2490\,\text{N}\,\text{m}^{-1}$, similar to the value $2290\,\text{N}\,\text{m}^{-1}$ for N_2.

Bond order can be misleading, however. If we apply a Lewis structure model to Na_2, we would assign it a bond order of 1, the same as Cl_2, and therefore might expect it to have a similar force constant. However, the alkali metal dimers form bonds primarily with weakly bound single valence electrons. The resulting

TABLE 8.2 Ground state molecular vibrational constants for selected diatomic molecules.

Molecule	μ (amu)	k (N m^{-1})	ω_e (cm^{-1})	$\omega_e x_e$ (cm^{-1})	$\omega_e y_e$ (cm^{-1})
^{1}H ^{1}H	0.50	570	4401.21	121.34	0.81
^{1}H ^{2}H	0.67	570	3811.92	90.71	0.48
^{2}H ^{2}H	1.01	580	3118.46	117.91	1.25
^{1}H ^{19}F	0.96	970	4138.32	89.88	0.90
^{1}H ^{35}Cl	0.98	520	2990.95	52.82	0.22
^{1}H ^{79}Br	1.00	410	2649.67	45.21	
^{1}H ^{127}I	1.00	310	2309.60	39.36	−0.09
^{2}H ^{19}F	1.82	960	2998.19	45.76	
^{12}C ^{16}O	6.86	1900	2169.82	13.29	0.01
^{14}N ^{14}N	7.00	2290	2358.07	14.19	−0.01
^{14}N ^{16}O^{+}	7.47	2490	2377.48	16.45	
^{14}N ^{16}O	7.47	1600	1904.41	14.19	0.02
^{14}N ^{16}O^{-}	7.47	830	1372	8	
^{16}O ^{16}O	8.00	1180	1580.36	12.07	0.05
^{19}F ^{19}F	9.50	440	891.2		
^{35}Cl ^{35}Cl	17.48	320	560.50	2.90	0.02
^{79}Br ^{79}Br	39.46	250	325.29	1.07	
^{127}I ^{79}Br	48.66	210	268.71	0.83	
^{127}I ^{127}I	63.45	170	214.52	0.61	0.00
^{23}Na ^{23}Na	11.49	17	159.13	0.73	0.00
^{133}Cs ^{133}Cs	66.45	7	42.02	0.08	0.00
^{197}Au ^{198}Pb	101.16	150	158.6	0.6	
$B^1\Sigma_u^+$ ^{1}H ^{1}H	0.50	50	1358.94	21.53	0.87
$a^3\Pi$ ^{12}C ^{16}O	6.86	1230	1743.41	14.36	−0.04

Values for the vibrational constants are based on varying numbers of anharmonic terms.

molecular bond in Na_2 is weak and has a very low force constant: 17 N m^{-1}. The more massive alkali metals form even weaker bonds, as shown by the extremely low force constant and vibrational constant of Cs_2.

Vibrational Energies

Could we have guessed these vibrational constants in Table 8.2, using what we have learned about electronic states? The spacing between the ground ($v = 0$) and first excited ($v = 1$) vibrational states is roughly $\hbar\sqrt{k/\mu}$, where k is positive and approximately equal to $\left(\frac{\partial^2 U_{\text{vib}}}{\partial R^2}\right)\Big|_{R_e}$. The reduced mass μ is a constant, found from the particle masses. The vibrational force constant k represents the curvature of the potential curve near the bottom of the attractive well. As k gets smaller, the potential gets flatter (less curvature), and the energy level spacing

BIOSKETCH Nathan Hammer

Nathan Hammer, a chemistry professor at the University of Mississippi, uses vibrational Raman spectroscopy to help us understand the effects of intermolecular forces on molecular structure and behavior. The vibrational spectrum provides a valuable probe of the electron distribution in the molecules as well. In the spectra shown, for example, a vibrational transition in normal pyrimidine shifts from roughly 1570 cm^{-1} (where it overlaps with another transition in the spectrum at left) to a clearly distinct peak at over 1580 cm^{-1} (the spectrum at right) when a water molecule attaches to one of the nitrogen atoms. This upward shift occurs because some electron density transfers from a non-bonding lone pair orbital on the pyrimidine over to the water molecule, and that shift strengthens several of the C—C and C—N bonds, raising the corresponding vibrational constants. In this experiment, Professor Hammer takes advantage of the way that the Raman signal depends on the polarization of the input and output radiation to determine the normal modes associated with the observed transitions. The interaction studied here, an OH group weakly bonded to an N atom, plays a role in many important biological processes, such as the proton transfer from a carboxylic acid group (COOH) to an amine group (NH_2) in proteins.

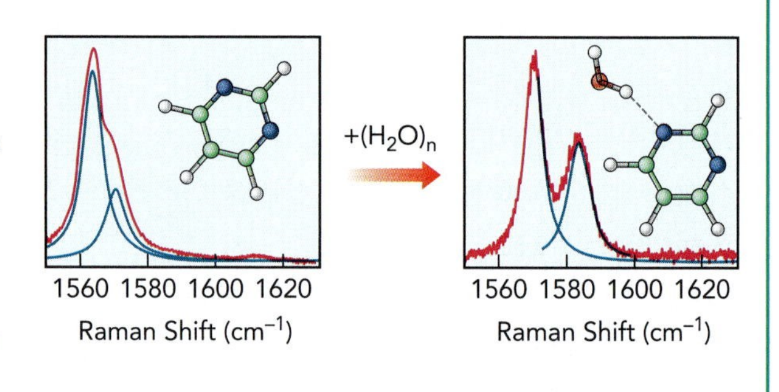

gets smaller as well. If we know the width of the potential well ΔR at some given potential energy $U_{\text{vib}}(R_1)$, then we can estimate k as follows:

$$U_{\text{vib}}(R_1) \approx \frac{1}{2}k(R_1 - R_e)^2 \approx \frac{1}{2}k(\Delta R/2)^2 = \frac{k}{8}\Delta R^2$$

$$k \approx \frac{8U_{\text{vib}}(R_1)}{\Delta R^2}. \tag{8.27}$$

(We use the width of the well ΔR rather than $R_1 - R_e$ for this estimate to average the slopes of the two sides of the well.)

Take a representative case. In the H_2 ground state potential energy curve drawn in Fig. 7.7, at an energy of 2 eV above the minimum, the potential well extends from about 0.6 Å to 1.2 Å, for a ΔR of 0.6 Å. The force constant k is therefore roughly

$$k \approx \frac{8(2\,\text{eV})(1.602 \times 10^{-19}\,\text{J/eV})}{(0.6\,\text{Å})^2(10^{-10}\text{m/Å})^2} = 710\,\text{N}\,\text{m}^{-1}.$$

For H_2, the reduced mass μ is 0.5 amu, and

$$\omega_e(\text{cm}^{-1}) = 130.27\sqrt{\frac{k(\text{N}\,\text{m}^{-1})}{\mu(\text{amu})}} \approx 130.27\sqrt{\frac{710\,\text{N}\,\text{m}^{-1}}{0.5\,\text{amu}}}$$
$$= 4900\,\text{cm}^{-1},$$

which is about 10% off from the actual value of 4401 cm^{-1}.

This crude calculation mimics what we do when we solve the Schrödinger equation: we start with a potential energy function and use it to derive the quantum state energies and wavefunctions. That general approach is a standard technique in theoretical studies of molecules. Experiments usually tackle the issue in the opposite direction, using the vibrational energies (which can be measured directly by spectroscopy) to determine the shape of the potential energy curve. This interplay was outlined in Fig. 2.3.

The force constant for H_2 is typical of the diatomics, but the very low reduced mass gives H_2 the highest ω_e value of any molecule. If we consider instead the O_2 molecule, with a more typical reduced mass of 8.0 amu and a force constant of 1180 N m^{-1}, we obtain a vibrational constant of 1580 cm^{-1}. This value is equal to $0.00720\,E_{\mathrm{h}}$, 0.196 eV, or a wavelength of 6.33 microns. Because the energy levels of the harmonic oscillator are evenly spaced, spectroscopic transitions between adjacent vibrational energy levels ($\Delta v = \pm 1$) occur at energies of roughly ω_e and involve radiation in the infrared region of the spectrum.

The Morse Potential

The harmonic approximation works well at low vibrational quantum number, but in many applications we need a better approximation, one that includes the effects of anharmonicity. And for chemistry, the critical drawback of the harmonic oscillator approximation is that there is no dissociation. The bond never breaks, it just gets harder and harder to stretch the more you try to pull it apart. We can't have any chemistry if we can't break bonds, so it is worthwhile to consider a more realistic (and, of course, more complicated) model potential for the vibrating molecule that includes this basic feature.

First, we pause for a word about dissociation energies. In real molecules, as in the ideal harmonic oscillator, there is zero-point energy; we can never stop the nuclei from vibrating. Consequently, there are two different reference points that are commonly used when vibrational energies are reported. Theoreticians often measure energy from the minimum of the potential energy curve ("the bottom of the well"), because that energy corresponds to a unique geometry of the molecule—the equilibrium geometry. For the same reason, this is also a convenient reference for the electronic energy, and we set the electronic energy equal to some constant E^0_{elec} when the molecule is at the equilibrium geometry and (we pretend) not vibrating at all. However, there is no quantum state at that energy that experimentalists can probe, so they measure vibrational energies starting from the ground vibrational state $v = 0$. The well depth or binding energy (the energy necessary to climb out of the potential well and break the chemical bond) is written D_e when measured from the bottom of the well, and it is written D_0 when measured from the ground vibrational state. Similarly the **term energy** is written T_e when it is the difference in energy between the minima of the potentials for two states, and T_0 when it is the difference in ground vibrational state energies. These definitions are illustrated in Fig. 8.4. We have to be careful now when saying "ground state" to specify whether we mean the lowest-energy electronic, vibrational, or rotational state (or some combination).

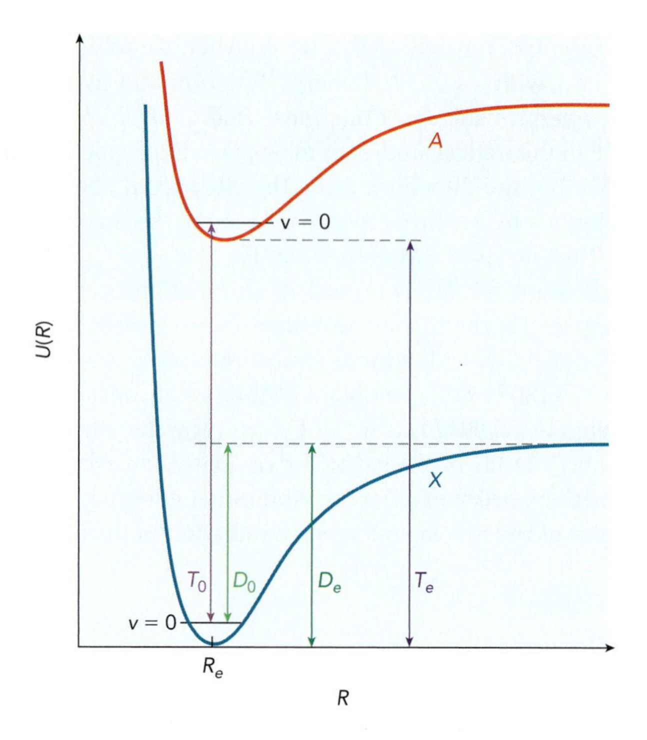

▶ **FIGURE 8.4 Distinctions between equilibrium properties (D_e and T_e) and experimentally measurable properties (D_0 and T_0),** for two electronic states X and A of a diatomic molecule.

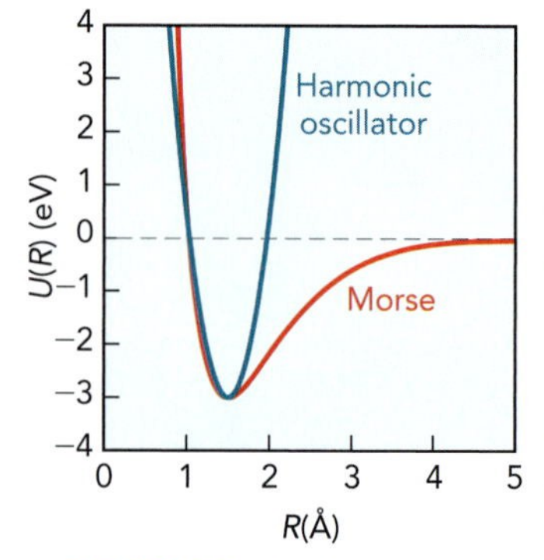

▲ **FIGURE 8.5 The Morse potential,** with $D_e = 3\text{eV}$, $R_e = 1.5\,\text{Å}$, and $\beta = 1.5\,\text{Å}^{-1}$.

Our improved model potential is the **Morse potential,** sketched in Fig. 8.5 and given by the formula

$$U_{\mathrm{M}}(R) = D_e\{[1 - e^{-\beta(R-R_e)}]^2 - 1\}, \tag{8.28}$$

where D_e and R_e are the equilibrium well depth and bond length, and where $\beta = \sqrt{k/(2D_e)}$ determines the width of the well. For this case, because we usually use the Morse potential while considering a single electronic state, we have set the potential energy to zero at the dissociation limit of large R. The Morse potential is an extremely useful curve for simple cases because it has the right general behavior: $U_{\mathrm{M}}(\infty) = 0$ and $U_{\mathrm{M}}(R_e) = -D_e$. The repulsive wall at low R is also quite large, $U_{\mathrm{M}}(0) = D_e(1 - e^{+\beta R_e})^2$, which is usually in the range of $10D_e$. Disagreement between experimental data and the Morse curve is usually large only at very high vibrational excitation, much higher than is easily accessible. Standard techniques in vibrational analysis take experimentally measured vibrational energies and search for a consistent model potential, such as the Morse potential, to obtain approximate vibrational wavefunctions as well as D_e, k, and R_e. Solving the Morse potential Schrödinger equation is not easy, and we will not do so here. Nevertheless, we can use the general concepts of potential surface analysis to guess at the form of the wavefunctions, and the energies exactly correspond to the first two terms in Eq. 8.26:

$$E_{\mathrm{Morse}} = \left(v + \frac{1}{2}\right)\omega_e - \left(v + \frac{1}{2}\right)^2 \omega_e x_e.$$

The Morse oscillator allows the first anharmonic correction to be related directly to the curvature and depth of the potential well:

$$\omega_e x_e = \frac{\hbar^2\beta^2}{2\mu} = \frac{\hbar^2\omega_e^2}{4D_e},$$

which shows that as the well gets deeper (greater D_e), the anharmonicity decreases.

8.3 Vibrations in Polyatomics

When we try to solve the vibrational Schrödinger equation for molecules with more than two atoms, we still focus first on the vibrational potential energy function. For the diatomic, we used the approximation that the potential energy curve looked like a parabola—a harmonic oscillator potential—at low vibrational energy. We will apply the same approximation to the polyatomic, but now the potential energy function is a surface that depends on each vibrational coordinate of the molecule. The point on this multi-coordinate surface where the potential energy reaches its lowest values corresponds to the molecule's **equilibrium geometry.** For the diatomic molecule, the equilibrium geometry was determined solely by the equilibrium bond length. For polyatomics, the equilibrium geometry depends on each bond length and bond angle in the molecule.

The molecule vibrates back and forth across that equilibrium geometry, and our next task is to describe what those motions look like. Which atoms are moving, and in what directions? And how many different kinds of motion do we need to consider?

Vibrational Modes

The vibrational states of the polyatomic molecule stretch each of those bond lengths and bend each of those bond angles. How many of these vibrational coordinates are there, really? For example, the BH_3 molecule has three B—H bond lengths, three H—B—H bond angles, and dihedral angles (see Fig. 8.6) that change if the molecule becomes non-planar. Right away, we can see that these are not all independent: if the molecule is planar, then once we know two of the H—B—H bond angles, the third angle is fixed at 360° minus the other two angles. We find the number of independent vibrational coordinates by counting the other coordinates.

For a molecule with N_{atom} atoms, there are X, Y, and Z coordinates for each atom, for a total of $3N_{\text{atom}}$ coordinates to describe any possible distribution of those atoms. (Recall that we measure X, Y, and Z along coordinate axes that are fixed in the laboratory.) Of those $3N_{\text{atom}}$ coordinates, three give the location of the center of mass, leaving $3N_{\text{atom}} - 3$ coordinates. Of those remaining, some correspond only to a change in the orientation of the molecule. If we fix all the bond lengths of a linear molecule, for example, we could still rotate the molecule by an angle Φ in the XY plane or by an angle Θ from the Z axis, without changing the location of the center of mass. For a linear molecule, these two angles are sufficient to describe any orientation of the molecule; for a non-linear molecule, we need three such angles, as shown in Fig. 8.7.

That means that for a linear molecule, $N_{\text{vib}} = 3N_{\text{atom}} - 3 - 2 = 3N_{\text{atom}} - 5$ coordinates remain apart from those that give the center of mass location and orientation angle. These coordinates are the vibrational coordinates, and they

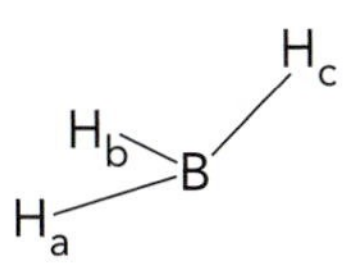

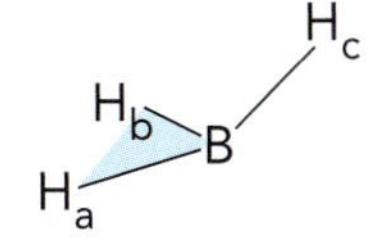

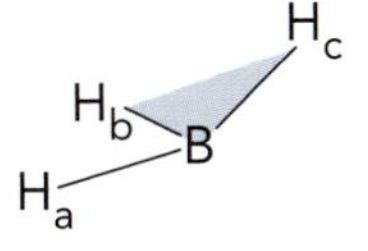

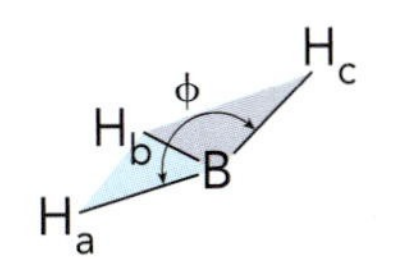

◀ FIGURE 8.6 **The dihedral angle,** shown using BH_3 as an example. Two planes are defined by the atom groups H_aBH_b and H_bBH_c. The dihedral angle ϕ is the angle between the two planes.

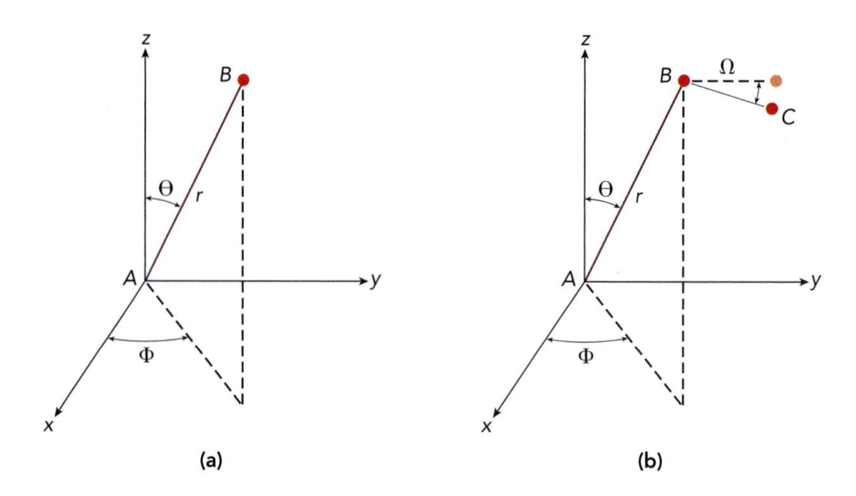

▶ FIGURE 8.7 **Rotational coordinates:** two (Θ and Φ) for a linear molecule AB (a) and three (Θ, Φ, and Ω) for a non-linear molecule ABC (b).

count the number of ways the atoms can move *relative* to one another. For a non-linear molecule, the number of vibrational coordinates N_{vib} is $3N_{atom} - 6$ because, of the $3N$ coordinates overall, three correspond to translation and three to rotation. These two expressions are worth using up an equation number:

$$\textbf{linear}: N_{vib} = 3N_{atom} - 5 \qquad \textbf{non-linear}: N_{vib} = 3N_{atom} - 6. \tag{8.29}$$

The formula works even for the diatomic, which must be linear:

$$3 \cdot 2 - 5 = 1.$$

There is one vibrational coordinate, and it is the bond length. Changing the bond length is the only way to change the *relative* position of the two atoms. In a non-linear triatomic molecule, there are $3 \cdot 3 - 6 = 3$ vibrational coordinates, or **vibrational modes.**

For most molecules, there are *many* different vibrational motions, and for each of them the harmonic oscillator remains our first approximation. Note that there is some zero-point energy in *each* vibrational mode, and a surprising amount of energy can be locked up in the zero-point vibration of a large molecule.

Normal Modes

We haven't yet considered what these vibrations might actually look like. In any system of vibrating objects, such as a molecule, there is a set of equations of motion (in classical physics) or vibrational wavefunctions (in quantum mechanics) called **normal modes** that describe the lowest-energy motions of the system. In the normal modes, each atom in the molecule oscillates (if it moves at all) back and forth across its equilibrium position at the same frequency and phase as every other atom in the molecule. At higher vibrational energy, the motions can be more complicated, but we can write those motions as a combination of different normal modes. Any vibration of the system can be expressed as a sum of the normal modes: they are one possible basis set of vibrational coordinates.

For example, in a linear triatomic molecule there are four vibrational coordinates: $3N_{atom} - 5 = 4$. Two of these are **stretching** coordinates, corresponding to changes in the bond lengths, and the other two are degenerate **bending** coordinates, corresponding to changes in the bond angle

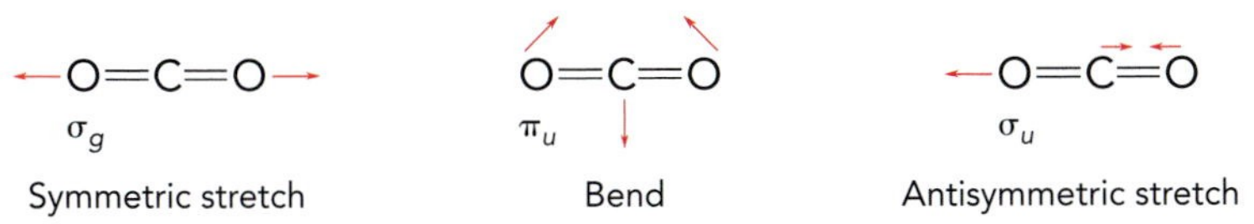

O=C=O σ_g Symmetric stretch

O=C=O π_u Bend

O=C=O σ_u Antisymmetric stretch

◀ **FIGURE 8.8 The normal modes of CO_2.** The arrows indicate the direction of displacement for each atom.

in the *xz* or *yz* planes, where the *z* axis is the internuclear axis. In CO_2, the normal mode stretches are the **symmetric stretch,** where both C=O bonds get longer simultaneously and get shorter simultaneously, and the **antisymmetric stretch,** where one bond lengthens as the other shortens. In both cases, the bond lengths reach their equilibrium values at the same time. Because the two oxygen atoms are indistinguishable, the two bond lengths reach the same maximum and minimum values during the vibration. These normal modes are illustrated in Fig. 8.8.

If we change one oxygen to a sulfur atom, we lose that symmetry. This molecule is carbonyl sulfide, OCS. The equilibrium bond length for the C=S is larger than for the C=O, since sulfur is a bigger atom and the displacement during the vibration is larger. Nonetheless, in the normal mode the equilibrium geometry is always one of the points reached during the vibration.

There is a numbering scheme for identifying the different normal modes based on symmetry. For triatomics that have two equivalent atoms (i.e., in the point groups $D_{\infty h}$ or C_{2v}), the convention is to use ν_1 to label the symmetric stretch, the bend is the ν_2 mode, and the antisymmetric stretch is labeled ν_3. To keep track of how much energy is in each mode, each mode gets its own quantum number: v_1 for the symmetric stretch, v_2 for the bend, and v_3 for the antisymmetric stretch.[3] For larger molecules, this system can be extended, but most chemists simply refer to the modes by the type of motion and the atoms moving, such as "the CO stretch" rather than using a specific mode number.

Normal modes treat motion of all the atoms at once, and they have the advantages and disadvantages of symmetry analysis. The normal modes of a molecule can be classified according to the representations of the point group, and the number of atoms and the point group of the molecule can predict exactly the number of normal modes given by each representation. How do we classify a vibration according to the point group? Instead of asking what happens to the phase of the electronic wavefunction under each of the symmetry operators, we find what happens to the direction of vibrational motion. For example, the symmetric stretch in CO_2 can be drawn with arrows indicating the direction and magnitude of motion for each oxygen atom. The carbon is at the center of mass and remains motionless due to the symmetry of this mode. The directions of the arrows are invariant whether we rotate, reflect, or invert the drawing by any of the operations of the $D_{\infty h}$ point group, and therefore the vibrational mode is represented by σ_g (lowercase letters are used for vibrational representations as well as one-electron orbitals). The antisymmetric stretch, on the other hand, has arrows that change direction when the drawing is inverted or reflected through the horizontal mirror plane, but not when rotated about the principal axis. Therefore, the antisymmetric stretch is represented by σ_u. Similarly, the doubly degenerate bending mode is represented by π_u.

[3]Unfortunately, the vibrational quantum number (italic "v," v) and the mode label (Greek letter "nu," ν) are difficult to distinguish in print. The context should make it clear which is being used.

Tools of the Trade | Fourier Transform Infrared (FTIR) Spectroscopy

Between 1880 and 1881, Albert Michelson developed an *interferometer,* which split a beam of coherent light into two beams traveling in different directions before reflecting them back to a common point on a detector. Coherent light has a single wavelength and phase, so that where the two beams recombine, they interfere with each other. If the two beams travel the same distance at the same speed, the phase of the waves will match at the detector and so interfere constructively with each other, giving a strong detector signal. If the two speeds are different, the wave peaks in one beam will arrive at different times from the peaks in the other beam, and some destructive interference will take place, cancelling part of the signal where the two beams overlap. With this device, Michelson and Edward Morley in 1887 demonstrated that to high precision, the speed of light was the same in all directions. Their measurements refuted a prevailing theory of how light traveled through space and provided an important foundation for the theory of special relativity. Michelson was awarded the 1907 Nobel Prize in Physics for this work.

Michelson continued to work with other applications of his interferometer, in particular how the interference pattern of the two beams could be used to determine the properties of the radiation when many wavelengths were present at the same time. By changing the position of the mirror that reflected one of the two beams, Michelson could change which wavelengths interfered destructively at the detector and which interfered constructively. If the mirror position was moved through a great enough distance, the detector signal contained all the information needed to calculate the intensity at each wavelength. This would make it possible to measure a spectrum without having to disperse the light through a prism and detect the signal at many locations.

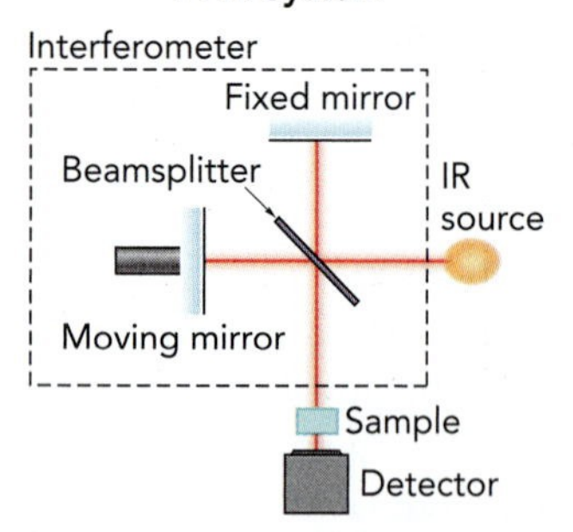

Schematic of an FTIR spectrometer, including the Michelson interferometer (the beamsplitter and pair of mirrors).

The interferometer produced an **interferogram,** a graph of signal versus mirror position. However, an interferogram is not the same as a spectrum (which is a graph of intensity versus radiation wavelength or frequency), because at any given mirror position, several wavelengths could contribute to the signal. The conversion of an interferogram to a spectrum can be accomplished by means of a Fourier transform, but Fourier transforms on complicated functions aren't easy, and the mathematics involved proved too demanding to be practical through the early 20th century. Interferometry became an established method in spectroscopy only with the development of the fast Fourier transform (FFT), a computational algorithm devised in 1965 by J. W. Cooley and John W. Tukey to obtain extremely efficient and accurate solutions to Fourier transforms. The first FTIR spectrometers were sold in the 1960s, and they rapidly grew in popularity, becoming a standard tool for chemical analysis by the late 1970s.

What is FTIR spectroscopy? Fourier transform infrared spectroscopy measures the infrared spectrum of a chemical sample, typically from about 400 cm^{-1} to 3500 cm^{-1}. In this region of the spectrum, all the observable vibrational transitions occur except those of the heaviest atoms or largest molecules. A Michelson interferometer initially measures an interferogram for the sample, and the FFT algorithm converts the interferogram to the spectrum. Vibrational spectroscopy in the 1940s employed dispersive infrared spectrometers, in which a diffraction grating, like a prism, would split infrared light into different wavelengths. The FTIR spectrometer measures the same spectrum with greater sensitivity, in part because the radiation does not need to be distributed over different positions to separate the wavelengths, and in part because the signal at many wavelengths can be acquired at the same time.

Why do we use FTIR spectroscopy? A vibrational spectrum serves two principal purposes in chemical analysis: (1) the spectrum unambiguously identifies a pure sample of almost any molecule of more than two atoms, as long as it is present in sufficient concentration, and (2) the spectrum provides the simplest direct probe of specific chemical bonds. The infrared (IR) spectrum of a molecule is usually more detailed than the UV-vis spectrum, and specific features can often be associated directly with specific arrangements of atoms (such as the carbonyl group C=O, which absorbs strongly at about 1760 cm^{-1}).

The IR spectrum does not give the same detailed structural information that an NMR spectrum does, so it is not used to solve molecular structures. However, the sample preparation is much simpler than for NMR, and the spectrometer itself is *much* less expensive and bulky, so the FTIR is a more common device for routine spectroscopic analysis, when you already know what you are looking for. It is particularly useful in laboratories that test large quantities of samples for known compounds, as in criminal forensics or water quality analysis. In research, we also use the FTIR to quantify the effects of forces such as intermolecular attractions on chemical bonds. For example, if a C=O bond is weakened by attraction of the O atom for a nearby hydrogen atom, then the C=O force constant k decreases. The vibrational constant $\omega_e = \hbar\sqrt{k/\mu}$ decreases as well, and the C=O transition appears at a lower wavenumber (cm^{-1}) value—corresponding to a lower transition energy—than when the oxygen–hydrogen interaction is absent.

How does it work? Although the FTIR does not directly measure absorbance as a function of wavelength, it has the design of the most fundamental absorption spectrometers: a radiation source, a sample, and a detector.

- *Radiation source.* Solids, when heated to temperatures around 1500 C, emit blackbody radiation (see *TK*, Section 5.1) throughout the mid-infrared. Rods of silicon carbide are commonly used for this purpose because the compound is strong and resistant to oxidation—a rare property at these high temperatures—and the heating is easily achieved by running an electrical current through the rod.
- *Sample preparation.* Samples in chemical analysis are usually most convenient to handle in solution. However, sample containers (such as NMR tubes or the cuvettes used in UV-vis spectroscopy) are problematic for this region of the spectrum because almost any substance absorbs *some* infrared light. The alkali halide salts are exceptions, and windows made of NaCl or KBr transmit IR radiation efficiently over the entire wavelength range of the spectrometer. Therefore, in FTIR spectroscopy the sample is usually prepared by pressing a few drops of the analyte between two NaCl or KBr windows.
- *Detector.* The radiation is detected by infrared-sensitive semiconductors, such as indium gallium arsenide (InGaAs) or mercury cadmium telluride (HgCdTe, but often abbreviated MCT). Infrared light striking these materials can promote electrons into conduction, allowing the radiation to be detected as an electrical current.

In addition to its function as a stand-alone analytical tool, FTIR may be used as the detector in various chromatography methods and combined with microscope optics to allow IR spectroscopy of small sample areas. This last technique has been applied to the analysis of the fine detail in historic paintings, using the FTIR to chemically analyze the paint and taking advantage of the greater penetrating power of long wavelengths to probe even below the surface of the painting.

Normal Modes and Local Modes

But must we treat the displacements of all the atoms at once if we are interested in a single bond? Any real motion of the molecule can also be described by a sum over **local vibrational modes.** These are the one-bond-at-a-time vibrations that are often the easiest to visualize: one bond length or bond angle changes while all the other bonds remain rigid. There are some cases in which the local mode picture is a useful approximation, in particular when there is a significant difference in atomic masses. For example, in HCN the hydrogen is considerably less massive than the carbon or nitrogen, and the H—C stretch operates almost independently of the C≡N stretch. At low vibrational excitation, the local mode picture is accurate in this molecule.

An important qualitative result available from the local mode description is that *bending modes tend to have lower vibrational constants than stretching modes.* Stretching a bond pulls atoms away from the prime bonding region that stabilizes the molecule, whereas pushing the atoms too close together rapidly increases the repulsion between the nuclei. In contrast, bending a bond angle doesn't directly affect the stability of the two bonds involved. Instead, the energy climbs as the bond angle strains the electron distribution governed by the orbital hybrid, or as bending bonds start to bump into other parts of the molecule. Therefore the ω_e values for bending modes are generally lower than for stretches. Similarly, torsional motions can change a dihedral bond angle while leaving the attached bond lengths and angles unchanged, so torsions often have even lower vibrational constants than bends.

The local mode picture is used extensively in infrared spectroscopy of organic compounds. Because a C—H or N—H or O—H bond behaves in a similar fashion in virtually any molecule, the roughly 3000–3600 cm^{-1} absorption signal is used to identify the presence of chemical bonds involving hydrogen in a compound being characterized. Similarly, the C=C and C≡C bonds have stretching transitions that absorb between 1800 and 2200 cm^{-1}. However, the picture falls apart in some limit, and the low frequency range of these IR spectra vary enormously from molecule to molecule. The local modes do not describe the molecule well at the lower frequencies typical of many-atom bending transitions, and a much more complex normal mode analysis becomes necessary in that regime. The spectrum below about 1200 cm^{-1} is often a complex, uninterpreted mass of overlapping bending or torsional transitions, as well as stretches of relatively weak single bonds and stretches involving heavy atoms or groups of atoms. Even though it is often impossible to identify the individual transitions at low wavenumber, the vibrational spectrum in that region is unique to the molecule and is therefore called the **fingerprint** region of the spectrum.

The local mode and normal mode models are related the same way the MO model and local bond model of electronic orbitals are related (Chapter 5). They are different basis sets for expressing the vibrational wavefunctions of the molecule. Any complete vibrational analysis of a particular molecule, whether according to the local mode or the normal mode scheme, should yield the same number of vibrational modes: $3N_{\text{atom}} - 5$ (linear) or $3N_{\text{atom}} - 6$ (non-linear).

Determination of Normal Modes

CHECKPOINT Here we are showing by example how the normal modes of a polyatomic molecule are derived from the atomic masses and the equilibrium geometry. The approach is essentially to calculate the force constants k for each vibrational mode by finding the second derivatives of the potential energy. Then we use matrix algebra to find the particular set of motions q_i such that those second derivatives $\partial^2 U/(\partial q_i \partial q_j)$ are all zero unless $i = j$. In other words, we are looking for the motions such that each force constant (second derivative) corresponds to exactly one motion without being mixed with any other motion in the set.

In a polyatomic molecule, the normal modes depend on the relative masses and bond strengths. Although the analysis for large molecules is an ideal problem to leave to a computer, we will work through one simple example here: the stretching modes of a $D_{\infty h}$ triatomic molecule such as CO_2, labeling the atoms $O_aC_bO_c$.

In order to find the normal modes and the vibrational frequencies of the molecule, we need to know how the potential energy of the nuclei varies with nuclear position, a function commonly called the vibrational **potential energy surface,** or PES. The PES may be estimated from quantum mechanical calculations or, more approximately, from classical arguments by treating the bonds as ideal springs. With the PES, we can set up and solve an eigenvalue equation for the force constants of the nuclear motions of the molecule. The idea is that the potential energy will depend on the relative positions of the nuclei (in this case, just on the two bond lengths R_{ab} and R_{bc}), and so if we can put all those dependencies into one equation, we can find the set of motions—the normal modes—that correspond to the fundamental vibrational frequencies of the molecule.

Let's start by placing the three atoms of our molecule on the z axis and labeling the positions z_a, z_b, and z_c from left to right, where each z value is the distance from the atom's equilibrium position. We take advantage of the molecular symmetry right away, recognizing that the potential energy must change equally with R_{ab} as with R_{bc}:

$$\frac{\partial^2 U}{\partial z_a^2} = \frac{\partial^2 U}{\partial z_c^2}$$

$$\frac{\partial^2 U}{\partial z_a \partial z_b} = \frac{\partial^2 U}{\partial z_b \partial z_c}.$$

We are neglecting all the motions along x and y, which would correspond to bending and rotational motions.

Now we set up what's called the mass-weighted **Hessian matrix,** a square, symmetric matrix whose elements are the second derivatives $\partial^2 U/(\partial z_i \partial z_j)$ divided by $\sqrt{m_i m_j}$:

$$\mathrm{H} = \begin{pmatrix} A & B & D \\ B & C & B \\ D & B & A \end{pmatrix}, \tag{8.30}$$

where

$$A = \frac{1}{m_a}\frac{\partial^2 U}{\partial z_a^2} = \frac{1}{m_c}\frac{\partial^2 U}{\partial z_c^2}$$

$$B = \frac{1}{\sqrt{m_a m_b}}\frac{\partial^2 U}{\partial z_a \partial z_b} = \frac{1}{\sqrt{m_b m_c}}\frac{\partial^2 U}{\partial z_b \partial z_c}$$

$$C = \frac{1}{m_b}\frac{\partial^2 U}{\partial z_b^2}$$

$$D = \frac{1}{\sqrt{m_a m_c}}\frac{\partial^2 U}{\partial z_a \partial z_c}.$$

Now we diagonalize this matrix, which is possible but tedious for a 3×3 matrix, unless you use a computer. In our case, the resulting cubic equation—found after removing zero-order terms that sum to zero (Problem 8.45)—is

$$\lambda^3 - (2A + C)\lambda^2 + (2AC + A^2 - 2B^2 - D^2)\lambda = 0, \tag{8.31}$$

which has solutions

$$\lambda = 0, A - D, A + C + D. \tag{8.32}$$

The value of zero corresponds to the motion when all the nuclei move along the z axis the same amount and in the same direction; in other words, the whole molecule just moves to the right or the left. This motion is a translation, not a vibration, and the potential energy does not depend on the location of the molecule, hence the zero eigenvalue.

The remaining eigenvalues are the squares of the fundamental vibrational frequencies, and the eigenvectors (which are linear combinations of z_a, z_b, and z_c) are the normal modes. Let's look at a simple case where the derivatives are easy to evaluate. If the vibrational potential energy for motion along z is just the sum of two independent, equivalent harmonic oscillators, then we may write

$$U = \frac{1}{2}k(R_{ab} - R_e)^2 + \frac{1}{2}k(R_{bc} - R_e)^2$$

$$= \frac{1}{2}k(z_b - z_a - R_e)^2 + \frac{1}{2}k(z_c - z_b - R_e)^2. \tag{8.33}$$

In this limit, the derivatives become

$$A = \frac{1}{m_a}\frac{\partial^2 U}{\partial z_a^2} = \frac{k}{m_a}$$
$$B = \frac{1}{\sqrt{m_a m_b}}\frac{\partial^2 U}{\partial z_a \partial z_b} = -\frac{k}{\sqrt{m_a m_b}}$$
$$C = \frac{1}{m_b}\frac{\partial^2 U}{\partial z_b^2} = \frac{2k}{m_b}$$
$$D = \frac{1}{\sqrt{m_a m_c}}\frac{\partial^2 U}{\partial z_a \partial z_c} = 0. \quad (8.34)$$

The non-zero eigenvalues from Eq. 8.32 simplify in this case to

$$\lambda_1 = A = \frac{k}{m_a}$$
$$\lambda_2 = A + C = \frac{k}{m_a} + \frac{2k}{m_b} = \frac{(2m_a + m_b)k}{m_a m_b}. \quad (8.35)$$

The corresponding vibrational constants are the square roots of these:

$$\omega_1 = \sqrt{\frac{k}{m_a}}$$
$$\omega_2 = \sqrt{\frac{(2m_a + m_b)k}{m_a m_b}}, \quad (8.36)$$

and the eigenvectors that accompany these solutions are the normal mode coordinates q:

$$q_1\colon z_a = -z_c, \qquad z_b = 0 \quad (8.37)$$
$$q_2\colon z_a = z_c, \qquad z_b = \frac{2B}{A} z_a = -2\sqrt{\frac{m_a}{m_b}}\, z_a. \quad (8.38)$$

The first solution, q_1, is the symmetric stretch, with the end atoms (a and c) moving in opposite directions while the central atom (b) remains still. The vibrational constant ω_1 therefore depends only on the mass m_a and in fact is essentially the same as would be measured for a diatomic molecule with the same potential energy curve, except that the mass of atom a is used instead of the reduced mass.

The second solution is for the antisymmetric stretch: atoms a and c move in one direction while the central atom moves in the opposite direction, and consequently the vibrational constant ω_2 depends on both masses m_a and m_b. The second solution predicts, for example, the symmetric and antisymmetric stretches of $^{12}C\,^{16}O_2$ have vibrational constants (1333 cm^{-1} and 2349 cm^{-1}, respectively) different by a factor of

$$\frac{\omega_2}{\omega_1} \approx \sqrt{\frac{(2m_{^{16}O} + m_{^{12}C})k/(m_{^{16}O}m_{^{12}C})}{(k/m_{^{16}O})}} = \sqrt{\frac{(44/192)}{(1/16)}} = 1.91, \quad (8.39)$$

which overshoots the actual ratio ω_2/ω_1 of 1.76. The discrepancy is primarily due to our assumption that the two bonds are independent, whereas stretching one C=O bond in CO_2 tends to strengthen the other bond. Better agreement is seen in molecules where this coupling is weaker, as, for example, in C_3, where the predicted ratio is 1.73 and the actual ratio is 1.67.

EXAMPLE 8.1 Isotope Dependence of Vibrational Constants

CONTEXT In 1971, Linus Pauling discovered that exhaled breath contained hundreds of volatile organic compounds (VOCs) in low concentrations. By the late 1970s, the abundances of these VOCs were found to contain signals for certain diseases. For example, lung cancer is associated with higher concentrations of alkanes in the breath, and elevated formaldehyde is found in cases of breast cancer. In many cases, these compounds are formed at various stages of metabolism of different nutrients and serve as indicators of how the metabolism is affected by disease or by the administration of drugs. These concentrations are normally measured by mass spectrometry and chromatography because the concentrations are so low and the mixtures so complex. More recently, however, we have learned that the ratio of $^{13}CO_2$ to $^{12}CO_2$ in exhaled breath may also be used as a diagnostic tool, specifically for evidence of liver malfunction and certain bacterial infections. For this measurement, infrared spectroscopy is a useful tool because the CO_2 concentrations are much greater than the VOC concentrations. Infrared spectroscopy has an advantage over the other techniques in offering a rapid, real-time analysis.

PROBLEM Estimate the vibrational constants for the symmetric and antisymmetric stretches of $^{13}C\,^{16}O_2$.

SOLUTION Changing the isotope of the carbon atom does not affect the solution to ω_1, so $^{13}C\,^{16}O_2$ has roughly the same vibrational constant for the symmetric stretch, 1333 cm^{-1}, as $^{12}C\,^{16}O_2$. The antisymmetric stretch, on the other hand, should have a vibrational constant lower than the $^{12}C\,^{16}O_2$ value of 2349 cm^{-1} by a factor of

$$\sqrt{\frac{(2m_{^{16}O} + m_{^{13}C})/(m_{^{16}O}m_{^{13}C})}{(2m_{^{16}O} + m_{^{12}C})/(m_{^{16}O}m_{^{12}C})}} \approx \sqrt{\frac{(45)/(208)}{(44)/(192)}} = 0.972.$$

This ratio predicts a value of 2283 cm^{-1}, which is correct to within 1 cm^{-1}.

The fundamental vibrational frequencies of several polyatomic molecules are given in Table 8.3, together with local mode descriptions and symmetry labels for the modes.[4] Despite the complications inherent in vibrating polyatomics, several features are predictable from our experience with diatomics and a fundamental knowledge of vibrational modes. For example, we know that the isotopes of H_2O, a non-linear triatomic ($N_{atom} = 3$) molecule, should each have $3N_{atom} - 6 = 3$ vibrational modes, and so they do: a bend and two stretches. The OCS molecule, a linear triatomic, should have $3N_{atom} - 5 = 4$ modes. Why are only three vibrational modes shown in the table? One of the modes, the bend at 520 cm^{-1}, has the symmetry representation π, and is therefore doubly degenerate. Similarly, NH_3 should have $3N_{atom} - 6 = 6$ modes, and it does: two non-degenerate a_1 modes and two doubly degenerate e modes.

The mass dependence of the vibrational frequencies is qualitatively what we would expect. If we increase the mass of one of the H atoms in H_2O, all the vibrational frequencies decrease, and the frequency of the mode most closely connected to the affected atom drops the most. We see the lower energy stretch of H_2O fall from 3656 cm^{-1} to 2726 cm^{-1} when one ^1H isotope is replaced by ^2H, as it changes from an antisymmetric stretch of both O—H bonds to a more

[4]The fundamental vibrational energy ν_0 includes contributions from the anharmonicity and therefore is usually approximately but not exactly equal to the harmonic vibrational constant ω_e.

TABLE 8.3 Fundamental vibrational frequencies for selected polyatomic molecules. Representations are assigned according to the character tables in Chapter 6, with planar molecules lying in the xz plane.

molecule	point group	ω_e (cm^{-1})	symmetry representation	local mode description
$^{1}H_2{}^{16}O$	C_{2v}	1594	a_1	bend
		3656	b_1	antisymmetric stretch
		3756	a_1	symmetric stretch
$^{1}H^{2}H^{16}O$ (HDO)	C_s	1402	a'	bend
		2726	a'	$O^{2}D$ stretch
		3703	a'	$O^{1}H$ stretch
$^{2}H_2{}^{16}O$ (D_2O)	C_{2v}	1178	a_1	bend
		2671	b_1	antisymmetric stretch
		2788	a_1	symmetric stretch
$^{14}N^{1}H_3$	C_{3v}	950	a_1	inversion
		1627	e	bends
		3337	a_1	symmetric stretch
		3444	e	antisymmetric stretch
$^{12}C^{1}H_4$	T_d	1306	t_2	bends
		1534	e	bends
		2917	a_1	breathing
		3019	t_2	asymmetric stretches
$^{16}O^{12}C^{16}O$	$D_{\infty h}$	667	π_u	bend
		1333	σ_g	symmetric stretch
		2349	σ_u	antisymmetric stretch
$^{16}O^{12}C^{32}S$	$C_{\infty v}$	520	π	bend
		859	σ	CS stretch
		2062	σ	CO stretch

local O—D stretch. The other stretching frequency remains roughly the same, because it is still primarily an O—H stretch. For this isotope of H_2O, the H atoms are no longer identical, and the local mode picture becomes somewhat appropriate. One mode can be clearly identified with stretching of the $O^{1}H$ bond, and another with the $O^{2}H$ stretch. When the second H atom is also substituted, the two stretching frequencies again become very similar, and neither motion is well confined to a single bond.

We could estimate these isotopic shifts by recalling that the reduced mass depends primarily on the lighter atom. Roughly doubling the mass of the H atoms to get $^{2}H_2O$ (D_2O) from H_2O should roughly double the reduced masses and decrease the vibrational constants by about $\sqrt{2}$. In fact, this predicts the D_2O constants to within about 100 cm^{-1}, and higher accuracy can be achieved using the correct two-atom reduced mass. When we move from H_2O to D_2O, the reduced mass of the O—H pair changes from 0.9482 to 1.789. If we multiply the H_2O vibrational constants in Table 8.3 by $\sqrt{0.9482/1.789}$, we get 1176 cm^{-1}, 2689 cm^{-1}, and 2767 cm^{-1}, all within 25 cm^{-1} of the analogous D_2O constants. This type of quantitative analysis of isotope shifts is often used to help identify a

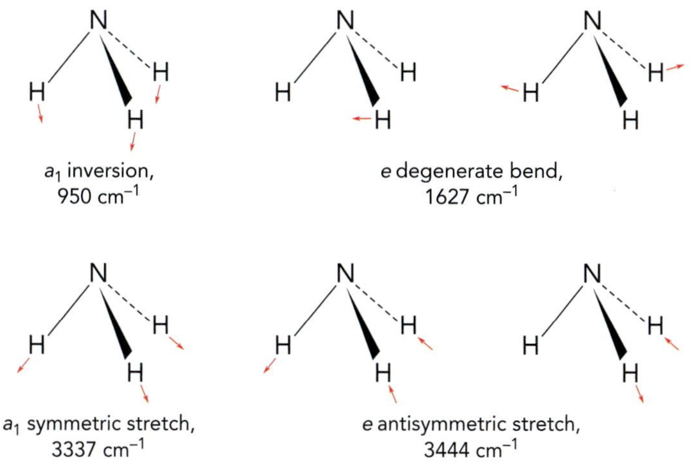

a_1 inversion, 950 cm^{-1}

e degenerate bend, 1627 cm^{-1}

a_1 symmetric stretch, 3337 cm^{-1}

e antisymmetric stretch, 3444 cm^{-1}

◀ **FIGURE 8.9 The normal vibrational modes of NH_3.**

molecule in an IR spectrum when the molecule cannot be analyzed by NMR or x-ray crystallography.

We can usually predict the number of vibrational constants over 2800 cm^{-1} by counting the number of H atoms in the molecule. In the local mode picture, each bond to a hydrogen atom would have its own stretching mode, with a reduced mass (1 amu) five or more times lower than most others, and therefore with a higher vibrational frequency than most others. Even though the stretching normal modes may have several H atoms moving at once, as shown for NH_3 in Fig. 8.9, they can be formed by sums of local mode vibrations, and they have comparable frequencies. We find, therefore, that NH_3 has three modes that correspond to NH stretches rather than bends, and they all have values of ω_e above 2800 cm^{-1}.

The bending modes have the lowest frequencies in every example in Table 8.3. The bending of H atom bonds has a lower reduced mass than other atoms, and therefore these bending modes may have ω_e values over 1600 cm^{-1}, greater than the stretching frequencies of relatively heavy atoms, such as the 860 cm^{-1} stretch of OCS.

Multiple Vibrational Excitation

Measurements of highly excited vibrational states of polyatomic molecules become extremely confusing to analyze, because vibrational energy can go into any mode or combination of modes any number of times, leading to nearly endless possibilities. For example, in CO_2, the fundamental vibrational wavenumbers are $\omega_2 = 667$ cm^{-1}, $\omega_1 = 1333$ cm^{-1}, and $\omega_3 = 2349$ cm^{-1}. However, the lowest nine vibrational states, indicated by the set of quantum numbers (v_1,v_2,v_3), lie at approximately the excitation energies (in cm^{-1}): 667 (0,1,0), 1334 (0,2,0), 1333 (1,0,0), 2001 (0,3,0), 2000 (1,1,0), 2349 (0,0,1), 2668 (0,4,0), 2667 (1,2,0), 3016 (0,1,1) (see Fig. 8.10). All of these states exist, and the number increases with energy because more combinations become possible and because the degeneracy of the excited bending states, equal to $v_2 + 1$, is climbing. In this example, there are only three states below 1000 cm^{-1}, including the ground state, four states between 1000 and 2000 cm^{-1}, and sixteen between 2000 and 3000 cm^{-1}. In larger

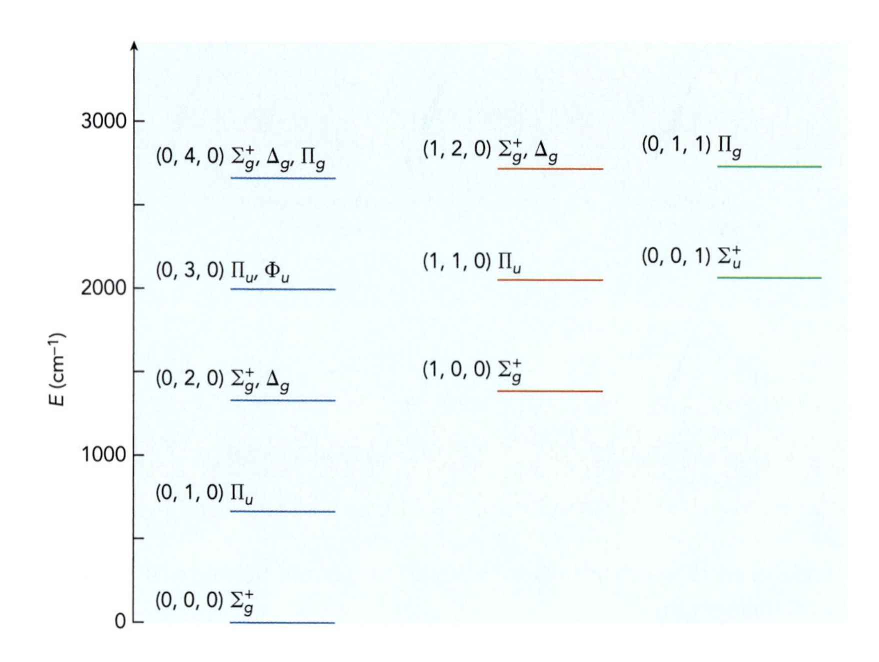

▶ **FIGURE 8.10 Energies of the lowest vibrational states of CO_2.**

polyatomics, the added number of modes causes the complexity to increase more rapidly.

For states in which more than one mode is excited, we can estimate the energies of these states relative to the ground state by summing over the excitation energies of all the vibrational modes:

$$E(v_1,v_2,v_3,\ldots) - E(0,0,0,\ldots) = v_1\omega_1 + v_2\omega_2 + v_3\omega_3 + \ldots \quad (8.40)$$

Equation 8.40 gives the energy relative to the ground state. In the harmonic limit, each mode also contributes a zero-point energy of $\omega_e/2$. Degenerate modes count separately here, so the total zero-point energy is given by

$$E_{zp} = \frac{1}{2}(g_1\omega_1 + g_2\omega_2 + g_3\omega_3 + \ldots), \quad (8.41)$$

where g_i is the degeneracy of mode i.

8.4 Spectroscopy of Vibrational States

To the extent that we can trust the harmonic approximation, each level of vibrational excitation (each increment in vibrational quantum number v) costs one vibrational constant ω_e in energy. As Table 8.2 shows, the vibrational constants of typical stretching motions place vibrational excitation energies in the infrared region of the spectrum. We can measure vibrational transitions that occur by absorption or emission or by scattering. Section 6.3 introduced the concept of Raman scattering, which in principle can be applied to the spectroscopy of any degree of freedom, but which is most commonly used for spectroscopy of vibrational states.

Some of the terminology of a vibrating string accompanies vibrational spectroscopy. The $v = 0 \rightarrow 1$ transition is called the **fundamental,** and the $v = 0 \rightarrow n + 1$ transition is called the ***n*th overtone.** Any absorption transition originating from a state where $v \neq 0$ is called a **hot band,** and transitions in polyatomic molecules that involve changes in more than one vibrational quantum number at the same time are called **combination bands.**

From the spacing of the vibrational energy levels, we can deduce the bond strength and the force constant. How do we determine the energy levels in the first place? Spectroscopy was the experimental method that led to the first quantitative tests of quantum mechanics, and it remains our major technique for determining molecular energy levels. For vibrational spectroscopy to succeed, however, we need to be able to induce a transition between vibrational energy levels. In the case of electronic transitions, we made the case that spectroscopic transitions occur because the electric field of the radiation pushes the electrons into a new state. For strong vibrational transitions, we need the electric field of the radiation to somehow coerce the molecule to change the relative motions of neutral atoms. How does that work?

The transition strength for the change from any initial quantum state i to final state f, by way of interaction with a photon's electric dipole field, is proportional to (Eq. 6.10)

$$\mathrm{I} \propto \left| \int \psi_f^* \hat{\mu}_t \psi_i d\tau \right|^2,$$

where $\hat{\mu}_t$ is the transition moment operator, proportional to the separation of positive and negative charges in the molecule. For vibrational transitions, the convenient way to look at the transition moment is to Taylor-expand the dipole moment μ around its equilibrium value:

CHECKPOINT The Taylor series expansion in Eq. A.29 is used here to show that any typical bond dipole moment will change roughly in proportion to the bond length. Therefore, the electric field of the radiation can drive changes in the bond length, inducing vibrational motion.

$$\mu = \mu_e + \left(\frac{\partial \mu}{\partial R}\right)_{R_e} (R - R_e) + \frac{1}{2}\left(\frac{\partial^2 \mu}{\partial R^2}\right)_{R_e} (R - R_e)^2 + \dots . \quad (8.42)$$

Let's assume that the dipole moment is given by small, fixed charges q and $-q$ on two atoms separated by a bond length R. If the difference q between the two charges stays constant, then μ is proportional to the bond length and the **dipole derivative** $\left(\frac{\partial \mu}{\partial R}\right)_{R_e}$ is a constant. Therefore the transition strength for a vibrational transition becomes

$$\mathrm{I} \propto \left| \int \psi_f^* \left[\mu_e + \left(\frac{\partial \mu}{\partial R}\right)_{R_e} (R - R_e) \right] \psi_i d\tau \right|^2 \qquad \text{combine Eqs. 6.10 and 8.42}$$

$$= \left| \int \psi_f^* \left[\mu_e - \left(\frac{\partial \mu}{\partial R}\right)_{R_e} R_e \right] \psi_i d\tau + \int \psi_f^* \left[\left(\frac{\partial \mu}{\partial R}\right)_{R_e} R \right] \psi_i d\tau \right|^2 \qquad \text{separate the } R \text{ from constants}$$

$$= \left| \left[\mu_e - \left(\frac{\partial \mu}{\partial R}\right)_{R_e} R_e \right] \int \psi_f^* \psi_i d\tau + \int \psi_f^* \left[\left(\frac{\partial \mu}{\partial R}\right)_{R_e} R \right] \psi_i d\tau \right|^2$$

$$= \left| \left[\mu_e - \left(\frac{\partial \mu}{\partial R}\right)_{R_e} R_e \right] \int \psi_f^* \psi_i d\tau + \left(\frac{\partial \mu}{\partial R}\right)_{R_e} \int \psi_f^* R \psi_i d\tau \right|^2 \qquad \text{factor out constants}$$

$$= \left| \left(\frac{\partial \mu}{\partial R}\right)_{R_e} \int \psi_f^* R \psi_i d\tau \right|^2 . \qquad \psi_i \text{ and } \psi_i \text{ are orthogonal}$$

The vibrational selection rules are determined by the fact that the interaction requires *a dipole moment for the molecule that changes as it vibrates.* This result does not mean that the molecule must have a permanent electric dipole moment, because vibrational motions in polyatomic molecules can disrupt the molecular symmetry. Bond dipoles that cancel in the equilibrium geometry do not necessarily cancel at each point of the vibrational motion. Carbon dioxide, for example, has no *permanent* dipole moment, but the bend and antisymmetric stretching vibrations temporarily break the symmetry of the two C=O bonds, giving a non-zero dipole *derivative*. However, diatomics have only one vibrational mode—the stretch—which does not change the symmetry of the molecule. A homonuclear diatomic such as N_2 has no permanent dipole moment, and stretching the bond will not give it a dipole moment. Therefore, the dipole derivative is zero for homonuclear diatomics, and vibrational absorption or emission transitions are simply all forbidden.

For all molecules, another approximate selection rule is $\Delta v = \pm 1$. When our transition skips over intervening vibrational states, as in $v = 1 \rightarrow 4$, it tends to be very weak. This selection rule is rigorous in the limiting case of a true harmonic oscillator and loses validity as the anharmonicities contribute more to the vibrational transition energy.

With a little group theory, we can determine whether or not the vibration has a dipole derivative. The same symmetry selection rules apply to vibrations as to electronic transitions: for a transition to be allowed, the direct product of the representations for the initial and final states must be one of the representations Γ_μ for the transition moment. The transition moments for electric dipole or infrared selection rules correspond to the functions x, y, and z. For Raman transitions, the transition moments correspond to any of the second-order functions of x, y, and z, such as xz or $x^2 + y^2$. The representation of the ground vibrational state is always the totally symmetric representation, so $\Gamma_i \otimes \Gamma_f$ is equal to Γ_f for fundamental transitions. Therefore, the selection rule for fundamental transitions is $\Gamma_i \otimes \Gamma_f = \Gamma_f = \Gamma_\mu$. For example, the group theory predicts that for CO_2 the transitions $v_2 = 0 \rightarrow 1$ and $v_3 = 0 \rightarrow 1$ are infrared-allowed, because those vibrational modes have π_u (x,y) and σ_u (z) symmetry, respectively. On the other hand, the symmetric stretch transition $v_1 = 0 \rightarrow 1$ is forbidden by infrared selection rules but allowed by Raman selection rules, because that vibrational mode has σ_g $(x^2 + y^2, z^2)$ symmetry. Here are the relevant rows from the character table in Table 6.4:

$D_{\infty h}$	$\hat{E}$	$\hat{C}_\infty$	$\infty\hat{\sigma}_v$	$\hat{i}$	$\infty\hat{C}_2$	Functions
$\Sigma_g^+(\sigma_g)$	1	1	1	1	1	$x^2 + y^2, z^2$
$\Sigma_u^+(\sigma_u)$	1	1	1	−1	−1	z
$\Pi_g(\pi_g)$	2	$2\cos\phi$	0	2	0	$xz, yz; R_x, R_y$
$\Pi_u(\pi_u)$	2	$2\cos\phi$	0	−2	0	x, y

One can justify these results physically by examining the change in dipole moment along the normal modes. If we look at the instantaneous dipole moment of the molecule during each of these vibrations, as in Fig. 8.11, we see that for the σ_g mode the dipole never changes; it is always zero. For the σ_u, the dipole becomes non-zero, pointing along the z axis, and that is why the representation for the z function makes this transition allowed. Similarly, the π_u mode has an instantaneous dipole in the xy plane, and the representation for x and y makes this transition allowed. When the inversion symmetry is lost, say by replacing one oxygen with sulfur to form OCS, both stretching transitions are allowed because the dipole moment is changing during transitions in either stretching mode.

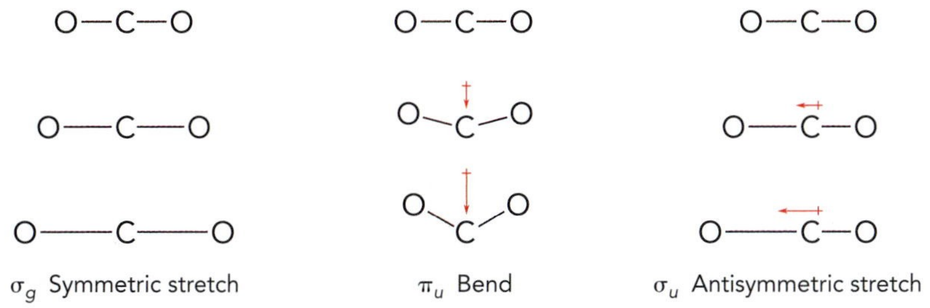

σ_g Symmetric stretch π_u Bend σ_u Antisymmetric stretch

◀ **FIGURE 8.11 Instantaneous dipole moments** for various points along the CO_2 vibrational coordinates. The dipole moment during the symmetric stretch is always zero, whereas motion in the bend or antisymmetric stretch changes the dipole moment.

This way of visualizing a vibrational mode is useful for seeing what excited vibrational states are accessible from the ground state in any molecule with symmetry. For example, in BF_3 there is an a_1 vibrational mode that corresponds to all the fluorines stretching away from the B atom at the same time. Such a mode, where all the bonds lengthen and contract in unison, is called a **breathing mode.** The dipole moment remains zero throughout the BF_3 breathing mode vibration, and the transition to this state from the ground state is electric dipole forbidden. Vibrational modes for which the $v = 0 \rightarrow 1$ transition is allowed by electric dipole selection rules are called **infrared-active.** Otherwise, as for this breathing mode of BF_3, the mode is **infrared-inactive.**

Infrared-inactive transitions may instead be observable by Raman spectroscopy. The spectra of these transitions are usually much harder to observe than those for electric dipole-allowed transitions, because most of the scattering is Rayleigh scattering, resulting in no net change in vibrational state. This is the only way, however, to directly observe vibrational transitions in homonuclear diatomics.

A very rough rule is that a vibrational mode is accessible from the ground state by a Raman transition if the vibrational motion changes the overall size of the molecule, because the size is roughly proportional to the polarizability. By this argument, the breathing mode stretch in BF_3 would be Raman-active, and it is.

In vibrational spectroscopy, one relevant conversion is to kelvin: 1000 cm^{-1} corresponds to about 1400 K, and 200 cm^{-1} is about 290 K, whereas the typical energy levels in our environment fall in the neighborhood of 300 K. Therefore, under ordinary terrestrial conditions, some heavy molecule vibrations are excited, but very few light molecule vibrations are, and virtually every molecule is in its ground electronic state. One way to make molecules heavier is to add atoms, so polyatomic molecules often have vibrational constants below 1000 cm^{-1} and are typically present in a wide range of vibrationally excited states. These excited states can emit, just as electronic states do (although the probability is weaker and the radiation harder to detect because each photon has less energy). Consequently, infrared detectors are good nighttime detectors for warm things, such as people or engine exhaust. These things can all emit infrared, which becomes detectable when other sources of heat (such as the sun) are hidden.

The air itself, being composed primarily of homonuclear diatomic molecules such as N_2 and O_2, does not compete because those vibrational transitions are infrared inactive. Sunlight, primarily in the visible and near-infrared region of the spectrum, warms the earth's surface. As the earth heats up, it releases some of that energy as infrared radiation. Atmospheric gases absorb some of that radiation, but regions of the infrared spectrum where the atmosphere is nearly transparent provide "windows" that allow much of the energy to be re-radiated into space. This balance between the amounts of incoming high-frequency (UV and visible) radiation and emitted low-frequency (IR) radiation largely determine Earth's average surface temperature. However, the increasing concentrations of more complex "greenhouse gases," such as CO_2 and hydrocarbons, absorb radiation in those windows, decreasing the amount of energy that would otherwise be lost back into space. This increased trapping of the sun's energy has led to an average warming of the atmosphere over the last century.

Another relevant conversion is to frequency; 1000 cm^{-1} is $3 \cdot 10^{13}\ \text{s}^{-1}$. The value of ω_e corresponds to the difference in classical vibrational frequency of the molecule between two states v and $v + 1$. If we think of the two nuclei as classical particles moving back and forth, we can characterize the vibration by the time it takes for the nuclei to move together from their point of farthest separation and then back again. This is the **vibrational period** τ, and $1/\tau$ is the classical vibrational frequency, the number of times the nuclei vibrate each second. The dipole of the molecule is changing at that frequency, and infrared radiation can induce a change in this vibrational frequency according to our semiclassical picture of Chapter 3. With a vibrational frequency of $3 \cdot 10^{13}\ \text{s}^{-1}$, the vibrational period is $33 \cdot 10^{-15}$ s, or 33 femtoseconds.

Vibrational transitions of H_2O have been cited as the source of water's intrinsic blue color. Seas and rivers and lakes often appear blue, largely because blue light from the sky reflects off the liquid surface. However, white light passing through a long sample of pure water also appears blue. The strength of the H_2O vibrational absorptions and the high densities and path lengths often associated with water cause water to strongly absorb radiation in the infrared, where the fundamental $v = 0 \rightarrow 1$ of the 3700 cm^{-1} stretching transitions are located. In fact, the combination of a significant dipole derivative and large anharmonicity allow even the very high $v = 0 \rightarrow 4$ overtone transition to absorb a noticeable amount of radiation in the 600–800 nm wavelength range. This range is in the red end of the visible spectrum, and so these $0 \rightarrow 4$ transitions suck in a little of the red light passing through the sample, leaving the shorter wavelength blue light to pass through unscathed. The effect is small, nonetheless, because the overtone transitions are indeed quite weak. This is a very unusual example of vibrational (rather than electronic) transitions determining color.

Vibronic Spectroscopy

A change in the electronic state of a molecule is normally accompanied by a change in the vibrational state as well. When we resolve different vibrational transitions within an electronic spectrum, we call this **vibronic** spectroscopy. Labeling the upper state by primes and lower state by double-primes, the energy difference between the two quantum states of the transition is given by

$$\begin{aligned} h\nu &= E' - E'' \qquad (8.43) \\ &= T_e + \left[\left(v' + \frac{1}{2}\right)\omega_{e'} - \left(v' + \frac{1}{2}\right)^2 \omega_e x_{e'}\right] - \left[\left(v'' + \frac{1}{2}\right)\omega_{e''} - \left(v'' + \frac{1}{2}\right)^2 \omega_e x_{e''}\right], \end{aligned}$$

where T_e is the term energy difference between the two electronic states, the difference in energies between the minima of the two potential wells. The harmonic vibrational selection rule, $\Delta v = \pm 1$, loses much of its validity in vibronic transitions. Spectroscopic transitions are drawn as vertical lines in the potential diagram, and R_e changes from one electronic state to the next. It is often the case that v has to change by several units for any transition to take place at all, as shown in Fig. 8.12. For a vibronic transition, the intensity integral can be divided into electronic and vibrational components:

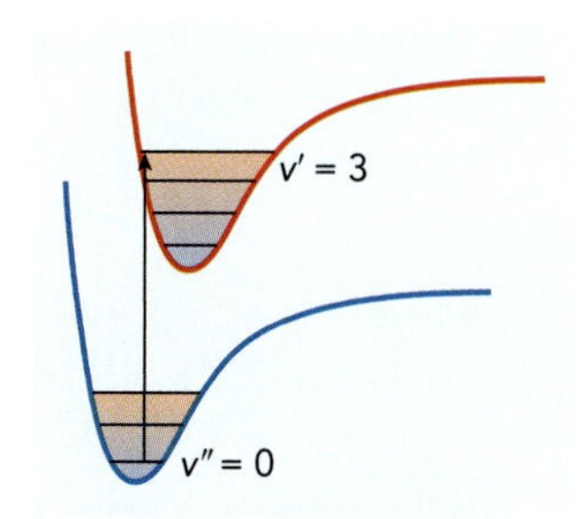

▲ **FIGURE 8.12 A vibronic transition.** Transitions between different electronic states often involve a large change in vibrational quantum number v, so that we can connect the states by a vertical line.

$$\begin{aligned} \mathcal{S} &= \int \psi_{\text{vibronic}}{}'^{*} \hat{\mu}_{\text{elec}} \psi_{\text{vibronic}}{}'' \, d\tau_{\text{elec}} \, d\tau_{\text{vib}} \\ &= \left[\int \psi_{\text{elec}}{}'^{*} \hat{\mu}_{\text{elec}} \psi_{\text{elec}}{}'' \, d\tau_{\text{elec}} \right] \left[\int \psi_{\text{vib}}{}'^{*} \psi_{\text{vib}}{}'' \, d\tau_{\text{vib}} \right]. \end{aligned} \tag{8.44}$$

Only the integral over the electronic coordinates depends on the transition moment operator, because the transition is being driven by the interaction between the electric fields of the radiation and the molecule's electron distribution. But within this transition, the intensity of any particular *vibrational* component $v'' \rightarrow v'$ is proportional to the value of the second integral. The second integral, over the upper and lower vibrational states, is an overlap integral between the two vibrational states called the **Franck-Condon factor,** which depends strongly on the difference between the equilibrium geometries of the two electronic states. If the equilibrium geometries are similar, then the Franck-Condon factors will approximate the harmonic oscillator selection rule $\Delta v = \pm 1$. But when the geometries differ substantially, the vibrational quantum number has to change by more than 1 just to find two states that are connected by a vertical arrow. Finally, for the transition to be intense, it must also satisfy the symmetry selection rule of Eq. 6.13: $\Gamma_i \otimes \Gamma_f \ni \Gamma_\mu$ for the electronic states.

The energetic processes in electronic spectroscopy discussed in Section 7.4—fluorescence, for example—also involve transitions among vibrational states. The level crossing between the potential energy curves in phosphorescence involves infrared emission as the vibrational levels of the excited states cool to lower levels, changing the electronic state in the process.

CONTEXT *Where Do We Go From Here?*

The vibrational motions of the molecule stretch the chemical bonds and bend the joints, all changing *relative* positions of the atoms. But even a rigid molecule—if such a thing existed—could store energy in the form of rotational motion. We complete our picture of internal molecular energy and structure with rotations in Chapter 9. From there, we move on to examine what happens when we have more than one molecule, which is the first step toward chemical reactions.

KEY CONCEPTS AND EQUATIONS

8.1 The vibrational Schrödinger equation.

a. The harmonic oscillator is often used to approximate molecular vibrations, with Hamiltonian

$$\hat{H} = \hat{K} + U = -\frac{\hbar^2}{2\mu}\frac{d^2}{dx^2} + \frac{1}{2}kx^2 \tag{8.14}$$

and wavefunctions

$$\eta_v(y) = A_v H_v(y) e^{-y^2/2} \tag{8.16}$$

b. The solution for the energies E_v as a function of the quantum number v is

$$E_v = \left(v + \frac{1}{2}\right)\omega_e, \tag{8.17}$$

where

$$\omega_e\,(\mathrm{J}) = \hbar\sqrt{\frac{k}{\mu}} \tag{8.18}$$

is the harmonic vibrational constant. To calculate ω_e in wavenumbers for a given vibration, we use

$$\omega_e(\mathrm{cm}^{-1}) = \frac{1}{2\pi c}\sqrt{\frac{k}{\mu}} = 130.28\sqrt{\frac{k(\mathrm{N\,m}^{-1})}{\mu(\mathrm{amu})}} \tag{8.19}$$

where in general the force constant is given by

$$k = \left(\frac{\partial^2 U_{\mathrm{vib}}}{\partial R^2}\right)\Bigg|_{R_e} \tag{8.20}$$

8.2 Vibrational energy levels in diatomics.

a. In spectroscopy, corrections are added to the harmonic energy to account for **anharmonicity:**

$$E_{\mathrm{vib}} = \left(v+\frac{1}{2}\right)\omega_e - \left(v+\frac{1}{2}\right)^2\omega_e x_e + \ldots \tag{8.26}$$

b. For the Morse potential,

$$U_{\mathrm{M}}(R) = D_e\{[1 - e^{-\beta(R-R_e)}]^2 - 1\}, \tag{8.28}$$

where D_e and R_e are the equilibrium well depth and bond length, and where $\beta \approx \sqrt{k/(2D_e)}$ determines the width of the well.

8.3 Vibrations in polyatomics. A molecule having N_{atom} atoms will have $3N_{\mathrm{atom}} - 6$ vibrational modes if the molecule is non-linear, $3N_{\mathrm{atom}} - 5$ if it is linear.

KEY TERMS

- The **harmonic oscillator** is an ideal system in physics where the potential energy increases as the square of a change in distance away from the center.
- The **force constant** is the second derivative of the harmonic oscillator potential energy.
- The **harmonic vibrational constant** is equal to the energy gap between any two adjacent quantum states in the harmonic oscillator.
- The **Morse potential** is an expression that yields a realistic potential energy curve for the interaction of two atoms in a chemical bond.
- The **equilibrium geometry** is the geometry of a molecule where the effective potential energy reaches its minimum value.
- Harmonic oscillator wavefunctions $\eta_v(y)$ consist of a Gaussian exponential multiplied by one of the tabulated **Hermite polynomials.**
- A **model potential** is a mathematical formula that attempts to mimic key aspects of a real potential energy function.
- **Normal modes** are a complete and orthogonal set of collective motions of the atoms in a vibrating molecule.

OBJECTIVES REVIEW

1. *Approximate vibrational energies and wavefunctions using the simple harmonic oscillator.*
   Find the total energy and wavefunction of the $v = 2$ state of a harmonic oscillator with vibrational constant $1450\,\mathrm{cm}^{-1}$.
2. *Sketch approximate solutions to a one-dimensional Schrödinger equation based on the shape of the potential energy curve and the reduced mass.*
   Sketch the lowest four energy levels and wavefunctions in a system with potential energy $U(x) = k_4 x^4$.
3. *Predict trends in the vibrational constant based on atomic mass and chemical bond strength.*
   Calculate the reduced mass of $^{39}K_2$ and estimate its force constant, based on values in Table 8.2. Then estimate the vibrational constant.
4. *Use group theory to predict the symmetries of a molecule's vibrational normal modes and the selection rules for transitions among the vibrational quantum states.*
   Find the symmetry representations of the C—H stretching vibrations in acetylene, HCCH, and determine whether any are IR active or Raman active.

PROBLEMS

Discussion Problems

8.1 For each of the following pairs, predict which bond will have the *lower* vibrational frequency, and give a reason.

a. $H^{79}Br$ or $D^{80}Br$
b. C=N or C≡N
c. $X^1\Sigma^+$ CO (bond length 1.128 Å) or $I^1\Sigma^-$ CO (bond length 1.391 Å)

8.2 Linear molecules can have $\lambda \neq 0$ MOs, such as the π orbitals, and these wavefunctions have non-zero angular momentum. Bending vibrations in linear molecules can also have π symmetry. Explain how, for example, the bending vibrational mode in CO_2 can have angular momentum.

8.3

a. Which molecule would you expect to have the *lowest* vibrational constant ω_e: MgO, CaO, or CaS?
b. Which molecule would you expect to have the *lowest* vibrational constant for the C—C stretch: H_3CCH_3 (ethane), H_2CCH_2 (ethene), or HCCH (ethyne)?

8.4 The Woods-Saxon potential $U(z) = -[\frac{U_0}{1 + e^{(|z|-a)/d}}]$, expressed here as a function of only one coordinate z, is in used in its three-dimensional form as a model for the potential energy experienced by neutrons inside a nucleus. Label the states using the quantum number v, and set the $v = 0$ ground state at an energy of $-0.8U_0$. Sketch as accurately as you can the potential energy curve and the energy levels and wavefunctions of the $v = 1$ and $v = 2$ states.

8.5 Describe an experiment that would measure the energy gap illustrated in the following figure, assuming that the sample is initially in the $X\ v = 0$ vibronic state and given all the equipment necessary to generate electromagnetic radiation at any frequency and to measure its intensity or energy. This is not an easy measurement, however, so you should point out where an experimentalist might run into difficulty using your method.

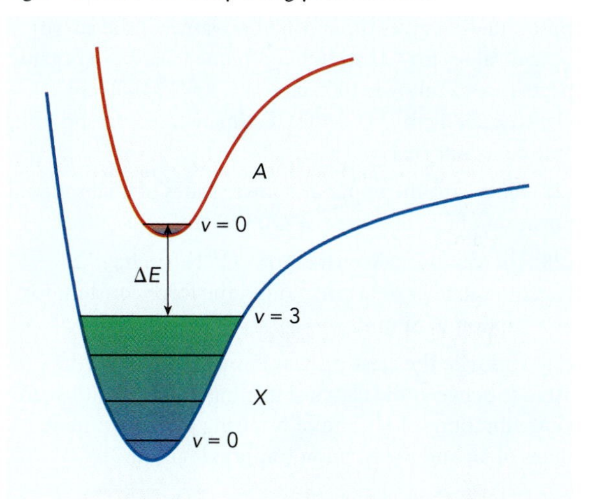

8.6 If we count *all* of the spatial coordinates that describe one H_2O molecule,

a. how many of these are translational coordinates?
b. how many of these are rotational coordinates?
c. how many of these are vibrational coordinates?
d. how many of these are electronic coordinates?

The Harmonic and Anharmonic Oscillators

8.7 In the Taylor series expansion used to obtain the harmonic approximation for vibrational energies, show whether the third-order term $(\partial^3 E_{elec}(R)/\partial R^3)|_{R_e}(R - R_e)^3$ should be positive or negative for real diatomic molecules.

8.8 For any state of a vibrating chemical bond, the two classical turning points are the positions at which $E = U(R)$. The distance ΔR between these points is one measure of the spatial extent of the wavefunction (i.e., how much the bond is stretching). In the following table, we will increase the value in the left-hand column, leaving everything else constant. In the right-hand column, write "+" if ΔR increases, "−" if ΔR decreases, and "0" if ΔR stays the same.

increase	change in ΔR
v	
μ	
R_e	
$\omega_e x_e$	
k	

8.9 The equations at the bottom of Table 8.1 are **recursion relations,** equations that allow us to determine the next term in a series from the preceding terms. Working from these relations, try to determine an expression for $H_{v=6}(y)$. Don't worry about the normalization constant.

8.10 Verify the normalization constant A_2 given in Table 8.1 for the harmonic oscillator wave function $\eta_2(x)$.

8.11 We can treat the anharmonic terms in Eq. 8.11 using perturbation theory (Section 4.2). However, it turns out that the most important anharmonic term, the term that depends on $\partial^3 U_{vib}/\partial R^3$, does not contribute to the energy in first-order perturbation theory. Show this.

8.12 For some bending vibrations, the potential energy surface is roughly given by the function

$$U(\theta) = a(\theta - \theta_e)^2 + b(\theta - \theta_e)^4,$$

where a and b are positive constants. If the vibrational states on such a potential surface are fit to Eq. 8.26, what will the signs of $\omega_e x_e$ and $\omega_e y_e$ be?

8.13 The square modulus of a particle's wavefunction at a given point is proportional to the probability of measuring the particle at that point. According to the correspondence principle, high energy states of molecules should act more classically than low energy states. Given these rules, explain how the amplitude of the wavefunction for high-energy nuclear vibration in a diatomic molecule should vary as a function of R between the classical turning points of the potential energy curve.

8.14 Use the de Broglie equation to *estimate* the momentum p in kg m s^{-1} of the hydrogen atom in the $v = 10$ state of HBr, with classical turning points at 1.20 and 1.80 Å.

8.15 The normalization constants A_v for the Hermite polynomials in Table 8.1 are for unitless wavefunctions $\eta_v(y) = A_v H(y) e^{-y^2/2}$ of the unitless coordinate $y = (R - R_e)(k\mu/\hbar^2)^{1/4}$. We haven't worried about it so far because we can always scale y to get distances again. The trouble is that if we want to find the probability of the bond being a certain length, then we'll need the wavefunction $\eta_v(R) = A'_v H(y) e^{-y^2/2}$ to have units of (distance)$^{-1/2}$, and therefore a different normalization constant A'_v is required. What is the ratio A'_v/A_v?

8.16 The expectation value of x^2 in the harmonic oscillator is given by

$$\langle x^2 \rangle = \frac{\omega_e}{k}\left(v + \tfrac{1}{2}\right).$$

Use this to find an expression for $\langle p^2 \rangle$ in terms of μ, ω_e and v.

8.17 Prove that the Morse potential energy curve

$$U_M(R) = D_e\{[1 - e^{-\beta(R-R_e)}]^2 - 1\}$$

has equilibrium bond length R_e and well depth D_e.

8.18 Prove that in the limit of low vibrational energy, the Morse potential with $\beta = \sqrt{k/(2D_e)}$, is equivalent to the harmonic oscillator potential $U(R) = \frac{1}{2}k(R - R_e)^2 + \text{constant}$.

8.19 The energy halfway between the bottom of the Morse oscillator well and the dissociation limit is $-D_e/2$, or $D_e/2$ when measured from the bottom of the well. Call the difference between minimum and maximum classically allowed R values at this energy δR (by "classically allowed" we mean that tunneling is neglected). At the same energy from the bottom of the well, $D_e/2$, a harmonic oscillator with the same force constant k, will predict a different value for δR, where the Morse oscillator parameter $\beta = \sqrt{k/2D_e}$. Find the ratio of δR values at $D_e/2$ from the bottom of the well for the Morse and simple harmonic oscillators having the same force constant.

8.20 A diatomic molecule AB is excited from the $v = 1$ vibrational level of the ground electronic state to a photodissociating electronic state that falls apart into A and B*, where B* is an excited state of atom B. Let $h\nu$ be the energy of the incident photon, ω_X be the ground state vibrational constant, D_X be the equilibrium dissociation energy of the ground state (measured from the bottom of the well), and E_B be the excitation energy of B*. Any excess energy left after the dissociation goes into the kinetic energy of the fragments K_{ex}. Write an expression for K_{ex} in terms of these other parameters.

Vibrations in Diatomics

8.21 The vibrational constant for $^{14}N\,^{1}H$ is about 3300 cm^{-1}. Rank the letters a–e for the following diatomic molecules in order of *increasing* vibrational constant, ω_e.

a. $^{15}N\,^{14}N^+$ b. $^{14}N_2^+$ c. $^{14}N\,^{1}H$
d. $^{14}N\,^{2}D$ e. $^{14}N_2$

8.22 If an H_2 molecule had a pure harmonic oscillator potential curve with force constant $k = 570$ N m^{-1} and equilibrium bond length R_e of 0.74 Å, estimate the probability of finding the ground state molecule with a bond length of $R = 2.00 \pm 0.01$ Å.

8.23 Estimate the force constants k (in N m^{-1}) for $^{90}Zr\,^{16}O$ ($\omega_e = 937.2$ cm^{-1}) and YF ($\omega_e = 630.9$ cm^{-1}).

8.24 The $v = 0 \rightarrow 4$ transition in a homonuclear diatomic is observed at 1353.788 cm^{-1}. Independent measurements find the force constant to have a value 10.31 N m^{-1}. Identify the molecule.

8.25 The first experimental observation of He_2 was reported in 1993 [*J. Chem. Phys.* **98**, 3564 (1993)]. Assuming the well depth is only 8 cm^{-1} (from dissociation energy measurements) and has essentially a harmonic oscillator form up to the dissociation energy, calculate an upper limit to the force constant k for this intermolecular potential energy curve.

8.26 We measure vibrational constants directly by spectroscopy. From those clues we can begin to infer what the effective potential energy curve looks like, for example, by calculating the force constants (which in the harmonic approximation gives the second derivative of the curve). Typical values for CC stretching frequencies ω_e in organic spectra are as follows: 1000 cm^{-1} (C—C); 1600 cm^{-1} (C=C); 2100 cm^{-1} (C≡C). Estimate the corresponding force constants in N m^{-1}.

8.27 Find v for the upper and lower states of a vibrational transition of CO observed at 4208.37 cm^{-1}.

8.28 The vibrational constant of $^{12}C\,^{14}N^+$ in its $a\,^1\Sigma^+$ electronic state is 2033 cm^{-1}. Find the force constant for the vibration in SI units.

8.29 Estimate the classical vibrational amplitude (the distance between the classical turning points) for the zero-point vibrations of H_2 and of N_2 using the experimental values of ω_e and the harmonic approximation.

8.30 Based on Table 8.2, estimate the vibrational constant ω_e for $^{37}Cl_2$ in cm^{-1}.

8.31 Calculate the vibrational energy of the $v = 2$ state of $^1H^{127}I$ relative to the ground state, including the anharmonic corrections.

8.32 Using only the information in Table 8.2, try to estimate ω_e and $\omega_e x_e$ for $^{127}I\,^{35}Cl$.

Vibrations in Polyatomics

8.33 How many vibrational modes are there in each of the following molecules: (a) H_2O; (b) acetylene, C_2H_2; (c) PF_5; and (d) glucose, $C_6H_{12}O_6$?

8.34 For *n*-butane ($CH_3CH_2CH_2CH_3$), identify the following:

a. the total number of vibrational modes
b. the number of hydrogen stretches
c. the total number of stretches
d. the total number of bends
e. the kind of motion associated with the lowest frequency vibrational constant

8.35 Estimate the reduced mass and force constant for any typical C—H stretching mode in a large organic compound, such as decane.

8.36 NH_3 has six vibrational modes as follows:

ν_1	a_1	3337
ν_2	a_1	950
ν_3	e	3444
ν_4	e	1627

List all the states and degeneracies with energies less than 4000 cm^{-1}.

8.37 For formaldehyde (H_2CO), do the following.

a. Give the symmetry representations for all the bending vibrations.
b. Give the symmetry representations for all the stretching vibrations.

8.38 Draw any reasonable vibrational motions (using arrows and/or plus and minus signs) of 1,3,5-trichlorobenzene that have the indicated symmetry representations. Try to draw the arrows in such a way that the center of mass and inertial axes are fixed (i.e., so that there is no rotation or translation mixed in).

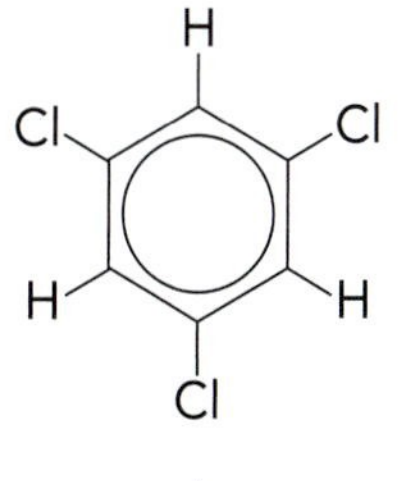

a'_2 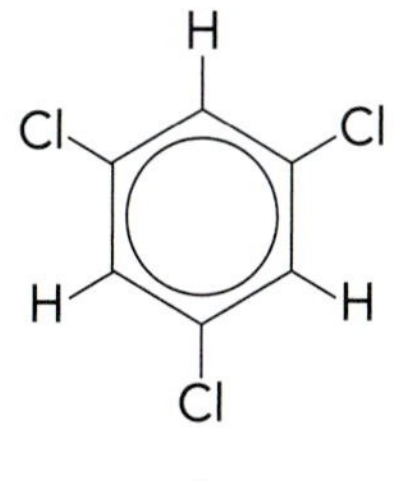 a''_1

8.39 The vibrational frequencies of the three modes in the linear triatomic C_3 are $\omega_1 = 1200\,cm^{-1}$, $\omega_2 = 63\ cm^{-1}$, and $\omega_3 = 2159\ cm^{-1}$. Predict the excitation energies (energies above the ground state) for the states $(v_1,v_2,v_3) = (1,0,0)$, $(1,1,0)$, and $(0,1,2)$.

8.40 For the CO_2 vibrational state (1,1,2), find (a) the energy relative to the ground state, (b) the symmetry representation, and (c) the degeneracy. The vibrational constants are $\omega_e^{(1)} = 1330.0\ cm^{-1}$, $\omega_e^{(2)} = 667.3\ cm^{-1}$, $\omega_e^{(3)} = 2349.3\ cm^{-1}$.

8.41 The vibrational constants for H_2O are: $\omega_e^{(1)} = 3657\ cm^{-1}$, $\omega_e^{(2)} = 1595\ cm^{-1}$, and $\omega_e^{(3)} = 3756\ cm^{-1}$. Give the energies, neglecting zero-point energy, for all the vibrational states of H_2O below 6000 cm^{-1}. What is the zero-point energy for this molecule?

8.42 Calculate the total zero-point energy for CO_2 in cm^{-1}, given the vibrational constants $\omega_1 = 1330.0\,cm^{-1}$, $\omega_2 = 667.3\,cm^{-1}$, and $\omega_3 = 2349.3\,cm^{-1}$. Note that ω_2 is the vibrational constant for the doubly degenerate bending mode.

8.43 Without referring to its measured molecular parameters, estimate the zero-point energy of acetylene (linear HCCH), showing your reasoning.

8.44 The highest vibrational constant in NH_3 is at 3444 cm^{-1}. Estimate the value of the highest vibrational constant in cm^{-1} for ND_3.

8.45 Diagonalize the Hessian matrix in Eq. 8.30 to verify the eigenvalues in Eq. 8.32. This involves proving that one of the terms in the determinant vanishes, which may be done by a qualitative argument.

8.46 The difference between the observed and calculated ω_1/ω_2 ratios in CO_2 (see Eq. 8.39) can be used to estimate the value of the second derivative $D \equiv \partial^2 U/(\partial z_a \partial z_c)$. Assume that the equations for A and C in terms of the single force constant k and the atomic masses remain valid, and solve for D in units of $N\,m^{-1}$. What is the significance of the sign of D (i.e., how would the sign of D appear in a laboratory experiment)?

8.47 We have kept the vibrational energy equation for a polyatomic molecule in simple form by assuming that motions along different modes are completely separable. The real world, of course, is more interesting. One commonly observed breakdown in the separation between vibrational modes occurs by means of the **Coriolis interaction,** and the principle can be crudely illustrated as follows. Draw the three nuclei of CO_2, and draw the displacement arrows for the in-plane bend; these will correspond to position vectors $\vec{r}$ for each nucleus relative to its equilibrium position. Now assume that at the same time one of the stretching modes of the molecule is also excited, resulting in a momentum vector $\vec{p}$ for each nucleus, parallel to the internuclear axis. Now show that for one of the excited stretches, the angular momentum vectors $\vec{r} \times \vec{p}$ do not cancel. (a) To what kind of motion does the resulting *net* angular momentum (if it could be applied to all the atoms) correspond? (b) For any molecule of $D_{\infty h}$ symmetry, what are the selection rules for this interaction (i.e., what vibrational modes can be coupled), assuming that this total angular momentum must be non-zero?

8.48 To calculate the magnitude of the Coriolis interaction described in Problem 8.47, one calculates integrals of the form

$$\left(\int \phi_{\text{bnd}}\,\hat{p}_{\text{bnd}}\,\phi'_{\text{bnd}}\,d\tau\right)\left(\int \phi_{\text{str}}\,\hat{q}_{\text{str}}\,\phi'_{\text{str}}\,d\tau\right) - \left(\int \phi_{\text{bnd}}\,\hat{q}_{\text{bnd}}\,\phi'_{\text{bnd}}\,d\tau\right)\left(\int \phi_{\text{str}}\,\hat{p}_{\text{str}}\,\phi'_{\text{str}}\,d\tau\right),$$

where the q's are the vibrational normal mode coordinates. These terms mix the bending and stretching vibrational modes, and if the harmonic approximation is used, the integrals may be very easy to calculate. Just to get an idea how this works, calculate instead the expectation value of $(yp_y - p_y y)$ for an arbitrary harmonic oscillator function $\phi_v(y)$, where $p_y = -i\hbar\frac{\partial}{\partial y}$. Use the equations beneath Table 8.1.

Symmetry and Vibrations

8.49 H_2O has vibrational *modes* of a_1 and b_1 symmetry. If we consider all possible combinations of stretching and bending, what are the possible symmetries of the vibrational *states*?

8.50 Find the point group of the molecule in the following figure, and label the vibrational motion by its symmetry representation.

8.51

a. PtF_2Cl_2 has the structure drawn as (a). Identify the symmetry representations for the two vibrational modes (b) and (c).

b. Identify the representation for the single vibrational state that results when $v = 1$ for both of the modes drawn next simultaneously.

(a) (b) (c)

8.52 Some of the normal vibrational modes of the $\hat{C}_{2v}$ molecule C_3H_2 are shown, with "⊕" and "⊖" indicating motion out of or into the page. Label each mode according to its point group representation.

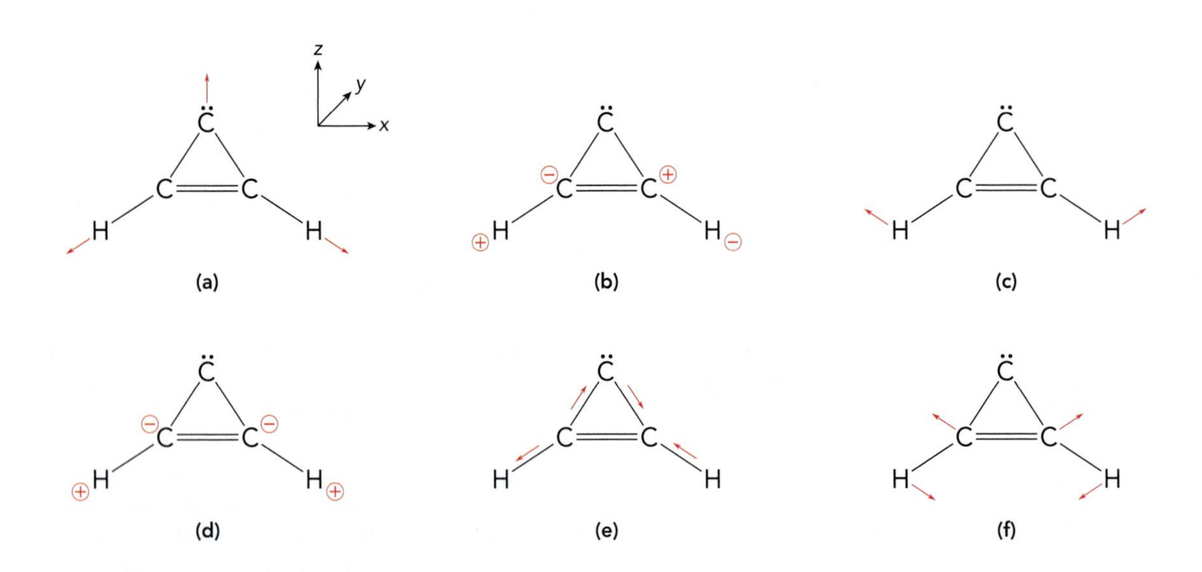

8.53 Rotations about a single bond, such as the methyl group torsions drawn here for propane, are technically vibrational motions, although their energy levels are often poorly predicted by the harmonic approximation. Give the representation for each of the two following vibrational modes.

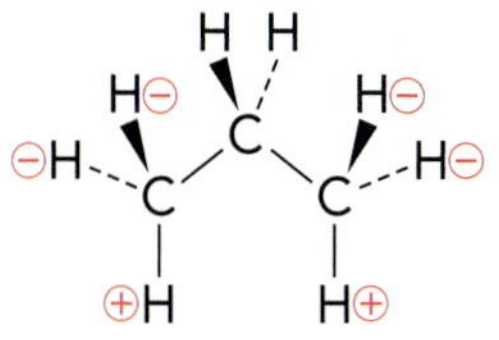

(a)

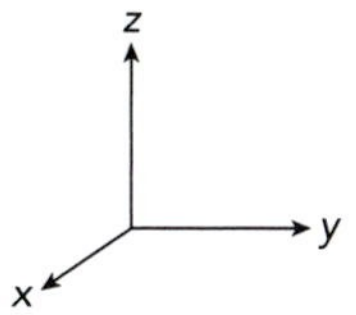

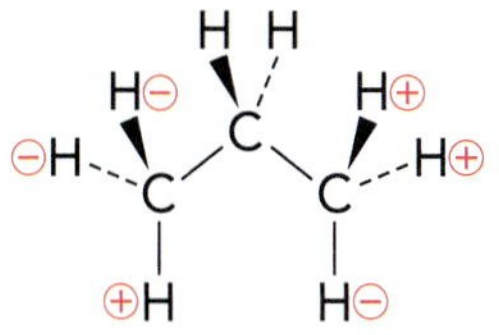

(b)

8.54 Hydrogen peroxide, HOOH, has a C_{2h} geometry.

a. How many vibrational modes does HOOH have?
b. Draw the structures and displacement arrows for each vibrational mode.
c. Label the representation for each mode.

8.55 The two vibrational modes drawn here are essentially the same motion for two different substitutions of benzene. Identify the symmetry representations for the two vibrational modes. Draw and label any axes or planes of symmetry as needed to clarify your assignment. If either mode is infrared active, write "IR."

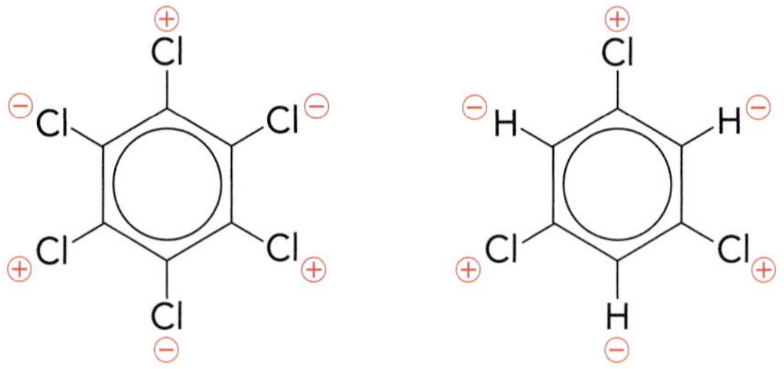

8.56 Hexamethylene triperoxide diamine (HMTD) was one of the explosive compounds identified at a property in Escondido, California, in November 2010, which led to the house being intentionally burned to the ground by authorities. Each atom is equivalent to the other atoms of same atomic number (so all six carbons are equivalent, both nitrogens, and so on). Assign HMTD to its point group, and identify the representation of the vibrational motion where the two nitrogens twist in opposite directions.

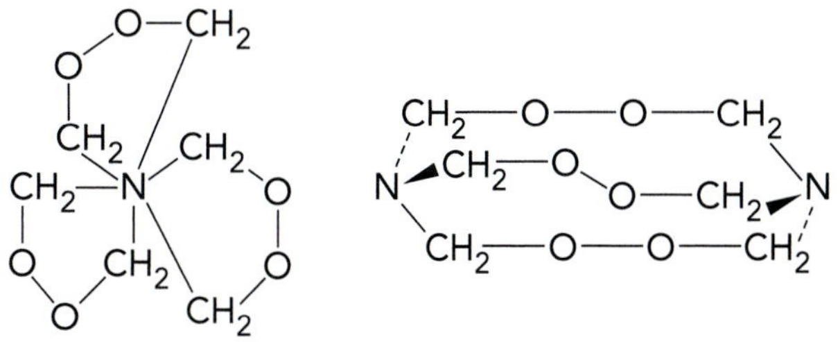

8.57 Consider the following normal motions of 1,3-difluoro-1,3-cyclohexadiene:

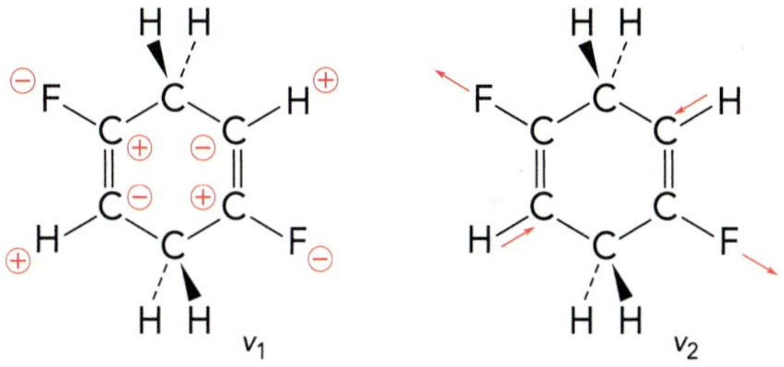

v_1 v_2

Determine the symmetry representations for these two modes, and whether the transition between the states $v_1 = 1$ and $v_2 = 1$ is allowed by either Raman or electric dipole selection rules.

8.58 Transitions between vibrational levels of different electronic states must obey the selection rules for the *vibronic* states, where $\Gamma_{\text{vibronic}} = \Gamma_{\text{electronic}} \otimes \Gamma_{\text{vibrational}}$. H_2O has three modes: the ν_1 symmetric stretch (a_1), the ν_2 bend (a_1), and the ν_3 antisymmetric stretch (b_2). Show if the transition $X^1A_1 \rightarrow A^1B_1$, $v_3 = 1 \rightarrow 2$ is allowed by electric dipole selection rules.

8.59 The (0,1,0) state of CO_2 has allowed transitions to which of the following states? (a) (1,0,0); (b) (0,0,1); (c) (0,2,0); (d) (1,0,1).

8.60 The $v = 0 \rightarrow 1$ transition in CH^+ is measured at 2046.3 cm^{-1}. If the force constant is 259.0 N m^{-1}, calculate the anharmonicity $\omega_e x_e$.

Potential Energy Functions

8.61 The accompanying figure shows a potential energy well for a particle trapped by opposing electric fields. The energy of the first excited state, $n = 2$, is drawn in.

a. Draw where you estimate the corresponding lines should appear for the $n = 1$ and $n = 3$ states. Include any *qualitative* features you know should be found, but don't try to calculate the answer.

b. On the line for the $n = 3$ state, sketch the shape of the $n = 3$ wavefunction.

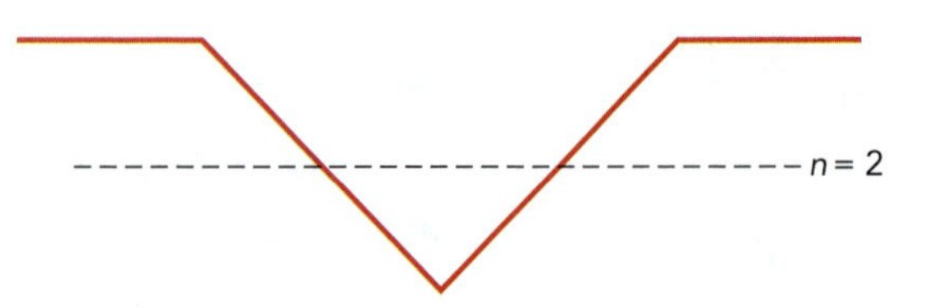

8.62 The lowest six energy levels are graphed and selected wavefunctions sketched for the solution to a one-dimensional Schrödinger equation. On this graph, draw in the potential energy curve that could account for these solutions.

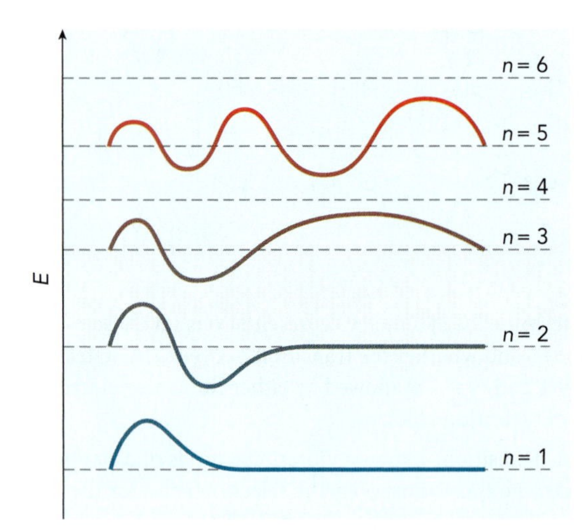

8.63 Describe as accurately as possible the potential energy curve and the state of a particle that has the normalized wavefunction

$$\psi = (2.0\ \text{Å})^{-1/2}\sin[8\pi x(\text{Å})].$$

8.64 The four lowest-energy wavefunctions of a one-dimensional system with their corresponding energy levels are shown in the accompanying figure. Sketch the potential energy curve that leads to these solutions.

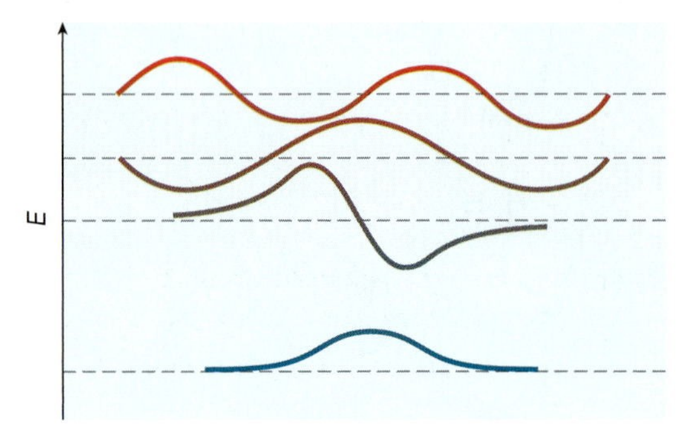

9 Rotational States and Spectroscopy

LEARNING OBJECTIVES

After this chapter, you will be able to do the following:

1. Calculate rotational constants for molecules based on their geometry.
2. Use those constants to predict or estimate the rotational energies for linear molecules and symmetric top molecules.
3. Predict trends in the rotational constant based on atomic masses and molecular geometry.
4. Predict spectroscopic transition energies involving rotation, including vibrational effects.

GOAL *Why Are We Here?*

The goal of this chapter is to analyze the quantum states and energy levels that result from solving the rotational part of the molecular Schrödinger equation.

CONTEXT *Where Are We Now?*

We've been through electronic and vibrational terms in the molecular Hamiltonian. At last we come to the final quantum contributor to the overall molecular energy: rotation. Unfettered rotational motion occurs only in the gas phase, limiting the importance of rotations to typical laboratory chemistry. However, our most precise measurements of molecular structure come from the spectroscopy of rotational states, and rotations play a significant role in the properties of gas phase compounds.

In Section 8.1, we obtained Eq. 8.10 for the partition of the molecular energy into different kinds of motion:

$$E = E^0_{\text{elec}} + E_{\text{vib}} + E_{\text{rot}} + E_{\text{trans}}. \tag{8.10}$$

Now our job is to find the energies and wavefunctions for that third part of the molecular Schrödinger equation, the rotational term. As usual, we start with the simplest case—which for molecules means diatomics.

SUPPORTING TEXT *How Did We Get Here?*

- Equation 3.18 gives the eigenvalue for the $\hat{L}^2$ operator,

$$L^2 = \hbar^2 l(l + 1),$$

which is the square of the electronic orbital angular momentum. The one-electron angular wavefunctions in Table 3.1 will also be helpful for describing the angular motion of molecular rotation.

- We can write the kinetic term of the molecular Hamiltonian in the form given by Eq. 5.11,

$$\hat{K}_{\text{nuc}} = -\frac{\hbar^2}{2\mu R^2}\frac{\partial}{\partial R}R^2\frac{\partial}{\partial R} + \frac{1}{2\mu R^2}\hat{J}(\Theta,\Phi)^2 - \frac{\hbar^2}{2M_{\text{nuc}}}\nabla^2_{\text{COM}},$$

where M_{nuc} is the combined mass of all the nuclei. The middle term will be our Hamiltonian for rotation in this chapter.
- Equation 6.13,

$$\Gamma_f \otimes \Gamma_i \ni \Gamma_\mu$$

shows that the direct product of the representations of two states gives the representations for the corresponding transition moments. If $\Gamma_f \otimes \Gamma_i$ does not correspond to the representation for any transition moment, the transition $i \rightarrow f$ is forbidden. This rule is true for rotational transitions as well as for electronic and vibrational transitions.
- Section 8.1 shows the separation of variables that allows us to (approximately) solve for the rotational energies independently of the vibrational part of the Schrödinger equation. The energy has likewise been divided among its electronic, vibrational, rotational, and translational contributions (Eq. 8.10):

$$E = E^0_{\text{elec}} + E_{\text{vib}} + E_{\text{rot}} + E_{\text{trans}}.$$

9.1 Rotations in Diatomics

Good news: we've already solved the rotational part of the Schrödinger equation, not knowing it at the time. If the Born-Oppenheimer and rigid rotor assumptions are valid, then the rotation of a molecule is just another example of angular motion in three dimensions—similar in key respects to the angular motion of the electron in the one-electron atom. Just as the classical equations for angular momentum are the same whether describing the orbit of a moon around a planet or a taxi turning a sharp corner, the basic shape of the solutions to the Schrödinger equation for angular motion depend little on the source of the motion. For the electron in an atom, we found that the angular wavefunctions are given by the spherical harmonics (Table 3.1), and the angular momentum L is equal to $\sqrt{l(l+1)}\hbar$.

We use the quantum number J to label the rotational energy levels of the molecule, and (as with the atomic quantum number l) it can have integer values from 0 to 1 and on up. The rotational wavefunctions are again the spherical harmonics, but now we index them by the rotational quantum numbers J and M_J: $Y_J^{M_J}(\Theta,\Phi)$, where M_J gives the projection of the rotational angular momentum $\overrightarrow{J}$ onto the lab-fixed Z axis. Like m_l and l for the electron in an atom, M_J can take any integer value from $-J$ to J, a total of $2J+1$ possible values for a given value of J. We will make little use of these wavefunctions explicitly, but among other uses they help determine the selection rules for transitions between these quantum states.

CHECKPOINT We only briefly encounter our color-coding for the Schrödinger equation in this chapter, because the math is shown in detail in Chapter 3. Again the blue Hamiltonian (here just for rotational kinetic energy) and the red energy expression emphasize that our fundamental approach to the quantum mechanics of these different degrees of freedom remains the same.

So we have the wavefunctions, but what are their energies? We single out the kinetic energy operator for rotation from Eq. 5.11:

$$\hat{K}_{\text{rot}} = \frac{1}{2\mu R^2}\hat{J}(\Theta,\Phi)^2$$

and take advantage of knowing the eigenvalue of the squared angular momentum operator (by analogy with Eq. 3.18, $\hat{L}^2 Y_l^{m_l} = l(l+1)\hbar^2 Y_l^{m_l}$) to get the rotational energy:

$$E_{\text{rot}} = \frac{\hbar^2}{2\mu R^2} J(J+1). \tag{9.1}$$

As with the particle in a box, there is no potential energy for the rotation, and therefore the energy level spacing increases as J increases, as shown in Fig. 9.1. We can improve on this equation for E_{rot} by recalling that the molecule is always vibrating, so R is not constant. Instead we write

$$E_{\text{rot}} = \frac{\hbar^2}{2\mu} \langle R^{-2} \rangle J(J+1), \tag{9.2}$$

where $\langle R^{-2} \rangle$ is the average value of R^{-2} integrated over a vibrational period. Near the vibrational ground state of the molecule, $R \approx R_e$, which allows us to define an equilibrium **rotational constant** B_e, such that

$$B_e(J) = \frac{\hbar^2}{2\mu R_e^2} \tag{9.3}$$

such that the rotational energy may be written

$$E_{\text{rot}} \approx \frac{\hbar^2}{2\mu R_e^2} J(J+1) \equiv B_e J(J+1). \tag{9.4}$$

CHECKPOINT We obtained the rotational energy in Eq. 9.4 by analogy with the orbital motion of the electron in Chapter 3. We don't need another derivation to solve the rotational Schrödinger equation. The detailed solution would be identical to all the work done between Eqs. 3.11 and 3.20, replacing L, r, and m_e for the electron by J, R, and μ for the molecule. Solving that differential equation once is enough!

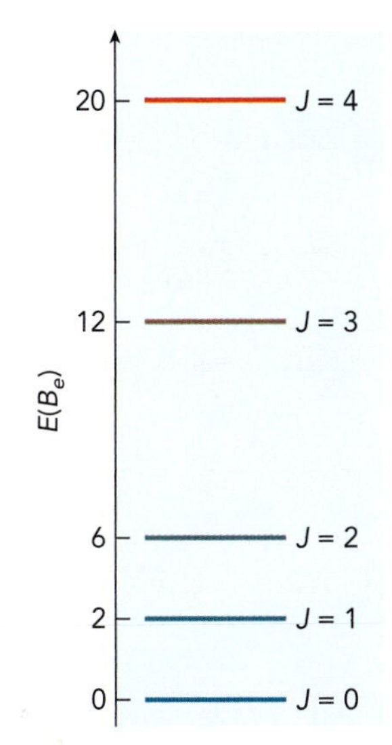

▲ **FIGURE 9.1 The rotational energy level diagram of a rigid linear molecule.**

Like electronic and vibrational energies, the rotational energy is most often expressed in non-SI units convenient to the experimentalist. The unit cm^{-1} may be used because rotational energies are often measured in conjunction with vibrational transitions. Pure rotational transitions, and the resulting constants, are instead usually given in MHz, because the transitions occur in the radio or microwave regions of the spectrum where the frequency of the radiation (rather than the wavelength) is measured directly by signal-counting electronics. The rotational constant can be calculated directly from μ and R_e in those units, written as follows:

$$B_e(\text{cm}^{-1}) = \frac{\hbar}{4\pi c \mu R_e^2} = \frac{16.858}{\mu(\text{amu}) R_e(\text{Å})^2}. \tag{9.5}$$

$$B_e(\text{MHz}) = \frac{\hbar}{4 \cdot 10^6 \pi \mu R_e^2} = \frac{5.0538 \cdot 10^5}{\mu(\text{amu})\, R_e^2}. \tag{9.6}$$

Representative values of the rotational constant for various diatomics are given in Table 9.1.

Even in the ground vibrational state, however, $\langle 1/R^2 \rangle$ is not quite the same as $1/R_e^2$, because the squared average of two numbers is never equal to the average of the squares unless the numbers are the same. For example, if a parameter f has two values, a and b, then the difference between the average of the squares and the square of the average is

$$\langle f^2 \rangle - \langle f \rangle^2 = \frac{a^2+b^2}{2} - \left(\frac{a+b}{2}\right)^2 \tag{9.7}$$
$$= \frac{a^2}{2} + \frac{b^2}{2} - \frac{a^2}{4} - \frac{2ab}{4} - \frac{b^2}{4} = \frac{a^2}{4} - \frac{2ab}{4} + \frac{b^2}{4} = \frac{(a-b)^2}{4},$$

TABLE 9.1 Ground state molecular rotational and vibrational constants for selected diatomic molecules. Values for the vibrational constants are based on varying numbers of anharmonic terms. In this table D_0 is the $v = 0$ rotational distortion constant. Missing values indicate the constant was not measured in the corresponding experiment.

Molecule	μ (amu)	R_e (Å)	B_e (cm^{-1})	α_e (cm^{-1})	D_0 (10^{-6} cm^{-1})	ω_e (cm^{-1})
$^1H\,^1H$	0.50	0.742	60.8536	3.0622	46660	4401.21
$^1H\,^2D$	0.67	0.742	45.6378	1.9500		3811.92
$^2D\,^2D$	1.01	0.742	30.442	1.0623		3118.46
$^1H\,^{19}F$	0.96	0.917	20.9557	0.798	2150	4138.32
$^1H\,^{35}Cl$	0.98	1.275	10.5934	0.3702	532	2990.95
$^1H\,^{79}Br$	1.00	1.414	8.3511	0.226	372	2649.67
$^1H\,^{127}I$	1.00	1.609	3.2535	0.0608	526	2309.60
$^2D\,^{19}F$	1.82	0.917	11.0000	0.2907	585	2998.19
$^{12}C\,^{16}O$	6.86	1.128	1.9313	0.0175	6	2169.82
$^{14}N\,^{14}N$	7.00	1.098	1.9987	0.0171	6	2358.07
$^{14}N\,^{16}O^+$	7.47	1.063	1.9982	0.0190		2377.48
$^{14}N\,^{16}O$	7.47	1.151	1.7043	0.0173	−37	1904.41
$^{14}N\,^{16}O^-$	7.47	1.286	1.427			1372
$^{16}O\,^{16}O$	8.00	1.207	1.4457	0.0158	5	1580.36
$^{19}F\,^{19}F$	9.50	1.418	0.8828			891.2
$^{35}Cl\,^{35}Cl$	17.48	1.988	0.2441	0.0017	0.2	560.50
$^{79}Br\,^{79}Br$	39.46	2.67	0.0821	0.0003	0.02	325.29
$^{127}I\,^{79}Br$	48.66	2.470	0.0559	0.0002	0.008	268.71
$^{127}I\,^{127}I$	63.45	2.664	0.0374	0.0001	−0.005	214.52
$^{23}Na\,^{23}Na$	11.49	3.077	0.1548	0.0009	0.7	159.13
$^{133}Cs\,^{133}Cs$	66.45	4.47	0.0127	0.00003	0.005	42.02
$B^1\Sigma_u^+\;^1H\,^1H$	0.50	0.742	60.864	3.0764		1358.94
$a^3\Pi\;^{12}C\,^{16}O$	6.86	1.206	1.6912	0.0019	6	1743.41

which is always greater than zero (as long as a and b are not equal). If the parameter f represents the values of $1/R$, this means that the average $\langle 1/R^2 \rangle$ is larger than $1/R_e^2$ for the harmonic oscillator, although the difference is often small.

The approximation in Eq. 9.4 runs into a different problem as we climb to higher vibrational states. Equations 9.3 and 9.4 assume that the vibrational motion of the molecule can be cleanly separated from its rotational motion. If we measure transition energies carefully, however, that assumption begins to break down. For example, the form of $U_{\text{vib}}(R)$ (Fig. 9.2) requires that at a higher vibrational quantum number v the average value of R will tend to increase, so $\langle 1/R^2 \rangle$ *decreases.* In other words, the rotational constant of a molecule tends to get smaller as it climbs to higher vibrational energies, because on average the chemical bonds get longer. In Eq. 8.26, we use a power series expansion in $(v + \frac{1}{2})$ to account for anharmonicity in the vibrational energy levels, and we use exactly the

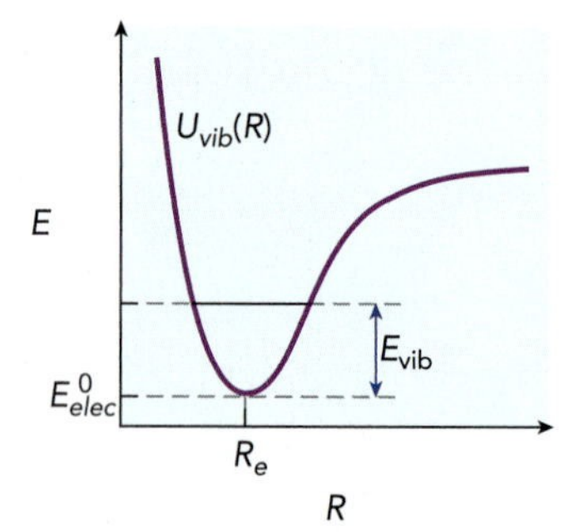

▲ **FIGURE 9.2 A typical vibrational potential energy curve.** A sample vibrational energy level is shown.

same power series to treat the effect of anharmonicity on the rotational constant. This time it takes the form of a vibration-dependent rotational constant B_v, where

$$B_v = B_e - \left(v + \frac{1}{2}\right)\alpha_e + \left(v + \frac{1}{2}\right)^2 \beta_e + \ldots . \tag{9.8}$$

Defined in this way, the **vibration-rotation coupling constant** α_e will usually be positive, because B_v, which includes the effect of the vibrational motion, will usually be smaller than B_e, which depends only on the equilibrium bond length. If rotational constants are measured in several vibrational states, then we add higher order terms with coupling constants such as β_e to fit Eq. 9.8 to the experimental data.

CHECKPOINT We obtained Eq. 9.3 for the rotational constant B_e from theory by solving the Schrödinger equation, assuming that the molecular structure was completely rigid. But the molecule is not rigid, and experimentalists measure the rotational energies too precisely to ignore the deviations from Eqs. 9.3 and 9.4. Here we introduce the vibration-rotation coupling and distortion parameters as empirical corrections to the energy of the rigid rotating molecule.

Another correction to our rotational energy expression in Eq. 9.4 results from **rotational distortion:** as the molecule rotates more and more, the apparent centrifugal force lengthens the chemical bond. This effect also increases the average bond length R and lowers the rotational energy that we would predict from the equilibrium bond length. But this time, the correction becomes more important as we increase the molecular *rotation,* rather than vibration. We treat the rotational distortion by applying a power series again, but this time the expansion is in powers of the rotational quantum number J instead of the vibrational quantum number v:

$$E_{\rm rot} = B_v J(J+1) - D_v[J(J+1)]^2 + H_v[J(J+1)]^3 + \ldots . \tag{9.9}$$

Here too, the minus sign on the right-hand side is chosen so that the **distortion constant** D_v will normally be a positive number. The higher order distortion constants such as H_v are less predictable. The value of each distortion constant depends on the vibrational quantum number v, but a formula based on second-order perturbation theory provides an estimate of the distortion constant in the vibrational ground state $v = 0$:

$$D_0 \approx \frac{4B_e^3}{\omega_e^2}. \tag{9.10}$$

Both α_e and D_v result from the breakdown of our assumption that the rotational and vibrational contributions to the energy are completely separable. Both terms provide a way for the non-rigid nature of the molecule to sneak into our rotational energy expression.

At the beginning of this section we introduced the rotational wavefunctions $Y_J^{M_J}(\Theta,\Phi)$, the same spherical harmonics as the angular wavefunctions of the one-electron atom except for how we label the angles and quantum numbers. Recall that the l quantum number in the atom determines the magnitude of the angular momentum vector, but the orientation of the vector is determined by the magnetic quantum number m_l. One value of l corresponds to $2l + 1$ values of m_l because the magnetic quantum number can take any integer value from $-l$ to $+l$. In the same way, the rotational angular momentum—and the rotational energy—depend only on J, not on M_J. Therefore, the rotational states are degenerate, with the degeneracy equal to the number of M_J states:

CHECKPOINT Note that the $v = 0$ distortion constant D_0 has exactly the same symbol but is *not* the dissociation energy D_0 defined in Section 8.2. The distortion constant will only appear in expressions related to the rotational energy, and has a magnitude less than 0.1 cm^{-1}. The dissociation energy gives the bond strength, so will usually be on the order of 100 kJ mol^{-1} (8000 cm^{-1}).

$$g_{\rm rot} = 2J + 1. \tag{9.11}$$

Let's take a look at the rotational constant in various units for a generic first-row diatomic with reduced mass 10 amu and $R_e = 1$ Å:

$$B_e = \frac{1.055 \cdot 10^{-34}\,\mathrm{J\,s}}{(4 \cdot 10^6 \pi)(10\,\mathrm{amu})(1.661 \cdot 10^{-27}\,\mathrm{kg\,amu^{-1}})(10^{-10}\,\mathrm{m\,\AA^{-1}})^2} = 5 \cdot 10^{10}\,\mathrm{s^{-1}} = 5 \cdot 10^4\,\mathrm{MHz}$$

$$= 1.7\,\mathrm{cm^{-1}} = 2.4\,\mathrm{K}.$$

As we found for the vibrational constant, writing this constant in units of frequency has a classical interpretation—it tells us the classical rotational frequency of the molecule. For example, in the $J = 1$ state, the rotational energy of this molecule will be roughly $2B_e$, which corresponds to a frequency of $1 \cdot 10^{11}$ rotations per second. In contrast, a vibrational constant of 1000 cm^{-1} corresponds to a vibrational frequency of $3 \cdot 10^{14}\,s^{-1}$. A typical molecule may vibrate 1000 times during a single rotation, which is why we average the effects of the vibration when we consider the rotational motions rather than the other way around. The very low excitation energy ($2B_e = 4.8$ K) relative to 300 K implies that most molecules will be rotating with high J value under typical laboratory conditions.

9.2 Rotations in Polyatomics

CHECKPOINT Most experimental studies that observe distinct rotational states are carried out on the vibrational ground state of the molecule. Therefore, it is common to drop the "v" subscript from the rotational constant symbol B_v unless different vibrational states are observed. From now on, if the rotational constant has no subscript, we mean it to apply to the vibrational ground state.

We study molecular rotations for several reasons, but chief among these is the close relationship between the rotational energy levels and the molecular geometry. As shown in Section 9.1, the rotational constant B for a diatomic molecule at any instant is inversely proportional to the bond length R, the one parameter that describes the molecular geometry. A polyatomic molecule has a more complex geometry than a diatomic, and a more complex relationship between the structure and the rotational energies. But tracing that relationship from the rotational Hamiltonian to the experimental spectrum rewards us with our most discriminating determination of how the molecule is put together.

The Rotational Hamiltonian

The expression μR^2 plays a key role in the rotational energies of the diatomic molecule, determining the value of the rotational constant $B = \hbar^2/(2\mu R^2)$. A physicist would recognize this as the **moment of inertia** for the diatomic. The moment of inertia, like the angular momentum, is a useful quantity not because we measure it directly, but because so many parameters that we *can* measure directly depend upon it. We place the origin of a Cartesian coordinate system at the molecule's center of mass and label the molecule-fixed axes x, y, and z, in no particular order. Two examples show how we then calculate the moments of inertia $I_{\alpha\beta}$ relative to any two coordinates α and β:

$$I_{zz} = \sum_{i=1}^{N} m_i(x_i^2 + y_i^2) \tag{9.12}$$

and

$$I_{xy} = -\sum_{i=1}^{N} m_i x_i y_i, \tag{9.13}$$

where N is the number of atoms in the molecule, m_i is the mass of atom i, and x_i and y_i are the coordinates of atom i along the two axes x and y. In general, the I_{xy} and I_{yz} and I_{zx} values are not zero, but we can always choose at least one set of axes a, b, and c such that $I_{ab} = I_{bc} = I_{ca} = 0$. Those coordinate axes are called the **principal inertial axes.**[1] Unless otherwise stated, moments of inertia are always

[1]These axes are not directly related to the principal *rotation* axis introduced in Chapter 6, which is the (often unique) rotation axis for the $\hat{C}_n$ proper rotation with the greatest value of n. However, the principal rotation axis often corresponds to *one* of the principal inertial axes, as we'll see shortly.

calculated using the principal inertial axes. These principal moments of inertia may then be written I_a, I_b, and I_c. The principal inertial axes are another set of molecule-fixed coordinate axes, much like the x, y, and z axes we specified when discussing molecular symmetry. The principal inertial axes are also mutually perpendicular, and they intersect at the molecular center of mass, but their labels are fixed by the mass distribution of the molecule, so the c axis has the largest moment of inertia and a has the smallest:

$$I_a \leq I_b \leq I_c. \tag{9.14}$$

For symmetric molecules, there may be more than one way to orient the principal inertial axes, as in the following example.

For the diatomic, the principal rotational axis is the a axis because the moment of inertia along this axis is zero, treating the nuclei as point masses.[2] Also, the symmetry of the system requires $I_b = I_c$, and both of these moments of inertia reduce to μR^2. For linear polyatomics, finding the moment of inertia I_b takes a little more work, but once it is obtained, the energy levels obey the same expression as for diatomics:

$$E_{\text{rot}} \approx BJ(J+1),$$

EXAMPLE 9.1 Moments of Inertia

CONTEXT Carbon suboxide, OCCCO, is a **quasilinear** molecule, meaning that the minimum energy on the potential energy surface corresponds to a bent structure, but the linear geometry lies only slightly higher in energy. As the bending motion is excited, carbon suboxide behaves more and more like a linear molecule. Does it matter then whether the molecule is linear or non-linear? It does, because if the barrier to the linear geometry is high, then the ground vibrational state of OCCCO may be bent like water, which would give the molecule a dipole moment, whereas if the molecule rapidly vibrates back and forth through the linear geometry, then its average geometry—like carbon dioxide, OCO—will be linear and the dipole moment will be zero. The polarity of OCCCO determines how it will interact with its surroundings in its liquid and crystal forms, so many of its properties hinge on the question. This remained an open question for years because the vibrational spectra resembled a nonlinear molecule in some respects, but it was finally determined that the zero-point motion in carbon suboxide is greater than the barrier to the linear structure. The molecule bends back and forth through the linear geometry, and on average it is linear, having no permanent dipole moment. One of the distinguishing features of a linear molecule is that the symmetry axis is always the a inertial axis, as we determine here.

PROBLEM Prove that the molecular axis in a linear molecule is the a principal axis.

SOLUTION Call the molecular axis z. All the atoms then lie at points $(0, 0, z_i)$. Since $x_i = y_i = 0$ for any atom i, then $I_{xy} = I_{yz} = I_{zx} = 0$, and $I_{zz} = 0$. However, $I_{xx} = I_{yy} = \sum_{i=1}^{N} m_i z_i^2$. Since $I_{xy} = I_{yz} = I_{zx} = 0$, the x, y, and z axes are the principal axes. Since $I_{zz} \leq I_{yy}$ and I_{xx}, we have proven $I_{zz} = I_a$. In other words, the z axis is the a axis, and all atoms lie on the a axis. The x and y axes correspond to the b and c axes. We do not have to specify the direction of the b and c axes because all directions perpendicular to a (or z) are equivalent.

---

[2]One argument against worrying about the rotation about the a axis in a linear molecule is that the moment of inertia of one nucleus amounts to a value on the order of $(10\text{ amu})(10^{-15}\text{ m})^2 = 1.66 \cdot 10^{-56}\text{ kg m}^2$ (using a nuclear radius estimated by Rutherford's scattering experiment), which gives a rotational constant of $3 \cdot 10^{-13}$ J or $2 \cdot 10^{8}$ kJ mol^{-1}, well out of the energy range per molecule that we associate with chemistry.

BIOSKETCH **Stewart Novick**

Stewart Novick is a professor in the Department of Chemistry at Wesleyan University, studying molecules that you cannot buy in a bottle from a chemical company: highly reactive free radicals and molecular ions and weakly bonded molecule–molecule pairs that are broken apart at room temperature. These species last no more than microseconds under typical experimental conditions, but they play central roles in chemistry. The radicals and ions are chemical intermediates, controlling the direction and speed of reactions in combustion, chemical synthesis, and other environments. The weakly bonded complexes are prototypes of liquids, perhaps the most important phase of matter in laboratory chemistry but also the toughest to understand. Novick's group generates these species in short bursts in a vacuum chamber and then immediately probes their rotational energy levels with a short pulse of microwave radiation. The Fourier transform of the time-dependent signal seen by a microwave detector gives the frequency-dependent rotational spectrum. The rotational spectroscopy of these molecules provides the highest precision measurements available on molecular structure, and Novick's work has determined the geometries of dozens of short-lived molecules and complexes.

neglecting vibration-rotation and distortion terms. Because the linear molecule can only rotate about some axis perpendicular to the a axis, and because the moments of inertia along the b and c axes are identical, the rotational energies are all functions of a single rotational constant B.

For nonlinear molecules, however, the situation is more complicated. The rotational Hamiltonian is

$$\hat{H}_{\text{rot}} = \frac{\hat{J}_a^2}{2I_a} + \frac{\hat{J}_b^2}{2I_b} + \frac{\hat{J}_c^2}{2I_c}, \tag{9.15}$$

where $\hat{J}_a$, $\hat{J}_b$, and $\hat{J}_c$ are the operators whose eigenvalues are the angular momenta for rotation *about* the a, b, and c axes, respectively. The principal moments of inertia may be worked out from Eq. 9.12 and used to define the A, B, and C rotational constants:

$$A \equiv \frac{\hbar^2}{2I_a} \qquad B \equiv \frac{\hbar^2}{2I_b} \qquad C \equiv \frac{\hbar^2}{2I_c}. \tag{9.16}$$

Our rotational Hamiltonian can then be written

$$\hat{H}_{\text{rot}} = \frac{[A\hat{J}_a^2 + B\hat{J}_b^2 + C\hat{J}_c^2]}{\hbar^2} \tag{9.17}$$

We run into an old dilemma here: the uncertainty intrinsic to quantum mechanics forbids us from knowing the exact orientation of an angular momentum vector. For example, we can specify a state of the molecule according to the quantum numbers J and J_a, but then we cannot specify J_b and J_c because they will not be constant for that state (see Section 2.1). This prevents us from finding one simple expression for the rotational energy that covers all polyatomic molecules. Thankfully, some cases are easy. (Make that, *relatively* easy).

Example 9.2 Rotational Constant of a Polyatomic

CONTEXT The rotational states of a molecule tell us about the molecular geometry, and few molecules have had their rotational spectrum more closely studied than ammonia. The geometry of ammonia plays a role in many of its applications, including its use in refrigeration. To refrigerate an object, we need to move energy from the object somewhere else, usually by transferring the energy into a refrigerant. The refrigerant evaporates (absorbing the heat) and is carried away to another point where it is re-condensed (releasing the heat). But transporting the energy in that refrigerant is a process that itself requires energy. Typical refrigerators use an electrically driven compressor to condense the refrigerant. In contrast, an **absorptive refrigerator** uses heat as its energy source, allowing it to run where electricity is inaccessible or expensive. Ammonia is the most efficient refrigerant used for these systems—it's slightly more efficient than the halocarbons used for compressor refrigeration—and operates within an appropriate temperature range. The boiling point of ammonia is higher than for most other small molecules (aside from water) because it has a large dipole moment, 1.42 D, which binds the molecules together in the liquid. The strength of that dipole moment is a direct result of the non-planar geometry, which sums together the contributions of the individual N—H bond dipoles rather than canceling them out.

PROBLEM Calculate the A, B, and C rotational constants of NH_3 in cm^{-1} based on the following Cartesian coordinates (given in Å) for the atoms, which place the center of mass at the origin:

atom	*a*	*b*	*c*
N	0.000	0.000	0.114
H	−0.938	0.000	−0.267
H	0.469	0.811	−0.267
H	0.469	−0.811	−0.267

SOLUTION The moments of inertia are

$$\begin{aligned} I_a &= [(14.00)(0^2+0.114^2) + (1.008)(0^2+0.267^2) + (1.008)(0.811^2+0.267^2) \\ &\quad + (1.008)(0.811^2+0.267^2)]\ \text{amu Å}^2 \\ &= 1.73\ \text{amu Å}^2 \\ I_b &= [(14.00)(0^2+0.114^2) + (1.008)(0.938^2+0.267^2) + (1.008)(0.469^2+0.267^2) \\ &\quad + (1.008)(0.469^2+0.267^2)]\ \text{amu Å}^2 \\ &= 1.73\ \text{amu Å}^2 \\ I_c &= [(14.00)(0^2+0^2) + (1.008)(0.938^2+0^2) + (1.008)(0.469^2+0.811^2) \\ &\quad + (1.008)(0.469^2+0.811^2)]\ \text{amu Å}^2 \\ &= 2.66\ \text{amu Å}^2. \end{aligned}$$

So the rotational constants are (using the conversion factor from Eq. 9.5)

$$A\,(\text{cm}^{-1}) = \frac{16.858}{1.73\,\text{amu Å}^2} = 9.74\ \text{cm}^{-1}$$

$$B\,(\text{cm}^{-1}) = \frac{16.858}{1.73\,\text{amu Å}^2} = 9.74\ \text{cm}^{-1}$$

$$C\,(\text{cm}^{-1}) = \frac{16.858}{2.66\,\text{amu Å}^2} = 6.34\ \text{cm}^{-1}.$$

These do not agree exactly with the values in Table 9.2 because vibration-rotation coupling is neglected in this calculation.

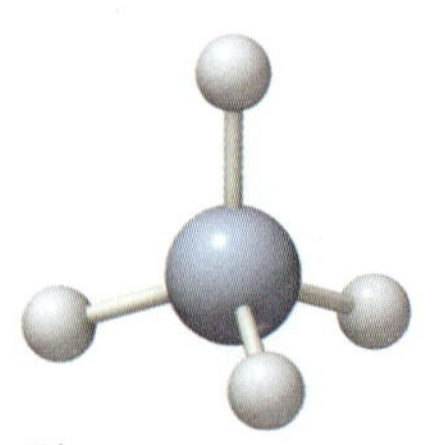

Silane
Spherical top
$A=B=C=2.85\ \text{cm}^{-1}$

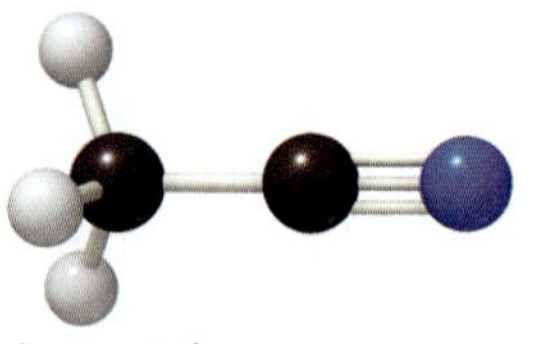

Acetonitrile
Prolate top
$A=5.33\ \text{cm}^{-1}$ $B=C=0.309\ \text{cm}^{-1}$

Chloroacetylene
Linear molecule
A undefined $B=C=0.190\ \text{cm}^{-1}$

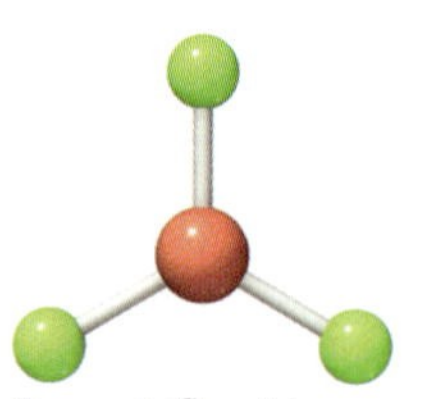

Boron trifluoride
Oblate top
$A=B=0.342\ \text{cm}^{-1}$ $C=0.171\ \text{cm}^{-1}$

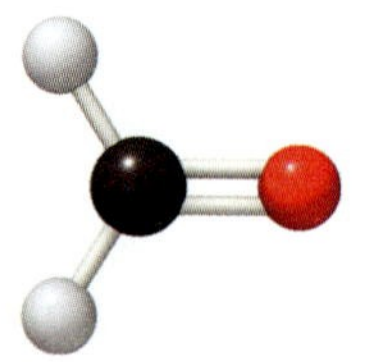

Formaldehyde
Near-prolate asymmetric top
$A=9.52\ \text{cm}^{-1}$ $B=1.31\ \text{cm}^{-1}$
$C=1.15\ \text{cm}^{-1}$

▲ **FIGURE 9.3 Different tops in rotational spectroscopy.** Examples of the different possible relationships among the three rotational constants.

Symmetric Tops

Take the case of a **symmetric top,** in which two of the moments of inertia are equal. If $I_a = I_b$, the object has a mass distribution like that of a dinner plate or a benzene molecule and is called an **oblate top.** If $I_b = I_c$, the distribution is similar to that of a pencil or a propyne molecule (H_3CCCH) and is called a **prolate top.** So $A = B$ for oblate tops, and $B = C$ for prolate tops. Examples are shown in Fig. 9.3. A molecule is a symmetric top whenever it has an n-fold principal axis where $n > 2$. Any molecule in the C_n, C_{nh}, C_{nv}, D_n, D_{nh}, or D_{nd} point groups with $n > 2$ is a symmetric top.

We can take advantage of the relation

$$\hat{J}^2 = \hat{J}_a^2 + \hat{J}_b^2 + \hat{J}_c^2 \tag{9.18}$$

to rewrite the Hamiltonian for prolate and oblate tops:

$$\text{prolate:}\quad \hat{H}_{\text{rot}} = \frac{[A\hat{J}_a^2 + B(\hat{J}^2 - \hat{J}_a^2)]}{\hbar^2} \tag{9.19}$$

$$\text{oblate:}\quad \hat{H}_{\text{rot}} = \frac{[C\hat{J}_c^2 + B(\hat{J}^2 - \hat{J}_c^2)]}{\hbar^2}. \tag{9.20}$$

The eigenvalue of the operator $\hat{J}_a$ is $K_a\hbar$, and for $\hat{J}_c$ it is $K_c\hbar$. These are projection operators like L_z or J_Z, but with quantum numbers K instead of Λ or M_J.[3] The eigenvalue equation is essentially the same, and K can take any integer value between $-J$ and J. The eigenvalue for $\hat{J}^2$ is still $J(J+1)\hbar^2$, so our energy expressions for symmetric tops become

$$\text{prolate:}\quad E_{\text{rot}} = K_a^2(A - B) + BJ(J+1) \tag{9.21}$$

$$\text{oblate:}\quad E_{\text{rot}} = K_c^2(C - B) + BJ(J+1). \tag{9.22}$$

For $K \neq 0$, each J state has a twofold degeneracy because the K rotation can be in either direction. Because $(A - B)$ is a positive number while $(C - B)$ is negative, the energy level diagrams of the two types of symmetric tops are qualitatively different: for a given J value, an increase in K lowers the energy in the oblate top, whereas in the prolate top the energy increases with K. The difference between the rotational energy levels of the two symmetric tops is shown in Fig. 9.4, where the rotational quantum states are drawn as a series of K-**stacks,** each with its own manifold of states that increase in energy with J.

Because the energy level expression for symmetric tops is so simple, the more complicated problem of **asymmetric tops**—molecules where no two moments of inertia are equal—is usually solved by using the symmetric top wavefunctions as a basis set. This approximation works well if the molecule is *nearly* a symmetric top, such as formaldehyde, H_2CO. The point group of formaldehyde is C_{2v}, so it is not a symmetric top, but the smallest moment of inertia I_a contains contributions only from the light hydrogen atoms and is much smaller than I_b and I_c. Therefore, this is a *nearly prolate* asymmetric top, and the rotational energy level diagram of formaldehyde closely mimics that of a true prolate top. The effect of the asymmetry is to break the degeneracy of the $K_a \neq 0$ states: the $K_a = \pm 1$ states, for example, now have slightly different energies.

[3]The difference between M_J and K is that M_J gives the projection of J onto a space-fixed axis Z, while K is the projection onto one of the molecule-fixed inertial axes. It matters because the rotational energy levels tend to be easily grouped by K values, whereas the influence of a lab-fixed magnetic field will break apart the energy levels according to their M_J values.

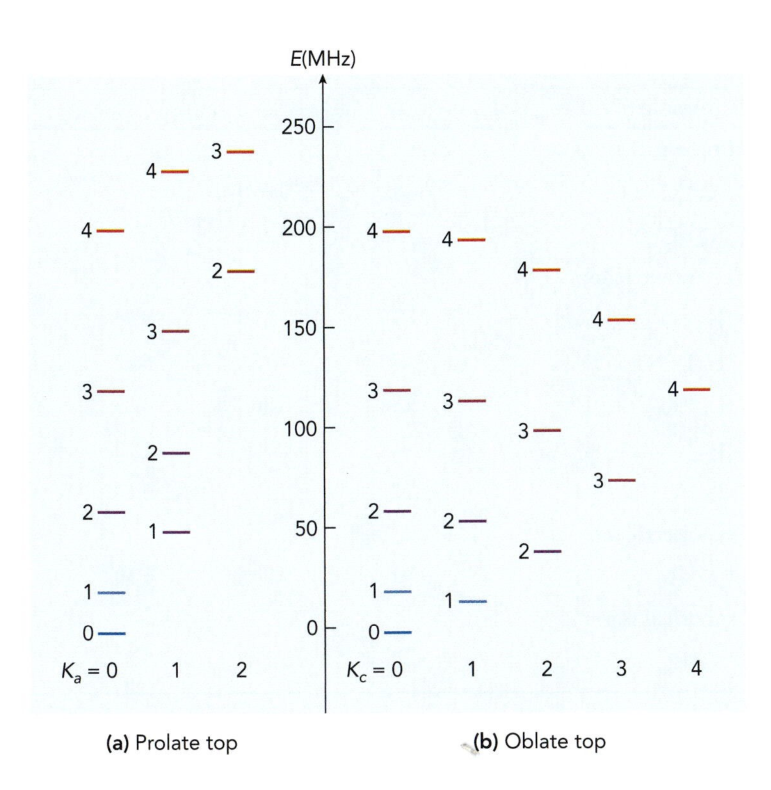

◀ **FIGURE 9.4 Rotational energy level diagrams for the symmetric tops: (a)** a prolate top ($A = 40$ MHz, $B = 10$ MHz) and **(b)** an oblate top ($B = 10$ MHz, $C = 5$ MHz). Energy levels are accompanied by their J values and are arranged in K-stacks.

Spherical Tops

Finally, there are the **spherical tops,** for which all three moments of inertia are equal. This condition is satisfied only for those molecules belonging to the cubic groups, having more than one C_n proper rotation axis with $n > 2$. This set includes the tetrahedral group T_d, so methane, carbon tetrachloride, and ammonium ion are all spherical tops. The spherical top Hamiltonian simplifies to

$$\hat{H}_{\text{rot}} = \frac{\hat{J}_a^2}{2I_a} + \frac{\hat{J}_b^2}{2I_b} + \frac{\hat{J}_c^2}{2I_c} = B(\hat{J}_a^2 + \hat{J}_b^2 + \hat{J}_c^2) = B\hat{J}^2 \tag{9.23}$$

and the energies to

$$E_{\text{rot}} = BJ(J + 1). \tag{9.24}$$

Equation 9.24 looks essentially the same as Eq. 9.4 for the rotational energy of a linear molecule, but the non-linearity of the molecule causes the degeneracy of a particular value of J to be higher. A given J value can correspond to any of a number of possible combinations of rotations about the a and c axes. Their spectra turn out to be more complicated than implied by this zero-order energy level expression, because molecular vibrations partially break the rotational state degeneracy.

The rotational constants for several polyatomic molecules are given in Table 9.2. As for the diatomics, we see that more massive compounds have lower rotational constants. Hydrogen atoms contribute very little to the moments of inertia, and so molecules such as H_2O, H_2CO, and NH_3 have very high A rotational constants because rotation about the a axis corresponds only to a rotation of very light H atoms. These A values are similar to the B values of the diatomic hydrides.

CHECKPOINT The K and M_J quantum numbers correspond to the projection of the rotational angular momentum vector onto different axes: for M_J, the projection is onto a space-fixed axis, and for K we project onto the molecule-fixed a or c axis. Therefore, the K value affects the energy whenever the molecule rotates, whereas M_J determines the energies only when we put the rotating molecule in an external magnetic field.

TABLE 9.2 **Rotational constants (cm^{-1}) for selected polyatomic molecules.**

molecule	*A*	*B*	*C*
linear			
$^{16}O\,^{12}C\,^{16}O$		0.390	
$^{16}O\,^{12}C\,^{32}S$		0.203	
asymmetric top			
$^{1}H_2\,^{16}O$	27.79	14.50	9.96
$^{1}H\,^{2}D\,^{16}O$	23.48	9.13	6.62
$^{2}D_2\,^{16}O$	15.36	7.26	4.85
$^{1}H_2\,^{12}C\,^{16}O$	9.41	1.30	1.13
$^{14}N\,^{12}C\,^{14}N_3$	1.27	0.11	0.10
symmetric top			
$^{14}N\,^{1}H_3$	9.94	9.94	6.30
spherical top			
$^{12}C\,^{1}H_4$	5.27	5.27	5.27

These are *observed* values for the rotational constants and therefore contain contributions from the zero-point vibrational motion of the atoms. The rigid equilibrium structure for water, for example, would have rotational constants $A = 31.2\ cm^{-1}$, $B = 13.7\ cm^{-1}$, and $C = 9.51\ cm^{-1}$. Zero-point bending tends to make the water molecule (on average) a little more linear than its equilibrium structure. Increased linearity reduces the moment of inertia along the *b* and *c* axes, which are largely determined by the position of the oxygen atom. In the linear limit, all three axes would pass through the oxygen atom and the oxygen would not contribute to any moment of inertia. Therefore, as the molecule becomes more linear, I_b and I_c decrease, so *B* and *C* increase. On the other hand, the *a* axis always contains the oxygen, and I_a is only determined by the positions of the hydrogens. Making the molecule more linear extends the hydrogens further from the *a* axis, making the observed values of I_a increase and *A* decrease compared to the equilibrium values.[4]

9.3 Spectroscopy of Rotational States

The initial and final states of any transition in a free molecule are characterized by the amount of energy in each of the electronic, vibrational, and rotational degrees of freedom. The type of transition is generally labeled by the highest-energy degree of freedom that changes; for example, if the electronic state changes, the transition is an **electronic transition;** if the vibrational state changes but not the electronic state, it is a **vibrational transition;** and if only the rotational state changes, it is a **rotational transition.** But in a gas-phase electronic transition, the vibrational and rotational states can also change; and in vibrational transitions the rotational state often changes.

[4]Another reason the observed *A* constant decreases from its equilibrium value is that the zero-point stretching motion tends to lengthen the OH bonds.

Pure Rotational Spectroscopy

In pure **rotational spectroscopy,** absorption or emission of radiation leads to changes only in the rotational state of the molecule, leaving the initial vibrational and electronic states intact. For the relatively strong electric dipole transitions, that means that the molecule must possess a charge distribution that changes somehow when the rotational state changes, providing a handle for the radiation's electric field to crank the rotation faster or slower. If the molecule is rigid, that charge distribution comes from the permanent electric dipole moment of the molecule. If the molecule rotates to change the direction of its permanent dipole moment, that gives the necessary interaction with the radiation. If the molecule is non-polar, or if the rotation is about the axis of the dipole moment (so that the dipole moment doesn't move), then pure rotational transitions are forbidden (Fig. 9.5). In fact, if we turn to our general expression for transition intensity, Eq. 6.10,

$$\mathrm{I} \propto \left| \int \psi_f^* \hat{\mu}_t \psi_i \, d\tau \right|^2,$$

then for an electric dipole allowed, pure rotational transition, the transition dipole moment operator $\hat{\mu}_t$ in this expression *is* the permanent dipole moment. More precisely, it is the component of the permanent dipole moment that is aligned with the electric field vector of the radiation.

Any homonuclear diatomic molecule has no permanent dipole moment and therefore no allowed rotational spectrum. Because the dipole moment remains zero as the bond stretches, homonuclear diatomics also have no allowed vibrational spectrum. Only electronic spectroscopy in homonuclear diatomics is allowed by electric dipole selection rules, and high precision measurements of the rotational and vibrational constants in molecules as simple as H_2 and N_2 can be quite difficult.

One way to directly measure rotational transitions in non-polar molecules such as these is by rotational Raman spectroscopy, which operates on the same principle as other Raman techniques (see Section 6.3). A rotational Raman transition connects initial and final rotational levels within the same vibrational state, so only the rotational quantum number changes. However, this technique is limited in precision by the uncertainties in the photon energies of the incident and scattered light. The scattering intensity increases dramatically with photon energy,

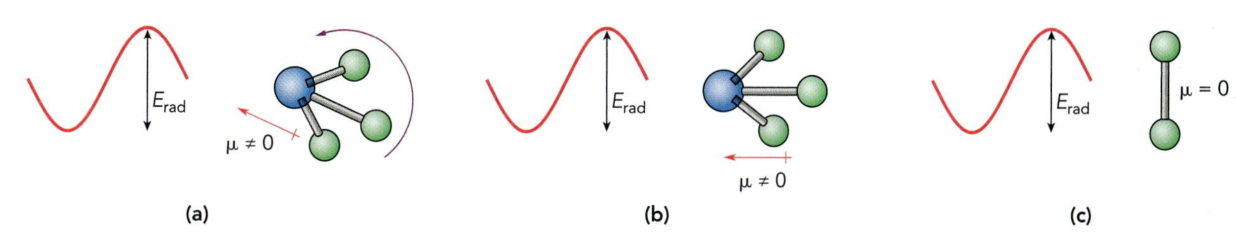

▲ FIGURE 9.5 **The dipole selection rule for rotational transitions.** A polar molecule can be induced to change its rotational state by interaction with the electric field of a photon, as long as the field vector has some component that is parallel to the dipole moment of the molecule **(a).** If the electric field vector of the photon is perpendicular to the dipole moment **(b),** or if the dipole moment is zero **(c),** there is no allowed interaction.

TOOLS OF THE TRADE | Radio Astronomy

Although rotational spectroscopy is carried out in many chemistry laboratories, the technique has also found a home in a seemingly distinct field of study: the astronomy of star-forming regions in our galaxy. Let's take advantage of that to look at the tools of physical chemistry at work in a setting outside the lab.

At the turn of the 20th century, our rapidly advancing understanding of electromagnetic radiation and the cosmos led many scientists to search for radiation from space at wavelengths other than the visible. Heinrich Hertz had already developed the radio transmitter and receiver in 1888, and with Guglielmo Marconi's subsequent improvements, the technology for manipulating and detecting radio frequency signals outdid our technology in any other region of the spectrum. However, early attempts to detect radio waves from the sun were fruitless. In the 1920s, we learned that our upper atmosphere consisted of ionized gases that would reflect radio waves. That appears to have ended public attempts to detect radio waves from space, until a lucky accident reawakened interest in the idea.

Karl Jansky worked for Bell Labs, and around 1930 he built an antenna to investigate radiation that could cause interference in radio communications. Over a period of months, he tracked down the origin of one mysterious source of static that appeared at regular intervals, rising and falling over the course of the day. After many measurements, he realized that the time period of the static was 4 minutes less than 24 hours, roughly 364/365 of a day. This interval is the same period of time it takes for Earth to rotate exactly once in relation *to the stars.* The source of the interference was the center of the Milky Way, in the direction of the constellation Sagittarius.

Jansky had found the hole in the atmosphere's shield against long-wavelength radiation. The ionosphere increasingly interacts with radio photons as wavelengths extend beyond about 20 cm, and by 10 m wavelengths the ionosphere completely reflects the radiation. The ionosphere does not impede shorter wavelengths, but gases in the lower atmosphere—especially water—absorb strongly at wavelengths less than about 1 cm. Radiation in that 1 to 20 cm "window" has taught us much about the nature of the universe, back to its earliest moments.

What is radio astronomy? Radio telescopes are dish antennas that measure the intensity of radiation from space at centimeter wavelengths. The goal of radio astronomy is to use these signals to learn about the source that emits them: stars, clouds of interstellar gas, and even objects that would be invisible to optical telescopes.

Why do we use radio astronomy? In the same way that optical telescopes can pinpoint stars and galaxies, radio telescopes can find astronomical objects. But because they operate at wavelengths more than 10,000 times longer than visible light, radio telescopes are sensitive to much colder objects and can also see through dust clouds that block shorter wavelengths. Furthermore, several important astronomical sources emit radiation specifically in the radio range of the spectrum. Two of these sources merit a brief introduction. (1) The **cosmic microwave background radiation** was first detected in 1965—also by accident—by Arno Penzias and Robert Wilson (both awarded the 1978 Nobel Prize in Physics). The cosmic background is radiation released early in the history of the universe, when electrons combined with protons to form the first neutral atoms. The particles that were at the outer edge of the Big Bang were accelerating away from the other particles at such incredible speeds that the radiation reaches us only now. (2) The two most famous spectroscopic transitions in astronomy are both hydrogen atom transitions, but they are near opposite ends of the electromagnetic spectrum. The ultraviolet **Lyman** α transition at 91 nm is the $n = 1$—2 transition, which can be seen in either emission or absorption from energetic sources such as stars, and is the most common signal used to measure the expansion of the universe. And at radio wavelengths, the **21-cm line** emission is caused by the flip of the proton spin relative to the electron spin in neutral hydrogen. Hydrogen is the most abundant element in the universe, and this transition has been used to map out the density of atomic hydrogen in this and other galaxies. Examples (1) and (2) relate to the simplest particles in chemistry, and also to vast distance scales. In contrast, a relatively small volume of our Galaxy consists of gases so dense that they block out the ultraviolet radiation emitted by hot stars. ("Dense" in this context corresponds to a million particles per cubic centimeter, lower than any attainable laboratory vacuum.) Molecules form in these interstellar clouds, and we can identify those molecules by their distinctive spectra. But the outside of these clouds shields the interior from radiation so effectively that the molecules near the center are *very* cold, too cold for electronic or vibrational excitation. Instead, the spectra we see are from rotational transitions of the molecules, at wavelengths that are observable in this 1–20 cm window. The analysis of these spectra has given us detailed information about the elemental and isotopic constitution of these clouds, velocities and angular momenta of different parts of the cloud, and the temperature.

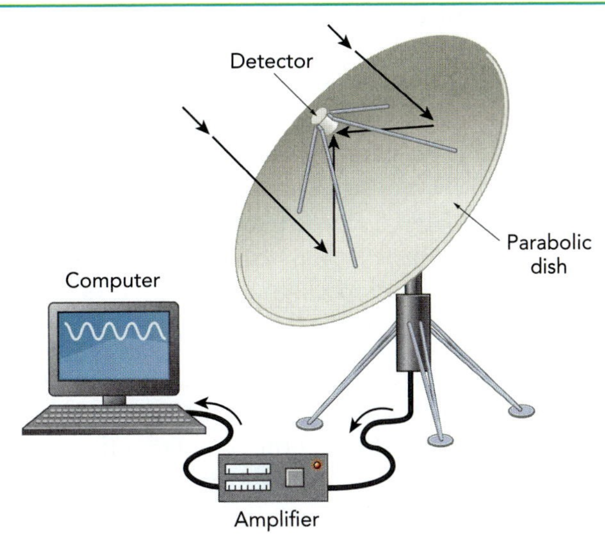

▲ **Diagram of a typical radio telescope.**

How does it work? Although all electromagnetic radiation can be classified as a single phenomenon, the technology we use to manipulate and detect radiation in different regions of the spectrum vary widely. Few examples make that point better than a comparison of optical and radio telescopes. Radio photons have much lower energy than visible photons, so radio telescopes rely on very large areas to capture enough radiation for sensitive detection. The size of optical telescopes is limited by the weight of the mirror and our ability to smoothly shape a large surface. But wavelengths of 1–20 cm are much less sensitive to surface imperfections and do not even need a continuous surface to be reflected. Therefore, the telescope dishes tend to be like nets, with lots of gaps to reduce the weight of the dish. The dish has a parabolic shape, which reflects the radiation into the horn of a receiver, usually suspended above the dish. The receiver amplifies the signal before sending it to a series of frequency-tuned filters and detectors.

In Fourier transform IR (FTIR) spectroscopy (see *Tools of the Trade* in Chapter 8), we use interferometry to generate a spectrum from time-dependent measurements. A different sort of interferometry provides the basis for a dramatic advance in radio astronomy. The resolution of a telescope—its ability to distinguish two stars, say, that appear close together—increases with the size of the telescope. The low frequency of radio waves makes it possible to measure precisely the relative phases of the same signal received at two separate telescopes, generating an interferogram of the signal. A Fourier transform of the interferogram yields an image of the source, but by combining the signals, the two telescopes take on the resolution of a telescope as big as the their separation distance. Our most detailed images of distant galaxies are achieved by this technique.

so rotational Raman spectroscopy is typically carried out using light at visible wavelengths. In the visible, even an uncertainty in wavelength of one part in 10^5 is equivalent to a 0.2 cm^{-1} uncertainty in the transition energy, as compared to a precision of 0.0005 cm^{-1} routinely achieved in the pure rotational spectroscopy of polar molecules. Therefore, rotational Raman is usually reserved for the study of non-polar molecules, which have few other options for the characterization of their rotational states.[5] One prominent application of this technique has been to

[5]Pure rotational transitions in H_2, following electric quadrupole selection rules, are now often observed in emission from interstellar gas, and the signals are one of many indicators that astronomers use to assess the energy distribution in those clouds. Those signals are too weak for routine observation in the laboratory.

atmospheric analysis, for example, where rotational transitions may be resolved for N_2, O_2, and CO_2, all invisible molecules in direct rotational spectroscopy. Figure 9.6 shows a sample rotational Raman spectrum of N_2.

When the transition is allowed, the electric dipole selection rules are $\Delta J = \pm 1$ for pure rotational transitions, obeying our general rule that the photon behaves like a particle with one unit of angular momentum. For a transition $J \rightarrow J + 1$, the photons absorbed or emitted have energy (neglecting distortion terms)

$$\begin{aligned} h\nu &= E' - E'' \\ &= B_v[(J'' + 1)(J'' + 2) - J''(J'' + 1)] = 2B_v(J'' + 1), \end{aligned} \tag{9.24}$$

where we have used a standard notation in which the upper state of the transition (not necessarily the final state) is identified by a single prime ($'$) and the lower state by a double prime ($''$). From the energies of pure rotational transitions, we can determine the rotational constants of the molecule and therefore its moments of inertia I. From these, in turn, we can derive the bond length of the molecule as a function of vibrational and electronic state. Thanks largely to the very high precision of radio and microwave frequency measurements, pure rotational spectroscopy remains the most precise means of measuring the bond lengths and bond angles of gas-phase molecules. This is not an easy determination for polyatomic molecules. Often the molecular symmetry forbids direct measurement of one of the rotational constants. And even if we can measure all three rotational constants, we only get three moments of inertia, from which we need to determine all the bond lengths and bond angles in the molecule. Additional measurements with isotopically substituted forms of the compound provide the other known values we need to determine the different geometric parameters, but it remains an arduous calculation because each isotope also changes the vibrational zero-point motion.

Symmetric tops will have zero dipole moment perpendicular to the principal axis, and for this reason pure rotational transitions that change K are forbidden. If there is a permanent dipole moment along the principal axis,

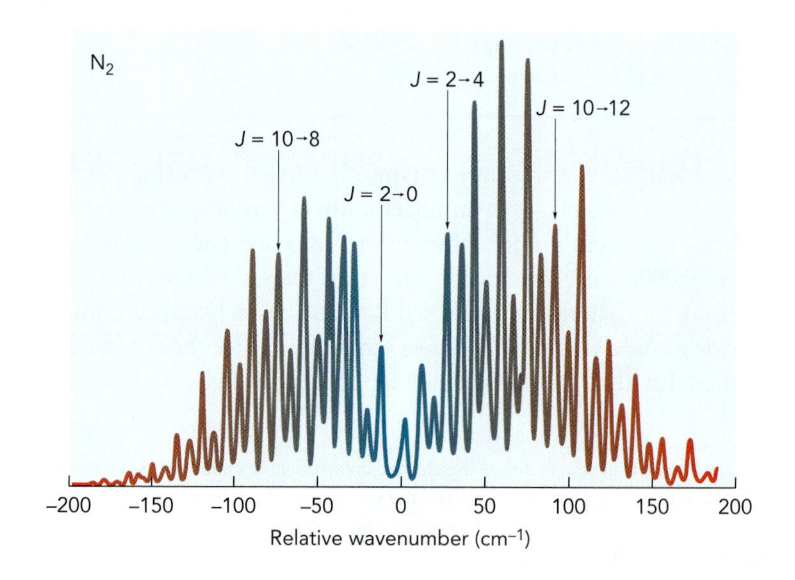

▶ **FIGURE 9.6 A rotational Raman spectrum of N_2.** The spectrum results from the scattering of a 532 nm YAG laser beam. The signals are plotted as a function of wavenumber difference between the excitation and scattered radiation. The angular momentum selection rules require J to change by two units for each transition, so the spacing between the transitions is roughly $4B$ or about 8 cm^{-1}. Transitions to lower energy rotational states are marked in blue, and red indicates transitions to higher energy states. [NASA]

then the pure rotational selection rules are $\Delta J = \pm 1$, $\Delta K = 0$. In vibrational or electronic transitions, there will be some allowed $\Delta K = \pm 1$ transitions, but that depends on the symmetry of the states involved, and the selection rules are determined the same way as for electronic and vibrational transitions, using the character table.

Rovibrational and Rovibronic Spectroscopy

The high symmetry of spherical tops and homonuclear diatomic molecules prevents their having any permanent dipole moment, so pure rotational spectroscopy of these molecules is forbidden. However, if we look high enough in energy, we will find electric dipole-allowed transitions, and differences among the rotational energy levels can often be measured at the same time. The spherical tops have allowed vibrational transitions, and all molecules, including the homonuclear diatomics, have allowed electronic transitions. As Table 9.1 shows, rotational constants typically range two to three orders of magnitude below the vibrational constants, and electronic transition energies are usually an order of magnitude greater than vibrational ones. As a result, the rotational contributions to a vibrational or electronic transition appear as small shifts of the overall transition energy. In a **rovibrational** spectrum, a vibrational transition is directly induced by the radiation, and the effect of different rotational states splits the signal into many distinct rotational components. A **rovibronic** spectrum directly probes an electronic transition but with sufficient resolution to observe the distinct vibrational and rotational contributions to the signal.

In the gas phase, the molecules in the initial state of an electronic or vibrational transition will be distributed among many rotational levels. Even if we don't allow the rotational state to change, the differences in the initial and final rotational constants (from parameters such as α_e) cause the transition energies from each of these initial rotational states to be a little different.

CHECKPOINT In this section, we bring together the equations we've assembled for the rotational and vibrational energies and see how they combine to predict the *transition* energies between different sets of vibrational and rotational levels. Although the Schrödinger equation solves for the energies of individual states, the transition energies are what we measure experimentally.

Let's examine the transition energies in some detail for a rovibrational spectrum. The transition energy is always given by the difference in energy of the final and initial states. Neglecting centrifugal distortion, and again using prime and double-prime to label upper and lower state quantum numbers respectively, the transition energy can be calculated by combining Eqs. 8.26 and 9.8:

$$\begin{aligned} h\nu &= E' - E'' \qquad (9.26) \\ &= \left[\left(v' + \frac{1}{2}\right)\omega_e - \left(v' + \frac{1}{2}\right)^2 \omega_e x_e + J'(J' + 1)B_e - J'(J' + 1)\left(v' + \frac{1}{2}\right)\alpha_e\right] \\ &\quad -\left[\left(v'' + \frac{1}{2}\right)\omega_e - \left(v'' + \frac{1}{2}\right)^2 \omega_e x_e + J''(J'' + 1)B_e - J''(J'' + 1)\left(v'' + \frac{1}{2}\right)\alpha_e\right]. \end{aligned}$$

If J'' and J' are the same, the terms proportional to B_e will cancel, but not the terms proportional to α_e.

In fact, the J value usually does change during a rovibrational or rovibronic transition, even though the rotational state has relatively little impact on the overall transition energy. In electric-dipole transitions, the photon transfers one unit of angular momentum to the molecule, and this must be absorbed by the molecule's electronic, vibrational, or rotational motion. Angular momentum of the electronic or vibrational state is present only if the symmetry of the state corresponds to a degenerate representation in the molecule's point group. On the other hand, all rotational states (except the $J = 0$ ground state) have angular momentum. In the rovibrational spectrum of a diatomic molecule, for example, no vibrational angular momentum is possible, because the only vibrational mode is a non-degenerate stretch with σ symmetry. Therefore, the photon's angular momentum must go into the rotational angular momentum, and the selection rule for the J quantum number remains $\Delta J = \pm 1$, as for pure rotational spectroscopy. The rovibrational spectrum of the diatomic then divides into two sets of transitions: (i) lines with $J' = J'' + 1$, which are labeled $R(J'')$, and (ii) lines with $J' = J'' - 1$, labeled $P(J'')$. Substituting these expressions for J' into Eq. 9.26, and also obeying the harmonic selection rule $\Delta v = \pm 1$, so that $v' = v'' + 1$, we obtain the transition energies:

$$R(J''): \quad h\nu = \omega_e - 2v'\omega_e x_e + 2(J'' + 1)B_e - (J'' + 1)(J'' + 2v' + 1)\alpha_e \quad (9.27)$$

$$P(J''): \quad h\nu = \omega_e - 2v'\omega_e x_e - 2J''B_e - J''(J'' - 2v')\alpha_e. \quad (9.28)$$

CHECKPOINT In linear molecules, Σ states are non-degenerate, and in non-linear molecules only A and B states are non-degenerate. Electronic or vibrational transitions involving any other representations (such as Π or E) allow changes in vibrational or electronic angular momentum that can account for the photon angular momentum. In those cases, it is possible to have a Q branch in which the rotational angular momentum does not change.

This family of rotational components that contribute to the overall vibrational or vibronic transition is called a **band,** and it is roughly centered on an energy called the **band origin ν_0.** The band origin would be the transition frequency if there were no rotational excitation at all, the energy difference between the $J' = 0$ and $J'' = 0$ states. For a fundamental vibrational transition ($v = 0 \rightarrow 1$), ν_0 is roughly equal to $\omega_e - 2\omega_e x_e$.

If we set α_e to zero for a moment, the spectrum predicted by Eqs. 9.27 and 9.28 becomes a set of evenly spaced transitions, separated by $2B_e$ around the center frequency of the transition. The ***P* branch** is the set of $P(J)$ transitions and starts at a transition energy just below the band origin, whereas the ***R* branch** transition energies start just above the band origin, as shown in Fig. 9.7.

If we turn α_e back on and give it a positive value, it partly cancels the $+2J'B_e$ contribution of the $R(J'')$ transition energy but adds to the $-2J''B_e$ contribution of the $P(J'')$ transitions. The result is that the rotational splitting gets smaller with J'' in the R branch, and it increases with J'' in the P branch.

This pattern of transitions in the rovibrational spectrum of the diatomic is roughly similar to the appearance of a rovibronic spectrum. Equations 9.27 and 9.28 need to be adjusted to include the equilibrium term energy T_e for the difference in energy between the potential minima for the initial and final electronic states and to separate the power series expansions in the vibrational and rotational energies. Parameters such as B_e, ω_e, $\omega_e x_e$, and α_e depend on the potential energy curve of each individual electronic state. No simple equation relates these potential energy curves for different electronic states, and therefore distinct values for each of these parameters are given to each electronic state:

$$\begin{aligned} h\nu &= E' - E'' \qquad (9.29) \\ &= T_e + \left[\left(v' + \frac{1}{2}\right)\omega_{e'} - \left(v' + \frac{1}{2}\right)^2 \omega_e x_{e'} + J'(J' + 1)B_{e'} - J'(J' + 1)\left(v' + \frac{1}{2}\right)\alpha_{e'}\right] \\ &\quad - \left[\left(v'' + \frac{1}{2}\right)\omega_{e''} - \left(v'' + \frac{1}{2}\right)^2 \omega_e x_{e''} + J''(J'' + 1)B_{e''} - J''(J'' + 1)\left(v'' + \frac{1}{2}\right)\alpha_{e''}\right]. \end{aligned}$$

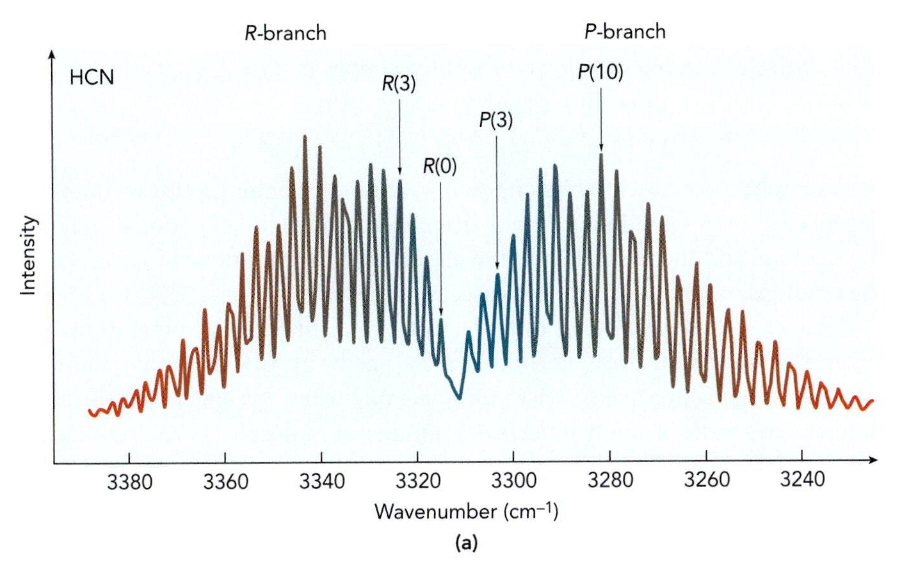

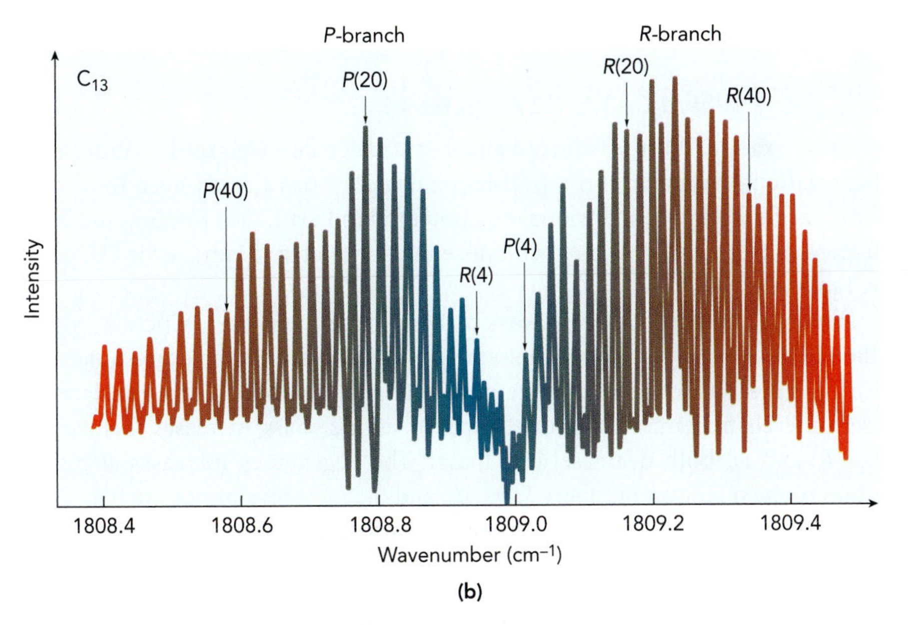

◀ **FIGURE 9.7 Rovibrational spectra of two linear molecules.** Both examples show the P and R branch transitions, centered around the band origin ν_0. **(a)** The HCN spectrum shows rotational components of the $v_3 = 0 \rightarrow 1$ transition of the CH stretch, with adjacent lines separated by roughly $2B$ ($B = 1.48\ \text{cm}^{-1}$). The spectrum is obtained with a relatively low-resolution FTIR. [*J. Chem. Ed.* **69** A296 (1992)] **(b)** The rovibrational spectrum of the linear C_{13} carbon cluster, obtained with a laser spectrometer capable of resolving the much smaller spacings ($<0.02\ \text{cm}^{-1}$) between transitions. [*Sci.* **265** 756 (1994)]

Equations 9.27 and 9.28 hold for a linear polyatomic molecule as well as a diatomic. There is an added complication, however, in that a linear polyatomic molecule has at least one degenerate bending mode. The bending mode makes the rovibrational selection rules more flexible, because the angular momentum of the photon may be absorbed into *either* the vibration or the rotation. A transition that changes the vibrational angular momentum is called a **perpendicular transition** (because a change in the bending state requires an electric field vector of the radiation perpendicular to the axis of the molecule), and the rotational selection rule extends to $\Delta J = 0, \pm 1$. The $\Delta J = 0$ transitions comprise the **Q branch** and are usually closely packed near the band origin. In the $v_2 = 0 \rightarrow 1$ fundamental transition of the bending mode in CO_2, for example, there appears a Q branch. In a **parallel transition,** the vibrational angular momentum does not change, and the selection rule remains $\Delta J = \pm 1$.

Extending these principles to nonlinear molecules, and to rovibronic transitions, the selection rules for J can be written simply as

$$\Delta J = \pm 1 \text{ when } \Delta l = \Delta\Lambda = 0 \qquad (9.30)$$
$$\Delta J = 0, \pm 1 \text{ when } \Delta l \neq 0 \text{ or } \Delta\Lambda \neq 0.$$

For example, molecules that belong to low-symmetry point groups without any degenerate representations cannot have electronic or vibrational angular momentum, and therefore the photon angular momentum must be absorbed by the rotations, so $\Delta J = \pm 1$ in that case.

Relative intensities of the different rotational lines in a rovibrational or rovibronic spectrum are determined only partly by the transition moment integral, which is most sensitive to the value of J when J is small. The variation in intensities more strongly reflects the number of molecules in each rotational level, which depends on the degeneracy of the level, the symmetry of the molecule, and the *temperature*. The temperature-dependence of the relative intensities is no longer a function of the properties of individual molecules, however, and so it is discussed instead in *Thermodynamics, Kinetics, and Statistical Mechanics*, Section 6.3.

CONTEXT *Where Do We Go From Here?*

We have examined three forms of energy in molecules: electronic, vibrational, and rotational. A fourth form, translational energy, has already been treated by the case of the three-dimensional box potential in Part I. Our findings for these degrees of freedom allow us to compose the parameter summary for HF given in Table 9.3.

Table 9.3 compares different ways of storing energy in one particular system (the HF molecule), but we see the same general trends that the correspondence principle predicts for all such systems. Vibrational, rotational, and translational energy levels become more closely spaced as the mass increases (remember that B_e and ω_e both decrease with mass). The degeneracy increases at higher values of the quantum numbers J, n_x, n_y, and n_z. For polyatomics, such as CO_2, we have seen that the vibrational degeneracy also climbs with v.

TABLE 9.3 Summary of for the four principal molecular degrees of freedom, applied to the case of the HF molecule. Energy expressions are approximate only.

degree of freedom	E	N_{dof}	U	g
electronic	$0.38\,E_h$	$3 \times 10 = 30$	$\sum_k^{10}[\sum_i^2(-Z_i e^2/r_{ik}) + \sum_{j\neq k}^{10}(e^2/r_{jk})]$	$(2S+1)g'_{elec}$
vibrational	$\omega_e(v_i + 1/2)$	1	$k(R - R_e)^2/2$	1
rotational	$B_e J(J+1)$	2	0	$2J + 1$
translational	$n^2\pi^2\hbar^2/(2ma^2)$	3	0	$\dfrac{4\pi V(2m^3E)^{1/2}dE}{h^3}$

This completes our look at the detailed structure of individual atoms and molecules and also concludes our introduction to quantum mechanics. The results from these chapters will determine the molecular properties and behavior as we allow our molecules to bump into each other and eventually to grow into the massive chemical networks that we know as liquids and solids.

KEY CONCEPTS AND EQUATIONS

9.1 Rotations in diatomics.

a. The rotational energy is given by

$$E_{\text{rot}} \approx \frac{\hbar^2}{2\mu R_e^2} J(J+1) \equiv B_e J(J+1), \quad (9.4)$$

where we have defined the rotational constant

$$B_e(J) = \frac{\hbar^2}{2\mu R_e^2}. \quad (9.3)$$

To calculate the rotational constant for a diatomic, the following formula is convenient:

$$B_e(\text{cm}^{-1}) = \frac{\hbar}{4\pi c\mu R_e^2} = \frac{16.858}{\mu(\text{amu})R_e(\text{Å})^2}. \quad (9.5)$$

b. Another effect is that as the molecule rotates more and more, the chemical bond lengthens due to the apparent centrifugal force, which means that the average R increases. This rotational distortion is also treated by a power series, this time in J:

$$E_{\text{rot}} = B_v J(J+1) - D_v[J(J+1)]^2 + H_v[J(J+1)]^3 + \dots \quad (9.9)$$

c. The rotational degeneracy is given by

$$g_{\text{rot}} = 2J + 1. \quad (9.11)$$

9.2 Rotations in polyatomics.

a. We define three rotational constants for nonlinear polyatomic molecules:

$$A \equiv \frac{\hbar^2}{2I_a} \quad B \equiv \frac{\hbar^2}{2I_b} \quad C \equiv \frac{\hbar^2}{2I_c}. \quad (9.16)$$

b. The rotational constants can be used to predict the rotational energies, but this is easy only for symmetric tops:

$$\text{prolate:} \quad E_{\text{rot}} = K_a^2(A - B) + BJ(J+1) \quad (9.21)$$

$$\text{oblate:} \quad E_{\text{rot}} = K_c^2(C - B) + BJ(J+1). \quad (9.22)$$

9.3 Spectroscopy of rotational states. If we obey the harmonic selection rule $\Delta v = \pm 1$, so $v' = v'' + 1$, we obtain the following transition energies for rovibrational spectra:

$$R(J''): \quad h\nu = \omega_e - 2v'\omega_e x_e + 2(J''+1)B_e - (J''+1)(J''+2v'+1)\alpha_e \quad (9.27)$$

$$P(J''): \quad h\nu = \omega_e - 2v'\omega_e x_e - 2J''B_e - J''(J''-2v')\alpha_e. \quad (9.28)$$

KEY TERMS

- **The moment of inertia** I is measure of the mass distribution in a structure, which happens to be convenient for calculating rotational properties.
- The **principal inertial axes** a, b, and c are the rotational axes that we choose for defining the rotational motion of a molecule.
- The **rotational constant** of a molecule is inversely proportional to one of the molecule's principal moments of inertia and—neglecting several small corrections—proportional to the rotational energies of the molecule.
- **Rotational distortion** is the effect that high rotational excitation tends to lengthen chemical bonds and decrease the rotational constant. Its effect on the energy levels is proportional to the **distortion constant D_V.**
- An **asymmetric top** has no two moments of inertia equal to one another.

- A **symmetric top** is any molecule that has two of its moments of inertia equal.
- A **spherical top** has all three moments of inertia equal.
- A **rovibrational spectrum** is a set of transitions in which both the rotational and vibrational quantum state may change.
- In a **rovibronic spectrum,** the electronic state, as well as vibrational and rotational states, is changing.

OBJECTIVES REVIEW

1. *Calculate rotational constants for molecules based on the geometry.*
   Calculate the rotational constant (in cm^{-1}) for $^{13}C\,^{18}O$, if the equilibrium bond length is 1.128 Å.
2. *Use those constants to predict or estimate the rotational energies for linear molecules and symmetric top molecules.*
   Using Table 9.2, estimate the energy of the $J, K_c = 2, 1$ state of NH_3.
3. *Predict trends in the rotational constant based on atomic masses and molecular geometry.*
   Rank the following in order of increasing rotational constant: $^{12}C\,^{16}O$, $^{12}C\,^{16}O_2$, $^{12}C\,^{18}O_2$.
4. *Predict spectroscopic transition energies involving rotation, including vibrational effects.*
   Calculate the $J = 2 \rightarrow 3$ transition energy for $^{1}H\,^{19}F$ in the $v = 1$ state.

PROBLEMS

Discussion Problems

9.1 Write the name of the molecule from the list on the right next to its rotational constant (in cm^{-1}) on the left:

0.0470	$^{85}Rb\,^{81}Br$
0.0324	$^{87}Rb\,^{79}Br$
3.020	$^{87}Rb\,^{19}F$
0.2098	$^{87}Rb\,^{1}H$
0.0469	$^{87}Rb\,^{127}I$

9.2 On which of the following do the rovibrational transition frequencies of a molecule depend?

a. the atomic isotopes
b. the electronic state
c. the density of the sample

9.3 Determine whether or not each of the following transitions is allowed by electric dipole selection rules:

a. $^{1}D_2 \rightarrow {}^{1}S_0$ in the ground $1s^2 2s^2 2p^2$ electron configuration of atomic carbon
b. $^{1}\Sigma_g \rightarrow {}^{3}\Pi_u$ in N_2
c. $v = 0 \rightarrow 1$ of the symmetric C—H stretch in acetylene
d. $J = 0 \rightarrow 2$ in CO
e. $J = 1 \rightarrow 2$ in N_2

9.4 Rovibrational perpendicular transitions change the vibrational angular momentum by one unit and have the selection rule $\Delta J = 0, \pm 1$. That appears to suggest that for a perpendicular transition with $\Delta J = \pm 1$, *both* the vibrational angular momentum and J change at the same time. If the photon acts like a particle with only one unit of angular momentum, how can that be?

Rotational Constants and Moments of Inertia

9.5 Calculate the equilibrium rotational constant in cm^{-1} for $^{32}S_2$, which has an equilibrium bond length of 1.892 Å.

9.6 Find the rotational constant for $^{37}Cl_2$ in cm^{-1}.

9.7 According to Table 9.1, H_2 has the same bond length in both the ground and $B^1\Sigma_u^+$ excited states. Use the experimentally determined B_e values and the precise H atom mass (1.0079 amu) to calculate more precise values of R_e for these two states.

9.8 Find the equilibrium bond length (in Å) of $^{58}Ni\,^{1}H$ if B_e is 7.700 cm^{-1}. Precise values for the atomic masses are given in the periodic table at the end of the notes.

9.9 The rotational constants B_0 for CN and I_2 are 1.9013 and 0.037389 cm^{-1} respectively. From their rotational constants, estimate the bond lengths for these two molecules.

9.10 Find the rotational constant B_e in cm^{-1} for the linear molecule cyanogen, NCCN, if the equilibrium bond lengths are 1.147 and 1.513 Å for the CN and CC bonds, respectively.

9.11 Estimate ω_e and B_e for FCl from the values in Table 9.1.

9.12 Each of the following molecules is drawn with accompanying coordinate axes that are parallel to the principal axes of the molecule. Label the a, b, and c axes in each case.

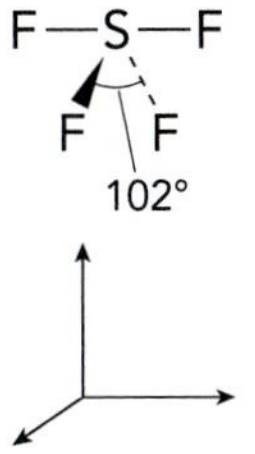

F—S—F
F F
102°

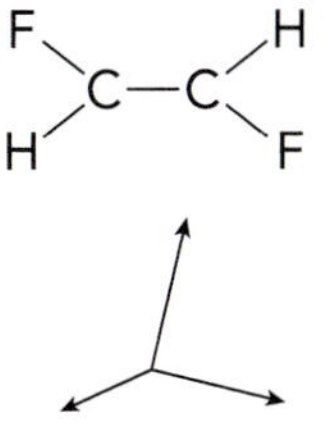

F H
C—C
H F

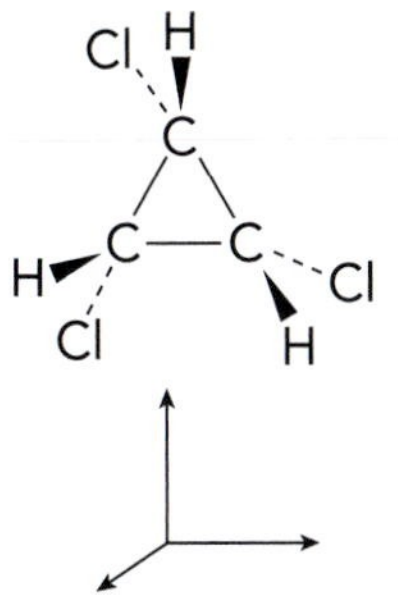

Cl H
C
H C—C Cl
Cl H

9.13 What homonuclear diatomic molecule has a bond length of 2.56 Å and an equilibrium rotational constant (for its most abundant isotope) of 0.0396 cm^{-1}?

9.14 Calculate the moments of inertia in amu Å^2 and the rotational constants in cm^{-1} for the D_{3h} molecule BH_3. The B—H bond length is 1.210 Å.

9.15 Calculate the A, B, and C rotational constants for $^{12}C\,^{35}Cl_4$ (point group T_d). The CCl bond lengths are all $R = 1.766$ Å, and the ClCCl bond angles are all $\theta = 109.47°$.

9.16 Lay the ethene molecule in the xy plane with the C—C bond along the x axis. First determine by inspection which Cartesian axes are the a and c inertial axes. Then use bond lengths estimated from Table 9.1 and 120° bond angles to predict the rotational constants in cm^{-1} of the molecule. Assume the most abundant isotopes.

9.17 The asymmetry parameter κ is used as a measure of how close an asymmetric top molecule is to one of the symmetric top limits. It is defined by the equation

$$\kappa = \frac{2B - A - C}{A - C}.$$

Calculate κ for the following molecules, and state whether they are prolate, near-prolate, oblate, near-oblate, spherical, or very asymmetric:

molecule	A(MHz)	B(MHz)	C(MHz)
CH_2Cl_2 (methylene chloride)	32001.8	3320.4	3065.2
$COCl_2$(phosgene)	7918.75	3474.99	2412.25
CH_3F (fluoromethane)	154000	25536.12	25536.12
C_4H_4O (furan)	9447.04	9246.76	4670.84
H_2O	833200	434700	298500

9.18 The HC_3O radical, believed to be formed during acetylene combustion, was originally predicted to have the structure shown in the following figure, which predicts rotational constants of 900.0 GHz, 4.42 GHz, and 4.40 GHz. However, subsequent experiments determined rotational constants of 262 GHz, 4.58 GHz, and 4.49 GHz. Draw a different Lewis-type structure for the molecule that could explain the discrepancy.

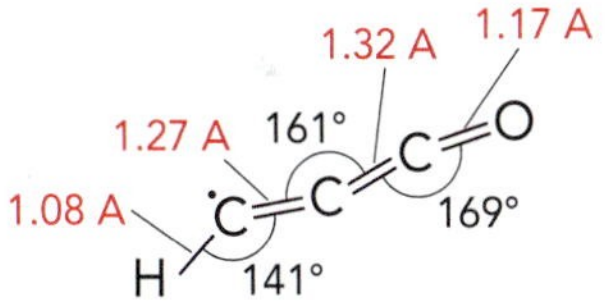

1.32 A 1.17 A
1.27 A 161°
1.08 A
C=C=C=O
169°
H 141°

9.19 The ammonia molecule NH_3 has C_{3v} symmetry, and therefore it is a symmetric top, with each NH bond at an angle θ measured from the symmetry axis (choose the direction so that $\theta \leq \pi/2$). The molecule is a prolate top for values of θ near 0 and an oblate top for values of θ near $\pi/2$. Given the actual NH length of 1.012 Å, at what value of θ does NH_3 become an "accidental" spherical top, with all three moments of inertia equal? (This one is much easier with a math program.)

Rotational, Rovibrational, Rovibronic Spectra

9.20 Derive an equation for the energy of an electric dipole-allowed pure rotational transition in the absorption spectrum of an oblate symmetric top molecule.

9.21 The rotational distortion constants D_0 for CN and I_2 are $2.6 \cdot 10^{-7}$ and $-4.54 \cdot 10^{-9}$ cm^{-1}, respectively. Predict the $J = 0 \rightarrow 1$ and $J = 10 \rightarrow 11$ pure rotational transition frequencies in MHz for these molecules.

9.22 Find the transition energy for the pure rotational transition $J, K_a = 2,0 \rightarrow 3,2$ in CH_3F. The rotational constants are $A = 154$ GHz and $B = 25.536$ GHz ($1 \text{ GHz} = 10^9$ Hz).

9.23 The lowest rotational transitions observed for a new diatomic molecule in its ground vibrational state are at 2481.384 MHz and 4950.312 MHz. Find the distortion constant D in MHz for this molecule.

9.24 In the pure rotational spectrum of PbF in its ground vibrational state, two transitions are measured at 14.4679 GHz and 28.9357 GHz. In a spectrum of the $v = 0 \rightarrow 1$ vibrational transition, the R(0) and R(1) are measured at 502.589 and 503.065 cm^{-1}. Find α_e.

9.25 Measuring the pure rotational spectrum of an unknown molecule between 9000 and 15000 MHz, we detect absorptions *only* at frequencies 12835.20 MHz and 9640.68 MHz. What are the rotational quantum numbers for these transitions, and what are B and D in MHz?

9.26 Predict the transition energies of the fundamental $P(1)$ and $R(0)$ transitions for $H^{35}Cl$ and IBr.

9.27 Find the transition energy in cm^{-1} for the $v = 0 \rightarrow 1$ $R(1)$ transition of HI.

9.28 Determine the wavelength in micrometers of light absorbed during the $R(4)$ fundamental transition of $^1H^{35}Cl$. Obtain the most precise value possible using the constants provided in this chapter.

9.29 Equations 9.27 and 9.28 give the transition frequencies for R and P branch transitions in $v = 0 \rightarrow 1$ rovibrational spectra in terms of the molecular constants ω_e, $\omega_e x_e$, B_e, and α_e. Similar expressions can be written for other kinds of transitions, including some that do not satisfy the electric dipole selection rules. Find the vibrational quantum numbers and ΔJ value for the transitions with frequencies given by the following equation:

$$h\nu = 2\omega_e - 10\omega_e x_e - (4J'' - 2)B_e - (2J''^2 - 12J'' + 7)\alpha_e.$$

9.30 For $^1H^{79}Br$, the rotational constant B_e is 8.3511 cm^{-1} and the vibration-rotation constant α_e is 0.226 cm^{-1}. The $J = 6 \rightarrow 7$ rotational transitions are observed in two different vibrational states of this molecule at 105.33 cm^{-1} and 102.17 cm^{-1} Find the quantum numbers for the two vibrational states.

9.31 Derive an equation in terms of B_v and D_v for the spacing between transitions in the R and P branches of a rovibrational spectrum.

9.32 In rovibrational spectra taken at high temperature, there usually occurs a point at which either the P branch or the R branch appears to "turn around"; for example, in the R branch there may occur a transition $R(J^\bullet)$ such that the R branch line frequencies increase with J for $J < J^\bullet$ but decrease with J for $J > J^\bullet$. This occurs primarily because the values of the upper and lower state rotational constants B' and B'' are different. The frequency of the $R(J^\bullet)$ transition in this example is called the **band head.** Estimate the value of $J^\bullet$ at the band head when it occurs in (a) the R and (b) the P branch, using only the simplest equation for the rotational energy:

$$E_{\text{rot}} = BJ(J + 1),$$

where B in the upper state is B' and in the lower state is B''.

9.33 Obtain an equation for the frequencies of the Q-branch transitions of the hot band transition $v = 1 \rightarrow 2$ in terms of B_e, α_e, D_1, and D_2.

9.34 Derive an expression in terms of the rotational constants B_0 and B_1 and the centrifugal distortion constants D_0 and D_1 for the spacing between two Q branch lines, $Q(J + 1) - Q(J)$, where B_1 and D_1 are the upper state constants and B_0 and D_0 the lower state constants.

9.35 In Raman transitions, the rotational selection rules differ from those for electric dipole transitions. Derive an equation for the rovibrational transition energy in terms of ω_e, $\omega_e x_e$, B_e, and α_e for the S branch ($\Delta J = +2$) of a $v = 0 \rightarrow 2$ Raman spectrum.

9.36 The $v = 0 \rightarrow 1$ rovibrational spectrum of potassium iodide (KI) is drawn in the following figure.

a. Label each transition by the appropriate $R(J)$ or $P(J)$ symbol.
b. Estimate ω_e and B in cm^{-1}. Neglect centrifugal distortion and vibration-rotation coupling.
c. Find the molecule's moment of inertia in amu Å^2.

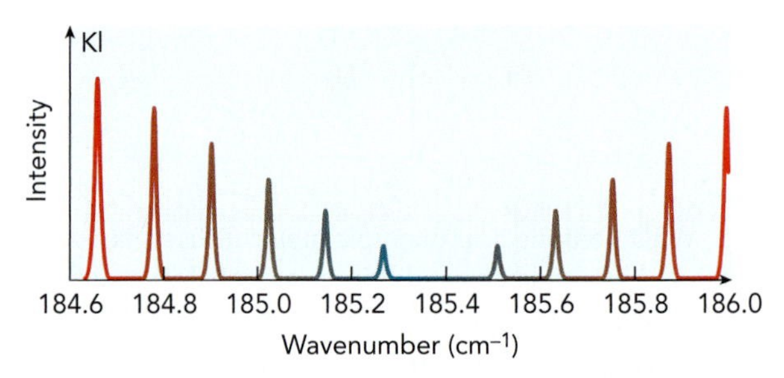

9.37 For the ground $^1\Sigma^+$ state of the molecule PN, $B_e = 0.786485\,\text{cm}^{-1}$ and $\alpha_e = 5.5364 \cdot 10^{-3}\,\text{cm}^{-1}$. Measuring the vibrational spectrum for the first time, we find that the $R(0)$ line for the $v = 0 \rightarrow 1$ transition occurs at 1324.830 cm^{-1} and the $R(0)$ for $v = 1 \rightarrow 2$ is at 1310.853 cm^{-1}. Rotational distortion is negligible for these transitions. Find ω_e and $\omega_e x_e$.

9.38 The transition frequencies for the $v = 0 \rightarrow 1$ $R(0)$ and $R(1)$ of SiO are 1231.04 cm^{-1} and 1232.47 cm^{-1}, respectively. Estimate the force constant in N m^{-1} and bond length in Å from these, using atomic masses of 28.00 amu for silicon and 16.00 amu for oxygen. Neglect centrifugal distortion and vibration-rotation coupling.

9.39 The selection rule for rovibrational transitions in linear molecules is $\Delta J = \pm 1$ if $\Delta\Lambda = 0$ (for example, $\Sigma \rightarrow \Sigma$ or $\Pi \rightarrow \Pi$). The spin selection rule is $\Delta S = 0$. Explain briefly what the Raman transition selection rules should be for ΔS and ΔJ.

9.40 Using Tables 8.2 and 9.1, calculate the total energy in rotation and vibration of the $v = 0$, $J = 10$ state of $^{12}C\,^{16}O$, including any corrections for which the data is available.

9.41 Calculate the bond length of $^{28}Si\,^{1}H$ if the $v = 0 \rightarrow 1$ $R(0)$ transition lies at 1985.34 cm^{-1} and the $v = 0 \rightarrow 1$ $R(1)$ transition lies at 1999.90 cm^{-1}. Neglect the vibration-rotation coupling constant.

9.42 A student compares two rovibrational spectra: one for the $v = 0 \rightarrow 1$ transition of $^{12}C\,^{16}O$ and one for the $v' - 1 \rightarrow v'$ transition of $^{14}N\,^{16}O^+$. Surprisingly, the spacing between the $R(0)$ and $R(1)$ lines in both spectra are equal to within 0.01 cm^{-1}. Use this fact and the following data to find the upper state v' of the $^{14}N\,^{16}O^+$ transition.

molecule	B_e	α_e
$^{14}N\,^{16}O^+$	1.9982 cm^{-1}	0.0190 cm^{-1}
$^{12}C\,^{16}O$	1.9313 cm^{-1}	0.0175 cm^{-1}

The Rotational Hamiltonian

9.43 For $^{12}C\,^{32}S$, $B = 24.58435$ GHz and $D = 0.040$ MHz. Calculate the energies in MHz of the $J = 2$ and $J = 20$ rotational levels.

9.44 For HF, $\omega_e = 4138.52$ cm^{-1}, $\omega_e x_e = 90.069$ cm^{-1}, $B_e = 20.939$ cm^{-1}, and $\alpha_e = 0.770$ cm^{-1}. Calculate the energies in cm^{-1} of the $v = 2$, $J = 0$, and $J = 1$ states.

9.45 The $^{14}N\,^{32}S$ radical has the constants $\omega_e = 1219.14$ cm^{-1}, $\omega_e x_e = 7.28$ cm^{-1}, $B_e = 0.7730$ cm^{-1}, and $\alpha_e = 0.0063$ cm^{-1}.

- Give the energy of the $v = 2$, $J = 3$ state in cm^{-1}.
- Estimate the bond length in Å.

9.46 The equation for the energy of a particle in a one-dimensional box is

$$E_n = \frac{n^2\pi^2\hbar^2}{2ma^2}.$$

What is the length a of a box for which the $n = 1 \rightarrow 2$ energy gap is equal to the $J = 1 \rightarrow 2$ rotational energy gap for $^{14}N_2$ ($B_e = 1.9987\,\text{cm}^{-1}$)? Neglect corrections such as centrifugal distortion.

9.47 There are some remarkable similarities between the quantum mechanics of the particle in a one-dimensional box and the rotating molecule; for example, the energies increase as n^2 or approximately as J^2. The wavefunctions for the particle in a one-dimensional box when the box extends along the x axis from $-\frac{a}{2}$ to $+\frac{a}{2}$ are $\phi_n = \sqrt{\frac{2}{a}}\cos\left(\frac{n\pi x}{a}\right)$.

a. For what value of a and what relation between x and Θ is the $n = 1$ particle in a box wavefunction identical to the $J = 1$, $M_J = 0$ rotational wavefunction?

b. Find the mathematical expression that relates the energy levels of the one-dimensional box and the three-dimensional rotator. As a hint, not all of the one-dimensional box states correspond to rotational wavefunctions, because there is a different rule for the allowed number of nodes as a molecule rotates through an angle of 2π.

9.48 The potential energy surface for twisting of the methyl group in ethane around the CC single bond is sketched in the following figure. Although this motion is called an "internal rotation," it is technically a vibrational mode, and the $v = 0$ energy level is indicated on the sketch by a dashed line. Add lines for $v = 1$ and 2, indicating the relative spacing between the levels.

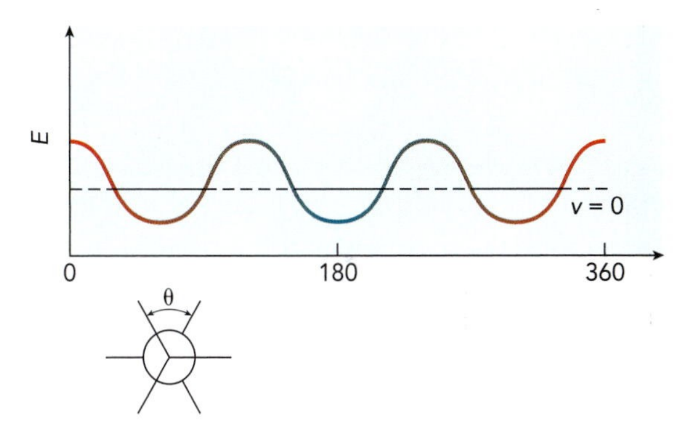

9.49 Consider a linear triatomic molecule in a $(\sigma^1\pi^1)\,^1\Pi$ excited electronic state. This state has one unit of electronic orbital angular momentum, with $\Lambda = \pm 1$. Unlike a closed-shell linear molecule, the electron density is not evenly distributed around the internuclear axis. Therefore, the potential energy curve for bending the molecule is different depending on whether the molecule bends in the plane of the π MO or perpendicular to that plane; this is the **Renner-Teller effect.** In the lowest excited bending state, the molecule also has one unit of vibrational angular momentum, with quantum number for projection onto the z axis of $l = +1$. The electronic and vibrational angular momenta can constructively add, giving a resultant projection $K \equiv \Lambda + l = \pm 2$, or they may cancel, giving $K = 0$. If the angle ϕ gives the angle of the electron measured from the x axis about the z axis, and the angle ξ defines the angle of the bending plane measured from the x axis about the z axis, then the rovibronic wavefunction has the general form

$$\psi_{\Lambda,l,K,N,M_N,p} = \psi_0 \, Y_N^{M_N}\left[e^{i(\Lambda\phi + l\xi)} + p(-1)^N e^{-i(\Lambda\phi + l\xi)} \right].$$

In this equation, Y is the spherical harmonic rotational wavefunction for rotational quantum number N and projection M_N, and p is the parity under inversion, equal to either $+1$ or -1 (see Problem 6.32). The rest of the vibrational and electronic terms are absorbed into ψ_0. Show that the $K = 0$ wavefunctions divide into two groups of states, one group corresponding to vibration in the plane of the π MO and one perpendicular to it.

9.50 Draw a rotational energy level diagram for NH_3, using the rotational constants $B = 298$ GHz and $C = 189$ GHz. Include all levels up to $J = 3$, $K_c = 2$.

PART I

ATOMIC STRUCTURE

PART II

MOLECULAR STRUCTURE

PART III

MOLECULAR INTERACTIONS

10 Intermolecular Forces

LEARNING OBJECTIVES

After reading this chapter, you will be able to do the following:

❶ Determine the strongest forces relevant to the interactions between a given pair of molecules, and determine the functional form of the resulting potential energy.

❷ Derive the functional form of various intermolecular interaction potential energies.

❸ Estimate the shape and magnitude of the potential energy functions for these interactions, given the appropriate molecular constants.

❹ Estimate the relative strength of forces between a given pair of molecules, based only on the molecular structures.

❺ Graph approximate potential energy curves based on any of several model potentials.

❻ Predict the relative importance of different modes of energy exchange between molecules.

GOAL *Why Are We Here?*

The principles presented in this chapter justify the distinct properties of liquids, solids, and non-ideal gases. The goal of the chapter is to familiarize ourselves with the forces that one molecule exerts on another, so that we will be able to predict the behavior of small groups of non-reacting molecules and later extend those predictions to *very* large groups.

CONTEXT *Where Are We Now?*

In Part II of this text, we examined the potential energy surfaces of chemical bonds, surfaces with wells running as deep as 30,000 cm^{-1} or 400 kJ mol^{-1}. These interactions are strong, and many weaker contributions to the potential energy are insignificant by comparison. Once we look *outside* the molecule, however, away from the chemical bonds, we find that those weaker interactions may become the dominant terms that control what happens when two or more of these molecules come into contact.

In this chapter, we use what we know about individual molecules to predict how molecules interact with one another. For now, we will assume that the chemical bonding stays intact—in other words, that the molecules do not react with one another. But we now expand the application of our chemical model from the structure of individual molecules to the structure of small groups of molecules. When we looked at the structure of the H atom and later the H_2^+ molecule, our first question in each case was "What forces shape the potential energy function for this system?" We pose that question here as well, this time examining not the forces *within* molecules, but the forces *between* molecules.

SUPPORTING TEXT *How Did We Get Here?*

The main *qualitative* preparation we need for the work ahead is a general chemistry-level understanding of the structure of individual molecules: each molecule is an assembly of small, dense, positively charged nuclei trapping a cloud of negatively charged electrons. For a more detailed background, we will draw on the following equations and sections of text to support the ideas developed in this chapter:

- The force that holds each molecule together is the Coulomb force (Eq. A.38),

$$\vec{F}_{\text{Coulomb}} = q_1\mathcal{E}.$$

  The same force will also turn out to control all the molecular interactions in this chapter.
- We can estimate the repulsion between molecules using the general exponential decay seen in the atomic orbitals (Table 3.2) and a rough application of Hartree theory (Eq. 4.26):

$$\hat{H}(i)\,\varphi(i) = \left[-\frac{\hbar^2}{2m_e}\nabla(i)^2 - \frac{Ze^2}{4\pi\varepsilon_0 r_i} + \sum_{j\neq i}\left(\int |\varphi(j)|^2 \frac{e^2}{4\pi\varepsilon_0 r_{ij}} d\tau_j\right)\right]\varphi(i) = \varepsilon_i\varphi(i).$$

- The polarizability α, discussed in Section 3.3, is defined by Eq. 3.36,

$$\mu_{\text{induced}} = \alpha\mathcal{E},$$

  and Problem 4.66 presents a way to estimate its value. This parameter measures how the electrons in one molecule will respond to the electric field of another molecule, so it plays a central role in several kinds of intermolecular interactions.
- Section 4.2 introduces perturbation theory, the basis of the one long derivation in this chapter.
- Two approximations that conveniently simplify some math for us in this chapter are the Taylor series (Eq. A.29),

$$f(x) \approx f(x_0) + (x - x_0)\left(\frac{df(x)}{dx}\right)\Bigg|_{x_0},$$

  and one particular application of the Taylor series, Eq. A.31:

$$\frac{1}{1+x} \approx 1 - x \quad \text{for } |x| < 1.$$

- For our treatment of dispersion, Section 5.2 describes two key ways to write wavefunctions for the molecules: MO theory, which for H_2 gives the wavefunction (Eq. 5.20)

$$\psi_{\text{MO}}(\text{ground}) = C[1s_A(1)1s_A(2) + 1s_A(1)1s_B(2) + 1s_B(1)1s_A(2) + 1s_B(1)1s_B(2)],$$

  and valence bond theory, which gives the wavefunction (Eq. 5.24)

$$\psi_{\text{VB}}(\text{ground}) = C[1s_A(1)1s_B(2) + 1s_B(1)1s_A(2)].$$

10.1 Intermolecular Potential Energy

The potential energy for a system combines all the force vectors into a single function. We can find the force along some direction by taking the derivative of the potential energy along that coordinate.

The forces between non-reacting molecules are known collectively as **van der Waals forces.** Any pair of atoms or molecules—or even different parts of the same molecule—will interact through van der Waals forces. Like the forces at work in chemical bonds, the van der Waals forces arise from either the Coulomb force or exchange force (the quantum-mechanical effect that separates electrons with the same value of m_s, Section 4.3). Also like the forces in chemical bonds, the result may be an attraction or a repulsion.

Intermolecular Repulsion

Paradoxically, we spend the least effort on the most important of the intermolecular forces: the repulsive force between the electron clouds on each molecule. The repulsion corresponds to a positive potential energy term that climbs steeply as the molecules get very close.

In Section 4.3, we show how the electron–electron repulsion in an atom can be separated into two components. We carry out the same separation here, expressing the repulsion between two molecules as the sum of

1. **Coulomb repulsion,** which results from the classical electron–electron and nucleus–nucleus interactions between the two molecules, as given by Coulomb's law (Eq. A.42)

$$U_{\text{Coulomb}} = \frac{q_1 q_2}{4\pi\varepsilon_0 r_{12}}, \text{ and}$$

2. **exchange repulsion,** which results from the Pauli exclusion principle forbidding two electrons of the same spin from occupying the same space simultaneously. The exchange repulsion has the same form as the exchange integral we derived for helium in Eqs. 4.48 and 4.49.[1]

To approximate both contributions to the intermolecular repulsion, we need the electron densities around each molecule. To estimate those, we borrow a result from the general form of the atomic orbitals found in Eq. 3.33: a few angstroms from the nucleus, *all* the atomic orbital wavefunctions ψ are dominated by an exponential decay:

$$\lim_{r\to\infty} \psi(r) \approx A_{nlm} e^{-Zr/(na_0)}, \tag{10.1}$$

[1]We are going to neglect two other components that appear when we no longer require the electron densities of the interacting molecules to stay the same. When the molecules affect one another, additional contributions to the repulsion come from *polarization* (the molecules may be neutral, but because they are made up of distinct charged particles they do carry an electric field) and from *charge transfer* (a fraction of the electron density on one molecule may become associated with the density on the other, as if to form a partial ionic bond). In groups of water molecules, the effects of polarization and charge transfer are each calculated to be about 25% of the contribution from exchange repulsion.

where A_{nlm} is the normalization constant, Z is the atomic number, n is the principal quantum number, and a_0 is the Bohr radius. If we also make the approximation that the effective size of the atom is determined primarily by the outermost, most easily ionized electron, we can write this exponential decay in terms of the effective atomic number Z_{eff} and the value of n for the highest energy electron, n_{max}.

$$\lim_{r \to \infty} \psi(r) \approx A_{nlm} e^{-Z_{\text{eff}}\, r/(n_{\text{max}} a_0)}, \tag{10.2}$$

where Z_{eff} is determined by Eq. 4.29,

$$Z_{\text{eff}} = \left(\frac{2(\text{IE}_1) n_{\text{max}}^2}{E_{\text{h}}} \right)^{1/2},$$

where we've replaced the orbital energy by the first ionization energy IE_1.

EXAMPLE 10.1 Effective Atomic Number from Ionization Energy

CONTEXT Our most precise method for measuring time uses the constant energy spacing of electronic quantum states in atoms to generate photons of very well-defined frequency ν. From ν, a fundamental unit of time, $\tau = 1/\nu$ can be defined with the same precision. The limitation of this precision has been fundamental uncertainties in ν, including fluctuations caused by the oscillating electric fields of thermal radiation.[2] One way to reduce the effect of these electric fields is to reduce the polarizability of the clock atom. The polarizability decreases as the electrons become more tightly bound to the nucleus, so we want atoms with *high effective atomic numbers* Z_{eff}. One way to achieve higher Z_{eff} is to use atomic cations, rather than neutral atoms.

PROBLEM Estimate the effective atomic number for the Al^+ ion, given that its next ionization energy (IE_2, the second ionization energy of Al) is 18.828 eV. For comparison, carry out the same calculation for cesium atom ($\text{IE}_1 = 3.894$ eV).

SOLUTION We get the n_{max} values from the period that each element occupies in the periodic table: $n_{\text{max}} = 3$ for Al^+ and $n_{\text{max}} = 6$ for Cs. Converting the IE_1s from eV to E_{h} to be consistent with our version of Eq. 4.29 above, we can then solve for Z_{eff}:

$$\text{IE}_2(\text{Al}^+) = (18.828\text{ eV})(0.03675\, E_{\text{h}}/\text{eV}) = 0.6919\, E_{\text{h}}$$

$$\text{IE}_1(\text{Cs}) = (3.894\text{ eV})(0.03675\, E_{\text{h}}/\text{eV}) = 0.1431\, E_{\text{h}}$$

$$Z_{\text{eff}} = \left(\frac{2(\text{IE}) n_{\text{max}}^2}{E_{\text{h}}} \right)^{1/2}$$

$$Z_{\text{eff}}(\text{Al}^+) = \left(\frac{2(0.6919\, E_{\text{h}})3^2}{E_{\text{h}}} \right)^{1/2} = 3.53$$

$$Z_{\text{eff}}(\text{Cs}) = \left(\frac{2(0.1431 E_{\text{h}})6^2}{E_{\text{h}}} \right)^{1/2} = 3.21.$$

Although it is harder to generate atomic ions under controlled conditions than neutral atoms, this principle led to the changing in 2010 of the atomic clock standard from cesium to Al^+.

[2] The matter that surrounds the clock is always emitting a spectrum of radiation corresponding to its temperature. This radiation contains oscillating electric fields, which polarize the atom in proportion to the atom's polarizability α. That in turn causes random shifts in the transition energy.

BIOSKETCH **Helen O. Leung and Mark D. Marshall**

At Amherst College, George H. Corey 1888 Professor of Chemistry **Helen O. Leung** and her husband Class of 1959 Professor of Chemistry **Mark D. Marshall** use molecular beams to study weakly bound molecular complexes. They form the complexes by sending pulses of a mixture of argon and 1—2% of another gas through a roughly 1 mm nozzle into a vacuum chamber. The expansion cools the particles to low enough energy that weak attractive forces such as dispersion and dipole–dipole interactions are sufficient to bind the molecules together as they enter the nearly collisionless molecular beam (see *Tools of the Trade* later in this chapter). From rotational spectra obtained in the microwave region of the spectrum, the Leung–Marshall team and their coworkers have been able to determine precise geometries for the complexes, often showing the relative influence of competing binding mechanisms. As one example, their recent studies of acetylene complexed to the dihaloethenes $FClC{=}CH_2$ and $F_2C{=}CH_2$ find that the most stable geometry is a planar arrangement that allows for two weak hydrogen bonds to coexist. As shown below for acteylene-chlorofluoroethene, one H atom on the acetylene leans towards one of the ethene's halogen atoms, while one of the ethene H atoms binds to the π system of the acetylene. Precise measurements such as these test our ability to calculate the effects of these weak forces, which shape the nature of liquids, atmospheric aerosols, and many other chemical systems.

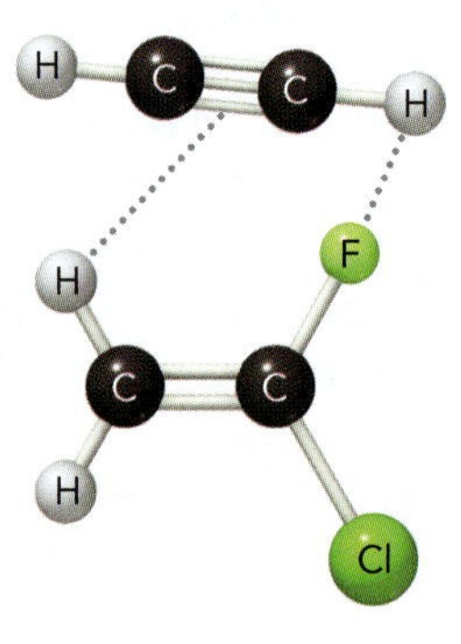

Let's use a lower case "u" to label the potential energy between two particular molecules (the **pair potential**), so that we can save "U" for the *intramolecular* potential energy and for the total potential energy of a large number of molecules. As we bring two non-bonding molecules A and B toward one another so that their centers of mass are separated by a distance R, the electronic wavefunctions begin to overlap. The higher density of electron charge in the overlap region raises the potential energy by an amount approximated by this expression, borrowed from the Hartree method (Eq. 4.26):

$$u_{\text{repulsion}} \approx \sum_{i(\text{A})} \sum_{j(\text{B})} \int |\varphi_i|^2 \left(\int |\varphi_j|^2 \frac{e^2}{4\pi\varepsilon_0 r_{ij}} d\tau_j \right) d\tau_i, \tag{10.3}$$

where the sums are over electrons i of molecule A and over electrons j of molecule B, and where the φ's are the MO wavefunctions. If the interaction is weak enough that the wavefunctions are unaltered, then we can factor $|\varphi_i|^2$ into the integral over the coordinates r_i and r_j of the electrons:

$$u_{\text{repulsion}} \approx \sum_{i(\text{A})} \sum_{j(\text{B})} \int \left(\int |\varphi_i|^2 |\varphi_j|^2 \frac{e^2}{4\pi\varepsilon_0 r_{ij}} d\tau_j \right) d\tau_i.$$

Furthermore, we simplify the wavefunctions at large R using Eq. 10.2 to rewrite them as $e^{-c_A r_i}$ and $e^{-c_B r_j}$, for molecules A and B respectively, where $c = Z_{\text{eff}}/(n_{\max} a_0)$ for each of A and B,[3] and see how the integral simplifies:

$$u_{\text{repulsion}} \approx \sum_{i(\text{A})} \sum_{j(\text{B})} \iint |e^{-c_A r_i}|^2 |e^{-c_B r_j}|^2 \frac{e^2}{4\pi\varepsilon_0 r_{ij}} d\tau_j d\tau_i \tag{10.4}$$

$$= \sum_{i(\text{A})} \sum_{j(\text{B})} \iint e^{-2c_A r_i} e^{-2c_B r_j} \frac{e^2}{4\pi\varepsilon_0 r_{ij}} d\tau_j d\tau_i \qquad (e^x)^2 = e^{2x}$$

Assume here for simplicity that c_A and c_B are equal. Let's also assume that the main overlap between the orbitals occurs close to the internuclear axis, so $r_i + r_{ij} + r_j = R$, where r_{ij} is the distance between the electrons along the internuclear axis. With R held constant for now, these simplifications yield

$$u_{\text{repulsion}} \approx \sum_{i(\text{A})} \sum_{j(\text{B})} \iint e^{-2cr_i} e^{-2cr_j} \frac{e^2}{4\pi\varepsilon_0 r_{ij}} d\tau_j d\tau_i \qquad \text{let } c_B = c_A = c$$

$$= \sum_{i(\text{A})} \sum_{j(\text{B})} \iint e^{-2c(r_i + r_j)} \frac{e^2}{4\pi\varepsilon_0 r_{ij}} d\tau_j d\tau_i \qquad \text{combine exponents}$$

$$= \sum_{i(\text{A})} \sum_{j(\text{B})} \iint e^{-2c(R - r_{ij})} \frac{e^2}{4\pi\varepsilon_0 r_{ij}} d\tau_j d\tau_i \qquad \text{let } r_i + r_{ij} + r_j = R$$

$$= e^{-2cR} \sum_{i(\text{A})} \sum_{j(\text{B})} \iint e^{2cr_{ij}} \frac{e^2}{4\pi\varepsilon_0 r_{ij}} d\tau_j d\tau_i \qquad \text{factor out constant}$$

We simplify this relation by combining the constants, including the integral, into a single value A so we can focus on the R-dependence:

$$u_{\text{repulsion}} \approx A e^{-2cR/a_0}. \tag{10.5}$$

The energy climbs exponentially as the molecules get closer.

It turns out that, except for hydrogen and helium, the value of c in the exponent ranges by less than a factor of 3 from atom to atom. That may seem surprising, but this is why, despite atomic numbers that vary over two orders of magnitude and n quantum numbers for the valence electrons from 2 to 6, the atomic radii (other than H and He) vary by less than a factor of 4. Electron shielding is a great equalizer among atoms, making them much more similar in their properties than the one-electron wavefunctions might suggest.

Because $u_{\text{repulsion}}$ dominates other contributions to the potential energy, it is often incorporated directly into the way we view molecules. When drawing molecules, it is usually easier to view the arrangement of the nuclei and the bonds by ball-and-stick models such as the one shown in Fig. 10.1a.

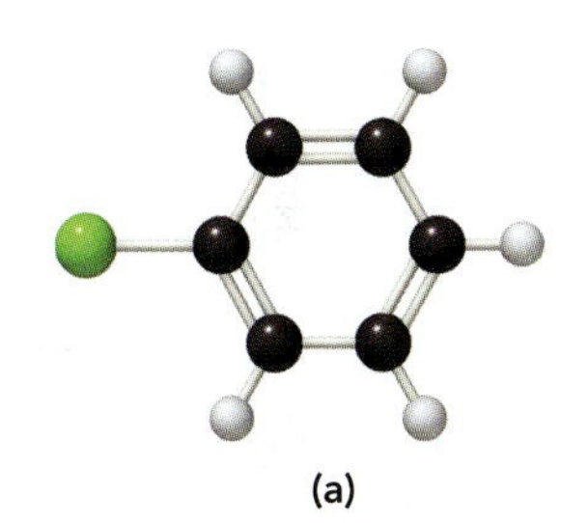

(a)

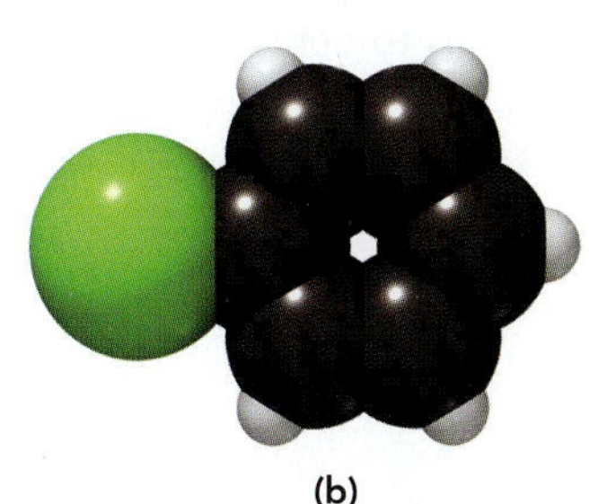

(b)

▲ FIGURE 10.1 **Molecular models of chlorobenzene.** The molecule is represented using **(a)** a ball-and-stick model, and **(b)** a space-filling model.

[3]Beware of the two definitions of e in these expressions: e in $e^2/(4\pi\varepsilon_0 r)$ is the fundamental charge (1.602×10^{-19} coulombs); the other e's are the base of the natural logarithm and are unitless.

EXAMPLE 10.2 How Soft Is the Repulsive Wall?

CONTEXT Consider the collision between two non-reacting, gas-phase atoms. As the temperature increases, the atoms can penetrate each other's electron clouds more and more. It's as though the atoms are *smaller* at high temperature than at low temperature, which affects properties such as the ability of the gas to transmit heat (its thermal conductivity) from a reaction to the cooled walls of the vessel. Smaller atoms slip past each other more often, so they can transmit heat more directly to the walls of the vessel than large atoms, which collide many times before transferring the energy out of the gas. Helium is a particularly interesting case, because it has the highest thermal conductivity of any monatomic gas.

PROBLEM To use typical values, let A in Eq. 10.5 be 1000.0 kJ mol^{-1} and $c = Z_{\text{eff}}/n_{\text{max}}$ be 1.34. The distance between the two nuclei is R. Assume that if the two atoms strike each other head on, the minimum value of R occurs at the point where the potential and kinetic energies are equal. Find the minimum values of R (in units of a_0) for this case when the kinetic energy is 2.50 kJ mol^{-1} (typical for room temperature) and when the kinetic energy is 16.6 kJ mol^{-1} (as in a high-temperature flame).

SOLUTION We set the potential energy $u_{\text{repulsion}}$ equal to the kinetic energies K given, replace $u_{\text{repulsion}}$ by the expression in Eq. 10.5, and solve for R:

$$u_{\text{repulsion}} \approx Ae^{-2cR/a_0} = K$$

$$R \approx -\frac{\ln(K/A)}{2c}a_0$$

$$R(K = 2.50\text{ kJ mol}^{-1}) = -\frac{\ln(2.50/1000.0)}{2(1.34)}a_0 = 2.24\,a_0$$

$$R(K = 16.6\text{ kJ mol}^{-1}) = -\frac{\ln(16.6/1000.0)}{2(1.34)}a_0 = 1.53\,a_0.$$

So the size of the atoms—which limits how close you can squish two of them together—is actually a function of the temperature. This result is exaggerated, however. The actual potential energy will climb faster than this exponential because the core electron density becomes increasingly important as R gets smaller. Still, the effective reduction in the size of the atom at high temperature is a measurable effect in high-temperature, high-density processes such as explosions. This effective reduction in size also leads real gases to approach the properties of an ideal gas—an imaginary system of non-interacting particles—at high temperature. As the collision energies increase, real atoms interact less because their speeds are so great that only head-on collision affect their trajectories.

However, to study the dynamics of several molecules it may be more helpful to draw the molecule using a **space-filling model,** as shown in Fig. 10.1b. In this case, the atoms are represented as spheres, each with a radius set to the distance at which the electron–electron repulsion will typically repel a non-bonding atom.[4] Although the space-filling model may conceal much of the chemical bonding framework, it more successfully conveys the effective

[4]The modeler has to decide just how that distance will be determined. In general, it depends on the particular atoms interacting and on the temperature.

size of the molecule—the volume that other molecules will not occupy because the repulsive forces will push them away. This potential energy term accounts for what chemists call the **steric effect,** the influence on chemical structure of the repulsion between non-bonded atoms. In sterics, the interaction is often between different parts of the same molecule, but the origin of the force remains the same: repulsion between the electron clouds of non-reacting atoms.

If we had to write a general equation for the electron density of a molecule as a function of position $\vec{R}$, it would look something like this:

$$u_{\text{repulsion}} = \sum_i c_i e^{-b|\vec{R} - \vec{R}_i|/a_0}, \tag{10.6}$$

where the vectors $\vec{R}_i$ give the positions of the atomic nuclei, and $|\vec{R} - \vec{R}_i|$ gives the distance of each nucleus in the molecule from the location $\vec{R}$. Equation 10.6 describes an exponential envelope that follows the shape of the molecule. To help visualize this envelope, we often graph in three dimensions an **isosurface** of the electron density around the molecule (Fig. 10.2). The surface shows where the electron density reaches some particular value. The greater the electron density, the greater the repulsion experienced by a second molecule trying to occupy that space.

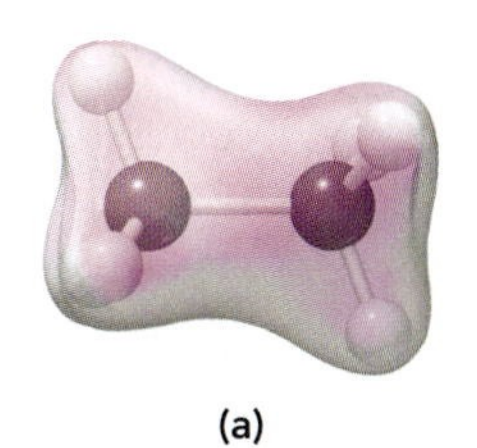

(a)

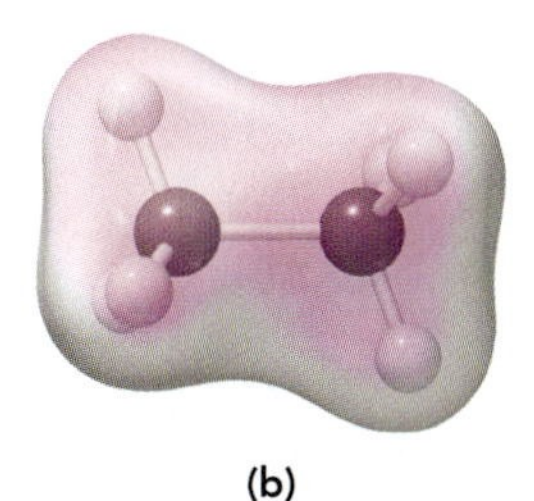

(b)

▲ FIGURE 10.2 **The electron density isosurface of ethane.** The surface is shown at two values of the density: **(a)** 0.08 and **(b)** 0.02. Because the electron density drops as we get further from the nuclei, the surface for (b) corresponds to a larger surface. The ball-and-stick model is shown in a cutaway.

Intermolecular Attraction

There are also attractive forces between molecules, which are feeble by comparison to the repulsive forces. In order to limit ourselves to intermolecular interactions for now, we assume throughout this chapter that when we bring our molecules together these attractive forces are too weak to significantly affect the existing chemical bonds. This assumption means that the features in our typical intermolecular potential energy curve are small compared to the well depth of our chemical bonds. In fact, intermolecular potentials between small, non-reacting molecules commonly have well depths of less than 1000 cm^{-1} or 12 kJ mol^{-1}, as does the argon-argon potential shown in Fig. 10.3.

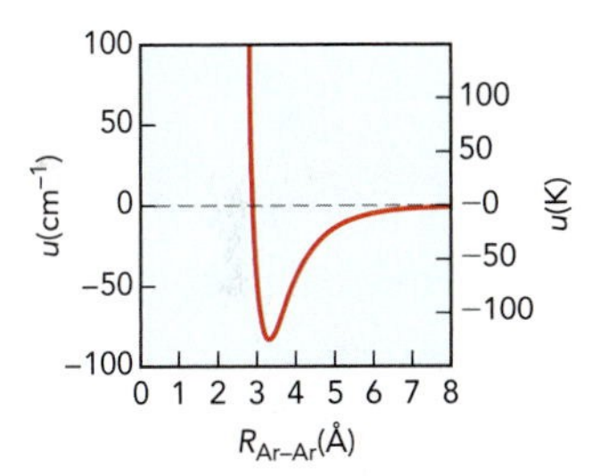

▲ FIGURE 10.3 **A typical intermolecular potential energy curve.** The potential energy in cm^{-1} and in kelvin is graphed as a function of the distance R between the centers of mass. The general features are (i) at large R, the molecules don't interact and the potential energy is zero; (ii) at intermediate values of R, there is usually a small attractive well; (iii) at very small values of R, the electron cloud of each molecule repels the other, leading to a rapidly rising Coulomb repulsion potential energy.

Any pair of atoms will experience an attraction from van der Waals forces. Compared to the forces in chemical bonds, however, the van der Waals attractions are too weak to have a noticeable effect except at large distances, where the repulsion potential drops below about 10 kJ mol^{-1}.

If we consider only the intermolecular forces, we can always find some distance (and perhaps some relative orientation) at which the attractions will dominate, holding the molecules together.[5] Two independent molecules, at rest and in the same vicinity, will eventually gravitate toward each other. These attractive forces may be strong enough to hold two or more molecules together in a stable (but weak) bond, creating a **van der Waals complex.** A van der Waals complex of two molecules is called a **dimer.**

[5]Small ions with like charge (two cations, or two anions) will generally repel at all distances and orientations, and Chapter 7 gives examples of certain electronic states that are dissociative at all distances. These involve stronger interactions than what we would classify as van der Waals forces.

Two argon atoms provide an example. Argon is a closed-shell atom, a noble gas like helium. The dimer Ar_2, like He_2, has zero bond order, and we might expect it to fall apart spontaneously. However, there is a very slight depression about 100 cm^{-1} or 1.2 kJ mol^{-1} deep in the potential surface that connects the two atoms (Fig. 10.3), with the minimum potential energy at a distance of about 3.8 Å and a zero-point vibrational energy of about 5 cm^{-1}. It is possible to form argon dimer, but the dissociation energy of this van der Waals bond is extremely low. At room temperature, molecules collide with an average energy equivalent to roughly 200 cm^{-1}. Under these conditions, even if two argon atoms combined into a dimer, chances are good that the bond would be broken as soon as the dimer collided with another argon atom (Fig. 10.4a).

However, if we cool the argon gas below 100 K, the average collision energies drop to less than 70 cm^{-1}, and this is low enough for argon dimer to form in measurable quantities (Fig. 10.4b). (Figure 8.4 shows that the experimental dissociation energy D_0 is always less than the equilibrium well depth D_e, by an amount equal to the zero-point energy, so we need to cool the system until the collision energies are considerably below D_e if we want many of these clusters to form.) Addition of a third argon atom leads to a trimer, Ar_3. If the temperature, and hence the collision energy, is low enough, this aggregation continues, leading to an argon droplet of increasing size and eventually to liquid argon.

What are the origins of the attractive intermolecular forces? Let's follow a common approach of breaking them into three principal forms:

1. **Electrostatic forces:** interactions between static electric fields
2. **Inductive forces:** interactions between a static field and an *induced* field

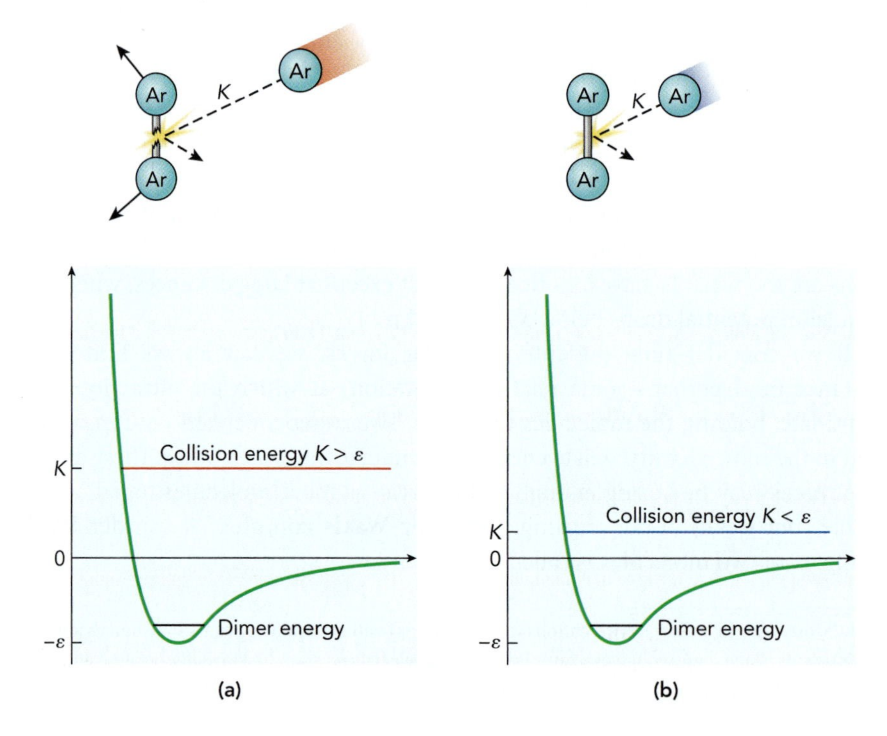

▶ FIGURE 10.4 **Formation of weakly bound complexes.** (**a**) In a room temperature gas, the well depth of the argon dimer is much less than the typical kinetic energy K of other atoms, and the dimer will be broken apart almost as soon as it is formed. (**b**) However, at very low temperatures, where the average kinetic energy K of the atoms is much less than the well depth ε, the argon dimer can survive because it can absorb the kinetic energy of a third atom without reaching the dimer's dissociation energy.

3. The **dispersion force:** the quantum-mechanical coupling of two electronic wavefunctions

Each of these is highlighted in the following sections.

Electrostatic Forces

Coulomb forces bind electrons to the nucleus to form an atom and bind atoms together to form molecules. And once molecules are formed, Coulomb forces also dictate how they interact. Molecules are assembled out of numerous charged particles, and those charges are responsible for virtually all the dynamics we study in chemistry. Let's take a look at the effect of Coulomb's law when two small particles, atoms or molecules, encounter one another.

Multipole fields. Even molecules with no net charge have some electric field. For example, in the polar covalent bond that binds the SiO molecule, the electronegative element (O) pulls electron density away from electropositive element (Si) in a polar covalent bond. One end of the bond gains a net positive charge while the other end has a net negative charge, although the molecular charge overall remains zero. If these are the only two atoms, there is a cylindrically symmetric electric field around the bond, a dipole field whose symmetry axis contains the chemical bond (Fig. 10.5a). A similar charge separation occurs in CO_2 (Fig. 10.5b), but in this case the dipoles of the two CO bonds cancel, so the dipole moment is zero. Nonetheless, the uneven distribution of charges generates an electric field, again cylindrically symmetric about the bond axis but (unlike a dipole) also symmetric under inversion.

For more complicated distributions of charge, we can define the dipole moment by either of these two equations:

$$\vec{\mu} = \sum_{i=1}^{N} \vec{r}_i q_i \qquad \text{for } N \text{ discrete charges } q_i \quad (10.7)$$

$$\vec{\mu} = \int \vec{r}\, \rho_q(\vec{r})\, d\tau \qquad \text{for continuous charge} \quad (10.8)$$

where $\rho_q(\vec{r})$ is the charge density (the charge per unit volume), which depends on the position $\vec{r}$. It is from the first of these general definitions that we get a result commonly used for chemical bond dipoles, such as the SiO bond in Fig. 10.5. For two point charges $+q$ and $-q$ separated by a distance d:

$$\vec{\mu} = \sum_{i=1}^{2} \vec{r}_i q_i = -q\vec{r}_1 + q\vec{r}_2 = q(\vec{r}_2 - \vec{r}_1). \quad (10.9)$$

The difference $\vec{r}_2 - \vec{r}_1$ is a vector of magnitude d pointing from the $-q$ charge to $+q$, so $\vec{\mu}$ is a vector that points from the negative charge to the positive charge[6] and has a magnitude of qd for two point charges $\pm q$.

$$\mu = qd \quad (10.10)$$

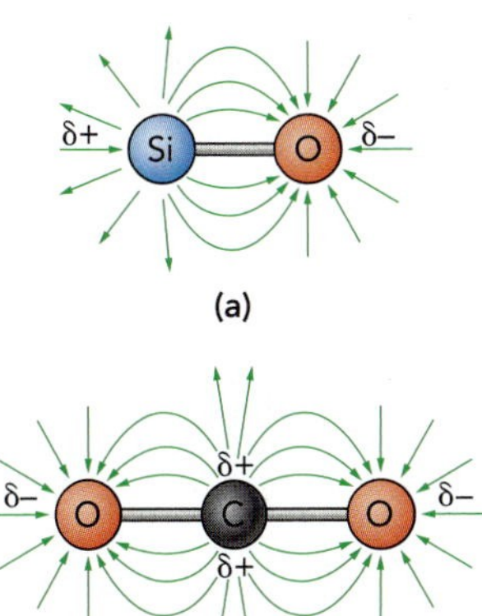

▲ **FIGURE 10.5 Electric fields near a molecule.** Neutral molecules such as **(a)** SiO, and even nonpolar molecules such as **(b)** CO_2, generate electric fields. The electric field lines drawn here show the direction of force that would be exerted on a positive charge near each molecule.

[6]This direction may be opposite the one you know from general chemistry (where we often draw the dipole from positive to negative), but it is the more common convention in physics. It can be justified by the simple form of Eq. 10.7.

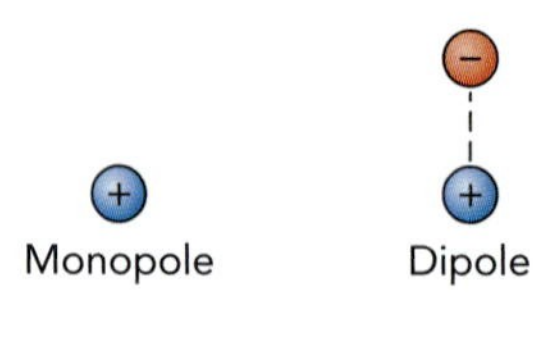

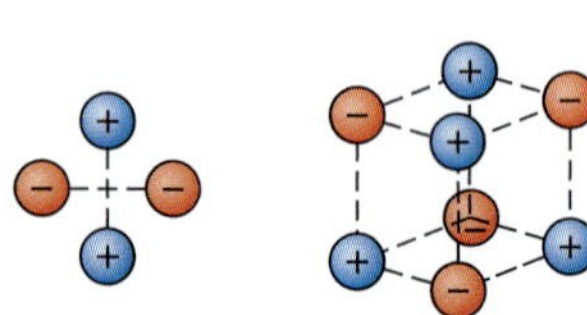

▲ **FIGURE 10.6 Arrangements of point electric charges that result in pure electric multipole fields.** These correspond to the spherical harmonics Y_0^0 (monopole), Y_1^0 (dipole), Y_2^1 (quadrupole), and Y_3^2 (octopole).

The inescapable (and emphatically non-SI) unit for the dipole moment is the debye (D), which is equal to 10^{-18} esu cm. When necessary, we can convert debye to the SI units of C m by the conversion

$$1\,\mathrm{D} = 3.3356 \cdot 10^{-30}\,\mathrm{C\,m}. \quad (10.11)$$

The debye unit has the advantage of being of the same order of magnitude as molecular dipole moments, but the disadvantage of running with the wrong crowd. Be careful of the units when calculating SI quantities such as energy or distance from dipole moment values.

The dipole is one example of a **multipole.** The lth multipole is the field generated by a symmetric arrangement of 2^l point sources or **poles** of the electric or magnetic field, where l can have integer values of 0, 1, 2, ... **Monopoles** (atomic ions, for example) have $l = 0$ and spherically symmetric electric fields. Higher order multipoles are the dipole ($l = 1$), **quadrupole** ($l = 2$), and **octopole** ($l = 3$). Intermolecular effects from multipoles of $l > 3$ have rarely been measured. Some examples of charge distributions corresponding to specific electric multipole fields are drawn in Fig. 10.6.

To be more specific, the multipole is the angular part of a potential energy function written using the spherical harmonics in Table 3.1. *Any* electric or

EXAMPLE 10.3 Dipole Moment of a Set of Point Charges

CONTEXT Carbonyl sulfide, OCS, is a polar molecule found in petroleum deposits. It is a gas under typical conditions and an irritant because it binds readily to water, where it may eventually react to form CO_2 and the toxic and reactive H_2S. The strong binding of OCS to water, compared to other gases with similar boiling points (such as propane), arises partly from its being a polar compound, allowing a favorable dipole–dipole interaction with water.

PROBLEM Use Eq. 10.7 to calculate the magnitude and direction of the dipole moment for OCS if the atoms are located at the points -1.69Å (O), -0.53Å (C), and $+1.04$Å (S) along the z axis, and have effective partial charges[7] $-0.172e$ (O), $+0.219e$ (C), and $-0.047e$ (S).

SOLUTION All the points lie along the z axis, so the x and y coordinates of $\vec{\mu}$ are zero. Thus, we carry out the sum in Eq. 10.7 just looking at the z coordinates:

$$\begin{aligned}\mu_z &= \sum_{i=1}^{N} z_i q_i \\ &= (-1.69\,\text{Å})(-0.172e) + (-0.53\,\text{Å})(0.219e) + (1.04\,\text{Å})(-0.047e) \\ &= (0.126\,\text{Å}\,e)(10^{-10}\mathrm{m}\,\text{Å}^{-1})(1.602 \cdot 10^{-19}\mathrm{C}/e)(3.3356 \cdot 10^{-30}\mathrm{C\,m/D})^{-1} = 0.60\,\mathrm{D}\end{aligned}$$

The actual dipole moment is 0.752 D.

[7]There are many ways to assign point charges to atoms, so these should not be considered a fixed set of numbers. In general, there are better ways to obtain dipole moments—whether by theory or experiment—than by adding partial charges, but this technique is useful, for example, in calculations of intermolecular bonding in large molecules, as described in Section 11.2.

magnetic potential energy function can be expressed as a linear combination of these multipole contributions:[8]

$$U(r,\theta,\phi) = \sum_{l=0}^{\infty} \sum_{m=-l}^{l} U_{l,m}(r)\, Y_l^m(\theta,\phi), \tag{10.12}$$

where all of the r-dependence of the field is absorbed into the function $U_{l,m}(r)$.

Multipole–multipole potential energies. As mentioned in Sections 3.3 and 3.4, the forces arising from magnetic fields tend to be orders of magnitude weaker than those from electric fields. For that reason, we will consider only electric field interactions in this section.

The force laws associated with the principal multipole–multipole interactions are straightforward to derive, in the limit that point charges generate the multipole fields. As examples, we shall find the R-dependence of the monopole–dipole and dipole–dipole interactions. Consider two molecules with electric dipoles, such as two H_2O molecules in the gas phase. They are

EXAMPLE 10.4 Multipole Fields in Water

CONTEXT In aqueous solutions of salts, we say that the ions manage to break the strong ionic bond because they are stabilized by favorable interactions with the water. But not all the interactions are favorable, and one interaction that tends to make the solution *less stable* is the interaction between the ion and the electric quadrupole field of the water. This effect turns out to be significant in concentrated solutions of chloride salts, where the water quadrupole reduces the ionic strength of chloride near the surface.

PROBLEM Drawing (a) in the accompanying figure represents a simplified version of the charge distribution in water. Show that this can be represented as the sum of a dipole field and a quadrupole field.

-0.2e
O
H H
+0.1e +0.1e
(a)

-0.2e
O
H H
+0.1e +0.1e
=
-0.1e
O
H H
+0.1e
+
-0.1e
O
H H
+0.1e +0.1e
-0.1e
(b)

SOLUTION The dipole moment vector passes through the O atom, between the two H atoms. The distribution of point charges necessary to create the dipole field must lie on that same axis. But this puts a positive charge below the O atom, where there is no atom. We can fold that charge into the quadrupole charge distribution, with positive point charges at the H atoms and negative charges at the O atom and below the O atom, as shown in part (b) of the figure.

[8]The lowest order *magnetic* multipole that has been observed is the dipole. The existence of magnetic monopoles remains an open question of particular importance to particle physicists and cosmologists.

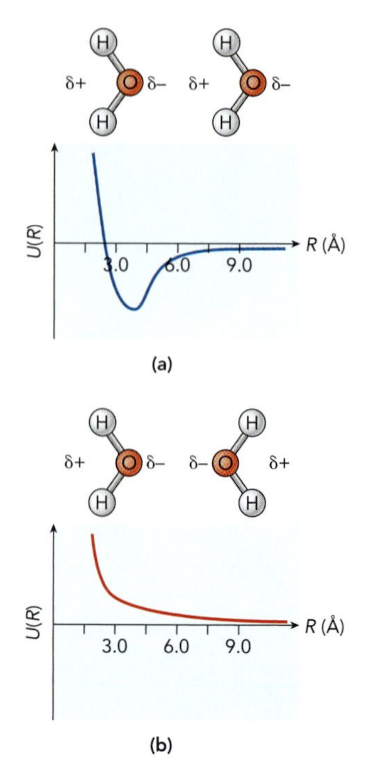

▲ **FIGURE 10.7 Sketched potential energy curves for H_2O-H_2O interactions.** **(a)** The molecular dipole moments are aligned end-to-end, creating an attractive well except at small values of R where the electron cores repel; **(b)** the dipoles oppose each other and the molecules repel at all R.

separated by a distance R and can be oriented in any direction. The dipole is the result of a charge separation: one part of the H_2O molecule (the oxygen atom) has a slight negative charge, and the other (the hydrogens) has a slight positive charge. Of all the possible relative orientations of these two molecules, the potential energy of the dimer will always be a little lower if the electropositive end of one H_2O molecule is closer to the electronegative end of the other molecule. Similarly, if the two molecules are arranged with like charges close together, there is a repulsive force, and the potential energy increases. In addition to this angular dependence, the potential energy function also depends on the distance R between the molecules, as discussed earlier. The distance dependence for two relative orientations of these H_2O molecules is illustrated in Fig. 10.7.

In the region where the repulsive wall has dropped to an energy low enough for the van der Waals forces to be significant, we shall assume the distance R between the molecules is large compared to the size of the molecule. This assumption lets us treat the electropositive and electronegative atoms as point charges. We concern ourselves only with finding the R dependence of the interaction and shall introduce the angular dependence later.

We now place a charge q_B at a distance R from the center of charge of a dipole moment μ_A (formed from charges q_A and $-q_A$ separated by distance d_A), such that the direction of the dipole moment is directly toward q_B (Fig. 10.8). If we evaluate the interaction of q_B with each of the charges in our dipole μ_A, we find two competing Coulomb potential terms:

$$u_{1-2}(R) = \frac{1}{4\pi\varepsilon_0}\left[\frac{q_A q_B}{R + (d_A/2)} - \frac{q_A q_B}{R - (d_A/2)}\right]. \qquad (10.13)$$

The subscript $1-2$ in u_{1-2} indicates that this is the contribution to the potential energy from the interaction between a monopole (1 charge) and a dipole (2 charges). In the limit that $R \gg d_A$, we can use the binomial series (Eq. A.32) to simplify the difference:

EXAMPLE 10.5 Deriving Distance from Dipole Moment

CONTEXT Carbon monoxide, a toxic byproduct of incomplete combustion, may be detected in the atmosphere using infrared absorption spectroscopy, which stimulates the C—O stretching motion. This transition is allowed because CO is a polar molecule, and stretching the bond changes the dipole moment (i .e., the dipole derivative is not zero, Section 8.4).

PROBLEM The dipole moment of CO (determined by Stark field spectroscopy) is small, only 0.112 D, and the bond length of CO (determined by rotational spectroscopy) is 1.128 Å. The dipole moment vector points toward the O atom (the O atom surprisingly has a partial positive charge). If the dipole moment is the product of the charge and the bond length, estimate the charges (a) in coulombs, and (b) in units of the fundamental charge e.

SOLUTION We solve Eq. 10.10 for q_A, using the bond length for d_A:

$$\mu_A = q_A d_A$$

$$q_A = \frac{\mu_A}{d_A} = \frac{0.112\text{D}}{1.128\text{Å}}$$

$$= \frac{(0.112\text{D})(3.3356 \cdot 10^{-30}\text{C m/D})}{(1.128\text{Å})(10^{-10}\,\text{m}^{-1})} = 3.31 \cdot 10^{-21}\text{C}$$

$$= \frac{3.31 \cdot 10^{-21}\text{C}}{1.602 \cdot 10^{-19}\text{C}/e} = 0.0207e.$$

In order to get such a small dipole moment, only a tiny fraction of the charge is redistributed from one atom toward the other. That said, keep in mind that the charges are not actually point charges, and the real electron distribution is more complicated and interesting.

$$u_{1-2}(R) = \frac{q_A q_B}{4\pi\varepsilon_0}\left[\frac{1}{R + (d_A/2)} - \frac{1}{R - (d_A/2)}\right] \qquad \text{factor out } q_A\, q_B$$

$$= \frac{q_A q_B}{(4\pi\varepsilon_0)R}\left[\left(1 + \frac{d_A}{2R}\right)^{-1} - \left(1 - \frac{d_A}{2R}\right)^{-1}\right] \qquad \text{factor out } 1/R$$

$$\approx \frac{q_A q_B}{(4\pi\varepsilon_0)R}\left[\left(1 - \frac{d_A}{2R}\right) - \left(1 + \frac{d_A}{2R}\right)\right] \qquad \text{by Eq. A.32}$$

$$= \frac{q_A q_B}{(4\pi\varepsilon_0)R}\left[-\frac{d_A}{R}\right] = -\frac{\mu_A q_B}{(4\pi\varepsilon_0)R^2} \qquad \mu_A = q_A d_A$$

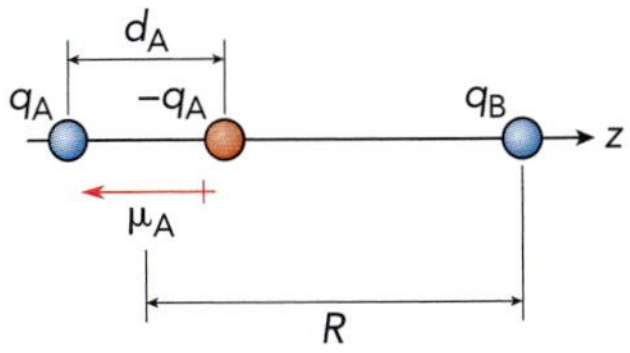

$$u_{1-2}(R) = -\frac{\mu_A q_B}{(4\pi\varepsilon_0)R^2}. \qquad (10.14)$$

▲ **FIGURE 10.8 Parameters appearing in the derivation of the dipole–monopole interaction potential.**

This result shows that at large distances the attractive energy between a monopole and a dipole varies as R^{-2}.

CHECKPOINT The interactions between molecules that we are studying here are weaker than chemical interactions, comparable to the average kinetic energy at room temperature, but they are so numerous that they play a significant role in shaping (literally as well as figuratively) the substances that chemistry seeks to understand. Here we find that an ion and a neutral molecule don't have to form a chemical bond to have a significant attraction for one another. As the interaction strengths decrease below the typical energy of molecular collisions, we will expect the effects to contribute less to the material's properties.

SAMPLE CALCULATION Monopole–Dipole Typical Magnitude. If we choose typical values of $\mu_A = 1.0\,\text{D}$, $q_B = e$, and $R = 3.0\,\text{Å}$, we find the monopole–dipole energy to be

$$-\frac{\mu_A q_B}{(4\pi\varepsilon_0)R^2} \approx -\frac{(1.0\text{D})(3.3356 \cdot 10^{-30}\,\text{C m/D})(1.602 \cdot 10^{-19}\,\text{C})}{(1.113 \cdot 10^{-10}\text{C}^2\text{J}^{-1}\text{m}^{-1})(3.0 \cdot 10^{-10}\,\text{m})^2}$$

$$= -5.3 \cdot 10^{-20}\,\text{J} = -32\,\text{kJ mol}^{-1}. \qquad (10.15)$$

Although a typical value for the attractive potential energy is about $-32\,\text{kJ mol}^{-1}$ (see the Sample Calculation), the potential well for this interaction would not be that deep. We have neglected the repulsion energy, which will partly cancel the attraction. Nevertheless, this interaction can lead to bond strengths between the ion and the polar neutral molecule that rival the bond strengths of traditional covalent or ionic chemical bonds, most of which lie in the range 100–800 kJ mol^{-1}.

If no ions are present, the lowest order multipole–multipole term we need to consider is for the interaction between two dipoles. To extend our results to the dipole–dipole case, we add a fourth charge $-q_B$ a distance d_B from the charge q_B (Fig. 10.9). Again we keep the charge separation d_B small compared to R, and we choose the dipole $\mu_B = q_B d_B$ to point in the same direction as the dipole μ_A.

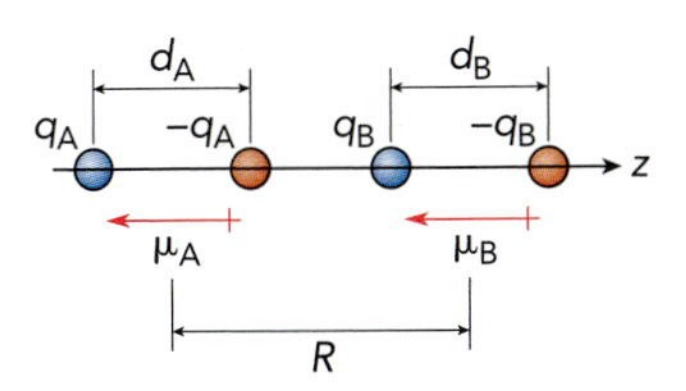

▲ **FIGURE 10.9 Parameters appearing in the derivation of the dipole–dipole interaction potential.**

EXAMPLE 10.6 Monopole–Dipole Interaction

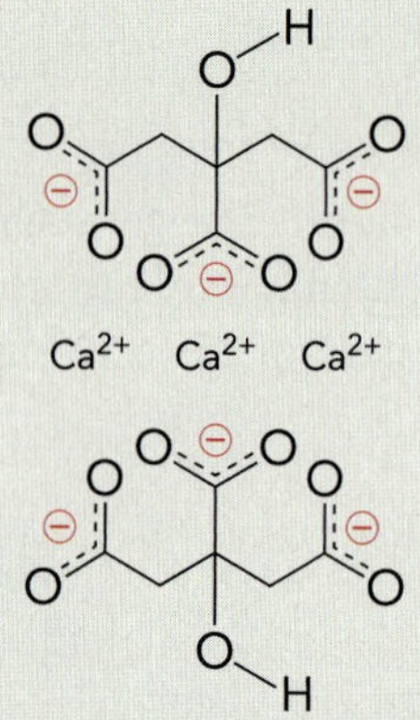

CONTEXT Calcium is present in the blood as the Ca^{2+} ion, stabilized by its monopole–dipole interaction with water. Calcium is one of the principal atomic ions critical for cell signaling: certain ion-channel proteins provide gateways through the cell membranes for Ca^{2+} to pass into or out of the cell interior, where they may stimulate muscle cells to contract or endocrine cells to secrete hormones. Blood platelet donors are administered small quantities of sodium citrate to slow coagulation, and the citrate can bind to calcium ion, decreasing the concentration of free Ca^{2+} in the blood. The drop in Ca^{2+} may trigger a small cell-signaling response in donors, which they experience as tingling around the jaw or fingers.

Ca^{2+} Ca^{2+} Ca^{2+}

PROBLEM (a) Estimate the monopole–dipole interaction energy between one Ca^{2+} ion and one water molecule ($\mu = 1.85$D) at a separation of 2.43 Å. (b) Ca^{2+} typically associates with seven water molecules, so multiply your answer to (a) by 7 to estimate the total binding energy of Ca^{2+} in water. (c) Finally, compare this to the ionic bond energy of Ca^{2+} and two O^- ions (from the carboxyl group of the citrate ion) using an average Ca—O distance from the crystal structure of 2.44 Å.

SOLUTION Use Eq. 10.14 for the monopole–dipole energy, and Coulomb's law for the attraction in the ionic bond:

$$\begin{aligned} u_{1-2}(R) &= -\frac{\mu_A q_B}{(4\pi\varepsilon_0)R^2} \\ &= -\frac{(1.85\text{D})(3.3356\cdot 10^{-30}\text{C m/D})(+2e)(1.602\cdot 10^{-19}\text{C}/e)}{(1.113\cdot 10^{-10}\text{C}^2\text{J}^{-1}\text{m}^{-1})[(2.43\text{Å})(10^{-10}\text{m Å}^{-1})]^2} \\ &= -3.01\cdot 10^{-19}\text{J} = -181\text{kJ mol}^{-1} \\ 7(u_{1-2}) &= -1270 \text{ kJ mol}^{-1}. \end{aligned}$$

Now use the monopole–monopole energy for the ionic bond: $u_{1-1} = \frac{q_{Ca^{2+}}q_{O^-}}{4\pi\varepsilon_0 R}$

$$\begin{aligned} &= \frac{(2+)(-1)(1.602\cdot 10^{-19}\text{C}/e)^2}{(1.113\cdot 10^{-10}\text{C}^2\text{J}^{-1}\text{m}^{-1})(2.44\text{Å})(10^{-10}\text{m Å}^{-1})} \\ &= -1.89\cdot 10^{-18}\text{J} = -1140\text{kJ mol}^{-1}, \end{aligned}$$

and multiply by 2 for the two O^- ions per Ca^{2+} ion:

$$2u_{1-1} = -2280 \text{ kJ mol}^{-1}.$$

So we find that, although the ion–dipole force is strong, the ionic bonding is probably stronger; thus Ca^{2+} will tend to bind to citrate ion.

We evaluate the interaction potential in the same way, this time looking at the sum of the potentials for the interaction of charges q_B and $-q_B$ with the dipole μ_A:

$$\begin{aligned} u_{2-2}(R) &= \left(\frac{1}{4\pi\varepsilon_0}\right)\left[\frac{\mu_A q_B}{[R + (d_B/2)]^2} - \frac{\mu_A q_B}{[R - (d_B/2)]^2}\right] && \text{by Eq. 10.14} \\ &= \frac{\mu_A q_B}{(4\pi\varepsilon_0)R^2}\left[\left(1+\frac{d_B}{2R}\right)^{-2} - \left(1-\frac{d_B}{2R}\right)^{-2}\right] && \text{factor out } \mu_A q_B/R^2 \\ &\approx \frac{\mu_A q_B}{(4\pi\varepsilon_0)R^2}\left[\left(1-\frac{2d_B}{2R}\right) - \left(1+\frac{2d_B}{2R}\right)\right] && \text{by Eq. A.32} \end{aligned}$$

$$= -\frac{2\mu_A q_B d_B}{(4\pi\varepsilon_0)R^3} = -\frac{2\mu_A\mu_B}{(4\pi\varepsilon_0)R^3} \qquad \mu_B = q_B d_B$$

$$u_{2-2}(R) = -\frac{2\mu_A\mu_B}{(4\pi\varepsilon_0)R^3}. \tag{10.16}$$

SAMPLE CALCULATION **Dipole–Dipole Typical Magnitude.** Using again rough values of $\mu_A = \mu_B = 1.0\,\text{D}$ and $R = 3.0\,\text{Å}$, a representative value for this interaction potential becomes

$$-\frac{2\mu_A\mu_B}{(4\pi\varepsilon_0)R^3} \approx -\frac{2(1.0\,\text{D})(1.0\,\text{D})(3.3356\cdot 10^{-30}\,\text{C m/D})^2}{(1.113\cdot 10^{-10}\text{C}^2\,\text{J}^{-1}\text{m}^{-1})(3.0\cdot 10^{-10}\text{m})^3}$$
$$= -7.4\cdot 10^{-21}\text{J} = -4.5\,\text{kJ mol}^{-1}.$$

This interaction is now roughly 100 times weaker than chemical bonding (see the Sample Calculation), and we don't expect it to be capable of affecting the overall *chemical* structure of our system. It makes a polar molecule sticky, but not too sticky. Forming and breaking this weak, intermolecular bond will have a nearly negligible effect on the molecule's shape.

Hydrogen bonds. An important class of intermolecular interactions, which we will group with the dipole–dipole attraction, are the **hydrogen bonds.** These are non-covalent bonds with a general form X—H· · · Y, where X and Y are highly electronegative atoms (particularly F, O, N, or Cl), X—H is a normal covalent bond, and the hydrogen bond is the H· · · Y interaction.[9] Hydrogen is an electropositive element, and it surrenders electron density across the covalent bond to X, resulting in a bond dipole, with a positive charge at the hydrogen. If Y, bonded to some other atoms, has acquired a negative partial charge, then we have all the ingredients for another dipole–dipole interaction.

This may not seem especially exciting news. But the significance of the hydrogen is that, unlike other electropositive atoms, *it has no core electrons.* When a hydrogen atom loses electron density to its bonding partner X, little Coulomb repulsion is left against the electrons of other atoms, even at very short distances. As a result, negatively charged atom Y sees a weakly shielded proton, and the two come quite close to each other before the negative partial charge of X begins to put up a fight. The dipole–dipole force becomes stronger as the separation decreases, so the relatively short H· · · Y distance stabilizes the interacting atoms. The electron density of X tends to smother one side of the H atom, so Y needs to approach from nearly a 180° angle to take advantage of the short hydrogen bonding distance. As an example, the H· · · O bond is only about 2 Å long in the $(H_2O)_2$ dimer, compared to typical dipole–dipole bond lengths of 3–5 Å.

[9]This adheres to Pauling's original definition of the hydrogen bond. The definition was greatly relaxed by Pimentel and McClellan to allow for the wide variation of electron-withdrawing power of elements among chemical species (in other words, in certain molecules some elements may act as thought they have higher electronegativities than given in the standard table). The later definition also allows for weaker and less rigid bonds to be incorporated under the same term, and debate on what is and isn't a hydrogen bond continues.

EXAMPLE 10.7 Dipole–Dipole Interaction

CONTEXT The distinction between plastics that bend *and* stretch (**elastomers** such as rubber) and plastics that bend but don't stretch (**fibers** such as nylon) is that the dipole–dipole attraction gives fibers great tensile strength without the rigidity that we associate with covalent bonds. Nylon is composed of polyamide chains, bound by the dipole–dipole attraction of the C=O bond ($\mu \approx 2.3$D) and the N—H bond ($\mu \approx 1.3$D), forming a hydrogen bond.

PROBLEM Estimate the potential energy of attraction in kJ/mol between a C=O bond ($\mu \approx 2.3$D) and N—H bond ($\mu \approx 1.3$D) separated by $R = 2.0\,\text{Å}$.

SOLUTION Using Eq. 10.16, we substitute in the values for μ_A, μ_B, and the separation R:

$$\begin{aligned} u_{2-2}(R) &= -\frac{2\mu_A\mu_B}{(4\pi\varepsilon_0)R^3} \\ &= -\frac{2(2.3\text{D})(1.3\text{D})(3.3356\cdot 10^{-30}\text{C m/D})^2}{(1.113\cdot 10^{-10}\,\text{C}^2\text{J}^{-1}\text{m}^{-1})(2.0\cdot 10^{-10}\,\text{m})^3} \\ &= -7.5\cdot 10^{-20}\,\text{J} = -45\ \text{kJ mol}^{-1}. \end{aligned}$$

This value indicates a strong intermolecular attraction, more than a tenth of typical chemical bond energies, which prevents the chains from pulling apart easily and gives nylon its tensile strength.

The close proximity of the electropositive H atom and the electronegative Y atom causes the hydrogen bonds to be stronger than most other van der Waals bonds, on the order of 1000 cm^{-1}.

Higher multipoles. Similar interactions occur between higher multipole fields, but these tend to be weaker than typical dipole–dipole interactions. Nevertheless, the higher multipole forces are often significant for interactions between molecules that have no permanent electric dipole moment. These include the homonuclear diatomics, such as N_2, and polyatomic molecules of high symmetry, such as BF_3 or CH_4.

The O_2 molecule provides an example of a different interaction. There is no *electric* dipole–dipole interaction, but each molecule in the ground $^3\Sigma_g^+$ electronic state has a magnetic moment due to the spin magnetic field of the two unpaired electrons. The magnetic dipole–magnetic dipole coupling between O_2 molecules is roughly 100 times weaker than typical electric dipole–electric dipole couplings, but it dominates the intermolecular potential in this case. This is a rare but important exception to our assumption that electric fields dominate the multipole forces.

Different multipole interactions also have different distance dependencies. To look at it simply, the interaction energy of two multipole fields depends on the product of the two multipole moments, the distance between them, and their relative angle. The electric n-pole moment has units of charge × distance^{k-1}, where $n = 2^k$, and the angle has no units. For the potential energy between an n-pole and an n'-pole,

$$u_{n-n'} = (a/4\pi\varepsilon_0)\,\mu_n\mu_{n'}\,R^{-k-k'-1}, \tag{10.17}$$

where a is a unitless coefficient that contains all the angular dependence of the potential. For example, the dipole–dipole interaction decreases as R^{-3}, as shown in Eq. 10.16, whereas the quadrupole–quadrupole interaction decreases much faster, as R^{-5}. The Coulomb force binding the electron to the nucleus is a monopole–monopole interaction and has a potential energy proportional to r^{-1}, as predicted by Eq. 10.17 (using little r for the electron distance).

TOOLS OF THE TRADE Molecular Beams

The French physicist Louis Dunoyer invented the molecular beam technique in 1911 for an elegant experiment to test the kinetic theory of gases. Dunoyer designed a glass tube, pinched in two places to separate it into three chambers, and connected to pumps so that the air could be removed. He placed solid sodium in the first chamber, evacuated the air with the pumps, and then heated the sodium to vaporize it. He discovered, as he had hoped, that the vaporized sodium reappeared as a thin film at the *opposite end* of the third chamber, rather than as an evenly distributed coating throughout the entire apparatus. The vapor had traveled through the first chamber into the second, where a vacuum pump had dropped the pressure to a point at which the atoms rarely collided with each other. The vapor had then traveled into the third chamber, where a second pump further reduced the pressure, which was now so low that the sodium atoms traveled in straight lines, hitting the opposite wall of the chamber, where they promptly solidified.

One of the great controversies in physics in the early years of the 20th century surrounded Peter Debye's attempt to explain the Zeeman effect, the splitting of certain atomic spectra into distinct transitions when exposed to a magnetic field (see Section 3.4). Debye employed the Bohr model of the atom to justify the possibility that the *orientation* of the magnetic fields in these atoms was quantized. This seemed much more mysterious than the notion that the energies of the atom would be quantized, and the proposed phenomenon was termed *space quantization.* In Frankfurt, Otto Stern, who did not believe Debye's space quantization, proposed to his colleague Walter Gerlach that they investigate the problem. Gerlach's prior experience with molecular beams established a basis for one part of the experiment. Other challenges lay in generating a sufficiently constant magnetic field and in attaining sufficiently high vacuum.

Overcoming these challenges, the two scientists carried out the stunning experiment now known by their names. The Stern-Gerlach experiment of 1922 proved that the magnetic moment of silver atoms is quantized, allowing a beam of the atoms to be separated by a magnetic field into two paths. The impact of this discovery on physics not only provided a crucial key to quantum mechanics, it also established the molecular beam as a prominent tool in atomic and molecular physics.

The molecular beam subsequently played a central role in the invention of nuclear magnetic resonance spectroscopy (see *Tools of the Trade,* Chapter 5) and formed the experimental basis for Nobel Prize–winning studies of chemical reactions. It has become a routine tool in many other forms of spectroscopy and is used in industry for molecular beam epitaxy, which takes advantage of the technology to grow carefully tailored semiconductor crystals.

What is a molecular beam? A molecular beam is a stream of neutral atoms or molecules, all traveling in nearly the same direction at nearly the same speed. It is not an analytical technique, like spectroscopy, but rather a way of preparing a chemical sample prior to analysis. There are also *ion beams,* but these tend to operate with different technology, with electric fields used to control the speeds and to oppose the natural repulsion between the ions in the beam.

How does it work? A schematic diagram of a typical molecular beam apparatus is shown in the accompanying figure. In retrospect, this design is remarkably similar to Dunoyer's original glass tube separated into three chambers: a nozzle, a vacuum chamber that a high-speed pump maintains at a much lower pressure than in the nozzle, and a third chamber in which the beam can be probed. The nozzle is a channel for delivering the gas through a hole typically less than 1 mm in diameter, small enough that the molecules have to collide several times on their way into the next chamber. These collisions force the molecules to exchange energy, with faster molecules speeding up the slower molecules and vice versa, until all the velocities are similar. Because all the random motion along X and Y and Z in the original gas has been converted into kinetic energy along a single axis (Z in the figure), the average energy in that one direction is greater in the vacuum chamber than in the nozzle. The molecules in the beam therefore travel slightly faster than the speed of sound in the nozzle.

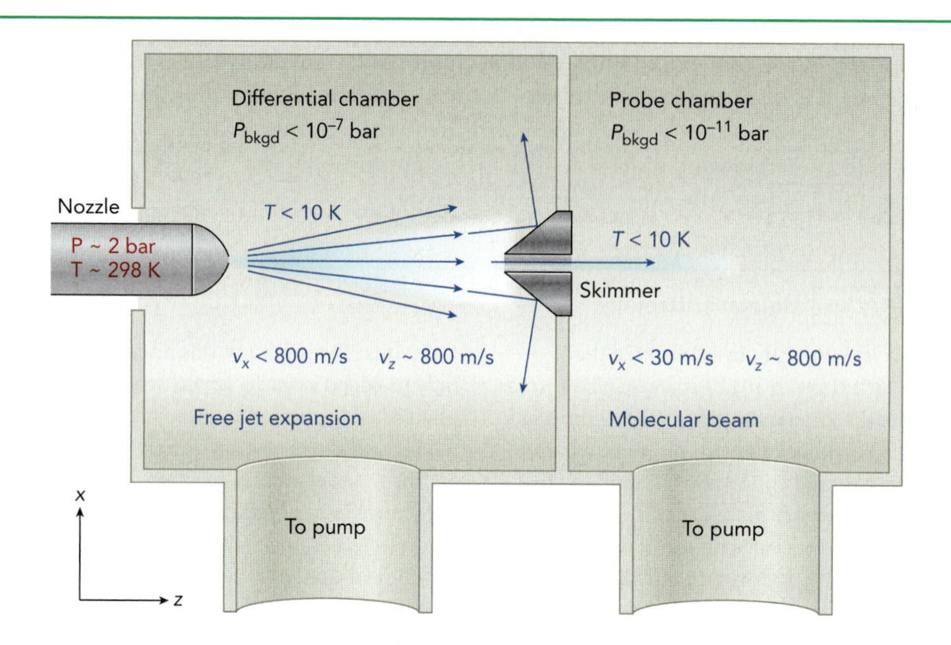

As they leave the nozzle, the molecules form a **free jet.** The speeds are supersonic, but the molecules are not all traveling in the same direction. To form the molecular beam, a *skimmer*—essentially a cone with a hole at the tip—lets molecules at the center of the free jet continue into the next chamber while it deflects other molecules away to prevent unnecessary collisions. In some experiments, an additional vacuum chamber with another aperture rejects more molecules to generate an especially narrow beam.

A typical pressure of the gas before it exits the nozzle is 1–2 bar, although any pressure inside the nozzle that is much greater than the pressure outside will generate a free jet. The pressures in the other two chambers must be low enough that the molecules in the jet or beam will not be easily scattered by the randomly moving background gas. Pressures less than 10^{-7} bar are required before the molecules can travel one meter on average without hitting another molecule in the chamber, and lower pressures are required for many of the detection techniques used.

Why do we use molecular beams? A supersonic molecular beam provides an environment that differs from an ordinary sample in several respects, any of which in a particular experiment may offer an important advantage. (1) *The forward speeds of the molecules cover a very narrow range, slightly higher than the speed of sound in the original sample.* The similar speeds of all the molecules in the beam allow numerous studies that would not be possible in a typical sample. For example, in studies of chemical reactions, this uniform motion greatly simplifies the analysis by reducing the range of initial energies that the particles have. Although the beams are accurately described as supersonic, researchers rarely take direct advantage of this feature, because the average speed in the beam is only about 20% faster than the speed of sound in the original sample. (2) *The molecules are confined to a very narrow region of space.* The molecules travel along nearly parallel trajectories. This property formed the basis for generations of molecular beam electric resonance experiments, which use electric fields and radiation to alter the path of molecules in specific quantum states before focusing them back on a detector. The ability to separate different quantum states in space hearkens back to the original Stern-Gerlach experiment. (3) *Very few collisions occur between the molecules in the beam.* When all the molecules in the beam travel in the same direction at the same speed, they don't run into each other as they would in a container full of randomly moving particles. Reactive molecules in this environment therefore have much less opportunity to react with another molecule, so they last longer, giving the experimenter more time to study them. (4) *Low temperatures can often be easily achieved.* When all the molecules in the beam travel in *nearly* the same direction at *nearly* the same speed, they do run into each other, but at very low relative speeds, so the collisions that do occur are quite soft. Temperature determines the amount of random motion in a gas, so these low collision speeds correspond to very low temperatures, typically 10 K or lower. The vast majority of studies on weakly bound molecular complexes have relied on the molecular beam to achieve these low temperatures so typical collisions at room temperature (occurring with energies at least 30 times higher) would not break apart the complex.

For a technique that offers such high selectivity, it is surprisingly easy to put into practice. No high-tech gadgets are required at all, just very good vacuum pumps. This simplicity is a hallmark of elegant design that researchers have recognized and appreciated now for over a century.

Rotating dipoles. When we derived Eqs. 10.13, 10.16, and 10.17, we assumed for simplicity that the charges all lay on the same axis. The assumption is okay for many environments, but in the gas phase the molecules are constantly rotating, and the relative orientations of the interacting molecules do matter. A more rigorous expression of the intermolecular potential for two dipoles includes the angles:

$$u_{2-2}(R,\theta_{\mathrm{A}},\theta_{\mathrm{B}},\phi) = -\frac{2\mu_{\mathrm{A}}\mu_{\mathrm{B}}}{(4\pi\varepsilon_0)R^3}\left[\cos\theta_{\mathrm{A}}\cos\theta_{\mathrm{B}} - \frac{1}{2}\sin\theta_{\mathrm{A}}\sin\theta_{\mathrm{B}}\cos(\phi_{\mathrm{B}}-\phi_{\mathrm{A}})\right], \quad (10.18)$$

and even this equation is valid only while R is large compared to the separation of charges d on the individual molecules. The angles θ_{A} and θ_{B} (Fig. 10.10) refer to the angle of each molecule relative to the z axis that connects the two centers of mass; the angles ϕ are the angles of rotation along the xy plane. This expression does not depend on the values of ϕ_{A} and ϕ_{B}, but only on the difference between them; therefore, we shall replace $\phi_{\mathrm{B}} - \phi_{\mathrm{A}}$ with ϕ in later equations.

The relative angles of interacting molecules are constantly changing in the gas phase because the molecules rotate and bounce around. The average potential energy experienced by a given polar molecule therefore usually includes partially canceling attractive and repulsive interactions with other polar molecules, with the shorter distance configurations being more heavily favored. When the dipoles are aligned in the same direction, the potential energy is negative and the molecules are pulled together. When they are aligned in opposite directions, the potential energy is positive and the molecules repel. Upon averaging these relative angles over a large number of molecules, we find that the distance dependence is squared, varying as R^{-6} for the dipole–dipole interaction instead of R^{-3}, and R^{-10} for the quadrupole–quadrupole interaction instead of R^{-5}. We derive this equation in Chapter 4 of *Thermodynamics, Statistical Mechanics, and Kinetics*; for now, we quote the resulting dipole–dipole *average* potential energy function:

$$\text{dipole–dipole: } \langle u_{2-2}\rangle_{N,\theta,\phi} = -\frac{2\mu_{\mathrm{A}}^2\mu_{\mathrm{B}}^2}{(4\pi\varepsilon_0)^2\, 3k_B T R^6}, \quad (10.19)$$

where k_{B} is $1.381 \cdot 10^{-23}$ J K^{-1}, where the subscript N indicates that this equation is obtained by averaging over some large number N of molecules, and where the subscripts θ and ϕ indicate that we are averaging over all angles.

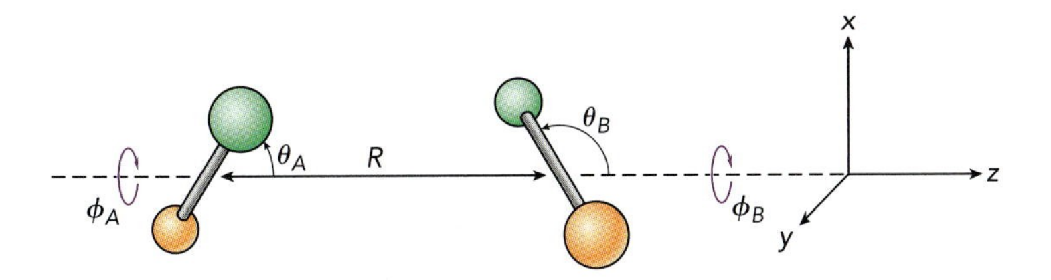

◀ FIGURE 10.10 **Definitions of the angles in the dipole–dipole interaction.**

SAMPLE CALCULATION **Averaged Dipole–Dipole Typical Magnitude.** Using the same values as before ($\mu_A = \mu_B = 1.0$ D and $R = 3.0$ Å) and assuming a temperature of 300 K, a typical value for the interaction energy is

$$-\frac{2\mu_A^2\mu_B^2}{(4\pi\varepsilon_0)^2\, 3k_BTR^6} \approx -\frac{2(1.0\text{D})^2(1.0\text{D})^2(3.3356\cdot 10^{-30}\text{ C m/D})^4}{(1.113\cdot 10^{-10}\text{ C}^2\text{ J}^{-1}\text{m}^{-1})^2(3)(1.381\cdot 10^{-23}\text{J K}^{-1})(300\text{K})(3.0\cdot 10^{-10})^6}$$
$$= -2.1\cdot 10^{-21}\text{ J} = -1.3\text{ kJ mol}^{-1}. \qquad (10.20)$$

This rotational averaging has a significant impact on the magnitude of the dipole–dipole interaction. At 3.0 Å, the interaction is about a factor of three weaker than for fixed dipoles (see the Sample Calculation), and it decreases now much more rapidly as R increases: at 5.0 Å the value drops to only 0.06 kJmol^{-1}. At this point, other attractive interactions may become important, and we take a look at those next.

Inductive Forces

What if one molecule has a dipole moment and the other does not? In that case, the electric field of the polar molecule can *induce* a dipole moment in the other. For example, N_2 has no permanent electric dipole field. When dissolved in H_2O, however, the dipole moment of the H_2O can cause the charges in the N_2 to separate a little bit, leading to an **induced dipole moment,** as illustrated in Fig. 10.11.

The induced dipole moment is an application of Eq. 3.36, which we first applied to electronic spectroscopy of atoms in Section 3.3:

$$\mu_{\text{induced}} = \alpha\mathcal{E}.$$

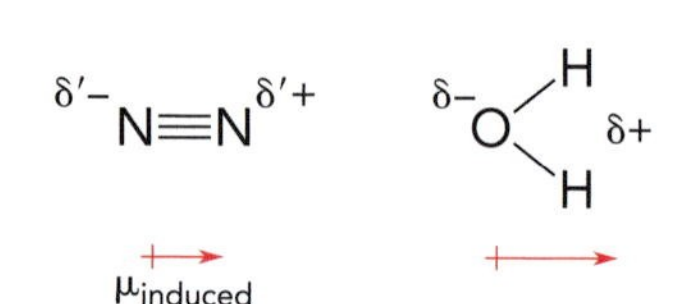

▲ **FIGURE 10.11** **A polar molecule (H_2O) can induce a dipole moment in a non-polar molecule (N_2) by forcing a small separation of charges δ'.**

The extent of the charge separation on the non-polar molecule depends on the polarizability of the molecule α, just as the ability of electromagnetic radiation to induce a Raman transition by inducing a dipole depends on the polarizability. How do we recognize a polarizable molecule? Table 10.1 lists some representative polarizability values. Polarizability increases with size (compare C_2H_2 and C_2H_6), with atomic number (compare H_2 and HI), and with number of electrons in π bonds (compare benzene and just about anything). The easier it is to distort the electronic wavefunction over a large volume, the higher the polarizability. It is easier to distort the electrons when there are more nuclei present to attract the electrons (large molecules) and when the electrons are not strongly bound to a particular nucleus (high atomic number and π bond electrons).

If we again neglect the angular terms, we can quickly obtain an approximate formula for the potential energy of the dipole-induced dipole interaction. According to Eq. 3.36, the induced dipole moment has a magnitude

$$\mu_{\text{induced}} = \alpha\mathcal{E} \qquad (10.21)$$

where $\mathcal{E}$ is the electric field due to the permanent dipole. The electric field due to a single point charge q_A has magnitude q_A/R^2. For a dipole consisting of two separated charges q_A and $-q_A$, the electric field at a distance R from the center of the dipole is given by

$$\mathcal{E}(R) = \frac{q_A}{[R-(d_A/2)]^2} - \frac{q_A}{[R+(d_A/2)]^2}$$
$$= \frac{q_A}{R^2}\left[\left(1-\frac{d_A}{2R}\right)^{-2} - \left(1+\frac{d_A}{2R}\right)^{-2}\right] \qquad \text{Factor out } \frac{q_A}{R^2}$$

$$\approx \frac{q_A}{R^2}\left[\left(1 + \frac{2d_A}{2R}\right) - \left(1 - \frac{2d_A}{2R}\right)\right] \qquad \text{by Eq. A.32}$$

$$= \frac{q_A}{R^2}\left[\frac{2d_A}{R}\right]$$

$$= \frac{2\mu_A}{R^3}, \qquad \mu_A = q_A d_A \quad (10.22)$$

TABLE 10.1 First ionization potentials (IE_1), permanent electric dipole moments (μ), and polarizabilities (α) of selected atoms and molecules. Polarizabilities for non-spherical tops vary with choice of axis, and only an average value, $(\alpha_{\|} - 2\alpha_{\perp})/3$, is given; $\alpha_{\|}$ and $\alpha_{\perp}$ are the polarizabilities measured parallel and perpendicular to the principal rotation axis, respectively. Lennard-Jones parameters (ε and R_{LJ}), when provided, are for the dimer of the indicated molecule.

Molecule	IE_1 (eV)	μ (D)	α (Å^3)	ε/k_B (K)	R_{LJ} (Å)
He	24.6	0.00	0.20	10	2.58
Ne	21.6	0.00	0.40	36	2.95
Ar	15.8	0.00	1.64	120	3.44
Kr	14.0	0.00	2.48	190	3.61
H_2	15.4	0.00	0.80	33	2.97
HF	16.0	1.83	2.46	2300	2.73
HCl	12.7	1.11	2.63	360	3.31
HBr	11.7	0.83	3.61		
HI	10.4	0.45	5.44	324	4.12
N_2	15.6	0.00	0.20	92	3.68
O_2	12.1	0.00	0.20	113	3.43
CO	14.0	0.11	1.95	110	3.59
CO_2	13.8	0.00	2.91	190	4.00
CH_4	12.5	0.00	2.59	137	3.82
C_2H_2	11.4	0.00	3.33	185	4.22
C_2H_4	10.5	0.00	4.25	205	4.23
C_2H_6	11.5	0.00	4.47	230	4.42
C_3H_6 (propane)	9.7	0.37	6.26	296	4.91
C_6H_6 (benzene)	9.2	0.00	10.3	440	5.27
C_8H_{18} (octane)		0.00	10.0	421	6.93
$C_{10}H_8$ (naphthalene)	8.1	0.00	17.5	554	6.45
H_2O	12.6	1.85	1.45	a	a
NH_3	10.2	1.47	2.81	a	a

[a] Lennard-Jones parameters not given because the strong dipole–dipole terms are treated separately (see text).

CHECKPOINT The series of derivations we've completed predict the attractive potential energy between two particles, where the particles have various charge-related characteristics: ionic, polar, and/or polarizable. These expressions are most useful at distances R of a few angstroms or more, both because we've been assuming that the distance is large compared to the size of the individual particle and because the repulsive contribution to the potential energy becomes the dominant effect when the molecules get very close.

where we have again taken advantage of the approximation $R \gg d$ and the equality $\mu_A = q_A d_A$. Returning to Eq. 10.16 for the interaction energy of two dipole moments, where now μ_B is our induced dipole moment $\mu_{induced}$, we find

$$u_{2-2^*}(R) = -\frac{2\mu_A \mu_{\text{induced}}}{(4\pi\varepsilon_0)R^3} \quad \text{by Eq. 10.16}$$

$$= -\frac{2\mu_A \alpha_B \mathcal{E}}{(4\pi\varepsilon_0)R^3} \quad \text{by Eq. 10.21}$$

$$= -\frac{4\mu_A^2 \alpha_B}{(4\pi\varepsilon_0)R^6} \quad \text{by Eq. 10.22}$$

$$u_{2-2^*}(R) = -\frac{4\mu_A^2 \alpha_B}{(4\pi\varepsilon_0)R^6}, \tag{10.23}$$

where the 2^* in the subscript for u indicates an induced dipole moment.

SAMPLE CALCULATION **Dipole-Induced Dipole Typical Magnitude.** To estimate a typical interaction strength, let's assume a polarizability of 2.00 Å^3, based on the values in Table 10.1. Combining this with our other assumed values of $\mu_A = 1.0$ D and $R = 3.0$ Å predicts a typical dipole-induced dipole interaction energy of about

$$-\frac{4(1.0\text{D})^2(3.3356 \cdot 10^{-30}\text{ C m/D})^2(2.00 \cdot 10^{-30}\text{m}^3)}{(1.113 \cdot 10^{-10}\text{ C}^2\text{ J}^{-1}\text{m}^{-1})(3.0 \cdot 10^{-10}\text{m})^6} = -1.1 \cdot 10^{-21}\text{J}$$

$$= -0.66 \text{ kJ mol}^{-1}.$$

For these values, the dipole-induced dipole coupling has about the same strength as the rotationally averaged dipole–dipole coupling (see the Sample Calculation), and it shares the rapid R^{-6} drop-off in strength as the separation increases. This may seem odd, since we can reasonably expect the induced dipole moment to be weaker than the permanent dipole moment. And in fact, the magnitude of the dipole-induced dipole is about a tenth the strength of the *fixed* dipole–dipole interaction strength in our Sample Calculations. However, unlike the dipole–dipole interaction, rotation of the molecules does not necessarily weaken the interaction or alter the R dependence. The dipole-induced dipole interaction is *always attractive*, because the induced dipole is oriented along the electric field lines of the permanent dipole moment. For this reason, the dipole-induced dipole interaction is often an important contributor to interactions between polar molecules in the gas phase, not just between polar and nonpolar molecules.

EXAMPLE 10.8 Dipole–Dipole and Dipole-Induced Dipole Interactions in HBr

CONTEXT Hydrogen bromide combines with alkenes to form bromoalkanes, which in turn provide useful starting materials for a wide range of reactions in organic synthesis (including Grignard reactions, nucleophilic substitutions, and radical-mediated reactions). Although quite toxic, HBr is easy to make and store in the laboratory, compared to HCl, because its boiling point (−66.4 °C) is above the temperature of dry ice (−78.5 °C). The boiling point is higher if the attractions between the molecules are stronger.

PROBLEM Using the values from Table 10.1, estimate the potential energies of the rotationally averaged dipole–dipole and the dipole-induced dipole interactions of HBr at 298 K and a separation of 2.5 Å.

SOLUTION The dipole moment is 0.83 D and the polarizability is 3.61 Å^3. Because we are looking at the interaction between one HBr and another HBr, we set $\mu_A = \mu_B = 0.83\text{D}$. Substituting these values into Eqs. 10.19 and 10.23, we find

$$\langle u_{2-2} \rangle_{N,\theta,\phi} = -\frac{2\mu_A^2\mu_B^2}{(4\pi\varepsilon_0)^2\, 3k_BTR^6}$$

$$= -\frac{2(0.83\text{D})^4(3.3356 \cdot 10^{-30}\text{C m/D})^4}{(1.113 \cdot 10^{-10}\text{C}^2\text{J}^{-1}\text{m}^{-1})^2(3)(1.381 \cdot 10^{-23}\text{JK}^{-1})(298\text{K})(2.5 \cdot 10^{-10}\text{m})^6}$$

$$= -3.1 \cdot 10^{-21}\text{J} = -1.9\text{kJ mol}^{-1}$$

$$u_{2-2^*}(R) = -\frac{4\mu_A^2\alpha_B}{(4\pi\varepsilon_0)R^6}$$

$$= -\frac{4(0.83\text{D})^2(3.3356 \cdot 10^{-30}\text{C m/D})^2(3.61 \cdot 10^{-30}\text{m}^3)}{(1.113 \cdot 10^{-10}\text{C}^2\text{J}^{-1}\text{m}^{-1})(2.5 \cdot 10^{-10}\text{m})^6}$$

$$= -4.1 \cdot 10^{-21}\text{J} = -2.5\text{kJ mol}^{-1}.$$

In this case we expect the dipole-induced dipole contribution to the attraction to be greater than the dipole–dipole contribution. How can the induced dipole effect be greater than the permanent dipole effect? This counter-intuitive result stems from the fact that while the dipole–dipole interaction is averaged over orientations that are repulsive as well as attractive, the dipole-induced dipole interaction is attractive at *all* orientations.

The Dispersion Force

A qualitative look at dispersion. What if the molecules have no dipole moments or hydrogen bonds at all? There always remains at least one significant van der Waals force, the **dispersion** or **London force,** which results from correlations between the wavefunctions of the two molecules. The dispersion force is often very weak for small molecules, but it rapidly becomes the dominant intermolecular attractive force for molecules of more than about five atoms. Unfortunately, while the dispersion force is the most widely applicable intermolecular attractive force, it is also the hardest to explain. Whereas the multipole–multipole and multipole-induced multipole forces can be treated classically, the origin of the dispersion force is rooted in quantum mechanics.

First, let's try a qualitative picture of what's going on. We find in Section 5.2 that the simplest MO wavefunction for the H_2 molecule is really very poor because it predicts that half the time the molecule dissociates into ions (Eq. 5.19). We fix this inadequacy in Section 5.2 by constructing a valence bond wavefunction (Eq. 5.24),

$$\psi_{VB}(\text{ground}) = [1s_A(1)1s_B(2) + 1s_B(1)1s_A(2)].$$

What does this wavefunction really mean? Electron 1 can be found on either nucleus A or nucleus B with equal probability, and the same goes for electron 2. What makes this different from the MO wavefunction is that there is no term in the wavefunction that represents both electrons on the same nucleus *simultaneously.* At any given time, we can't say whether the electron we measure near nucleus A is electron 1 or electron 2, but we know that there will only be one electron there. By separating the two electrons this way, the valence bond wavefunction is able to reduce the electron–electron repulsion energy and yields a more accurate and lower-energy solution than the MO wavefunction.

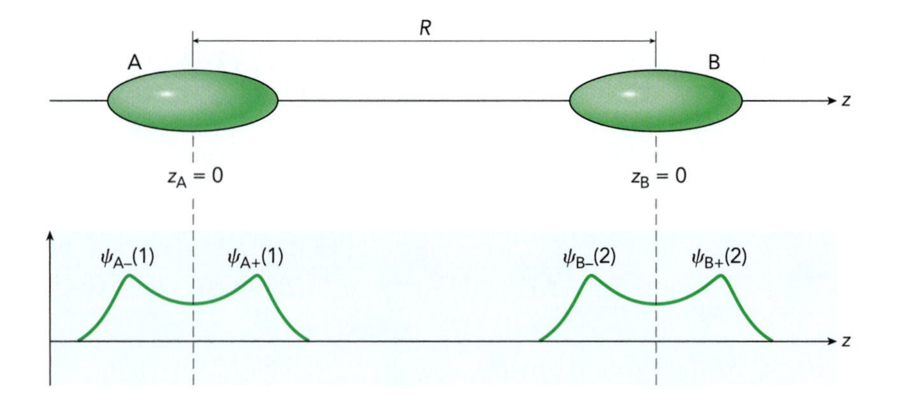

▶ FIGURE 10.12 **Two molecules A and B, with centers of mass separated by a distance *R*.** We divide the wavefunction for electron 1 on molecule A into two pieces on either side of the center of A and do the same for electron 2 on molecule B.

The dispersion force works in much the same way. Two non-polar molecules A and B approach each other, like the two nuclei in H_2, along the z axis. Let's put the origin for the electrons in molecule A at the center of molecule A, and the origin for the B electrons at the center of B (Fig. 10.12). To simplify things more, let's consider just one valence electron on each molecule: electron 1 on molecule A and electron 2 on molecule B. Electrons 1 and 2 spend an equal time on the $-z$ and $+z$ sides of their respective molecules. We can write a wavefunction for each electron that's divided into its $-z$ and $+z$ parts:

$$\psi(1) = \psi_{A-}(1) + \psi_{A+}(1), \psi(2) = \psi_{B-}(2) + \psi_{B+}(2), \qquad (10.24)$$

where, for example, $\psi_{A-}(1)$ is the wavefunction for electron 1 in the region $z_A < 0$. When the two molecules come together, we can write a combined wavefunction for both electrons. And—here's the key step—*we reduce the electron–electron repulsion if each term in the overall wavefunction places each of the two electrons on the same side of its nucleus*:

$$\psi(1,2) \approx \psi_{A-}(1)\psi_{B-}(2) + \psi_{A+}(1)\psi_{B+}(2). \qquad (10.25)$$

This wavefunction has a lower energy than we would obtain by multiplying $\psi(1)$ and $\psi(2)$ because the two terms in the wavefunction are both stabilized by an effective dipole–dipole interaction energy. The first term, $\psi_{A-}(1)\psi_{B-}(2)$, has two molecular dipole moments pointing in the $+z$ direction, while the second term has two dipole moments pointing in the $-z$ direction. Both are attractive orientations, so the two molecules are *more stable next to each other than separated.* However, because both terms are equally likely, there is still *no net molecular dipole moment on either molecule* (Fig. 10.13).

There's no reason that the molecules must be non-polar; the same argument applies to polar molecules as well. We only used non-polar molecules here so that we could neglect the dipole–dipole and dipole-induced dipole contributions to the potential energy, and to show that dispersion forces work even when no other binding mechanism is present.

Dispersion force: a derivation. Now let's try a more rigorous approach so that we can predict how properties of the molecule affect the dispersion force. The perturbation theory that we discussed in Section 4.2 provides a useful way to approach this. We'll use a simplified picture of molecules A and B: each molecule consists of a single nuclear point charge $+e$, at position z_A or z_B,

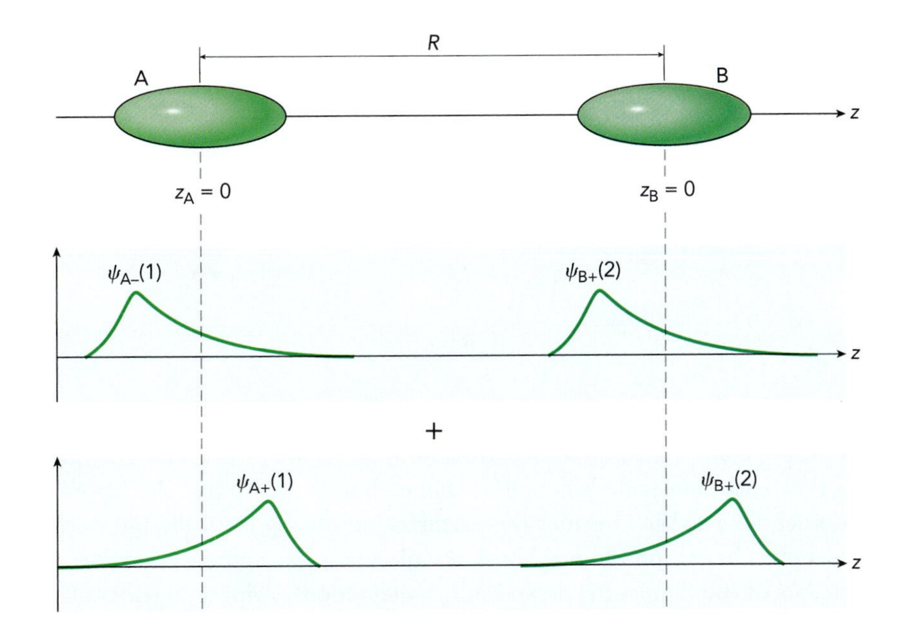

FIGURE 10.13 In-phase molecular wavefunctions. The overall wavefunction for the two molecules may be stabilized by combining the A and B wavefunctions so that the positions of electrons 1 and 2 are correlated.

surrounded by an electron cloud with charge $-e$ (Fig. 10.14). To eliminate contributions from dipole–dipole and dipole-induced dipole interactions, we make each electron distribution symmetric about the positive charges so that the permanent dipole moment is zero. We will further simplify the system by treating it as a strictly one-dimensional function of the coordinate z.

Parameters Key: the dispersion force

$\hat{H}'$	dispersion Hamiltonian
$\hat{H}_{x(A)-y(B)}$	Hamiltonian for interaction between particle x on molecule A and particle y on molecule B
R	distance between the centers of molecules A and B
z_i	z axis position of the center of molecule i
e	the fundamental charge
$\Psi^{(0)}_{A+B}$	the zeroth order wavefunction for the combined system A + B
$\psi^{(i)}_{A1}$	the ith order wavefunction for molecule A in its ground state
$E^{(i)}_j$	the ith order energy of quantum state j for the combined system A + B
A	normalization constant for $\psi^{(1)}$
E	electric field
μ_{induced}	the molecular dipole moment induced by an electric field outside the molecule
α	polarizability
u_{disp}	the potential energy of dispersion; the goal of the derivation

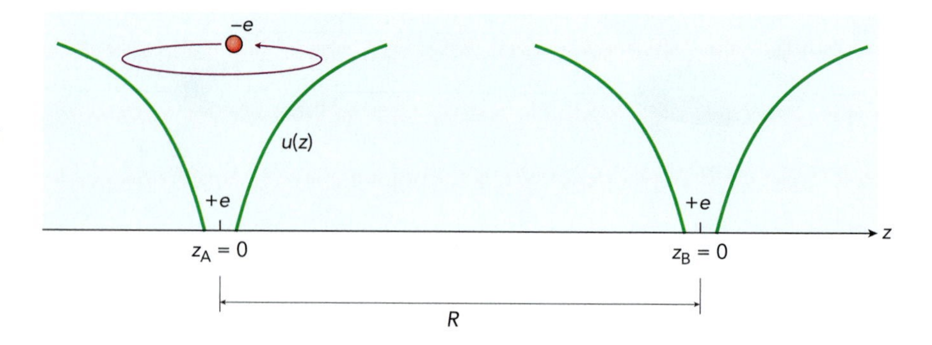

▶ FIGURE 10.14 **The coordinates for our derivation of the dispersion potential energy.** The curve $u(z)$ illustrates the potential energy of the electron–nucleus attraction.

There is a complicated molecular Hamiltonian, $\hat{H}_0$, that we can use to find the energy of each molecule when it is alone. But as the molecules approach each other, we need to consider the perturbation arising from the interactions between the charges on the neighboring molecules. This approach parallels our treatments of the dipole–dipole and other interactions. Adding together all the terms, we have a perturbation Hamiltonian of

$$\hat{H}' = \hat{H}_{\text{nuc(A)}-\text{nuc(B)}} + \hat{H}_{\text{elec(A)}-\text{elec(B)}} + \hat{H}_{\text{elec(A)}-\text{nuc(B)}} + \hat{H}_{\text{nuc(A)}-\text{elec(B)}}$$

$$= \frac{1}{4\pi\varepsilon_0}\left(\frac{e^2}{R} + \frac{e^2}{R - z_A + z_B} - \frac{e^2}{R - z_A} - \frac{e^2}{R + z_B}\right)$$

$$= \frac{e^2}{(4\pi\varepsilon_0)R}\left[1 + \left(1 + \frac{z_B - z_A}{R}\right)^{-1} - \left(1 - \frac{z_A}{R}\right)^{-1} - \left(1 + \frac{z_B}{R}\right)^{-1}\right] \quad \text{factor out } e^2/R$$

$$\approx \frac{e^2}{(4\pi\varepsilon_0)R}\left[1 + \left(1 - \frac{z_B - z_A}{R} + \left(\frac{z_B - z_A}{R}\right)^2\right) \right. \quad \text{Taylor series, Eq. A.31}$$

$$\left. - \left(1 + \frac{z_A}{R} + \left(\frac{z_A}{R}\right)^2\right) - \left(1 - \frac{z_B}{R} + \left(\frac{z_B}{R}\right)^2\right)\right]$$

$$= \frac{e^2}{(4\pi\varepsilon_0)R}\left[-\frac{2z_Az_B}{R^2}\right] = -\frac{2e^2z_Az_B}{(4\pi\varepsilon_0)R^3}. \quad \text{combine terms} \tag{10.26}$$

We use the Taylor series expansion $(1 + x)^{-1} \approx 1 - x + x^2$ to simplify adding the terms together, and we have to go at least to second order in this expansion (out to the x^2 term) to find terms that don't exactly cancel.

In the simplest picture, zero-order in perturbation theory, we neglect this perturbation and there is no interaction between the two non-polar molecules. The overall wavefunction of the system A + B is just the product of the individual A and B wavefunctions:

$$\Psi^{(0)}_{A+B} = \psi^{(0)}_{A1}\,\psi^{(0)}_{B1}, \tag{10.27}$$

where the subscript 1 indicates that the electronic wavefunction is evaluated for the ground state (electronic energy level number 1) of each molecule. The overall energy is then the sum of the individual ground state energies for A and B:

$$E^{(0)} = E^{(0)}_{A1} + E^{(0)}_{B1}. \tag{10.28}$$

To get the first-order correction to this energy, we apply the average value theorem to get the average perturbation energy using the zero-order wavefunction:

$$E^{(1)} = \int \Psi^{(0)}_{\mathrm{A+B}} \hat{H}' \Psi^{(0)}_{\mathrm{A+B}} \, d\tau = -\int \Psi^{(0)}_{\mathrm{A+B}} \frac{2e^2 z_{\mathrm{A}} z_{\mathrm{B}}}{(4\pi\varepsilon_0)R^3} \Psi^{(0)}_{\mathrm{A+B}} \, dz_{\mathrm{A}} dz_{\mathrm{B}}$$

$$= -\frac{2e^2}{(4\pi\varepsilon_0)R^3}\left(\int \psi^{(0)}_{\mathrm{A}1} z_{\mathrm{A}} \psi^{(0)}_{\mathrm{A}1} dz_{\mathrm{A}}\right)\left(\int \psi^{(0)}_{\mathrm{B}1} z_{\mathrm{B}} \psi^{(0)}_{\mathrm{B}1} dz_{\mathrm{B}}\right)$$

$$= -\frac{2e^2}{(4\pi\varepsilon_0)R^3}\left(\int (\psi^{(0)}_{\mathrm{A}1})^2 z_{\mathrm{A}} dz_{\mathrm{A}}\right)\left(\int (\psi^{(0)}_{\mathrm{B}1})^2 z_{\mathrm{B}} dz_{\mathrm{B}}\right) = 0. \quad \text{by symmetry}$$

You see, for the molecule to be strictly non-polar, the zero-order electron density distribution $\psi^2_{\mathrm{A}1}$ must be symmetric about $z_{\mathrm{A}} = 0$, whereas z_{A} is antisymmetric. Therefore, the integrand $\psi^2_{\mathrm{A}1} z_{\mathrm{A}}$ is antisymmetric, and it integrates to zero (Fig. 10.15). To first order in perturbation theory, there is no dispersion force.

To expose the interaction, we have to go to the second-order energy correction, which now depends on excited states i and j (both greater than 1) of the individual molecules:

$$E^{(2)} = \sum_{\text{excited states}} \frac{\left[\int \Psi_{\mathrm{A+B}}(\text{ground}) \hat{H}' \Psi_{\mathrm{A+B}}(\text{excited}) \, dz_{\mathrm{A}} \, dz_{\mathrm{B}}\right]^2}{E^{(0)}(\text{ground}) - E^{(0)}(\text{excited})}$$

$$= \sum_{i>1}\sum_{j>1} \frac{\left[\int (\psi_{\mathrm{A}1}\psi_{\mathrm{B}1}) \hat{H}' (\psi_{\mathrm{A}i}\psi_{\mathrm{B}j}) \, dz_{\mathrm{A}} \, dz_{\mathrm{B}}\right]^2}{(E_{\mathrm{A}1} + E_{\mathrm{B}1}) - (E_{\mathrm{A}i} + E_{\mathrm{B}j})}$$

$$= \frac{4e^4}{(4\pi\varepsilon_0)^2 R^6} \sum_{i>1}\sum_{j>1} \frac{\left[\int \psi_{\mathrm{A}1} z_{\mathrm{A}} \psi_{\mathrm{A}i} \, dz_{\mathrm{A}}\right]^2 \left[\int \psi_{\mathrm{B}1} z_{\mathrm{B}} \psi_{\mathrm{B}j} \, dz_{\mathrm{B}}\right]^2}{(E_{\mathrm{A}1} + E_{\mathrm{B}1}) - (E_{\mathrm{A}i} + E_{\mathrm{B}j})}. \quad (10.29)$$

So far, despite a few simplifications, we haven't had to make any assumptions about the energy levels of the molecules. The equation just shown is widely applicable, but the sums do make it rather difficult to read and visualize. Now let's make it much simpler by including only the lowest excited state of each molecule, with i and j equal to 2:

$$E^{(2)} \approx \frac{4e^4}{(4\pi\varepsilon_0)^2 R^6} \frac{\left[\int \psi_{\mathrm{A}1} z_{\mathrm{A}} \psi_{\mathrm{A}2} \, dz_{\mathrm{A}}\right]^2 \left[\int \psi_{\mathrm{B}1} z_{\mathrm{B}} \psi_{\mathrm{B}2} \, dz_{\mathrm{B}}\right]^2}{(E_{\mathrm{A}1} + E_{\mathrm{B}1}) - (E_{\mathrm{A}2} + E_{\mathrm{B}2})}. \quad (10.30)$$

This energy will be our dispersion energy, so we set $u_{\text{disp}} = E^{(2)}$.

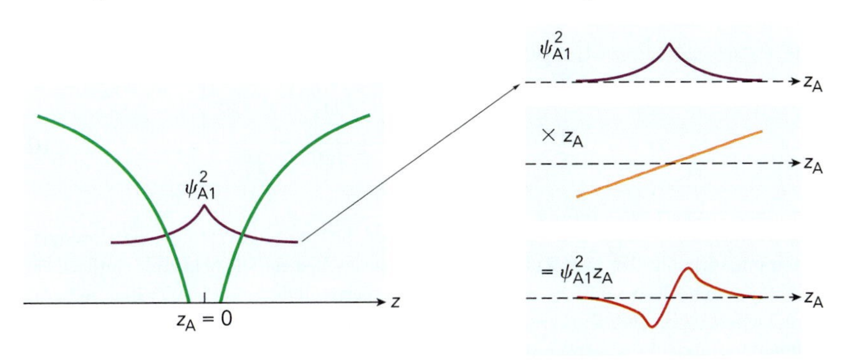

FIGURE 10.15 Symmetry of the integrals. The ground state probability density, $\psi^2_{\mathrm{A}1}$, times z_{A} yields an antisymmetric function, which integrates to zero.

You may have encountered the integrals in Eq. 10.30 previously in Problem 4.67, where they give the impact of an applied electric field on the electronic energy.[10] The polarizability, for our simplified molecules, can be solved to first order in perturbation theory as follows. We need the first-order wavefunction of one of our molecules when subjected to an applied external field ε. The interaction of the electron with this field yields a potential energy function $e\varepsilon z/(4\pi\varepsilon_0)$, which we can use to find the first-order wavefunction:

$$\psi^{(1)} = A\left[\psi_1^{(0)} + \sum_{i\neq 1}\frac{\int \psi_1^{(0)}(e\mathcal{E}z)\psi_i^{(0)}\,dz}{(4\pi\varepsilon_0)(E_1^{(0)} - E_i^{(0)})}\psi_i^{(0)}\right]. \tag{10.31}$$

We will set the normalization constant A to 1, under the assumption that only a very weak electric field is applied and therefore $\psi^{(1)}$ is almost equal to $\psi^{(0)}$.

Now we use this wavefunction to estimate how much the electron distribution deforms when a weak field is applied. The induced dipole moment will then be given by the following equation:

$$\begin{aligned}
\mu_{\text{induced}} &= \int \psi^{(1)}(ez)\psi^{(1)}\,dz \\
&= e\int\left\{\left[\psi_1^{(0)} + \sum_{i>1}\frac{\int \psi_1^{(0)}(e\mathcal{E}z')\psi_i^{(0)}\,dz'}{(4\pi\varepsilon_0)(E_1^{(0)} - E_i^{(0)})}\psi_i^{(0)}\right](z)\right.\\
&\quad\left.\left[\psi_1^{(0)} + \sum_{j>1}\frac{\int \psi_1^{(0)}(e\mathcal{E}z')\psi_j^{(0)}\,dz'}{(4\pi\varepsilon_0)(E_1^{(0)} - E_j^{(0)})}\psi_j^{(0)}\right]dz\right\} \qquad \text{by Eq. 10.31}\\
&= e\int(\psi_1^{(0)})^2 z\,dz + 2e\sum_{i>1}\left[\frac{\int \psi_1^{(0)}(e\mathcal{E}z)\psi_i^{(0)}\,dz}{(4\pi\varepsilon_0)(E_1^{(0)} - E_i^{(0)})}\right]\int \psi_1^{(0)} z\psi_i^{(0)}\,dz \\
&\quad + e\sum_{i>1}\sum_{j>1}\left\{\frac{\left[\int \psi_1^{(0)}(e\mathcal{E}z)\psi_i^{(0)}\,dz\right]\left[\int \psi_1^{(0)}(e\mathcal{E}z)\psi_j^{(0)}\,dz\right]}{(4\pi\varepsilon_0)^2(E_1^{(0)} - E_i^{(0)})(E_1^{(0)} - E_j^{(0)})}\right\}\int \psi_i^{(0)} z\,\psi_j^{(0)}\,dz.
\end{aligned} \tag{10.32}$$

Again the $\int(\psi_1^{(0)})^2 z\,dz$ integral vanishes, and we neglect the higher order terms that have large energy denominators $(E_1^{(0)} - E_i^{(0)})(E_1^{(0)} - E_j^{(0)})$, leaving

$$\begin{aligned}
\mu_{\text{induced}} &\approx 2e\sum_{i>1}\left[\frac{\int \psi_1^{(0)}(e\mathcal{E}z)\psi_i^{(0)}\,dz}{(4\pi\varepsilon_0)(E_1^{(0)} - E_i^{(0)})}\right]\int \psi_1^{(0)} z\psi_i^{(0)}\,dz \\
&= \frac{2e^2\mathcal{E}}{4\pi\varepsilon_0}\sum_{i>1}\frac{\left[\int \psi_1^{(0)} z\psi_i^{(0)}\,dz\right]^2}{E_1^{(0)} - E_i^{(0)}}.
\end{aligned} \tag{10.33}$$

This expression gives the polarizability:

$$\alpha \equiv \frac{\mu_{\text{induced}}}{\mathcal{E}} = \frac{2e^2}{4\pi\varepsilon_0}\sum_{i>1}\frac{\left[\int \psi_1^{(0)} z\psi_i^{(0)}\,dz\right]^2}{E_1^{(0)} - E_i^{(0)}}. \tag{10.34}$$

[10]The same integrals also appear in Problem 3.61, where they are needed to find spectroscopic transition intensities. In that case, a photon provides the applied electric field. This similarity allows polarizabilities to be determined in some applications (particularly for molecular ions and other molecules where dipole moments are hard to measure directly) by adding together transition intensities determined by spectroscopy.

Our final form for α looks roughly like what we could use in our equation for the dispersion energy, but again let's simplify by including only the lowest excited state, $i = 2$:

$$\alpha \approx \frac{2e^2}{4\pi\varepsilon_0} \frac{\left[\int \psi_1^{(0)} z \psi_2^{(0)}\, dz\right]^2}{E_1^{(0)} - E_2^{(0)}}. \tag{10.35}$$

This equation can now be applied separately to molecule A and to molecule B.

We can use Eq. 10.35 to solve for the integrals in Eq. 10.30 in terms of the polarizability α:

$$\left[\int \psi_1^{(0)} z \psi_2^{(0)}\, dz\right]^2 dz \approx \frac{4\pi\varepsilon_0}{2e^2} \alpha(E_1^{(0)} - E_2^{(0)}). \tag{10.36}$$

Finally, we can use this to obtain a fairly clean expression for the dispersion energy:

$$\begin{aligned} u_{\text{disp}} &\approx \frac{4e^4}{(4\pi\varepsilon_0)^2 R^6} \frac{\left[\int \psi_{A1} z_A \psi_{A2}\, d\tau_A\right]^2 \left[\int \psi_{B1} z_B \psi_{B2}\, d\tau_B\right]^2}{(E_{A1}^{(0)} + E_{B1}^{(0)}) - (E_{A2}^{(0)} + E_{B2}^{(0)})} \\ &\approx \frac{4e^4}{(4\pi\varepsilon_0)^2 R^6} \frac{\left[\frac{4\pi\varepsilon_0}{2e^2}\alpha_A(E_{A1}^{(0)} - E_{A2}^{(0)})\right]\left[\frac{4\pi\varepsilon_0}{2e^2}\alpha_B(E_{B1}^{(0)} - E_{B2}^{(0)})\right]}{(E_{A1}^{(0)} + E_{B1}^{(0)}) - (E_{A2}^{(0)} + E_{B2}^{(0)})} && \text{by Eq. 10.36} \\ &= \frac{4e^4}{(4\pi\varepsilon_0)^2 R^6} \frac{\left[\frac{4\pi\varepsilon_0}{2e^2}\alpha_A(E_{A1}^{(0)} - E_{A2}^{(0)})\right]\left[\frac{4\pi\varepsilon_0}{2e^2}\alpha_B(E_{B1}^{(0)} - E_{B2}^{(0)})\right]}{(E_{A1}^{(0)} - E_{A2}^{(0)}) + (E_{B1}^{(0)} - E_{B2}^{(0)})} && \text{rearrange denominator} \\ &= -\frac{\alpha_A \alpha_B}{R^6} \frac{\Delta E_A \Delta E_B}{\Delta E_A + \Delta E_B}, \end{aligned} \tag{10.37}$$

where ΔE is the zero-order energy difference between the ground and first excited states, $E_2^{(0)} - E_1^{(0)}$, in each molecule. In the case that A and B are chemically identical—for example, if they are both water molecules—then $\alpha_A = \alpha_B$ and $\Delta E_A = \Delta E_B$, and our estimate of the dispersion energy simplifies further to

$$u_{disp} \approx -\frac{\alpha^2 \Delta E}{2R^6}. \tag{10.38}$$

A more elegant derivation, known as the **Drude model,** approximates the molecule as an electron harmonically oscillating about a positively charged core, and it results in a similar equation:

$$u_{\text{disp}} \approx -\frac{3\alpha^2 \omega_{\text{elec}}}{4R^6}, \tag{10.39}$$

where ω_{elec} is the harmonic oscillation frequency of the electron in the model. The precision of this equation is not great; it is reliable to within about a factor of two for most typical molecules. An accurate treatment of the polarizability would draw from many excited states, not just the lowest excited state (see, for example, Problem 4.67). The larger energy difference $E_i - E_1$ that would reduce the importance of these terms is countered by the larger contribution the excited states can make to the polarizability, because excited, loosely bound electrons are much more polarizable.

CHECKPOINT At this point in our derivation of the potential energy for dispersion, an expression for the polarizability α drops out of the math. The dispersion and the polarizability are two manifestations of the same phenomenon, and scattering (including Raman transitions) can be said to be a third. Our construction of electronic wavefunctions in Chapters 4 through 7 has been valuable for understanding how molecules are put together, but we find that by itself it is not sufficient to explain how molecules *behave.*

SAMPLE CALCULATION Dispersion Energy Typical Magnitude for Small Molecule. For $\alpha = 2.00\,\text{Å}^3$ and $R = 3.0\,\text{Å}$ as before, and using an excitation energy ΔE of $0.2E_h$ (roughly 5 eV), we estimate a dispersion energy of

$$-\frac{(2.00 \cdot 10^{-30}\text{m}^3)^2 (0.2)(4.36 \cdot 10^{-18}\text{J}/E_h)}{2(3.0 \cdot 10^{-10}\text{m})^6} = -2.4 \cdot 10^{-21}\,\text{J} = -1.4\,\text{kJ mol}^{-1}.$$

The dispersion energy in our Sample Calculation is the smallest of the interaction energies we've seen so far. But, unlike dipole moments, the polarizability α tends to increase reliably with the size of the molecule. For example, if we choose ethane, with polarizability of $4.47\,\text{Å}^3$ and ionization energy of 11.5 eV ($1.84 \cdot 10^{-18}$J), then the dispersion energy at a distance of $3.0\,\text{Å}$ is roughly $3.8\,\text{kJ mol}^{-1}$. Dispersion usually dominates the intermolecular attraction among molecules of ten or more atoms.

The important features of Eqs. 10.38 and 10.39 are that the attraction between molecules again increases inversely with R^6 and that it increases as the product of the polarizabilities. Perhaps of equal importance, the preceding derivation *demonstrates* that the basis of the dispersion force is still the Coulomb force, as in all our other chemical interactions, but it is manifested in a very quantum-mechanical fashion. Two non-polar molecules can attract each other because it is possible for the wavefunction of one electron to borrow some character from one of its own excited electronic states. The wavefunction therefore polarizes itself in phase with the wavefunction of a neighboring molecule, and then the two stabilize each other. The electron remains wavelike throughout this interaction, and the average dipole moment of both molecules can remain zero.

There is an important lesson for computational chemistry in this derivation itself: we have shown that in order to calculate the effects of dispersion forces quantitatively, it is not enough to know the electron distribution in the ground state. For this reason, Hartree-Fock calculations, which in many respects are powerful predictors of molecular properties, cannot calculate dispersion forces. This limitation severely restricts our ability to predict properties of large molecules and large groups of molecules by quantum-mechanical methods.

EXAMPLE 10.9 Relative Dispersion Energies and Boiling Points

CONTEXT In commercial chemical reactors, one of the first design questions is whether the compounds present will be gases or liquids at the reactor's operating temperature. Although both the liquid and gas may be present at the same time, the boiling point lets us know which is likely to dominate at a particular temperature. For reactions that require a solvent, choosing a solvent that has the right temperature characteristics for the reaction is one of the key decisions. For non-polar molecules, the relative boiling points are largely determined by the different strengths of the dispersion forces that bind the molecules together in the liquid.

PROBLEM Carbon tetrachloride (CCl_4, $\alpha = 10.5\text{Å}^3$), carbon tetrafluoride (CF_4, $\alpha = 3.21\text{Å}^3$), and propane (C_3H_8, $\alpha = 6.26\text{Å}^3$) are all non-polar molecules. Rank these in order of increasing boiling point.

SOLUTION The strongest attractive forces will most resist separation of the molecules to form the gas, so the boiling point increases as the dispersion force goes up. The higher the polarizability, the higher the expected dispersion force. Thus, the expected ranking would be (1) CF_4 (lowest α, lowest boiling point); (2) C_3H_8; (3) CCl_4 (highest α, highest boiling point). And sure enough, the boiling points are -128 °C (CF_4), -42 °C (C_3H_8), and 76.5 °C (CCl_4).

Model Potentials

We approximated the potential energy of the vibrating molecule by a harmonic oscillator in Section 8.1, making it our first example of a model potential. For low levels of vibrational energy, the harmonic oscillator works well. However, the harmonic oscillator doesn't have a dissociation limit; there's no energy at which you can climb out of the well, because the walls on both sides keep rising forever. For a more detailed treatment, we may use the Morse potential, Eq. 8.28, which does include dissociation at the expense of more complicated solutions to the wavefunctions and energies. The weaker interactions *between* molecules do not allow us much leeway to use the harmonic oscillator, because the dissociation energy is so low that we can't ignore that part of the potential energy curve.

The trouble is that problems that deal with intermolecular interactions are often *big* problems, with many molecules, or large molecules, or both. That often limits our ability to use the most accurate available intermolecular potential energy function. Instead, intermolecular forces are calculated using approximate model potentials, and often very crude ones.

Developing an accurate intermolecular potential just for two molecules is a formidable task to begin with. The total potential energy at any given point is the sum of the various contributions we've examined in preceding sections. Only the electric monopole field due to an ionic charge is isotropic (angle-independent). All other contributions to the potential energy are angle-dependent, or anisotropic. And all of the intermolecular forces are distance-dependent, weakening as the separation between the molecules becomes large. For atoms and linear molecules it may be possible to simplify the angle dependence, but normally there are five angles that must be specified, along with the center of mass separation R, to completely define the relative orientations of the components of a van der Waals dimer. In Fig. 10.16, six angles are shown: the three necessary to specify the orientation of each molecule in the coordinate system defined for the two molecules. However, two of these angles, ϕ_A and ϕ_B, can be reduced to a single relative angle $\phi = \phi_A - \phi_B$, because the potential energy for the interaction

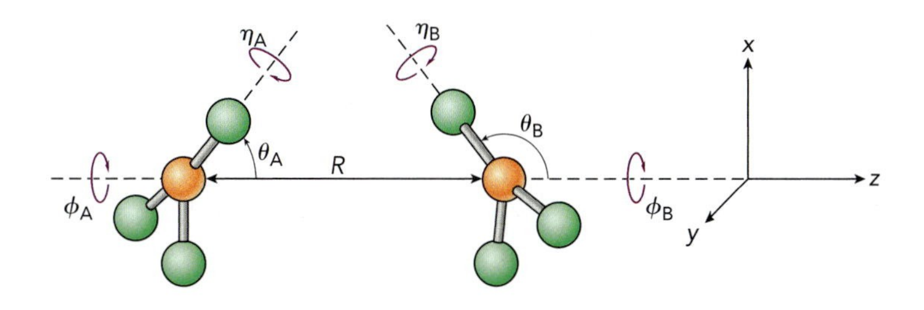

◀ **FIGURE 10.16 The six coordinates necessary for the general bimolecular potential energy surface.** The center of mass separation is R; θ_A and θ_B give the angles of each molecule's a inertial axis with respect to the intermolecular z axis; ϕ_A and ϕ_B give the angle of each molecular a axis projected into the xy plane; and η_A and η_B give the angle of rotation of each molecule about its own a axis. There are only six coordinates, because the potential energy depends on $\phi_A - \phi_B$ and not on the individual values of ϕ_A and ϕ_B.

between the molecules does not change if ϕ_A and ϕ_B are changed by the same amount (this amounts simply to a rotation of the entire dimer by ϕ_A about the z axis). Counting R, therefore, the bimolecular potential surface generally has six coordinates.

There is usually also an isotropic component to the attractive terms in the potential energy. The potential is often treated using a separation of variables:

$$u(R,\theta_A,\theta_B,\phi,\eta_A,\eta_B) = u_{\text{isotropic}}(R) + u_{\text{anisotropic}}(\theta_A,\theta_B,\phi,\eta_A,\eta_B). \qquad (10.40)$$

The dispersion force in particular often has a large isotropic component, and in large or non-polar molecules, the isotropic terms may dominate the attractive forces. On the other hand, the repulsive force is determined by the shape of the molecule and tends to be very anisotropic. Even so, the repulsive wall is most important at low values of R, just as in the diatomic potential energy function. At high R, the anisotropic terms can often be ignored.

If we consider only the isotropic terms, this leaves us with a potential that depends only on R. This potential has the same qualitative appearance as the Morse oscillator for intramolecular vibrations: a repulsive wall at low R that drops to an attractive well, this time at longer values of R roughly between 2 and 5 Å, and then an asymptotic approach to zero interaction energy as R increases to infinity. The depth of the well depends on the sum of the repulsion term and the various attractive contributions to the potential energy. In H_2O dimer, for example, there are attractive contributions to the overall potential energy from dipole–dipole, dipole-induced dipole, and dispersion forces, among other weaker ones. These distinct contributions are tabulated for selected molecules in Table 10.2. The resulting binding energy is given by the sum of all these components:

$$u(R) = u_{\text{repulsion}}(R) + u_{2-2}(R) + u_{2-2^*}(R) + u_{\text{disp}}(R) + \ldots. \qquad (10.41)$$

In any treatment of the intermolecular potential energy, the most important term is the repulsion between the molecules at low R. If we wish our simplest model potential to reflect the single dominant term, we employ the **hard sphere potential.** In this model, the molecules are similar to billiard balls: the molecules bounce off one another, but there is no attractive potential:

$$u_{\text{hs}}(R) = \begin{cases} \infty & \text{if } R \le R_{\text{hs}} \\ 0 & \text{if } R > R_{\text{hs}} \end{cases} \qquad \text{hard sphere} \qquad (10.42)$$

There is only one parameter that changes from molecule to molecule: R_{hs}, the molecular diameter. When we graph this potential energy curve (Fig. 10.17a), we find that the centers of mass of two hard sphere molecules cannot approach closer

TABLE 10.2 Contributions from attractive forces to the total intermolecular potential energy of selected molecules, calculated at the equilibrium center of mass separation.

Molecule	R_e (Å)	u_{2-2}/k_B (K)	u_{2-2^*}/k_B (K)	u_{disp}/k_B (K)
Ar_2	3.7	0	0	−130
$(CO)_2$	4.0	−0.005	−0.10	−160
$(NH_3)_2$	2.9	−750	−110	−1550
$(H_2O)_2$	3.0	−1940	−110	−640

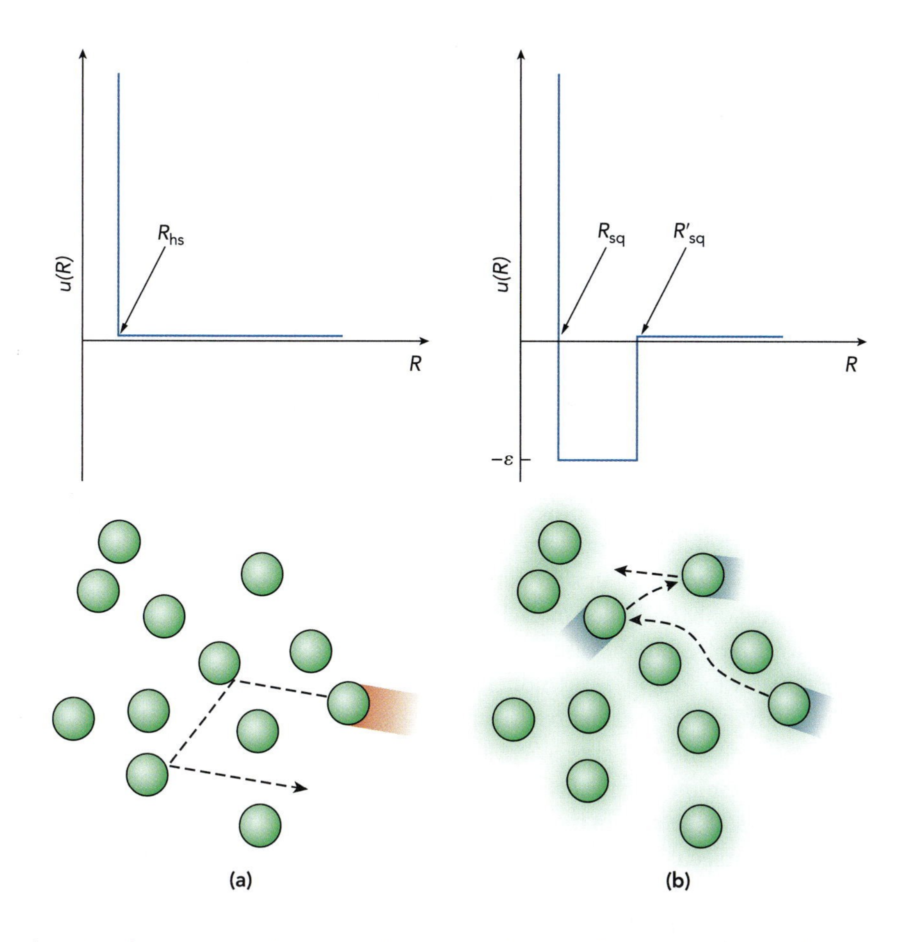

◀ **FIGURE 10.17 Motions of particles according to the simplest model potentials.** **(a)** Hard spheres bounce off one another like balls on a pool table. One hard sphere does not affect the motion of another unless the repulsive walls come in contact (here only one sphere is shown in motion). **(b)** A spherical particle with a square well is a hard sphere (solid circles) surrounded by a spherical shell (shaded area around each circle) that pulls on its neighbors. When one sphere passes near another, the two are accelerated toward each other, but the repulsive walls still cause particles to bounce away if their centers approach too closely.

than R_{hs}, because past that point the potential energy becomes infinite. Therefore, the *radius* of each molecule is $R_{hs}/2$, and R_{hs} is the molecular *diameter.*

The motions of hard spheres follow the same laws of mechanics that work for billiard balls or marbles. Collisions may be head-on or glancing, but the total kinetic energy, angular momentum, and linear momentum are all unchanged (Fig. 10.17a).

The hard sphere potential allows molecules in a calculation to occupy space, rather than behave like point particles. This potential is sufficient to predict the qualitative features of many bulk properties of molecular motion, including diffusion and viscosity. However, a gas of hard spheres will not suddenly condense into clumps as we lower the kinetic energy, so it lacks a key feature of real molecules, which have attractive forces. Schematically, the next simplest model potential is the **square well potential,** a three-parameter model that includes a straightforward attractive term (Fig. 10.17b):[11]

$$u_{sq}(R) = \begin{cases} \infty & \text{if } R \leq R_{sq} \\ -\varepsilon & \text{if } R_{sq} < R \leq R_{sq}' \\ 0 & \text{if } R > R_{sq}' \end{cases} \qquad \text{square well} \quad (10.43)$$

[11]In physics, "square well" often means a potential energy function that is symmetric about the middle of the well, with *both* walls on either side being either finite or infinite (the same as our particle in a box potential in Chapter 2). To distinguish, the potential we are using now may be called the *intermolecular square well potential.*

The molecular diameter R_{sq} represents the impenetrable distance between the centers of mass, equivalent to R_{hs} in the hard sphere potential. There is now also an attractive well that extends to a diameter R_{sq}' with a well-depth of ε_{sq}. Because this well represents a region of *lower* potential energy than the surrounding space, particles with a square well potential will *accelerate* toward one another when they enter this region (Fig. 10.17b).

There is no strict equilibrium bond length R_e for the hard sphere or square well potentials, because there is no unique value of R at which the potential energy reaches a minimum. The square well, like the Morse potential, does have a dissociation energy. For intermolecular bonds, we shall label this dissociation energy (the well-depth) by ε, rather than by D_e, which we use for dissociation of a chemical bond.

The hard sphere and square well potentials are both discontinuous functions, and therefore integrations over these functions are broken up into discrete pieces, so the correct value of the potential energy can be substituted for each range of integration. We took advantage of this property in Chapters 1, 2, and 3. Because the potential energy has only one value in each of these regions (∞, $-\varepsilon$, or 0), the integrals may be easy to solve, but the discontinuities remind us that these are not very realistic functions. A more appealing model potential would be some sort of continuous function with the right qualitative shape, something like the Morse potential used for the potential energy between covalently bound atoms in Eq. 8.28. The most common of these more realistic model potentials takes advantage of one remarkable fact: the leading terms in the three principal mechanisms for attraction between two neutral molecules—dipole–dipole (for rotating dipoles, Eq. 10.19), dipole-induced dipole (Eq. 10.23), and dispersion (Eq. 10.39)—*all* vary as R^{-6}.[12]

The Morse potential itself is not a good model here, because the exponential dependence of the curve at large R is not very similar to the R^{-6} power law. Instead, the **Lennard-Jones 6-12 potential** tends to be the model potential of choice for intermolecular interactions:

$$u_{LJ}(R) = \varepsilon\left[\left(\frac{R_e}{R}\right)^{12} - 2\left(\frac{R_e}{R}\right)^{6}\right]. \tag{10.44}$$

This two-parameter model potential has ε again representing the well-depth and R_e the equilibrium well-depth. In most references, the equation is written in terms of R_{LJ}, the R value where $u_{LJ}(R)$ crosses zero,[13] rather than R_e:

$$u_{LJ}(R) = 4\varepsilon\left[\left(\frac{R_{LJ}}{R}\right)^{12} - \left(\frac{R_{LJ}}{R}\right)^{6}\right], \tag{10.45}$$

[12]Symmetry in an unexpected place often signifies a hidden relationship. When it appears, whether in art or in science, it's worth examining. In this case, it can appear that random coincidence lends the same R^{-6} dependence to three distinct forms of intermolecular attraction. However, we've seen that these are not completely independent mechanisms, because each arises from the Coulomb interactions between two groups of charged particles, and each can be analyzed as a form of dipole–dipole interaction. The argument can also be made that the dipole-induced dipole term, which is only rarely the dominant contributor to intermolecular attractions, gets as much attention as it does precisely because it has the right R-dependence.

[13]The symbol used for the bond length at which the intermolecular potential energy crosses zero is usually σ. As indicated in Chapter 3, we have too many uses for σ already, and the traditional notation is sacrificed here for clarity.

which we can minimize with respect to R to show that

$$R_{\mathrm{LJ}} = 2^{-1/6} R_e. \tag{10.46}$$

This function has the advantage of being easily integrated and of correctly modeling the R^{-6} power law at large R. In the limit of large R, the R^{-12} term becomes insignificant, and the R^{-6} dependence dominates. A sample Lennard-Jones potential is sketched in Fig. 10.18. Values for the Lennard-Jones parameters in homogeneous interactions (interactions between like molecules) are given in Table 10.1. To estimate the Lennard-Jones potential for a heterogeneous interaction, when the compositions of two small molecules are different, the following equations are often successful:

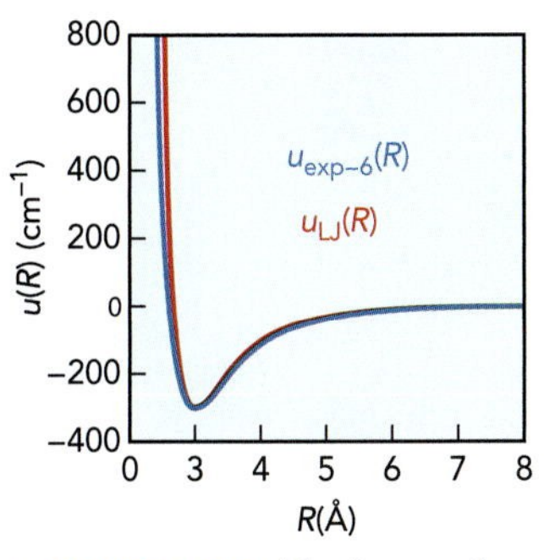

▲ **FIGURE 10.18 The Lennard-Jones 6-12 and exp-6 potential energy curves,** with $R_e = 3$Å, $\varepsilon = 300\,\mathrm{cm}^{-1}$, and (for the exp-6) $\zeta = 12$.

$$R_{\mathrm{LJ,\,AB}} \approx \frac{R_{\mathrm{LJ,A}} + R_{\mathrm{LJ,B}}}{2} \tag{10.47}$$

$$\varepsilon_{\mathrm{AB}} \approx 2\left(\frac{1}{\varepsilon_{\mathrm{AA}}} + \frac{1}{\varepsilon_{\mathrm{BB}}}\right)^{-1}. \tag{10.48}$$

SAMPLE CALCULATION **Combined Lennard-Jones Parameters.** We use Eqs. 10.47 and 10.48 to estimate the properties of neon ($R_{\mathrm{LJ,Ne}} = 2.95$Å, $\varepsilon_{\mathrm{Ne}} = 36$K) interacting with krypton $R_{\mathrm{LJ,Kr}} = 3.61$Å, $\varepsilon_{\mathrm{Kr}} = 190$ K),

$$R_{\mathrm{LJ,AB}} = \frac{R_{\mathrm{LJ,Ne}} + R_{\mathrm{LJ,Kr}}}{2}$$
$$= \frac{(2.95\text{Å} + 3.61\text{Å})}{2} = 3.28\text{Å}.$$
$$\varepsilon_{\mathrm{AB}} = 2\left(\frac{1}{\varepsilon_{\mathrm{Ne}}} + \frac{1}{\varepsilon_{\mathrm{Kr}}}\right)^{-1}$$
$$= 2\left(\frac{1}{36\mathrm{K}} + \frac{1}{190\mathrm{K}}\right)^{-1} = 61\mathrm{K}.$$

The final answers should (and do) lie between the values we substituted.

In more sophisticated treatments of polar molecules, only the repulsion and polarizability-dependent terms are included the Lennard-Jones potential. The dipole–dipole interaction is modeled by a separate Coulomb potential, using the partial charges q_i on the atoms for molecule A and q_j for molecule B. For this reason, no Lennard-Jones parameters are given for H_2O and NH_3 in Table 10.1. For example, one intermolecular potential function for water uses $R_{\mathrm{LJ}} = 3.165$Å, $\varepsilon = 78$K, and partial charges $-0.82e$ on the oxygen and $+0.41e$ on each of the hydrogens. Because the strong dipole–dipole (in this case, hydrogen-bonding) term has been shifted into a separate Coulomb term, the 78 K value of ε is much smaller than the values tabulated for very polarizable molecules such as CO.

Such treatments move beyond treating the molecules as spherically symmetric and starts the fans humming on your computers. To represent a large molecule, the Lennard-Jones potential itself may be broken up into several terms, corresponding to different regions of the molecule. In general, therefore, a combined function

$$u = \sum_{iA}\sum_{jB}\left\{4\varepsilon_{ij}\left[\left(\frac{R_{\mathrm{LJ},ij}}{R_{ij}}\right)^{12} - \left(\frac{R_{\mathrm{LJ},ij}}{R_{ij}}\right)^{6}\right] + \frac{q_i q_j}{(4\pi\varepsilon_0)R_{ij}}\right\} \tag{10.49}$$

represents the total potential energy of interaction between two molecules, with the charges q and the Lennard-Jones parameters ε and R_{LJ} adjusted for each pair of interacting regions on molecules A and B.

For applications that are sensitive to the repulsive part of the potential, the Lennard-Jones 6-12 potential may be unsatisfactory because the R^{-12} term does not very accurately represent the exponential $u_{\mathrm{repulsion}} \approx Ae^{-2cR/a_0}$ of Eq. 10.5. At the expense of including another adjustable parameter, ζ, a more adaptable model of the intermolecular potential energy is available using the **exp-6 potential** (Fig. 10.18):

$$u_{e6}(R) = \frac{\varepsilon\zeta}{\zeta - 6}\left[\left(\frac{6}{\zeta}\right)e^{\zeta[1-(R/R_e)]} - \left(\frac{R_e}{R}\right)^6\right]. \tag{10.50}$$

In this equation, ε is the well-depth at the equilibrium bond length R_e, and ζ is a parameter controlling the steepness of the repulsive wall. A higher value of ζ corresponds to a steeper wall, with $\zeta = 15$ yielding a well that is very similar to the Lennard-Jones 6–12 potential well. Equation 10.50 does fail one criterion: as R approaches zero from R_e, $u(R)$ reaches a positive maximum and then rapidly diverges toward $-\infty$, whereas the actual potential energy should continuously approach a very high positive value as R approaches zero. Normally this flaw can be overlooked because $u(R)$ takes on unrealistic values only at much smaller values of R than the molecules can actually reach, but it means being careful with the lower limit of any integral that uses the exp-6 expression.

Like the other model potentials we have considered, the exp-6 potential only accounts for the R-dependent term. Model potentials that include angular contributions are also available, but they require more (often *many* more) adjustable parameters. The well depth is so small in these intermolecular potentials that, for light molecules, only a few vibrational energy levels may exist within the well. The He_2 molecule, for example, has such a small binding energy (about 8 cm^{-1}) that only a single bound vibrational state is believed to be present. The wavefunctions that are obtained for the intermolecular potential correspond to vibrational states in which the van der Waals bonds are stretching and bending. If the anisotropic part of the potential is small compared to the isotropic part, it is possible to have bending vibrational states in which one molecule spins completely around on its own axis while still weakly bound to the second molecule. For example, vibrational spectra have been observed in which the HeHCN complex is excited from its hydrogen-bound ground state (He-HCN) to a state in which the HCN is essentially freely rotating and the helium atom happens to be nearby.

10.2 Molecular Collisions

With the masses of two particles and the potential surface for their interaction, we can in principle determine all the possible results of any encounter between the particles. From the atomic masses and the harmonic force constant, for example, we can describe all the vibrational states of a harmonically oscillating diatomic molecule made from those two atoms. Now we apply this principle to a different kind of state: an **unbound** or **free** state. These are states of two or more particles for which the energy is greater than any potential energy barriers that would hold the particles together (Fig. 10.19). Even though many of the features of the potential energy surface are below the energy of the state, these features may still play a critical role in determining the nature of an interaction between the particles.

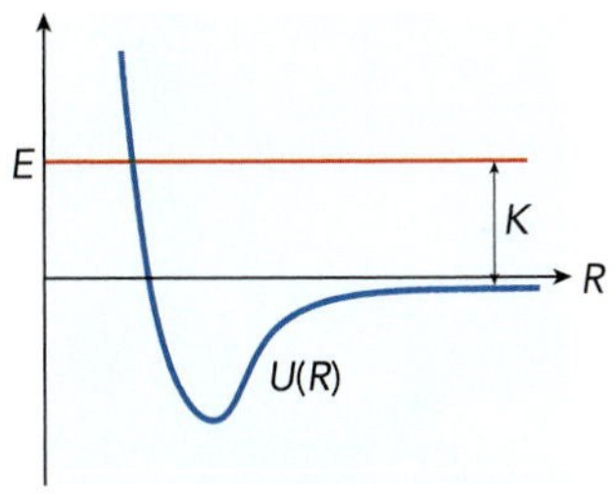

▲ FIGURE 10.19 **The energy level of an unbound state of two atoms.** The translational kinetic energy of the collision K is the difference between the energy E and the potential energy u.

Our attention will be focused through much of the remaining text on a dramatic class of interactions between free particles: collisions. In a gas, the molecules are nearly independent, but they do collide with one another and with the walls of their container. If the collisions do not change the total kinetic energy of the centers of mass—the translational kinetic energy—then then the collision is said to be **elastic** (Fig. 10.20). Collisions can also be **inelastic,** in which case some of the translational energy is transferred into one or more of the **internal coordinates** or **internal degrees of freedom,** especially the rotational, vibrational, or electronic coordinates. During an inelastic collision, the translational kinetic energy is converted into one of these other forms of energy. In our most extreme interactions, so much energy may be injected into one of the vibrational coordinates that a chemical bond breaks, or enough energy may be released that a new bond forms. This kind of energy transfer has a special relevance to chemistry, and we shall call the energy associated with bond cleavage or formation **chemical energy.** Please note that this energy does not depend on any new degrees of freedom. Depending on the initial coordinate system, the chemical energy is an extension of either the vibrational energy, involving vibrational states above the dissociation limit, or of the translational energy, involving translational states trapped in a potential well.

Elastic Collisions

Elastic collisions are chemically inert; no chemical bonds are broken or formed. Our interest in elastic collisions will usually be limited to problems involving molecular transport (the movement of molecules from one place to another, without any chemical transformation). Elastic collisions are also useful for illustrating the following important point about chemical bond formation.

Consider a container of non-interacting hydrogen atoms moving at some average translational kinetic energy K_{trans}. This kinetic energy is added to the potential energy of the dissociated atoms to get the total energy, as in Fig. 10.19. Therefore, if the collision is elastic (so that the vibrational energy remains constant), any pair of atoms colliding has an energy greater than the dissociation energy for the H_2 molecule that would be made. *The chemical bond cannot form* if energy is not transferred from the translational kinetic energy. This rule also holds for the formation of van der Waals bonds; a dimer almost never forms from simply bringing two molecules next to each other. To form a new bond between two free particles, a third particle may be introduced, so the energy released by the bond formation can be carried away in the translational energy of the third particle.

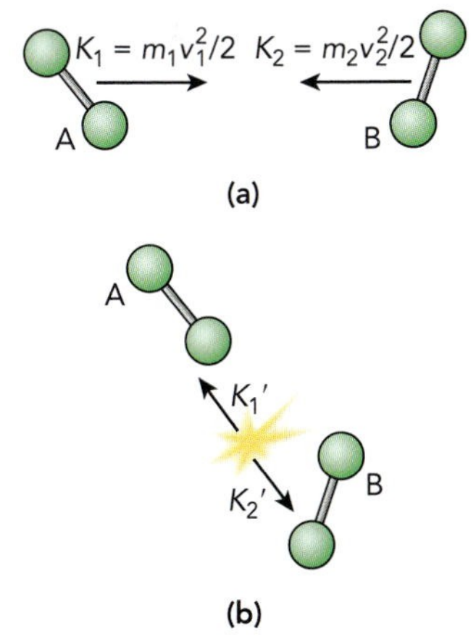

▲ FIGURE 10.20 **Elastic collisions between two molecules A and B.** The total translational kinetic energy is not altered, so $K_1 + K_2 = K_1' + K_2'$ in the illustration.

Even if no bond forms, the motion of the particles is still affected by the potential surface. As R decreases and approaches the equilibrium bond length (where the potential energy is lowest), the kinetic energy rises and the particles accelerate toward one another. Below R_e, the potential energy increases and the particles decelerate, bouncing off one another at the classical turning point. The direction of motion changes, but the total kinetic energy stays the same.

Inelastic Collisions

You will not be shocked to learn that inelastic collisions, in which energy is redistributed among different degrees of freedom, can be extremely complicated. Elastic collisions involve only a few *external* coordinates of the potential, which in the center of mass frame can be reduced to the distance between the molecules. Inelastic collisions involve the transfer of energy between these external coordinates and any combination of the internal coordinates: $3N_{atom} - 5$ or $3N_{atom} - 6$ vibrational coordinates, 2 or 3 rotational coordinates for each molecule, and perhaps even the electronic coordinates. The *elastic* collision of two CO_2 molecules, for example, at its worst is a function of only two independent coordinates: the distance between the molecules R and the scattering angle θ, which measures any change in the axis of motion for the molecules before and after the collision. On the other hand, the *inelastic* collision must also include the $3N_{atom} - 5 = 4$ vibrational coordinates and two rotational coordinates of each CO_2, a total of 13 coordinates. As a result, to model the dynamics of all the possible inelastic collisions, we would need in principle a 13-dimensional potential surface that takes all of those coordinates into account.

How does this energy transfer take place? The energy is conserved within the collision system, so whatever energy is lost from one coordinate in one molecule is gained by another coordinate somewhere else. However, some additional considerations apply. To transfer energy from translation into vibration, for example, there must be enough translational energy available to overcome the relatively large gap between vibrational states. Table 10.3 illustrates the approximate energy gaps typical of each degree of freedom,

TABLE 10.3 Energy gaps between ground and first excited states and dissociation energy D_0 for N_2 in a cubical container of volume 1 L.

Degree of freedom	$\Delta E/k_B$ (K)
translation	$5.52 \cdot 10^{-19}$
rotation	5.73
vibration	3350
electronic	71,600
dissociation	113,700

using the example of N_2 in a 1 L container. The dissociation energy D_0 gives an idea of how much energy is necessary for chemical changes to take place. With a typical available energy corresponding to about 300 K, it is clear from this table that translations in this system will be very highly excited, rotations will be moderately excited, and the molecules will almost always be in their ground vibrational and electronic states. Two molecules in this system colliding have enough energy to change the rotational state but usually not enough to change the vibrational or electronic states.

A general rule of coupling is that degrees of freedom exchange character most easily when they have similar energy level densities (Fig. 10.21). The energy level density is the number of quantum states per unit energy, and it depends on the degree of freedom and the total energy available. For example, at $E = 1000\,\text{cm}^{-1}$, H_2 does not have enough energy to reach its lowest excited electronic state at 36,000 cm^{-1} or its lowest excited vibrational state at 4400 cm^{-1}, but it could be in its $J = 4$ rotational state (which has a degeneracy of $2J + 1 = 9$) and in any of a large number of translational states. We would say that at this energy, H_2 has an electronic energy level density of only about $1/(36{,}000\,\text{cm}^{-1}) = 3 \cdot 10^{-5}/\text{cm}^{-1}$, a vibrational energy level density of about $1/(4400\,\text{cm}^{-1}) = 2 \cdot 10^{-4}/\text{cm}^{-1}$, and a rotational energy level density of about $9/(500\,\text{cm}^{-1}) = 0.018/\text{cm}^{-1}$ (because the $J = 5$ state is 500 cm^{-1} higher than the $J = 4$). The translational energy level density depends on the size of the container for the H_2 molecules, but it will be a much larger number (over 10^{17} for a 10 cm^3 volume). We predict, for example, that in most molecules the electronic and rotational wavefunctions interact very weakly, because electronic energy level densities are usually orders of magnitude smaller than rotational energy level densities. In compounds of the transition metals, where the density of electronic states is often much higher than for other atoms, the electronic states may couple more easily to vibrations and other degrees of freedom.

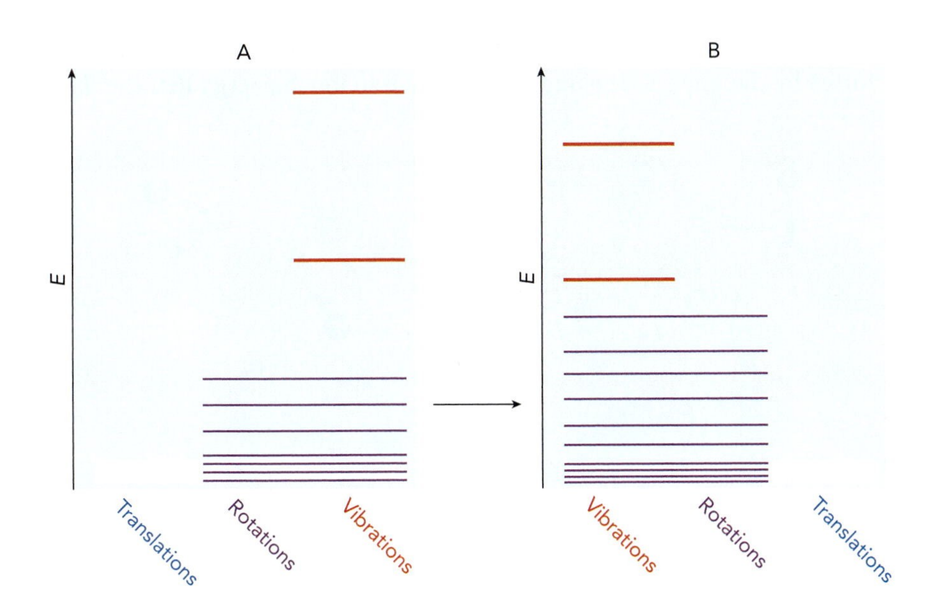

◀ FIGURE 10.21 **Qualitative comparison of the energy gaps in different degrees of freedom for two molecules, A and B.** Energy is transferred most easily between kinds of motion that have similar energy gaps. In this case, the energy transfer is represented by an arrow between rotational states of A and B. Often energy is transferred between the same kinds of motion in two different molecules, such as rotation in A to rotation in B as shown. Energy can also be transferred between rotation and translation, or between vibrational and electronic energy, more easily than between motions where the quantum states have very different energy gaps.

We can pull a visual justification for this out of quantum mechanics. Energy transfer between quantum states during a collision is a problem in time-dependent quantum mechanics. From the initial quantum states of two particles, we can calculate the probability per collision that they will arrive in a particular configuration of states after the collision. The probability is generally higher for cases where the initial and final wavefunctions are more similar. In particular, a rapidly oscillating wavefunction (corresponding to high excitation and a high density of states) does not couple easily to a slowly oscillating wavefunction (low excitation, relatively low density of states).

Rotational Energy Transfer

When rotational energy is transferred, the *total* angular momentum before and after the collision must be the same. The conservation of angular momentum can be tricky to recognize. For example, when two non-rotating molecules approach each other and their centers of mass do not quite have intersecting trajectories, there is a net angular momentum, even though the molecules are not rotating. In such an off-axis collision, the one molecule may "trip" the other, resulting in both molecules entering excited rotational states (Fig. 10.22a–c). In order to conserve energy, the combined translational energy of the two centers of mass must decrease.

Rotational energies are usually the lowest energies treated quantum-mechanically in collisions. Translational energy gaps are so small that they are treated classically, that is, as though there is no gap and the energy levels are continuous. These two degrees of freedom, having the smallest scale energies, often couple very well; in other words, rotational and translational energy are exchanged very readily during intermolecular collisions.

Vibrational Energy Transfer

The translational energy of two colliding molecules can be converted to vibrational energy if one of the molecules is strongly compressed or otherwise distorted by the impact. The impact must have so much energy that the *shape*

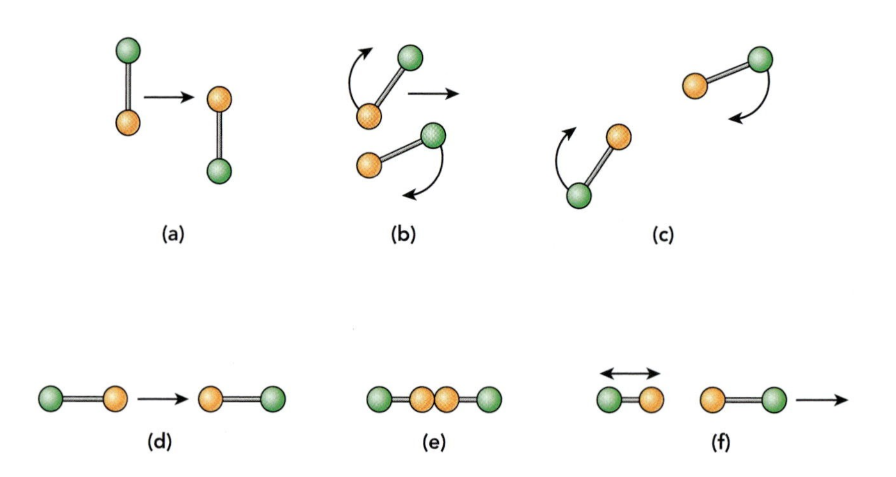

▶ FIGURE 10.22 **Rotational and vibrational energy transfer.** Two molecules approaching off-axis **(a)** may convert translational energy into rotational energy during the collision **(b)** by transferring some of the angular momentum of the off-axis motion into rotational angular momentum, **(c)** leaving one or both of the molecules in a new rotational state. Two molecules approaching with sufficient energy **(d)** may convert translational energy into vibrational energy during the collision **(e)** by deforming the bond lengths or bond angles, leaving one of the molecules (or possibly both) in an excited vibrational state **(f)**.

of the molecule actually changes (Fig. 10.22d–f)—the bond lengths or bond angles have been affected by the collision. Changes in the bond lengths or bond angles correspond to excitation of one or more of the molecule's vibrational coordinates. In quantum-mechanical terms, the molecule has been excited into a virtual state from which it will relax into one of the lower energy stationary states. The relaxation may leave the molecule vibrationally excited, in which case the total translational energy is diminished by the amount lost to vibration.

However, vibrational excitation energies are generally much higher than room temperature translational energies. There must be at least enough translational energy to promote one of the molecules from its initial vibrational state to the final state, and for small molecules, this energy typically corresponds to more than 700 K (assuming $\omega_e \geq 500\ \text{cm}^{-1}$). Although the molecules have all sorts of vibrational states available, the collision is similar to one between two very stiff springs; it takes a lot of force in the collision to bend the springs. At higher temperatures or for larger molecules (which have lower vibrational force constants), collisions are much more likely to change the vibrational state of the molecule.

Electronic and Chemical Energy Transfer

The conversion of translational energy into electronic energy is unlikely, unless there is already some strong impetus for interaction between the two molecules. For example, if chemical bonds are broken or formed, the electronic states of the products may be very different from those of the reactants. A chemical reaction results from a chain of events, beginning for example with the conversion of translational energy into vibrational energy (existing bonds distort), which is then converted into chemical energy (bonds are broken or formed). During this process the electron distribution of the atoms changes substantially, so some of the energy may be converted to electronic energy. Since these collisions are so complex and so central to chemistry, we will deal with them separately.

A common example of a strong interaction between translational and electronic degrees of freedom is the interconversion between kinetic energy of electrons and electronic energy of atoms or molecules in electrical discharges and other devices. Free electrons, formed near the cathode of a gas discharge tube or at a hot filament, typically accelerate to kinetic energies of at least 10 eV, comparable to molecular ionization or dissociation energies. On collision with a molecule, one of these electrons causes a substantial distortion in the valence molecular orbitals and may break bonds or liberate additional electrons. Electron impact ionization, in which electrons from a filament are directed by charged plates and meshes into a gas sample, has long been a staple ionization method in mass spectrometers.

CONTEXT ***Where Do We Go From Here?***

This chapter is the first baby step from one individual, fully quantum-mechanical atom or molecule toward the teeming masses of molecules that we are normally faced with anywhere in our common experience. Molecules are constantly bumping into one another, exchanging energy. This understanding of the interactions between molecules will justify why large molecules and clusters of molecules behave the way they do (Chapter 11), why solids are different from liquids, and why both are so very different from gases (Chapters 12 and 13). In *Thermodynamics, Statistical Mechanics, and Kinetics* Chapter 3, we use the model potentials in this chapter to simplify our predictions of the properties of non-ideal gases. That extrapolation from the behavior of individual molecules to the behavior of mole quantities is the great accomplishment of modern chemistry, and the results from this baby-step chapter are the key to making that *giant* step.

KEY CONCEPTS AND EQUATIONS

10.1 Intermolecular potential energy. Intermolecular repulsion. The dominant intermolecular force is repulsion. When non-reacting molecules are pushed very close together (typically to separations R of less than about 2 Å between the closest nuclei), they repel one another strongly. The repulsion potential energy $u_{\text{repulsion}}$ climbs exponentially as R approaches zero:

$$u_{\text{repulsion}} \approx Ae^{-2cR/a_0}. \tag{10.5}$$

Intermolecular attraction. Any pair of molecules also exhibits attractive forces. Although these are always weaker than the repulsion at short distances, at intermediate distances (roughly $R \approx 2 - 5$Å) the attractions may take over. There are three principal forms of intermolecular attractions:

a. Electrostatic forces are Coulomb forces between fixed charges, such as the partial charges on the atoms if we imagine the molecule to be rigid. Except for ion–ion interactions, these are direction-dependent and may be repulsive at certain relative angles. But in a gas, the rotations may average the orientations to give a net attractive potential energy.
   - Every molecule generates an electric field, which we can break down into pieces: the total charge (monopole), the charge separation along a single axis (dipole), and so on.
   - The strongest of these (non-chemical bonding) interactions is the monopole–dipole attraction between an ion and a polar molecule, which is proportional to the dipole moment μ_A and the ion charge q_B:

$$u_{1-2}(R) = -\frac{\mu_A q_B}{(4\pi\varepsilon_0)R^2}. \tag{10.14}$$

   - The strongest of these interactions between neutral molecules is the dipole–dipole interaction. Rotations in the gas phase average the dipole–dipole potential energy to the form

$$\langle u_{2-2}\rangle_{N,\theta,\phi} = -\frac{2\mu_A^2\mu_B^2}{(4\pi\varepsilon_0)^2\, 3k_B TR^6}. \tag{10.19}$$

b. Inductive forces arise from the electric field of one molecule polarizing the electron distribution of a neighboring molecule. The case of greatest interest is the induction of a dipole moment in a molecule by its polar neighbor, the dipole-induced dipole interaction. In this case the potential energy is proportional to the dipole moment μ_A of the polar molecule and the polarizability α_B of the second molecule:

$$u_{2-2'}(R) = -\frac{4\mu_A^2\alpha_B}{(4\pi\varepsilon_0)R^6}. \tag{10.23}$$

c. The dispersion force is an attractive force between *any* two polarizable particles. Even non-polar molecules are pulled together by dispersion. Furthermore, because the polarizability α tends to increase with the size of a molecule, dispersion is often the most significant attractive force between molecules with more than a few atoms, whether they are polar or not. An approximate expression for the dispersion potential energy when two identical molecules interact shows that it has the same $-1/R^6$ dependence that the rotationally averaged dipole–dipole and dipole-induced dipole interactions exhibit:

$$u_{\text{disp}} \approx -\frac{\alpha^2 \Delta E}{2R^6}. \tag{10.38}$$

Model potentials. We often need to represent the potential energy for the interaction between two molecules using simplified models. The most common are the following:

a. The hard sphere potential, which stands in for only the repulsive part of the potential.
b. The square well, which adds a flat-bottomed attractive region to the hard sphere potential.
c. The Lennard-Jones potential, which uses a very convenient power law to give a continuous curve (unlike the hard sphere and square well potentials) and even includes the $-1/R^6$ dependence shared by the dispersion, dipole–dipole, and dipole-induced dipole potentials:

$$u_{\text{LJ}}(R) = 4\varepsilon\left[\left(\frac{R_{\text{LJ}}}{R}\right)^{12} - \left(\frac{R_{\text{LJ}}}{R}\right)^{6}\right]. \tag{10.45}$$

10.2 Molecular collisions. In an elastic collision between two molecules, the total translational kinetic energy does not change, although the molecules may exchange energy and the directions of motion may change. When a collision transfers some of the translational energy into rotation, vibration, or electronic energy, then the collision is inelastic. Energy is most easily exchanged between types of motion that have similar energy gaps. So, for example, energy can normally move more easily between translational and rotational degrees of freedom (small energy gaps between quantum states) than between translation and electronic motions (small and large energy gaps, respectively).

KEY TERMS

- **Intermolecular forces** or **van der Waals forces** are the forces between non-reacting molecules.
- **Coulomb repulsion** is the repulsion between like charges, as between electron clouds around molecules separated by small distances.
- **Exchange repulsion** is the repulsion between electrons in the same volume arising from the Pauli exclusion principle.
- **Electrostatic forces** are interactions between static electric fields charge distributions.
- **Inductive forces** result from the polarization of the charge distribution by a static field, creating a net attraction between the two particles.
- The **dispersion force** or **London force** is the quantum-mechanical coupling of two electronic wavefunctions that attracts the two particles.
- **Hydrogen bonds** are relatively strong dipole–dipole attractions in which a hydrogen atom pulls together two electronegative atoms (usually one is chemically bonded to the hydrogen, one not).
- Electric **multipoles** are a way of breaking up an electric field into a sum of components from a **monopole** (a single charge), dipole (two equal and opposite charges), and higher order components (quadrupole, octopole, and so on).
- The **hard sphere potential** is a simplified intermolecular potential energy function that represents the particles as balls with an impenetrable shell.
- The **square well potential** is another simplified intermolecular potential energy function that adds a shell with a constant, negative potential energy value to the hard sphere potential.
- The **Lennard-Jones 6-12 potential** is a continuous intermolecular potential energy function, depending on two parameters and reproducing the $-1/R^6$ dependence common to many intermolecular attractions.
- An **elastic** collision is an encounter between two particles in which the total translational kinetic energy is unchanged.
- In an **inelastic** collision, some of the translational energy is converted to or from the molecule's **internal degrees of freedom,** such as the rotational, vibrational, or electronic motions.

OBJECTIVES REVIEW

1. *Determine the strongest forces relevant to the interactions between a given pair of molecules, and the functional form of the resulting potential energy.*
Write the name of the principal contributions to the intermolecular potential energy for the following pairs of molecules, and for each contribution write an expression that shows how that term in the potential energy depends on (if it depends on) R, q, μ, and/or α. Assume that the molecules are in the gas phase unless otherwise indicated.
   a. CO_2 and C_6H_6 (benzene)
   b. H_2CO (formaldehyde) and NH_3
   c. CO and HCCF freely rotating
   d. CO and HCCF fixed so all atoms share the same axis

2. *Derive the functional form of various intermolecular interaction potential energies.*
What will be the R-dependence of the electric quadrupole–quadrupole interaction, assuming the molecules are not rotating?

3. *Estimate the shape and magnitude of the potential energy functions for these interactions, given the appropriate molecular constants.*
Use Eq. 10.23 and the values in Table 10.1 to estimate the dipole-induced dipole interaction energy between HI and N_2, when they are separated by 3.90 Å (the average of their R_{LJ} values).

4. *Estimate the relative strength of forces between a given pair of molecules, based only on the molecular structures.*
For the interaction between two molecules of CH_3CH_2Br (bromoethane) in the gas phase, rank the following in the expected order of increasing strength at a distance of about 4.0 Å: (a) dipole–dipole, (b) dipole-induced dipole, (c) dispersion.

5. *Graph approximate potential energy curves based on any of several model potentials.*
Graph the square well potential for Kr, using the Lennard-Jones parameters in Table 10.1 for the well-depth and R_{sq}, and assuming $R_{sq}' = (3/2)R_{sq}$.

6. *Predict the relative importance of different modes of energy exchange between molecules.*
Two N_2 molecules A and B are each initially in the vibrational state $v = 2$, and then they collide, leaving A in the state $v = 1$. What degree of freedom in which molecule is most likely to have acquired the missing energy?

PROBLEMS

Discussion Problems

10.1 The polarizability of Ar is 1.64 Å^3, of HCl is 2.63 Å^3, and of CO_2 is 2.91 Å^3. Assume that the excitation energies of all three are similar. Identify the following as true or false:

a. Bonds between HCl and CO_2 will probably be stronger than bonds between CO_2 and CO_2.

b. The strongest dispersion forces are for $CO_2 + CO_2$.

c. If HCl bonds to HCl, there is no induced dipole moment contribution to the bonding.

10.2 For HF, $\mu = 1.83$ D and $\alpha = 2.46$ Å^3. For HCl, $\mu = 1.11$ D and $\alpha = 2.63$ Å^3. At $R = 3.0$ Å, which dimer—$(HF)_2$ or $(HCl)_2$—will have the strongest attraction due to the following?

a. the dipole–dipole force

b. the dipole-induced dipole force

c. the dispersion force

10.3 Compared to a diatomic molecule like Cl_2, a dimer (two molecules bound by van der Waals forces) typically has which of the following (ω_e here refers to the intermolecular bond stretch)?

a. a large ω_e and large B_e

b. a large ω_e and small B_e

c. a small ω_e and large B_e

d. a small ω_e and small B_e

10.4 We have three identical polar molecules AB. The three molecules are arranged in a line with equal separations between the first and second and between the second and third. For each of these arrangements, is the overall potential energy for the intermolecular dipole–dipole interactions positive or negative?

a. AB BA AB

b. AB AB BA

10.5 Consider a perfectly tetrahedral distribution of four point charges: $-q$, $-q$, $+q$, and $+q$, drawn schematically in the following figure. Does this arrangement result in a pure quadrupole field? If not, what other multipoles contribute to the electric field?

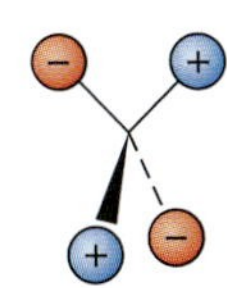

10.6 The electric field generated by the distribution of charges shown in the following figure has which of the following as its principal multipolar term: monopole, dipole, quadrupole, neither?

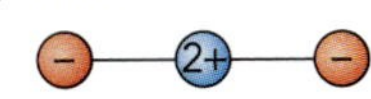

10.7 An argon atom and NH_3 molecule are weakly bound in the gas phase. Write "+" if the bond energy of this dimer increases, "–" if it decreases, and "0" if it stays the same for each of the following changes:

a. The Ar is ionized to form Ar^+.

b. The temperature is raised by 50 K.

c. The hydrogen atoms are changed from ^1H to ^3H (tritium).

d. The NH_3 is changed to NF_3.

e. The NH_3 is changed to N_2.

THINKING AHEAD ▶ [What sort of attractive interaction do we have at the start?]

10.8 State which two of the following pairs will form the strongest bonds, and sketch the most stable geometry you would expect for each of those systems:

a. Ar and H^+

b. Ar and CO_2

c. Ar and Ar

d. Ar and HCN

10.9 Pressures below 10^{-10} bar are not usually obtained quickly in the laboratory, because molecules can cling to the walls of the vacuum chamber by van der Waals forces. Of the molecules typically found in air (N_2, O_2, H_2O, CO_2) explain which is most likely to be observed in the vacuum chamber at these very low pressures.

10.10 Oils are long-chain, unsaturated hydrocarbon esters and are liquids at room temperature. How can you explain the fact that many oils have higher viscosity and surface tension than H_2O, but smaller dipole moments?

10.11 Consider the intermolecular potential curve of a molecule represented on the one hand by the hard sphere model (with repulsive wall $R_{hs} = 2.4$ Å) and on the other by the square well model (with well between $R_{sq} = 2.4$ Å and $R_{sq}' = 3.5$ Å). What is the difference, if any, between the two models when predicting the trajectory of one molecule passing at a close distance between 2.4 and 3.5 Å of another molecule?

Multipole Potential Energy Functions

10.12 Derive an equation for $u_{1-2'}(R)$, the monopole-induced dipole interaction energy, which results from the interaction of an ion and a non-polar molecule. THINKING AHEAD ▶ [If it helps, draw a picture similar to Fig. 10.8. What field is going to induce the induced dipole?]

10.13 The interaction energy for two dipoles μ_A and μ_B is given approximately by Eq. 10.18:

$$u_{2-2}(R,\theta_A,\theta_B,\phi) = -\frac{2\mu_A\mu_B}{(4\pi\varepsilon_0)R^3}\left[\cos\theta_A\cos\theta_B - \frac{1}{2}\sin\theta_A\sin\theta_B\cos(\phi_B - \phi_A)\right].$$

Calculate the interaction energy in cm^{-1} for $\mu_A = \mu_B = 1.0$ debye and $\phi_A = \phi_B$ at the coordinates $(R\,(\text{Å}), \theta_A, \theta_B) =$

a. (2.0, 0°, 0°)

b. (3.0, 0°, 0°)

c. (3.0, 90°, 0°)

d. (3.0, 180°, 0°)

e. (3.0, 0°, 90°)

f. (3.0, 90°, 90°)

g. (3.0, 180°, 90°)

h. (3.0, 180°, 180°)

i. (4.0, 0°, 0°)

10.14 Prove that the potential energy u_{1-3} for the force between an electric monopole and an electric quadrupole varies as R^{-3}. Let the distance between the monopole and the center of the quadrupole be R, and assume that the quadrupole is made of the charge distribution drawn in Fig. 10.6, with the spacing between the positive charges the same as the spacing between the negative charges, d. THINKING AHEAD ▶ [Can you break this down into a combination of systems we've looked at in the chapter?]

10.15 A molecular ion A^+ and a neutral molecule B are represented by the formal charges (in units of e) on the atoms and their *fixed* positions as drawn in the following figure. Average both charge distributions into dipole and monopole components, and then use the appropriate equations to calculate these interaction energies between A^+ and B.

a. monopole–dipole energy

b. dipole–dipole energy

Leave the energies in units of $e^2/[(4\pi\varepsilon_0)a_0]$.

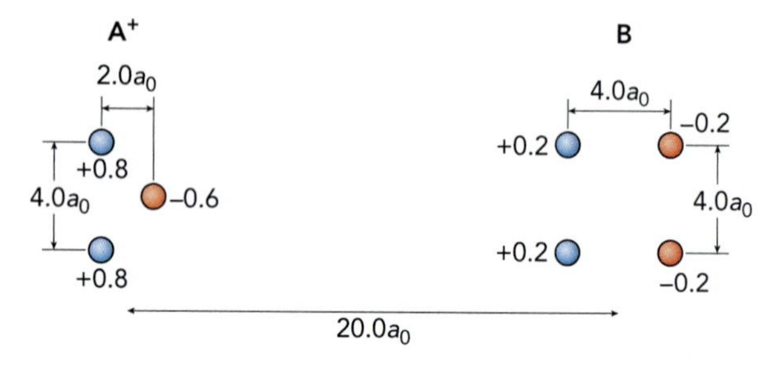

10.16 We could write the dipole–dipole interaction energy briefly as

$$u_{2-2}(R) = -C\frac{\mu_A\mu_B}{R^3}, \qquad (10.51)$$

where the constant C takes care of several constants. The system sketched in the following figure consists of a neutral atom with polarizability α_A suspended halfway between two ions of equal and opposite charges q_B and $-q_B$. The electric field generated by an ion with charge q at a distance R is q/R^2. Write an expression in the form of Eq. 10.51 showing how the interaction potential energy u between the atom and the two ions depends on R, q_B, and α_A, but don't worry about any other constants.

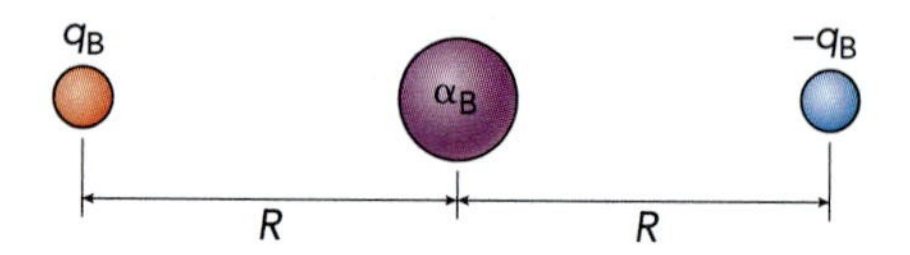

10.17 We found a formula for the potential energy of interaction for two fixed dipoles that were co-aligned. Use the same approach to find the potential energy as a function of R and in terms of the dipole moments μ_A and μ_B for the case in which the two dipole moments are parallel, as shown for the charges drawn in the following figure.

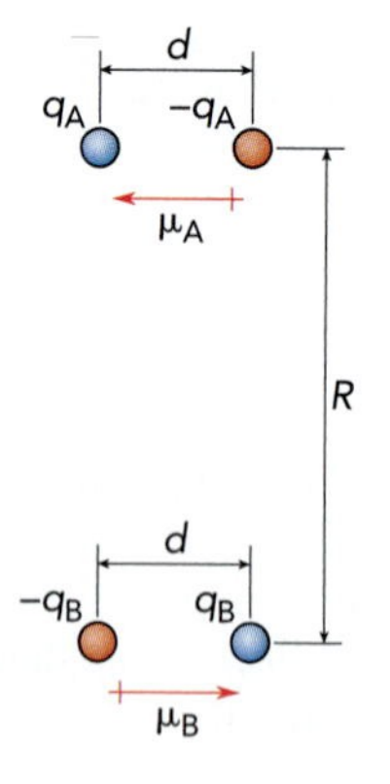

10.18 Using two point charges for a dipole, and four point charges for a quadrupole, draw an orientation in the xz plane of a dipole and a quadrupole such that there is no net attraction or repulsion.

10.19 Find the potential energy due to dispersion between two electrons trapped in one-dimensional boxes of length a and separated by a distance R, under the same approximations we used to obtain our (more general) expression for the dispersion energy. Do *not* assume we know the polarizabilities of the boxes.

10.20 For the charge distribution drawn in the following figure, write the monopole component and the dipole component (in units of ea_0). Draw the dipole moment vector (use the convention minus to plus) on the picture.

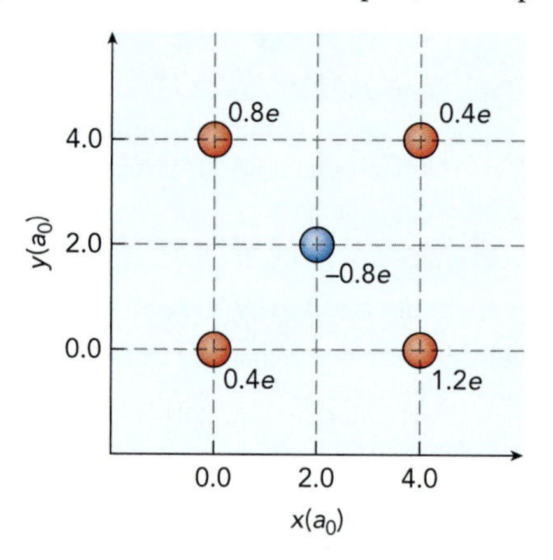

General Intermolecular Potential Energies

10.21 What is the *minimum* number of coordinates necessary to describe completely the intermolecular potential energy function of two linear molecules? Assume that you can neglect any coordinates for the internal vibrations of each molecule. Give an example of such a set of coordinates. THINKING AHEAD ▶ [If we assume that the two molecules are both diatomics, how many coordinates are there altogether? How many are for just the internal vibrations?]

10.22 The interactions among N non-linear molecules, L linear molecules, and A atoms are dictated by a multi-dimensional intermolecular potential energy surface. Find a general equation for the minimum number of coordinates necessary to completely describe this surface. Assume the molecules are completely rigid (so coordinates for the internal vibrations should be neglected), but they may rotate and translate relative to each other. THINKING AHEAD ▶ [How many coordinates are there for each of the different components (non-linear, linear, atomic), if we leave out the internal molecular vibrations?]

10.23 The potential energy, averaged over angle, between two H_2O molecules at a separation of $R = 3$ Å is roughly $-300\ \text{cm}^{-1}$, relative to the potential energy at infinite R. Assume that the same measurement for the interaction between one H_2O molecule and one *trans*-1,3-butadiene molecule is roughly $-200\ \text{cm}^{-1}$. Compare the potential energies for these two interactions at $R = 6$ Å and at $R = 12$ Å. What does this imply for the solubility of butadiene in H_2O?

10.24 Draw a qualitative potential energy curve for the interaction between two helium atoms. Give approximate values for the distances and energies of any features on the curve. Then on the same graph, plot a qualitative curve for the interaction between two UF_6 molecules, showing how UF_6 differs from He_2.

10.25 The HCl dimer, $(HCl)_2$, has a well-depth of 173 cm^{-1} and an equilibrium bond length of 4.2 Å for the van der Waals bond at 300 K. The dipole moment for HCl is 1.109 D and the polarizability is $2.7 \cdot 10^{-24}\ \text{cm}^3$. Estimate the contributions (in cm^{-1}) to the well-depth from (a) the dipole–dipole term (averaged over rotations), (b) the dipole-induced dipole term, and (c) the dispersion term.

10.26 Based on the following parameters, estimate in kJ mol^{-1} the total intermolecular potential energy for attraction between freely rotating HI and CO at a separation of 4.0 Å and a temperature of 298 K:

	ΔE (eV)	μ (D)	α (Å^3)
HI	7.72	0.45	5.44
CO	8.07	0.11	1.95

10.27 Draw what you think will be the most stable structure for the van der Waals molecules (a) Ar-HCl and (b) Ar-benzene.

Model Potentials

10.28 Draw approximate energy levels and wavefunctions for the bound states ($E < 0$) of the square well potential. THINKING AHEAD ▶ [How is this like other cases in quantum mechanics that you may know from earlier chapters?]

10.29 Prove that for the Lennard-Jones 6-12 potential curve, $u_{LJ}(R)$ has its minimum at R_e and has a value $-\varepsilon$ at that point. THINKING AHEAD ▶ [How can we find the minimum of any function from its algebraic form?]

10.30 Use the Lennard-Jones potential to estimate the energy in K at which the centers of mass of two Kr atoms colliding head-on will reach a minimum separation of 3.00 Å. THINKING AHEAD ▶ [Sketch this interaction on a graph showing the potential energy. What is the mathematical condition that holds when the two atoms reach their minimum separation?]

10.31 A possible form of the intermolecular potential for R^{-3} interactions (such as monopole–quadrupole) is the following:

$$u(R) = \beta\varepsilon\left[e^{-\alpha R} - \left(\frac{\gamma R_e}{R}\right)^3\right].$$

Find β and γ in terms of ε, R_e, and α so that ε is the well-depth and R_e is the equilibrium bond length.

10.32 Prove that for the exp-6 potential (Eq. 10.50),

$$u(R) = \frac{\varepsilon\zeta}{\zeta - 6}\left[\left(\frac{6}{\zeta}\right)e^{\zeta[1-(R/R_{e6})]} - \left(\frac{R_{e6}}{R}\right)^6\right],$$

R_{e6} is the equilibrium bond length and ε is the well-depth. THINKING AHEAD ▶ [How do we define the equilibrium bond length and well depth mathematically?]

10.33 The Lennard-Jones parameters for three van der Waals complexes are as follows:

a. LiAr: $\varepsilon = 0.92 \cdot 10^{-21}$ J, $R_{LJ} = 3.7$ Å
b. Ar_2: $\varepsilon = 2.0 \cdot 10^{-21}$ J, $R_{LJ} = 3.8$ Å
c. Xe_2: $\varepsilon = 3.9 \cdot 10^{-21}$ J, $R_{LJ} = 3.9$ Å

Express the dissociation energies D_e in degrees K, equilibrium bond lengths R_e in Å, and explain the observed trends in D_e and R_e. THINKING AHEAD ▶ [What force or forces are at work, and what atomic parameters are relevant?]

10.34 The Lennard-Jones parameters for Ar_2 are $\varepsilon = 2.0 \cdot 10^{-21}$ J and $R_{LJ} = 3.8$ Å. At what values of R are the classical turning points (where $E = u(R)$) for the lowest-energy state, which is at about $-70\ \text{cm}^{-1}$? THINKING AHEAD ▶ [How can we solve for more than one value of R?]

10.35 If the zero-point energy in a Lennard-Jones potential is $9\varepsilon/100$, what are the minimum and maximum classically allowed values of R in the $v = 0$ state, in terms of R_e?

10.36 One model potential used for intermolecular forces combines the square well and an exponential attractive term:

$$u_{sq}(R) = \begin{cases} \infty & \text{if } R \le R_0 \\ -\varepsilon e^{-R/R_0} & \text{if } R > R_0 \end{cases}$$

Assume that this potential function has only three bound states. Estimate the energy levels and sketch the wavefunctions for those states.

Molecular Collisions

10.37 The potential curve for the Ar_2 dimer has a dissociation energy of about 85 cm^{-1}. Let one Ar_2 collide with a third Ar atom so that the total kinetic energy of the system is 200 cm^{-1}. If the Ar_2 bond is broken and all three atoms are left traveling at equal speeds v, find v.

THINKING AHEAD ▶ [What relation allows us to link the energy of the system to the speeds?]

10.38 Two N_2 molecules are each initially in the $v, J = 0, 0$ state, traveling toward each other with the speed $5.15 \cdot 10^2 \text{ ms}^{-1}$. They collide. What are the possible values of v and J in which they are left after the collision?
THINKING AHEAD ▶ [What conservation law is important here?]

10.39 Two H_2O molecules in their ground rotational and vibrational states approach each other along trajectories that do not intersect but approach within 0.5 Å of each other. The total translational kinetic energy is about 50 cm^{-1}. Predict their behavior (a) just before the collision and (b) their condition after the collision.

10.40 Find the second derivative of the exp-6 potential at the equilibrium separation, and show that the value of ζ must be greater than 7 or less than 6 for this expression to correctly model the intermolecular potential energy.
THINKING AHEAD ▶ [If the exp-6 potential does behave correctly, what do we know about its second derivative at R_e?]

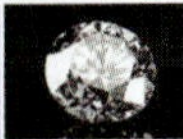

11 Nanoscale Chemical Structure

LEARNING OBJECTIVES

After reading this chapter, you will be able to do the following:

❶ Predict the major bonding forces and trends in various physical properties in homogeneous molecular clusters.

❷ Identify the chief terms in the force fields used for molecular mechanics calculations.

GOAL *Why Are We Here?*

The goal of this chapter is to extend our understanding of chemical bonding and intermolecular forces to distance scales of several nanometers. This is an area of enormous enthusiasm and effort at present, and it is relatively new to the physical chemistry curriculum.

CONTEXT *Where Are We Now?*

In Chapter 10, we treated intermolecular interactions in a series of examples, two particles at a time. As we expand our perspective from the microscopic view of individual molecules out toward the macroscopic limit, we rapidly reach a point where the particles are too numerous to treat their interactions with that same attention to detail. But this point is critical to our understanding of chemistry, because here is also where we first begin to glimpse the properties of bulk materials.

SUPPORTING TEXT *How Did We Get Here?*

Chapters 1–4 of this text deal with individual atoms, and we collect a few of those together to make the molecules studied in Chapters 5–9. The focus in Chapters 10–13 is on what happens when we continue this process, collecting a few molecules together or even larger numbers of atoms to make enormous molecules. Therefore, we will draw principally on the results of Chapter 5 (chemical bonding) and Chapter 10 (intermolecular forces) to describe how these components interact to form large clusters or molecules. Specific equations and sections of text information we will use include the following:

- Equation A.42 defines the Coulomb potential energy, which is at the heart of nearly all the attractions and repulsions between particles that we consider:

$$U_{\text{Coulomb}} = \int_0^{U(r_{12})} dU = -\int_\infty^{r_{12}} F_{\text{Coulomb}}\, dr = -\int_\infty^{r_{12}} \frac{q_1 q_2}{4\pi\varepsilon_0 r^2}\, dr = \frac{q_1 q_2}{4\pi\varepsilon_0 r_{12}}.$$

- The de Broglie wavelength (Eq. 1.3), $\lambda_{\text{dB}} = h/(mv)$, is the key parameter determining whether these clusters and macromolecules reach the limit where we can abandon quantum mechanics. (Don't get your hopes up.)
- Equation 3.21 gives one form of the Schrödinger equation used for one-electron atoms:

$$\left[-\frac{\hbar^2}{2m_e}\frac{1}{r^2}\frac{\partial}{\partial r}r^2\frac{\partial}{\partial r} + \frac{\hbar^2 l(l+1)}{2m_e r^2} - \frac{Ze^2}{4\pi\varepsilon_0 r}\right]R(r)\, Y_l^{m_l}(\theta,\phi) = ER(r)Y_l^{m_l}(\theta,\phi),$$

  where $\psi(r,\theta,\phi)$ has been divided into a radial wavefunction $R(r)$ and an angular function $Y_l^{m_l}(\theta,\phi)$. This relationship turns out to be helpful in considering loosely bound electrons in metal clusters, as does Eq. 3.57 for the degeneracy of the electronic states in a one-electron atom,

$$g = 2\sum_{l=0}^{n-1}(2l+1) = 2n^2.$$

- Section 5.4 describes some of the non-covalent interactions that can lead to cluster formation. Specifically, (1) ionic bonding occurs when electron density is largely transferred from one atom to another, rather than filling out the middle of the binding region; (2) metallic bonding occurs when metal cations are bound by the negative charge of valence electrons that have become delocalized throughout the structure. Both of these examples are consistent with the binding force defined in Section 5.1, so the force at work in these chemical bonds is fundamentally the same as in covalent bonds: the attraction of the nucleus of one atom for the valence electrons of another atom.
- Section 10.1 shows us how to estimate the potential energy functions relevant to the clusters and intramolecular interactions in macromolecules that we consider in this chapter. Table 10.1 provides some sample parameters for molecules that we can use for numerical estimates.
- The **Lennard-Jones 6-12 potential** tends to be the model potential of choice for intermolecular interactions (Eq. 10.44):

$$u_{\text{LJ}}(R) = \varepsilon\left[\left(\frac{R_e}{R}\right)^{12} - 2\left(\frac{R_e}{R}\right)^{6}\right].$$

11.1 The Nanoscopic Scale

From the microscopic perspective on individual molecules and molecule–molecule interactions, the path toward bulk properties of matter may seem obvious: just add more molecules. From Chapter 10, we know how to estimate any relevant potential energy functions for the interactions, and with this we can write the Hamiltonian for the system and integrate the Schrödinger equation to find the wavefunctions and energies.

Quantum mechanics successfully models the structure and dynamics of microscopic chemical systems and in theory continues to work for macroscopic systems as well. But quantum mechanics has practical limitations. For large systems, solving the Schrödinger equation becomes impossibly demanding. A single molecule of glucose has 96 electrons, and the simplest Hartree-Fock calculation (Section 7.3) for that molecule requires over 4000 individual electron-pair integrals. Even if we *can* solve the system's quantum mechanics, we may not want all the information we'd get. For a collection of 100 atoms or 50 molecules, we tend to lose interest in the dynamics of the individual particles. Instead we want the big picture, the overall trends and features of the system.

There's a middle ground, which we'll call the **nanoscopic scale.** As methods improve for manipulating and probing molecules at the microscopic scale and in bulk quantities, an intermediate molecular regime has become the focus of intense study. In this regime, the systems are just becoming too complicated for detailed quantum mechanical analysis, but the effects of quantum mechanics are not completely hidden. Typical system sizes in the nanoscale regime run to the hundreds of angstroms or tens of nanometers, encompassing up to several thousand atoms.

Clusters and Macromolecules

Just as an appreciation of the interatomic potential energy helped us to describe the structure of molecules in Chapter 5, the potential energy functions for intermolecular interactions discussed in Chapter 10 let us address the structure of very large *groups* of atoms. Let's divide these groups into two principal forms:

- groups of many identical atoms or molecules, called **clusters**
- single molecules of roughly a thousand atoms or more, called **macromolecules**

Because a cluster is composed from one building block, such as silicon atoms or H_2O molecules, its characteristics vary rather smoothly as these blocks are added or removed. A convenient reference parameter for a particular cluster is its **cluster size,** the number of building blocks present.

A macromolecule, on the other hand, is composed of many distinct chemical constituents, and its properties may change substantially with changes in a few critical atoms.

Why are the intermolecular forces so important for these systems? Many clusters are held together solely by these weak forces, and macromolecules are often flexible enough to bend and fold upon themselves under the influence of these forces. Long carbon chain macromolecules are typically more stable rolled up into balls than strung out like strings; when dispersion and hydrogen bonds are formed among some fifty atoms out of a thousand, the resulting stabilization energy may dictate the overall shape of the molecule.

Much of the current research in physical chemistry is directed toward understanding these systems. Clusters and macromolecules form the bridge between the microscopic world, successfully described by quantum mechanics, and the macroscopic, classical realm of bulk liquids and solids. Although quantum mechanics works well in the microscopic limit, it rapidly becomes too cumbersome as the molecules expand into the nanoscopic realm. On the other hand, the chemistry of bulk fluids and solids is still described largely by empirical laws, which are not only theoretically unsatisfying but often incapable of guiding our ventures into new chemical territory. To this end, experimentalists and theoreticians are struggling to develop methods and, more importantly, insights in the examination of the compounds in this middle ground.

Pragmatic interests as well as the desire for theoretical insights drive our fascination with nanoscopic chemistry. In particular,

- biochemical research increasingly capitalizes on our ability to design and manipulate macromolecular proteins and nucleic acids,
- metal and semiconductor clusters are being pursued for applications to molecular-scale electronic circuitry and for their unique optical properties, and
- polymers, which are macromolecules composed of identical or nearly identical subunits, form a vast class of materials whose properties depend on the structure at these small-distance scales.

11.2 Clusters

In keeping with our overall strategy of moving from the small, non-interacting systems to large, strongly-interacting systems, let's start by looking at clusters held together by the weakest forces. As we turn up the strength of the binding, we'll see accompanying changes in the resulting structure and properties.

Weakly Bound Clusters

The simplest weakly-bound clusters are those built from noble gas atoms, such as neon and argon.[1] What do we expect to be the optimal structure for one of these noble gas clusters, and how does the structure depend on the cluster size? For that matter, can we even describe these clusters in terms of just one structure?

With all the electrons paired and a spherically symmetric charge distribution, the noble gas atoms (to a first approximation) have no electric or magnetic multipole moments whatever. That leaves the dispersion forces to account for the binding of the cluster. The dispersion force depends on the polarizability, and since these single atoms are equally polarizable in all directions, the potential energy curve for each atom is completely isotropic; it initially depends only on R.

Starting from a single atom, as we increase the cluster size by adding more atoms, the structure of the cluster tends to reflect this isotropy—the atoms form roughly globular clusters. The dispersion potential energy disappears quickly with distance, in proportion to R^{-6}, so it is only effective near R_e, where the atoms are essentially in contact. Each atom that is added to the cluster is therefore stabilized by a net dispersion energy roughly in proportion to the number of nearest neighbor atoms, so it will be surrounded by as many nearest neighbors as possible. If we pick one atom and name it "A," the number of atoms in contact

[1]Helium turns out to be not so simple. Its low polarizability binds its clusters in very shallow potential wells, where minimal vibrational excitation leads to dissociation. That, coupled with the low atomic mass, makes helium clusters peculiarly quantum-mechanical in character compared to clusters of heavier atoms. We follow up on this property of helium in *TK* Section 4.4.

with it is called the **coordination number** of A, $\mathcal{C}$. According to this model, the coordination number tends toward a maximum, which minimizes the number of exposed atoms near the surface of the cluster. For a given cluster size N, the volume of the cluster is roughly NV_{atom}, regardless of the arrangement of the atoms. The atoms adopt the geometry with the smallest ratio of volume to surface area, and that geometry is the sphere. This is a fairly general feature of chemical structure: when the attractive forces within a system are stronger than the attractions for its surroundings, the system tends to roll up into a ball.

To maximize the coordination number $\mathcal{C}$, what's the best we can do? The question is an old one, known as the Gregory-Newton or "kissing number" problem, and it has a well-known answer. But let's carry out a crude geometrical analysis just to see that the answer makes sense.

Take one atom of radius r and surround it with several other identical atoms; we don't know how many yet (Fig. 11.1a). If we pack the surrounding atoms around the central atom as closely as possible, we can hope that they will all contact one another (Fig. 11.1b). The points of contact will occur at a distance $2r$ from the middle of the central atom and will define a sphere with a surface area of $4\pi(2r)^2 = 16\pi r^2 = 50.3r^2$ (Fig. 11.1c). Call that the "cage sphere." Next, imagine that we've unwrapped that sphere of surrounding atoms into a flat sheet. If we arrange the cage atoms in a rectangular grid over an area $50.3r^2$, giving each atom a square area $(2r)^2$, then there's room for 12 atoms. (Although we could crowd the atoms more closely together on the sheet, when we curve the sheet back into a sphere we'll find that the atoms bump into one another too much if we have more than 12.)

Simplistic though this analysis is, we arrive at the correct value, and the result is worth remembering:

The maximum coordination number of a set of identical spheres is 12.

We didn't worry about exactly how the atoms were spread out, and there are in fact *two* distinct arrangements of the cage atoms that attain $\mathcal{C} = 12$: the cuboctahedral and the icosahedral forms pictured in Fig. 11.2. Both of these forms stabilize the central atom with 12 attractive interactions, but they differ in the number of interactions among the atoms in the cage.

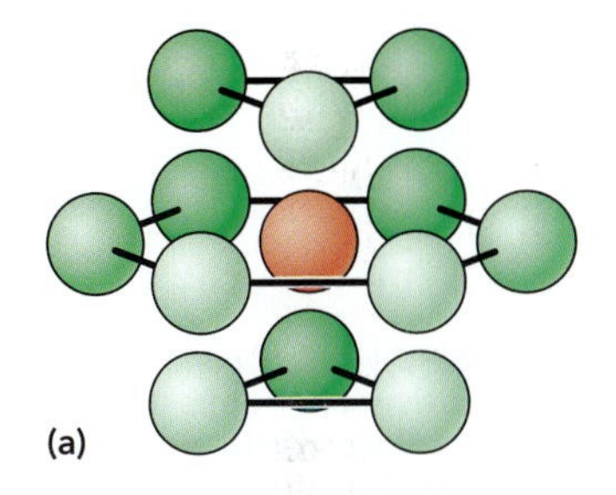

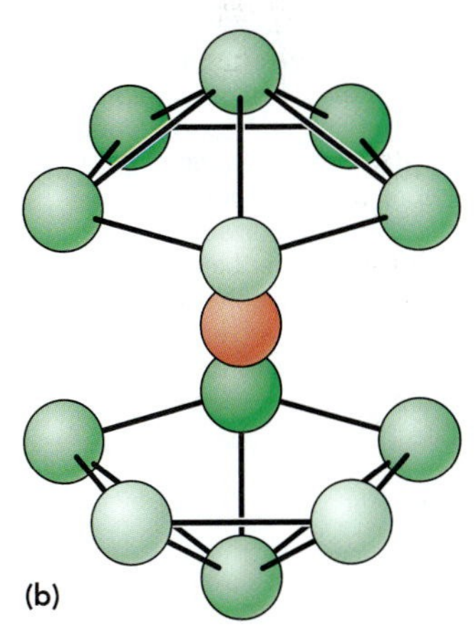

▲ FIGURE 11.2 **Two structures of the 12 cage atoms in a noble gas complex with $\mathcal{C} = 12$.** **(a)** The cuboctahedron, which can be divided into three layers of three, six, and three cage atoms; and **(b)** the icosahedron, which can be separated into four layers of one, five, five, and one.

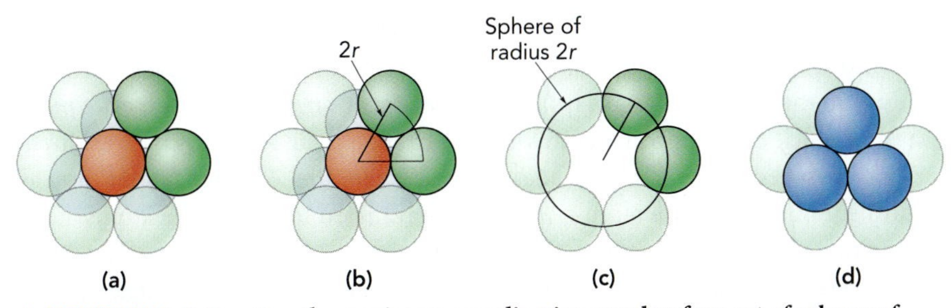

▲ FIGURE 11.1 **Estimating the maximum coordination number for a set of spheres of radius r.** **(a)** We first pack the cage atoms together as tightly as we can so that each cage atom is pressed up against other cage atoms and against the central atom. **(b)** For any cage atom that touches our central atom, the line between the centers of the two atoms is of length $2r$. **(c)** The center of every cage atom therefore lies on a circle of radius $2r$. **(d)** We can place three atoms on top and three below to complete the arrangement.

Let's assume that each pair of adjacent atoms lowers the overall energy by ε. To figure out the cluster stabilization energy, we can count all the atoms involved in each interaction and then divide by 2 at the end (because we want the number of interactions *between* each pair of atoms, not the number of atoms). For example, in the cuboctahedral arrangement, dividing the arrangement into three layers as shown in Fig. 11.2, each atom in the top and bottom layers interacts with four other atoms: the two in the same layer and two in the middle layer. So we take the number of atoms in the top and bottom layers (6), multiply by the number of atoms in contact (4), and divide by 2 to get the number of interactions: $(6 \times 4)/2 = 12$. Next, if we look at the interactions in the middle layer, we see that each of those six atoms interacts with four neighbors: two in the same layer, one atom on the top, and one on the bottom. This arrangement contributes an additional $(6 \times 4)/2 = 12$ interactions. We have a total interaction energy of $(12 + 12)(-\varepsilon) = -24\varepsilon$ in the cage. Adding this to -12ε for the interactions with the central atom gives a total stabilization of -36ε for the cuboctahedral arrangement.

In the icosahedral arrangement, there are five interactions in each of the two middle layers, five interactions between the top layer and the layer below, five interactions between the bottom layer and the layer above, and ten interactions between the two middle layers, giving an overall stabilization in the cage of $(5 + 5 + 5 + 5 + 10)(-\varepsilon) = -30\varepsilon$. Including the stabilization of the central atom gives a total stabilization of -42ε, so the icosahedral arrangement is more stable.

CHECKPOINT A general hypothesis, which forms the basis for this analysis, is that when direction-independent (*isotropic*) attractive forces determine the structure of a cluster or any other grouping, the resulting structure tends to be roughly spherical. A sphere has the highest ratio of volume to surface area of any three-dimensional shape, so it maximizes the number of particles that are in the middle, surrounded in all directions by the attractive forces of other particles.

SAMPLE CALCULATION Cluster Stabilization Energy. Let's estimate the maximum stabilization energy of the Ar_4 cluster if the interaction energy for each pair of atoms in contact is $-\varepsilon$. We can arrange the four atoms in a trigonal pyramid so that each of the four atoms is simultaneously in contact with the other three. This geometry gives $(4 \times 3)/2 = 6$ distinct interactions for a stabilization energy of -6ε.

We can't go much farther with a structural analysis of these clusters, because their geometries shift too easily. If the bonding mechanisms are weak, then a little intramolecular vibrational energy can easily be transferred into reorientation or relocation of the particles in the cluster. With cluster sizes of 10 or more, there will be competing geometries with similar energies. The dynamics are even harder to study than the structure (as always), because even for one structure there are *many* ways in which the individual atoms may rearrange their locations. For example, in the icosahedral Ar_{13} cluster described earlier, the 12 cage atoms are all equivalent but the central atom—which is coordinated to all 12 cage atoms at the same time—is unique. Therefore, there are 12 more structures that have exactly the same energy but with a different atom in the middle. We could investigate such a system, say, by using a mixture of isotopically labeled atoms (ignoring the effect those masses would have on the zero-point energies).

For large cluster sizes N, we expect the volume of the cluster to increase in proportion to N. A very crude estimate of the volume would come from assuming that each atom occupies a cubical space roughly R_e on a side (keep in mind that R_e is equilibrium separation between two atomic centers of mass and is therefore roughly equal to the atomic diameter, not the radius). This arrangement is not the most efficient way to squeeze the atoms together, and a better estimate takes 71% of that value. (We'll explore the geometric basis for this estimate in Section 13.2.) The volume of one of these clusters is then estimated to be roughly $(0.71)NR_e^3$.

EXAMPLE 11.1 Argon Cluster Volume

CONTEXT Although the thermodynamics of evaporation has been extensively studied, the actual path of a molecule from the liquid into the gas phase is—perhaps surprisingly—poorly understood, because evaporation is a relatively rare event. For example, when water evaporates under typical conditions, only about one molecule per 10 nm^2 of surface enters the gas phase from the liquid every 10 nanoseconds. Although the 10 nm^2 and 10 ns may sound at first like small quantities, this is roughly one molecule per 100 molecules on the surface every *million* vibrations. One approach to studying this process experimentally is to form clusters of simple species such as argon at temperatures consistent with the liquid near its boiling point and to look for evidence of single units escaping the cluster. The overall evaporation rate depends on the cluster size and is expected to be roughly proportional to the surface area.

PROBLEM Estimate the volume of the Ar_{30} cluster in nm^3, using the R_{LJ} value from Table 10.1 for argon of 3.44 Å. Then estimate the radius and surface area of the cluster, assuming it is roughly spherical.

SOLUTION The equilibrium separation R_e is related to R_{LJ} by Eq. 10.46:

$$R_e = 2^{1/6}R_{LJ} = 2^{1/6}\,(3.44\text{ Å}) = 3.86\text{ Å},$$

or 0.386 nm. Therefore the Ar_{30} cluster should run up a volume of about

$$V = (0.71)NR_e^3 = (0.71)(30)(0.386\text{ nm})^3 = 1.2\text{ nm}^3.$$

The volume of a sphere is $4\pi R^3/3$, so the radius of a sphere with volume 1.2 nm^3 is

$$R \approx \left(\frac{3V}{4\pi}\right)^{1/3} = 0.66\text{ nm}.$$

That diameter is 13 Å, or 3.4 argon atom diameters, which seems reasonable. The surface area is then

$$A = 4\pi R^3 = 4\pi\,(0.66\text{ nm})^2 = 5.5\text{ nm}^2.$$

Turning Up the Binding Energy

Moving beyond the noble gas clusters to clusters of small, non-polar molecules such as H_2 and N_2, the potential energy function becomes less isotropic. The repulsive part of the potential (determined mainly by the shape of the molecule) is not the same in all directions and neither is the polarizability (which is normally larger along the longer axis of the molecule). Making the molecule large, providing sites for hydrogen bonding, and giving it a strong dipole moment all further decrease the isotropy and make the geometry of the cluster more sophisticated.

The most extensively studied case of this type is water. By the year 2000, the structures of the water clusters up to $N = 6$ had been experimentally determined. The geometries are governed largely by hydrogen bonding. The dimer forms only one hydrogen bond at a time, but the hydrogen atoms rapidly trade places to be involved in the bond by tunneling. The trimer, tetramer, and pentamer are all cyclic structures, with each water molecule on the receiving end of one hydrogen bond and the donating end of another. With the hexamer, the three-dimensional networking observable in bulk water begins to appear, with four of the molecules each participating simultaneously in three hydrogen bonds.

Many dimers and trimers have been studied by rotationally resolved spectroscopy, allowing accurate determinations of the structures by the means described in Chapter 9. At larger cluster sizes, however, there have been few direct measurements of structure, weakly bound or otherwise, because it becomes increasingly difficult to select a specific cluster size. The conditions that lead to effective formation of a dimer in the laboratory may not yet be cold or dense enough to form trimers effectively. But the same conditions that form a lot of tetramers will also form pentamers and hexamers effectively. Therefore, with larger clusters, many studies settle for a more qualitative look at the trends in various properties over a range of cluster sizes.

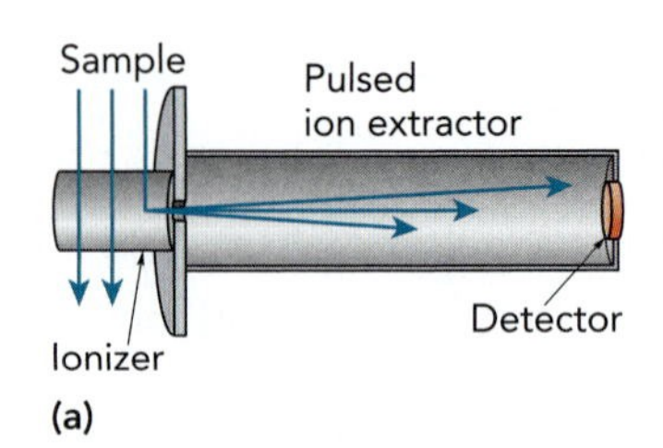

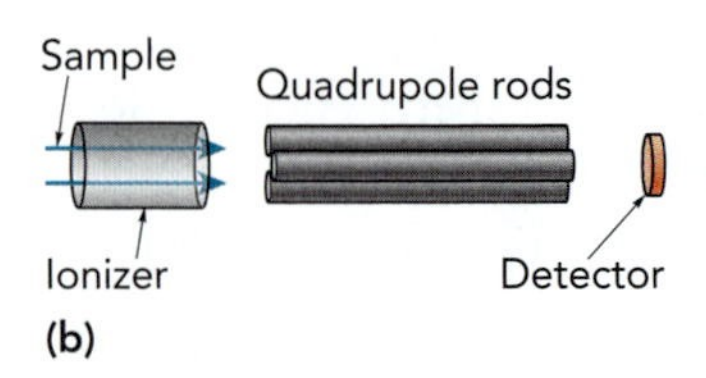

▲ FIGURE 11.3 **Two common mass spectrometric (MS) techniques.** (a) A **time-of-flight MS** pulses an electric field near the ion source that briefly accelerates the ions down the path of a drift tube. Because the acceleration is proportional to the ion's charge-to-mass ratio, q/m, the lightest molecules are accelerated to the highest speeds and reach the other end of the drift tube first. **(b)** A **quadrupole MS** sends the ions down the central axis of a set of four long rods charged with oscillating voltages, establishing a varying quadrupole electric field that separates the ions according to their q/m ratios. Most of the ions are forced out of the quadrupole field, but an adjustable constant voltage difference between the rods focuses a specific q/m ratio at the opposite end of the quadrupole field. This quadrupole MS can monitor the number of ions continuously (unlike the pulsed time-of-flight method), but it suffers lower sensitivity by losing a significant fraction of the ions in the path through the rods.

Weakly Bound Cluster Ions

One way around the size-selectivity issue has been to hide a charge among the molecules in the cluster. This allows a specific cluster size to be identified, or even isolated, using mass spectrometric methods. In order to cleanly separate molecules of similar chemical behavior but different composition in the gas phase, we need a force that acts differently on different masses. Gravity is too weak a force for all but the largest molecules, so mass spectrometers instead use electric or magnetic fields. Because these forces (F) acting on a charged particle in these fields depend on the charge but not the mass (m) of the particle, the acceleration, F/m, is mass-dependent. These methods therefore allow us to isolate molecules with a particular charge-to-mass ratio. The most common methods of mass spectrometry are summarized in Fig. 11.3 (and more detail is given in *Tools of the Trade, TK* Chapter 5).

Two applications motivate many studies of these weakly bound cluster ions: (a) direct studies of the cluster ions probe the behavior of ions in solution, and (b) the ionic charge allows clusters of a specific size to be selected by electric fields for subsequent study of the *neutral* clusters. The neutral, weakly bound clusters provide small-scale models of the structure and interactions in the liquid state. In one method of this type, a negatively charged cluster ion is formed, selected for size, and then irradiated to detach an electron and form the neutral cluster (Fig. 11.4). By measuring simultaneously the photon energy of the ionizing radiation and the kinetic energy of the expelled electron, we determine the transition energies between the quantum states of the neutral cluster and the cluster anion. Both kinds of clusters can be studied at the same time in one experiment.

As with the neutral weakly bonded clusters, the binding energy in these clusters generally increases with cluster size. However, the first few molecules added to the ionic cluster occupy the best sites to take advantage of the strong monopole–dipole or monopole-induced dipole attraction to the ion. As coordination sites around the ion get more crowded, the binding energy of additional

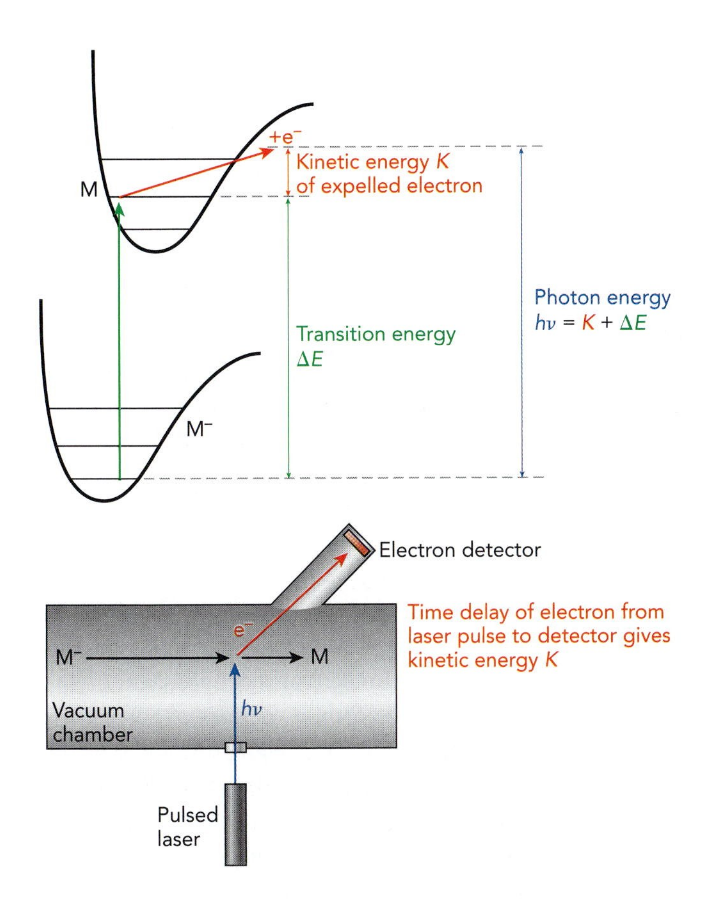

◄ **FIGURE 11.4** **Photodetachment spectroscopy.** This technique measures the kinetic energy of an electron ejected from a negative ion as a function of incident photon energy. From the electron kinetic energies, the energy levels of the anion and the corresponding neutral species can be determined.

molecules does not climb as fast. Consequently, these clusters may be relatively rigid and structured near the ion, with the outer molecules more mobile as in the weakly bonded cluster.

Strongly Bound Clusters

When the binding between components of the cluster approaches chemical bond energies, the cluster becomes highly structured (although the number of distinct, stable geometries still increases rapidly with cluster size). The available cluster structures depend strongly on the specific components, but we can make a few generalizations based on the principal bonding mechanism.

One of these generalizations connects the bonding mechanism to the *electrical properties* of the cluster rather than its structure. **Electricity** is the net flow of charge from one location to another, for example, by the opposing motion of anions and cations in an electrochemical cell. The electricity we're going to examine in this section is simply the flow of electrons from one end of the cluster to the other. In the more weakly bound clusters that we've considered in the previous sections, electrons stay close to the individual subunits and do not easily flow through the cluster. For the present examples, strong bonds extend throughout the cluster and electrons may be able to move through the bonds. Let's start with clusters where this flow is restricted and progressively free up the electrons.

Covalently Bound Clusters: Carbon Clusters

Covalent bonds tend to keep the electrons more localized than do other chemical bonding mechanisms. Among the covalently bound atoms, carbon poses a unique problem because several different bulk forms of pure carbon exist. We can define the first set of carbon clusters according to the standard molecular orbital hybrids we encounter in carbon–carbon bonds:

- sp^3 hybridized carbon atoms bond into the three-dimensional network that makes **diamond** one of the hardest minerals.
- sp^2 hybridized carbons combine to form the **graphite** structure, a collection of aromatic planes held together by dispersion forces.
- *sp* hybrid orbitals can extend into long, chemically reactive chains of carbons known collectively as **polyynes** if the bonding alternates as $C{\equiv}C{-}C{\equiv}C$ or **cumulenes** if the bonding more closely resembles a consistent $C{=}C{=}C{=}C$.

These forms are drawn in Fig. 11.5a–c.

The small clusters of carbon, with a few added elements, can resemble any of these forms. Pure carbon clusters, up to ten atoms or so, are either linear or cyclic, most strongly resembling the graphite or polyyne forms. Larger cluster sizes, or impurity atoms, stabilize the diamond form. The graphite and polyyne forms have in common a conjugated π electron system, so these small carbon clusters and their relatives may be cultivated into wires for nanometer-scale electronics.

However, the age of nanometer-scale technology, or **nanoengineering,** was ushered in by the discovery of an example of a *fourth* form of carbon cluster, having hybridization intermediate between sp^2 and sp^3: the C_{60} cluster, an elegant soccer ball–shaped structure (Fig. 11.5d). This molecule, other similarly structured carbon clusters, and their derivatives are all collectively known as **fullerenes,** in recognition of their resemblance to Buckminster

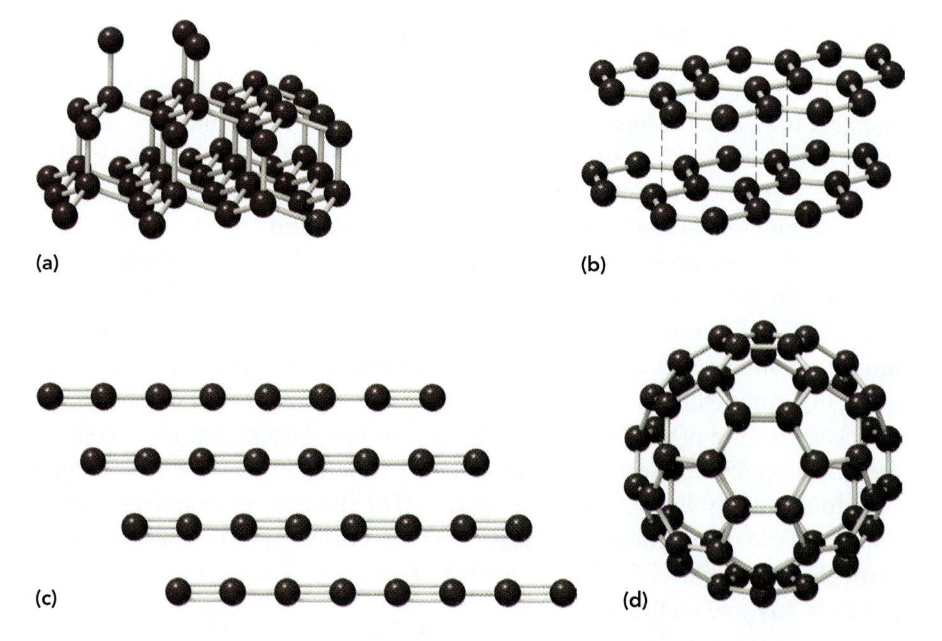

▶ FIGURE 11.5 **Four forms of essentially pure carbon.** Pure carbon can be found as **(a)** diamond, **(b)** graphite, **(c)** polyacetylene, and **(d)** fullerene. The first three examples are each dominated by a different orbital hybridization (sp^3, sp^2, and *sp*, respectively), while orbital hybridization in the fullerenes lies between the sp^2 and sp^3 cases.

Fuller's geodesic dome. The fullerenes are based on any of several three-dimensional carbon structures with twelve 5-membered carbon rings and a variable number of 6-membered rings. The fullerenes share some properties with graphite, which is also built from 6-membered rings, but lack the extended delocalization of the π electrons that graphite has. As a result, the electrical conductivity across the fullerene surface is much lower. If h (for "hexagons") is the number of 6-membered carbon rings in the structure, the chemical formula of a fullerene is C_{20+2h}. Up to large values of h, the fullerenes are found to be very stable as pure carbon-containing molecules. The most stable forms, consistent with the experimentally observed quantities, are those that avoid the strain energy of two adjacent 5-membered rings.

The C_{60} structure, the most stable of this family, is a perfect **truncated icosahedron,** a geometric solid having 20 6-membered rings and the considerable symmetry of the point group I_h: 6 $\hat{C}_5$ axes, 10 $\hat{C}_3$ axes, and 15 $\hat{C}_2$ axes. The symmetry was first established by measuring the ^{13}C NMR spectrum of $^{13}C_{60}$ in which only one peak was observed, verifying that all 60 carbon atoms were chemically equivalent.

The fullerenes are easily isolated in large quantities. They are stable and only mildly reactive, because each carbon atom is bound to three other carbons in a rigid geometry that reaction would disrupt. Although the 6-membered rings would be most stable in a planar arrangement, each C—C bond on the surface lies at an angle only 11.6° from the plane of the 5- or 6-membered ring it joins to, resulting in relatively little strain. However, the large density of π electrons results in a large polarizability and dispersion force, so the fullerenes are sticky molecules. Just as graphite has carbon p orbitals perpendicular to the sheet, C_{60} has **normal orbitals** that extend perpendicular to the surface of the sphere. These normal orbitals are nearly pure p orbitals, with only 8% s-character. The C_{60} derivatives that have atoms bound to the surface by these normal orbitals may have up to 20% s-character at the binding site, and the carbon atom at that site approaches the tetrahedral bonding of a pure sp^3 hybrid orbital, which has 25% s-character.

At first glance, C_{60} would appear to be an ideal aromatic molecule, consisting of a three-dimensional network of aromatic benzene rings. It turns out, however, that C_{60} does not fit that description well. Unlike benzene, the bonds in the 6-membered rings of C_{60} are not all equivalent, because half lie between two 6-membered rings while the other half form the border between one 5- and one 6-membered ring. As a result, the two types of bonds are not the same length and they do not share π electrons equally. This reduces the degree of resonance stabilization in C_{60} compared to benzene sufficiently that this and other fullerenes are not generally considered to be aromatic.

Carbon Nanotubes

The discovery of fullerenes led quickly to the development of techniques for manufacturing **carbon nanotubes.** These are cylindrically rolled, single sheets of graphite that yield extremely strong tubes of very narrow diameter. Of particular interest is the manner in which their electronic properties can be tuned by adjusting the parameters for their construction.

Let us start with a single, covalently bound layer from a sheet of graphite, known as **graphene,** a piece of which is shown in Fig. 11.6. If the sheet is infinite

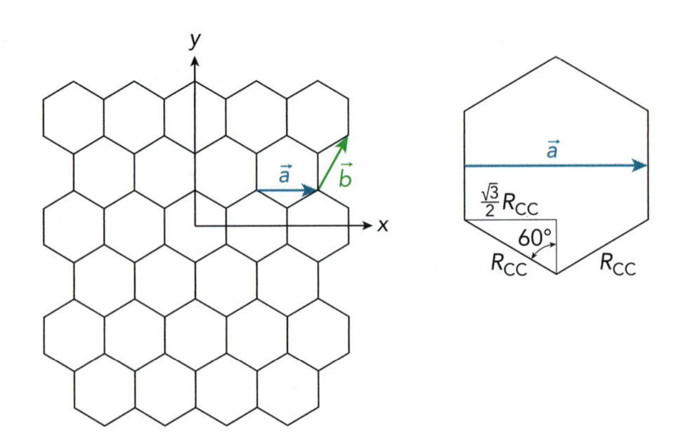

▶ FIGURE 11.6 **Graphene, a single sheet of sp^2 covalently bonded carbon atoms.** The circumferential coordinate vectors $\vec{a}$ and $\vec{b}$ shown.

initially, all the carbons are equivalent and a $\hat{C}_6$ symmetry axis passes through the center of each hexagon and a $\hat{C}_3$ axis passes through the center of each carbon atom (the vertices of the hexagons in Fig. 11.6). To stay consistent with our axis definitions from Chapter 6, we'll define our coordinate system so that z is the axis perpendicular to the sheet and parallel to the principal rotation axes. The x axis points perpendicular to one set of bonds separating adjacent hexagons, and the y axis is perpendicular to the x axis (Fig. 11.6). Next, we cut out a long strip of the graphene sheet and roll it into a long cylinder, bringing the atoms along one edge of the strip up to the atoms of the other edge (Fig. 11.7). But in cutting out the original strip, we have chosen from among many possible diameters and orientations. To describe the structure of the nanotube we have rolled, we use two vectors that reach between two non-adjacent carbon atoms within a hexagon, either parallel to the x axis ($\vec{a}$) or at a $\pi/6$ (60°) angle to the x axis ($\vec{b}$). If the hexagonal symmetry is not distorted by formation of the nanotube, these vectors are each of length $\sqrt{3}R_{CC}$, where R_{CC} is the carbon–carbon bond length, equal to 1.42 Å in the original graphene.

We define a vector $\vec{C}$, with magnitude equal to the circumference of the tube and which points along the direction that the sheet will be rolled up. Rather than giving the vector in units of distance, we give it in units of $\vec{a}$ and $\vec{b}$:

$$\vec{C} = n\vec{a} + m\vec{b}, \tag{11.1}$$

and each nanotube can then be characterized by its circumferential vector coordinates (n,m). These are convenient indices for the nanotube because they describe at once the possible tube diameters *and* the possible orientations of the hexagonal rings within the nanotube with a pair of integers. Non-integer values of m or n would correspond to structures in which the atoms at each edge of the sheet fail to line up properly to close the sheet, and these are not observed.

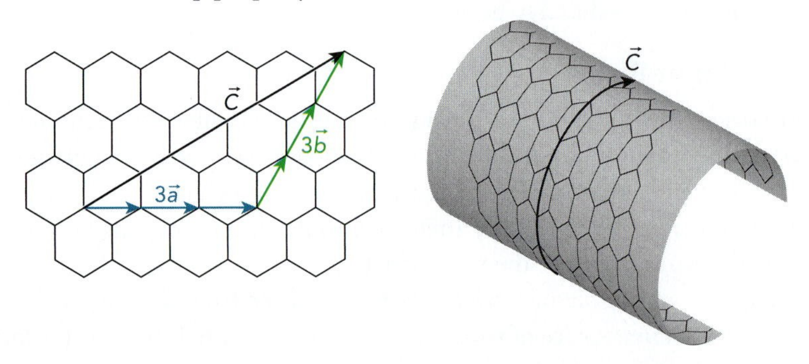

▶ FIGURE 11.7 **The circumferential vector $\vec{C}$ for a (3,3) nanotube.**

For example, the (3, 3) nanotube in Fig. 11.7 would be rolled along the $\vec{C}$ axis shown, resulting in a tube that has C—C bonds parallel to the circumferential vectors and therefore perpendicular to the axis of the tube. Similarly, any $(n, 0)$ tube—or equivalently any $(0, m)$ tube—will have the C—C bonds oriented *parallel* to the axis of the tube.

The electronic energy distribution depends strongly on the direction of the circumferential vector. Of particular interest, the electrons in the π system formed from the p_z orbitals, perpendicular to the surface of the nanotube, find themselves in a nearly metallic environment only when $n - m$ is an integer multiple of 3. At these orientations, the lowest excited π-electron MO in the nanotube drops to the energy of the highest occupied MO, allowing the electrons to be excited with the tiniest push. These excited state orbitals connect adjacent hexagons, and the electrons can move freely from one end of the nanotube to the other. The same holds for the π electrons in ordinary graphite, so this is not terribly surprising. However, as we change the value of $n - m$, we separate the HOMO and LUMO energies so that the nanotube becomes a semiconductor requiring some additional energy to push the localized electrons into the excited state MOs. As we expect, the nanotube returns toward the electrical conductance of graphite as the diameter of the tube increases and the surface of the tube becomes flatter.

CHECKPOINT The vectors $\vec{a}$ and $\vec{b}$ are like unit vectors in geometry, pointing in the right directions and having the right length so that any integer combination m and n of $\vec{a}$ and $\vec{b}$ guarantees that the nanotube can form with minimal strain on the carbon–carbon bonds. Like the x and y axes, $\vec{a}$ and $\vec{b}$ are equivalent—the hexagonal pattern looks exactly the same along one as along the other—and *any possible* angle of twist of the hexagonal pattern with respect to the axis of the nanotube is represented by some combination of m and n.

Because $\vec{a}$ and $\vec{b}$ are not perpendicular, the length of $\vec{C}$ must be calculated using the angle $\pi/6$ between the two coordinate vectors:

$$\begin{aligned}
|\vec{C}|^2 &= |n\vec{a} + m\vec{b}|^2 \\
&= |n\vec{a}|^2 + |m\vec{b}|^2 + 2(n\vec{a})\cdot(m\vec{b}) \\
&= n^2|\vec{a}|^2 + m^2|\vec{b}|^2 + 2nm|\vec{a}||\vec{b}|\cos\left(\frac{\pi}{6}\right) \\
&= [n^2 + m^2 + 2nm(\tfrac{1}{2})](\sqrt{3}R_{CC})^2 \\
|\vec{C}| &= \sqrt{3}[n^2 + m^2 + nm]^{1/2}R_{CC}.
\end{aligned} \tag{11.2}$$

The diameter D of the nanotube can then be calculated as

$$D = \frac{|\vec{C}|}{\pi} = \frac{\sqrt{3}}{\pi}[n^2 + m^2 + nm]^{1/2}R_{CC}. \tag{11.3}$$

SAMPLE CALCULATION **Carbon Nanotube Diameter.** We will estimate the diameter of the (3, 3) carbon nanotube to be prepared in Fig. 11.7, assuming a carbon–carbon bond length of 1.42 Å. The diameter works out to only

$$\begin{aligned}
D &= \frac{\sqrt{3}}{\pi}[n^2 + m^2 + nm]^{1/2}R_{CC} \\
&= \frac{\sqrt{3}}{\pi}[3^2 + 3^2 + 9]^{1/2}(1.42\,\text{Å}) \\
&= \frac{9}{\pi}(1.42\,\text{Å}) = 4.07\,\text{Å}.
\end{aligned}$$

This diameter is among the smallest possible for carbon nanotubes, but it has been observed in experiments.

Although the rolled sheet is a convenient way to describe the structure of a carbon nanotube in terms of plane geometry, this is not the way the tubes are normally formed. Several different methods exist, but one common way to make carbon nanotubes currently is to grow them by plasma-enhanced chemical vapor deposition (PECVD) from metal nanoparticles attached to a solid surface in a vacuum chamber. An electric discharge runs through a combination of gases—a noble gas to provide the ions that maintain the discharge current, a hydrocarbon to provide the carbon atoms, and a process gas such as hydrogen or ammonia that helps suppress the formation of more stable graphite.

Individual carbon nanotubes have an unusually high ratio of strength to weight, and at present this is the basis of their principal application, which is in the development of new lightweight construction materials. Their small size has also led to their use in probe tips for atomic force microscopes (see *Tools of the Trade*). Attracting interest more recently is the unrolled graphene sheet, the single layer of graphite described at the beginning of this section. The 2010 Nobel Prize in Physics was awarded to Andre Geim and Konstantin Novoselov for discovering a means of isolating graphene and opening the door on the virtually unexplored realm of two-dimensional solids.

Except for the sp^3 hybridized diamond structure, all of these forms of carbon conduct electricity (and the fullerenes, with help from the alkali metals, can even become superconductors, with zero electrical resistance). The ease with which carbon forms conjugated π electron systems leads to several geometries that can carry electricity, including conducting polymers. Carbon continues to draw our attention more than any other element, as research continues to search for ways to efficiently capitalize on the electronic and structural properties of these novel materials.

TOOLS OF THE TRADE | Scanning Tunneling and Atomic Force Microscopy

Microscopy is the use of technology to observe images too small for human eyesight alone to resolve. A traditional *optical microscope* magnifies the light bouncing off or transmitted through the sample being studied. But when the distance between two particles in the sample approaches the wavelength of the light being observed, beams of light from the two particles to the observer will interfere with each other. At a limit of about 200 nm, the interference fringes become too strong for the particles themselves to be distinguished.

Working in the years immediately after Davisson and Germer verified the wave character of electrons, Max Knoll and Ernst Ruska invented the **electron microscope,** a device that works on the same principle as the optical microscope but that manipulates the paths of electron beams to image the sample. At easily attainable speeds, the de Broglie wavelength of electrons drops to less than an angstrom, and the resolution limit of the electron microscope is instead determined by our ability to control the electric fields used for focusing the electron beams. By 1950, the resolution of the electron microscope was 2 nm, far beyond the capabilities of the optical microscope.

However, electron microscopy must be carried out in a vacuum, and single-atom resolution is achieved only for electrons accelerated to over 100 kV. In 1955, Erwin Müller and his student Kanwar Bahadur imaged single atoms for the first time using a *field ion microscope,* a clever technique but one limited to samples that can be sharpened to a fine tip. The search for general techniques that could routinely achieve single-atom resolution continued until the invention of *scanning probe microscopy* in the 1980s. In scanning probe microscopy, the image is acquired one position at a time. A fine-tipped probe tests the sample at points along a rectangular grid, forming a *raster image* rather than detecting the whole image simultaneously. The most common tools of this technique are the scanning tunneling microscope (STM) and its close relation, the atomic force microscope (AFM).

The STM was developed at IBM by Gerd Binnig and Heinrich Rohrer in 1981, earning them the Nobel Prize in Physics just five years later. (Sharing that Nobel Prize was Ernst Ruska for his part in developing the electron microscope.) Binnig went on to develop the more general atomic force microscopy (AFM) in 1986, working with Calvin Quate and Christoph Gerber.

What are STM and AFM? The scanning tunneling microscope measures the electrical current of electrons that tunnel across a small gap between sample and probe. The atomic force microscope measures the force arising from intermolecular interactions between the probe tip and sample surface. Both methods are used to map—or even to *alter*—the structure of a surface with single-atom precision.

Why do we use STM and AFM? When carrying out research on a chemical system, the better we know our sample, the more confident we can be of the conclusions we draw. The LEED technique (*Tools of the Trade*, Chapter 1) characterizes the structure over a large area of a surface and allows the experimentalist to determine, for example, the geometry of the atoms on the exposed face of a crystal. But LEED is not sensitive to the *local* structure of the crystal. The AFM and STM techniques provide that local information, measuring point by point the distribution of atoms across the surface. For example, the distance between circuit elements in a typical microcircuit cannot be determined by LEED, because the microcircuit does not form a repeating pattern like the atoms in a crystal, but an AFM can measure the distance directly. Interest in AFM and STM is also motivated by numerous potential applications to industry, such as the capability of AFM and STM to write and then read back specific arrangements of individual atoms, perhaps pointing the way to a new generation of high-density data storage devices.

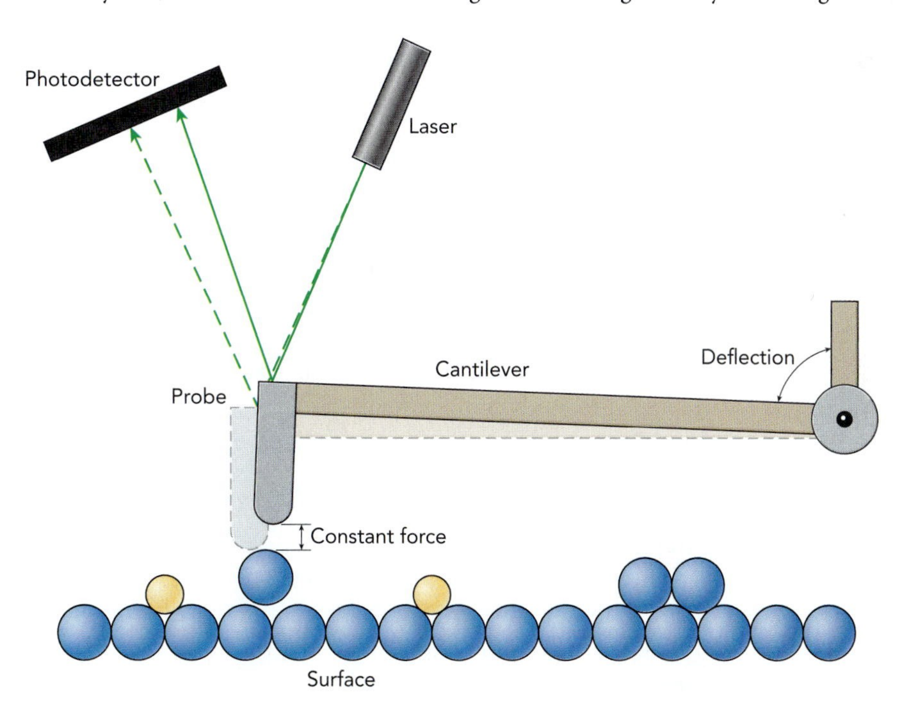

▲ **Atomic force microscopy.** In this design, the probe is maintained at a constant distance from a surface by a feedback circuit that maintains a constant intermolecular attractive force for AFM. The surface structure is mapped by measuring the deflection of the probe arm (the cantilever) as the surface is moved beneath it. The STM uses a similar principle, but a different configuration for supporting the probe.

How do they work? Optical and electron microscopes image the photons or electrons that have interacted with the sample. In contrast, the interaction of a tiny probe tip with the surface of the sample, point by point, generates STM and AFM images. When a charged needle (the probe) is brought to within 10 Å of a conductive surface, electrons can tunnel across the potential energy barrier of the intervening space (as described in Section 3.2). If the probe is given a positive voltage, a very small current of electrons will flow from the surface to the probe. The probes may be formed with tip diameters of only a few Å, and the tunneling current is extremely sensitive to the distance between the probe and the surface. Therefore, by moving the probe along the surface and measuring the current, an STM map of the conducting surface can be generated. One application of these detailed maps is to improve the catalysis of reactions on a solid surface. STM can measure directly the roughness of the surface and help to quantify how the reaction rates depend on surface imperfections. STM can also be used to move individual atoms about on the surface. The current of tunneling electrons repels atoms with high electron densities, such as metal atoms adhered to the surface. By this method, STM has been used to form a ring of iron atoms on a copper surface, enabling the predictions of quantum mechanics to be tested for a simple two-dimensional particle-in-a-box potential.

STM is specifically applied to the study and manipulation of conductive surfaces and their attached molecules. A more general technique, using similar equipment, is atomic force microscopy (AFM), which takes advantage of the intermolecular attractive forces to provide a high-resolution map of solid surfaces. The probe is supported in this case at the end of a delicate cantilever (see figure). As the tip of the probe is brought near the surface, the long-range van der Waals forces pull the tip closer. An extremely sensitive detection system measures this force and uses electrical feedback to maintain a constant attractive force between the probe and the surface. Changes in the vertical position of the probe, measured to a precision of a few Å using the interference pattern of laser light reflected off the arm, then record the shape of the surface. Like STM, AFM may function as an engineering tool. By changing the force setting, it is possible to use the tip to scratch surfaces.

Although the technology has existed since the 1980s, we continue to find new applications of these techniques—the first engineering tools to be applied at scale of individual atoms.

Ionically Bound Clusters

Molecules that have very strong ionic bonds can be difficult to get into the gas phase. At high enough temperatures to evaporate the compound, a salt tends to separate into its component elements, rather than retain the ionic bonds in the vapor phase. However, it is possible to simultaneously evaporate the atomic components of a salt using high-powered lasers, which will vaporize anything, and then let the atoms coalesce into small gas-phase molecules. This method has been used to study **nanocrystals**—ionically bound molecules containing fewer than 100 atoms, measuring on the order of nanometers (10^{-9} m or 10 Å). These molecules are still too complicated to be studied spectroscopically, but with mass spectrometry it has been shown that certain combinations of atoms $X_m Y_n$ are extremely stable with respect to others. These are presumed to correspond to structures that resemble the bulk solid, in which the positions of the atoms optimize the binding energy. With continuing improvements in atomic-scale microscopy and diffraction, the structures of nanocrystals may soon be determinable by non-spectroscopic means.

Metal Clusters

Section 5.4 briefly introduces the concept of metallic bonds, and in some regards these are less complicated than covalent bonds. Metallic bonding, like the dispersion forces that hold our noble gas clusters together, is relatively non-directional. However, these clusters have typical chemical bond strengths: it takes a few eV or a few hundred kJ mol^{-1} to remove one of the atoms. The electrons, instead of being localized into distinct chemical bonds, are bound to the volume of the cluster by the charges of the dispersed metal cations. The potential energy generated by the cations in these clusters has been approximated using the **jellium model.** The jellium model assumes that the electronic energies are independent of the precise locations of the nuclei. The potential energy experienced by the electrons may be approximated instead by a smooth function such as the following:

$$U(r) = \begin{cases} -U_0\left[\dfrac{N}{2r_0}\left(3 - \dfrac{r^2}{r_0^2}\right)\right] & r \leq r_0 \\ -U_0\dfrac{N}{r} & r > r_0 \end{cases}, \tag{11.4}$$

where $r_0 = 3N^{1/3}/(4\pi\rho_0^{1/3})$, ρ_0 is the number of electrons per unit volume (the electron density) at the center of the cluster, and N is the cluster size. Like the $U(r) = -Ze^2/(4\pi\varepsilon_0 r(r))$ curve that we used for the one-electron atom, this potential has spherical symmetry and decreases in strength with distance from the center of the system but replaces the singularity at the origin with a finite potential energy $U(r = 0) = -U_0$.

It makes sense, therefore, that the electronic states of small metal clusters roughly resemble what we would expect for atoms with an extended nucleus. As in Chapter 3, we can again break the wavefunction into a spherical harmonic containing all of the angle-dependence and described by quantum numbers l and m_l, and a radial part, again with a long-range exponential decay and indexed by the principal quantum number n. The effect of the distributed positive charges, compared to the point nuclear charge that we used for the single atom, is that the quantum numbers n and l are no longer so closely connected. For this potential, we arrive at a Schrödinger equation analogous to Eq. 3.21:

$$\left[-\frac{\hbar^2}{2m_e}\frac{1}{r^2}\frac{\partial}{\partial r}r^2\frac{\partial}{\partial r} + \frac{\hbar^2 l(l+1)}{2m_e r^2} - U(r)\right]R_{n,l}(r)\,Y_l^{m_l}(\theta,\phi) = E_{n,l}\,R_{n,l}(r)Y_l^{m_l}(\theta,\phi). \tag{11.5}$$

For the one-electron atom, the solutions to this differential equation generated terms that exactly canceled the angular momentum contribution $\hbar^2 l(l+1)/(2m_e r^2)$. For the jellium potential, that cancellation does not occur, so the energy depends on l as well as on n. In effect, the angular and radial terms become less dependent on each other. As a result, n is used only to indicate the degree of excitation along r, and l is no longer limited to values less than n. Subshell labels such as $1p$ and $2f$, forbidden in the atomic case, become valid. The energy increases faster with n (excitation along r) than with l (angular excitation), because the potential energy still rises quickly as the electron tries to move away from the field of the metal cations. This orbital arrangement, based on a more sophisticated jellium model, is shown for the Na_{20} cluster in Fig. 11.8. Incidentally, in this respect the jellium resembles the **nuclear shell model,** which is used to explain the structure of atomic nuclei. Nuclei are similar in that each particle experiences forces from several other particles simultaneously over distances comparable to the separation between the

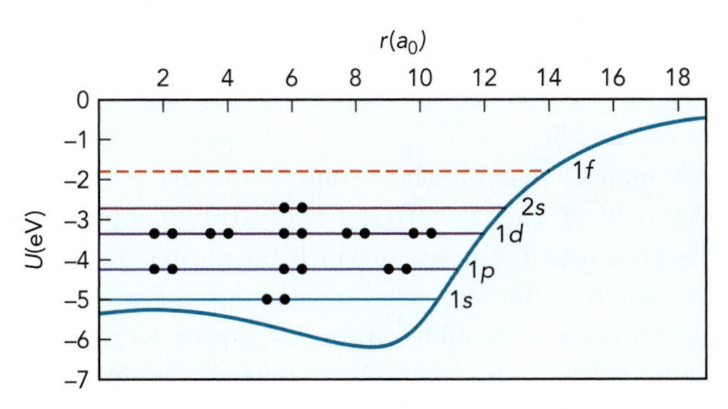

◀ FIGURE 11.8 **The jellium potential energy curve for Na_{20}.** The potential energy curve of the 20 valence electrons (shown as dots) is calculated using a jellium model to average the effects of the nuclei [*Phys. Rev. B* **29,** 1558 (1984)].

CHECKPOINT The jellium model replaces the potential energy term in the atomic Hamiltonian by another function, which reflects the roughly constant stabilizing influence of the nuclei throughout the center of the cluster. The new potential energy curve means that the allowed radial kinetic energies are different from those in the single-atom Schrödinger equation. But the angular component of the kinetic energy is unchanged. Therefore, the balance between the angular and radial kinetic energies is different, and that is why the total energy in the jellium model depends on both n and l, rather than on n alone.

particles. The radial and angular motions again are not as strongly coupled as in the atomic Schrödinger equation, and states with $l \geq n$ (such as $1p$ and $2f$) are allowed for nuclei as well.

This model works best for the one-electron metals, particularly the alkali metals and copper. Among the metals, the alkali metals are the simplest because they lack the high number of valence electrons and the density of low-lying electronic states that typify the transition metals and (to a lesser extent) the alkaline earth metals. Copper often behaves in a similar fashion because its crowded $4s^1 3d^{10}$ electron configuration leaves only one unpaired electron and relatively few nearby electronic states.

The valence electrons of the atoms contribute to a cluster-wide electronic wavefunction that obeys shell-filling rules of its own. Filled electron shells stabilize some of these clusters more than others, in the same way that the noble gas atoms are stabilized by filled electron shells at atomic numbers 2, 10, 18, 36, and 54. This additional stability causes those clusters to appear in relatively high abundance in experimental studies. In a series of clusters of any composition, the cluster sizes that correspond to relatively high stability are known as **magic numbers.** The magic numbers for the alkali metal clusters are predicted to fall where a subshell becomes filled. Depending on the values of U_0 and r_0 (which also determine to some extent the energy shifts arising from the electron–electron repulsion) the early magic numbers typically fall at cluster sizes of

- two (two electrons to fill the $1s$),
- eight (add six more to fill the $1p$), and
- ten (two more electrons fill the $2s$).

One series of magic numbers is given by the expression

$$N_{\text{magic}} = \sum_{i=0}^{n-1}\left[\sum_{j=0}^{i} 2(2j + 1)\right]. \tag{11.6}$$

The second sum, $\sum_{j=0}^{i} 2(2j + 1)$, gives the degeneracy of the jth shell in the one-electron atom, from Eq. 3.57. The magic number N_{magic} is the same as the number of electrons necessary to fill the atomic orbitals up to the principal quantum number n: $2(n = 1)$, $10(n = 2)$, $28(n = 3)$, and so on. The energy levels are grouped the way the electron energy levels would be grouped in an atom if the shielding were not important.

The average binding energies per atom of the small lithium and sodium clusters are graphed in Figure 11.9. For both elements, the binding energy generally increases with size, as metal ions increasingly become surrounded by the diffuse sea of delocalized valence electrons. However, when one of the subshells fills up, the addition of the very next atom is somewhat destabilizing and the binding energy drops. This effect explains the dips in these curves after the magic number values of 2 and 8.

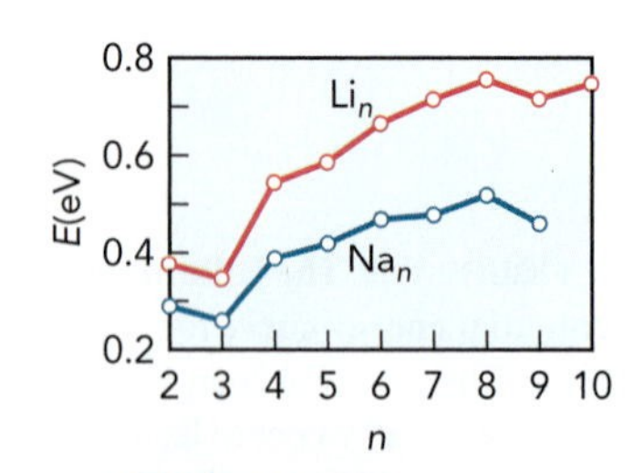

▲ FIGURE 11.9 **Average binding energy per atom for small clusters of lithium and sodium.**

These clusters quickly begin to show some of the characteristic properties of bulk metals. As a general rule, the properties of a cluster should become similar to those of the bulk near the point at which the cluster diameter exceeds the de Broglie wavelength of the associated wavefunction. For example, one of the hallmarks of a bulk metal is its ability to conduct electricity. It turns out that this ability can be loosely tied to the de Broglie wavelengths of the electrons.

The de Broglie wavelength of a particle in a box wavefunction is straightforward to determine. Each state in the particle in a box is a perfect sine wave that lasts $n/2$ wavelengths (so the $n = 1$ ground state is half a wavelength, the $n = 2$ is a full wavelength, etc.). The de Broglie wavelength is therefore $\lambda_{dB} = 2a/n$, where a is the length of the box. Using the particle in a box as a primitive model for the valence electrons in a metal cluster, we could then estimate that the de Broglie wavelength of the highest energy electrons would be roughly twice the cluster diameter, $2D$, divided by the principal quantum number n of those electrons. We can estimate the number of electrons needed to reach a particular value of n by integrating the magic number equation to get

$$N_{\text{electrons}} \approx \int_0^n 2(2j+1)\,dj = \frac{4n^2}{2} + 2n = 2n(n+1) \approx 2n^2. \quad (11.7)$$

For an alkali metal cluster, in which each atom contributes one electron, we can set $N_{\text{electrons}}$ equal to the cluster size N, and the value of n for the valence electrons is then approximately $\sqrt{N/2}$. So we expect to see conductivity similar to the bulk appearing in cluster sizes where

$$D \gg \lambda_{dB} \approx \frac{2D}{n} \approx \frac{2D}{\sqrt{N/2}} = \frac{2\sqrt{2}D}{\sqrt{N}}$$

$$N \gg 8.$$

A cluster size of 8 is quite low, and the conductivity of such small clusters is difficult to define, let alone measure. However, we can examine a related electronic property directly. The energy required to remove an electron from the metal is given for the neutral atom (the first ionization energy) and for the bulk metal (the **work function**) in Table 11.1 for several metals. The energy ordering is about the same for the elements whether considering the single atoms or the bulk metal, but the work function tends to be about half of the ionization potential. The electron delocalization of the bulk metal results in a much lower ionization potential (2—3 eV) than in the bare atoms, and much of that reduction is accomplished before the clusters reach ten atoms.

TABLE 11.1 First ionization energies (IE_1) and work functions of selected metals.

		IE_1 (eV)		work
metal	**atom**	**M_2**	**M_6**	**function (eV)**
Rb	4.177			2.16
K	4.341	4.3	3.5	2.30
Na	5.139	5.1	4.6	2.75
Sr	5.695			2.59
Ca	6.113			2.87
Mg	7.646			3.66
Fe	7.870	7.2	6.8	4.70

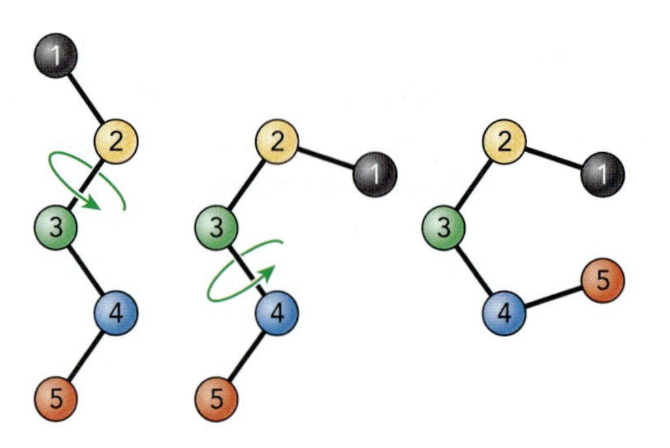

▶ FIGURE 11.10 **Torsion in macromolecules.** The five-carbon chain drawn here may fold from its denatured (unfolded) state to a compact form by internal rotation about single bonds.

11.3 Macromolecules

The macromolecules of greatest interest in biochemistry are proteins and nucleic acids, although numerous other biological molecules may fall into this category as well. These biomolecules are extensions of the polyatomic molecules we have already discussed but with a crucial complication: chemical forces bind adjacent atoms, but the overall structure is also determined by van der Waals forces and electrostatic interactions between different parts of the same molecule.

The biomolecules can usually be represented as lengthy single-bonded chains of carbon and nitrogen atoms, with attached hydrogens and functional groups. These are *very* flexible molecules. Internal rotations or **torsions** about the C—C and C—N single bonds can drastically change the shape of the molecule, often without having to overcome potential energy barriers much greater than 500 cm^{-1} (Fig. 11.10). Molecular geometries that differ from one another by some sequence of internal rotations are called **conformers.**

How many different conformers do we need to worry about? For a frightening upper-limit estimate, picture a substituted alkane chain as shown in Fig. 11.11. Each 120° rotation about any single bond gives a new conformer with distinct orientations among the R groups. A chain of N carbon atoms has $N - 1$ carbon–carbon bonds, each of which can be twisted by 120° to get a new conformer. The maximum number of possible conformers is therefore 3^{N-1} for the chain. A carbon chain of 50 atoms would therefore have roughly $3^{49} = 10^{23}$ distinct conformers. This figure is a significant overestimate as it turns out, because a large fraction of the conformers obtained this way place some atoms too close together for the conformer to be stable. Still, it gives us an idea of the magnitude of the problem. A popular perspective points out that if we tested every one of these geometries at roughly the rate of a chemical bond vibration ($10^{12}\ \text{s}^{-1}$), it would take 10,000 years.

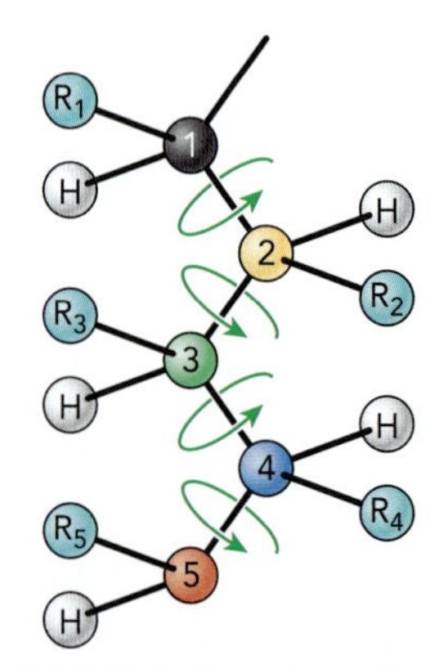

▲ FIGURE 11.11 **Alkane chain with functional groups attached to each carbon.** A new conformer may be generated for each twist of 120° about any C—C bond.

The large molecular masses cause some of the vibrational bending modes to be of extremely low frequency and easily excited because the vibrational constant is roughly proportional to $\sqrt{1/\mu}$, where the reduced mass μ now may be of the order 10^4 amu or more. Unlike bends and stretches, torsions do not necessarily push atoms against one another, and the force constants are therefore low, usually less than 10 $\mathrm{N\,m^{-1}}$. This value suggests that the excitation energies for these torsions are on the order of (using Eq. 8.19)

$$\omega_{\text{torsion}} \approx 130.28\sqrt{\frac{10}{10^4}} \approx 0.1 \text{ cm}^{-1}.$$

The energy levels are almost continuous, and the energy can be treated as a classical variable to a good approximation.

This marks a major divide in our development of the chemical model. Up to this point, even with the clusters, we justified the structures of molecules using a foundation of quantum mechanics. *We're all done with that now.* The size of the molecules and the lack of the fundamental symmetry that occurs in clusters of identical subunits forbid treating these systems with the excruciating detail of quantum mechanics.

BIOSKETCH **James V. Coe**

James V. Coe is a professor of chemistry at the Ohio State University, where he and his students are taking the next step beyond the nanoscale and working at distance scales of a few micrometers. By electrochemically depositing copper onto a commercially available nickel mesh, Coe and his research group are able to form meshes with uniformly distributed square apertures less than 6μm on a side, comparable to infrared wavelengths. They have used these meshes to demonstrate *extraordinary optical transmission* (EOT) at these wavelengths. In EOT, the fraction of light transmitted through a mesh is *more than* the fraction of the total surface area containing holes—as though the light were concentrated at the holes. The radiation excites waves called *plasmon resonances* along the surface of the metal. Plasmon resonances are effectively synchronized oscillations of large numbers of valence electrons, and these motions have the effect of transferring the energy of the radiation selectively toward the holes in the surface. At the hole, the energy of the oscillating electric field is released again as photons, which now propagate through the hole to the other side of the mesh. The effect is highly sensitive to the wavelength of the light, so the mesh can be designed to selectively enhance the transmission of radiation that corresponds to specific vibrational transitions. Coupled with the ability of these holes to trap micrometer-scale particles (such as individual dust grains), the wavelength discrimination makes these meshes promising devices for chemical analysis. As with many applications of nanotechnology, this work capitalizes on the distinct properties of the interaction of radiation with matter at distances less than or equal to the radiation wavelength.

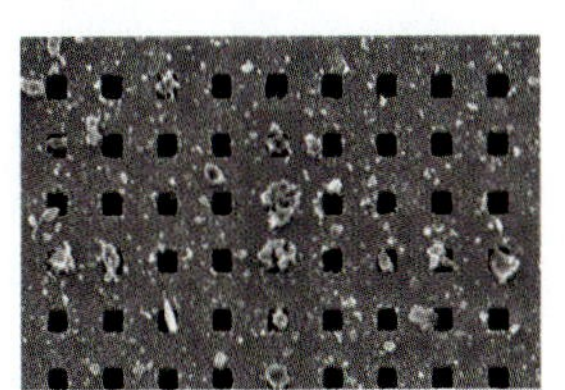

We can appreciate the challenge of determining the geometry of a complex molecule with thousands of atoms, and structural measurements of macromolecules are conducted by a variety of means, many of them clever extensions of techniques that were initially developed to characterize small molecules.

- X-ray crystallography (see *Tools of the Trade* in Chapter 13) was the earliest technique able to determine macromolecular structures, with the precision and range of the method gradually improving over the decades since its first use in 1912. By the early 1950s, x-ray diffraction was able to identify the double helix structure of DNA, and it remains the primary method for determining structures of proteins and other biomolecules.
- Nuclear magnetic resonance (NMR, see *Tools of the Trade* in Chapter 5) measures the local structure in molecules, and a widening array of experimental methods now allow NMR to tie these local structure determinations together for macromolecules into a cohesive structure for the whole. It remains a challenging approach, limited to smaller molecules than can be studied by x-rays, but it has the advantage of allowing the molecules to be studied in solution and of allowing dynamic properties such as isomerization rates to be measured.
- Mass spectrometric techniques—briefly described in Section 11.2 in relation to clusters—are increasingly available for the study of macromolecules. Mass spectrometry was long restricted from very large molecules, first by the need to get these delicate molecules into the gas phase without decomposition, and secondly by the need to ionize the sample. Ionization energies lie well above the dissociation energies of chemical bonds, and traditional means of ionizing molecules would break a macromolecule into too many fragments for analysis to be meaningful. A method invented in the 1980s, matrix-assisted laser desorption and ionization (MALDI), is now used to gently vaporize and ionize the molecule by channeling the energy from a laser through a smaller molecule that can absorb the radiation without damage. Although mass spectrometry directly measures only mass-to-charge ratios, extensions of the technique make it possible to determine amino acid sequences and other structural properties much more easily than by x-ray diffraction or NMR.

Molecular Mechanics

Because quantum mechanical calculations would be too demanding and yield too much detail we usually study the motions of these large molecules by means of an informed but approximate classical method. Ideally, this method takes advantage of the central results of our quantum-mechanical treatment of atoms and molecules but dispenses with the most difficult mathematics and the least interesting information. For example, we don't really need the MO wavefunctions here. Methods that predict molecular properties by applying classical mechanics to molecular structure are known collectively as **molecular mechanics.**

Typical assumptions and procedures in molecular mechanics include the following:

1. Atoms are treated as single, spherical particles with radius determined by the atomic number; there are no explicit electrons. Each atom may be assigned a partial ionic charge based either on quantum-mechanical calculations or experimental data.
2. Chemical bonds between adjacent atoms are treated using the harmonic oscillator potential for stretching and bending:

$$U_{\text{stretch}} = \frac{1}{2}k_R(R - R_e)^2 \tag{11.8}$$

$$U_{\text{bend}} = \frac{1}{2}k_\theta(\theta - \theta_e)^2, \tag{11.9}$$

   here the equilibrium bond lengths and bond angles are R_e and θ_e, respectively. A table of empirical data uses the bond order and the atomic numbers of the atoms to set values for the force constants k_R and k_θ.
3. Internal rotations are modeled by a potential energy term of the form

$$U_{\text{torsion}}(\phi) = \frac{U_{t0}}{2}\{1 - \cos[n(\phi - \phi_e)]\}, \tag{11.10}$$

   where U_{t0} is the energy barrier to the internal rotation, ϕ is the internal rotation angle, n is the number of identical conformations in a 360° internal rotation, and ϕ_e is the equilibrium torsion angle. The most stable conformations occur when $\phi - \phi_e$ is any even multiple of π/n, because then

$$U_{\text{torsion}}(\phi) = \frac{U_{t0}}{2}\{1 - \cos[2\pi]\} = \frac{U_{t0}}{2}\{1-1\} = 0. \tag{11.11}$$

4. Atoms that are not chemically bound to one another can still interact via their van der Waals forces, provided the atoms are close enough for those forces to be significant. A model potential, such as the Lennard-Jones 6-12 potential, is used (Eq. 10.45) for each interaction:

$$U_{\text{vdW}} \approx u_{\text{LJ}}(R) = 4\varepsilon\left[\left(\frac{R_{\text{LJ}}}{R}\right)^{12} - \left(\frac{R_{\text{LJ}}}{R}\right)^{6}\right]. \tag{11.12}$$

   Empirical data tables again provide values for ε and R_e based on the atomic numbers or polyatomic functional groups involved. The repulsive wall in this potential accounts for the steric effect, which increases the energy of conformations that place non-bonded atoms too close together.
5. Electrostatic interactions between different sites on the molecule are calculated using the Coulomb potential energy (Eq. A.42):

$$U_{\text{Coulomb}} = \frac{q_1 q_2}{4\pi\varepsilon_0 R_{12}}. \tag{11.13}$$

   It is assumed that electrostatic forces between chemically bonded atoms are already folded into the chemical bond model. Therefore, as with the van der Waals interactions, these terms are calculated for every pair of atoms that are *not* bonded to each other, and they are added to get the total electrostatic potential energy:

$$U_{\text{electrostatic}} = \sum_{i=1}^{j-1} \sum_{j=2}^{N} \frac{q_i q_j}{4\pi\varepsilon_0 R_{ij}}, \tag{11.14}$$

where j is not bonded to i.

CHECKPOINT The force field is a sum over all the terms that we think will contribute measurably to the potential energy of a particular molecular structure. If we can calculate that potential energy at any structure of the molecule, then we can compare the stabilities of different conformers. Furthermore, because the force can be found from the derivative of the potential energy, we can simulate the motions of the atoms in response to any stress or change in conformation.

The total potential energy of the molecule is then written as the sum of these various interaction potential energies:

$$U_{\text{tot}} = U_{\text{stretch}} + U_{\text{bend}} + U_{\text{torsion}} + U_{\text{vdW}} + U_{\text{electrostatic}}. \tag{11.15}$$

Although written in terms of the potential energies, Eq. 11.15 is called the **force field** of the molecular mechanics method. The typical choice for a reference energy is the potential energy of the denatured (completely unfolded) structure at equilibrium, with all atomic charges and van der Waals interactions set to zero.

What's the result of all this work? A model of a very complex system (typically thousands of atoms), which is so simple that experimentally measurable parameters—such as conformer geometry—can be successfully predicted. That doesn't mean that the problem is trivial. All of the vibrational modes are still treated explicitly, and there are plenty of those to worry about.

SAMPLE CALCULATION **Vibrational Modes of a Single Amino Acid.** How many vibrational modes are there in the amino acid tryptophan ($C_{11}H_{12}O_3N_2$)? The atoms add up to a total of 28, and the molecule is non-linear. Therefore it has $3(28) - 6 = 78$ vibrational modes. And this is just *one* amino acid. Proteins typically consist of hundreds of amino acids.

These calculations quickly establish that, if we consider only the isolated molecule, proteins and other macromolecules tend to be much more stable coiled up than stretched out. Despite the greater steric interaction, this folding allows the van der Waals forces between different parts of the molecule to contribute to the overall stabilization. As a particularly important case, if an amine group R—NH—R in the peptide chain of a protein folds to a position near a carboxyl group R—CO—R, the hydrogen bond formed between the NH group and oxygen stabilizes this conformation. With vibrational and internal rotational energies of a few hundred cm^{-1} or less, the van der Waals forces significantly influence the motions in these coordinates. Each dispersion interaction that stabilizes the molecule by 100 cm^{-1} more than makes up for a 50 cm^{-1} increase in energy from climbing a torsional energy barrier.

EXAMPLE 11.2 Molecular Mechanics and Folding

CONTEXT The secondary structure of a protein is crucial to its biological activity, whether for good or ill. Snake venoms consist of saliva with a high proportion of toxic proteins. Cobra venom in particular acts on the victim by means of *neurotoxins,* proteins that attack the nervous system, in this case by blocking the acetylcholine receptors on the membranes of muscle cells. Normally, when acetylcholine binds to one of these receptors, it triggers a flow of calcium ions that stimulate muscle contraction. Once the job is done, the acetylcholine is broken down by acetylcholinesterase. Cobra venom proteins bind to the same receptor but are not deactivated by the acetylcholinesterase, paralyzing the muscle cell. The venom targets diaphragm muscles, causing respiratory failure and death. The venom neurotoxins are folded in a way that complements the acetylcholine receptor, allowing strong binding. The forces that cause the neurotoxin to adopt this conformation are the same forces that contribute to the terms in Eq. 11.15.

PROBLEM Calculate the change in potential energy in kJ mol^{-1} when conformer I (drawn in the following figure) folds to conformer II, according to the mechanical model described earlier and using the following table of data. Let R_e for the Lennard-Jones potential of two atoms i and j be 2.2 Å and let $\varepsilon = 2.0\,\text{kJ mol}^{-1}$ for all van der Waals interactions. For torsional energies set $n = 3$ and $U_{t0} = 4.0\,\text{kJ mol}^{-1}$, and set the equilibrium torsional angles to those shown in conformer I. Assume that all bond lengths and bond angles are at their equilibrium values in both conformers.

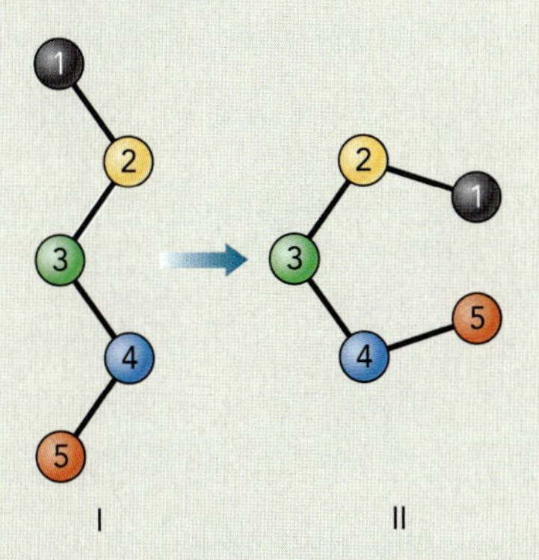

atom i	q_i	R_{1i}	R_{2i}	R_{3i}	R_{4i}	R_{5i}
	(e)	(Å)	(Å)	(Å)	(Å)	(Å)
				conformer I		
1	+0.30	0	1.5	2.5	3.8	5.0
2	−0.15	1.5	0	1.5	2.5	3.8
3	0.00	2.5	1.5	0	1.5	2.5
4	+0.15	3.8	2.5	1.5	0	1.5
5	−0.30	5.0	3.8	2.5	1.5	0
				conformer II		
1	+0.30	0	1.5	2.5	2.6	1.8
2	−0.15	1.5	0	1.5	2.6	2.6
3	0.00	2.5	1.5	0	1.5	2.5
4	+0.15	2.6	2.6	1.5	0	1.5
5	−0.30	1.8	2.6	2.5	1.5	0

SOLUTION We can first neglect contributions from stretching and bending because the bond lengths and angles are assumed to be at their equilibrium values. However, we do need to evaluate the torsional, van der Waals, and electrostatic contributions.

1. The torsional potential energy of conformer I is 0 because the ϕ's are at their equilibrium values there. The torsional energy for conformer II has only two contributions because we arrive at configuration II by twisting the molecule 180° about two bonds: 2—3 and 3—4. The torsional energy then contributes a total of

$$U_{\text{torsion}}(\text{II}) = (4\,\text{kJ mol}^{-1})\{1 - \cos[3(\pi)]\} = 8\,\text{kJ mol}^{-1}.$$

2. The van der Waals and electrostatic contributions are each the sum of the potential energies of all the interacting atom pairs, neglecting any chemically bonded atoms:

$$\begin{aligned} U_{\text{vdW}}(\text{I}) &= \sum_i \sum_j \varepsilon[R_{\text{eLJ}}^{12} R_{ij}^{-12} - 2R_{\text{eLJ}}^{6} R_{ij}^{-6}] \\ &= (2\,\text{kJ mol}^{-1})\{2.2^{12}[2.5^{-12} + 3.8^{-12} + 5.0^{-12} + 2.5^{-12} + 3.8^{-12} + 2.5^{-12}] \\ &\quad -2(2.8)^6[2.5^{-6} + 3.8^{-6} + 5.0^{-6} + 2.5^{-6} + 3.8^{-6} + 2.5^{-6}]\} = -4.5\,\text{kJ mol}^{-1} \\ U_{\text{vdW}}(\text{II}) &= (2\,\text{kJ mol}^{-1})\{2.2^{12}[2.5^{-12} + 2.6^{-12} + 1.8^{-12} + 2.6^{-12} + 2.6^{-12} + 2.5^{-12}] \\ &\quad -2(2.8)^6[2.5^{-6} + 2.6^{-6} + 1.8^{-6} + 2.6^{-6} + 2.6^{-6} + 2.5^{-6}]\} = 3.9\,\text{kJ mol}^{-1}. \end{aligned}$$

Because this potential energy function describes only the weakly interacting atoms, the sum omits the bonded atom pairs (1—2, 2—3, 3—4, and 4—5).

3. For the electrostatic energy, we do not need to include atom 3 in the sum because it has no charge:

$$\begin{aligned} U_{\text{electrostatic}}(\text{I}) &= \sum_i \sum_j \frac{q_i q_j}{4\pi\varepsilon_0 R_{ij}} \\ &= -4.6\,\text{kJ mol}^{-1}. \\ U_{\text{electrostatic}}(\text{II}) &= -33.4\,\text{kJ mol}^{-1}. \end{aligned}$$

The electrostatic energy significantly stabilizes the folded conformer II.

The overall change in potential energy is the sum of these contributions:

$$\begin{aligned} U_{\text{tot}}(\text{I}) &= U_{\text{torsion}}(\text{I}) + U_{\text{vdW}}(\text{I}) + U_{\text{electrostatic}}(\text{I}) = -9.1\,\text{kJ mol}^{-1} \\ U_{\text{tot}}(\text{II}) &= U_{\text{torsion}}(\text{II}) + U_{\text{vdW}}(\text{II}) + U_{\text{electrostatic}}(\text{II}) = -21.5\,\text{kJ mol}^{-1} \\ \Delta U_{\text{tot}} &= U_{\text{tot}}(\text{II}) - U_{\text{tot}}(\text{I}) = -12.4\,\text{kJ mol}^{-1}. \end{aligned}$$

The molecule favors conformer II over conformer I in this case.

When the alkane chains are much longer than the one shown in Example 11.2, it becomes much easier to find conformations that keep the atoms far enough away from one another to avoid severe steric repulsions while lowering the energy by virtue of Coulomb and van der Waals interactions. Such structures are central to the function of nucleic acids, proteins, and lipids.

For example, globular proteins regulate many transport and biochemical reaction processes in organisms, and their ability to function is determined by their solubility and by the chemical nature and structure of specific active sites. The overall shape of these proteins generally results from peptide chains folded back on themselves, orienting the **hydrophobic groups** (which do not bond well to water) toward the center of the structure, where they are bound by van der Waals forces. This leaves the **hydrophilic groups** free to hydrogen-bond with the water of their environment, and it permits the molecule to travel where the water takes it. In the same way, the critical active sites can be exposed to the substrates or buried within a web of peptide chains. The α helix and β-pleated sheet in proteins and the double-helical structure of DNA are the result of hydrogen-bonding between different functional groups of very large, flexible molecules. In DNA, these are (N—H---O) and (N—H---H) hydrogen bonds between the nucleotides. A similar structure, the β-pleated sheet in proteins, is illustrated in Fig. 11.12.

Hydrogen-bonded conformers are often very pH-dependent, because H^+ ions protonate the exposed electronegative atoms (for example, the carboxyl oxygens in a peptide), robbing those atoms of the opportunity to form intramolecular hydrogen bonds. Without the stabilizing influence of the hydrogen bonds, these molecules form disordered structures called **random coils** at low pH.

Because pH and temperature can change the relative stability of different structures of the same protein, there is a great deal of interest in this aspect of protein dynamics. Macromolecular structure can be studied by diffraction techniques or by spectroscopy of particular sites of the molecule. For example, biomolecules with aromatic rings are susceptible to strong fluorescence when the π electrons of the ring are electronically excited by a laser, but the transition energy depends on the environment of the ring. Fluorescence spectroscopy can therefore be used to infer the location or conformation of the molecule by providing data on a specific site.

Molecular mechanics serves to probe the dynamics of macromolecules as well as their structure. From the potential energy barrier that lies between two stable conformers, molecular mechanics can predict the temperature at which the molecule twists easily between the two forms. Specific folding pathways can be tested in principle, but the enormous range of possible conformations often limits these studies to statistical methods, described in *TK* Chapter 7.

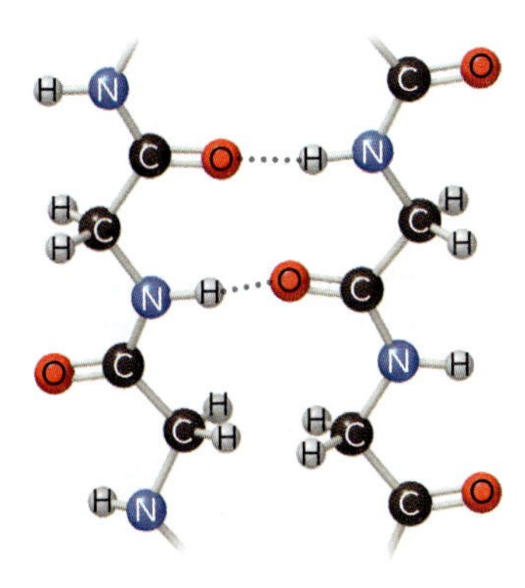

▲ FIGURE 11.12 **Hydrogen bonding in the β-pleated sheet of proteins.** The C=O groups participating in the hydrogen bonding are shown oriented toward N—H groups.

CONTEXT *Where Do We Go From Here?*

We can extrapolate from what we've learned here about molecular structure over distances of several nanometers to the structure of molecular interactions over longer distances. To keep things as simple as possible, we can continue to focus just on structures and energies, without letting our system change while we watch it. That way we can get familiar with the microscopic structure of liquids (Chapter 12) and solids (Chapter 13) which will complete our picture of matter at the molecular level.

KEY CONCEPTS AND EQUATIONS

11.2 Clusters. Any attractive interaction between molecules can lead to formation of clusters, so there is no single equation that describes the binding of the particles. One that we don't run into elsewhere, however, is this jellium potential, derived for small clusters of metal atoms:

$$U(r) = \begin{cases} -U_0\left[\frac{N}{2r_0}\left(3 - \frac{r^2}{r_0^2}\right)\right] & r \leq r_0 \\ -U_0\frac{N}{r} & r > r_0, \end{cases} \quad (11.4)$$

where $r_0 = 3N^{1/3}/(4\pi\rho_0^{1/3})$, ρ_0 is the the electron density at the center of the cluster, and N is the cluster size.

11.3 Macromolecules. In molecular mechanics, the total potential energy of the molecule is written as the sum of these various interaction potential energies:

$$U_{\text{tot}} = U_{\text{stretch}} + U_{\text{bend}} + U_{\text{torsion}} + U_{\text{vdW}} + U_{\text{electrostatic}}. \quad (11.15)$$

Many of these terms must be based on parameters measured experimentally, but the resulting method is *much* faster at predicting molecular properties than quantum mechanical calculations.

KEY TERMS

- **Clusters** are bound groups of identical atoms or molecules. The number of those units is the **cluster size.**
- **Macromolecules** are very large, single molecules.
- The **coordination number** is the number of particles in direct contact with some reference particle.
- The **jellium model** is a model potential for the binding of clusters of metal atoms.
- **Magic numbers** are cluster sizes that have significant stability relative to other similar cluster sizes.
- **Molecular mechanics** is a method of predicting molecular properties by applying classical physics to the motions of the atoms in (usually large) molecules.
- A **force field** is the sum of mathematical expressions defining the contributions to the potential energy of a molecule, and it is used to optimize the geometry and calculate conformational energies in a molecular mechanics calculation.
- **Scanning tunneling microscopy** maps a sample by measuring the current of electrons tunneling from the surface to the microscope probe.
- **Atomic force microscopy** maps the intermolecular forces between the microscope probe and the sample surface.

OBJECTIVES REVIEW

1. *Predict the major bonding forces and trends in various physical properties in homogeneous molecular clusters.* What forces will be primarily responsible for binding a small cluster of CO_2 molecules? Will the average binding energy per molecule increase or decrease as the cluster size grows?
2. *Identify the chief terms in the force fields used for molecular mechanics calculations.* List the terms in a typical molecular mechanics force field that account for non-chemical bonding interactions.

PROBLEMS

Discussion Problems

11.1 Of K_3 and K_4, which will have the higher ionization energy, and why?

11.2 Silicon is in the same group of the periodic table as carbon. Would you expect silicon to easily cluster in a form similar to any of the following: (a) diamond, (b) graphite, (c) polyacetylene, or (d) fullerene?

11.3 Predict the qualitative results when two water clusters $(H_2O)_{100}$ collide at typical velocities for a temperature of 300 K.

Weakly Bound Clusters

11.4 Calculate the *change* in energy when two tetrahedral Ar_4 clusters combine to form a cubic Ar_8 cluster, assuming that each nearest-neighbor interaction is $-\varepsilon$. THINKING AHEAD ▶ [Should the change be positive or negative, or are there competing effects?]

11.5 Four geometries for the Ar_4 cluster are sketched in the following figure. Number them from 1 to 4 in order of decreasing stability so that 1 is the *most* stable and 4 is the *least* stable.

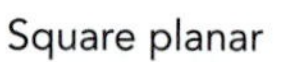

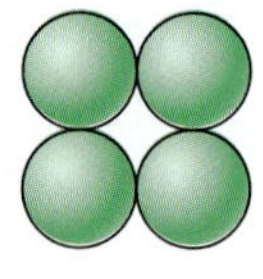

Rhombohedral

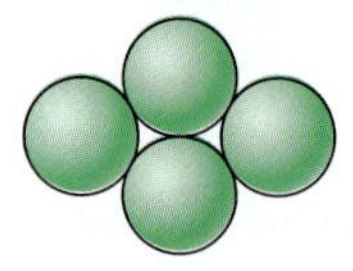

Linear

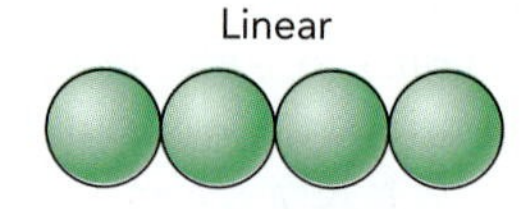

Tetrahedral

11.6 According to calculations using a Lennard-Jones potential, there are four competing structures for the cluster of seven noble gas atoms, all differing slightly in energy. Try to guess any of these stable arrangements of this cluster. THINKING AHEAD ▶ [What *approximate* shape is likely for any arrangement that maximizes the stability?]

11.7 Section 11.2 includes a rough calculation to estimate the maximum coordination number of a set of identical spheres. The result is that one sphere may be immediately surrounded by up to 12 spheres, giving a magic number cluster size of 13. Estimate the next magic number cluster size by this method. THINKING AHEAD ▶ [As the cluster grows from 13, where will the next spheres be added?]

11.8 The equilibrium geometry of the water hexamer is pictured in the following figure. Each water molecule in this cluster belongs to one of three classes, based on the hydrogen bonding. Two water molecules (1 and 2 in the sample configuration drawn) participate in only two hydrogen bonds, donating a hydrogen to one bond and accepting the hydrogen in the other bond; two other waters (3 and 4) serve as hydrogen donors in two bonds and as hydrogen acceptors in a third bond; and the remaining two waters (5 and 6) function as hydrogen donors in one bond and hydrogen acceptors in two bonds. Find the number of distinct conformations of this structure that can be formed if we keep the oxygens at *roughly* the same position (i.e., the water molecules do not change location in the cluster), but we allow all of the hydrogens to tunnel or rotate to new positions.

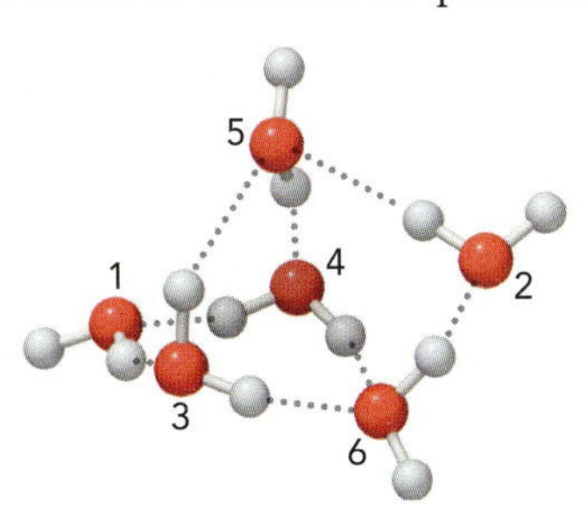

11.9 From the value of ε in Table 10.1 for argon dimer, Ar_2, and the 6.43 kJ mol^{-1} required to vaporize bulk argon from liquid to gas, estimate the total number of intermolecular bonds that must be broken to free one argon atom from the bulk liquid.

11.10 Repeat the derivation of Eq. 10.38, this time using three one-dimensional boxes centered at x values of $-R$, 0, and $+R$. Compare the energy of the three-box system to the energy of three independent boxes to estimate how the binding energy of a linear noble gas trimer compares with that of the dimer. Does the binding energy increase or decrease with cluster size? THINKING AHEAD ▶ [Why might the binding energy increase in comparison to the dimer? Why might it decrease?]

Strongly Bound Clusters

11.11 One critical spectroscopic measurement of C_{60} was the determination of the gap between the HOMO to LUMO by electron loss spectroscopy in 1991.

a. How many electrons are in the π system of C_{60}?

b. How many nodes perpendicular to the surface of each 6-membered ring do you expect for the HOMO? Based on this, try to identify just the *degeneracy* of the symmetry representation of the HOMO.

c. The LUMO has t_{1u} symmetry, and the HOMO has u inversion symmetry. Is the HOMO–LUMO transition allowed by electric dipole selection rules?

11.12 Among many candidates for quantum electronic devices are the *tolanes*, two benzene rings connected by -C≡C- groups. The electrical properties can be controlled by twisting the benzene rings so that the planes differ by a torsion angle ϕ.

a. Sketch the shape of the potential energy curve as a function of the torsion angle ϕ.

b. Sketch the expected trends in the HOMO and LUMO energies of the π electron system as ϕ varies from 0 to $\pi/2$.

11.13 Sketch the energy levels and radial wavefunctions for the 1*s*, 2*s*, and 3*s* states of the jellium potential energy model. Point out the qualitative differences when compared to the same states of the one-electron atom. THINKING AHEAD ▶ [What is the shape of the potential energy curve, and how will this affect the energies and wavefunctions?]

11.14 The carbon network in a cylindrical nanotube can be thought of as a plane of sp^2 hybridized carbons strained by the rolling of the plane to form the tube. This strain can increase the sp^3 character of the carbon atom hybrids.

a. Find an equation for the CCC bond angles in an ideal carbon nanotube as a function of the radius R of the tube, assuming that all the bond lengths are 1.40 Å and that the axis of the tube lines up with one set of carbon–carbon bonds.

b. Use this equation to estimate the fraction of $p_\perp$ character in the carbon orbital hybrids when $R = 5.0$ Å and 3.1 Å, where $p_\perp$ is the atomic p orbital perpendicular to the plane sp^2 hybrid. (For example, a $p_\perp$ fraction of 0 would correspond to a pure sp^2 hybrid, while a fraction of 0.25 would indicate a pure sp^3 hybrid.)

11.15 For a single electron in a metal cluster obeying the jellium model, do the following.

a. Write the Schrödinger equation that must be solved to obtain the radial wavefunction.

b. Write an angular wavefunction for the lowest energy excited state.

11.16 Assume that clusters of Mg atoms obey the jellium model. Plot the ionization potential IE_1 versus cluster size n for $n = 1 - 6$, showing *qualitatively* how you expect the IE_1 values to vary as a function of n. THINKING AHEAD ▶ [The ionization energies depend on the energy or energies of what quantum states?]

11.17 Estimate the volume inside a C_{60} molecule, based on Fig. 11.5, if the average C—C bond length is 1.43 Å. Which of the following noble gas atoms could be contained *inside* a C_{60} molecule: Ne ($R = 2.4$ Å), Ar ($R = 2.9$ Å), Xe ($R = 3.4$ Å)?

11.18 Assume that the van der Waals radius of the lanthanum atom is about 3.5 Å. Use the carbon–carbon bond lengths in Table 5.1 to estimate the minimum number of interconnected, sp^2 hybridized, and aromatic carbon atoms that would be required to completely encase one lanthanum atom in a roughly spherical shell.

11.19 Verify that the 11.6° non-planarity of the "sp^2" hybrids on the surface of pure C_{60} comes from donating 8.4% s character to the out-of-plane orbitals. Start with a normalized sp^2 hybrid in the xy plane that gives three equivalent bonds separated by 120°. This leaves a pure p_z orbital. Add 0.084 of the s orbital to the p_z orbital, and transfer enough of the p_z into the sp^2 hybrid to keep all the orbitals normalized. Then show that this shifts the sp^2 bonds below the xy plane by 11.6°.

11.20 The density of graphite is 2.25 g cm^{-3}; the density of polyethylene is roughly 0.95 g cm^{-3}.

a. Use these values to estimate the spacing between adjacent sheets of graphite and between adjacent polymer chains in polyethylene.
b. Estimate the density of bulk C_{60}, assuming each molecule is a sphere with a radius of 2.8 Å.

Macromolecules

11.21 Consider a protein made up of 50 amino acids in a chain. If each amino acid can be oriented in three different ways, how many conformations does the protein have? If the protein moves from one conformation to the next in a typical vibrational period (10^{-13} s), what is the minimum necessary time required for the protein to sample only 1% of these conformations?

11.22 A protein possesses a backbone chain of 200 sp^3-hybridized carbon atoms with bond angles of 109 degrees. If we stretch the backbone out as much as possible, allowing the atoms to retain their tetrahedral bonding, we obtain a chain about 230 Å long. For a typical zero-point motion of ± 0.06 Å per bond, how much could the overall length of the molecule change from these zero-point vibrations, assuming they are all in phase? Can we assume that they are all in phase?

11.23 Consider only the bending and electrostatic contributions to the potential energy for the system drawn in the following figure. If only the center bond angle is allowed to change, find the value of θ that minimizes the potential energy if $\theta_e = 109°$ for all three bond angles, $R = 1.5$ Å, $k_\theta = 0.5$ kJ mol^{-1} rad^{-2}, and $q_1 = -q_2 = 0.03e$.

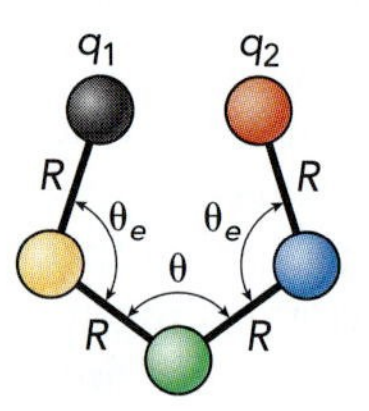

11.24 Estimate the de Broglie wavelength for a protein of roughly 100 amino acids, the amino acids having an average mass of about 80 amu, if the average atomic speed in solution is 540 cm s^{-1}. One of the major fields in current biochemical research is the folding of proteins into globular shapes by torsions around the bonds of the amino acids. Is it likely that quantum mechanics is necessary to adequately describe this process?

11.25 The work function of a metal is the energy necessary to release an electron from its surface. Gold has a work function of about 5.4 eV; aluminum has a work function of about 4.2 eV. Sketch the potential energy for electrons in an STM experiment as a function of z (the distance above the surface) for each of these metals, assuming that the current flows from the metal surface to the STM tip and fixing probe voltage and the position of the tip for each metal. State which of the two metals has the higher tunneling current and how this is proven by your graphs.

12 The Structure of Liquids

LEARNING OBJECTIVES

After reading this chapter, you will be able to do the following:

❶ Use the pair correlation function $\mathcal{G}$ as a measure of long-range order in liquids, and predict trends in $\mathcal{G}$ as a function of density, temperature, and intermolecular forces.

❷ Qualitatively predict the miscibility of two substances based on the relevant intermolecular forces, and describe the microscopic structure of the resulting solution.

GOAL *Why Are We Here?*

The goal of this chapter is to describe the structure of liquids on small-distance scales, examining how the balance among diverse forces influences the properties of the substance.

CONTEXT *Where Are We Now?*

As we've progressed through this text, our chemical system has been growing in complexity, from single atoms to single molecules, from pairs of molecules to clusters. Eventually, we aspire to extend these methods to even bigger systems so we can describe chemical systems in bulk quantities like grams and milliliters. To make this extension manageable, we begin by studying only the *microscopic* structure of the bulk system.

First, let's stop assuming that the system is small. Consider as many molecules as you like. More than that, actually—consider as many molecules as makes you uncomfortable. Come to think of it, we will need still more.

Our examination of isolated molecules in Part II and two-molecule interactions in Chapter 10 mimicked the gas phase, where one molecule rarely encounters another and has little influence on the structure and energy of its neighbors. In the microscopic picture, it doesn't matter whether we have two gas molecules or a thousand. The neighborhood around our molecule is quiet, with just the occasional visitor. For the nature of our system to be influenced by the huge number of molecules we've just added, we squeeze the molecules together, into such a small volume that they bump up against each other constantly. The neighborhood becomes crowded, and—if we haven't reduced the energy of the molecules by much—it's *noisy.* We've formed a liquid.

SUPPORTING TEXT **How Did We Get Here?**

The main *qualitative* preparation we need for the topics in this chapter is grasp of the forces exerted by one molecule on another, as described in Chapter 10. For a more detailed background, we will draw on the following equations and sections of text to support the ideas developed in this chapter:

- Table 10.1 gives the parameters for calculating intermolecular interactions involving a number of molecules. One of the parameters we find there is the polarizability, α, which we can convert to SI units using Eq. 6.15:

$$\alpha(\mathrm{Cm^2V^{-1}}) = (10^{-30})4\pi\varepsilon_0\,\alpha(\text{Å}^3).$$

- We often approximate these intermolecular potentials using the Lennard-Jones potential (Eqs. 10.44 and 10.45),

$$u_{\mathrm{LJ}}(R) = \varepsilon\left[\left(\frac{R_e}{R}\right)^{12} - 2\left(\frac{R_e}{R}\right)^6\right] = 4\varepsilon\left[\left(\frac{R_{\mathrm{LJ}}}{R}\right)^{12} - \left(\frac{R_{\mathrm{LJ}}}{R}\right)^6\right].$$

To estimate the Lennard-Jones potential for a heterogeneous interaction when the compositions of the two molecules are different, the following approximation is often successful (Eq. 10.47):

$$R_{\mathrm{LJ,AB}} \approx \frac{R_{\mathrm{LJ,A}} + R_{\mathrm{LJ,B}}}{2}$$

(We will replace Eq. 10.48 for the well depth by a simpler expression more appropriate for large molecules.)

12.1 The Qualitative Nature of Liquids

A material is said to flow when the distribution of its molecules changes freely under the influence of very weak forces, such as Earth's gravity, and any such substance is called a **fluid.** A gas is a fluid that expands or contracts to occupy the full volume of its container. A **liquid** is also a fluid, but unlike a gas a liquid maintains a constant volume, within limits. (All liquids are compressible under some circumstances, binding forces can limit a liquid's ability to change shape, and the time scale for flowing can cover a wide range.)

Bonding Mechanisms

For a given mass of liquid to be of constant volume, it must retain a constant average density: the number of particles per unit volume must be constant. If the density is a fixed value, then the average distance R between neighboring particles in the liquid must also be fixed. What keeps the average distance R the same? In a liquid, attractive forces bind the particles together, with the average separation being roughly the equilibrium distance R_e where the potential energy reaches a minimum. Liquids and solids share this characteristic of their particles being bound together to maintain a constant density, making these two phases of matter the **condensed phases.** In order for liquids to have freedom of shape, however, the binding force must allow the bonds to change orientation and to break and reform. This means that the directional (anisotropic) character to the binding force is small enough compared to the non-directional (isotropic) character that it can be overcome by weak external forces or intermolecular collisions.

Covalent chemical bonds tend to form stiff structures, with little isotropic contribution to the potential energy. Major changes in the shape of a covalently bound molecule correspond to enormous vibrational excitation, often approaching the bond dissociation energy. Some other, less directional bonding mechanism is responsible for the cohesiveness of a liquid, allowing it to maintain a constant volume. Three general classes of bonding fill this bill:

1. **Van der Waals liquids.** Even the direction-dependent dipole–dipole force has much less angle dependence, allowing greater angular motion, than do chemical bonds, and the dipole-induced dipole and dispersion forces are largely isotropic. An isotropic potential allows the molecules in a liquid to be bound together, but at the same time to arrange themselves in any fashion. As a result, weaker forces, such as gravity, can determine the shape of the liquid as long as they do not have an effect on the density. In the same way, a jar of marbles has an average density determined by the size of the jar and number of marbles, but after those constraints are met, gravity and the shape of the jar determine the orientation of the marbles.
2. **Ionic liquids.** Another form of isotropic force is the monopole–monopole force. In ionic liquids (also called molten salts), the monopoles are molecular anions and cations. Because these monopole–monopole forces are much stronger than the van der Waals forces, salts are normally liquids only at much higher temperature than the more familiar van der Waals liquids. At temperatures closer to 300 K, ionic liquids *usually* solidify as the very strong binding forces pull the particles together into a dense, rigid arrangement. However, there are several room temperature ionic liquids that use cyclic organo-nitrogen cations to distribute the ionic charge over a relatively large volume. By increasing the average distance between the ionic charges, we reduce the strength of the attractions, so the attractions between the ions are comparable in strength to the dispersion forces in smaller molecules. Bulky substituents may also reduce the chance of the substance freezing into an ordered structure, even at temperatures below its freezing point, leading to a *supercooled* ionic liquid.
3. **Molten metals.** Metals are also bound by a monopole–monopole force, but in this case the monopoles are metal cations and electrons. Mercury is the only pure metal that is liquid at room temperature, but a few other metals (cesium, gallium, rubidium) as well as several alloys have melting points under 40 °C. What these metals have in common is a single unpaired valence electron, which limits the number of electrons that can be donated into the delocalized electron sea binding the metal cations. The large atomic radii of mercury, rubidium, and cesium further reduce the binding force between the atoms. Gallium is smaller but has too much nonmetal character to form strong metallic bonds.

Surface Tension

One probe of the strength of the forces that hold the liquid together is the liquid's ability to deform under the influence of weak forces, such as gravity. The attractive forces between the particles tend to reduce the surface area of the liquid because molecules at the surface experience the attractions of fewer neighbors and therefore raise the average potential energy of liquid. For any given volume, the sphere has the smallest surface area, so the attractive forces in the liquid tend to pull the liquid into a sphere. In the absence of gravity, water droplets are spherical because this shape maximizes the number of intermolecular bonds. However, on a tabletop in Earth's gravity, a water droplet will spread out because the influence of gravity is sufficient to break a few intermolecular bonds. As the droplet spreads out, more molecules are brought to the surface, which requires breaking some of the bonds between the particles that hold the liquid together. Beyond a certain point, the droplet stops spreading, and the intermolecular forces yield no further to gravity. This limit is determined by the **surface tension** γ of the fluid, where γ gives the ratio of the energy change dE to the change in surface area of the liquid dA:

$$\gamma = \frac{dE}{dA}. \tag{12.1}$$

Although given here as an energy divided by an area, the units of surface tension (J m^{-2}) are more often given in the equivalent units of force per unit length (N m^{-1}). Several values for the surface tension are given in Table 12.1. The greater the surface tension, the greater the energy required to increase the liquid's surface area. The surface tension increases roughly in proportion to the strength of the bonds between the particles and the number of those bonds per particle and decreases roughly in proportion to the size of the particle. Liquids with high

TABLE 12.1 Surface tensions for selected liquids. The bonding mechanisms listed are between the particles that move independently in the liquid, as given in column 1. For example, each water molecule is held together by covalent bonds, but the liquid consists of water molecules held together by *hydrogen* bonds. The measurement temperature is *T*.

substance		Inter-particle bonding (liq)	T(K)	surface tension γ (N m^{-1})
He	helium	dispersion	4.2	0.00037
N_2	nitrogen	dispersion	79.2	0.00844
$C_{10}H_{22}$	decane	dispersion	298.15	0.0234
NH_3	ammonia	H-bonding	298.15	0.0248
CCl_4	carbon tetrachloride	dispersion	298.15	0.0264
$CHCl_3$	chloroform	dipole–dipole	298.15	0.0265
C_6H_6	benzene	dispersion	298.15	0.0281
Br_2	bromine	dispersion	298.15	0.0409
$C_8H_{15}N_2BF_4$	bmim[a] tetrafluoroborate	ionic	293	0.0448
S_8	sulfur	dispersion	433	0.0567
H_2O	water	H-bonding	298.15	0.0720
SiO_2	silicon dioxide	ionic	1773	0.2982
AgCl	silver chloride	ionic	733	0.1783
Hg	mercury	metallic	298.15	0.4855
Si	silicon	metallic	1683	0.720
Cu	copper	metallic	1560	1.216

[a]1-Butyl-3-methylimidazolium

intermolecular binding energies (e.g., molten metals, ionic liquids, hydrogen-bonding molecules) have greater surface tensions than liquids with low intermolecular bonding energies (e.g., nonpolar or weakly polar molecules).

Silicon and silicon dioxide provide interesting examples in Table 12.1. The particles form strong covalent bonds in the solid, but covalent bonds are too highly directional to allow the freedom of motion that characterizes the liquid. The forces need to be more isotropic. As a result, these substances become liquids only at temperatures high enough to allow the nature of the bonding to change. Liquid silicon dioxide is an *ionic* liquid, and liquid silicon is *metallic.* In both cases, the elevated temperature provides the energy needed to redistribute the electrons in a manner that allows more freedom of motion to the atoms.

The liquid lies in a regime between the two extremes of the ideal gas, in which the particles don't interact at all, and the rigid structure of the ideal solid. Without the advantage of these limiting cases, the liquid is a difficult phase of matter to describe. So for now we'll assume a relatively weak, purely isotropic potential energy function initially, so that the attractions between the particles depend only on R. We will also assume that all the molecules in our sample have the same chemical composition. In other words, the liquid is a *pure substance.*

12.2 Weakly Bonded Pure Liquids

The key to the structure of the liquid is the nature of the potential energy function that describes the attractions and repulsions between the particles. Chapter 10 deals with the kind of weak attractions that can bind molecules together without locking them into the rigid structure of a solid. But in the liquid we have many molecules interacting simultaneously. How do we take that into account?

Potential Energy in the Condensed Phase

We could naively estimate the potential energy experienced by one molecule (call it molecule A) in the liquid by adding together the potential energies for all the interactions involving that molecule with the other $N-1$ molecules:

$$u_{\mathrm{A}} \approx \sum_{i=2}^{N} u(R_{\mathrm{A}i}) \approx -\varepsilon \sum_{i=2}^{N}\left[\left(\frac{R_e}{R_{\mathrm{A}i}}\right)^{12} - 2\left(\frac{R_e}{R_{\mathrm{A}i}}\right)^{6}\right], \tag{12.2}$$

where we've assumed Lennard-Jones potential (Eq. 10.44). Energy is an additive parameter, so what could be wrong with this?

The trouble is that, without saying so, we have assumed that only molecule A feels the effects of other molecules. But we know better. Each molecule polarizes its neighbor, if only a little bit, and that affects how the neighbor molecule will then interact with the next one. The **Axilrod-Teller correction** estimates the effect of this indirect interaction by applying third-order perturbation theory to the dispersion energy of three spherical molecules. In Section 10.1 we see how second-order perturbation theory predicts the attraction between two polarizable particles. It takes third-order terms to treat the effect of a third particle, and the derivation is lengthy, so let's just jump to the result this time:

$$u'_{\mathrm{ABC}} = \frac{3\alpha\varepsilon}{4(4\pi\varepsilon_0)} \frac{3\cos\theta_{\mathrm{A}}\cos\theta_{\mathrm{B}}\cos\theta_{\mathrm{C}} + 1}{(R_{\mathrm{AB}}R_{\mathrm{BC}}R_{\mathrm{CA}})^3}. \tag{12.3}$$

CHECKPOINT The Axilrod-Teller correction provides a slightly more sophisticated and accurate description of interactions in a liquid than we can get from the pair interactions in Chapter 10 alone. But even this is only an approximation, enough to see how challenging it is to describe the liquid with algebraic formulas.

In this equation, α is the polarizability of each molecule, and the θ values are the angles measured between each pair of molecules, as shown in Fig. 12.1a. This correction can then be added to the pairwise potential energies to get the overall potential energy for all three molecules:

$$U_{ABC} = u_{AB} + u_{BC} + u_{AC} + u'_{ABC}. \tag{12.4}$$

Negative values of u'_{ABC}, which further stabilize the three-molecule system, appear when $\cos\theta_A\cos\theta_B\cos\theta_C < -1/3$. This occurs, for example, when the three molecules fall roughly in a line, because then two of the cosines are close to $+1$ and the other close to -1. In that case, the correlation of the electronic wavefunctions on two of the molecules reinforces the correlation with the third molecule (Fig. 12.1b). As one molecule, say C, moves toward a point perpendicular to the line connecting A and B (so $\theta_B = \pi/2$ in Fig. 12.1c), C impairs the ability of A and B to stabilize each other. If A polarizes molecule B along one axis, while at the same time C polarizes B along a perpendicular axis, then molecule B compromises and polarizes along an intermediate axis. The A—B and B—C interactions are weaker than they would be without the three-body interaction, and therefore the Axilrod-Teller correction becomes a positive contribution to the potential energy, destabilizing the three-molecule system.

SAMPLE CALCULATION **Axilrod-Teller Correction.** We will compute the Axilrod-Teller correction in kJ mol^{-1} for three argon atoms positioned at the points of an equilateral triangle with sides of length R_e, using the values (from Table 10.1) $R_e = 2^{1/6} R_{LJ} = 3.86\,\text{Å}$, $\alpha = 1.64\,\text{Å}^3$, and $\varepsilon = 120\,\text{K}$. An equilateral triangle sets the three values of θ each at 60°, so the cosines are all $\frac{1}{2}$. Don't forget to convert the polarizability to SI units using Eq. 6.16. The correction adds up to

$$u'_{ABC} = \frac{3\alpha\varepsilon}{4(4\pi\varepsilon_0)} \frac{3\cos\theta_1\cos\theta_2\cos\theta_3 + 1}{(R_{12}R_{23}R_{13})^3}$$

$$= \left(\frac{3(1.64\,\text{Å}^3)(10^{-30})(4\pi\varepsilon_0)(120\,\text{K})(1.381\cdot10^{-23}\,\text{J K}^{-1})}{4(4\pi\varepsilon_0)}\right)\left(\frac{3(\frac{1}{2})^3 + 1}{(3.86\cdot10^{-10}\,\text{m})^3}\right)$$

$$= 4.87\cdot10^{-23}\,\text{J} = 0.029\,\text{kJ mol}^{-1}.$$

Compare this to the sum of the three pair potentials, $-3\varepsilon = -2.99\,\text{kJ mol}^{-1}$.

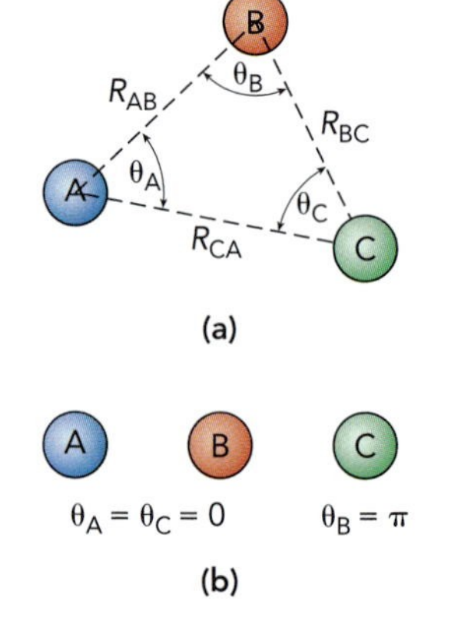

(a)

A B C

$\theta_A = \theta_C = 0$ $\theta_B = \pi$

(b)

C

$\theta_B = \pi/2$

(c)

▲ **FIGURE 12.1 Three spherical molecules sharing dispersion interactions.** **(a)** Definition of parameters; **(b)** a mutually stabilizing orientation; **(c)** a destabilizing orientation.

The Pair Correlation Function

Having just emphasized the importance of isotropic bonding in liquids, we now turn around and look for structure in liquids. The idea of a liquid having structure may sound weird, but over distances that cover a few molecules liquids do have *some* regular structure. More and more randomness becomes evident as we look over larger distances.

A non-random or ordered structure is one for which we can successfully predict the location of one molecule if we know the locations of its neighbors. In the case of a liquid, our predictions will not always be correct—there is always some randomness. However, if we select a single reference molecule, we can predict the location of its nearest neighbor molecules with greater success than we could if the distribution were purely random. The farther we get from our reference, the less successful our predictions become.

We describe this very tenuous structure using a function $\mathcal{G}(R)$ called the **pair correlation function.** The pair correlation function provides a way to graph the average structure of a liquid, quantifying the delicate balance between the kinetic and potential energies that defines a liquid. Because liquids are generally difficult to parameterize, $\mathcal{G}(R)$ is an important function. We use it to describe the properties of bulk liquids as well as their microscopic structure. But the precise definition of the pair correlation function takes some careful thinking. To define the structure of the liquid, we will select a reference molecule at random and then count the average number of molecules $\overline{N}(R)$ that we find within a radius R from the origin (the center of mass of the reference molecule). We express $\overline{N}(R)$ in terms of the **number density** ρ, the average number of molecules per unit volume of the liquid, as follows. If the molecules are evenly distributed, then $\overline{N}(R)$ is given by integrating the number density ρ over the radius R:

Even distribution:

$$\overline{N}(R) = \int_0^{2\pi}\int_0^{\pi}\int_0^{R} \rho\, R'^2 dR' \sin\theta d\theta d\varphi = 4\pi \int_0^{R} \rho R'^2 dR' = \frac{4\pi R^3}{3}\rho.$$

Now, consider a real liquid, where the molecules tend to cluster around each other rather than spreading themselves randomly. In that case, $\mathcal{G}(R)$ is the fraction of the sphere's surface area that intersects other molecules (Fig. 12.2), and $\rho\mathcal{G}(R)$ gives the probability per unit volume of finding a molecule at a distance R from the reference molecule. The number of neighboring molecules $\overline{N}(R)$ within a radius R in that case is

$$\overline{N}(R) = \int_0^{2\pi}\int_0^{\pi}\int_0^{R} \rho\mathcal{G}(R')\, R'^2 dR' \sin\theta d\theta d\phi = 4\pi\rho \int_0^{R} \mathcal{G}(R') R'^2 dR'. \quad (12.5)$$

Without knowing the exact form of the pair correlation function $\mathcal{G}(R)$, we cannot simplify the integral over R, but the pair correlation function can be measured experimentally by x-ray diffraction, and it is relatively simple to obtain from computational models.

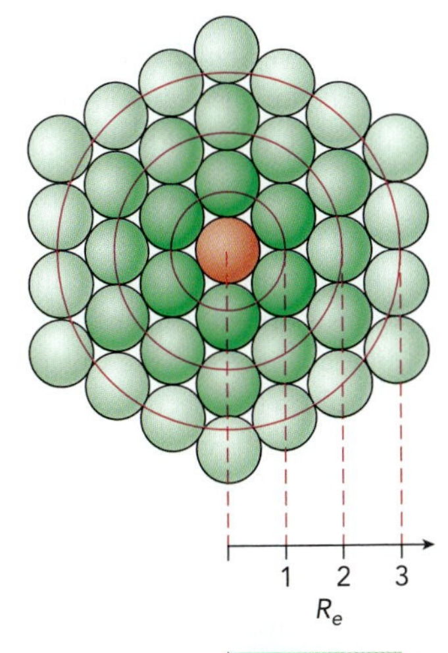

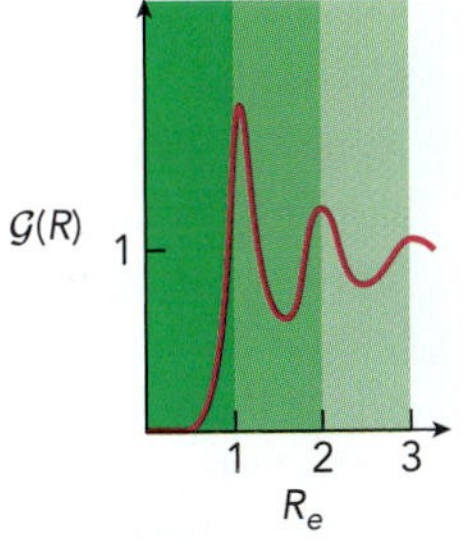

▲ **FIGURE 12.2 The correlation function.** The correlation function $\mathcal{G}(R)$ reflects the degree of order in a substance by measuring the probability of intermolecular distances favoring certain average values. If the substance is arranged so that molecules group together in solvation shells, then periodic peaks or spikes appear in the correlation function.

TOOLS OF THE TRADE Neutron Diffraction

In 1928, it was found that beryllium released an unknown form of radiation when struck by high-energy alpha particles (helium-4 nuclei). The radiation was difficult to characterize, and for a time it was mistaken for γ radiation. In 1932, James Chadwick showed that this new radiation was in fact a beam of neutrons—particles with a mass similar to the proton's but with no charge.

At the time of Chadwick's discovery, the Davisson-Germer experiment was five years old, and the wavelike nature of subatomic particles was already widely accepted. Scientists at once began to look for evidence of neutron diffraction and were able to observe the phenomenon by 1936. However, neutrons were hard to make and hard to detect. The polonium-beryllium source that Chadwick and others had used delivered too few neutrons to be of use for practical purposes. This changed with the rapid development of nuclear reactors during World War II. By 1949, Ernest O. Wollen and

Clifford G. Shull had developed a practical design based on existing x-ray diffractometers, and neutron diffractometers today share the same essential components.

What is neutron diffraction? In neutron diffraction, a beam of neutrons of uniform kinetic energy is directed at a target, and the resulting diffraction pattern is observed by measuring the intensity of the neutrons scattered at different angles by the target. Carrying out a Fourier transform (see Section A.1) of the diffraction pattern, coupled with careful analysis of the scattering intensities, leads to a determination of the structure of the sample.

Why do we use neutron diffraction? By the late 1940s, diffraction patterns of materials could already be obtained using beams of electrons (see *Tools of the Trade,* Chapter 1) and x-rays (see *Tools of the Trade,* Chapter 13). Electron beams are scattered by their strong Coulomb repulsion of the electrons near the surface of the sample, allowing electron diffraction to provide detailed information about the exterior structure of the sample but not its interior. Similarly, x-ray scattering occurs when the electric field of the radiation distorts the electron distribution, but x-rays have greater penetrating power than electrons and can probe the interior structure of the sample.

Neutron beams differ from both of these alternatives in that the neutron has no charge and generates no measurable electric field, so the neutron beam is usually not greatly affected by the electron distribution in the sample. Instead, the neutrons are scattered principally by the atomic nuclei. Like x-rays, the neutrons penetrate the interior of the sample. In fact, one of the disadvantages of neutron scattering is that the interaction with the sample is often much weaker than with x-rays, since much of the neutron beam passes through the sample without being scattered at all. However, the sensitivity of neutron beams to the *nuclear* distribution provides an important advantage in some studies over x-ray diffraction: neutrons are much more sensitive to the positions of protons in the sample.

Hydrogen atoms, which are small to begin with, contribute their single electron to any covalent bond that they form with a larger atom. The electron distribution around the hydrogen atom then merges so closely with the electron distribution of the larger atom that the position of the proton becomes very difficult to determine from electrons alone. However, the protons—like other nuclei—scatter neutrons. In one of the classic experiments carried out in 1956, the locations of the deuterium atoms in D_2O ice were determined for the first time by neutron scattering, verifying Linus Pauling's theory of hydrogen bonding. Neutrons are also capable of easily distinguishing nuclei of similar atomic number, such as Fe and Mn, which may have similar electron distributions and therefore appear indistinguishable in x-ray diffraction.

One of the other chief applications of neutron scattering has been to the investigation of magnetic materials. Like the electron, the neutron has a spin magnetic moment, but magnetic interactions of an electron beam with a sample are swamped by the much stronger effects of Coulomb repulsion. Because neutrons have no charge, they do not experience Coulomb repulsion, and magnetic interactions become detectable. As a result, neutrons that are not scattered by electrons in filled atomic subshells (where the magnetic fields all cancel) *are* scattered by unpaired electrons. Neutron scattering in paramagnetic materials therefore allows a direct assessment of, for example, the size of the partly filled *d* orbitals in a paramagnetic transition metal compound.

Of particular interest to the study of liquids is the use of neutron diffraction to obtain experimental measurements of the pair correlation function (Eq. 12.6). Because neutrons respond to the nuclei in the sample rather than the electron clouds, neutron diffractometry produces higher resolution maps of the distribution of the atoms than can be obtained using x-rays. This feature becomes especially valuable when studying liquids, where the distances between the molecules are comparable to the distances *within* the molecules, and where there is no regular pattern to the orientations of the molecules,

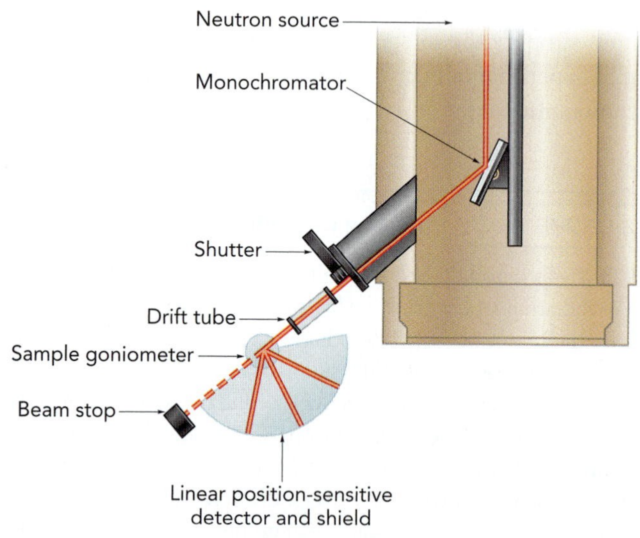

▲ **Schematic of a neutron diffractometer.**

as there is in crystals. Without the benefit of a repeating pattern, the higher resolution of neutron scattering data and the unique sensitivity to distinct nuclei are generally required to distinguish between intramolecular and intermolecular positions of the atoms. From these measurements, the pair correlation function of the liquid can be directly determined, essentially by taking the Fourier transform of the diffraction pattern.

How does it work? The accompanying figure is a schematic drawing of a typical neutron diffractometer. In addition to the sample, there are three essential components: the neutron source, an energy selector, and a detector.

Neutron source. Neutrons are most commonly generated by a nuclear reactor or a radioactive isotope. Use of radioactive isotopes, similar to Chadwick's original polonium-beryllium neutron source, allows for a more compact design but still suffers from very low output intensities. More recently, intense bursts of high-energy neutrons have been generated by *spallation,* in which pulses of accelerated protons strike a mercury target, stimulating the neutron emission.

Energy selector. In order to generate an interpretable interference pattern, the neutrons must all be of roughly equal de Broglie wavelength—in other words, traveling with the same momentum. A nuclear reactor emits neutrons continuously with a wide range of momentum values, so a monochromator selects a single momentum component by diffraction. Bouncing the neutron beam off a metal crystal splits the beam so that each value of the momentum corresponds to a particular angle of deflection. Adjusting the angle of the monochromator directs neutrons of a specific momentum value onto the sample. Energy selection in spallation sources differs because the neutrons in that case are generated in short bursts. As the neutrons fly down a *drift tube,* they separate themselves according to their speeds, and a series of timed beam stops block all the neutrons except those of a particular speed.

Detector. Electrons and ions are relatively easy to detect by the current they generate at a detector. Because neutrons carry no charge, they are usually detected by indirect means. One common method is to measure ions or photons generated when the neutron combines with an atomic nucleus, such as ^3He (to form $^3H + {}^1H$) or ^{10}B (to form $^4He + {}^7Li$). These processes release charged particles as well as radiation, which can then be detected by standard current-measuring or photon-measuring devices.

Although a much more cumbersome technique than x-ray or electron diffraction, neutron diffraction continues to find new applications in science and engineering. In addition to the advantages listed earlier, neutron diffraction is relatively non-destructive to biological compounds and is increasingly used for studies of structure and hydrogen bonding in drug discovery.

Graphing $\mathcal{G}(R)$ against R is a useful way to view the stability of any structural regularity in the system over distance. Let's build that graph a piece at a time. We start at our origin, the center of mass of our reference molecule. Two molecules will repel each other once the separation between their centers of mass comes within about R_{LJ}. Therefore the pair correlation function is zero at values of R less than this, because these values of R are effectively inside our reference molecule, and other molecules are excluded. However, several neighboring molecules will normally be attracted to our reference molecule by van der Waals forces, finding themselves most stable at separations of roughly the equilibrium distance R_e, which is equal to $1.12R_{LJ}$ in the Lennard-Jones potential.

Whatever the bonding mechanism between molecules, the reduction of one molecule's potential energy—its stabilization—by interaction with its neighbors is called **solvation.** The first group of molecules attracted to our reference molecule form the first **solvation shell,** leading to a peak in $\mathcal{G}(R)$ at about R_e. Then a second solvation shell of molecules is attracted by the first shell, and we expect to see a second peak in $\mathcal{G}(R)$ at about $2R_e$. But the arrangement of these molecules is not so dependent on the position of the reference molecule, and their distribution appears more random. Some are closer to the reference than $2R_e$, and some are farther away. What's happening is that this second shell of molecules is not directly influenced by our reference molecule, but only by the molecules immediately next

CHECKPOINT The pair correlation function $\mathcal{G}(R)$ demonstrates a solution to the problem of describing the liquid. Instead of trying to represent the liquid exactly, we use an average distribution of the relative positions of the particles. This varies sufficiently from one liquid to another and from one set of conditions to another that we can learn a great deal about the nature of the liquid from $\mathcal{G}(R)$. In *Thermodynamics, Kinetics, and Statistical mechanics,* we show how the pair correlation function can be used to predict many bulk properties of liquids.

to the second shell. Therefore, the second peak in $\mathcal{G}(R)$ is much weaker and more spread out than the first. The structure seems to disappear as we look farther from the reference. Within one or two molecular diameters, $\mathcal{G}(R)$ converges to 1, as shown in Fig. 12.3a, meaning that the distribution of molecules relative to the reference has effectively become random. By converging to 1, the pair correlation function becomes normalized, showing that if we set R large enough to reach the edge of the system, then we count all N molecules (except our reference):

$$\overline{N}(R_{\max}) = 4\pi\rho \int_0^{R_{\max}} \mathcal{G}(R')R'^2 dR' = 4\pi\rho \int_{R_{\mathrm{LJ}}}^{R_{\max}} (1)R'^2 dR' = N - 1. \tag{12.6}$$

One useful parameter of the structure derived from the pair correlation function is the coordination number $\mathcal{C}$—the average number of molecules directly in contact with the molecule at the origin:

$$\mathcal{C} \approx \overline{N}(R_{\mathrm{LJ}}) = 4\pi\rho \int_0^{R_e} \mathcal{G}(R)R^2 dR. \tag{12.7}$$

In this equation we have used the equilibrium distance R_e as determined for the Lennard-Jones potential, because at this distance the nearest neighbors will be most strongly stabilized.

For a gas, $\mathcal{G}(R)$ will remain near 1 for any distance larger than R_{LJ} because the probability of finding another molecule is roughly the same anywhere in space. In solids, on the other hand, the rigid structure limits neighboring molecules to fixed distances, often to form a regular pattern. In this case, the pair correlation function is commonly redefined to point along a particular Cartesian axis, so the regular structure of the solid may be visible as a predictable series of spikes, as sketched in Fig. 12.3b. For liquids, the pair distribution function lies between these two extremes.

At higher density, the ordering in the liquid becomes longer range, but the distance to the nearest neighbor atoms remains fixed by the repulsive electron cloud of the reference atom. The pair correlation function is relatively insensitive to temperature, assuming that the density is constant and the substance remains liquid, although there may be a smearing of the structure at higher temperature.

The average radius of the first solvation shell is usually less than the equilibrium distance R_e in the Lennard-Jones potential. Each additional intermolecular bond further stabilizes the reference molecule, so an increased density can improve the stability of the system by increasing the average coordination number.

The time-dependence of structure in the liquid further complicates our grasp of liquid structure. In the solid, the molecule at a given position at a particular time is likely to be the same molecule an hour later. In the liquid, however, the molecules responsible for a peak in $\mathcal{G}(R)$ are constantly being replaced by other molecules. The timescale for this replacement is roughly the period for one cycle of the intermolecular vibrations, on the order of a picosecond (10^{-12} s) in liquid argon. Therefore, even the short range structure in the liquid fluctuates rapidly.

With polyatomic molecules in a liquid, anisotropic attractions such as the dipole–dipole force may play a major role. The repulsive force also becomes anisotropic (although spherical top molecules, as the name suggests, may be

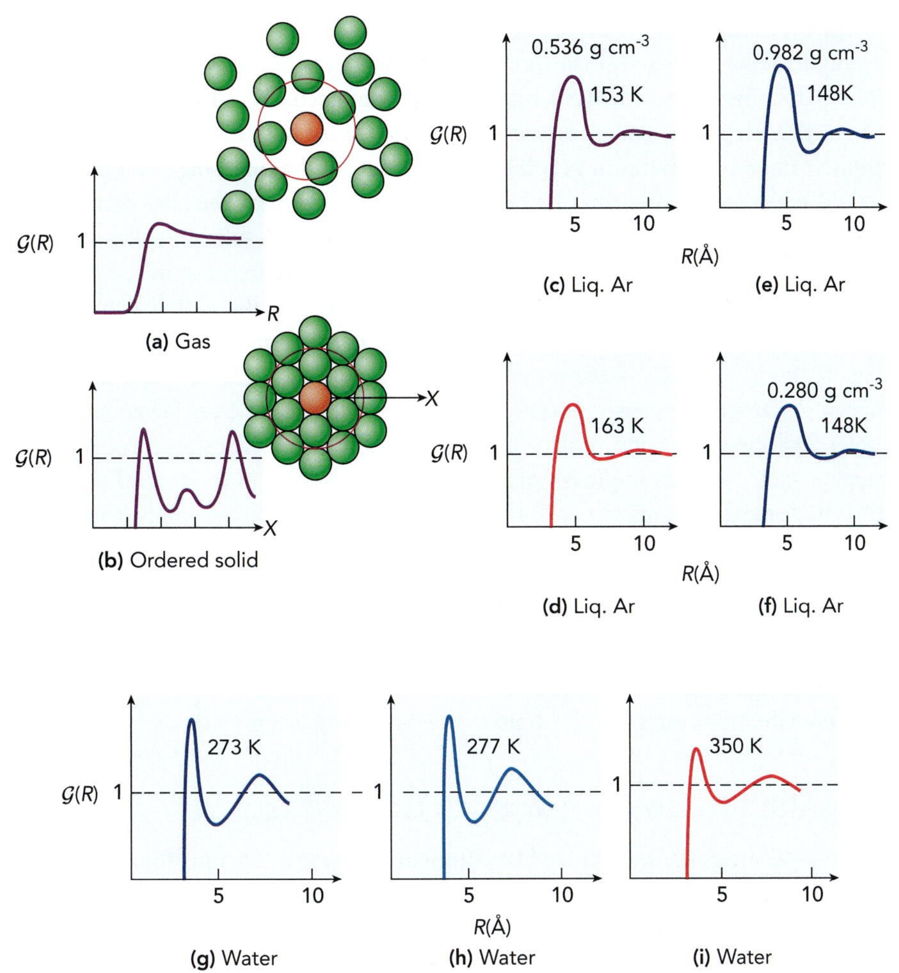

▲ FIGURE 12.3 **Trends in the pair correlation function.** In a diffuse, random environment such as a gas, **(a)** $\mathcal{G}(R)$ will rapidly converge to one, whereas in a rigid, ordered substance such as many solids, **(b)** $\mathcal{G}(X)$ will have periodic spikes. In liquid argon, we see that the degree of order decreases as we raise the temperature from **(c)** 153 K to **(d)** 163 K, and as we decrease the density from **(e)** 0.982 g cm^{-3} to **(f)** 0.280 g cm^{-3}. In the case of liquid water, there is a very slight narrowing of the first peak from **(g)** 273 K to **(h)** 277 K, indicative of a tiny increase in density, but **(i)** at 350 K the peaks broaden, reflecting much lower order.

nearly sphere-like as far as their interactions are concerned). As a result, the structure of the liquid over small distance scales may be much more sophisticated than for the monatomic liquids. Water is a convenient example because it has been very closely studied, and although it has a strong dipole–dipole intermolecular force, the repulsive force is still fairly isotropic (the hydrogens contribute little to the shape of the overall electron distribution).

In this case, as seen in Fig. 12.3g–i, there is noticeably decreasing order as the temperature increases. The anisotropic part of the potential energy surface makes an arrangement with aligned dipoles more stable, but at higher temperatures there is enough energy in the van der Waals vibrational modes that the dipoles can oscillate substantially around their most stable configuration. At low temperatures, that energy is no longer available and the structure becomes more rigid.

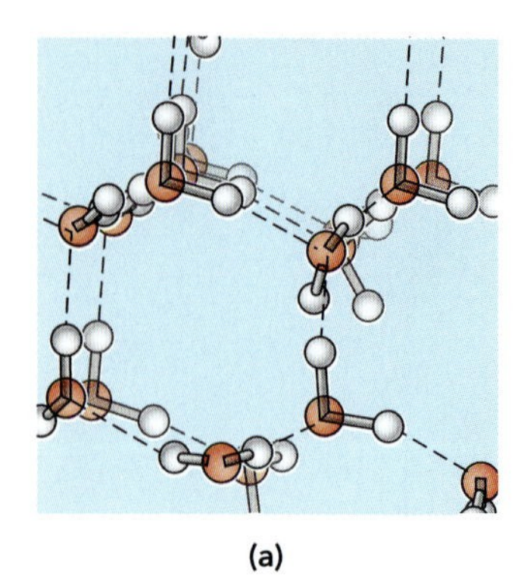

(a)

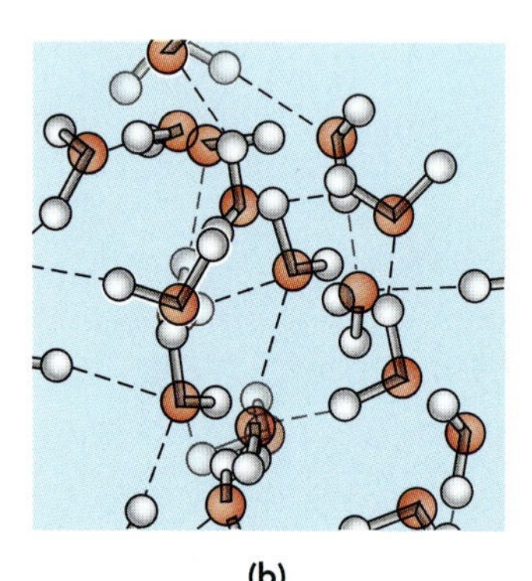

(b)

▲ FIGURE 12.4 **Schematic structures of water in solid and liquid phases.** The less dense structure **(a)** is more stable. As energy is added to form the liquid, the increased bending motion of the molecules makes the structure more dense **(b)**.

As the temperature decreases, a typical liquid approaches one of its solid configurations. Water begins to form tetrahedral structures in which the oxygen lone pairs bind with hydrogens on an adjacent molecule. At still lower temperatures, this structure becomes the rigid structure of water **ice.**[1] Although the transformation from liquid to solid comes rapidly in the final stages of cooling, it would be wrong to describe the liquid as formless up until the last instant. It is estimated that liquid water, near the point at which it solidifies, has already formed 87% of the intermolecular bonds that it will have in the solid.

One of the remarkable features of liquid water is that it does not always expand when warmed. This anomaly appears in the pair correlation functions sketched in Fig. 12.3g–h. From the freezing point of 273 K up to 277 K, the liquid actually becomes denser. The cause appears to be the following: around any given reference water molecule, the water molecules in the first solvation shell stay at roughly the same distance as the sample is warmed, but the molecules in the second shell can turn their oxygen atoms closer to the reference at higher temperatures (Fig. 12.4). The closer arrangement of second-nearest neighbors is less stable, so it occurs only at higher temperatures, and the average distribution of molecules becomes denser. At 277 K, the molecules have reached their closest average distance under typical pressures. Beyond 277 K, the second shell has enough energy for the oxygen atoms to vibrate significantly *away* from the reference molecule as well as toward it, and as these vibrations increase, the density begins to drop again.

Quantum States in a Weakly Bonded Liquid

We can separate the molecular Hamiltonian in Section 5.1 into four contributions to the energy: electronic, vibrational, rotational, and translational. Each of these degrees of freedom is affected differently when the molecule moves from the gas phase to the liquid.

Electronic states. A liquid is described by a wavefunction just as any other kind of matter is, but as the number of atoms in the liquid increases, the wavefunction becomes more complicated. However, because the molecules in the liquid must still be independent enough to flow, the energy level diagram of a liquid often resembles the energy level diagram of the same substance as a gas.[2] For example, the electronic wavefunction is relatively insensitive to the formation of van der Waals bonds, because the changes in electron distribution due to polarization are usually tiny compared to the change induced by an electronic transition. As a result, the liquid still has electronic energy levels essentially the same as those of the gas phase.

---

[1]Apparently, only the solid form of water is properly called "ice," but the term is also commonly applied to the solid form of any substance if it is glassy in appearance and melts or vaporizes at temperatures at or below about 300 K. Hence, frozen CO_2 is "dry ice," and the clouds on Saturn's largest moon, Titan, carry tiny crystals of methane ice.

[2]Glaring exceptions include the molten metals and molten salts, both of which have electrical conductivity entirely unlike the gaseous forms of the same substances.

Vibrational states. The **intramolecular** vibrations—the vibrations of the chemical bonds—of a polyatomic molecule are still present in the liquid, but they are more strongly perturbed than the electronic levels. Weak chemical bonds or those that participate strongly in the intermolecular forces may have much different stretching and bending frequencies in the liquid than in the gas. A good example is water. In the gas phase, the OH stretching transitions occur at about 3700 cm^{-1}. In liquid water, there is a great deal of hydrogen bonding, and the hydrogen experiences an attractive potential from both oxygen atoms: the one it is bound to chemically, and the one it is bound to by the hydrogen bond. This means that the potential surface for the OH stretch is flatter than in the gas phase, and the vibrational frequency drops to about 3300 cm^{-1}. Furthermore, because this shift depends on the precise orientation and environment of each molecule in the liquid, which change from molecule to molecule, the vibrational frequency can no longer be as precisely defined as in the gas phase. Many molecules in the liquid will have formed strong hydrogen bonds, resulting in a particularly low absorption energy for the OH stretch, while others will be in transit from one hydrogen-bonding site to another and will have a transition energy for the stretch more similar to the gas-phase value. Therefore, these transitions in the liquid become very broad.

Rotational states. Condensation of polyatomic molecules from the gas phase to the liquid phase has a much more drastic effect on the rotational motions. Once in the liquid, the molecule is usually no longer free to rotate about its center of mass, because it will tend to run into an adjoining molecule. It bounces back in the opposite direction until it runs into a different molecule, and so on (Fig. 12.5). What was a rotational motion has transformed into something better described as an intermolecular bending vibration. This rotation-like motion in a liquid is called **libration.** The excitation energy of these motions is higher than for the corresponding rotations—shifting to the low-frequency infrared from the microwave region—because now there is a constraining potential energy, which causes the energy levels to shift upward.

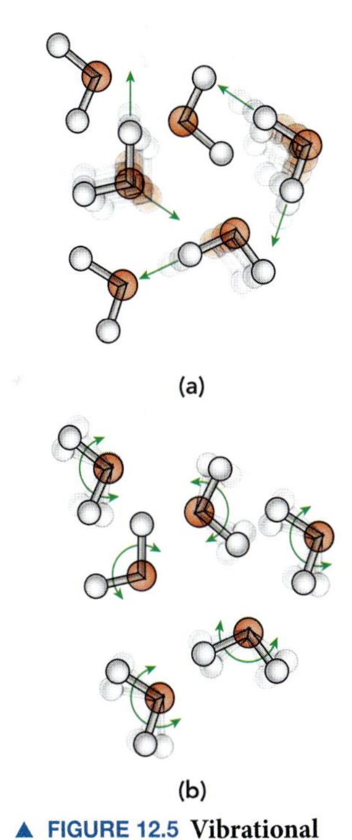

▲ FIGURE 12.5 **Vibrational motions in liquids.** **(a)** Internal vibrations exist as in the gas phase, but with vibrational frequencies affected by the neighboring molecules. **(b)** Rotations are also affected, no longer being free as in the gas phase. The hindered rotations are called librations.

Translational states. Translational motions are the motions of entire molecules. We have largely ignored them in previous chapters, because translations rarely affect (or tell us much about) the molecular structure. Translational motion in a gas is normally a classical degree of freedom—the domain (the distance over which the particles travel) is too large for quantum properties to be measured. For example, the translational energy levels are so close together that we cannot measure the gaps between them. However, in a liquid the particles no longer travel freely, because each particle is trapped in a potential energy well by the attractive forces that loosely hold the liquid together. For liquids near room temperature, the potential energy well is of the sort described in Chapter 10, much shallower than the well for chemical bonding, governed instead by dipole–dipole or dispersion forces or forces of similar magnitude. These resemble the potential energy curves for vibration shown in Chapter 8, but they have much

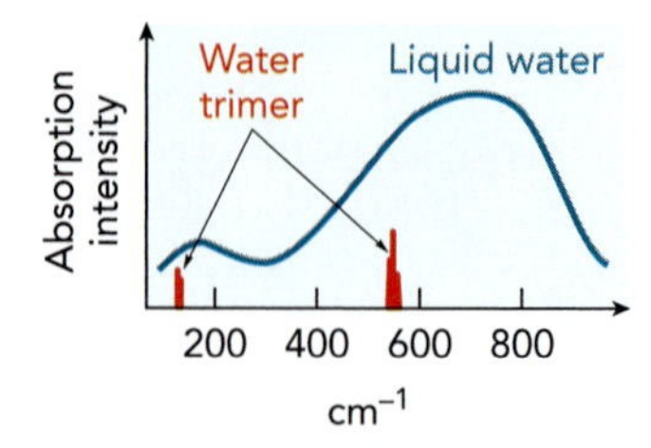

▲ FIGURE 12.6 **The spectrum of liquid water between 200 and 800 cm^{-1}.** The spectrum includes transitions that correspond to librations in the water trimer. (After a figure in F. N. Keutsch and R. J. Saykally, *Proc. Natl. Acad. Sci.* **98,** 10533, 2001.)

lower force constants k, and therefore much lower vibrational constants $\omega_e \approx \sqrt{k/\mu}$ (Eq. 8.18). Another difference between these motions and chemical bond vibrations is that they have none of the consistent structure of a chemical bond. The particles are constantly reorienting and shifting. The result is that the distribution of *intermolecular* vibrational energy levels becomes more dense and complex than the *intramolecular* vibrations.

If the liquid is composed of a polar molecule, librations and the translational intermolecular vibrations can be excited by radiation near the border between microwave and infrared regions of the spectrum, above about 100 cm^{-1}. The high density of a polar liquid makes these transitions opaque in that region; the sample absorbs essentially all the radiation at those frequencies. Liquid water, for example, appears transparent to visible light, but it strongly absorbs infrared radiation across the entire range 100–5000 cm^{-1} (2–100 μ), at least 10^4 times more effectively than the visible. The absorption at the low-frequency end of the infrared region corresponds to excitation of the intermolecular motions, and the absorptions above about 1000 cm^{-1} excite the intramolecular vibrational frequencies of water (at about 1600 cm^{-1} and 3600 cm^{-1} in the gas phase) alone and in combination with the intermolecular motions. Figure 12.6 shows that the observed spectrum of liquid water broadly overlaps the spectrum of the water trimer, $(H_2O)_3$. In the water trimer, the three water molecules are hydrogen bonded together, and the bending of one water molecule is affected by its bond to the next one. Whereas a single gas-phase water molecule has rotational transition energies below 100 cm^{-1}, the crowding of other molecules in the liquid restricts that motion and pushes the transition energies up into the hundreds of cm^{-1}.

12.3 Solvation

If we mix two different chemicals, A and B, to form a gas, they mix down to the molecular level. The molecules of A, for example, do not clump together. In the gas phase, molecules interact too weakly to influence one another's motions to that extent.

But what happens if we combine A and B and end up with a *liquid,* where the intermolecular forces are stronger by definition? There are two possibilities for such an interaction: either the attraction between different molecules (A and B) is sufficient to stabilize the mixture, or it isn't.[3] In the second case, the mixture is not stable and A and B separate. We will take a quick look at that case.

Because liquids can have different bonding mechanisms, a mixture of different liquids can give rise to two or more distinct **liquid phases** of the sample. Water and **oil** provide the proverbial example. Oil is a glyceride, an ester of glycerol and a carboxylic acid, having a large number of π bonds. In the liquid form the internuclear attractive force is dominated by dispersion forces arising from the molecule's large volume and highly polarizable π electrons. Water,

[3]Although the intermolecular potential energy figures heavily in determining whether or not two liquids mix, the role of another major factor—the entropy—is discussed in *TK* Chapter 11.

by contrast, binds via hydrogen bonding and dipole–dipole forces, and it is not very polarizable over its tiny three-atom volume. Therefore, water and oil do not bind well to each other, and they normally coexist in separate regions of the container, forming two distinct *liquid phases*, with the denser phase on the bottom. As for the case of the pure liquid, the greater the surface tension, the greater tendency of each phase to distribute itself in a way that minimizes its surface area. If we *force* oil and water to mix, say by shaking the container violently, the two substances spontaneously separate, first into an **emulsion,** in which tiny droplets of one substance (such as the oil) are suspended in the other substance. The emulsion increases the surface area of each phase, because small droplets have relatively low ratios of volume to area. Having different densities, these droplets respond differently to gravity, and they gradually combine with other droplets of the same compound, driven partly by the need to minimize the surface area. Eventually the two liquids are again completely separated, with the denser liquid on the bottom. This property is routinely exploited for the separation of organic compounds during a synthesis. An organic reaction using the chlorinating agent $AlCl_3$, for example, may be carried out in a non-polar solvent such as dichloroethane. Rinsing with water then removes the ionic $AlCl_3$, leaving the organic products behind.

It is the other case, when A and B stay mixed, that has been the most heavily exploited in laboratory chemistry. When compounds mix to form a liquid and the mixing is homogeneous down to the molecular scale (i.e., there is little or no clumping of like molecules surrounded by unlike molecules), the mixture is called a **solution.** When one of these molecules is present in much greater abundance than any other, we call that substance the **solvent,** and the substances in low abundance are called the **solutes.** Solutions are crucial to chemistry primarily for these reasons:

1. Chemical reactions usually require reactants to mix. Solutions allow substances to mix at much higher densities than they could attain in the gas phase. The typical liquid has a density a thousand times greater than the corresponding gas at 1 bar.
2. Many reactive species may be stable and easily formed in solution but not in the gas phase. The sodium ion Na^+, for example, is formed in the gas phase by ionization of the neutral sodium atom at an energy of 5.14 eV, which corresponds to a temperature of over 12,000 K. Yet Na^+ is commonly formed at room temperature by adding NaCl to water.
3. A diverse range of chemical environments is possible in solution: electron-donating or electron-accepting, polar or non-polar, high-dispersion forces or low-dispersion forces, and transparent to radiation or opaque.

The molecular structure of a solution, like that of a pure liquid, can be described in terms of the pair correlation function. We may specify, for example, a function $\mathcal{G}_{B:A}(R)$ that describes the pair correlation for a solute molecule B surrounded by molecules of the solvent A. As in a pure liquid, the amount of order seen in the function depends on the strength and directionality of the binding forces involved. Some qualitative conclusions can often be drawn from the nature

BIOSKETCH Martina Havenith-Newen

Martina Havenith-Newen is a chair of physical chemistry at the Ruhr-Universität Bochum in Germany, where she uses **terahertz spectroscopy** to probe the ever-changing structure of liquid water. One terahertz (THz) is equal to $10^{12}\,s^{-1}$, and photons in this frequency range lie at wavelengths of roughly 0.1–1.0 mm, at the long wavelength end of the infrared. In this region of the spectrum, Professor Havenith is able to detect transitions between *librational* states of water, the hindered rotational motions that correspond to the exchange of hydrogen bonds. These motions are especially sensitive to the presence of other compounds that alter or disrupt the hydrogen bonding. Professor Havenith's research group has recently taken advantage of this characteristic to apply terahertz technology to the study of water molecules in biological systems. Shifts in the spectrum distinguish between waters bonded to a solute and water molecules that interact only with each other. In particular, the Havenith group has been able to analyze proteins in solution and determine how many water molecules directly stabilize a protein by hydrogen bonding. Knowing the number of water molecules involved in hydrating the protein provides valuable information needed to understand the protein's thermodynamics, which in turn leads us to a better understanding of how biological molecules function in living organisms.

of the solvent A and solute B, however, without resorting to the more quantitative correlation function:

1. When the bonding force between A and B is very strong, the structure of the pure liquid A becomes subordinate to the A—B bond. Molecules of A will move to increase the number of A—B bonds. (Fig. 12.7a). This is often the case for ionic solutes (**electrolytes**) in polar solvents; the monopole–dipole bond is much stronger than the typical A—A liquid bond. Liquid water is a highly ordered liquid, with hydrogen bonds orienting adjacent water molecules to obtain a coordination number of about four. Yet even this stable structure is sacrificed to make room for cations and anions of considerable size. Solubilities in water of more than 2 mol L^{-1} are common among ionic compounds, and (since each solute molecule decomposes into at least two ions in solution) this corresponds to at least (4 mol ions)/(56 mol H_2O) or 1 ion for every 14 water molecules. A typical polyatomic ion has a volume several times that of 1 water molecule, so the tetrahedral structure of the water must be substantially altered to allow such high concentrations. Because the monopole electric field is isotropic, the structure of the solvent near one of the solute ions need not be highly ordered except in one respect: the positive end of the dipolar solvent molecules will tend to point toward the dissolved anions and away from the cations. The ions become surrounded by neutral, polar molecules.

These stabilizing interactions, together with the electron affinity to form the anion, must be sufficiently strong to make up for the energy required to ionize the cation.

2. If the A — B attractive force is comparable to or weaker than the A — A attractive force, and B is of comparable size or larger than A, the solvent tends to reject high concentrations of B. High concentrations disturb the original solvent structure more than low concentrations do. If compound B forms a weak A — B bond, it may be tolerated in a dilute solution, but addition of more B will not disrupt the solvent structure any further, and it does not dissolve. The solvent tends to retain its original structure, with occasional cavities opened to allow room for solute molecules (Fig. 12.7b). As the strength of the A — B bond increases, the solvent becomes more able to make adjustments. A common manifestation of this is the formation of a **solvent cage,** an arrangement of solvent molecules that encloses the solute in a shell, or series of shells. The arrangement of molecules within a shell may be reminiscent of the structure of the pure solvent liquid, or it may be a new structure. In either case, the solvent cage does not have a definite boundary; it blends smoothly into the liquid solvent structure at large distances from the solute.
3. If B is much smaller than A, regardless of the strength of the A — B bond, B may have very high solubility in A because it can occupy gaps in the solvent structure without disturbing the solvent (Fig. 12.7c). Helium is an example of a solute with very weak interactions but such a low atomic volume that it still has about half the solubility in water of the much more polarizable CO molecule.

Simple solutions, in which the solvent and solute are of similar size and the intermolecular interactions remain relatively weak, have been modeled theoretically by treating the solution as a single fluid of the same molecule. This allows one Lennard-Jones potential to predict the experimental observables.

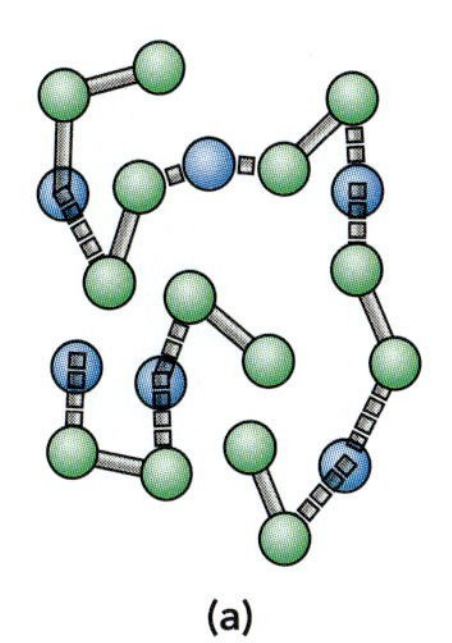

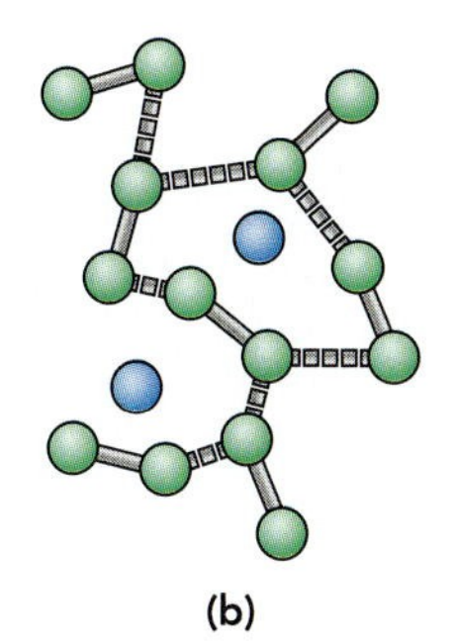

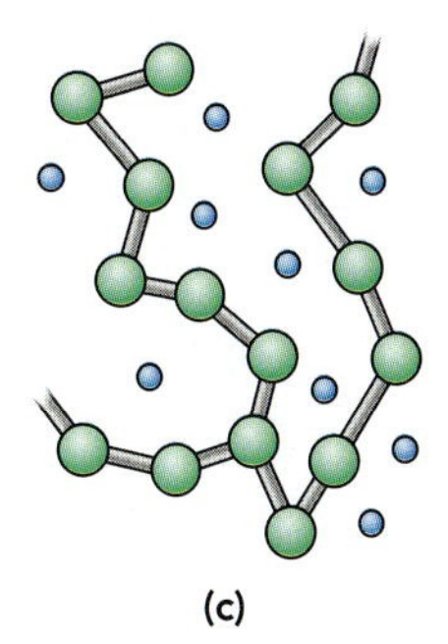

▲ FIGURE 12.7 **Solvation schemes.** **(a)** The attractive forces between the solute (filled circles) and the solvent are very strong, and the concentration can be very high. **(b)** The solvent–solute forces are weak relative to the solvent–solvent bonding forces, and solvent cages form around solute molecules. **(c)** The solvent–solute forces are weak, but the solute can occupy cavities in the solvent structure without disturbing the solvent–solvent bonds, thereby keeping the solution stable even at high concentrations.

EXAMPLE 12.1 Solvation of an Ionic Compound

CONTEXT Electrochemical cells take advantage of differences in the electrical potential energy between various elements and compounds to create solutions that carry electrical current. The process can drive chemical reactions and can simultaneously yield information about the mechanism for the reaction. When an electrical current runs through two adjoining electrolyte solutions, there are two challenges: (1) maintaining the overall charge neutrality of the two solutions (because a buildup of charge will eventually halt the flow of current) and (2) preventing the ions from one solution from mixing freely (and perhaps reacting) with the ions of the other solution. A common way to address both challenges is to connect the two solutions by *salt bridge,* a tube containing a solution of KCl (and often other salts) at maximum concentration. The K^+ and Cl^- ions carry the charge across the bridge, rather than the reactive ions, maintaining the charge balance while allowing current to flow. Keeping the salt bridge solution saturated ensures maximum conductivity and reduces the likelihood of reactive ions crossing the bridge into the other solution.
To solve this problem, we need to convert concentration in moles solute per kg solution to the ratio of solvent to solute molecules. The molar masses of the solvent and solute allow us to change masses to moles, and the ratio of moles solvent to moles solute will be the answer we want.

PROBLEM The maximum molality of KCl in water at 25 °C is 4.76 mol (kg solvent)$^{-1}$. (For concentrated solutions, molality is a more convenient concentration unit than molarity, because the volume of the solution can be a sensitive function of the amount of solute.) Calculate the average number of water molecules per ion at this concentration.

SOLUTION We can convert from molality m to mole fraction X with a little algebra:

$$m_{\text{KCl}} = \frac{n_{\text{KCl}}}{M_{\text{water}}} = \frac{n_{\text{KCl}}}{n_{\text{water}}\mathcal{M}_{\text{water}}},$$

where $\mathcal{M}_{\text{water}}$ is the molar mass of water. Now we can solve for the ratio $n_{\text{water}}/n_{\text{KCl}}$.

$$\frac{n_{\text{water}}}{n_{\text{KCl}}} = \frac{1}{m_{\text{KCl}}\mathcal{M}_{\text{water}}} = \frac{1}{(4.76\ \text{mol kg}^{-1})(0.0180\ \text{kg mol}^{-1})} = 11.7,$$

or 11.7 molecules of water per molecule of KCl in solution. Because KCl dissociates into two ions, K^+ and Cl^-, this indicates that there are on average only about six molecules of water per ion.

The Lennard-Jones parameters for this **one-fluid model** are estimated from the values for the various A and B pair Lennard-Jones potentials as follows:

$$R_{\text{soln}} = (X_A^2 R_{\text{LJ,A}}^3 + 2X_A X_B R_{\text{LJ,AB}}^3 + X_B^2 R_{\text{LJ,B}}^3)^{1/3} \tag{12.8}$$

$$\varepsilon_{\text{soln}} = \left(\frac{1}{R_{\text{soln}}}\right)^3 (X_A^2 \varepsilon_{\text{AA}} R_{\text{LJ,A}}^3 + 2X_A X_B \varepsilon_{\text{AB}} R_{\text{LJ,AB}}^3 + X_B^2 \varepsilon_{\text{BB}} R_{\text{LJ,B}}^3), \tag{12.9}$$

where the Xs are the mole fractions of A and B. This formula appears to successfully predict, for example, the change in energy upon combining two pure liquids to form a solution.

EXAMPLE 12.2 The One-Fluid Model

CONTEXT Gasoline is a mixture of alkanes and other hydrocarbons having roughly between four and ten carbon atoms in a chain. The properties of gasoline depend on the relative amounts of these various compounds and are crucial to its use and safe handling, but a wide range of mixtures may be sold as gasoline. In the 1940s, a *congruency relationship* was discovered that allowed several properties of liquid alkane mixtures to be predicted based on only the chain lengths of the alkanes. The basis for the congruency relationship is the additive nature of the attractive wells and separation distances between the compounds, as implied by the one-fluid model. One of the original tests of the congruency relationship was based on mixtures of propane and octane, where it was found that the principle successfully predicts molar volumes and other properties of the mixture.

PROBLEM Calculate the value of $\varepsilon_{\text{soln}}$ for the interaction between octane ($\varepsilon = 421$ K, $R_{\text{LJ}} = 6.93$ Å) and propane ($\varepsilon = 296$ K, $R_{\text{LJ}} = 4.91$ Å) in a 50:50 mixture (by mole) of the two liquids. Estimate the heterogeneous interaction parameters (for the interaction between octane and propane) using Eq. 10.47 and using a geometric mean $\sqrt{(\varepsilon_{\text{AA}}\,\varepsilon_{\text{BB}})}$ to represent the average well depth.

SOLUTION First we need the parameters for the heterogeneous interactions:

$$R_{\text{LJ,o-p}} = \frac{R_{\text{LJ,A}} + R_{\text{LJ,B}}}{2} = 5.92\,\text{Å}$$

$$\varepsilon_{\text{o-p}} = \sqrt{\varepsilon_{\text{oo}}\,\varepsilon_{\text{pp}}} = \sqrt{(421\text{ K})(296\text{ K})}$$
$$= 353\,\text{K}.$$

A 50:50 mixture has mole fractions $x_{\text{o}} = x_{\text{p}} = 0.50$, so we can factor the x^2 out of the sums. The characteristic Lennard-Jones distance becomes

$$R_{\text{soln}} = \left[(0.50)^2\left((6.93\,\text{Å})^3 + 2(5.92\,\text{Å})^3 + (4.91\,\text{Å})^3\right)\right]^{1/3} = 6.00\,\text{Å},$$

which is the same as $R_{\text{LJ,o-p}}$. The characteristic well depth is predicted to be

$$\varepsilon_{\text{soln}} = \left(\frac{1}{6.00\,\text{Å}}\right)^3\left(\frac{1}{2}\right)^2\left((421\,\text{K})(6.93\,\text{Å})^3 + 2(353\,\text{K})(5.92\,\text{Å})^3 + (296\,\text{K})(4.91\,\text{Å})^3\right) = 372\,\text{K}.$$

These are corrections of only a few percent compared to the values for the two-molecule interactions, but they predict that the interaction among many molecules will tend to stabilize the system (because $\varepsilon_{\text{soln}} > \varepsilon_{\text{o-p}}$).

CONTEXT *Where Do We Go From Here?*

Liquids are the hardest phase of matter to understand, because they lie between the limits of independent molecules in the gas phase and the systematic structure typical of the solid phase. The strong bonding in solids (Chapter 13) is a relief by comparison, and it is the last stage of our microscopic examination of chemical structure.

KEY CONCEPTS AND EQUATIONS

12.1 The qualitative nature of liquids. Three major classes of liquids are the following:

- Van der Waals liquids, which are bound by the weak intermolecular forces described in Chapter 10.
- Ionic liquids, which are bound by the Coulomb force between cations and anions.
- Molten metals, which are bound by the Coulomb force between metal cations and delocalized electrons.

  In each case, the attractive forces are largely isotropic—angle independent—which allows the forces to hold the liquid together at a constant density while allowing enough freedom of movement that the distribution of particles can change so the liquid can flow.

12.2 Weakly bonded pure liquids. The average number of molecules with their centers of mass at distances within a distance R from the origin (the center of mass of the reference molecule) can be calculated from the pair correlation function $\mathcal{G}(R)$, which is defined by the relation

$$\overline{N}(R) = \int_0^{2\pi}\int_0^{\pi}\int_0^{R} \rho\mathcal{G}(R')\, R'^2 dR' \sin\theta d\,\theta d\phi'$$

$$= 4\pi\rho\int_0^{R} \mathcal{G}(R')R'^2 dR'. \qquad (12.5)$$

The pair correlation function $\mathcal{G}(R)$ always approaches 1 at large distances.

12.3 Solvation. A liquid normally incorporates a solute in one of three ways to form a solution:

- If the solvent–solute attractive forces are stronger than the solvent–solvent attractions, the solvent structure is disrupted in order to increase the interaction with the solute.
- If the solvent–solute attractions are weaker than solvent–solvent, then the solvent retains its structure as much as possible while forming small solvent cages around the solute molecules.
- Regardless of the relative attractive forces, if the solute molecule is much smaller than the solvent, then the solute fills gaps in the solvent structure.

KEY TERMS

- A **fluid** is a substance that flows, in other words, where the intermolecular attractions are too weak to keep two molecules permanently bound to each other.
- The **Axilrod-Teller correction** extends the dispersion interaction to three particles.
- The **pair correlation function** shows on average how many molecules are found at some distance R from a reference molecule.
- The **number density** is the number of molecules per unit volume.
- **Surface tension** is the tendency for a sample of liquid to minimize its surface area by approaching a spherical shape.
- A **solution** is any mixture of two or more substances in which the mixing is roughly even down to the molecular scale. When one substance is much more abundant than the others, dominating the structure of the solution, we call that substance the **solvent** and the others the **solutes.**

OBJECTIVES REVIEW

1. *Use the pair correlation function $\mathcal{G}$ as a measure of long-range order in liquids, and predict trends in $\mathcal{G}$ as a function of density, temperature, and intermolecular forces.* Sketch a graph of $\mathcal{G}(R)$ for liquid ammonia, and show how the curve will change as the temperature approaches the boiling point.
2. *Qualitatively predict the miscibility of two substances based on the relevant intermolecular forces, and describe the microscopic structure of the solution.* Is hexane likely to be soluble in benzene?

PROBLEMS

Discussion Problems

12.1 The following table lists the range of temperatures over which some selected compounds are stable as liquids at a pressure of 1 bar. What accounts for the difference between the very small values at the left and the large values at the right?

compound	$T_{bp} - T_{mp}$(K)	compound	$T_{bp} - T_{mp}$(K)
CO_2	0.0	H_2O	100.0
H_2	6.3	HF	102.6
CO	7.5	acetone: $(CH_3)_2O$	150.5
N_2	14.1	heptane	189.0

12.2 Identify any of the following liquids that you expect to be transparent in the infrared: benzene (C_6H_6), ammonia (NH_3), carbon tetrachloride (CCl_4), liquid nitrogen (N_2).

12.3 Identify the intermolecular attractive forces likely to be dominant in the liquids (a) water, (b) n-pentane ($CH_3CH_2CH_2CH_2CH_3$), (c) mercury, and (d) benzene. Based on this, list the compounds in order of increasing surface tension.

12.4 In what ways would the pair correlation function for CCl_4 differ qualitatively from that of H_2O?

12.5 Will surface tension generally increase or decrease as temperature increases?

12.6 For which of the following would an infrared detector probably be useless?

a. detecting volcanic activity on one of Jupiter's moons from orbit

b. locating deep-sea divers

c. detecting volcanic activity at the ocean floor

d. detecting jet engine emissions in the day sky

General Liquid Structure and Dynamics

12.7 Regarding the Axilrod-Teller correction u'_{ABC} in Eq. 12.2, answer the following.

a. What geometry (what values of the θs) gives the biggest negative value for u'_{ABC}?

b. What geometry gives the biggest positive value?

c. For what value of $\theta_A = \theta_B$ is the correction zero?

12.8 Simplify the equation for the Axilrod-Teller correction when molecules A, B, and C form an isosceles triangle with $\theta_A = \theta_B$. Find the maximum and minimum values for the correction, and the values of θ_C at which they occur, if R_{AB} is fixed at $2R_e$.

12.9 The vibrational energy of argon liquid at 80 K is roughly the same as the translational kinetic energy. The time that one argon atom spends in the vicinity of another is about 10^{-12} s. Approximately how many times can a pair of argon atoms vibrate at this temperature during the time they are next to each other? THINKING AHEAD ▶ [How is the vibrational energy related to the number of vibrations per second, the vibrational *frequency*?]

12.10 How do you expect the vibrational frequencies of HCN to differ between the gas phase and liquid phase?

12.11 Based on the information for NH_3 in Tables 8.3 and 9.2, and the effect of librations shown for water in Fig. 12.5, make a rough sketch of the infrared spectrum of liquid ammonia.

12.12 List the following liquids in order of *increasing* surface tension:

a. H_2O

b. CH_3OH (methanol)

c. $C_6H_{13}OH$ (1-hexanol)

d. C_6H_{14} (hexane)

Pair Correlation Functions

12.13 The following equation is proposed as a model pair correlation function:

$$\mathcal{G}(R) \approx \left[1 - \cos(\pi R/R_0)\right] e^{-R/R_0}.$$

In what limit of R and in what way does this function have the wrong behavior?

12.14 For an infinite square array of atoms, a section of which is drawn in the following figure, find the first five values of R at which spikes appear in the pair correlation function.

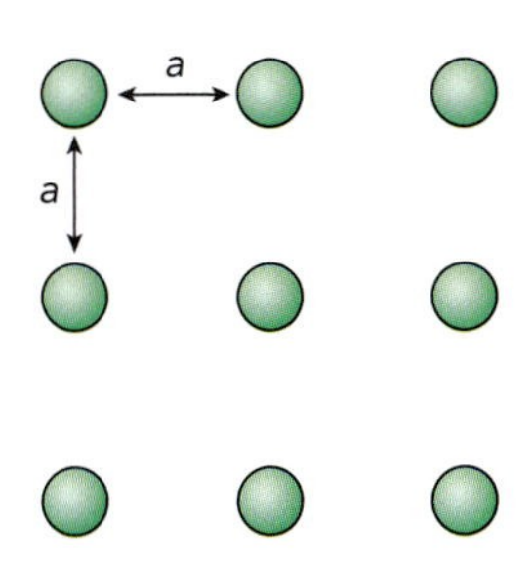

12.15 An approximate model for the pair correlation function is drawn in the following figure. Find the coordination number C for this correlation function when $\rho = 1.1 \cdot 10^{22}\ \text{cm}^{-3}$, $N = 2.3$, and $R_{LJ} = 3.0$ Å.

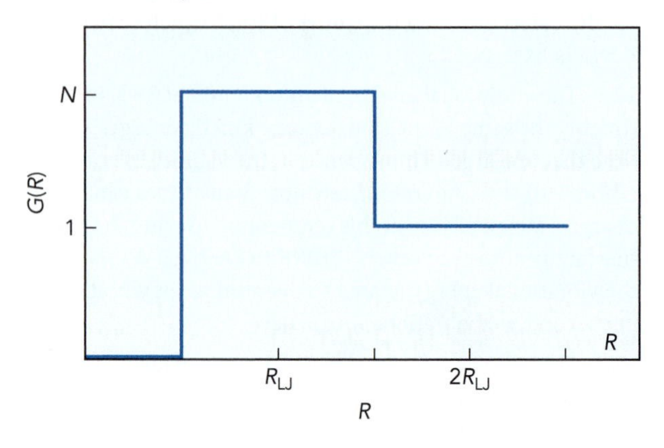

12.16 A model pair correlation function is given by

$$\mathcal{G}(R) = e^{-R/R_0}\cos\left(\frac{2\pi R}{R_{LJ}}\right) + 1, \quad R > R_{LJ}/2$$
$$\mathcal{G}(R) = 0, \quad R \leq R_{LJ}/2.$$

Estimate the coordination number for water using this model function, with values $R_{LJ} = 2$ Å, $R_0 = 8$ Å. The indefinite integral solution is

$$\int x^2 e^{ax}\cos(bx)dx = e^{ax}\left\{\left[\frac{ax^2}{(a^2+b^2)} - \frac{2(a^2-b^2)x}{(a^2+b^2)^2} + \frac{2a(a^2-3b^2)}{(a^2+b^2)^3}\right]\cos bx + \left[\frac{bx^2}{(a^2+b^2)} - \frac{4abx}{(a^2+b^2)^2} + \frac{2b(3a^2-b^2)}{(a^2+b^2)^3}\right]\sin bx\right\}.$$

12.17 This question examines a limitation of the pair correlation function. Imagine an experiment in which we excite molecule A in the gas phase into a high-energy state by the absorption of two photons and then transfer that energy to a different molecule B during a collision. For the first step, molecule A must not be next to any other molecules because then the energy of the first photon will be lost during collision before the second photon can be absorbed. We want to know the best density of molecules so that molecule A has a small probability of being near any other molecules over the time required for the two-photon absorption but still has a good chance of colliding with B over the longer period of time available before A releases the energy by fluorescence. However, it is not possible to write an expression that gives the fractional probability that *no other atoms* lie within a distance R_1 of a randomly selected reference atom in the gas, in terms of only the pair correlation function $\mathcal{G}(R)$ and the density ρ. Why not?

Solvation

12.18 Make a rough sketch of a likely arrangement of solute and solvent molecules for each of the following systems.
a. propane in hexane
b. argon in water
c. F_2 in water

12.19 Iron(III) chloride, $FeCl_3$, has a maximum solubility in water of 84.6% by mass at 60°C. Find the mole fraction $x = n_{FeCl_3}/n_{tot}$ of $FeCl_3$ in the solution and the average number of water molecules per solvated ion (assuming the $FeCl_3$ dissociates completely). (The result should be rather surprising.)

12.20 The maximum mole fraction X_{KCl} of KCl dissolved in water is 0.079. Assume that this is limited by having a minimum number of waters per unit surface area of the K^+ and Cl^- ions. For example, replacing one ion by an ion with the same charge but twice the surface area reduces the solubility because it takes twice as many water molecules to stabilize the larger surface area. Then, given the following ionic radii, estimate the maximum mole fraction of NaCl in water at 25°C.

Cl^- 1.81 Å Na^+ 1.02 Å K^+ 1.38 Å

12.21 Describe the result of mixing three liquids A, B, and C if the interaction is dominated by the following trends in the attractive forces:

$$F_{AB} \ll F_{AC} \ll F_{CC} \approx F_{BB} \approx F_{BC} \ll F_{AA}.$$

13 The Structure of Solids

LEARNING OBJECTIVES

After reading this chapter, you will be able to do the following:

❶ Identify the lattice system, Bravais lattice, and crystallographic point group of a unit cell.

❷ Qualitatively predict properties such as piezoelectricity from the crystal symmetry.

GOAL *Why Are We Here?*

The goal of this chapter is to describe the structure of solids, especially crystals, which allow us to take advantage of symmetry principles in order to predict properties of various substances.

CONTEXT *Where Are We Now?*

Liquids form when we strengthen the intermolecular interactions enough to bring all the molecules together while still preserving some of the fluidity—the freedom of motion—that characterizes the gas phase. The molecular environment in a liquid is crowded and boisterous.

A **solid** is what we get when we turn the intermolecular attractions up enough to stifle the freedom of motion. Like the liquid, the solid is a condensed phase, but it is not fluid. Both the volume and shape of a solid substance are essentially independent of weak forces, such as gravity. Our goal for now is to study the microscopic properties and structure of solids.

Gases are relatively simple to treat mathematically because the molecular interactions are negligible. What makes liquids so complicated is that the molecular interactions compete with the translational energy; any given molecule stays put one instant and bounces around the next. Some simplicity returns when we consider solids, in which the translational energy is too weak to fight the intermolecular forces. The molecular arrangement becomes fairly rigid.

Another, more quantitative way to view the distinction among these phases of matter is provided by the time-dependence of translational and rotational motions. Let's briefly consider the rotational motion of a diatomic molecule as an example. As the molecule rotates freely in the gas phase, the path followed by one of the nuclei describes a circle around the center of mass. If the molecule collides with another, the orientation of the rotational motion is likely to change, which breaks the circular pattern of the motion. As the gas becomes denser, this disruption occurs more often, and in the liquid the path of the nucleus becomes virtually random. In the solid, unlike either the liquid or the gas, the rotational motion is quenched and the nucleus simply oscillates back and forth about an average position. The relationship of these paths to the state of matter is quantified by a *time correlation function*. Various experimental methods probe the time correlation function as a means of studying the molecular dynamics in detail over a range of conditions.

Some solids may be viewed as very large molecules, with chemical bonds providing the apparent rigidity of size and shape. These are extrapolations of the polyatomic molecules to macroscopic sizes, involving, say, 10^{23} atoms or so. This kind of solid illustrates the relevance of chemical bond strength to bulk materials. Cutting through a 10 cm^2 cross-section of a stiff piece of polyethylene, for example, may require breaking 10^{16} bonds at roughly 500 kJ mol^{-1} each, an expenditure of about 0.01 J. It takes about the same energy to make an easy toss of a softball. While this is a crude picture (the actual energy necessary to cut solids does not solely reflect the bond strength), the molecular-scale structure of the solid always plays a role in its macroscopic properties.

SUPPORTING TEXT **How Did We Get Here?**

The main *qualitative* preparation we need for the material in this chapter is a familiarity with the forces responsible for chemical bonding (Chapter 5). We will also draw on basic principles of molecular symmetry (Chapter 6) to help us understand the role of symmetry in crystals. For a more detailed background, we will draw on the following equations and sections of text to support the ideas developed in this chapter:

- Section 2.2 described the quantum-mechanical properties of the free particle, which are similar to those of conducting electrons in solids.
- In ionic bonds, we can approximate the potential energy by treating the ions as two point charges (Eq. 5.31):

$$U_{\text{ionic attr}} = \frac{q_A q_B}{4\pi\varepsilon_0 R},$$

  where q_A and q_B are the charges on the two ions and R is the distance between them.
- Section 6.1 defines the point group symmetry operators and Section 6.2 gives additional detail about the point groups themselves.
- Section 10.2 describes energy transfer in inelastic collisions.
- Descriptions of small clusters may provide some insight into the structure of the bulk solids, as in the cases of weakly bound molecules and metals (Section 11.2).

13.1 Amorphous Solids, Polymers, and Crystals

The molecular structures of solids range between two extremes: nearly random like the liquid, and highly ordered. We begin our exploration of solids by dividing them into these two classes, and also briefly describing a class that tends to lie between these two extremes.

Amorphous Solids

Amorphous solids are solids that do *not* have long-range order. They include all forms of glass, which is the solid formed when a liquid cools to rigidity without adopting any ordered arrangement. The amorphous liquid structure is preserved in a glass, but too little kinetic energy remains for the molecules to flow freely, even over long periods.[1] The short-range order of an amorphous solid may be characterized by determining the pair correlation function $\mathcal{G}(R)$. At values of R equal to two or three molecular diameters, $\mathcal{G}(R)$ will converge rapidly to one, just as it does for the liquid. The only difference between a glass and a liquid is that the liquid will keep shifting in time.

Not all amorphous solids are glasses. For example, liquid silicon dioxide is structurally identical to window glass: the bonding mechanisms, short-range order, electrical properties, density, and so forth, are the same. Pure elemental silicon, on the other hand, has a liquid state that is very different from the amorphous solid state. The liquid is a metal, and correspondingly dense, whereas the amorphous solid has a much less dense, tetrahedrally bound structure akin to that of diamond and is not metallic. If silicon could be made to form a glass, it would be a dense metal, like the liquid.

Polymers

Polymers are generally any molecule made up of several identical subunits, but not necessarily possessing the long-range symmetry of crystals. The term has come to be closely associated with the particular case of a long-chain hydrocarbon composed chiefly of a repeating link. Among the best known are the polyvinyl plastics, in which the repeating sequence is of the form—CR'R"—CR'R"—. Although "vinyl" actually refers to the functional group $H_2C{=}CH$, the polyvinyl plastics are single-bond carbon chains, so named because they are derived from vinyl precursors. Similarly, polyesters, polyamides (the nylons), and polyisoprenes (rubbers) are made from ester, amine, and butadiene (isoprene) precursors.

Copolymers, polymers made from more than one different monomeric subunit, are also possible. These may appear in any of a variety of forms, each with different macroscopic properties and different synthetic procedures. **Random copolymers** have no order to the appearance of one type of monomeric unit over another. **Block copolymers** have large isolated groups of each monomer, and the groups are then bound together. **Alternating copolymers** are formed with each monomer bonded to a different monomer.

Polymers encompass a wide range of physical properties, including:

- **plastics:** Compounds that distort under stress, such as polyethylene.
- **elastomers:** Compounds that distort under stress but return to their original structure when the stress is removed, such as rubber.
- **fibers:** Essentially one-dimensional polymers, forming long, thin strands, such as nylon.

[1]The notion persists that windows in ancient buildings are distorted from the slow action of gravity on the glass over centuries, but evidence suggests that any such distortion was present when the windows were made.

Bulk polymers may be either crystalline or amorphous, or a mixture of both. The crystalline regions within the bulk are called *crystallites*. The fraction of the bulk that exists in crystallites depends on the regularity of the polymer chain, but can also be controlled in some cases by the rate of solidification. Stressing a polymer tends to increase the size and number of crystalline regions, because external forces tend to push or pull the polymer units into line with one another. The amorphous regions allow the polymer to yield to stress, giving the substance *plasticity*. Crystallites, on the other hand, will resist any stress that opposes alignment of the molecules, and may fracture if the stress force exceeds the bonding force.

Crystals

In the same way that an individual molecule has an equilibrium geometry, in which the potential energy reaches its lowest value,[2] a pair of molecules joined by intermolecular forces also has an equilibrium geometry. As we add more and more molecules, the equilibrium arrangement may change, but it will always be the same for each molecule, assuming our solid has only one type of molecule. Therefore, an arrangement of the molecules exists that best stabilizes a pure solid, having all the molecules in the same environment, oriented the same way with respect to all their neighbors. The randomness of the amorphous solid requires that it occasionally deviate from the most stable positioning of the molecules, so some ordered form of the solid is always more stable than the amorphous solid.

We call such structures **crystals:** solid structures composed of a single, repeating molecular arrangement, such that the immediate environment of any molecule is exactly like that of any other molecule at any other location in the crystal. All crystals share this long-range order, which requires a strong direction-dependence of the forces between the particles, but those forces encompass many different bonding mechanisms. Whatever its origin, this direction-dependence (or *anisotropy*) arranges the molecules into a pattern, resulting in a solid with considerable symmetry.

This anisotropy also poses a challenge to the analysis of crystals by nuclear magnetic resonance spectroscopy, which is covered in Section 5.5. Each change in orientation of a crystal in the spectrometer corresponds to a different angle between the external magnetic field and the chemical bond axes, and a different spectrum. The rapid and random motions of molecules in a liquid average over these differences to yield narrow lines in the spectrum. To duplicate the same effect in crystals, the crystal is powdered and the sample is then rotated at rates of 10 kHz or faster.

We can imagine making any crystal by starting from one small component, the **unit cell,** that contains all the information necessary to reconstruct the

---

[2]That equilibrium geometry—a minimum in the effective potential energy function—doesn't have to be unique, in that there may be equivalent minima at other geometries. Perhaps the most common example occurs when we rotate the methyl group in a molecule. This changes the places of the hydrogen atoms but arrives at exactly the same effective potential energy.

entire crystal. Each unit cell corresponds to some arrangement of molecules that is then replicated throughout the crystal. The crystal can be constructed from the unit cell as follows (Fig. 13.1):

1. Place a single unit cell with one corner at the origin.
2. Place an infinite number of identical unit cells next to the first one along one axis—call it the *a* axis—to make a line of unit cells. These unit cells differ from the original *only by a translational operation*; they are not rotated, reflected, or inverted with respect to the original unit cell.
3. Repeat this line of unit cells infinitely along a second axis, the *b* axis, to produce a two-dimensional layer of unit cells. Again, the difference between each new line of unit cells and the original is only the result of translation.
4. Finally, repeat these layers infinitely along a third axis, the *c* axis, to make the three-dimensional crystal. Each layer resembles every other, except for the translation.

This similarity over large distances in all directions defines the crystal.

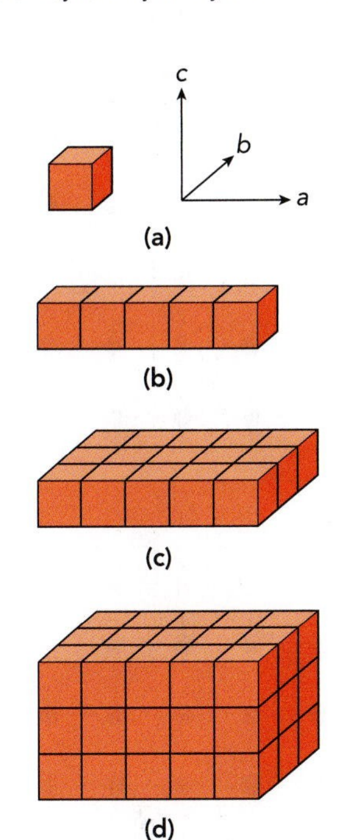

▲ FIGURE 13.1 **Repeating the unit cell structure along the *a*, *b*, and *c* lattice coordinates generates an entire crystal.**

A substance is **crystalline** if it displays long-range order over hundreds of molecules or more, a characteristic that can be determined by the interference pattern of electrons or other particles diffracted by the substance (see *Tools of the Trade* in Chapter 1 and this chapter). Crystalline substances do not have to be true crystals. Liquid crystals, for example, display some structural order along one or two dimensions, but not all three, so the environment of one molecule is not exactly like that of another (as it must be in a true crystal).

Materials may have both amorphous and crystalline solid phases. The most stable form will be a crystal, but the amorphous form may be more easily obtained by cooling the liquid. The glassy ice that makes up the nucleus of a comet is formed by water vapor freezing at very low pressure. The gas is a disordered substance, and the disorder is carried through into the solid as the gas freezes. Although the crystalline ice with its regular arrangement of the molecules would be a more stable form of the solid, once the molecules are frozen into the glass they no longer have enough kinetic energy to rearrange themselves into the crystal.

13.2 Symmetry in Crystals

In an isolated molecule, symmetry plays a role at many levels: there is symmetry in the structure of the atomic orbitals, in the interaction between any two electrons, and in the distribution of electrons into *s*- and *p*-bonding orbitals. However, there is no rule that the structure of the entire molecule be symmetric. A crystal, however, is *defined to have symmetry* by its repeating structure, and symmetry is therefore where we begin our look at crystals.

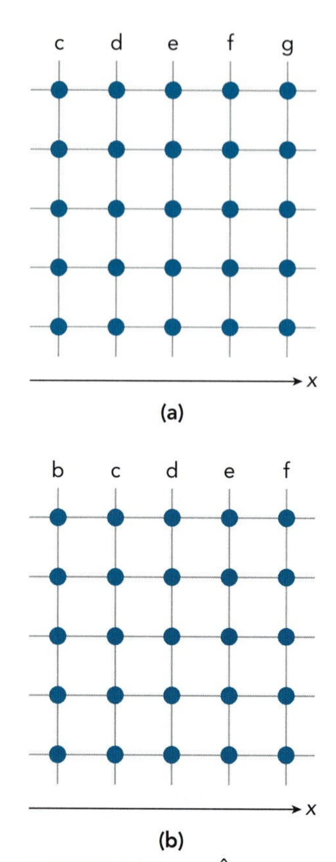

▲ FIGURE 13.2 **The $\hat{X}$ symmetry operation on an infinite two-dimensional square lattice.** This operation advances the lattice **(a)** along the x axis by one unit **(b).** The columns shown have been labeled b–g in the figure to show the motion of the lattice, but an ideal lattice would not be perceptibly altered by the operation.

Group Theory

The ideal crystal is a rigid, three-dimensional array of molecules extending infinitely in all directions. This is the model used to evaluate the symmetry of a group of real atoms. The infinite extent of this array allows us to add new symmetry operations to our list of point group symmetry elements (Section 6.1). Previously, we counted only operations that leave the center of mass unchanged. However, the center of mass is not defined for an infinite number of atoms, so we can ignore that constraint now by adding **translational symmetry elements** to the list.

Start with a **crystal lattice,** an infinite array of points in space arranged such that each point is exactly identical to every other point. For now, we will choose a cubic lattice, placing each lattice point at a location (x,y,z), spaced by 1 Å along x, y, or z from the previous point. At each lattice point, we place one atom, identical to every other atom in the crystal. The operator $\hat{X}$ that moves the entire lattice 1 Å along the x axis is now a symmetry operation. Because the lattice is infinite, there is no point (x,y,z) to which an atom has moved that did not have an atom previously. The lattice after operation is indistinguishable from the original lattice (Fig. 13.2). The minimum distance for which this translation is a symmetry element is called the **unit distance** and usually has different values for each of the three coordinate axes. Similarly, translations along the y and z axes by an integer number of angstroms are also symmetry operations ($\hat{Y}$ and $\hat{Z}$), as are an infinite number of combinations of these operations. Like the point group operators, successive or combined operation with the translation operators must be identical to operation by another symmetry element of the group. Examples of such operations include the **glide reflection** (translation by half the unit distance, followed by reflection through a plane containing the axis of translation, for example, $\hat{\sigma}_{XY}\,\frac{1}{2}\hat{X}$) and the **screw rotation** (translation followed by proper rotation about the translation axis, for example, $\hat{C}_2(Z)\hat{Z}$).

The lattice does not have to be cubic. The appearance of the lattice along one coordinate axis does not have to be the same as along the other axes, and the translational axes of symmetry need not even be perpendicular to each other like the Cartesian axes. The essential property is simply that the pattern along each axis continues to repeat in all three dimensions.

The symmetry elements of the point groups—proper rotation, reflection, inversion, and improper rotation—all still apply. For example, the lattice described by Fig. 13.2 has the same point group operations of the octahedral group—several $\hat{C}_3$ and $\hat{C}_4$ rotation axes, various mirror planes, improper rotation axes, and inversion. These can also be combined with the translation operators. We call the groups that contain the point group *and* translation operations the **space groups** and use them to classify different crystal structures.

Here's a surprise: the number of possible point groups is infinite, because n in the C_n groups can be as large as we want. To form the space groups, we keep all the same symmetry elements and then add the translations on top of those. You would think the possibilities would only get more complicated, but in fact there are only 230 possible space groups, and fewer than 20 of these account for nearly all observed crystal structures. The translational symmetry of crystals is an added degree of freedom, but it is also a constraint because only certain combinations of point group operations can coexist with translations in the same group. In particular, only proper rotations by 60°, 90°, and their multiples are possible.

Once we've identified the 230 space groups, we can eliminate the translation operations from our list of symmetry elements and classify those groups just looking at the point group symmetry operators. We find 32 distinct groups, called the **crystallographic point groups** and listed in Table 13.1. The crystallographic point groups can be labeled by either the Schönflies point group symbols or by the **Hermann-Mauguin crystal group symbols.**

The Schönflies symbols are the same point group labels we used in Chapter 6, but scientists studying crystals prefer the Hermann-Mauguin symbols. Let's go over that notation briefly. The Hermann-Mauguin symbols give, in shorthand, enough of the symmetry elements to uniquely identify the crystallographic point group:

1. The numbers 1, 2, 3, 4, and 6 indicate the order of the principal rotation axis (the symmetry elements $\hat{C}_1$, $\hat{C}_2$, $\hat{C}_3$, $\hat{C}_4$, and $\hat{C}_6$).
2. The barred numbers $\overline{1}$, $\overline{3}$, $\overline{4}$, and $\overline{6}$ indicate a rotation followed by inversion and correspond to the operations $\hat{I}$, $\hat{S}_6$, $\hat{S}_4$, and $\hat{S}_3$, respectively.
3. Mirror reflection planes are indicated by m, with $/m$ used for horizontal mirror planes.

The choice of symmetry elements used to construct the Hermann-Mauguin group symbol is sometimes a matter of convention. The T_d group, for example, is denoted by $\overline{4}3m$ in the Hermann-Mauguin system, in view of the $\hat{S}_4$, $\hat{C}_3$, and $\hat{\sigma}$ symmetry elements of T_d (see the Appendix).

TABLE 13.1 The 32 crystallographic point groups. For each group, the symbols, lattice system, and selected symmetry properties are given. Each crystallographic point group corresponds to just one lattice system, but it may correspond to any of the Bravais lattices listed for that lattice system.

Lattice system	Bravais lattices	Crystallographic point group: Schönflies symbol	Crystallographic point group: Hermann-Mauguin	Properties
triclinic	primitive	C_1	1	polar
		C_i	$\overline{1}$	centrosymmetric
monoclinic	primitive C-centered	C_2	2	polar
		C_s	m	polar
		C_{2h}	2/m	centrosymmetric
orthorhombic	primitive C-centered face-centered body-centered	D_2	222	
		C_{2v}	$mm2$	polar
		D_{2h}	mmm	centrosymmetric

(continued)

TABLE 13.1 continued The 32 crystallographic point groups. For each group, the symbols, lattice system, and selected symmetry properties are given. Each crystallographic point group corresponds to just one lattice system, but it may correspond to any of the Bravais lattices listed for that lattice system.

Lattice system	Bravais lattices	Crystallographic point group: Schönflies symbol	Crystallographic point group: Hermann-Mauguin	Properties
tetragonal	primitive body-centered	C_4	4	polar
		S_4	$\bar{4}$	
		C_{4h}	$4/m$	centrosymmetric
		D_4	422	
		C_{4v}	$4mm$	polar
		D_{2d}	$\bar{4}2m$	
		D_{4h}	$4/mmm$	centrosymmetric
rhombohedral	primitive	C_3	3	polar
		S_6	$\bar{3}$	centrosymmetric
		D_3	32	
		C_{3v}	$3m$	polar
		D_{3d}	$\bar{3}m$	centrosymmetric
hexagonal	primitive	C_6	6	polar
		C_{3h}	$\bar{6}$	
		C_{6h}	$6/m$	centrosymmetric
		D_6	622	
		C_{6v}	$6mm$	polar
		D_{3h}	$\bar{6}m2$	
		D_{6h}	$6/mmm$	centrosymmetric
cubic	primitive body-centered face-centered	T	23	
		T_h	$m3$	centrosymmetric
		O	432	
		T_d	$\bar{4}3m$	
		O_h	$m3m$	centrosymmetric

Bravais Lattices

CHECKPOINT Crystals are analyzed according to a dizzying number of different classification schemes because different levels of structure are relevant to different fields. For example, certain optical properties depend on the *point group* symmetry of the crystal, whereas the identification of a particular crystalline compound under a microscope may rely partly on the shape given by the *lattice system*.

Crystals are also commonly classified according to a less restrictive scheme based on the structure of the unit cell. The size and shape of the unit cell are described by the **lattice constants**—the lengths a, b, and c of the sides along each of the corresponding axes—and by the angles α (between axes b and c), β (between axes c and a), and γ (between axes a and b). How a, b, and c are assigned to the three translation axes depends on the lattice system, and in some cases the conventions have not been firmly established. The a, b, and c axes need *not* be perpendicular to one another.

As a simpler example, we can construct all the distinct lattices that have translational symmetry in *two* dimensions. There are only five, shown in Fig. 13.3a. Only three of these have the a and b axes perpendicular.

Figure 13.3b demonstrates that each of these lattices can be drawn with a different representation for the unit cell; the fundamental difference between these lattices lies in the *symmetry elements of the structure*, not in the symmetry elements of the unit cell we happen to choose. For example, the centered rectangular lattice is drawn in Fig. 13.3a with a rectangular unit cell. In Fig. 13.3b, however, we find that the hexagonal lattice can also be drawn with a centered rectangular unit cell. Why are the two lattices given different labels? The reason is because these five lattices are separated not according to their unit cells, but according to their symmetry elements, which include the following:

1. $\hat{C}_4$ for the square lattice
2. $\hat{C}_2$ and mirror reflections for the primitive rectangular lattice, but no $\hat{C}_6$
3. $\hat{C}_2$, mirror reflections, and a glide reflection along both the a and b axes for the centered rectangular lattice
4. $\hat{C}_6$ for the hexagonal lattice
5. $\hat{C}_2$ but no mirror reflections for the oblique lattice

There are more symmetry elements than these, but this list tells the difference among the different lattices.

The same principles hold in three dimensions. The unit cell provides a convenient representation of the entire crystal, but its boundaries are based on historical convention.[3] For any given crystal, there will be choices for where to position the vertices of the unit cell. Wherever we choose, we call the location of these vertices the **lattice points** of the unit cell. Any other position in the crystal with exactly the same structure is also a lattice point. There may not actually be any atoms at these sites, because sometimes the symmetry of the unit cell is more evident if we place the lattice points between interacting atoms or molecules. In fact, it is not even required that there be a one-to-one correspondence between the lattice points and the molecules in the unit cell. Each lattice point may represent a group of molecules, and this group is called the **basis** of the lattice. Figure 13.4 illustrates a unit cell with a pair of identical atoms for each lattice point.

As with our two-dimensional lattices, the shape of the unit cell alone, determined by the relative lengths of the sides and the angles at the corners, can be used to classify the crystal structure. In three dimensions, we end up with the seven **lattice systems** given in Table 13.2.

For each of these shapes, there may be choices as to where the lattice points may be placed in the unit cell while preserving the symmetry. If the lattice points are located only at the vertices of the unit cell, we have a **primitive lattice.** For some lattice systems, there may also be lattice points located in the centers of two opposite faces (**end-centered lattice**), in the center of each face (**face-centered lattice**), or in the middle of the unit cell (**body-centered lattice**). There are 14 distinct possibilities, and these are known as the **Bravais lattices.** The conventional unit cells for these lattices are shown in Fig. 13.5.

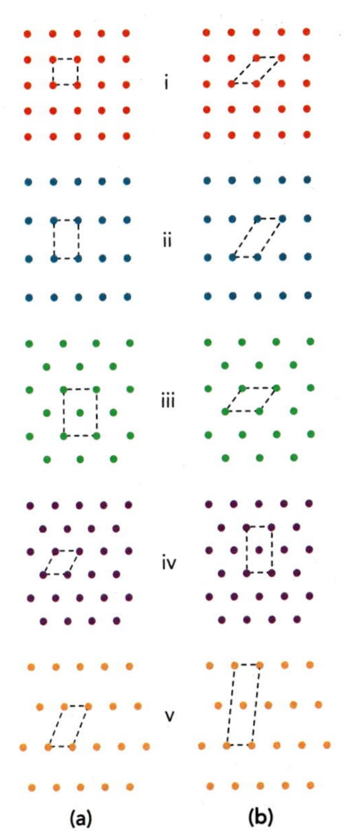

▲ FIGURE 13.3 **The five fundamental two-dimensional lattices.** The lattices are drawn with the customary choice of unit cell **(a)** and with different unit cells **(b).** (i) the square lattice ($a = b, \gamma = 90°$); (ii) the primitive rectangular lattice ($a \neq b, \gamma = 90°$); (iii) the centered rectangular lattice ($a \neq b, \gamma = 90°$); (iv) the hexagonal lattice ($a = b$, $\gamma = 120°$); (v) the oblique lattice ($a \neq b$, $\gamma \neq 90°$).

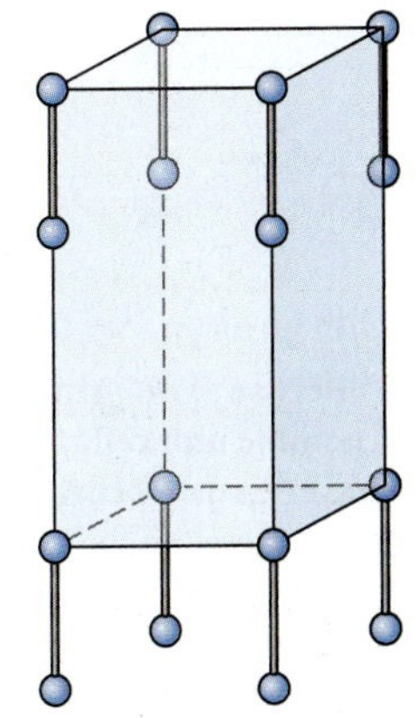

▲ FIGURE 13.4 **Orthorhombic unit cell with a basis of two atoms.**

[3]The conventional unit cells are usually not the smallest repeating subunit of the crystal, but are instead chosen to emphasize the principal symmetry elements of the crystal. Other choices for the unit cell will be available.

TABLE 13.2 The lattice systems. The angles and shapes of the faces describe a single unit cell in each system.

Name	Sides	Angles	*ab* face	*bc* face	*ac* face
cubic or isometric	$a = b = c$	$\alpha = \beta = \gamma = 90°$	square	square	square
hexagonal	$a = b \neq c$	$\alpha = \beta = 90°; \gamma = 120°$	rhombus	rectangle	rectangle
rhombohedral	$a = b = c$	$\alpha = \beta = \gamma \neq 90°$	rhombus	rhombus	rhombus
tetragonal	$a = b \neq c$	$\alpha = \beta = \gamma = 90°$	square	rectangle	rectangle
orthorhombic	$a \neq b \neq c$	$\alpha = \beta = \gamma = 90°$	rectangle	rectangle	rectangle
monoclinic	$a \neq b \neq c$	$\alpha = \beta = 90° \neq \gamma$	parallelogram	rectangle	rectangle
triclinic	$a \neq b \neq c$	$\alpha \neq \beta \neq \gamma$	parallelogram	parallelogram	parallelogram

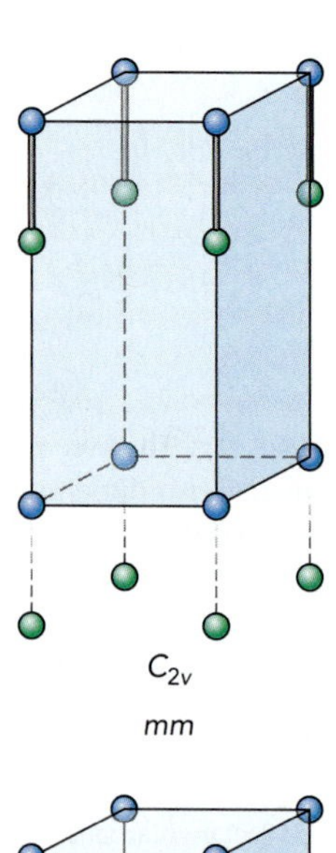

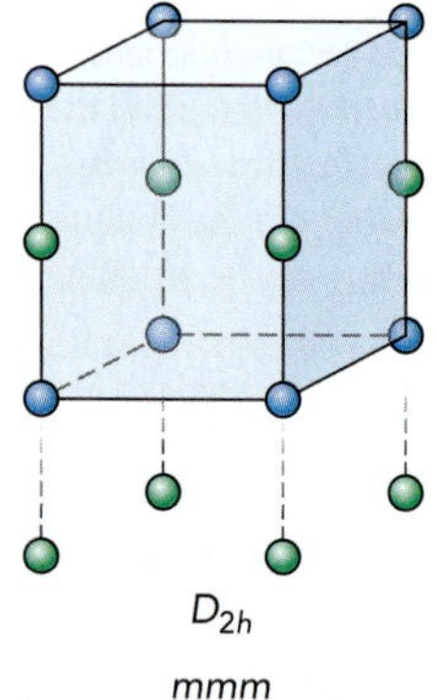

▲ **FIGURE 13.6 Two primitive orthorhombic unit cells for a diatomic crystal.** By changing the distance between the component atoms, it is possible in this case to increase the symmetry of the crystal from C_{2v} to D_{2h} without altering the Bravais lattice.

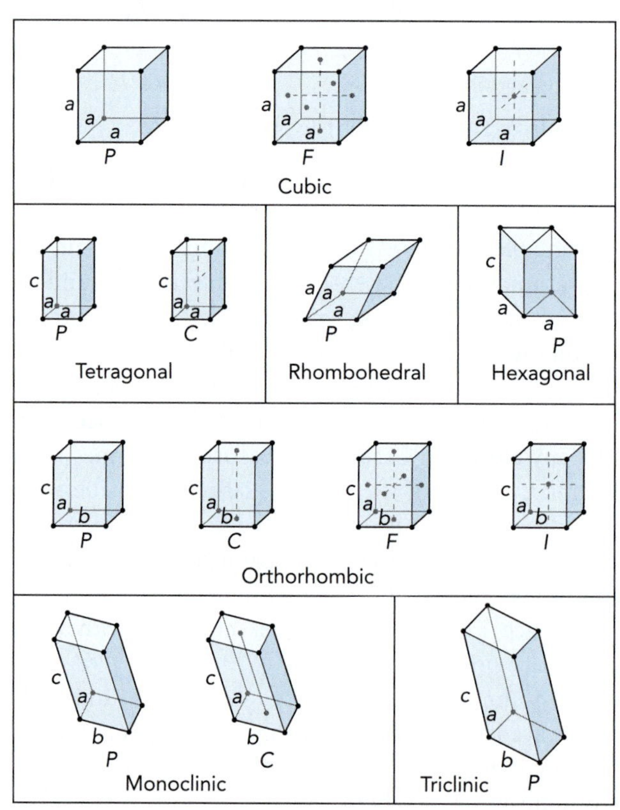

▲ **FIGURE 13.5 The Bravais lattices.** Within a set, *P* denotes the primitive lattice, *F* is face-centered, *I* is body-centered, and *C* is C-centered.

The 14 Bravais lattices do not uniquely determine the crystallographic point groups. For example, the primitive orthorhombic lattice can have inversion as one of its symmetry elements (point group C_i) or lack it (point group C_1). This depends on the symmetry properties and relative orientations of the molecules that make up the crystal. Figure 13.6 illustrates how this can be the case for an example using a diatomic molecule in a primitive orthorhombic lattice. For one arrangement of the molecules, the unit cell belongs to the C_{2v} point group; for another orientation, the unit cell belongs to the D_{2h} point group. In both cases, the lattice points are occupied by atoms in identical environments, in arrangements consistent with the primitive orthorhombic Bravais lattice. The Bravais lattices only specify the symmetry of the lattice points in the crystal, not the symmetry of the crystal itself.

These various schemes for classifying crystal structures are summarized in Table 13.3.

TABLE 13.3 Summary of the classification schemes for crystals.

Classification	Number	How it's defined	What it's used for
lattice system	7	Describes the geometry of the unit cell itself, no matter what the arrangement of atoms inside	Largely determines the shape of the macroscopic crystal
Bravais lattice	14	Gives the distribution of lattice points within the unit cell	Together with the basis, gives the number of particles per unit cell
crystallographic point group	32	Gives the point group symmetry elements of the unit cell	Determines whether or not the crystal possesses symmetry-dependent properties such as polarity
space group	230	Combines the point group symmetry with the translational symmetry elements	Normally needs to be determined in order to solve the crystal structure from a diffraction pattern

EXAMPLE 13.1 Bulk Properties from the Unit Cell

CONTEXT The nearly perfect symmetry of a crystal makes possible perhaps the most direct connection between microscopic and macroscopic properties that we encounter in chemistry. For example, the density of the unit cell is the same as the density of the bulk. It takes some understanding of the geometry of the different unit cells to make that connection, however. In this case, we use the structure of the sodium chloride unit cell to work our way up to the bulk density, a shift of roughly 23 orders of magnitude.

The density is the mass divided by volume, whether for the unit cell or the bulk crystal. For the unit cell, we use the volume of a cube with side of length a, and the mass is the molecular mass times the number of molecules (on average) in each unit cell.

PROBLEM Calculate the density in g cm^{-3} of the face-centered cubic lattice for ^{23}Na^{35}Cl, given the lattice constant $a = 5.640$ Å.

SOLUTION The face-centered cubic unit cell contains (8 vertex NaCl) $\left(\frac{1}{8}\text{ vertex/unit cell}\right)$ + (6 face NaCl) $\left(\frac{1}{2}\text{ face/unit cell}\right)$ = 4NaCl/unit cell.

$$m_{\text{NaCl}} = 22.99 + 35.45 = 58.44\text{ amu} = 9.704 \cdot 10^{-23}\text{ g}$$

The density is therefore

$$\text{density} = \frac{4m_{\text{NaCl}}}{V} = \frac{4(9.704 \cdot 10^{-23}\text{ g})}{(5.640 \cdot 10^{-8}\text{ cm})^3} = 2.164\text{ g cm}^{-3}.$$

TOOLS OF THE TRADE X-Ray Crystallography

In 1912, at the University of Munich, a theoretical physics student named Paul Peter Ewald developed the equations to describe the interaction of light of any wavelength with an orthorhombic lattice. At the time, it was assumed (but could not be proven) that the ordered structure observed for crystals under microscopes extended down to the atomic level. Ewald shared his work with Max von Laue, who was then working with Wilhelm Röntgen, discoverer of x-rays. Von Laue recognized immediately that the tiny spacings in crystals could form a diffraction grating on the same distance scales as the wavelengths of x-rays. Within the year, von Laue had generated the first successful x-ray diffraction pattern of a copper sulfate crystal, validating the hypothesis that crystals were indeed composed of a highly ordered array of atoms and simultaneously proving to a few remaining skeptics that x-rays were indeed a form of electromagnetic radiation. Von Laue was awarded the Nobel Prize in Physics in 1914 for this work. The very next year, the father-son team of William Henry Bragg and William Lawrence Bragg shared another Nobel Prize in Physics for showing how the precise geometry of the crystal could be predicted from the diffraction pattern. Together, in just two years, the Braggs had determined the structures of diamond and numerous ionic solids, providing the *first* direct measurements of chemical bond lengths and atomic sizes.

What is x-ray crystallography? X-ray crystallography is the measurement of the diffraction pattern formed by the interaction of x-rays with a crystalline sample. The x-rays are scattered by regions of electron density, so the interference pattern formed by the x-rays scattered in the same direction from different atoms is determined by the spacings between the atoms.

Why do we use x-ray crystallography? X-ray crystallography is our primary means for determining precise molecular structures. Chemists have long recognized that the properties of a substance depend not only on its composition, but also on its molecular structure. The challenge is often to find out just what that structure is. Rotational spectroscopy (Chapter 9) yields the most precise measurements of molecular geometries but only works for gas-phase samples of relatively small molecules. When the attractive forces between the particles make it hard to get the substance into the gas phase, it is usually possible to form crystals instead. Crystallography is therefore capable of determining structures for the strongly bound ionic solids and metals, for example, but can also determine geometries for molecular crystals formed from organic compounds and biochemicals. One of the greatest accomplishments of the technique was the determination of the double helix structure of DNA by James Watson and Francis Crick from x-ray diffraction patterns obtained by Rosalind Franklin.

How does it work? The design of an x-ray diffractometer is similar to that of a neutron diffractometer (*Tools of the Trade*, Chapter 12) and consists of an x-ray source, a sample holder, and a detector. To obtain the most precise geometries, a monochromatic source of x-rays is used. The x-rays are generated initially by bombarding a metal target with high-energy electrons to ionize core electrons. The ejected core electrons are replaced by higher energy electrons, which emit high energy photons as they drop into the core orbital. In order to determine the three-dimensional structure of the crystal, the sample is rotated on a device called a *goniometer,* which allows diffraction patterns to be measured for all the distinct orientations of the crystal. Thousands of images may be gathered for a single structure determination. The diffraction patterns are acquired electronically, for example by a charge-coupled device (CCD) detector.

Perhaps the most crucial advance in recent decades has been the rapid evolution of computational methods for determining the structure from the diffraction patterns. The quality of the raw data is affected by the number of atoms in the unit cell, the temperature of the crystal, and of course imperfections in the crystal. Often this makes the pattern ambiguous enough that there is no unique solution: more than one structure—often only slightly different—would result in the same pattern, to within some experimental error. Modern software solves for the set of possible solutions, using the known chemical composition of the crystal and basic rules of chemical bonding to arrive at the likeliest geometry.

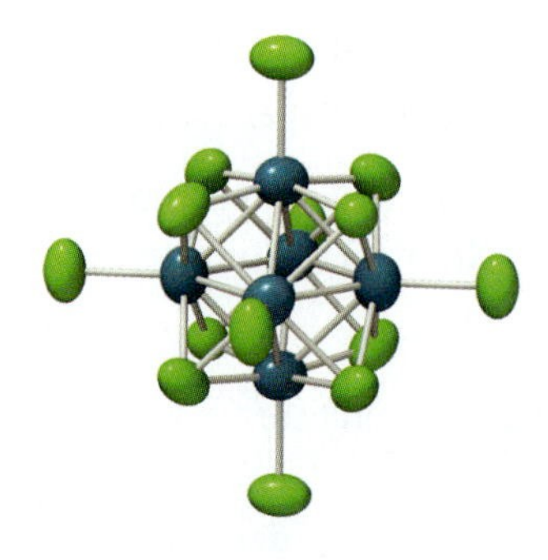

Molecular geometries obtained from x-ray diffraction are averaged over the vibrational motions of the atoms. (In fact, Peter Debye showed that the intensity of the x-ray scattering would depend so much on the vibrational motions that the temperature dependence of the signal could be used to demonstrate the existence of zero-point motion.) To show the vibrational averaging, the structures resulting from x-ray diffraction patterns are represented by models in which the each atom is represented by a *thermal ellipsoid,* proportional to the extent of its vibrational motion, as shown in the figure. For the $Mo_6Cl_{14}^{2-}$ cluster anion shown on previous page, notice that the chlorine atoms that extend farthest from the center of the cluster have the largest ellipsoids because they have the most freedom to move, and those ellipsoids appear to be flattened along the bond axis because the stretching motion is more constrained (has a higher force constant) than the bending.

Now a century old, few techniques have shown the same breadth of application in chemistry as x-ray diffraction, from questions of fundamental quantum mechanics and chemical bonding to the structures of proteins and nucleic acids.

Electrical Properties of Crystals

The symmetry elements of a molecule tell us much about its properties, and the symmetry elements of a unit cell tell us much about the properties of the crystal. For example, the unit cell of a crystal has no dipole moment if it meets either or both of these criteria (from Section 6.2):

1. All the bond dipoles lie perpendicular to either a $\hat{C}_n$ proper rotation axis or to a $\hat{\sigma}$ mirror plane.
2. The molecule has a $\hat{S}_{2n}$ symmetry element (including inversion, because $\hat{I} = \hat{S}_2$).

As a result, only certain crystallographic point groups yield polar unit cells, as indicated in Table 13.1.

If the unit cell is polar, the charge density at one end of the unit cell is greater than at the other end, and this pattern is repeated throughout the crystal. Polar crystals, whatever the bonding mechanism, may exhibit some useful electrical characteristics that arise from the high degree of order not found in amorphous solids. A **ferroelectric crystal** has a net charge separation from one end to the other due to a large number of co-aligned dipole moments. Such a crystal, if left to move freely, will align itself under the influence of an electric field.

The aligned dipole moments in the ferroelectric crystal correspond to a net charge separation throughout the crystal and therefore correspond to an electrostatic potential energy—what we call voltage—from one end of the crystal to the other. This voltage is sensitive to the spacing between the molecular dipoles, so when the crystal is compressed, the voltage changes. Looking at it from the other perspective, applying a voltage to such a crystal can compress it or expand it. Crystals that deform under the influence of applied voltages are called **piezoelectric crystals.** Piezoelectricity is possible in crystals belonging to any of the crystallographic point groups in Table 13.1 except O and those with centers of inversion (called **centrosymmetric crystals**). Of the 20 remaining crystallographic point groups, some (such as C_4) are polar and will compress or expand along one axis when a voltage is applied. Some others (such as D_{2d}) are *not* normally polar (and therefore not ferroelectric) but may become polar when twisted.

The piezoelectric crystal is a common component in applications that require very fast and precise distance adjustments. The crystal dimensions are usually changed by a factor of only 10^{-6} or less using the piezoelectric effect, but for small distance measurements, such as those in atomic force microscopy, this is ideal.

A related effect is **pyroelectricity,** which is the generation of voltage by the heating or cooling of a crystal. Because substances tend to expand when heated and contract when cooled, changes in temperature of polar crystals tend to change the overall dipole moment and voltage of the crystal.

Many polar crystals also have valuable optical properties resulting from the interaction of the electrons with the electric field of the radiation. For example, the frequency of a beam of electromagnetic radiation may be doubled by sending the beam through a non-centrosymmetric crystal. The incident radiation carries an oscillating electric field through the crystal, a field that deforms the electronic wavefunction, and therefore the electric field, of the crystal. The extent to which the electric field of the crystal is changed increases with the polarizability of the electrons and with the intensity of the radiation.

In any of the non-centrosymmetric crystals, the electron distribution is more easily pushed in one direction than in the opposite direction by the radiation's electric field. This is equivalent to multiplying a fraction of the incident radiation by a damping factor that varies at the frequency of the radiation's electric field. If the radiation enters the crystal with an electric field given by

$$\mathcal{E}_{\text{in}}(t) = \mathcal{E}_0\cos(\omega_0 t), \tag{13.1}$$

the direction dependence of the polarizability of the crystal causes a small fraction b of the incident electric field to be multiplied by a factor $a\cos(\omega_0 t)$. One of our more useful trigonometric identities permits us to write the resulting electric field $\mathcal{E}_{\text{mix}}$ as

$$\begin{aligned}\mathcal{E}_{\text{mix}}(t) &= [a\cos(\omega_0 t)][b\mathcal{E}_0\cos(\omega_0 t)] \\ &= ab\mathcal{E}_0\cos^2(\omega_0 t) \\ &= \frac{ab\mathcal{E}_0}{2}[1+\cos(2\omega_0 t)].\end{aligned} \tag{13.2}$$

The radiation leaves the crystal primarily at the same frequency ω_0 as the incident radiation, but a small component of the radiation has acquired a frequency of $2\omega_0$. We call the process **frequency doubling** or **second harmonic generation.**

The electric field has been labeled $\mathcal{E}_{\text{mix}}$ because this is a single example of several techniques for "mixing" oscillating electric fields to obtain radiation with a new frequency. In general, the damping factor has a more complicated time dependence than $\cos(\omega_0 t)$ and leads to weak output signals at three and four times the input frequency. One application has been the extension of the operating range of visible-light lasers into the ultraviolet. A popular crystal for this application has been KH_2PO_4 (potassium dihydrogen phosphate or KDP), which belongs to the D_2 (or 222) crystallographic point group.

If the damping factor were a constant of value a, there would be no effect on the frequency of the radiation, only on the amplitude. This is called a linear response (because the output signal depends on $\cos(\omega_0 t)$ to the first power).

The interactions that mix frequencies require a nonlinear response, and the investigation of these properties is the field of **nonlinear optics.**

Crystal Planes and Miller Indices

A useful way to think of the distinction between crystals and amorphous solids is that amorphous solids are isotropic; the properties of the amorphous solid, such as density and average bond strength, are the same along any line we draw through the sample. Crystals are anisotropic, and this characteristic is crucial to many of the unique properties of crystals.

For example, many crystals form stronger bonds along two axes than along the third. The strong bonds form more easily, and remain stable over a longer time than the weak bonds. Such crystals tend to develop in layers stacked along an axis perpendicular to the plane of the strong bonds. These crystals will resist breaking or cutting along that axis but are easily cut along planes parallel to the layers. This is an important consideration in some manufacturing processes, because crystals of this sort, such as brass, tend to cleave along the weakly bound planes rather than cutting easily along the intended plane. Similarly, different crystal faces may grow at different rates, accounting for the needle-like formations of some crystals.

In addition to these mechanical properties, the anisotropy of the crystal implies that electrical, thermal, and optical properties of the crystal depend on the orientation of the crystal with respect to the direction of motion of the electricity, heat, or radiation. Therefore, we not only have to understand the structure of the crystal, we must also have a standard method for describing orientation in the crystal. The same method is used to identify unambiguously the type of surface exposed on a real crystal.

The symmetry of the crystal often makes it possible to sever the chemical bonds along one of the planes of the crystal, or for the crystal to develop naturally along these planes in order to maximize the number of strong bonds. Cuts along different planes can expose different numbers of atoms per unit area as well as different geometries of the atoms. In consequence, the chemistry at the surface of a crystal often strongly depends on the angles of the surface plane relative to the lattice, and cutting is easier along a crystal face than through it.

The planes are described by the **Miller indices.** Locate one of the unit cell vertices at the point $(0, 0, 0)$, and define the unit cell dimensions as a, b, c along each of the three axes. Then let the surface plane of the crystal cut through the unit cell and intersect each of the three axes at points $(a/h,0,0)$, $(0, b/k,0)$, and $(0, 0, c/l)$. The exact location of the unit cell with respect to the surface is *chosen* such that h, k, and l are whole numbers. If the face is parallel to the a axis, convention is to set h equal to zero (i.e., the surface intersects the a axis only at infinity; $a/\infty = 0$). The face of the crystal is then reported as the (h,k,l) face. For example, the surface presented when the crystal is cut parallel to the ab plane intersects the axes of our strategically placed unit cell at $a = \infty$, $b = \infty$, and $c = 1$; this is the $(1/\infty, 1/\infty, 1/1) = (0, 0, 1)$ face. More examples are shown in Fig. 13.7.

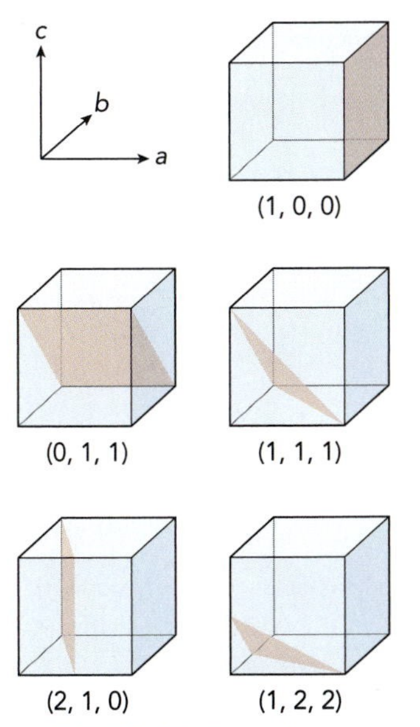

▲ FIGURE 13.7 **Examples of crystal planes labeled by the Miller indices.**

13.3 Bonding Mechanisms and Properties of Crystals

Translational symmetry is the one unifying feature of crystals. Aside from that, they vary over a huge range of chemical or physical properties. We group them, somewhat loosely, according to the bonding mechanism that holds the components together:

1. **Ionic crystals:** Atomic or molecular anions and cations are bound by the electric monopole–monopole attraction (e.g., NaCl, NH_4ClO_4).
2. **Metal crystals:** Atomic metal cations are bound together by a "sea" of delocalized electrons (copper, bronze).
3. **Covalent crystals:** Covalent bonds join the atoms of the crystal (diamond, quartz).
4. **Molecular crystals:** Van der Waals forces bind the atoms or molecules to form the crystal. Bond strengths are small and melting points are low compared to the previous examples (water ice, solid argon).

CHECKPOINT While the *structure* of the crystal is critical to determining the composition, optical properties, and electrical properties of the crystal, its chemical and material properties are often more dependent simply on what holds the crystal together. Whether or not the crystal is soluble in water and the crystal's hardness are more functions of the bonding mechanism than the lattice structure.

This is an appealingly chemistry-oriented approach, but classifying crystals according to their bonding mechanism can be problematic. For example, some crystals have different bonding mechanisms along different directions: graphite combines covalent bonding (in the aromatic plane) with van der Waals bonds (perpendicular to the aromatic plane). **Ceramics** are materials formed from inorganic compounds, often mixing ionic and covalent bonds. However, most crystals can be clearly identified with a single bonding mechanism from the list just given, and we now briefly examine each of these examples in turn.

Ionic Crystals

The bonding mechanism is the electric monopole–monopole attraction between atomic or molecular anions and cations, Eq. 5.31:

$$U_{\text{ionic attr}} = \frac{q_A q_B}{4\pi\varepsilon_0 R}.$$

Frequently the ionization results when electrons are transferred in order to fill atomic or molecular orbitals containing unpaired electrons. The stabilization energy acquired by this transfer must more than make up for the ionization potential of the cation precursor, although some of that energy will be provided by the energy released when the anion is formed.

Predicting crystal structure is a tough job. The monopole–monopole attraction is isotropic, and the arrangement of the atoms in the crystal is largely determined by the most efficient packing arrangement that balances the favorable interaction between unlike ions against the repulsive interaction between like ions. The optimum arrangement is generally not obvious, given that the cations and anions may have different sizes and may be polyatomic (HCO_3^- or NH_4^+, for example), and the stoichiometric ratio of the ions need not be 1:1 (for example, BaF_2). If the stoichiometric ratio is 1:1 and the ions are of similar size, as is the case in CsCl, the unit cell of the crystal is usually primitive cubic, with the cations at the vertices of the cube and the anion in the center or vice versa.[4] Since each of the eight corner atoms is shared by eight other unit cells, the stoichiometric ratio is preserved. The coordination number C for this structure is 8: each Cs^+ ion is a neighbor to 8 Cl^- ions. In the case of NaCl, the chlorine atom has a much more diffuse electron cloud than the sodium, and there would be a lot of empty space around the sodium atom in a primitive cubic structure. Instead the atoms are arranged in a face-centered cubic cell, which gives each atom a coordination number of only 6 but reduces the Na^+Cl^- distance.

One might easily forget that all common ions, whatever their total charge, have positively charged nuclei that repel the nuclei of other ions at short range. This repulsion eventually pushes Na^+ away from Cl^- if it gets too close. The Coulomb attraction between the ionic charges, $(q_A q_B)/(4\pi\varepsilon_0 R)$, is an approximation that works well at large R but neglects the complexity of the electron charge distribution at short distances. The balance between the ionic attraction and the electron–electron repulsion determines the ionic radii of the component ions. Values for the ionic radii indicate the distance from the atomic nucleus to the "edge" of the ion's electron distribution. But the ionic radius cannot be strictly defined to begin with, because the electron density fades away as the distance from the nucleus increases, rather than simply coming to an end. Furthermore, the effective radius can shift in response to the electron distribution of the neighboring ion, especially for the more weakly bound anionic electrons. Nevertheless, the ionic radii often provide valuable clues to the structure of an ionic solid when experimental results might be misinterpreted. Values of the ionic radii for several common atomic ions are given in Table 13.4. This table applies only to monatomic ions; polyatomic ions (such as NH_4^+ or SO_4^{-2}) are not easily treated so generally.

TABLE 13.4 Ionic radii (Å) for selected atomic ions. These values assume the ion is in a crystal with coordination number 6. Values for coordination numbers 8, 4, 3, and 2 can be estimated by multiplying these values by 1.015, 0.985, 0.97, and 0.965, respectively.

Ag^+	1.15	Co^{3+}	0.61	K^+	1.38	Pd^{2+}	0.86	Ti^{2+}	0.86	I^-	2.20
Al^{3+}	0.54	Cr^{3+}	0.62	La^{3+}	1.03	Pt^{2+}	0.80	Ti^{4+}	0.61	O^{2-}	1.40
Ba^{2+}	1.35	Cs^+	1.67	Li^+	0.76	Pt^{4+}	0.63	Tl^+	1.50	S^{2-}	1.84
Ca^{2+}	1.00	Cu^+	0.77	Mg^{2+}	0.72	Ra^{2+}	1.43	Zn^{2+}	0.74	Se^{2-}	1.98
Cd^{2+}	0.95	Cu^{2+}	0.73	Mn^{2+}	0.83	Rb^+	1.52	NH_4^+	1.48	Te^{2-}	2.21
Ce^{3+}	1.01	Fe^{2+}	0.78	Na^+	1.02	Sn^{2+}	0.93	Br^-	1.96		
Ce^{4+}	0.87	Fe^{3+}	0.65	Ni^{2+}	0.69	Sr^{2+}	1.16	Cl^-	1.81		
Co^{2+}	0.75	Hg^{2+}	1.02	Pb^{2+}	1.19	Th^{4+}	0.94	F^-	1.33		

[4]Neutral Cs is a much larger atom than Cl (with radii of 2.65 Å and 1.00 Å, respectively), but the *ionic* radii of Cs^+ (1.67 Å) and Cl^- (1.81 Å) are fairly similar.

Metal Crystals

Metal crystals, like ionic crystals, rely on the attraction between opposite charges that works equally well in all directions, but this time, the negative charge carriers are electrons instead of anions. Bulk metals resemble the tiny metal clusters of Chapter 11 in that the valence electrons are highly delocalized, not tied to any particular atom or group of atoms. The packing problem therefore becomes one of how we can best pack identical, spherical cations (for a pure metal) into a given space, an easier question than how best to pack the ionic crystals, which contain more than one kind of atom. The ratio of the volume occupied by the atoms to the total volume is called the **packing efficiency,** where the atoms are assumed to be hard spheres for this purpose. Because we can often reduce the problem to optimizing the packing efficiency, we have a fairly successful, general description of crystal structures in metals.

Treating the individual atoms as spheres of the same radius, we would find that the most efficient packing, filling 74% of the available space, occurs for two patterns, shown in Fig. 13.8: the **face-centered cubic** (fcc) or **cubic close-packed** (ccp) lattice, and the **hexagonal close-packed** (hcp) lattice. The hcp lattice can be visualized by arranging one set of atoms in staggered rows on a plane, then adding an identical plane of atoms on top of the first one

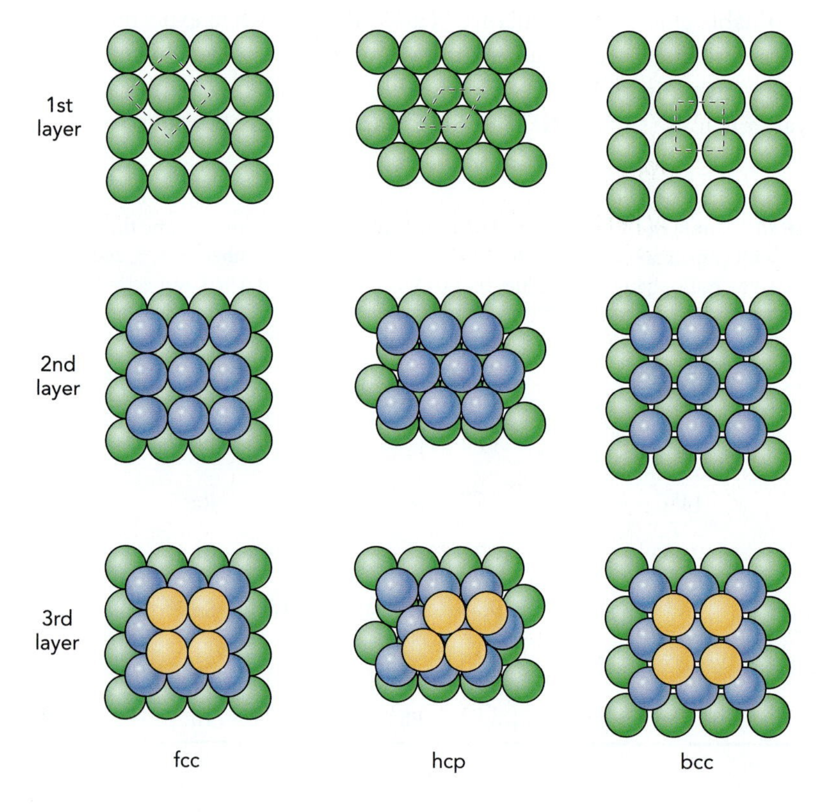

▶ FIGURE 13.8 **The structure of the face-centered cubic, hexagonal closest packing, and body-centered cubic lattices.** To form each lattice, start with the first layer (green atoms), add the second layer as shown (blue atoms), and then the third layer (yellow atoms). The cycle repeats, with every other layer being identical in each lattice. One face of one unit cell is outlined on the first layer of each lattice. This representation illustrates the (100) surface of each lattice.

so that these atoms lie in the gaps left by the atoms of the first plane, and then adding a third plane with atoms positioned directly above those of the first plane, and so on. The pattern repeats, with each second plane of atoms identical. The unit cell of this lattice turns out to be primitive hexagonal, and the crystallographic point group is D_{3h} $(6/mmm)$. The fcc lattice is constructed in Fig. 13.8 in such a way as to make the face-centered cubic Bravais lattice more apparent. However, if we examine the (111) surface of this lattice, we find that the lattice can also be constructed in the same way as the hcp lattice in Fig. 13.8, except the third plane is positioned over an equivalent set of gaps in the second plane, a set that does *not* lie exactly above the atoms in the first plane. The pattern then repeats, with each third plane of atoms being identically positioned. For this reason, fcc is often described as a close-packed arrangement with ABCABC stacking, while hcp (where every other layer is positioned identically) has ABABAB stacking.

EXAMPLE 13.2 Packing Efficiency of the fcc Lattice

CONTEXT The greatest packing efficiency minimizes the empty space in a system and therefore also maximizes the density. This is relevant not only to the density of metals and other crystalline solids, but to numerous other problems in structure from cellular biology to engineering. Taking a cue from the close-packed lattices of metals at the atomic scale, engineers have designed high-strength, lightweight foams using hollow metal spheres in bcc and fcc arrangements. The fcc foam, which is slightly heavier than the bcc arrangement, has superior strength. The density of the fcc foam is determined by the packing efficiency of the lattice, which is calculated in the same way that we calculate it for the atomic packing case.

PROBLEM Prove that the packing efficiency of the close-packed fcc lattice is 74%.

SOLUTION Take the unit cell as a representative sample of the whole crystal. Let each atom have a radius r, and therefore a volume $4\pi r^3/3$. What volume of the fcc unit cell is actually occupied by the atoms? We can figure that out if we know how many atoms (net) actually occupy the unit cell. There are atoms at each of the eight corners, with only 1/8 of the atom actually lying within our unit cell, and in the middle of each of the six faces, each with half their volume within our unit cell. The total is

$$(8 \text{ corners})\left(\frac{1}{8} \text{ per corner}\right) + (6 \text{ faces})\left(\frac{1}{2} \text{ per face}\right) = 4 \text{ atoms net.}$$

The packing efficiency is the ratio of the occupied volume (equal to the volume of four whole atoms) divided by the total volume. To get the total volume, we cube the length of one side of the unit cell. From Fig. 13.8 we see that the diagonal of one face is equal to $4r$, so the length of one side is $4r/\sqrt{2}$. The packing efficiency is therefore

$$\frac{\text{occupied volume}}{\text{total volume}} = \frac{4(4\pi r^3/3)}{(4r/\sqrt{2})^3} = \frac{\sqrt{2}\pi}{6} = 0.7405.$$

This result is among the many contributions to scientific and pure mathematics by Carl Friedrich Gauss.

EXAMPLE 13.3 The fcc Crystallographic Point Groups

CONTEXT X-ray diffraction is the most common method for determining molecular structures within a crystal, but other methods are capable of faster, less detailed information about the crystal. For example, *electron backscatter diffraction* (also called *backscatter Kikuchi diffraction*), from a scanning electron microscope, measures the diffraction patterns of electrons that scatter off more than one plane in the crystal. From the patterns, the crystallographic point group, the orientation of the crystal, and the exposed Miller indices of the surface can be determined. Copper crystals, which have the advantage of simple structure, have been used to test the strengths and limitations of this method.

To determine the crystallographic point group, we isolate the smallest part of the crystal that shares all the symmetry properties of the crystal. For all lattice systems except the hexagonal, this part is the unit cell. Once we've identified and isolated this subunit of the crystal, we analyze it according to the flow chart in Fig. 6.7.

PROBLEM By examining the symmetry elements of the lattice, determine the crystallographic point group for the fcc lattice of a pure metal, such as copper, having a basis of one atom.

SOLUTION As one of the cubic lattice systems, the fcc lattice must belong to one of the cubic crystallographic point groups listed in Table 13.1: T, T_h, O, or O_h. We may first examine the face-centered cubic unit cell for any of the symmetry elements of these point groups; if they are symmetry elements of the unit cell, they will also be symmetry elements of the crystal. The face-centered cubic unit cell with identical spherical atoms at each lattice point has all the symmetry elements of the perfect cube: $\hat{E}$, $\hat{I}$, six $\hat{C}_2$ axes, four $\hat{C}_3$ axes, three $\hat{C}_4$ axes, and nine mirror planes. These are sufficient to identify the point group of the unit cell, and the lattice, as O_h.

The hcp and fcc lattices each have a coordination number of 12, the maximum that was found in Section 11.1 and the highest density possible for identical spheres. Not all metals can take advantage of these arrangements, however, because a high density of metal cations needs to be stabilized by a high density of valence electrons. Metals such as the alkali metals, having a low number of valence electrons, often tend instead to adopt the **body-centered cubic** (bcc) lattice shown in Fig. 13.8c. In this pattern, the first plane of atoms consists of identical rows of atoms with a spacing of $2/\sqrt{3}$ diameters between the centers of the nearest neighbor atoms. The second plane is identical to the first but placed with the centers of its atoms over the gaps in the first plane. The third plane is identical to the first, and each subsequent plane is identical to the plane two levels below it. The coordination number is 8, and the packing efficiency is 68%.

Ultimately, many factors determine whether a particular metal adopts the fcc, hcp, or bcc lattice, or some other configuration. One is our conclusion that the bcc lattice is better suited to metals with few valence electrons. A more subtle argument allows us to distinguish between the hcp and fcc lattices. The crystallographic point group for the fcc lattice is O_h, significantly higher in symmetry than the D_{3h} point group of the hcp lattice. The greater symmetry allows electronic energy levels in the fcc lattice to have higher degeneracy than in the hcp. This means that metals with a high density of electrons are more likely to adopt

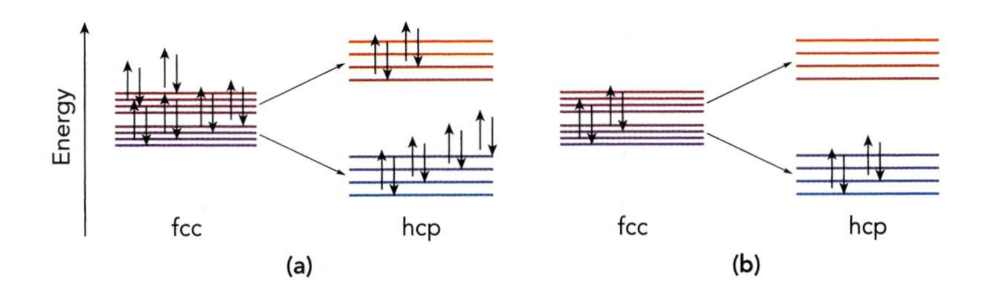

◀ **FIGURE 13.9 Schematic energy level diagrams of electronic states in the fcc and hcp lattices.** The diagrams illustrate how the higher degeneracy of the fcc lattice better accommodates electron-rich metals **(a)**, whereas electron-poor metals will tend to prefer the lower symmetry hcp lattice **(b)**.

the fcc than the hcp lattice, because highly degenerate levels can support more electrons without having to occupy high-energy states. This tendency is shown schematically in Fig. 13.9.

As a general rule, therefore, we expect that metals in the lower group numbers of the periodic table should tend to adopt either bcc or hcp lattices, and the fcc lattice will be more stable for metals with high group number. The lattice structures listed in Table 13.5 follow this qualitative rule fairly well, but there are several exceptions, including lattice structures we haven't even discussed. The exceptions arise from less obvious contributions of the electronic wavefunctions to the stability of the solid, as well as other effects. For example, as we heat or cool any particular metal, we often observe that the most stable crystal structure *changes*. Iron, heated above 912 K, changes from its bcc structure (called α-Fe) to fcc (γ-Fe). Tin, when cooled below 286 K, is more stable in a lattice with the structure of diamond, rather than the tetragonal crystal listed in Table 13.5. These different crystal structures of the same material are called **allotropes.**

One rule applies fairly well to both pure and alloyed metal crystals: the bond energy tends to increase with the number of unpaired electrons n_u available for bonding. This accounts for the peak in melting points of the transition metals in Group 6 of the periodic table (Fig. 13.10), because these atoms may have as many as six unpaired electrons.

TABLE 13.5 Most stable crystal structures of selected pure metals at 298 K and 1 bar. The entries are organized according to group number in the periodic table. In addition to the fcc, hcp, and bcc lattices defined in the text, the following appear: cI58 is a body-centered cubic lattice, different from the standard bcc; oC8 is C-centered orthorhombic; diam is the diamond structure; tI2 and tI4 are body-centered tetragonal lattices; and hP4 is a hexagonal lattice.

1	2	3	4	5	6	7	8	9	10	11	12	13	14
Li bcc	Be hcp												
Na bcc	Mg hcp											Al fcc	
K bcc	Ca fcc	Sc hcp	Ti hcp	V bcc	Cr bcc	Mn cI58	Fe bcc	Co Hcp	Ni fcc	Cu fcc	Zn hcp	Ga oC8	Ge diam
Rb bcc	Sr fcc	Y hcp	Zr hcp	Nb bcc	Mo bcc	Tc hcp	Ru hcp	Rh Fcc	Pd fcc	Ag fcc	Cd hcp	In tI2	Sn tI4
Cs bcc	Ba bcc	La hP4	Hf hcp	Ta bcc	W bcc	Re hcp	Os hcp	Ir Fcc	Pt fcc	Au fcc	Hg (liq)	Tl hcp	Pb fcc

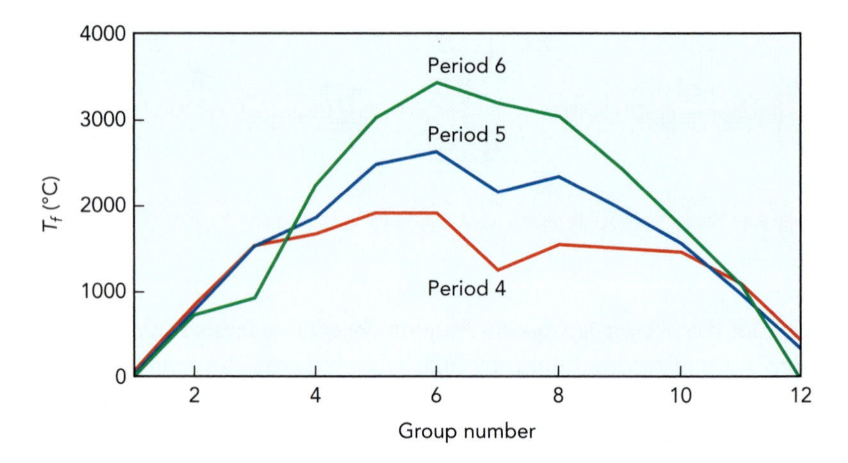

▶ FIGURE 13.10 **Melting points (°C) of the metals as a function of periodic table group number.**

Figure 13.10 partly explains the historic uses of several metals. For example, some early human civilizations improved upon tools made of rocks and wood (the Stone Age) by forging tools and utensils from bronze and pewter, durable alloys of copper (group 11) and tin (group 14), thus entering the Bronze Age. These were among the easiest metals to work with because their melting points are relatively low (1085 °C for copper, 232 °C for tin). The age of iron (group 8) and steel tools came later, only after the makers could achieve temperatures above iron's melting point of 1538 °C. Similarly, uses of lead for plumbing and tool making date from antiquity, at least in part because its low melting point of 328 °C allowed it to be easily separated from other minerals.

Covalent Crystals

Covalent crystals encompass too diverse a spectrum of molecules and bond structures to allow generalization of their crystal structures. The covalent bonds tend to be much more highly directional than the metal or ionic bonds,

BIOSKETCH Susan Kauzlarich

Susan Kauzlarich is a professor of chemistry at the University of California at Davis. Among other projects, Professor Kauzlarich has been studying crystals in the *Zintl phase,* an arrangement of compounds that combine electropositive metals (alkali metal, alkaline earth, or rare earth) with *relatively* electronegative *p*-electron metals or metalloids (groups 13–16). In the Zintl phase, the bonding becomes ionic (rather than metallic) and the crystal has magnetic properties distinct from other crystal forms of the substance. The Kauzlarich group has been particularly interested in the *magnetoresistance* of these crystals, the property that an external magnetic field changes the crystal's electrical resistance. From the crystal structures, Kauzlarich is able to identify how magnetic interactions between layers of the ions contribute to the magnetoresistance. These novel materials may play a role in the development of new molecular-scale electromagnetic devices.

so packing efficiency is not the dominant concern in determining the overall lattice structure. As with covalent bonds in general, the covalent crystals are typically formed from the non-metals of the periodic table, particularly the first- and second-row elements.

These crystals may be exceptionally *hard*, meaning impervious to damage. This is not the same as having a great bond strength, although that helps. Ionic salts such as NaCl also tend to have great bond strengths, but they can easily be damaged by pressure along one of the crystal planes. The bonding is non-directional, so the attractive forces can easily be redistributed. To cut through diamond, on the other hand, one must break a large number of highly directional covalent bonds, and these are not then stabilized by the remaining bonds.

Molecular Crystals

Molecular crystals are also too diverse in their chemical and electronic properties to allow generalization of their crystal structures. They tend to be held together by the same intermolecular bonding mechanism as the liquid state of the compound but now with all the translational energy removed, so there can be long-range order to the system. The crystalline structure is therefore usually very similar to the short-range structure in the liquid. Although these intermolecular forces are much weaker than the binding forces of the covalent crystals, they are still strong enough to supersede packing efficiency as the determining factor in the lattice structure.

Molecular crystals include the crystals of many very large, covalently bound molecules, such as proteins and nucleic acids. The complexity of the molecule causes complexity in the crystal. Large molecules tend to have many possible conformational isomers and several very low vibrational constants. Similarly, their crystals tend to have a greater tendency to disorder, particularly when the crystal structure has a great deal of open space between adjacent molecules. These crystals are therefore likely to be measurably disordered. One unit cell is not exactly identical to the next if one of the molecules has been frozen into the structure with a slightly different conformation than the others, or if one molecule is in a distinct vibrational quantum state. Crystals formed during the evaporation of a solvent, a common technique in synthetic chemistry, are prone to an additional source of disorder: small solvent molecules may become randomly trapped in the spaces between the large molecules. Disorder of these types does not alter the principal characteristic of crystals—the predictable arrangement of the molecules over large distance scales—but it may make the determination of crystal structures quite difficult.

A different class of crystals bound by the van der Waals forces is the group of frozen noble gases. Noble gas atoms are closed-shell and have no net bond order, and since they have no net electric or magnetic moments, the only bonding mechanism available is the dispersion force. This is non-directional and very short-range (dropping off as R^{-6}). Like metal crystals, which are also composed of single atoms bound by non-directional forces, the packing efficiency determines the lattice structure. The most stable lattice structures for the noble gas ices are close packed structures.

13.4 Wavefunctions and Energies of Solids

So far in this chapter we have considered only the structure of solids, but this structure dictates to a large extent the behavior of the substance as well. In using the Born-Oppenheimer approximation (Section 5.1) to help solve the quantum mechanics of a molecule, we place the nuclei in the molecule where we want them, and the Schrödinger equation then determines how the electrons are distributed and what their energy levels are. Similarly, now that we have an idea how to arrange the atomic nuclei in crystals, we can analyze (in a qualitative fashion, at least) the electron distribution and the energy levels of the solid.

Vibrational Conductivity

In a solid, molecules cannot rotate, but they do vibrate. The vibrational motions on the microscopic scale range from the low frequency, floppy motions between weakly bonded components of a molecular crystal to the high-frequency stretches of covalent bonds. The covalent bonds can be found either within individual molecules of a weakly bonded crystal (such as the carbon–carbon bonds in sugar) or in the bonds that make up the latticework of a covalent crystal (as in the carbon–carbon bonds in diamond).

In addition to these microscopic motions, the joining of the molecules over long distances leads to vibrational motions known as **lattice modes** or **phonon modes** because they carry sound. These modes involve the concerted motion of many adjacent lattice points, as illustrated in Fig. 13.11. The simultaneous motion of so many atoms leads to a large reduced mass and therefore very low vibrational frequencies. The audio frequency range, roughly 20—20,000 s^{-1}, becomes accessible if the solid is large enough. Furthermore, because the reduced mass varies slowly as we add one lattice point or drop one away, the available vibrational frequencies may be nearly continuous. In principal, therefore, a solid can efficiently transmit this low-frequency vibrational energy from one place to another.

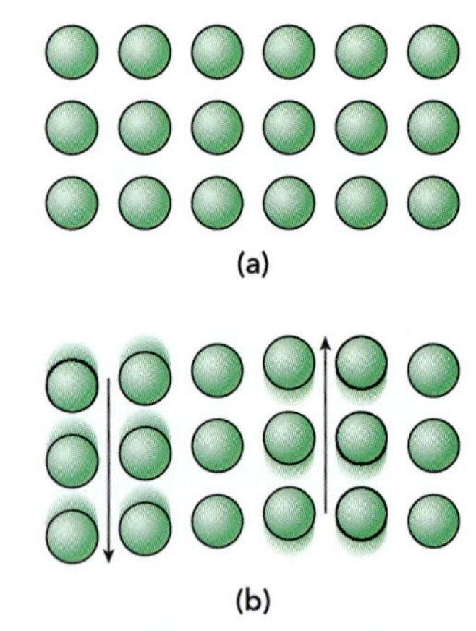

▲ FIGURE 13.11 **Schematic illustration of a lattice mode.** Such modes involve the coupled motion of several lattice points over distances much larger than the separation between the lattice points.

However, that energy could instead dissipate itself into other vibrational motions that send the energy in other directions. Or, if the solid is a molecular crystal or a glass, vibrational energy in a strongly bound subunit could find it difficult to move across the weak bonds that hold the solid together, because degrees of freedom with very different energy gaps couple poorly (Section 10.2). Consequently, those solids with relatively simple structures and strong bonding can rapidly channel that energy throughout the solid, making them the best conductors of vibrational energy. The audio frequency vibrational motions are slow enough that they can be excited and detected by mechanical means, and the ability of the solid to transmit the energy is its **acoustic conductance.** These motions rarely contribute substantially to the overall vibrational energy of the solid. The higher frequency vibrations, on the other hand, require higher energy excitation and can make the solid hot to the touch. If this vibrational energy is effectively transmitted through the solid, then the solid has a high **thermal conductivity.** Metals in particular tend to have the high symmetry and strong bonding that promotes both thermal and acoustic conductivity.

Amorphous solids and molecular crystals, in contrast, tend to be good thermal and acoustic **insulators.** Vibrational energy couples poorly from one region of the solid to another if the bonds are weak or if their organization is haphazard.

Table 13.6 lists thermal conductivities of some representative materials, covering five orders of magnitude. Close packed solids, such as the metals, and covalent crystals, such as diamond and graphite (which has covalent bonds parallel to the aromatic plane), tend to have very high thermal conductivity because the molecules cannot vibrate far without bumping into one another. At the opposite end of the range lie the gases, which transport heat only through the relatively inefficient mechanism of independent molecules occasionally bumping into one another. In between these extremes lie the molecular crystals, amorphous solids, and plastics (e.g., water ice, amorphous SiO_2, nylon). These tend to have relatively low thermal conductivity because the molecular subunits are held together by weak forces, which do not easily couple the vibrational energy in one region to another region.

Electronic Energies and Electrical Conduction

A continuous distribution of energy levels, such as the translational energies of the free particle (Section 2.2), is called a **continuum.** A macroscopic solid has enough atoms that the $3N_{\text{atom}} - 6$ vibrational modes lead to a continuum of vibrational energies. The same solid also has so many electrons and so complex a distribution of nuclei that a continuum of molecular orbital energies may also result. The core electrons remain at roughly the same energies they would have in the free atoms, but the valence electron energy levels may be very different.

Materials are transparent when they do not absorb or reflect incident radiation. In order to absorb radiation, the substance must have some accessible energy level available at $E'' + h\nu$, where E'' is the energy of the initial state and $h\nu$ is the energy of the photon. Reflection is the result of Rayleigh scattering—scattering in which the emitted radiation is of the same frequency as the incident radiation—and requires a polarizable medium, such as the diffuse

TABLE 13.6 Thermal conductivities k for selected materials. In these units, the thermal conductivity is the power in mW transmitted through a cross-sectional area of 1 cm^2 when the temperature gradient is 1 K cm^{-1} at 273 K. For graphite, the $\parallel$ and $\perp$ symbols indicate conduction parallel and perpendicular to the graphite covalent planes, respectively. Many of these values are highly sensitive to structure, orientation, and the presence of impurities.

	k (mW cm^{-1} K^{-1})		k (mW cm^{-1} K^{-1})		k (mW cm^{-1} K^{-1})
H_2O (g)	0.158	CO_2 (g)	1.50	NaCl (s)	64
N_2 (g)	0.240	oak	1.6	C (graphite, $\perp$)	64
cork	0.3	nylon	3.0	Al (s)	243
asbestos	0.9	H_2O (l)	5.61	Fe (s)	830
CCl_4 (l)	1.04	SiO_2(amorphous)	13	Cu (s)	4010
CH_2Cl_2 (g)	1.22	C (amorphous)	15	Ag (s)	4290
toluene (l)	1.39	H_2O (s)	22	C (graphite, $\parallel$)	$2.13 \cdot 10^3$
He (g)	1.41	$CaCO_3$ (s)	50	C (diamond)	$2.30 \cdot 10^3$

electron sea of a metal.[5] Electronic excitation energies correspond typically to visible or ultraviolet radiation, and therefore we reason that the optical properties of a solid depend principally on its electronic structure.

Crystals again, having high symmetry, are more likely to transmit radiation over large regions of the spectrum. The metals are an exception. Metals exhibit two competing effects: (*i*) because they have many electrons in partly filled shells, their lowest excited states occur very low in energy and in great quantity; and (*ii*) the polarizability of the electrons is so high that reflection is very probable. In some metals, such as copper and gold, electronic absorption becomes more probable at the high frequency end of the visible spectrum. This causes some of the blue and green light hitting those metals to be absorbed, increasing the fraction of reds and yellows in the reflected light. Many glasses and plastics are composed of covalently bonded subunits that require ultraviolet light to access the lowest excited electronic states, leaving those solids transparent in the visible.

The electrical properties of solids are also determined by the electronic energy level structure. To simplify the complicated and dense distribution of electronic energy levels, we often break the continuum of molecular orbitals into two groups, shown in Fig. 13.12. In the ground electronic state of the solid, the occupied molecular orbitals make up the **valence band** of the solid. The unoccupied orbitals, if they extend over large distances in the solid, comprise the **conduction band.** If the valence band ends just where the conduction band begins, the slightest electronic energy pushes the electrons into orbitals where they have more mobility. The formation of the conduction band from individual atomic orbitals is illustrated in Fig. 13.13, which shows approximate molecular orbital energies of a cluster of eight sodium atoms as they simultaneously approach one another.

In conductors, the gap between the valence and conduction bands is zero. In insulators, the gap is large. In **semiconductors,** the gap is small enough that we can drive electrons into the conduction band only after providing a push. Usually that push is an external electric field. In this case, the electric field adds a slope to the potential energy function of the electrons, effectively "tipping" them in one direction (Fig. 13.14).

In photo-conductors, the gap between the valence and conduction bands is small enough that electrons can be excited into the conduction band by absorption of a visible photon. Some polymers, such as polyacetylene, and covalent crystals, such as graphite, may also have accessible conduction bands if they have partly filled π-bond systems. Graphite has a much lower electrical and thermal conductance along the axis perpendicular to the plane of the sheets; in this direction the atoms are connected only by the weak van der Waals forces.

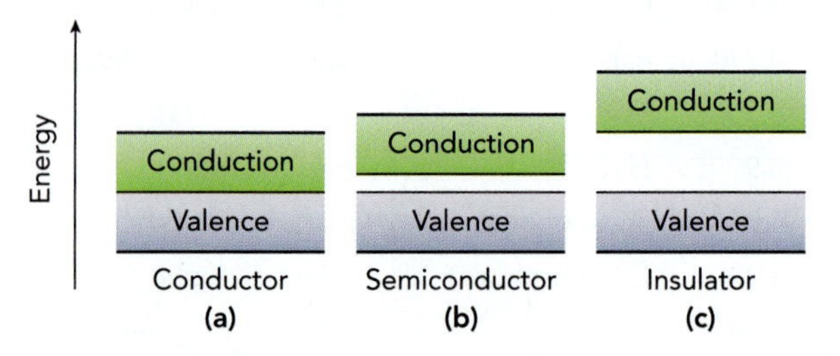

▶ FIGURE 13.12 **Examples of continuum electronic state structure in solids.**

[5]Reflection is more complicated in detail, depending on the change in the index of refraction—a function partly of the polarizability—when the radiation passes from one medium into another.

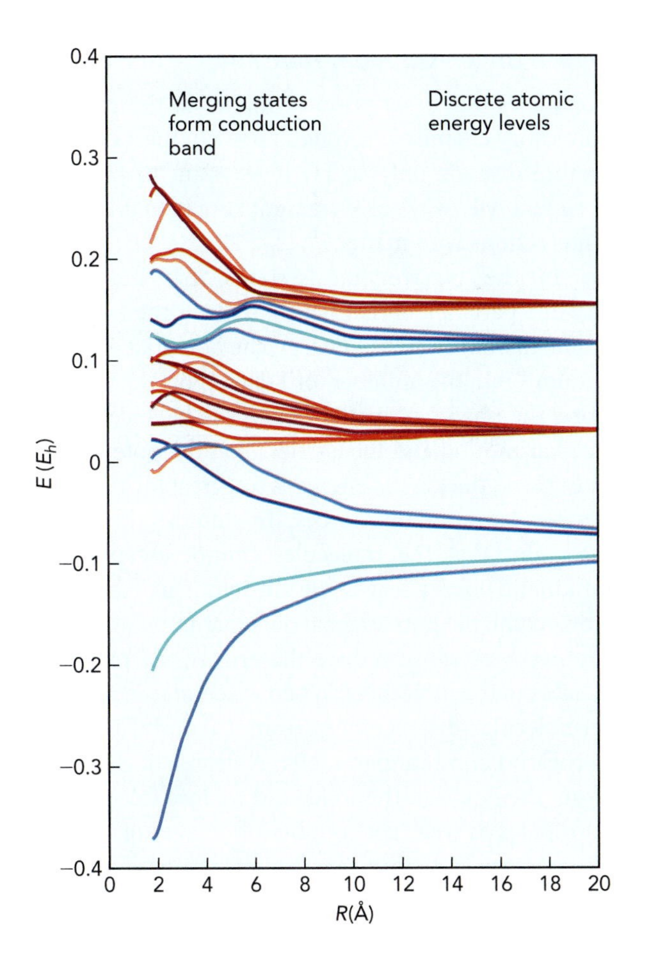

◀ **FIGURE 13.13 Approximate MO energies of a cubic Na_8 cluster as a function of nearest neighbor distance.**

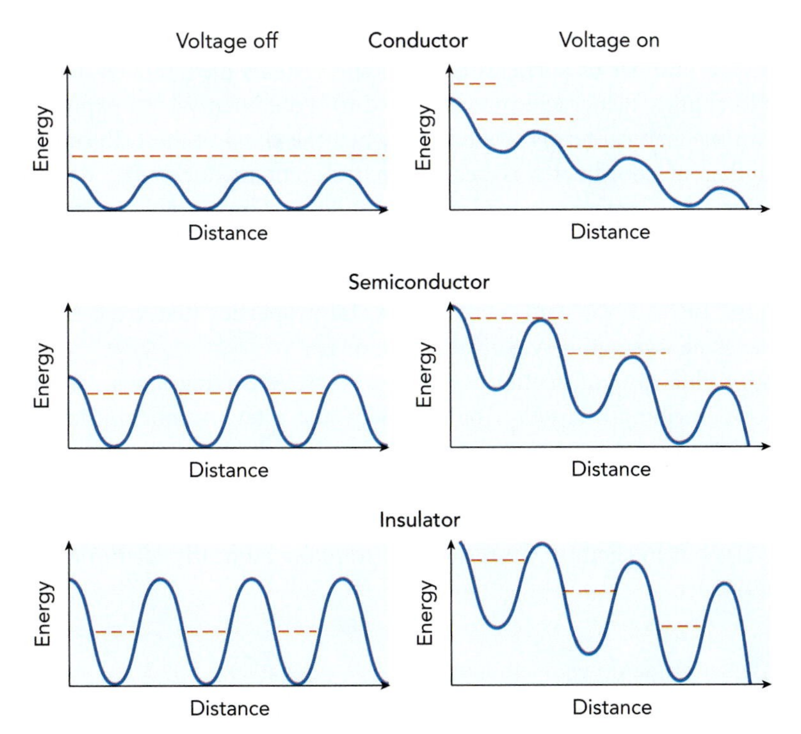

▶ **FIGURE 13.14 The effect on the electronic potential energy function of an applied voltage across a conductor, a semiconductor, and an insulator.** Given sufficiently high voltages, even insulators will allow electricity to flow, but often with damage to the material.

CONTEXT *Where Do We Go From Here?*

This chapter completes our description of matter at the microscopic level. Of course, most laboratory chemistry involves huge numbers of molecules, and our microscopic picture extends only so far if we want to predict how a typical laboratory experiment will work, or if we want simply to understand more about familiar, daily interactions with matter.

Nevertheless, in the macroscopic systems that make up our routine experience, the same parameters that have governed the behavior of individual molecules continue to hold sway: the energy, the potential energy, the number of degrees of freedom, and the number of states. For example, the key to the distinction among the phases of matter lies in the balance between the kinetic energy for molecular motion and the barriers that the potential energy function puts in the way. In gases, the kinetic energy is too great for molecules to be bound by their intermolecular attractive forces. In liquids, the kinetic energy has dropped to the point that the molecules cannot escape one another, and removing more kinetic energy leaves the substance in the solid state with too little energy to overcome the potential barriers that imprison the molecules.

To put it another way, as we reduce the energy, we reduce the number of coordinates available to the molecules. When we condense matter from a gas to a liquid, the intermolecular distance goes from a (nearly) limitless translational coordinate to a (nearly) constant parameter. A degree of freedom has been lost. Then we cool from the liquid to the solid, and we lose freedom of motion along the angular coordinates as well. The number of degrees of freedom N_{dof} for the entire system decreases with energy, as we expect.

So what really changes when we go from one or two molecules to billions? For a macroscopic system, even at the lowest attainable energies, we never return to the simplicity of the individual molecules analyzed in Chapters 1–9 of this text. The number of coordinates stays huge when we get to bulk matter, simply because the number of particles is huge. The systems have moved beyond our ability to count degeneracies precisely and to write energy level expressions in quantum-mechanical terms. We have reached the classical limit. Even counting the number of coordinates is becoming irrelevant in one sense, because the properties of the larger scale material do not depend measurably on the number of molecules. So, as our interest migrates to the large-scale properties of the system, it would be pointless to try to gather information about the individual molecules. Instead, we shall focus on general properties that are the same for each molecule, averaging over any fluctuations.

That's a clue to the direction we need to take. Averaging is a statistical tool, used for describing a large set of data with just a few numbers. Macroscopic chemistry is the perfect application for statistics. To the best of our measurement abilities, it appears that we cannot distinguish between two ^{1}H atoms, for example, on the basis of any fixed properties, such as rest mass or charge.[6] Where else but in chemistry can one find so *many*, so nearly *identical* members of a set?

Statistics will help us reduce the number of coordinates we follow. However, we must not forget the lessons of the molecular perspective. The potential energy function between two molecules is just as important when there are 10^{23} molecules as when there are only two, because *every pair of molecules will experience this same potential energy function.* The goal of *Physical Chemistry: Thermodynamics, Kinetics, and Statistical Mechanics* is to develop these methods and apply them to obtain the common laws of chemistry that describe how the macroscopic properties of matter are influenced by its microscopic structure.

KEY CONCEPTS AND EQUATIONS

13.1 Amorphous Solids, Polymers, and Crystals. Solids are grouped first according to the degree and type of structural order they exhibit. In amorphous solids the particles are distributed almost randomly, whereas in crystals the distribution forms a regular pattern.

13.2 Symmetry in Crystals. The structures of crystals are classified using several different but related considerations: the shape of the unit cell, the distribution of identical particles within the unit cell, the point group symmetry of the unit cell, and the space group symmetry of the entire lattice.

13.3 Bonding Mechanisms and Properties of Crystals. Crystals are also grouped according to the forces that hold them together into ionic, metal, covalent, and molecular crystals.

13.4 Wavefunctions and Energies of Solids. As we combine particles to form the solid, the vibrational motions become increasingly connected. In metals, the electronic states also become especially strongly linked, forming a conduction band that effectively allows electrons to move over large distances within the substance.

KEY TERMS

- **Amorphous solids** have no long-range order.
- **Crystals** are solids in which the atoms are distributed in regular patterns over large distances.
- A **unit cell** is the smallest subunit of a crystal such that copies laid side by side in all three dimensions will regenerate the complete crystal structure.
- **Translational symmetry elements** are symmetry operators that shift the entire function a fixed amount along a specified axis.
- **Space groups** are groups that combine certain point group operations, translational symmetry elements, and their products.
- **Crystallographic point groups** are the set of point groups that can give the symmetry elements of ideal crystals. Because crystals are limited to only a few rotational symmetries, this is a subset of only 32 point groups.
- A unit cell has three distinct sides, and the lengths of those sides are the **lattice constants.**
- The **face-centered cubic** (fcc) or **cubic close-packed** (ccp) lattice is a crystal lattice geometry based on a face-centered cubic unit cell, and it is capable of the highest possible packing efficiency for perfect spheres.
- The **hexagonal close-packed** (hcp) lattice is a crystal lattice geometry based on a hexagonal unit cell, and it is capable of the same packing efficiency as the fcc lattice.
- The **body-centered cubic** (bcc) lattice is based on the body-centered cubic unit cell, and it gives a packing efficiency slightly lower than the fcc or hcp lattices.

[6]There is a theoretical basis for the assumption that all electrons have the same values for these properties, as do all protons and all neutrons. Perhaps the best current evidence in support of the indistinguishability of like particles is the 12-digit agreement between experimental and theoretical values of the electron spin gyromagnetic ratio, g_s.

OBJECTIVES REVIEW

1. *Identify the lattice system, Bravais lattice, and crystallographic point group of a unit cell.*
Find the crystallographic point group of a primitive tetragonal lattice, assuming the basis can be treated as one atom per lattice point.

2. *Qualitatively predict properties such as piezoelectricity from the crystal symmetry.*
Will any monatomic crystal, such as pure iron, exhibit piezoelectricity?

PROBLEMS

Discussion Problems

13.1 Explain why the ionic radii in Table 13.4 are larger for larger coordination numbers.

13.2 Some crystals polarize radiation: electromagnetic waves transmitted through the crystal have electric field vectors that all lie in the same plane. What property of these crystals permits this?

Bravais Lattices and Lattice Systems

13.3 Verify that the Hermann-Mauguin symbol $\overline{3}$, indicating a $\hat{C}_3$ rotation followed by inversion, is equivalent to the operator $\hat{S}_6$. THINKING AHEAD ▶ [How would you write these operations mathematically?]

13.4 If $\hat{X}$, $\hat{Y}$, and $\hat{Z}$ are the operators that translate a crystal lattice one unit along the x, y, or z axis respectively, what combination of operations could result in the transformation illustrated in the following figure? (For simplicity, let any reflection plane, rotation axis, or inversion center contain the center of the unit cell, rather than the origin of the coordinate system.)

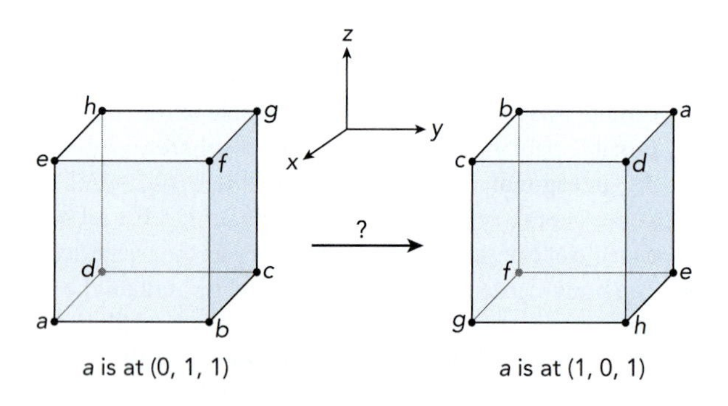

13.5 What point group operations are symmetry elements of the primitive monoclinic Bravais lattice?

13.6 Why is there no C-centered cubic unit cell?

13.7 How many molecules are there per unit cell of the C-centered monoclinic Bravais lattice if there is only one molecule in the basis? THINKING AHEAD ▶ [The corners are not all equivalent, but what happens when this unit cell is placed next to its neighbors in the crystal?]

13.8 Identify (a) the lattice system, (b) the Bravais lattice, and (c) the crystallographic point group for the CO_2 crystal with the unit cell drawn in the following figure. The dark atoms are the carbons, and the arrangement of the carbon atoms alone is given in the figure. For simplicity, molecules on the rear walls of the unit cell are not drawn.

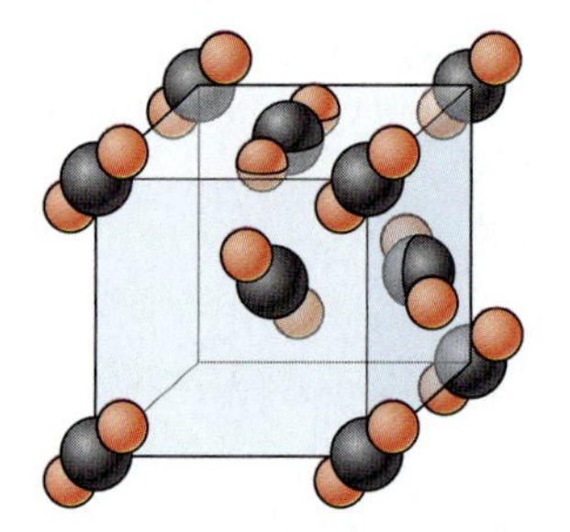

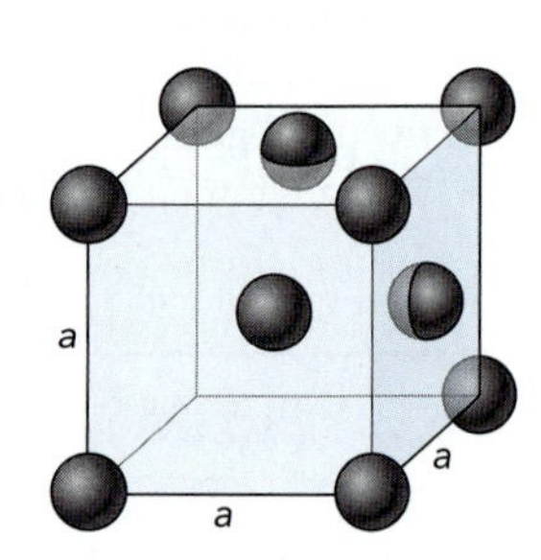

13.9 The density of diamond is 3.513 g cm^{-3}. Given that the molecular structure of the crystal consists of carbon atoms all tetrahedrally bonded to each other, find the CC bond length in diamond. To simplify the geometry, use the fact that a tetrahedron can be inscribed in a cube, but be careful: this is not the unit cell of diamond.

13.10 What is the maximum percentage of the occupied space in the primitive cubic lattice of a monatomic solid?

13.11 Drawn in the following figure is a proposed lattice for one of the Alnico alloys, magnetic alloys composed of nickel (A) cobalt (B), and aluminum (C).

a. Write the molecular formula for this alloy.

b. Calculate the density of the alloy.

c. On the following figure, draw the edges for one of the unit cells. The unit cell should be as small as possible while still conveying the proper symmetry and stoichiometry of the crystal. Even so, there is more than one possible solution.

d. Find the Bravais lattice and crystallographic point group for the structure.

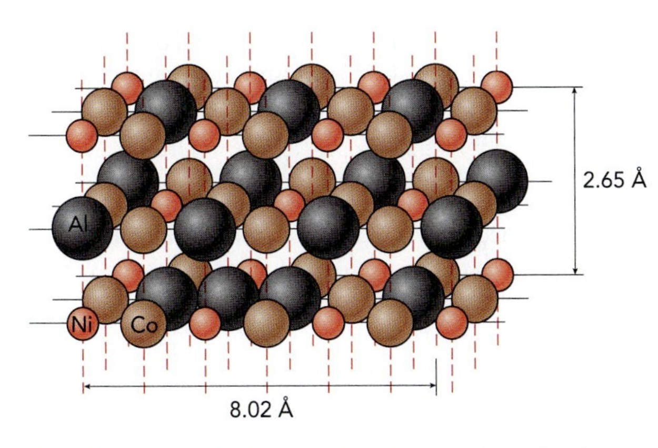

13.12 The unit cell drawn in the following figure has been determined for a high-temperature superconductor of the form $YBa_xCu_yO_z$.

a. Find the values of x, y, and z in the molecular formula.

b. Find the Bravais lattice if $\alpha = \beta = \gamma = 90^\circ$ (assume that sides with equal lengths must be exactly equivalent).

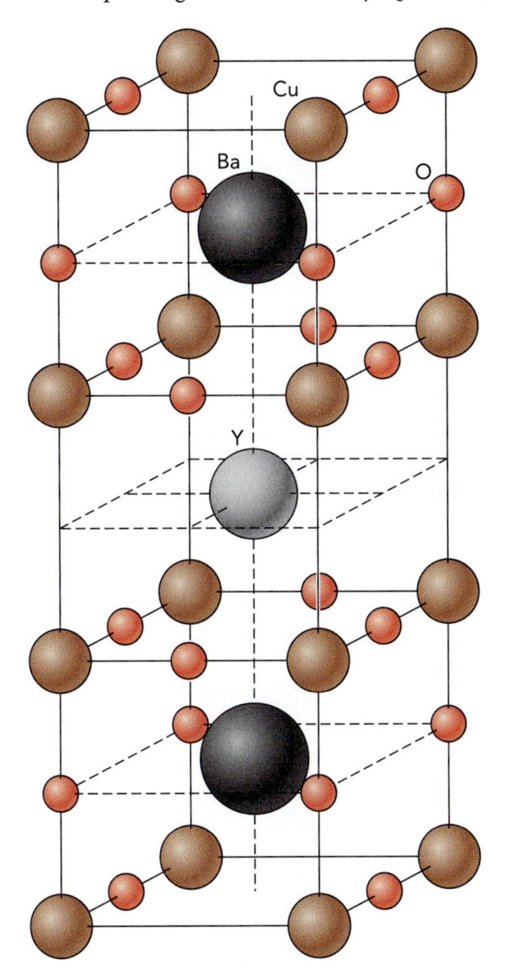

13.13 Consider the fourteen Bravais unit cells in Fig. 13.5, with $a = 1$ Å, $b = 2$ Å, and $c = 3$ Å. If the coordination number C is the number of atoms within 1 Å of any given reference atom in the lattice, find the largest value of C and the corresponding Bravais lattice or lattices.

Surface Characteristics

13.14 Of the (001), (110), and (111) faces in a body-centered tetragonal Bravais lattice with $a = 2.0$ Å and $c = 3.0$ Å, which has the highest surface density of molecules?

13.15 Identify by the Miller indices the surface plane parallel to the *ab* plane of a face-centered orthorhombic lattice, and bisecting the *b* edges.

13.16 In the primitive cubic lattice of a monatomic crystal, identify the Miller indices of the surfaces that are structurally identical to (a) the (1,0,0) face, and (b) the (1,1,0) face.

13.17 Give the number of atoms per unit area (in atoms Å^{-2}) on the *surface* of (a) the 100 face and (b) the 110 face of a primitive cube lattice with lattice constant $a = 3.0$ Å.

General Structure and Energetics

13.18 The density of an alkali chloride fcc lattice is 2.167 g cm^{-3}. The lattice constant is $a = 5.623$ Å. Identify the molecule. THINKING AHEAD ▶ [What property of the unit cell will tell us which alkali metal is present?]

13.19 Diodes are semiconductor devices commonly used to restrict electrical current to flow in a single direction. The typical diode is composed of Si or Ge (atoms having four valence electrons) divided into two parts:

a. an n-terminal doped with small quantities of As or Sb, which (having five valence electrons) increase the density of electrons

b. a p-terminal doped with small quantities of Ga or In, which (having three valence electrons) decrease the density of electrons

Electrons flow easily from the n terminal to the p terminal, provided that a voltage of about 0.6 V is applied across the terminals. Explain why this voltage is necessary and why the diode behaves as an insulator if the voltage is reversed.

13.20 The CsCl crystal has a primitive cubic Bravais lattice. Use the tabulated ionic radii to estimate the density (in g cm^{-3}) of crystalline CsCl.

13.21 Alumina, Al_2O_3, has a rhombohedral unit cell and a density of 3.965 g cm^{-3}. Calculate the volume of a single unit cell and the number of unit cells in one mole of alumina.

13.22 Assume a crystal with four molecules per unit cell is prepared with a molar purity of 99.999%. If the lattice constants are all 5.68 Å, what is the number of impurity molecules in one cm^3 sample?

13.23 Show that for the bcc lattice, 68% of the available space is filled.

13.24 Given the atomic radii 0.95 Å for Na^+ and 1.81 Å for Cl^-, estimate the density in g cm^{-3} for the densest structures of $Na^{35}Cl$ using the primitive cubic and the face-centered cubic Bravais lattices.

13.25 The **radius ratio rule** predicts that diatomic salts will tend to form primitive cubic lattices when $r_{small}/r_{large} > 0.73$ and face-centered cubic lattices when $0.73 \geq r_{small}/r_{large} > 0.41$, where r_{small} and r_{large} are the ionic radii of the smallest and largest of the two ions, respectively. Assume that these ions pack as closely as possible in the geometries shown in the following figure, and explain the significance of the values 0.73 and 0.41.

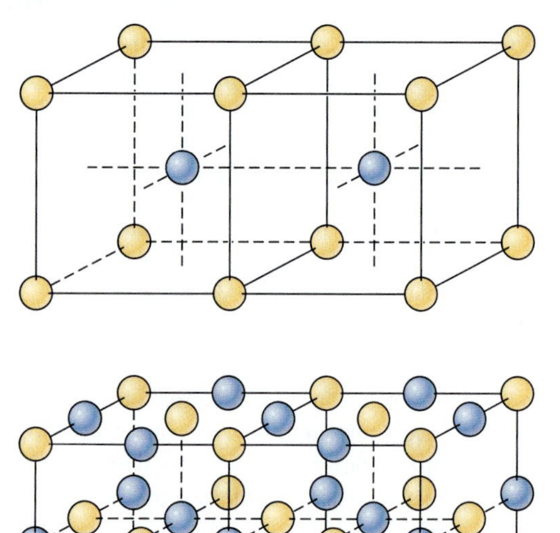

13.26 Suppose that we have a monolayer of noble gas atoms adsorbed to the surface of a solid, and we want to know of any efficient packing schemes. In Fig. 13.3 *iv* and *v*, two of the possible solutions are shown: the hexagonal lattice with a packing efficiency of 90.7%, and the square lattice, at 78.5%. We can think of these lattices as being connected by a sliding motion, such that the angle η is 60° in the hexagonal lattice and 90° in the square lattice, and where the circles along the sliding axis are always connected to each other.

a. What is the name of the lattice structure for values of η between 60° and 90°?

b. Find an equation for the packing efficiency of the lattice as a function of η, assuming that the rows of atoms are pressed as closely together as possible while they slide relative to one another. Is 78.5% a local minimum, a maximum, or not a stationary point on this curve?

13.27 Benzene crystallizes at 1 bar and 278.5 K to an orthorhombic lattice with lattice constants $a = 7.49$ Å, $b = 9.71$ Å, and $c = 7.07$ Å. The density of the unit cell is 0.606 amu/Å^3. If we put one molecule at each lattice point, where in the unit cell will we find the molecules?

APPENDIX

Character Tables for Common Point Groups

C_1	$\hat{E}$	Functions
$A(a)$	1	all

C_s	$\hat{E}$	$\hat{\sigma}$	Functions
$A'(a')$	1	1	$x, y; x^2, y^2, z^2, xy; R_z$
$A''(a'')$	1	−1	$z; yz, xz; R_x, R_y$

C_i	$\hat{E}$	$\hat{I}$	Functions
$A_g(a_g)$	1	1	$x^2, y^2, z^2, xy, xz, yz; R_x, R_y, R_z$
$A_u(a_u)$	1	−1	x, y, z

C_2	$\hat{E}$	$\hat{C}_2$	Functions
$A(a)$	1	1	$z; x^2, y^2, z^2, xy; R_z$
$B(b)$	1	−1	$x, y; yz, xz; R_x, R_y$

C_3	$\hat{E}$	$\hat{C}_3$	$\hat{C}_3^2$	Functions
$A(a)$	1	1	1	$z; x^2 + y^2, z^2; R_z$
$E(e)$	$\left\{\begin{matrix} 1 \\ 1 \end{matrix}\right.$	$\begin{matrix} \exp(2\pi i/3) \\ \exp(-2\pi i/3) \end{matrix}$	$\left.\begin{matrix} \exp(-2\pi i/3) \\ \exp(2\pi i/3) \end{matrix}\right\}$	$(x, y); (x^2 - y^2, xy), (yz, xz); (R_x, R_y)$

C_{2h}	$\hat{E}$	$\hat{C}_2$	$\hat{I}$	$\hat{\sigma}_h$	Functions
$A_g(a_g)$	1	1	1	1	$x^2, y^2, z^2, xy; R_z$
$B_g(b_g)$	1	−1	1	−1	$xz, yz; R_x, R_y$
$A_u(a_u)$	1	1	−1	−1	z
$B_u(b_u)$	1	−1	−1	1	x, y

C_{2v}	$\hat{E}$	$\hat{C}_2(z)$	$\hat{\sigma}_v(xz)$	$\hat{\sigma}'_v(yz)$	Functions
$A_1(a_1)$	1	1	1	1	$z; x^2, y^2, z^2$
$A_2(a_2)$	1	1	-1	-1	$xy; R_z$
$B_1(b_1)$	1	-1	1	-1	$x; xz; R_y$
$B_2(b_2)$	1	-1	-1	1	$y; yz; R_x$

C_{3v}	$\hat{E}$	$2\hat{C}_3$	$3\hat{\sigma}_v$	Functions
$A_1(a_1)$	1	1	1	$z; x^2 + y^2, z^2$
$A_2(a_2)$	1	1	-1	R_z
$E(e)$	2	-1	0	$(x, y); (x^2 - y^2, xy), (xz, yz); (R_x, R_y)$

C_{4v}	$\hat{E}$	$2\hat{C}_4$	$\hat{C}_2$	$2\hat{\sigma}_v$	$2\hat{\sigma}_d$	Functions
$A_1(a_1)$	1	1	1	1	1	$z; x^2 + y^2, z^2$
$A_2(a_2)$	1	1	1	-1	-1	R_z
$B_1(b_1)$	1	-1	1	1	-1	$x^2 - y^2$
$B_2(b_2)$	1	-1	1	-1	1	xy
$E(e)$	2	0	-2	0	0	$(x, y); (xz, yz); (R_x, R_y)$

D_2	$\hat{E}$	$\hat{C}_2(z)$	$\hat{C}_2(y)$	$\hat{C}_2(x)$	Functions
$A(a)$	1	1	1	1	x^2, y^2, z^2
$B_1(b_1)$	1	1	-1	-1	$z; xy; R_z$
$B_2(b_2)$	1	-1	1	-1	$y; xz; R_y$
$B_3(b_3)$	1	-1	-1	1	$x; yz; R_x$

D_3	$\hat{E}$	$2\hat{C}_3$	$3\hat{C}_2$	Functions
$A_1(a_1)$	1	1	1	$x^2 + y^2, z^2$
$A_2(a_2)$	1	1	-1	$z; R_z$
$E(e)$	2	-1	0	$(x, y); (x^2 - y^2, xy), (xz, yz); (R_x, R_y)$

D_{2d}	$\hat{E}$	$2\hat{S}_4$	$\hat{C}_2$	$2\hat{C}'_2$	$2\hat{\sigma}_d$	Functions
$A_1(a_1)$	1	1	1	1	1	$x^2 + y^2, z^2$
$A_2(a_2)$	1	1	1	-1	-1	R_z
$B_1(b_1)$	1	-1	1	1	-1	$x^2 - y^2$
$B_2(b_2)$	1	-1	1	-1	1	$z; xy$
$E(e)$	2	0	-2	0	0	$(x, y); (xz, yz); (R_x, R_y)$

D_{3d}	$\hat{E}$	$2\hat{C}_3$	$3\hat{C}_2$	$\hat{I}$	$2\hat{S}_6$	$3\hat{\sigma}_d$	Functions
$A_{1g}(a_{1g})$	1	1	1	1	1	1	$x^2 + y^2, z^2$
$A_{2g}(a_{2g})$	1	1	−1	1	1	−1	R_z
$E_g(e_g)$	2	−1	0	2	−1	0	$(x^2 - y^2, xy), (xz, yz); (R_x, R_y)$
$A_{1u}(a_{1u})$	1	1	1	−1	−1	−1	
$A_{2u}(a_{2u})$	1	1	−1	−1	−1	1	z
$E_u(e_u)$	2	−1	0	−2	1	0	(x, y)

S_4	$\hat{E}$	$\hat{S}_4$	$\hat{C}_2$	$\hat{S}_4^3$	Functions
$A(a)$	1	1	1	1	$x^2 + y^2, z^2; R_z$
$B(b)$	1	−1	1	−1	$z; x^2 - y^2, xy$
$E(e)$	$\left\{\begin{matrix}1 \\ 1\end{matrix}\right.$	$\begin{matrix}i \\ -i\end{matrix}$	$\begin{matrix}-1 \\ -1\end{matrix}$	$\left.\begin{matrix}-i \\ i\end{matrix}\right\}$	$(x, y); (xz, yz); (R_x, R_y)$

D_{2h}	$\hat{E}$	$\hat{C}_2(z)$	$\hat{C}_2(y)$	$\hat{C}_2(x)$	$\hat{I}$	$\hat{\sigma}(xy)$	$\hat{\sigma}(xz)$	$\hat{\sigma}(yz)$	Functions
$A_g(a_g)$	1	1	1	1	1	1	1	1	x^2, y^2, z^2
$B_{1g}(b_{1g})$	1	1	−1	−1	1	1	−1	−1	$xy; R_z$
$B_{2g}(b_{2g})$	1	−1	1	−1	1	−1	1	−1	$xz; R_y$
$B_{3g}(b_{3g})$	1	−1	−1	1	1	−1	−1	1	$yz; R_x$
$A_u(a_u)$	1	1	1	1	−1	−1	−1	−1	
$B_{1u}(b_{1u})$	1	1	−1	−1	−1	−1	1	1	z
$B_{2u}(b_{2u})$	1	−1	1	−1	−1	1	−1	1	y
$B_{3u}(b_{3u})$	1	−1	−1	1	−1	1	1	−1	x

D_{3h}	$\hat{E}$	$2\hat{C}_3$	$3\hat{C}_2$	$\hat{\sigma}_h$	$2\hat{S}_3$	$3\hat{\sigma}_v$	Functions
$A_1'(a_1')$	1	1	1	1	1	1	$z^2, x^2 + y^2$
$A_2'(a_2')$	1	1	−1	1	1	−1	R_z
$E'(e')$	2	−1	0	2	−1	0	$(x, y); (x^2 - y^2, xy)$
$A_1''(a_1'')$	1	1	1	−1	−1	−1	
$A_2''(a_2'')$	1	1	−1	−1	−1	1	z
$E''(e'')$	2	−1	0	−2	1	0	$(xz, yz); (R_x, R_y)$

D_{4h}	$\hat{E}$	$2\hat{C}_4$	$\hat{C}_2$	$2\hat{C}_2'$	$2\hat{C}_2''$	$\hat{i}$	$2\hat{S}_4$	$\hat{\sigma}_h$	$2\hat{\sigma}_v$	$2\hat{\sigma}_d$	Functions
$A_{1g}(a_{1g})$	1	1	1	1	1	1	1	1	1	1	$x^2 + y^2, z^2$
$A_{2g}(a_{2g})$	1	1	1	−1	−1	1	1	1	−1	−1	R_z
$B_{1g}(b_{1g})$	1	−1	1	1	−1	1	−1	1	1	−1	$x^2 - y^2$
$B_{2g}(b_{2g})$	1	−1	1	−1	1	1	−1	1	−1	1	xy
$E_g(e_g)$	2	0	−2	0	0	2	0	−2	0	0	$(xz, yz); (R_x, R_y)$
$A_{1u}(a_{1u})$	1	1	1	1	1	−1	−1	−1	−1	−1	
$A_{2u}(a_{2u})$	1	1	1	−1	−1	−1	−1	−1	1	1	z
$B_{1u}(b_{1u})$	1	−1	1	1	−1	−1	1	−1	−1	1	
$B_{2u}(b_{2u})$	1	−1	1	−1	1	−1	1	−1	1	−1	
$E_u(e_u)$	2	0	−2	0	0	−2	0	2	0	0	(x, y)

D_{6h}	$\hat{E}$	$2\hat{C}_6$	$2\hat{C}_3$	$\hat{C}_2$	$3\hat{C}_2'$	$3\hat{C}_2''$	$\hat{i}$	$2\hat{S}_3$	$2\hat{S}_6$	$\hat{\sigma}_h$	$3\hat{\sigma}_d$	$3\hat{\sigma}_v$	Functions
$A_{1g}(a_{1g})$	1	1	1	1	1	1	1	1	1	1	1	1	$x^2 + y^2, z^2$
$A_{2g}(a_{2g})$	1	1	1	1	−1	−1	1	1	1	1	−1	−1	R_z
$B_{1g}(b_{1g})$	1	−1	1	−1	1	−1	1	−1	1	−1	1	−1	
$B_{2g}(b_{2g})$	1	−1	1	−1	−1	1	1	−1	1	−1	−1	1	
$E_{1g}(e_{1g})$	2	1	−1	−2	0	0	2	1	−1	−2	0	0	$(xz, yz); (R_x, R_y)$
$E_{2g}(e_{2g})$	2	−1	−1	2	0	0	2	−1	−1	2	0	0	$(x^2 - y^2, xy)$
$A_{1u}(a_{1u})$	1	1	1	1	1	1	−1	−1	−1	−1	−1	−1	
$A_{2u}(a_{2u})$	1	1	1	1	−1	−1	−1	−1	−1	−1	1	1	z
$B_{1u}(b_{1u})$	1	−1	1	−1	1	−1	−1	1	−1	1	−1	1	
$B_{2u}(b_{2u})$	1	−1	1	−1	−1	1	−1	1	−1	1	1	−1	
$E_{1u}(e_{1u})$	2	1	−1	−2	0	0	−2	−1	1	2	0	0	(x, y)
$E_{2u}(e_{2u})$	2	−1	−1	2	0	0	−2	1	1	−2	0	0	

T	$\hat{E}$	$4\hat{C}_3$	$4\hat{C}_3^2$	$3\hat{C}_2$	Functions
$A(a)$	1	1	1	1	$x^2 + y^2 + z^2$
$E(e)$	$\begin{cases} 1 \\ 1 \end{cases}$	$\exp(2\pi i/3)$ $\exp(-2\pi i/3)$	$\exp(-2\pi i/3)$ $\exp(2\pi i/3)$	$\left.\begin{matrix} 1 \\ 1 \end{matrix}\right\}$	$(2z^2 - x^2 - y^2, x^2 - y^2)$
$T(t)$	3	0	0	−1	$(x, y, z); (xy, xz, yz); (R_x, R_y, R_z)$

T_d	$\hat{E}$	$8\hat{C}_3$	$3\hat{C}_2$	$6\hat{S}_4$	$6\hat{\sigma}_d$	Functions
$A_1(a_1)$	1	1	1	1	1	$x^2 + y^2 + z^2$
$A_2(a_2)$	1	1	1	−1	−1	
$E(e)$	2	−1	2	0	0	$(2z^2 - x^2 - y^2, x^2 - y^2)$
$T_1(t_1)$	3	0	−1	1	−1	(R_x, R_y, R_z)
$T_2(t_2)$	3	0	−1	−1	1	$(x, y, z); (xy, xz, yz)$

O	$\hat{E}$	$8\hat{C}_3$	$3\hat{C}_2(=\hat{C}_4^2)$	$6\hat{C}_4$	$6\hat{C}_2'$	Functions
$A_1(a_1)$	1	1	1	1	1	$x^2+y^2+z^2$
$A_2(a_2)$	1	1	1	-1	-1	
$E(e)$	2	-1	2	0	0	$(2z^2-x^2-y^2, x^2-y^2)$
$T_1(t_1)$	3	0	-1	1	-1	(x, y, z); (R_x, R_y, R_z)
$T_2(t_2)$	3	0	-1	-1	1	(xy, xz, yz)

O_h	$\hat{E}$	$8\hat{C}_3$	$6\hat{C}_2$	$6\hat{C}_4$	$3\hat{C}_2(=\hat{C}_4^2)$	$\hat{I}$	$6\hat{S}_4$	$8\hat{S}_6$	$3\hat{\sigma}_h$	$6\hat{\sigma}_d$	Functions
$A_{1g}(a_{1g})$	1	1	1	1	1	1	1	1	1	1	$x^2+y^2+z^2$
$A_{2g}(a_{2g})$	1	1	-1	-1	1	1	-1	1	1	-1	
$E_g(e_g)$	2	-1	0	0	2	2	0	-1	2	0	$(2z^2-x^2-y^2, x^2-y^2)$
$T_{1g}(t_{1g})$	3	0	-1	1	-1	3	1	0	-1	-1	(R_x, R_y, R_z)
$T_{2g}(t_{2g})$	3	0	1	-1	-1	3	-1	0	-1	1	(xz, yz, xy)
$A_{1u}(a_{1u})$	1	1	1	1	1	-1	-1	-1	-1	-1	
$A_{2u}(a_{2u})$	1	1	-1	-1	1	-1	1	-1	-1	1	
$E_u(e_u)$	2	-1	0	0	2	-2	0	1	-2	0	
$T_{1u}(t_{1u})$	3	0	-1	1	-1	-3	-1	0	1	1	(x, y, z)
$T_{2u}(t_{2u})$	3	0	1	-1	-1	-3	1	0	1	-1	

I_h	$\hat{E}$	$12\hat{C}_5$	$12\hat{C}_5^2$	$20\hat{C}_3$	$15\hat{C}_2$	$\hat{I}$	$12\hat{S}_{10}$	$12\hat{S}_{10}^3$	$20\hat{S}_6$	$15\hat{\sigma}$	Functions
$A_g(a_g)$	1	1	1	1	1	1	1	1	1	1	$x^2+y^2+z^2$
$T_{1g}(t_{1g})$	3	$\frac{1}{2}(1+\sqrt{5})$	$\frac{1}{2}(1-\sqrt{5})$	0	-1	3	$\frac{1}{2}(1-\sqrt{5})$	$\frac{1}{2}(1+\sqrt{5})$	0	-1	(R_x, R_y, R_z)
$T_{2g}(t_{2g})$	3	$\frac{1}{2}(1-\sqrt{5})$	$\frac{1}{2}(1+\sqrt{5})$	0	-1	3	$\frac{1}{2}(1+\sqrt{5})$	$\frac{1}{2}(1-\sqrt{5})$	0	-1	
$G_g(g_g)$	4	-1	-1	1	0	4	-1	-1	1	0	
$H_g(h_g)$	5	0	0	-1	1	5	0	0	-1	1	$(2z^2-x^2-y^2,$ $x^2-y^2, xy, yz, zx)$
$A_u(a_u)$	1	1	1	1	1	-1	-1	-1	-1	-1	
$T_{1u}(t_{1u})$	3	$\frac{1}{2}(1+\sqrt{5})$	$\frac{1}{2}(1-\sqrt{5})$	0	-1	-3	$-\frac{1}{2}(1-\sqrt{5})$	$-\frac{1}{2}(1+\sqrt{5})$	0	1	(x, y, z)
$T_{2u}(t_{2u})$	3	$\frac{1}{2}(1-\sqrt{5})$	$\frac{1}{2}(1+\sqrt{5})$	0	-1	-3	$-\frac{1}{2}(1+\sqrt{5})$	$-\frac{1}{2}(1-\sqrt{5})$	0	1	
$G_u(g_u)$	4	-1	-1	1	0	-4	1	1	-1	0	
$H_u(h_u)$	5	0	0	-1	1	-5	0	0	1	-1	

$C_{\infty v}$	$\hat{E}$	$\infty \hat{C}_{\infty}$	$\infty \hat{\sigma}_v$	Functions
$\Sigma^+(\sigma)$	1	1	1	$z; x^2 + y^2, z^2$
Σ^-	1	1	-1	R_z
$\Pi(\pi)$	2	$2 \cos \phi$	0	$(x, y); (xz, yz); (R_x, R_y)$
$\Delta(\delta)$	2	$2 \cos 2\phi$	0	$(xy, x^2 - y^2)$
$\Phi(\varphi)$	2	$2 \cos 3\phi$	0	
$\vdots$	$\vdots$	$\vdots$	$\vdots$	

$D_{\infty h}$	$\hat{E}$	$\infty \hat{C}_{\infty}$	$\infty \hat{\sigma}_v$	$\hat{I}$	$\infty \hat{S}_{\infty}$	$\infty \hat{C}_2'$	Functions
$\Sigma_g^+(\sigma_g)$	1	1	1	1	1	1	$x^2 + y^2, z^2$
Σ_g^-	1	1	-1	1	1	-1	R_z
$\Sigma_u^+(\sigma_u)$	1	1	1	-1	-1	-1	z
Σ_u^-	1	1	-1	-1	-1	1	
$\Pi_g(\pi_g)$	2	$2 \cos \phi$	0	2	$-2 \cos \phi$	0	$(xz, yz); (R_x, R_y)$
$\Pi_u(\pi_u)$	2	$2 \cos \phi$	0	-2	$2 \cos \phi$	0	(x, y)
$\Delta_g(\delta_g)$	2	$2 \cos 2\phi$	0	2	$2 \cos 2\phi$	0	$(xy, x^2 - y^2)$
$\Delta_u(\delta_u)$	2	$2 \cos 2\phi$	0	-2	$-2 \cos 2\phi$	0	
$\vdots$	$\vdots$	$\vdots$	$\vdots$	$\vdots$	$\vdots$	$\vdots$	

Solutions to Objectives Review Questions

Numerical answers to problems are included here. Complete solutions to selected problems can be found in the *Student's Solutions Manual.*

Chapter 1

P1.1 $3.0 \cdot 10^{11}$ s, $2.0 \cdot 10^{-22}$ J

P1.2 $\lambda_{\mathrm{dB}} = 1.8\,\text{Å} \ll 1.0\,\mu\text{m}$: no

P1.3 $-0.5\,E_{\mathrm{h}}, -1.0\,E_{\mathrm{h}}$

Chapter 2

P2.1 (a.) no; (b) yes (eigenvalue = 4)

P2.2 $0.176a^4$

P2.3 $\left(-\dfrac{\hbar^2}{2m}\dfrac{\partial^2}{\partial x^2} + U_0 x\right)\psi = E\psi$

P2.4 $3.28 \cdot 10^{-25}$ J

Chapter 3

P3.1 $$\frac{4\sqrt{2}}{27\sqrt{3}}\left(\frac{3}{a_0}\right)^{3/2}\left(\frac{3r}{a_0}\right)\left(1 - \frac{r}{2a_0}\right)e^{-r/a_0}$$
$$\sqrt{\frac{3}{8\pi}}\sin\theta\, e^{-i\phi}$$

P3.2 $$\frac{32}{3^7}\left(\frac{3}{a_0}\right)^3\int_0^\infty\left(\frac{3r}{a_0}\right)^2\left(1 - \frac{r}{2a_0}\right)^2 e^{-2r/a_0}r^3\,dr$$
$$= 25a_0/3$$

P3.3 1 angular, 1 radial

Chapter 4

P4.1 $$-\frac{\hbar^2}{2m_e}(\nabla(1)^2 + \nabla(2)^2 + \nabla(3)^2) - \frac{4e^2}{4\pi\varepsilon_0 r_1}$$
$$-\frac{4e^2}{4\pi\varepsilon_0 r_2} - \frac{4e^2}{4\pi\varepsilon_0 r_3} + \frac{e^2}{4\pi\varepsilon_0}\left(\frac{1}{r_{12}} + \frac{1}{r_{23}} + \frac{1}{r_{13}}\right)$$

P4.2 $1s^2 2s^1, -18\,E_{\mathrm{h}}$

P4.3 1.81

P4.4 antisymmetric

P4.5 ${}^3P_0, {}^3P_1, {}^3P_2, {}^1P_1$

Chapter 5

P5.1 $$\hat{H} = -\frac{\hbar^2}{2m_e}(\nabla(1)^2 + \nabla(2)^2 + \nabla(3)^2)$$
$$+ \frac{e^2}{4\pi\varepsilon_0}\left[-\frac{3}{r_{\mathrm{Li1}}} - \frac{1}{r_{\mathrm{H1}}} - \frac{3}{r_{\mathrm{Li2}}} - \frac{1}{r_{\mathrm{H2}}} - \frac{3}{r_{\mathrm{Li3}}} - \frac{1}{r_{\mathrm{H3}}}\right.$$
$$\left. + \frac{1}{r_{12}} + \frac{1}{r_{23}} + \frac{1}{r_{13}} + \frac{3}{R_{\mathrm{AB}}}\right]$$
$$- \frac{\hbar^2}{2m_{\mathrm{Li}}}\nabla(\mathrm{Li})^2 - \frac{\hbar^2}{2m_{\mathrm{H}}}\nabla(\mathrm{H})^2$$

P5.2 overlap between nuclei A and B, node near B but outside bonding region

P5.3 Curve such as in Fig. 5.14, but with minimum at $R = 1.5\,\text{Å}$, $U = -200$ kJ mol^{-1}

P5.4 The angle 2-3 increases, the angles 1-2 and 1-3 decrease

P5.5 ^{1}H: 2 triplets, equal intensity, $\delta \approx 3.55, 3.75$
13C: 2 singlets, equal intensity, d ≈ 25,35

Chapter 6

P6.1 $\hat{C}_2(z)$

P6.2 C_{2v}: $\hat{E}, \hat{C}_2, \hat{\sigma}_{xz}, \hat{\sigma}_{yz}$

P6.3 upper states for electric dipole: ${}^1A_1, {}^1B_1, {}^1B_2$; for Raman: ${}^1A_1, {}^1B_1, {}^1B_2, {}^1A_2$

P6.4 a_u

Chapter 7

P7.1 1/2

P7.2 $1\sigma_g^2\, 1\sigma_u^2\, 2\sigma_g^2\, 2\sigma_u^1$

P7.3 forbidden by spin

P7.4 phosphorescence

Chapter 8

P8.1 $3625\,\text{cm}^{-1}$, $\left(\frac{k\mu}{\hbar^2}\right)^{1/8}\left(\frac{1}{8\sqrt{\pi}}\right)^{1/2}(4y^2 - 2)e^{-y^2/2}$ with $y = (R - R_e)(k\mu/\hbar^2)^{1/4}$

P8.2 Spacing increases with E but more slowly than particle in box

P8.3 $\mu = 19.55$ amu, $k \approx 12\,\text{Nm}^{-1}$, $\omega_e \approx 102\,\text{cm}^{-1}$

P8.4 σ_g (Raman active), σ_u (IR active)

Chapter 9

P9.1 $1.755\ \text{cm}^{-1}$

P9.2 $56.0\ \text{cm}^{-1}$

P9.3 $^{12}C\,^{18}O_2$, $^{12}C\,^{16}O_2$, $^{12}C\,^{16}O$

P9.4 $118.32\ \text{cm}^{-1}$

Chapter 10

P10.1 Repulsion applies to all examples. (a) dispersion: $U(R) \propto \alpha_A\alpha_B/R^6$. (b) H-bonding: $U(R) \propto \mu_A\mu_B/R^6$. (c) dipole–dipole, dipole-induced dipole: $U(R) \propto \mu_A\mu_B/R^6$, $\mu_A\alpha_B/R^6$. (d) dipole–dipole (non-rotating): $U(R) \propto \mu_A\alpha_B/R^3$.

P10.2 R^{-5}

P10.3 $4.60 \cdot 10^{-24}$ J or 0.00277 kJ mol^{-1}

P10.4 b < a < c

P10.5 The well should extend from 3.61 Å to 5.42 Å and be 190 K (132 cm^{-1}) deep

P10.6 vibrations in B

Chapter 11

P11.1 dispersion, decrease

P11.2 U_{vdW} and $U_{\text{electrostatic}}$

Chapter 12

P12.1 Like curves in Figs. 12.3h and 12.3i, but more pronounced oscillation at large R; the oscillations smooth out near the boiling point (figure in solutions manual)

P12.2 yes (both bond principally through dispersion)

Chapter 13

P13.1 D_{4h}

P13.2 no

CREDITS

Cover Tony Jackson/Getty Images.

Front Matter Page v: Nathaniel Hawthorne (1804–1864) The Birthmark.

Chapter A Page 8: Arthur J. Stephson, "Mars Climate Orbiter Mishap Investigation Board Phase I Report" NASA, 11-10-1999.

Part I Page 37: Karramba Production/Fotolia.

Chapter 4 Page 179: Sylvia Ceyer.

Part II Page 205: Sportlibrary/Fotolia.

Chapter 5 Page 225: Douglas B. Grotjahn.

Chapter 6 Page 280: James Rondinelli. **Page 306:** Andrew Cooksy.

Chapter 7 Page 319: Arthur Suits.

Chapter 8 Page 374: Steven Vaughn, William Vaughn, FTIR Analytical Systems: Part II—Experimental Design, Gases & Instrumentation Magazine, January/February 2009.

Chapter 9 Page 402: Stewart E. Novick.

Part III Page 421: Vlad Ageshin/Shutterstock.com.

Chapter 10 Page 426: Helen Leung; Mark Marshall.

Chapter 11 Page 485: IBM characterizes single-atom DRAM, Courtesy IBM. **Page 491:** James V. Coe.

Chapter 12 Page 516: Martina Havenith.

Chapter 13 Page 534: Bragg WL (1913). "The Diffraction of Short Electromagnetic Waves by a Crystal." Proceedings of the Cambridge Philosophical Society 17: 43; J. Am. Chem. Soc. (2008) 130, 10038–10039. **Page 535:** Davies, M. (1970). "Peter Joseph Wilhelm Debye. 1884–1966." Biographical Memoirs of Fellows of the Royal Society 16. **Page 544:** Susan M. Kauzlarich.

INDEX

Note: Page numbers followed by n indicate footnotes.

I

J

K

L

Q

R